Barrier Systems for Waste Disposal

2nd edition

R. Kerry Rowe,
Queen's University Kingston, Ontario, Canada

Robert M. Quigley,

Richard W.I. Brachman
Queen's University Kingston, Ontario, Canada

and

John R. Booker

Spon Press
Taylor & Francis Group
LONDON AND NEW YORK

First edition published 1995 by E & FN Spon

This edition published 2004
by Spon Press
11 New Fetter Lane, London EC4P 4EE

Simultaneously published in the USA and Canada
by Spon Press
29 West 35th Street, New York, NY 10001

Spon Press is an imprint of the Taylor & Francis Group

Typeset in Times New Roman by
Integra Software Services Pvt. Ltd, Pondicherry, India
Printed and bound in Great Britain by
St Edmundsbury Press, Bury St Edmunds, Suffolk

British Library Cataloguing in Publication Data
A catalogue record for this book is available from the British Library

Library of Congress Cataloging in Publication Data
A catalog record for this book has been requested

ISBN 0-419-22630-3

Contents

Preface

The design of waste disposal facilities typically involves some form of "barrier" that separates the "waste" from the general groundwater system and a cover (also a barrier) that separates the waste from the above ground environment. The bottom barrier is intended to minimize the migration of contaminants from the facility and hence the environmental impact of the facility is intimately related to its design and long-term performance. Natural clayey deposits, recompacted clayey liners or geosynthetic clay liners frequently represent a key component of these barriers. However, there are many practical situations where neither a low-permeability clay nor plastic liner (geomembrane) alone is sufficient to prevent unacceptable long-term environmental impact. In these circumstances, the barrier is a designed system which may involve numerous components (e.g., composite liners, multiple leachate collection systems, etc.) which work together to provide protection to the groundwater environment. The cover may also involve drainage layers, geomembranes and/or clay liners, and may serve both to limit inward migration of water and outward migration of both dissolved contaminants and gases to the surface.

This book deals with the design and performance of barrier systems which may include natural clayey deposits, recompacted clayey liners or even porous rock as part of the barrier system. Considerable attention is also given to other components of the barrier system, such as the leachate collection system and geomembrane liners. The first edition of the book was written as the basis for short courses which the authors have given for educators, practising engineers and hydrogeologists as well as a text for a senior or graduate level course. This second edition addresses the same audience. We have attempted to explain all key concepts assuming little background knowledge other than introductory courses in chemistry, soil mechanics and/or hydrogeology. In particular, Chapter 1 is essential to the reader not familiar with the basic concepts. Appendix A also presents a glossary of terms which should assist the reader not familiar with some of the terminology.

The objectives of this book are threefold. First, to examine the clayey component of the barrier. Contaminants can potentially pass through a barrier by advection (e.g., the movement of contaminants due to groundwater movement) and diffusion (i.e., the movement of contaminants from locations of high concentration to locations of lower concentration). The potential impact of a given contaminant will generally depend on both the hydraulic conductivity and diffusion coefficient for the barrier. Thus the book deals with the construction/compaction of clayey liners, the significance of clay mineralogy in the design of barriers, geosynthetic clay liners and the determination of appropriate hydraulic conductivity values and diffusion coefficients for clayey liners. In particular, the field studies discussed in Chapter 9 demonstrate conclusively that diffusive contaminant transport can and does occur even in the absence of significant advection. The chapters dealing with these issues from the original book have been updated to reflect advances since the first book was published in 1995 and a new chapter on geosynthetic clay liners has been added.

The second objective is to provide the reader with the current state-of-the-art with respect to aspects of barrier system design other than the clayey component. Thus this second edition has a chapter devoted to leachate characteristics and the design and performance of leachate collection systems, a new chapter on geomembrane liners, a new chapter on covers and a new chapter on geotechnical and related design issues. In addition many of the chapters in the previous book have been extensively revised to address issues related to the use of geosynthetics, the interaction between geosynthetics and clay liners, the service life of engineered components of the system and the equivalency of different liner systems.

The third and final objective is to highlight the role which theoretical modelling can play in the design of barrier systems. This role may include the determination of some relevant design parameters from laboratory tests and the evaluation of the potential impact of a proposed design on groundwater quality both under working conditions and in the event of a failure of part of the engineered system (e.g., the leachate collection system and a geomembrane when its service life is reached).

The state-of-the-practice has now advanced to a point where modelling is frequently used as an aid to engineering judgement in landfill designs. However, the results of an analysis are only as good as the assumptions and parameters on which it is based. There will always be some uncertainty associated with parameters and design assumptions and for this reason analysis will be most useful for performing sensitivity studies to examine the implications of different design scenarios and for determining the potential significance of uncertainty regarding key parameters.

The first chapter deals with basic concepts including the types of barrier being considered, the nature and relative importance of different transport mechanisms, the choice of suitable boundary conditions, complicating factors, and design considerations. It also includes a discussion of methods of analysis ranging from simple hand methods to simple finite layer techniques which can be readily implemented on a personal computer.

The second chapter discusses leachate characteristics, leachate mounding, the design of collection systems, clogging and service life issues for leachate collection systems and the relationship between leachate mounding and liner temperature.

Chapters 3 and 4 examine issues of permeability and clay-leachate compatibility for clayey liners while Chapter 5 discusses the modelling of flow through liner systems.

Chapter 6 provides a discussion on the process of molecular diffusion and highlights the many factors involved in the migration of contaminants through both soil and geomembranes due to a difference in chemical potential (concentration).

Chapter 7 describes a finite layer technique for the analysis of contaminant migration in intact and fractured porous media, and provides the tools for modelling composite liner systems and describes the behaviour of the system while all components are functioning as designed as well as when the service lives of various components are reached.

Chapter 8 discusses the evaluation of a number of key parameters such as the diffusion coefficient and sorption parameters (i.e., distribution or partitioning coefficient), and shows that both parameters can often be estimated from a single test using a simple finite layer analysis to match calculated and observed migration of a chemical species of interest through a clayey soil or geomembrane. Chapter 9 then illustrates the important role of molecular diffusion with reference to a number of field case histories.

Chapter 10 discusses the importance of the finite mass of contaminant available for transport into the groundwater system together with factors such as the effect of landfill size, sorption, liner thickness and advective velocity. Examples will show how analysis may improve the designer's "feel" for the effectiveness of potential contaminant attenuation mechanisms and proposed design both for natural and engineered systems involving intact clayey barriers.

Chapter 11 examines the migration of contaminants in fractured porous media and will highlight the important role of matrix diffusion (i.e., diffusion from the fractures into the intact matrix adjacent to the fractures). Consideration is given to both fractured porous rock and fractured clayey deposits and to the benefits to be derived from installing an intact (e.g., compacted clay) liner over fractured media.

Chapters 12 and 13 provide a summary of the key considerations relating to the use of geosynthetic clay liners and geomembranes as part of the barrier system.

Chapter 14 addresses issues related to the design of covers and Chapter 15 addresses a number of geotechnical issues as well as issues related to the design of geotextiles, leachate collection pipes and manholes.

Finally, Chapter 16 discusses the integration of hydrogeology and engineering in the design of barrier systems and the assessment of potential contaminant impact. Topics discussed include the effect of the natural hydrogeology on engineering design, the service life of engineered systems, the contaminating lifespan of a landfill, the implications of failure of leachate collection systems, and the assessment of the equivalency of different liner systems.

Much of the research reported in this book was supported by operating and strategic funds from the Natural Sciences and Engineering Research Council of Canada. In addition, funding was received from the Ontario Ministry of the Environment, the Centre for Research in Earth and Space Technology (CRESTech), Terrafix Geosynthetics Inc. and Naue Fasertechnik GmbH & Co. KG.

The authors are very grateful to many individuals who have made suggestions or contributions to both the research and the preparation of this book. In particular we wish to acknowledge Dr A. Bouazza, Dr C. Lake, Dr T. Iryo, Mrs K. Rowe, Dr D. Smith, Mr J. Southen, Dr J. VanGulck, Mr K. von Maubeuge and Ms K. Lange for their suggestions and help. Many graduate students, postdoctoral fellows, research associates and collaborators have also been involved in generating the research that forms the basis for this book. They are: Drs K. Badv, F.S. Barone, A.J. Cooke, D.R. Cullimore, M. Darwish, F. Fernandez, I.R. Fleming, L. Hrapovic, D. Goodall, A. do Lago, C. Lake, H.M. Li, F. Longstaffe, R. McIsaac, T. Mukunoki, P. Nadarajah, C. Ohikere, B. Rittmann, H.P. Sangam, J. Southen, N. Touze-Foltz, J. VanGulck, E.K. Yanful, Y. Yoshida, Y. Zhou; Messrs M.D. Armstrong, P.J. Bennett, C.J. Caers, M.J. Fraser, S. Gudina, A. Hammoud, T. Helgason, R. Krushelnitzky, J. Mucklow, R.J. Petrov, E. San, J.D. Smith, A. Tognon; Ms L. Eggleston, M. Krol, P.C. MacKay, S.C. Millward, C. Orsini and V.E. Crooks. G. Lusk and W. Logan have expedited the laboratory work.

Two of the authors of the first edition, Drs J.R. Booker and R.M. Quigley passed away shortly after publication of the first edition. Their counsel and experience has been sadly missed in the preparation of this second edition – the undersigned take full responsibility for any errors or omissions that almost certainly would not have escaped their attention and request that any such errors or omissions be brought to our attention.

Finally, we owe a deep debt of gratitude to our families, and especially to our wives, Kathy and Dara, for their patience and understanding over the past three years as this edition has taken shape.

R. Kerry Rowe and Richard W.I. Brachman
Queen's University, Kingston, Canada.

Chapter one

Basic concepts

1.1 Introduction

This book was written as a reference book for professional engineers and hydrogeologists and as a text for senior level and graduate level courses dealing with the design, construction and performance of barrier systems. The objective of this chapter is to describe the basic concepts that will be expanded on in later chapters. Different types of barrier systems will be discussed. The different factors associated with contaminant transport through barrier systems are explained and the relative importance of different transport mechanisms is examined. Finally, some of the basic concepts associated with modelling of contaminant movement through barrier systems and assessing its impact on groundwater quality are introduced and illustrated for some simple cases.

1.2 Overview of barrier systems

The impact of a waste disposal facility on groundwater quality will depend on the nature of the site, the climate, the type of waste, the local hydrogeology and the presence of a dominant flow path, and, perhaps most importantly, on the nature of the barrier which is intended to limit and control contaminant migration. These days, barriers will usually include one or more of the following components:

(i) natural clayey soils such as lacustrine clay or clayey till;
(ii) re-compacted clayey liners;
(iii) cut-off walls;
(iv) natural bedrock; and
(v) geosynthetics either alone or as part of a composite liner system.

The following subsections provide an introduction to some of the design considerations associated with these components. In this discussion terms such as "diffusion", "advection" and "sorption" are used. The reader unfamiliar with these terms may wish to review Section 1.3 prior to reading the following. A glossary defining key terminology is given in Appendix A; key notation is defined in Appendix B.

1.2.1 Natural clayey deposits

(a) Diffusion-controlled systems

The simplest barrier systems involve thick natural deposits of clayey soil where the water table is near the surface. For the unusual case of a landfill which does not mound above the original ground surface, it may be possible to design the landfill such that, after closure, the liquid (leachate) levels in the landfill rise to the natural water table, as shown schematically in Figure 1.1, and there is no significant gradient or advective

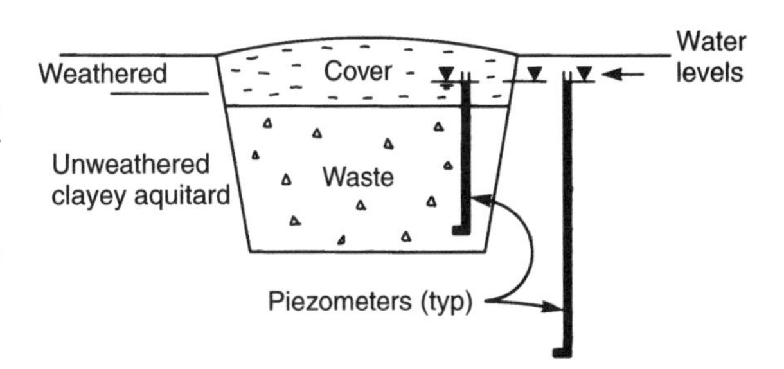

Figure 1.1 Natural barrier design involving diffusion transport (but no advection) in deep clayey deposit (after Rowe, 1988; reproduced with permission of the *Canadian Geotechnical Journal*).

transport. The primary consideration in the evaluation of this type of barrier design is to ensure that the hydrogeologic environment is such that one can reasonably expect low hydraulic gradients over the long term. Under these circumstances the movement of contaminants will be slow and controlled primarily by diffusion. The clay itself can also act as an important medium for the attenuation of some contaminants due to processes such as sorption, precipitation and biodegradation. Nevertheless, migration will occur and it is important to check that contaminant impact at the site boundary, due to molecular diffusion from the landfill, will meet regulatory requirements.

The diffusive movement of contaminants through saturated clayey deposits is relatively well understood and, as discussed in Chapter 9, research has shown that natural, diffusion-controlled, chemical profiles established over thousands of years are consistent with simple theoretical predictions.

(b) Advective–diffusive transport

While situations such as that shown in Figure 1.1 do exist, they are not common and a more typical situation involving mounded waste is shown in Figure 1.2. Prior to landfilling, there are downward gradients from the water table to the underlying aquifer. The clayey soil provides a natural barrier; however in order to design a suitable landfill, careful consideration must be given to the selection of the landfill base elevations and to the design of the leachate collection system.

Figure 1.2 shows a situation where the base elevations are below the groundwater level but above the potentiometric surface (i.e., water levels) in the aquifer and hence there will be downward advective flow from the landfill to the aquifer. Figure 1.3 shows a situation where the base elevations are such that the design leachate mound is below the potentiometric surface in the aquifer and hence there is upward advective flow from the aquifer to the landfill; this design relies on hydraulic containment of contaminants and

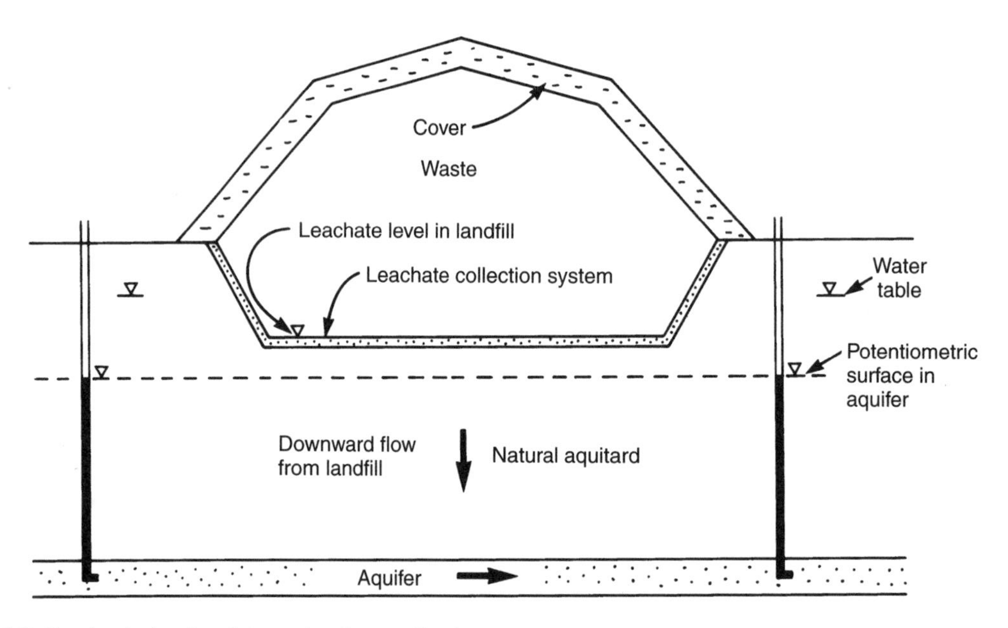

Figure 1.2 Barrier design involving a leachate collection system, a natural clayey deposit and downward advective–diffusive transport.

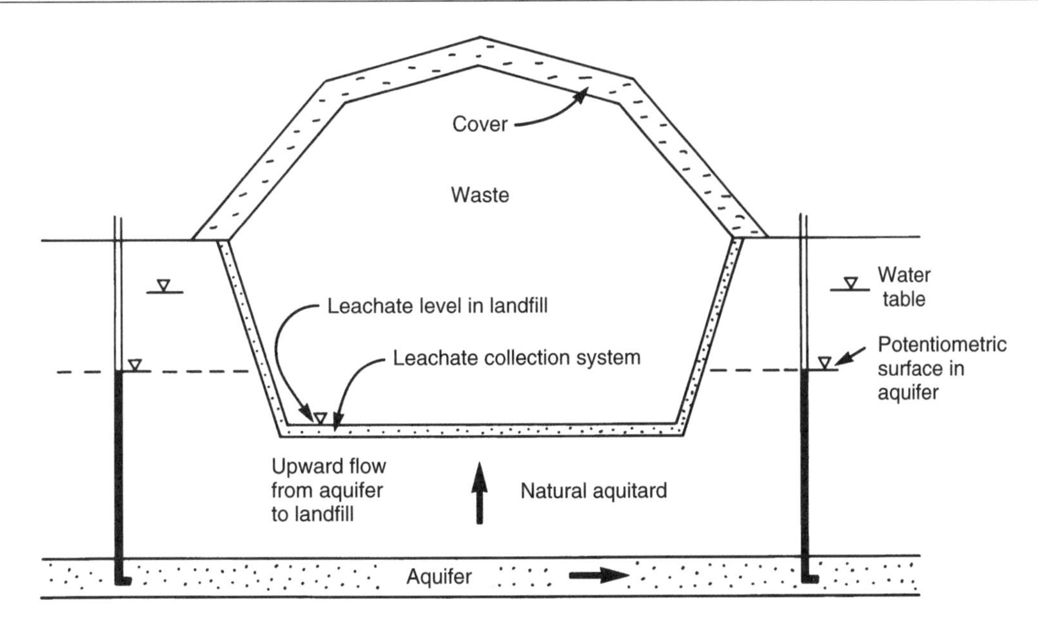

Figure 1.3 Barrier design involving a leachate collection system, a natural clayey deposit, upward advection and downward diffusion – a hydraulic trap.

may be referred to as a "hydraulic trap" (as discussed in Section 1.2.1(d)). A discussion of modelling of these types of situations for intact soil is given in Chapters 10 and 16.

(c) Fractured clays

A major consideration for a landfill design such as that shown in Figure 1.2 is whether the clayey soil or till is fractured. Research by D'Astous *et al.* (1989), Herzog *et al.* (1989), Rowe *et al.* (2000c) and others has provided evidence to suggest that clayey till beneath the obviously weathered and fractured zone may be fractured to depths of 6–15 m (e.g., see Table 1.1).

It is important to recognize that fractures with very thin (e.g., 10–15 μm) openings at relatively wide spacings of 1–3 m can still have a significant effect in terms of increasing the bulk hydraulic conductivity (compared to that of the intact soil). Furthermore, even if the fractures are closed prior to construction of the landfill, stress relief due to excavation of the landfill cell may, under some circumstances, cause opening of fractures in the underlying clays.

If present, fractures are likely to control the hydraulic conductivity of the clay and hence the downward advective velocity and it may be necessary to construct a clay liner by reworking the existing soil (if it is suitable) to reduce outward advective transport from the landfill to acceptable levels. Relatively simple semi-analytical models have now been developed which will allow the designer (see Chapter 11) to estimate the potential impact associated with fractured clay, or a liner overlying fractured clay, for situations similar to that shown in Figure 1.2.

(d) Hydraulic containment

The hydraulic containment (often called a "hydraulic trap") as shown in Figure 1.3 is very attractive from a contaminant impact perspective since the inward advective flow of groundwater from the aquifer tends to inhibit the outward diffusion of contaminants. However, as illustrated in Chapter 10, there may still be a potential impact on the underlying aquifer even with an operating hydraulic trap. Appropriate contaminant transport calculations are required

Table 1.1 Summary of a number of cases where fracturing was encountered below the obviously weathered zone in clayey till deposits

Site	*Approximate depth of weathered zone* (m)	*Approximate observed depth of fractures* (m)	*Was deep fracturing evident in borehole samples?*	*Were deep fractures evident from borehole tests?*	*Was fracturing implied by field pumping tests?*
Southern Ontario 1	4–5	10.5–11	No	No	Yes
Southern Ontario 2	4–5	7	No	No	Yes
Southern Ontario 3	4–5	10	No	No	NA
Southern Ontario 4	4	7	No	Maybe	Yes
Southern Ontario 5	4–5	>13	No	No	NA
Southern Ontario 6	7	>12	Yes	NA	NA
Southern Ontario 7	4–6	9	No	No	NA
Southern Ontario 8	4–6	8–8.5	No	Maybe	Yes
Southern Ontario 9	4–6	10	No	No	NA
Illinois	5–6	~15	No	Maybe	NA

NA, not applicable (either not performed or not performed at an elevation to detect fractures).

to assess whether the impact will be significant or negligible.

The design of landfills intended to operate as a hydraulic trap is far from straightforward. Very careful consideration must be given to a number of conflicting criteria. For example, lowering the landfill base elevations increases the inward hydraulic gradient and hence inward advective flow. This reduces the amount of outward diffusion (which is good); however by lowering the base contours, one also reduces the thickness of the barrier which is separating the waste from the underlying aquifer (which is not so good). Thus, if there were to be a failure of the leachate collection system and mounding of leachate which reversed the hydraulic gradient, the attenuation afforded by the barrier would be reduced. Furthermore, the lower the base of the landfill, the greater the potential for opening of fractures in the clay due to uplift pressure and the greater the potential for blowout of the base of the landfill. One can reduce the likelihood of blowout by depressurizing the aquifer during construction; however, this option requires careful evaluation of the number of wells required to gain adequate drawdown over large areas and the potential effect on off-site water users.

The operation of the hydraulic trap shown schematically in Figure 1.3 implicitly assumes that the construction of the landfill will not significantly affect the water levels in the underlying aquifer. It cannot be assumed that this will necessarily be the case. Construction of a landfill as shown schematically in Figure 1.3 would eliminate any recharge of the aquifer that had previously occurred over the area of the landfill footprint. In addition, the leachate collection system is removing groundwater from the aquifer for as long as the hydraulic trap is maintained. The combination of these two effects can be expected to result in a reduction in head (water pressure) in the aquifer unless the aquifer is very permeable or the site is over a natural groundwater discharge area. The extent to which the head in the aquifer is reduced will depend on the hydrogeological characteristics of the aquifer and overlying clayey soils as well as on the size of the landfill (i.e., the larger the landfill, the greater the shadow it casts on the aquifer and hence the greater the potential for a decrease in head within the aquifer). The implications of this shadow effect should be considered carefully in the design of a hydraulic trap and are discussed with respect to a specific landfill by Rowe *et al.* (2000c) and in more general terms in Chapter 16.

Consideration should also be given to the impact of potential pumping of the aquifer for use as water supply and the impact of potential long-term changes in climate over the contaminating lifespan of the landfill (i.e., the period of time during which the landfill contains contaminants which could have an unacceptable impact if released to the environment). In some cases these factors alone may necessitate a more elaborate design such as that discussed subsequently with respect to Figure 1.8 (Section 1.2.2(b)).

1.2.2 Compacted clay liners

Compacted clay liners have been the subject of debate with respect to both the hydraulic conductivity which can be achieved in the field (e.g., Day and Daniel, 1985 and related discussion) and the potential impact of soil–leachate interaction on hydraulic conductivity (e.g., Green *et al.*, 1981; Brown and Anderson, 1983; Brown *et al.*, 1983, 1984; Fernandez and Quigley, 1985; Bowders and Daniel, 1987). However, experience to date would suggest that with good engineering practice and quality control, good quality, low hydraulic conductivity liners can be constructed of compacted inactive clays as discussed in Chapter 9. Furthermore, it would appear that while some very concentrated organic wastes may increase the hydraulic conductivity of clay, provided that one considers clay–leachate compatibility in the selection of the liner material, large increases in hydraulic conductivity can be avoided as discussed in Chapter 4.

Inactive clayey liners which are compacted at a water content higher than the Standard Proctor optimum moisture content and not allowed to subsequently dry out will be nearly saturated, compressible on loading and should behave in a satisfactory manner similar to natural unfractured clayey barriers. Provided that the design minimizes outward gradients, the primary transport mechanism through a well-designed compacted clayey barrier will be diffusion.

The present state-of-the-art has not clarified the susceptibility of sodium bentonite/sand mixtures to damage by saline and organic leachates and particular care should be taken in the design of this type of compacted liner.

(a) Single liner

Figure 1.4 shows a compacted clay barrier constructed on an existing soil deposit. The liner may be required for one of two reasons. First, if the *in-situ* soil is fractured then the liner may be required to retard movement (or potential movement) of contaminant along the fractures. For situations where there is outward flow from the landfill (e.g., water level (a) in Figure 1.4), the assessment of potential impact is likely to consider advective–diffusive migration through the liner and the potential for attenuation due to matrix diffusion as contaminant migrates along the fractures, as discussed in Chapter 11. For situations where there is a hydraulic trap (e.g., water level (b) in Figure 1.4), the assessment of potential impact will require consideration of outward diffusion. The effect of the inward advective flow needs to be considered carefully. For example, if the inward flow is primarily through the fractures then the outward diffusion of contaminant through the matrix between fractures may not be significantly reduced by

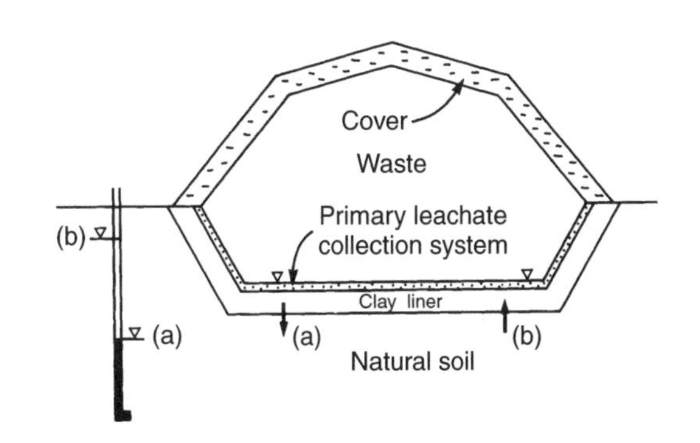

Figure 1.4 A compacted clayey liner used as the primary barrier.

the inward flow if the fracture spacing is relatively wide. On the other hand, if the fractures are close-spaced and/or non-conductive then the situation may be similar to that discussed in the previous section where the outward diffusion is resisted by inward advective flow. Clearly, each specific case should be carefully examined on its own merits.

The second reason for constructing a clay liner is that, while intact, the surrounding natural soil does not have a low enough hydraulic conductivity (permeability) to provide an adequate barrier. A typical example is Metropolitan Toronto's Keele Valley landfill in Ontario, Canada. The available data would suggest that, to date, the liner is performing well. Based on the results of the shallow lysimeters and the majority of conductivity sensor sets, it would appear that it is performing better than would be expected for the specified 1.2 m thick liner with a 1×10^{-10} m/s hydraulic conductivity (Reades *et al.*, 1989; King *et al.*, 1993; and Section 9.4). Even though the outward hydraulic gradient through the liner exceeds unity, the data suggest that the rate of contaminant migration through the liner is governed by diffusion. No evidence has been found for bulk field hydraulic conductivities higher than those measured on undisturbed samples in the laboratory at an effective consolidation pressure of about 160 kPa (Reades *et al.*, 1989).

It is noted that Reades *et al.* (1989) report that the diffusion coefficient inferred from the observed diffusion profile through the Keele Valley compacted clay liner of 6.5×10^{-10} m^2/s (0.02 m^2/a) is very close to that of 7×10^{-10} m^2/s (0.022 m^2/a) obtained earlier in laboratory testing. Furthermore, as discussed in Chapter 9, the data would appear to suggest that the 0.3 m thick sand blanket constructed over the liner is acting more like part of the barrier than as a drain, based on the observed change in the upper portion (5–10 cm) of the sand (i.e., the presence of a biologically produced black slime) (King *et al.*, 1993) and the fact that a very good diffusion profile can be traced through the sand and clay liner. In the Keele Valley landfill the sand blanket is primarily intended as protection for the liner, and so clogging of the sand is desirable since it increases the effective thickness of the liner with respect to diffusion. However, the clogging of the sand observed at Keele Valley is also a warning for other projects where one might be considering using a sand blanket as a drainage layer, since it raises concerns regarding the effectiveness of sand blankets as an essential part of a leachate collection system (as discussed in Section 2.4).

(b) Liner with secondary leachate collection/hydraulic control system

There are some situations where the conceptual designs indicated in Figures 1.2–1.4 may not provide sufficient confidence that there will be a negligible effect on groundwater quality. Under these circumstances, an additional level of engineering in the form of a secondary leachate collection system or hydraulic control layer (HCL) may be appropriate as shown in Figures 1.5–1.8.

For situations where the potentiometric surface in the underlying aquifer is at or below the base of the landfill (water level (a) in Figure 1.5), the construction of a permeable drainage system, which is located beneath the compacted clay liner, serves two purposes. First, the drainage layer functions as a secondary leachate collection system which can remove a portion of the leachate which passes through the liner

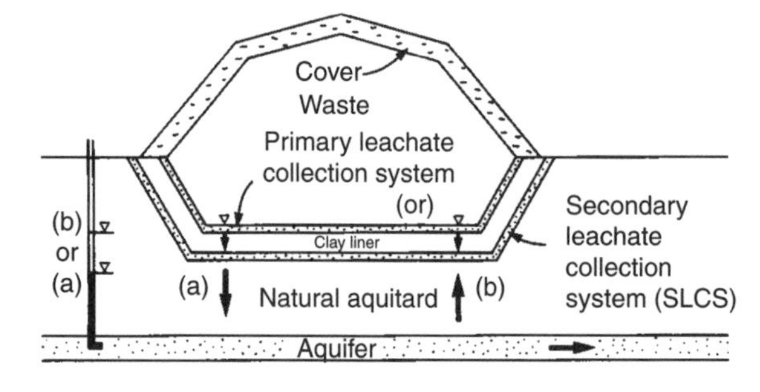

Figure 1.5 A compacted clayey liner used in conjunction with a secondary leachate collection/detection system.

(and some leakage is to be expected through any liner system where there are outward gradients). Second, this layer serves to reduce the hydraulic gradient through the underlying soil for aquifer water levels below the landfill base (case (a) in Figure 1.5) or to induce a hydraulic trap to the secondary system when the potentiometric surface in the aquifer is above the level of the secondary leachate collection system (water level (b) in Figure 1.5).

If the natural soil has a relatively low hydraulic conductivity (e.g., 1×10^{-9} m/s or less), then the permeable layer beneath the primary liner may be used as an HCL, as shown schematically in Figure 1.6. In this concept the permeable layer is saturated and maintained at a pressure above that in the landfill. This creates an inward gradient and an engineered hydraulic trap. The inward advective flow will resist outward diffusion of contaminants. Some fluid from the HCL will also move downward (i.e., there are downward gradients from the HCL for the case shown in Figure 1.6). The volume of fluid required to maintain the hydraulic trap can be reduced by introducing a geomembrane liner (and/or a second clay liner) below the HCL as shown in Figure 1.7.

In environments where the water table is at or near the ground surface, or where the potentiometric surface in an underlying aquifer is near the surface, a design such as that shown in Figure 1.8 may be warranted. In this case, there is both a natural hydraulic trap (i.e., water flows from the natural soil into the HCL) and an engineered hydraulic trap (i.e., water flows from the HCL into the landfill). Where practical, this design has the following advantages. First, since there is inward flow to the HCL and a relatively impermeable clay liner, it may be possible to design the system such that the engineered hydraulic trap is entirely passive. That is, all water required to maintain an inward gradient from the HCL to the landfill is provided by the natural hydrogeologic system and no injection of water to the HCL is required. Second, because of the two-level hydraulic trap, there will be substantially greater attenuation of any contaminants that do migrate through the primary liner. Thus, the impact at the site boundary can be expected to be substantially less than that which would be calculated for the case shown in Figure 1.6. Third, since fluid can be injected and withdrawn from the HCL it is possible to control the concentration of contaminant in the layer, and hence the impact at the boundary, in the event of a major failure of either the liner or primary leachate collection system. The designs shown in Figure 1.7 (water level (a)) and Figure 1.8 have the same advantage.

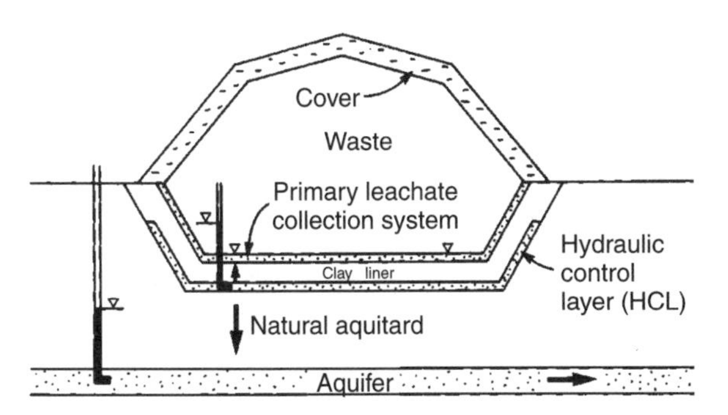

Figure 1.6 A compacted clayey liner used in conjunction with a hydraulic control layer (HCL) and engineered hydraulic trap above the HCL.

The design of all landfills should involve careful consideration of the design life of the engineered components of the system. The conceptual designs shown in Figures 1.6–1.8 provide a high level of redundancy in the event that there is a major failure of the primary leachate collection system followed by an increase in leachate levels to above the potentiometric surface in the underlying aquifer.

When designing hydraulic traps involving an engineered HCL, consideration should be given to the implications of failure of the primary leachate collection system, the effect of the landfill shadow on the long-term performance of the hydraulic trap and the potential for blowout. A more detailed discussion of the design of HCLs is given in Chapter 16.

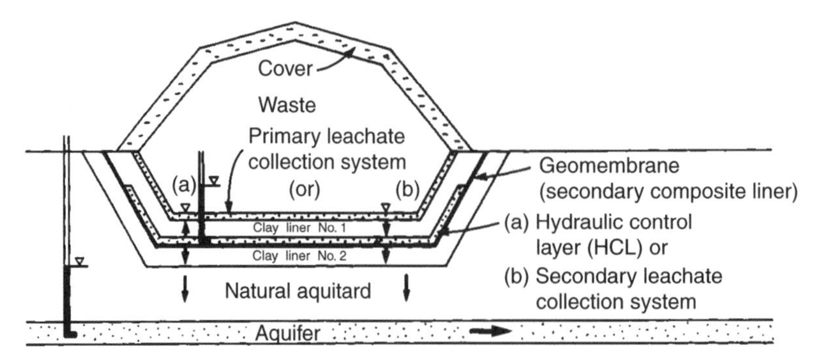

Figure 1.7 A compacted clayey primary liner used in conjunction with an engineered hydraulic control layer and hydraulic trap to minimize contaminant impact together with a composite secondary liner geomembrane (and clayey liner) used to minimize volume of fluid needed to maintain the hydraulic trap. By pumping the hydraulic control layer, this can also be used as a secondary leachate collection system. Note that second compacted clay liners could potentially be replaced by a GCL and foundation layer as discussed in Chapter 16.

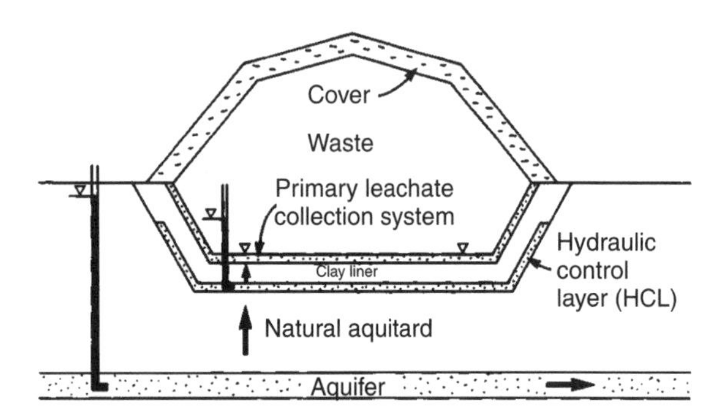

Figure 1.8 A compacted clayey liner used in conjunction with a primary leachate collection system and a hydraulic control layer to create a "natural" hydraulic trap.

1.2.3 Cut-off walls and permeable surrounds

Cut-off walls are most commonly used to limit contaminant migration from existing sites which have not been adequately designed; however, they can also be used in controlling migration from new sites where it may be desirable to isolate the (potentially contaminated) groundwater in a relatively thin and shallow aquifer beneath the landfill. For example, in the case shown schematically in Figure 1.9a, the thickness of the natural clay barrier may not be enough to prevent potential contamination of water flowing along the underlying minor aquifer. By constructing cut-off walls around the site and hence reducing the flow in the aquifer locally, it is possible to change an advection-controlled system beneath the landfill into a diffusion-controlled system thereby substantially reducing the impact on off-site groundwater quality. It is, of course, still necessary to consider diffusive migration through the cut-off wall and into the aquifer. This can be achieved using techniques similar to those which will be discussed for natural or compacted clayey barriers in Chapter 10.

The containment of contaminated land by the construction of a vertical cut-off wall around part or the entire contaminated zone is growing in popularity (Pankow and Cherry, 1996). For example, these walls may be used to control the migration and spreading of the dense non-aqueous phase liquids (DNAPLs) and allow time to implement other remediation technologies. The walls may consist of steel, polyethylene or soil (soil-bentonite or soil-bentonite and cement). However, containment of DNAPL spills creates a situation unlike that in other remediation applications because the dissolved concentrations associated with such pools can potentially be as high as the solubility limit of the DNAPL spilled. This gives rise to a very large concentration gradient that has the potential to have a

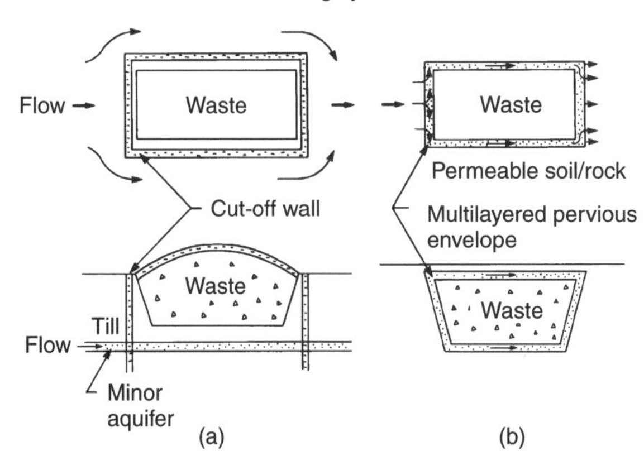

Figure 1.9 (a) A cut-off wall is used to divert groundwater flow from beneath a natural clayey barrier. (b) Pervious material is placed outside the waste to divert groundwater flow around (rather than through) the waste (after Rowe, 1988; reproduced with permission of the *Canadian Geotechnical Journal*).

substantial impact on the surrounding aquifer due to diffusion and it is important to select the appropriate barrier for the contaminated site.

An interesting alternative to construct a low-permeability cut-off wall is the "pervious surround" concept developed by Matich and Tao (1984), which involves minimizing advective transport through a waste pit by surrounding it with a multilayered pervious envelope with less permeable material adjacent to the waste and more permeable material outside of this as shown schematically in Figure 1.9b. In this way, water flow is directed around the outside of the pit rather than through the pit, and contaminant migration would be predominantly by diffusion from the waste through the less permeable material, together with advective–dispersive transport within the more permeable outer zone. Thus, from the standpoint of modelling, determination of contaminant loading of the groundwater for this case is also very similar to that for waste sites separated from an underlying aquifer or drainage system by a clayey barrier as shown in Figure 1.2.

1.2.4 Bedrock

A topic of particular interest in some regions is the migration of contaminants from existing or proposed landfills excavated into, or sitting on top of, fractured rock. Typically, the intact rock has a very low hydraulic conductivity and contaminant migration will primarily involve advective–dispersive transport along the fractures in the rock (see Figure 1.10). In these cases, the primary mechanism limiting the movement of contaminant is the process of matrix diffusion whereby contaminant is removed from the fracture as it diffuses into the matrix of the rock. For example, monitoring of an existing landfill at Burlington, Ontario (Gartner Lee and Associates Ltd, 1986), suggests that after 15 years migration, contaminant movement in fractured shale downgradient of the Burlington landfill is probably not more than 25 m and that substantial attenuation has occurred. The migration of contaminants through fractured porous rock is discussed in Chapter 11.

These days landfills proposed for old worked out quarries in fractured rock will typically have a liner system, where the major challenge is constructing a suitable liner along the side walls of the quarry. However, landfills also have been

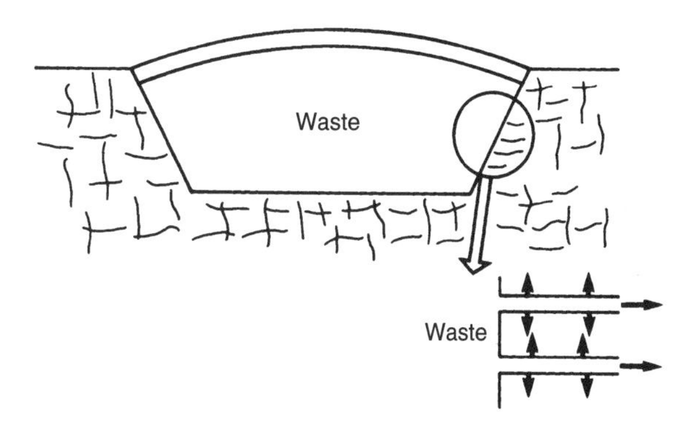

Figure 1.10 Landfill located in fractured shale. Contaminant transport along the fractures is attenuated by diffusion into the matrix of the shale adjacent to the fractures (after Rowe, 1988; reproduced with permission of the *Canadian Geotechnical Journal*).

proposed for fractured rock on the basis of their operation as hydraulic containment landfills. Here the concept is to have a drainage layer around the waste that is pumped to both collect leachate and induce inward flow through the fractures. This type of landfill may or may not have any low-permeability liner. In principle, no liner is required if the collection layer has a service life, and will continue to be operated, for a period greater than or equal to the contaminating lifespan of the landfill. However, these designs need to consider the issues raised in Section 1.2.1(d) and Chapter 16.

1.2.5 Geosynthetics as part of composite liner systems

There are situations where *in-situ* clays or compacted clays can provide a cost-effective and safe primary barrier for landfills. However, there are also situations where the type of waste, landfill size, local hydrogeology and geotechnical conditions are such that the natural soil liner alone is not sufficient to prevent unacceptable contaminant impact at some time in the future. In these cases, an appropriate design involving a geomembrane used in conjunction with either *in-situ* or compacted clay, or a geosynthetic clay liner as a part of a primary (and, if necessary, secondary) composite liner may provide a cost-effective means of gaining the environmental protection required.

Figure 1.11 is a schematic detail showing the use of a 1–2 mm thick geomembrane as the primary liner in a barrier system. This detail could be coupled with the scenario shown in Figure 1.4 (water level (b)) to provide a composite liner. This type of system might be necessary if the clay liner alone could not meet regulatory requirements in terms of contaminant impact. For water table conditions (a) shown in Figure 1.4 or the situations shown in Figures 1.2 and 1.5(a), there is an outward hydraulic gradient from the landfill. The installation of a geomembrane above the clay liner would be expected to reduce substantially the outward movement of contaminant by advective processes; however, some migration would still be expected due to leakage through the geomembrane (primarily due to holes, as discussed in Chapters 5 and 13). Furthermore, diffusion can also occur through the geomembrane (as discussed in Chapters 6, 8, 13 and 16). Consideration should be given to both mechanisms.

Groundwater conditions shown in Figure 1.3 or as implied by water level (b) in Figure 1.4 would pose a somewhat different problem if a geomembrane were proposed as part of the primary liner. Without the geomembrane, the landfill would operate as a hydraulic trap with small quantities of groundwater flowing into the landfill and being collected by a leachate collection system. If a geomembrane is used as part of the primary liner, careful consideration must be

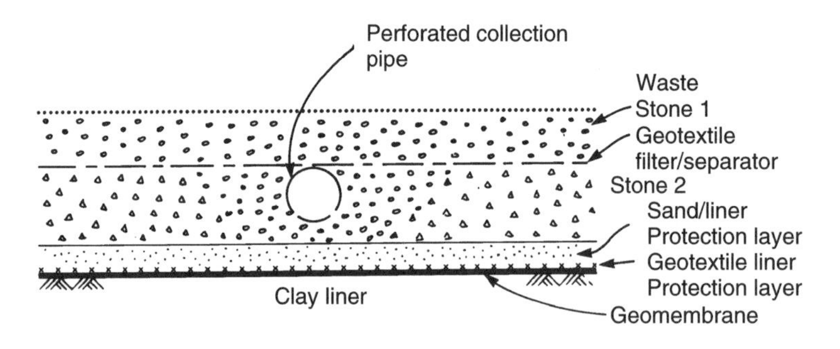

Figure 1.11 Schematic of a primary leachate collection system and geomembrane primary liner.

given to the effect of uplift pressure on the geomembrane and it may be necessary either to install a granular groundwater control layer beneath a liner (as discussed in the next paragraph) or to depressurize the underlying aquifer during construction in order to prevent problems due to uplift. Also, with the geomembrane in place, one cannot assume that there is any inward advective velocity, and contaminant migration by pure diffusion through the geomembrane and soil must be considered.

A design such as that shown in Figure 1.8 could be used to limit pressures on the liner by managing the pumping of the hydraulic (groundwater) control layer. In the limit where there is full drawdown of fluid levels at the pump, the HCL would act as a secondary leachate collection system (like that shown in Figure 1.5b) but would also collect groundwater. This would be beneficial from the perspective of contaminant transport since there would be a hydraulic trap to resist outward movement of any contaminant which did escape through the composite (geomembrane–clay) liner. The disadvantage would be the difficulty in distinguishing the level of leakage through the composite liner from the groundwater and hence monitoring of volumes of fluid would probably not provide a good indication of failure of the primary leachate collection system. Monitoring of the chemistry of fluid collected from the secondary collection system would provide an indication of a failure of the primary system but there is the potential for considerable dilution.

Geomembranes have also found application in double liner systems such as that shown in Figure 1.12. Here the leachate collection system overlies the primary geomembrane liner which, in turn, overlies a leak detection system (secondary leachate collection system) and secondary composite geomembrane and clay liner.

The leachate collection system above the geomembrane is primarily intended to minimize the head drop (driving force) across the geomembrane. Recognizing that it is almost impossible to construct a liner which does not allow some contaminant to escape, the leak detection system (which may be a granular layer (Figure 1.12) and/or a geosynthetic drainage layer (geonet) for some situations (Figure 1.13)) is intended to detect and collect most of the leachate which does escape through the primary geomembrane. A clay liner is often used as a backup in these systems to provide additional containment of any leachate that does escape through the geomembrane. The clayey liner also provides long-term attenuation of contaminant which may still remain after the landfill has been decommissioned and gone beyond the post-closure maintenance period (which is often as little as 30

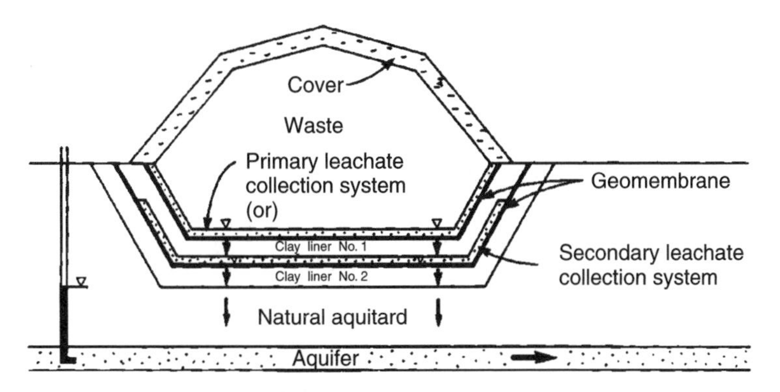

Figure 1.12 Geosynthetic liner system with primary and secondary composite liners each involving a geomembrane and compacted clayey liner. Note that one (or both) compacted clay liners could potentially be replaced by a GCL and foundation/attenuation layer as discussed in Chapter 16.

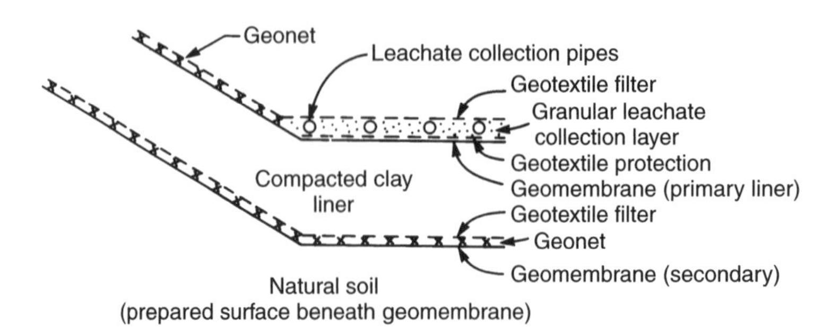

Figure 1.13 Schematic of a primary leachate collection system, composite liner and secondary geosynthetic leachate collection/detection system.

years post-closure but may vary considerably depending on local regulations). This may, in fact, be the most important role of the clayey barrier for systems such as that shown in Figure 1.12, and calculations can be performed to estimate the likely impact following the post-closure maintenance period.

For composite liner systems (e.g., Figure 1.12), particular care is required to prevent dessication of the underlying clay liner (especially on side slopes) due to potentially high temperatures at the underside of the geomembrane while it is exposed (Basnett and Brungard, 1992; Corser *et al.*, 1992; Rowe *et al.*, 2003a). The risk of this dessication can be reduced by covering the geomembrane with some insulating material (e.g., a protection layer and/or the leachate collection system) as quickly as possible. Consideration should also be given to the potential for clay liner dessication due to high temperatures in landfill leachate (e.g., see Collins, 1993). However, it would appear that in a number of practical situations in Canada neither the temperature nor the temperature gradient has been as high as that examined by Collins. Furthermore, the soil tested by Collins was particularly susceptible to dessication cracking. Nevertheless, consideration should be given to the potential for dessication cracking due to landfill-induced temperature effects when designing composite liners (see Döll, 1997; Southen and Rowe, 2002; Zhou and Rowe, 2003).

1.2.6 Significance of the cover

While the primary focus of this book is on the migration of contaminants in the groundwater system, it should be noted that the generation of mobile contaminants is often related to the movement of fluid or gas into the contaminated zone (e.g., waste). For example, the volume of leachate generated in a landfill is directly related to the movement of water through the cover. As will be discussed in subsequent chapters, the leachate concentrations and the contaminating lifespan of a landfill (defined fully in Section 1.9.5) may be related to the infiltration through the cover and this, in turn, may influence the performance of the underlying natural clayey deposit as a barrier.

The generation of contaminants may also be related to the movement of gas through a cover. For example, sulphide-rich mine tailings may pose little problem if kept in a reduced state but may generate considerable acidified leachate if oxygen is allowed to migrate through the cover to the waste. In these instances, the primary design criterion for a cover may be to minimize the diffusion of oxygen, from the atmosphere, through the cover and into the mine tailings (Yanful, 1993; Bonaparte and Yanful, 2001).

1.3 Transport mechanisms and governing equations

Evaluation of the design for a contaminant containment (e.g., waste disposal) facility involves making a quantitative prediction of potential impact of the facility on groundwater quality, keeping in mind that under most circumstances involving contaminant movement through a barrier and into an aquifer, the best one can expect to do is to predict trends and a likely range of concentrations at any given point in space and time.

There are four aspects of any attempt to make quantitative predictions, namely the need to:

1. identify the controlling mechanisms;
2. formulate or select a theoretical model;
3. determine the relevant parameters; and
4. solve the governing equations.

The controlling mechanisms for contaminant transport are discussed in the following subsections together with the development of the governing differential equation. The relevant boundary conditions are discussed in Sections 1.5 and 1.6. The evaluation of relevant parameters is discussed in Chapters 3, 4, 5, 8, 12, 13 and 16. Section 1.7 illustrates some simple "hand solutions" to the governing equations. More elaborate solution techniques are examined in Chapter 7 and their application is illustrated in Chapters 10, 11 and 16.

When dealing with contaminant transport through saturated clayey barriers or the matrix of an intact rock (e.g., shale), the primary transport mechanisms are advection and diffusion. When dealing with transport through aquifers the key transport mechanisms are usually advection and dispersion. In fractured materials, transport is generally controlled by advection and dispersion along the fractures and diffusion from the fractures into the matrix for the adjacent porous media.

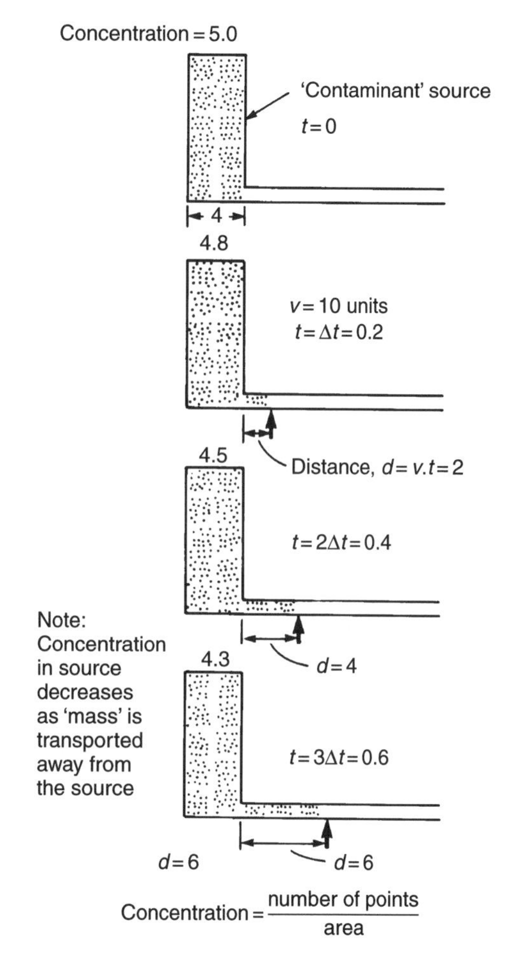

Figure 1.14 Schematic showing advective transport (e.g., people standing on a moving walkway) and a decrease in source concentration with time (e.g., in an airport holding bay) as individuals move out of the holding bay.

1.3.1 Advective transport

Advection involves the movement of contaminant with flowing water. This may be thought of as being analogous to a moving walkway at an airport (see Figure 1.14). If it is assumed that once people step onto the walkway they remain standing, then they will be transported along at the speed of the walkway. In this analogy, the speed of the walkway corresponds to the groundwater (seepage) velocity, v. The movement of contaminants at a speed corresponding to the groundwater velocity is often referred to as plug flow. The time required for a plug of contaminant to move a given distance, d, is equivalent to the distance, d, divided by the groundwater velocity, v (see Figure 1.14). The concentration, c, of people on the walkway is the number of people per unit area and the total number of people being transported past any given checkpoint in a unit of time is defined as the flux.

Thus, when dealing with contaminants in groundwater (rather than people in airports) the mass of contaminant transported by advection per unit area per unit time (i.e., the mass flux, f) is given by

$$f = nvc \tag{1.1a}$$

$$f = v_a c \tag{1.1b}$$

where n is the effective porosity of the soil [–]; v is the groundwater velocity (seepage velocity) [LT^{-1}]; v_a is the Darcy velocity (also called the Darcy flux) $= nv$ [LT^{-1}]; c is the concentration of contaminant at the point and time of interest; and concentration is defined as the mass of contaminant per unit volume of fluid [ML^{-3}]. L, T and M in square brackets indicate the basic dimensions of the quantity, L being length, T being time and M being mass.

Thus, just as the total number of people passing out of the holding bay of a terminal on a moving walkway (see Figure 1.14) is obtained by summing (counting) the number of people passing the checkpoint at the door of the holding bay, so too the total mass transported from a contaminant source into a barrier up to some specific time, t, is obtained by summing (integrating) the mass flux (equation 1.1) with respect to time, τ, viz.

$$m_a = A_0 \int_0^t nvc \, d\tau \tag{1.2}$$

where m_a is the mass of contaminant transported from the landfill by advection [M]; A_0 is the cross-sectional area of the landfill through which contaminant is passing [L^2]; and all other terms are as defined above.

If there were no flow, then there would be no movement of contaminant into the barrier **by advection**. It should be noted that the velocity at which contaminant moves **by advection** is the groundwater velocity. In the analogy, this is the velocity of the moving walkway since people are assumed to remain standing; if people walk on the walkway then they can move still faster and this is analogous to diffusion as discussed in the next section.

1.3.2 Diffusive transport

Diffusion involves the movement of contaminant from points of high chemical potential (concentration) to points of low chemical potential (concentration). In terms of analogy, this corresponds to the desire of people who have been locked up in a hot, crowded, airport holding bay to spread out once the doors of the holding bay are opened. Even if the moving walkway is stopped, people will walk at different rates away from the holding bay with the fastest walkers out in front having the most space around them (i.e., the lowest concentration).

The mass flux transported by diffusion alone can be written as

$$f = -nD_e \frac{\partial c}{\partial z} \tag{1.3}$$

where n is the effective porosity of the soil [–] (as previously defined); D_e is the effective diffusion coefficient [L^2T^{-1}] and $\partial c/\partial z$ is the concentration gradient (i.e., the change in concentration with distance). The negative sign arises from the fact that contaminants move from high to low concentrations and hence the gradient $\partial c/\partial z$ will usually be negative.

The total mass of contaminant, m_d, transported out of a landfill **by diffusion** up to some specific time, t, is obtained by integrating equation 1.3 with respect to time τ, to give

$$m_d = A_0 \int_0^t \left(-nD_e \frac{\partial c}{\partial z}\right) d\tau \tag{1.4}$$

where all terms are as previously defined.

1.3.3 Advective–diffusive transport

As discussed in Section 1.3.1, in the absence of diffusion, contaminant would be transported out of a landfill at the groundwater (seepage) velocity. Again, in the absence of diffusion, there would be no outward contaminant transport through a landfill barrier if the advective flow were into the landfill. However, diffusion cannot be neglected. If the direction of diffusive transport is the same as the direction of advective flow then it will increase the amount of contaminant transport and decrease the time it takes for contaminant to move to a given point away from the source. In terms of analogy, this corresponds to people spreading out from our airport holding bay by walking on the moving walkway. Suppose that the first and most eager person to leave the lounge drops yellow paint on the moving walkway as soon as he/she steps onto it. If the walkway is moving at a velocity v, then after some time Δt the yellow paint will have moved by advection a distance $d = v\Delta t$ (see Figure 1.15). However, if that individual was also walking at a speed v relative to the walkway then in the same time Δt they would have moved a distance $2d$. Thus the location of the first person out front marks the location of the contaminant plume. Other slower walkers will be spread out behind our lead walker. Thus we get a change in concentration (i.e., number of people per unit area) starting at zero ahead of the location of our lead walker, with the concentration increasing as we move back to behind the point where our lead walker first stepped onto the walkway (i.e., where there is yellow paint). Meanwhile, back in the holding bay, there had originally been five people per unit area (see Figure 1.15). The fact that many of them have moved out of the holding bay by stepping onto the walkway has reduced the number of people in the holding bay and so the concentration (number of people per unit area of holding bay) drops from the original value of five people per unit area as time passes and more people leave the lounge.

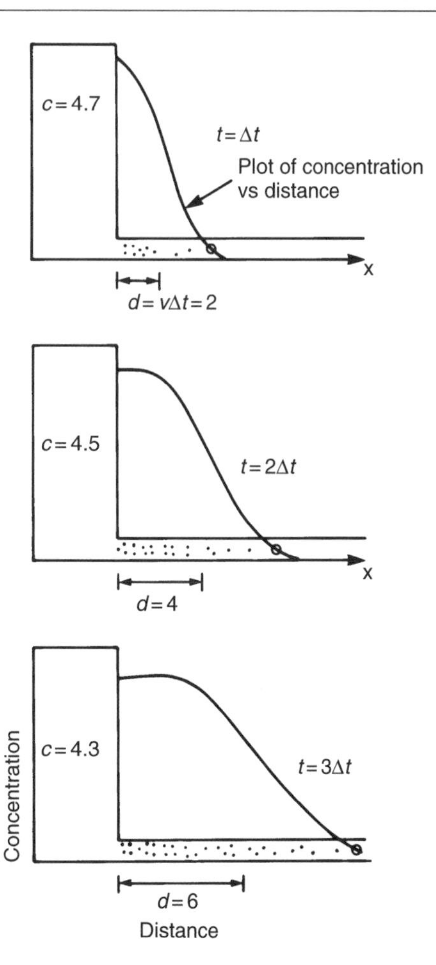

Figure 1.15 Schematic showing advective–diffusive transport (e.g., people walking at different rates on a moving walkway) and a decrease in source concentration with time (e.g., an airport holding bay) as individuals move out of the holding bay.

Diffusion can also occur in the direction opposite to advective transport and so, as will be demonstrated in Chapter 10, it is possible for contaminant to escape from a landfill even though the groundwater flow is directed into the landfill. The level of impact can be calculated and often the landfill can be designed to maintain an acceptable level of impact (which may not be significant). This situation is again analogous to our airport holding bay but where the moving walkway is moving into the holding bay; for people to escape they must walk along the walkway in the opposite direction to the walkway's movement. This is the hydraulic trap discussed in Section 1.2. If the walkway is moving slowly (i.e., low advective velocity into the holding bay/landfill) then it is not difficult for people to walk out of the holding bay. As the inward speed of the walkway is increased it becomes more and more difficult for people to escape and eventually one could set the inward velocity of the walkway so high that even the fastest runner cannot move outward along the inward moving walkway.

For the case of advective–diffusive transport the mass flux, f, is given by

$$f = nvc - nD_e \frac{\partial c}{\partial z} \tag{1.5}$$

and the total mass, m, transported from the landfill up to a specific time, t, is given by

$$m = A_0 \int_0^t \left(nvc - nD_e \frac{\partial c}{\partial z} \right) d\tau \tag{1.6}$$

where the velocity, v, is positive if it is out of the landfill and negative if it is into the landfill and all other terms are as previously defined.

1.3.4 Dispersion

When contaminant migration is associated with relatively high flows (as in many aquifers), there is a third transport mechanism to be considered, namely mechanical dispersion. This process involves mixing that occurs due to local variations in the flow velocity of the groundwater. It too can be illustrated in terms of analogy. Picture a busload of school children wearing bright red uniforms arriving at a crowded fair. In the bus they have a high concentration, i.e., there are a large number of children per unit area; once let out of the bus the children will disperse or spread out in the crowd and before long an aerial view of the fairground would show that the red uniforms were widely distributed in the crowd, not necessarily because the students wanted to be separated but rather because of mixing that takes place when trying to move through a large crowd.

Mechanical dispersion of contaminants involves a mixing and spreading of the contaminant caused by the convoluted pathways that water and contaminants (solutes, molecules, particles) follow while flowing through porous or fractured media (see Domenico and Schwartz, 1998). Mechanical dispersion is mixing caused by local variations in velocity around some mean velocity of flow (average linear groundwater velocity). Thus, it is purely an advective process related to the statistical variability in the grain size distribution in the aquifer and is not driven by a chemical potential like diffusion. The more heterogeneous the aquifer is, the greater the irregularity of the flow paths and the deviation about the mean velocity will be, and the greater the mechanical dispersion. Although this mechanism is totally different from the diffusion process, for most practical purposes, it can be mathematically modelled in the same way and hence the two processes are often lumped together as a composite parameter, D, called the coefficient of hydrodynamic dispersion:

$$D = D_e + D_{md} \tag{1.7}$$

where D_e is the effective (molecular) diffusion coefficient for the contaminant species of interest $[L^2T^{-1}]$ and D_{md} is the coefficient of mechanical dispersion $[L^2T^{-1}]$.

The mass flux is then given by

$$f = nvc - nD \frac{\partial c}{\partial z} \tag{1.8}$$

where D is defined by equations 1.7 and 1.9a (discussed below) and all other terms are as previously defined.

When dealing with transport through intact clayey soil, diffusion will usually control the parameter D and dispersion is negligible

(Gillham and Cherry, 1982; Rowe, 1987; and Section 1.3.8). In aquifers, the opposite tends to be true and dispersion tends to dominate. It is often convenient to model the dispersive process as a linear function of velocity (Bear, 1979; Domenico and Schwartz, 1998):

$$D_{md} = \alpha v \tag{1.9a}$$

where α is dispersivity [L].

The dispersivity, α, tends to be scale-dependent and is not a true material property (see Domenico and Schwartz, 1998). A number of values of dispersivity backfigured for a number of different cases and summarized by Anderson (1979) are given in Table 1.2. This table shows the estimated longitudinal dispersivity α_L in the direction of groundwater flow and the ratio of transverse dispersivity α_T (in a direction perpendicular to the direction of flow) to longitudinal dispersivity. Typically, the longitudinal and transverse directions relate to two horizontal directions (see Figure 1.16). For sand and gravel deposits the transverse dispersivity is commonly less than 30% of the longitudinal value (i.e., $\alpha_T/\alpha_L \leq 0.3$ in many cases) and is often of the order of 10% of the longitudinal value ($\alpha_T/\alpha_L \sim 0.1$) (Domenico and Schwartz, 1998). When there is pronounced horizontal stratification the vertical mechanical dispersivity may be small and the coefficient of hydrodynamic dispersivity may be similar to the diffusion coefficient (Sudicky, 1986).

Gelhar *et al.* (1985) examined field experiments at 55 sites and subsequently Gelhar *et al.* (1992) undertook a critical review of field experiments from 59 sites around the world. Of approximately 106 values for longitudinal dispersivity only 14 values were highly reliable. Based on the data, a number of approximate relationships can be established. A reasonable fit to the available data gives

$$\begin{aligned} \alpha_L &\simeq \frac{x^2}{100}\,\text{m} \qquad x \leq 100\,\text{m} \\ \alpha_L &= 100\,\text{m} \qquad x > 100\,\text{m} \end{aligned} \tag{1.9b}$$

where α_L is the longitudinal dispersivity in metres and x is the scale of the observation in metres.

In experiments with high-quality data (all of which involved scales of 100 m or less), the dispersivity was less than 3 m and typically lay between the dispersivity given by equation 1.9b and that given by equation 1.9c.

Table 1.2 Regional dispersivities (adapted from Anderson, 1979)

Aquifer	*Location*	*Porosity*	*Longitudinal dispersivity* (m)	α_T/α_L
Alluvial	Rocky Mountain	0.30	30.5	1.0
	Colorado	0.20	30.5	0.3
	California	NR	30.5	0.3
	Lyons, France	0.2	12	0.33
	Barstow, CA	0.40	61	0.3
	Sutter Basin, CA	0.5–0.2	80–200	0.1
Glacial	Long Island, NY	0.35	21.3	0.2
Limestone	Brunswick, GA	0.35	61	0.3
Fractured basalt	Idaho	0.10	91	1.5
	Hanford Site, WA	NR	30.5	0.6
Alluvial	Barstow, CA	0.40	61	0.003
	Alsace, France	NR	15	0.067
Glacial till over shale	Alberta, Canada	0.001 and 0.053	3.0 and 6.1	0.2
Limestone	Cutler Area, FL	0.25	6.7	0.1

Note: α_L, longitudinal dispersivity; α_T, transverse dispersivity.

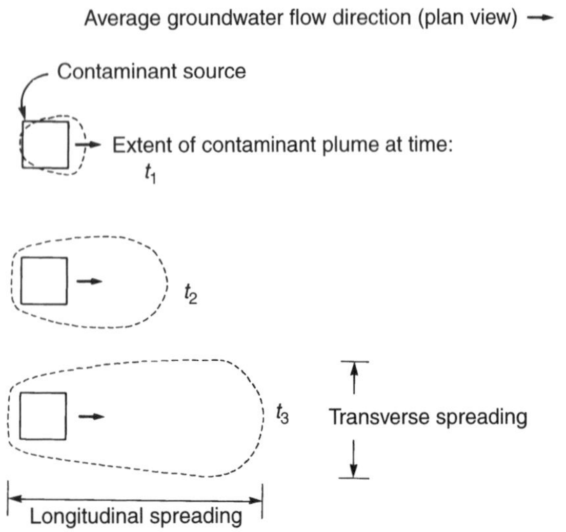

Average groundwater flow direction (plan view)

Contaminant source

Extent of contaminant plume at time:

t_1

t_2

t_3

Transverse spreading

Longitudinal spreading

Figure 1.16 Schematic showing plan view of longitudinal spreading of a contaminant plume due to advection and longitudinal dispersion parallel to the average direction of groundwater flow and transverse spreading perpendicular to groundwater flow due to transverse dispersion.

Looking at all the available data it would also appear that some of the high-quality data defines a typical lower bound which is given by

$$\alpha_L \simeq \frac{x}{100}\,\text{m} \qquad x \leq 1{,}000\,\text{m}$$
$$\alpha_L = 10\,\text{m} \qquad x > 1{,}000\,\text{m} \tag{1.9c}$$

Examining all the available data studied by Gelhar *et al.* (1992), it would appear that a reasonable upper estimate of dispersivity is given by equation 1.9d:

$$\alpha_L \simeq \frac{x^4}{100} \qquad x \leq 10\,\text{m}$$
$$\alpha_L \simeq 200\,\text{m} \qquad x > 10\,\text{m} \tag{1.9d}$$

It should be noted that for scales exceeding 1,000 m a limited number of dispersivities have been reported which are less than that given by equation 1.9c and a limited number exceed that given by equation 1.9d. The range for data for lengths exceeding 1,000 m is 5–5,600 m; none of this is high-quality data.

Any consideration of published dispersivity values (including those used to obtain the correlations presented above) must recognize that many values are unrealistically large due to errors related to the collection and interpretation of field data (Domenico and Schwartz, 1998). Furthermore, many published estimates of dispersivity are no more than numerical model calibration or curve fitting values (Anderson, 1984), and can be subject to a variety of other errors.

Since mechanical dispersion is related to local changes in groundwater velocity, and since this is related to local changes in hydraulic conductivity, it seems reasonable that there should be some relationship between dispersivity and a geostatistical description of the hydraulic conductivities in an aquifer and such relationships have been developed (e.g., Gelhar and Axness, 1983). These approaches may be useful when there are sufficient data to characterize the geostatistics of the hydraulic conductivities meaningfully; frequently, this is not possible.

When dealing with fractured porous media, the apparent dispersion inferred from tracer tests can be affected significantly by the phenomenon of matrix diffusion. Thus when considering migration of contaminants in this medium, the effects of matrix diffusion (to be discussed in Section 1.4.2) and mechanical dispersion (discussed above) should be examined separately. The modelling of matrix diffusion from fractures is discussed in Chapter 7 and illustrated in Chapter 11.

In modelling, the coefficient of hydrodynamic dispersion is typically considered to be a constant; however, in reality, it is likely to be spatially varying (as implied by the dependence of dispersivity on the scale being considered) because of the greater variability in detailed flow conditions that are likely to be encountered over larger distance. Peters and Smith (1999) have shown that there can be significant differences between the predicted migration obtained for constant and spatially varying mechanical dispersion. The potential effect of spatially varying dispersion warrants consideration in situations where failure to consider it could have significant practical implication. As in the case of modelling with constant dispersion, the biggest challenge is establishing the dispersion parameters.

It is evident from the foregoing (and especially by the spread of estimates of α implied by equations 1.9b–d) that there will always be considerable uncertainty regarding dispersivity in granular or fractured media and hence in the design it will usually be necessary to perform sensitivity studies to evaluate the effects of this uncertainty on predicted impact.

1.3.5 Sorption

In the previous sections, various transport mechanisms have been discussed. These represent means by which contaminant species in solution move in the groundwater. Equally important, however, are the mechanisms which remove contaminant from solution. These processes may include cation exchange whereby cations such as K^{+}, Na^{+}, Pb^{2+}, Cd^{2+}, Fe^{2+}, Cu^{2+}, etc. replace other cations (e.g., Ca^{2+}, Mg^{2+}) on the surface of the clay and hence are removed from solution (this is usually accompanied by an increase in the pore fluid concentration of the desorbed cations Ca^{2+}, Mg^{2+}, etc.). Other mechanisms include precipitation of heavy metals (e.g., see Yanful *et al.*, 1988b) and removal of organic contaminants from solution by interaction with solid organic matter in the soil (field examples are given in Chapter 9).

In the simplest case, the sorption processes can be modelled as being linear and reversible and so the mass of contaminant removed from solution, S, is proportional to the concentration in solution, c:

$$S = K_d c \qquad (1.10)$$

where S is the mass of solute removed from solution per unit mass of solid [–]; K_d is the partitioning or distribution coefficient [$M^{-1}L^{3}$] and c is the equilibrium concentration of solute (mass of solute per unit volume of pore fluid) [ML^{-3}].

A plot of the variation in solid-phase concentration, S, versus the solution-phase concentration, c, under equilibrium is called an isopleth or an isotherm, with the former term being preferred here. The case represented by equation 1.10 is a linear isopleth (see Figure 1.17) and is often regarded as a reasonable approximation for low concentrations of contaminant; the assumption that the process giving rise to sorption is reversible is generally safe even if the process is not reversible. At high concentrations sorption is non-linear and more complex relationships between the solid-phase concentration, S, and the solution concentration have been devised. Two that are more commonly used are the Langmuir and Freundlich isopleths.

The Langmuir isopleth is given by

$$S = \frac{S_m bc}{1 + bc} \qquad (1.11a)$$

where S_m is the solid-phase concentration corresponding to all available sorption sites being

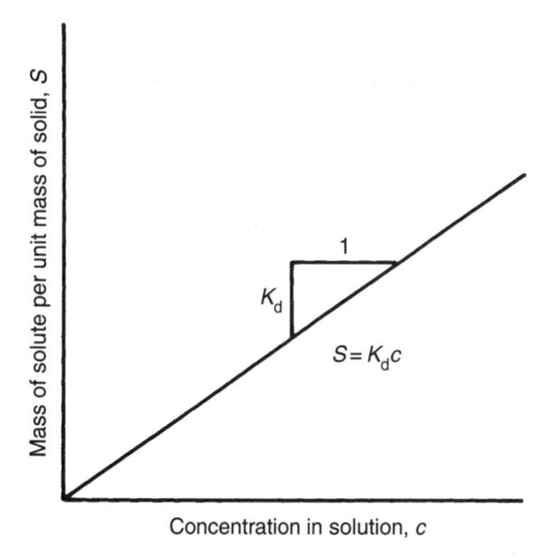

Figure 1.17 Linear isopleth.

occupied; b is a parameter representing the rate of sorption; S is the mass of solute removed from solution (sorbed) per unit mass of solid and c is the equilibrium concentration of solute (mass per unit volume of pore fluid).

The Langmuir isopleth is plotted in Figure 1.18a. The parameters S_m and b are best obtained by performing batch tests in which a known volume of contaminant at a known concentration is mixed with a known volume of dry soil. The solute concentration of species which interact with the soil will decrease to an equilibrium value c. From this test, the equilibrium solid-phase and solute concentrations (S and c) can be deduced for a range of solute concentrations. The parameters S_m and b can then be estimated by plotting the data in the form $1/S$ versus $1/c$ as shown in Figure 1.18b.

The Langmuir isopleth has a good theoretical basis but, unfortunately, is not always adequate for describing sorption processes. The Freundlich isopleth provides an empirically based alternative model which sometimes provides a better quantitative description of sorption. It can be written as

$$S = K_f c^{\varepsilon} \tag{1.12a}$$

where K_f and ε are empirically determined constants and the relationship between S and c is plotted as in Figure 1.19a for $\varepsilon > 1$ and $\varepsilon < 1$. As was the case for the Langmuir parameters S_m

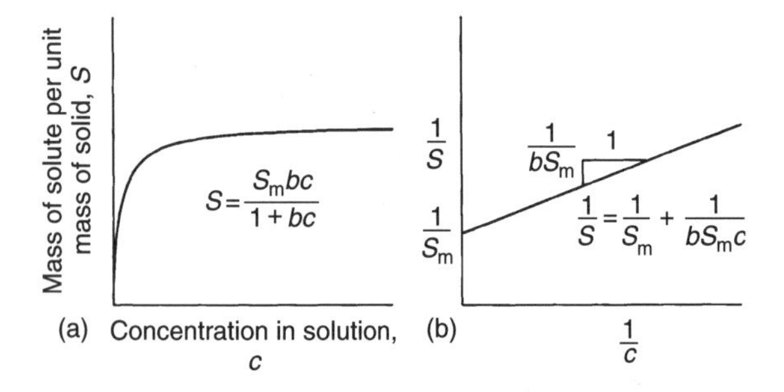

Figure 1.18 Langmuir isopleth: (a) plot showing sorption as a function of concentration and (b) plot used to determine parameters.

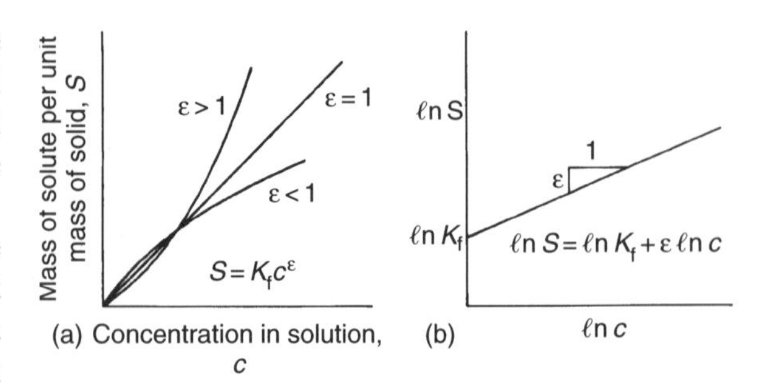

Figure 1.19 Freundlich isopleth: (a) plot showing sorption as a function of concentration and (b) log plot used to determine parameters.

and b, the Freundlich parameters are also best obtained by performing batch tests. Since, in this case, equation 1.12a can be rewritten as

$$\ell n\, S = \ell n\, K_f + \varepsilon\, \ell n\, c \tag{1.12b}$$

it is most convenient to plot $\ell n\, S$ versus $\ell n\, c$ as shown in Figure 1.19b and this readily allows the evaluation of K_f and ε.

At the beginning of this section it was indicated that the linear isopleth is the simplest representation of sorption, and indeed it is. It should not be inferred, however, that this is simply a mathematically convenient representation. Examination of equation 1.11a and Figure 1.18a shows that at low concentrations the Langmuir isopleth is essentially linear. Clearly, also, the Freundlich isopleth reduces to equation 1.10 for $\varepsilon = 1$. The linear isopleth is, in fact, often a good representation for the sorption of contaminants found in leachate from municipal waste disposal sites where the concentrations are often relatively low. Unfortunately this is not always the case. Furthermore, it should be noted that the sorption of a given species is often dependent on the presence of other competing species and hence cannot be determined in isolation.

Analyses of contaminant transport assuming linear sorption (equation 1.10) can readily be performed using analytic, semi-analytic (e.g., finite layer) or numerical (e.g., finite element or finite difference) techniques. Non-linear sorption poses more of a problem for analysis.

(a) Modelling non-linear sorption

Numerical techniques (finite element) can be used in a time marching algorithm in which the concentration at time $t_{i+1} = t_i + \Delta t$ is equal to the concentration, c_i, at the previous time, t_i, plus the change in concentration, Δc, due to contaminant migration over the time interval Δt. With this approach, non-linear sorption can be modelled using tangent approximations to the isotherm over the time period Δt. Thus, in the analysis, the increment in sorption, ΔS, over this period, Δt, is given by

$$\Delta S = \frac{S_{\mathrm{m}} b}{(1 + b c_{\mathrm{r}})^2} \Delta c \qquad (1.11\mathrm{b})$$

for the Langmuir isopleth and

$$\Delta S = (\varepsilon K_{\mathrm{f}} c_{\mathrm{r}}^{\varepsilon - 1}) \Delta c \qquad (1.12\mathrm{c})$$

for the Freundlich isopleth, where ΔS is the increment in mass removed from solution in the time Δt; Δc is the increase in solute concentration in the time Δt; c_{r} is the reference concentration which may be the concentration of solute at time t_i or, if an iterative procedure is adopted, c_{r} may be taken at the average of c_i and c_{i+1}.

Equation 1.11b or 1.12c can be implemented readily in standard finite element contaminant transport codes (essentially the distribution coefficient K_{d} varies with time and, for any time increment, is given by equations 1.11b and 1.12c since $K_{\mathrm{d}} = \Delta S / \Delta c$). However, it is not a trivial matter to obtain accurate results. Since the sorption is dependent on the reference concentration c_{r} and since the procedure is incremental (viz. $c_{i+1} = c_i + \Delta c$), significant errors can accumulate unless a small time step and fine finite element mesh are used to capture the high concentration gradient at small times.

An alternative approach is to adopt semi-analytic formulations (e.g., the finite layer technique) where an iterative approach is adopted to determine an appropriate secant approximation to the isotherm. Using semi-analytic techniques (to be discussed in detail in Chapter 7), the concentration at time t_{i+1} can be calculated directly (i.e., without determining the concentration at time t_i) provided that the non-linear isopleth can be approximated by an appropriate secant distribution coefficient. For a Langmuir isopleth, the secant K_{d} is given by

$$K_{\mathrm{d}} = \frac{S_{\mathrm{m}} b}{1 + b c_*} \qquad (1.11\mathrm{c})$$

For the Freundlich isopleth, the secant K_{d} is given by

$$K_{\mathrm{d}} = K_{\mathrm{f}} c_*^{\varepsilon - 1} \qquad (1.12\mathrm{d})$$

where c_* is the estimated concentration at the point and time of interest.

In order to implement this approach in finite layer codes it is necessary to split the deposit into sublayers and use the following iterative technique to obtain the secant K_{d} for each sublayer:

(a) calculate the concentrations at the top and bottom of each sublayer based on an estimated K_{d} value for each sublayer (an initial value of 0 may be used or a value of some user-specific starting estimate c_{r} may be used);
(b) calculate a new secant value of K_{d} for each sublayer using equation 1.11c or equation 1.12d, where c_* is the average concentration in the sublayer determined from the concentrations at the top and bottom of the sublayer in (a); and
(c) repeat step (a) using the new estimate of K_{d} for each sublayer until the process converges (i.e., the value of K_{d} used in step (a) is essentially the same as the new value of K_{d} determined in step (b)).

With this approach, the accuracy of the solution will depend on the number of sublayers. However, the number of sublayers required is not particularly large (compared with the number of finite elements required in a numerical analysis), provided that they are arranged to provide a reasonable cover in the zone where there is a significant concentration profile. There is no accumulation of error in the approach (i.e., the accuracy at time t_2 is not in any way dependent on the accuracy at time t_1) and one can directly calculate the concentrations at any particular time of interest without determining them at earlier times.

(b) Sorption for variable charged soils

There is a considerable amount of published information related to contaminant sorption on constant charged soils. For these soils, negative permanent charges on mineral particle surface are primarily formed by isomorphic substitution between ions from the crystal lattice and ions from pore solution. For 2:1 clay minerals (e.g., smectite and vermiculite – see Chapter 3), the charge density is constant and can be calculated using well-known models (Singh and Uehara, 1986; Mitchell, 1993). Variable charges are mostly found in Al–Fe–Mn–Ti–Si oxides and hydroxides, formed by the dissociation of H^+ and OH^- ions from the particle surface and the recombination of these surfaces with H^+ ions from pore solution (Zhang and Zhao, 1997). For these soils the surface charges are directly influenced by the pH of the solution. In addition to these pH-dependent charges, other variable charges are generated by the interaction among the soil functional groups and ions from the solid/solution interface (Sposito, 1984, 1989). In contrast to permanent negatively charged soils, where the electrostatic adsorption of cations is the main sorption process (and hence ions like Cl^- are considered conservative since they are not sorbed), in variable charged soils the electrostatic and specific sorption of anions can also play an important role (and hence anions like Cl^- may not be conservative).

The sorption sites in tropical soils are preferentially provided by the mineral clay-size fraction (e.g., kaolinite), organic matter and Fe–Al oxides and hydroxides. For these soils, the influence of pH on the sorption of heavy metals is well documented (e.g., Matos *et al.*, 2000; Sarkar *et al.*, 2000; Sauvé *et al.*, 2000; Gomes *et al.*, 2001), as is the effect of the ion competition for the sorption sites (e.g., Matos *et al.*, 1999, 2000; Fontes *et al.*, 2000; Gomes *et al.*, 2001).

As an example, a study of the sorption of Cd^{2+}, K^+, Cl^- and F^- on residual latosols from the south-eastern part of Brazil by Leite *et al.* (2003) showed that for a single salt solution of $CdCl_2$ at $pH < pH_0$, the Cl^- anion cannot be used as a tracer (non-reactive anion). However, Cl^- was not sorbed for $pH > pH_0$ or for any of the multiple salt solutions examined, suggesting that the sorption of Cl^- is related to the pH conditions and the dominate attraction of F^- for the soil sorption sites when both are present. This is consistent with previous findings (e.g., Ji, 1997) that the sorption of anions decreases for pH above the zero point charge (pH_0). The importance of pH with respect to sorption of F^- has also been noted by others (e.g., in a study of sorption onto kaolinite by Kau *et al.*, 1997).

(c) Summary

Sorption is an important process affecting the attenuation of contaminants migrating through soils. It can be quite complex and depends on both the mineralogy of soil and the detailed chemistry of the permeant (including both the concentration of contaminants and other factors such as pH). However, for many practical situations it can also be dealt with in a reasonably straightforward manner. Thus, unless otherwise noted, it will be assumed in this book that the sorption processes are linear and can be represented in terms of the partitioning or distribution coefficient K_d. The product ρK_d (where ρ is the dry density of the soil) is a dimensionless measure of the amount of sorption which is likely to occur. Notwithstanding the recognition that anions, like Cl^-, can be sorbed under some circumstances (as discussed in Section 1.3.5(b)), unless otherwise noted, it will also be assumed that the chloride ion (Cl^-) is a typical example of a conservative contaminant. Typical examples of contaminants whose migration and impact may be greatly retarded by sorption processes are the heavy metals Pb^{2+}, Cd^{2+}, Fe^{2+}, Cu^{2+} and, in the presence of soil organics, hydrocarbons such as benzene, toluene and halogenated aliphatic compounds such as dichloromethane, etc.

1.3.6 Radioactive and biological decay

Some elements undergo radioactive decay to lighter elements and many organic compounds undergo biological decay into other simpler compounds. The time it takes for the concentration of the particular species to be reduced to half of the original concentration may be referred to as its half-life. For example, phenol is a benchmark chemical for biodegradability studies and there is a large body of information on its time rates of degradation in sewage, soil and freshwater (Howard, 1989).

For substances which undergo first-order decay, the rate of reduction of concentration is proportional to the current concentration so that

$$\frac{\partial c}{\partial t} = -\lambda c \tag{1.13a}$$

where λ is the first-order decay constant $[T^{-1}]$, which has three components due to radioactive decay, biological decay and fluid withdrawal:

$$\lambda = \Gamma_R + \Gamma_B + \Gamma_S$$

Γ_R is the radioactive decay constant $= \ell n\, 2/$(radioactive decay half-life); Γ_B is the biological decay constant $= \ell n\, 2/$(biological decay half-life); and Γ_S is the volume of fluid removed per unit volume of soil per unit time from beneath a landfill (to be discussed below).

This equation has the solution

$$c(t) = c_0\, e^{-\lambda t} \tag{1.13b}$$

where c_0 is the initial concentration $[ML^{-3}]$.

For radioactive decay which is controlled by an element's atomic structure and is essentially independent of environment, there is substantial available data that can be used to estimate the decay constant Γ_R. Since biological decay depends on many factors such as the presence of appropriate bacteria, substrate, temperature, etc., the rate of decay will be specific to a given environment. Much more research is required to allow quantification of the decay, but lack of clear quantification should not mean that the process is overlooked. For example, examination of data from an Ontario landfill, over a period of one decade, suggests that phenol concentrations have degraded with a half-life of less than 1 year. However, because of potential differences from one landfill to another, care is required in extrapolating this experience.

In addition to the decay of naturally occurring organics (such as phenol, benzene, toluene), there is evidence to indicate that synthetic chemicals found in waste (e.g., halogenated aliphatic compounds) can undergo biologically mediated breakdown. Table 1.3 summarizes published half-lives of various halogenated aliphatic compounds in water (at 20 °C) in the absence of bacteria. As indicated by Vogel *et al.* (1987), these processes can be accelerated by microorganisms which mediate (i.e., enhance/promote) substitution reactions (e.g., for monohalogenated or dihalogenated aliphatic compounds such as dichloromethane, chloroethane, 1,1-dichloroethane, 1,2-dichloroethane, etc.). Thus the half-lives given in Table 1.3 should represent upper bounds to half-life in a biologically active environment and there is growing evidence that there can be significant biological breakdown of at least some of these contaminants under anaerobic conditions as discussed in Section 2.3.6.

A decrease in concentration may also occur if fluid is removed from the soil by some mechanism (as discussed below). In this case $\lambda = \Gamma_S$ where Γ_S equals the volume of fluid removed

Table 1.3 Environmental half-lives and products from abiotic hydrolysis or dehydrohalogenation of halogenated aliphatic compounds at 20 °C (adapted from Vogel *et al.*, 1987)

Compound	*Half-life (years)*
Methanes	
Dichloromethane	1.5, 704[a]
Trichloromethane	1.3, 3,500[a]
Tetrachloromethane (carbon tetrachloride)	7,000
Bromomethane	0.10
Dibromomethane	183
Tribromomethane	686
Bromochloromethane	44
Bromodichloromethane	137
Dibromochloromethane	274
Ethanes	
Chloroethane	0.12
1,2-Dichloroethane	50
1,1,1-Trichloroethane	0.5, 2.5
1,1,2-Trichloroethane	170
1,1,1,2-Tetrachloroethane	384
1,1,2,2-Tetrachloroethane	0.8
1,1,2,2,2-Pentachloroethane	0.01
Bromoethane	0.08
1,2-Dibromoethane	2.5
Ethenes	
Trichloroethene	0.9, 2.5
Tetrachloroethene	0.7, 6
Propanes	
1-Bromopropane	0.07
1,2-Dibromopropane	0.88
1,3-Dibromopropane	0.13
1,2-Dibromo-3-chloropropane	35

[a] Very large and questionable range of reported values.

from unit volume of soil per unit time in the region where fluid is being removed from below the landfill. There are a number of possible ways in which this may occur. For example, it may arise in a landfill barrier system because leachate is being withdrawn at a constant rate from the landfill by a collection system (e.g., a secondary leachate collection system or active HCL as discussed in Section 1.2.2(b)). Alternatively, it may occur if there is a horizontal flow in a soil deposit where the transport of contaminant is predominantly vertical. In this case, water entering below the upgradient edge of the overlaying landfill will generally bring negligible contaminant into the aquifer beneath the landfill while on the downgradient edge of the landfill contaminant will be removed from beneath the landfill by the flowing water.

1.3.7 Governing differential equations

As previously noted (e.g., see equation 1.8) the mass flux, f, which is transported per unit area per unit time due to advective–diffusive–dispersive transport can be given by

$$f = nvc - nD\frac{\partial c}{\partial z} \tag{1.8}$$

Thus, by considering conservation of mass within any small region, the change in concentration with time is given by

$$n\frac{\partial c}{\partial t} = -\frac{\partial f}{\partial z} - \rho\frac{\partial S}{\partial t} - n\lambda c \tag{1.14a}$$

which simply says that the increase in contaminant concentration within a small volume is equal to the increase in mass due to advective–diffusive transport ($-\partial f/\partial z$, where the negative sign implies a net increase in mass within the element) minus the decrease in mass due to sorption ($-\partial S/\partial t$) minus the decrease in mass due to first-order decay processes ($-n\lambda c$).

Substituting equation 1.8 for the mass flux f and equation 1.10 for the sorbed concentration S into equation 1.14a then gives

$$n\frac{\partial c}{\partial t} = \left(nD\frac{\partial^2 c}{\partial z^2} - nv\frac{\partial c}{\partial z}\right) - \rho K_{\mathrm{d}}\frac{\partial c}{\partial t} - n\lambda c \tag{1.14b}$$

where n is the effective porosity of the soil [–]; c is the concentration at depth z and time t [$\mathrm{ML^{-3}}$]; $D = D_{\mathrm{e}} + D_{\mathrm{md}}$ is the coefficient of hydrodynamic dispersion [$\mathrm{L^2T^{-1}}$]; v is the average linearized groundwater (seepage) velocity [$\mathrm{LT^{-1}}$]; ρ is the dry density of the soil [$\mathrm{ML^{-3}}$]; K_{d} is the distribution or partitioning coefficient [$\mathrm{M^{-1}L^3}$]; $v_{\mathrm{a}} = nv$ is the Darcy or discharge velocity [$\mathrm{LT^{-1}}$]; $\lambda = \Gamma_{\mathrm{R}} + \Gamma_{\mathrm{B}} + \Gamma_{\mathrm{S}}$ is the first-order decay constant [$\mathrm{T^{-1}}$] defined in Section 1.3.6.

It is worth noting that in some texts equation 1.14b is rearranged in the form

$$(n + \rho K_{\mathrm{d}})\frac{\partial c}{\partial t} = nD\frac{\partial^2 c}{\partial z^2} - nv\frac{\partial c}{\partial z} - n\lambda c \tag{1.14c}$$

Neglecting first-order decay and dividing throughout by $(n + \rho K_{\mathrm{d}})$, this becomes

$$\frac{\partial c}{\partial t} = D_{\mathrm{R}}\frac{\partial^2 c}{\partial z^2} - v_{\mathrm{R}}\frac{\partial c}{\partial z} \tag{1.14d}$$

where $D_{\mathrm{R}} = D/R$; $v_{\mathrm{R}} = v/R$; and $R = 1 + \rho K_{\mathrm{d}}/n$ is the retardation coefficient.

This is a mathematically correct procedure but can lead to severe difficulties if the boundary conditions are flux-controlled (e.g., boundary conditions like those discussed in Sections 1.5 and 1.6). For this reason, the use of parameters D_{R} and v_{R} and equation 1.14d is to be discouraged.

Equation 1.14b describes advective–diffusive–dispersive transport under one-dimensional (1D) conditions. This can be generalized readily to three-dimensional (3D) conditions and rewritten as

$$n\frac{\partial c}{\partial t} = nD_x\frac{\partial^2 c}{\partial x^2} + nD_y\frac{\partial^2 c}{\partial y^2} + nD_z\frac{\partial^2 c}{\partial z^2} - nv_x\frac{\partial c}{\partial x} - nv_y\frac{\partial c}{\partial y} - nv_z\frac{\partial c}{\partial z} - \rho K_{\mathrm{d}}\frac{\partial c}{\partial t} - n\lambda c \tag{1.15}$$

where D_x, D_y, D_z are the coefficients of hydrodynamic dispersion [$\mathrm{L^2T^{-1}}$] in the x, y and z Cartesian directions, respectively, and v_x, v_y, v_z are the components at groundwater velocity [$\mathrm{LT^{-1}}$] in the three Cartesian directions.

1.3.8 Relative importance of different transport mechanisms

A considerable number of laboratory tests have been performed to verify the applicability of the advection–dispersion model (equations 1.7, 1.8 and 1.14). The available data suggest that for the majority of cases the model was quite adequate for practical purposes (e.g., see Fried, 1976). The laboratory tests indicate that at relatively low velocities the dispersion coefficient is equal to the effective diffusion coefficient while at high velocities the dispersion coefficient increases as a linear function of velocity.

Perkins and Johnston (1963) published an empirical relationship which provides some insight as to what constitutes low and high velocities. Based on the results of a number of tests on homogeneous samples, the coefficient of hydrodynamic dispersion, D, was given by

$$D = D_e + 1.75dv \quad (\mathrm{m}^2/\mathrm{a}) \tag{1.16}$$

where d is the mean grain diameter of the soil (m).

The effective (molecular) diffusion coefficient D_e often lies in the range 0.005–0.05 m^2/a. Adopting these two values, equation 1.16 can be plotted to give the variation in the coefficient of hydrodynamic dispersion, D, with velocity as shown in Figure 1.20 for the case where the mean grain size is 200 μm.

Assuming that equation 1.16 is applicable to saturated, homogeneous, unfractured silts, silty clays or clayey soils with hydraulic conductivity (permeability, k) less than 10^{-7} m/s, mechanical dispersion may be neglected for hydraulic gradients less than 1 (i.e., in most such cases). For a saturated homogeneous sand with hydraulic conductivity of 10^{-5} m/s or less (and $d \leq 200$ μm), these results also suggest that diffusion will generally dominate over mechanical dispersion for hydraulic gradients less than 0.01. For coarser sands or where hydraulic conductivities or gradients are higher, mechanical dispersion may be significant.

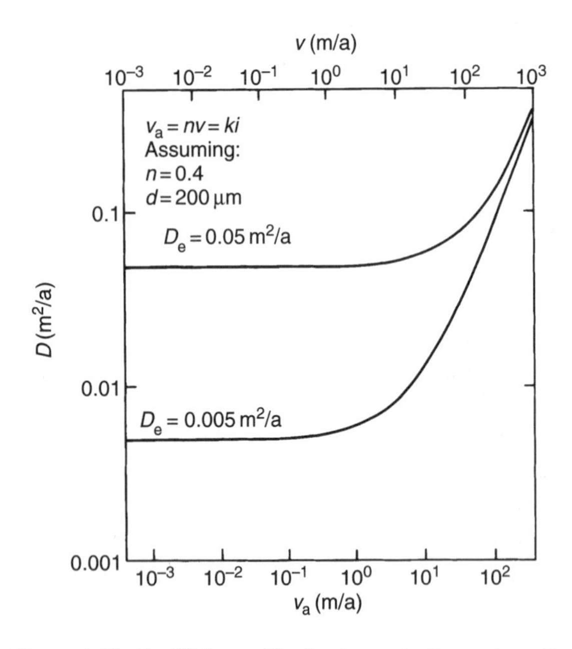

Figure 1.20 Coefficient of hydrodynamic dispersion, D, as a function of Darcy velocity, v_a, and seepage velocity, v, based on data from Perkins and Johnston, 1963 (modified from Rowe, 1987).

It is evident from Figure 1.20 that for $d \leq 200$ μm, the coefficient of hydrodynamic dispersion is quite independent of the groundwater velocity for Darcy velocities v_a less than 10^{-1} m/a. This then raises the question as to how important advective transport is for these velocities (i.e., $v_a < 10^{-1}$ m/a). To provide some answer to this question analyses were performed to determine the peak chemical flux exiting from beneath a 1.2 m thick clay liner ($n = 0.4$) as a function of the advective velocity v_a. For purposes of illustration the diffusion coefficient was taken to be 0.018 m^2/a (see Chapter 8 for a detailed discussion of diffusion coefficients). Unless otherwise noted the leachate concentration was assumed to be constant at $c_0 = 1$ g/l (i.e., $c_0 = 1{,}000$ $\mathrm{g/m}^3$) and the liner was assumed to be totally washed such that the base concentration $c_b = 0$.

Figure 1.21a shows the final (steady-state) variation in concentration with depth beneath the landfill for the case of pure diffusion ($v_a = 0$) and for a Darcy velocity of 0.006 m/a. Notice that the concentrations for $v_a > 0$ are greater than those for $v_a = 0$ throughout the layer. Furthermore, the concentration gradient ($\partial c/\partial z$) at the bottom of the liner for $v_a > 0$ is also greater than that for $v_a = 0$. Thus it follows that for a given soil and contaminant (i.e., given value of D), the mass of contaminant (per unit area per unit time) passing through the barrier and into the underlying aquifer (i.e., the exiting chemical flux or exit flux) will increase with increasing Darcy velocity v_a.

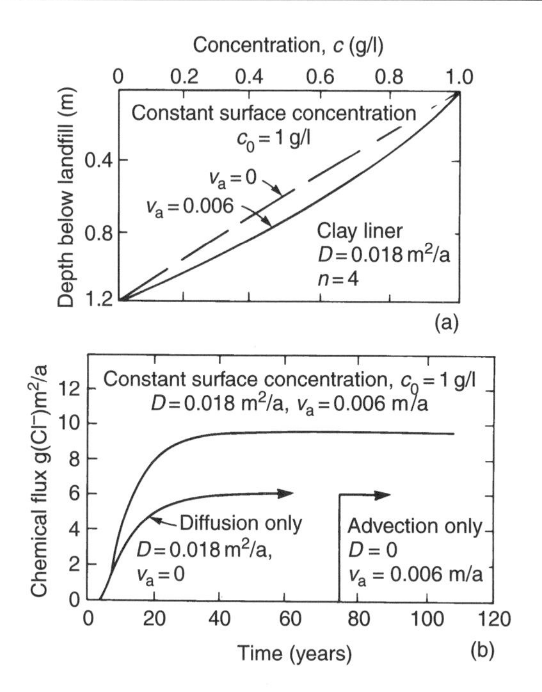

Figure 1.21 Effect of assumptions used to model contaminant migration through a liner: (a) final (steady-state) concentration profile through the liner and (b) chemical flux passing out of the liner (exit flux) assuming constant contaminant concentration above the liner.

It may be tempting to estimate the peak flux loading, f, on the aquifer by performing two simple hand calculations, viz.

$$f = -nD\frac{\partial c}{\partial z} = nD\frac{c_0}{H} \quad (\text{g/m}^2/\text{a}) \qquad \text{(assuming } v_a << nD) \tag{1.17a}$$

$$f = nvc_0 = v_a c_0 \quad (\text{g/m}^2/\text{a}) \qquad \text{(assuming } v_a >> nD) \tag{1.17b}$$

where H is the thickness of the liner (m) and c_0 is the constant leachate concentration ($1\,\text{g/l} = 1000\,\text{g/m}^3$). However, for cases where neither v_a nor D dominates, this approach can give misleading results. For example, Figure 1.21b shows the variation in exit flux with time for the case of pure diffusion ($D = 0.018\,\text{m}^2/\text{a}$, $v_a = 0$), pure advection ($D = 0$, $v_a = 0.006\,\text{m/a}$) and advective–diffusive transport ($D = 0.018\,\text{m}^2/\text{a}$, $v_a = 0.006\,\text{m/a}$). Coincidentally, in this example the maximum flux of $6\,\text{g/m}^2/\text{a}$ is identical for both the pure diffusion and pure advection cases. Conventional calculations performed neglecting diffusion and assuming plug flow ($D = 0$, $v_a = 0.006\,\text{m/a}$) suggest that no contaminant would escape into the aquifer until the seepage front arrived at the base after 75 years. However, diffusion is important and due to diffusion alone an exit flux exceeding 10% of the maximum flux would be expected after only 5 years. Indeed, the maximum flux of $6\,\text{g/m}^2/\text{a}$ would be attained after only 50 years (compared to 75 years for plug flow). Consideration of both advection and diffusion gives a substantially higher flux at any time, with the peak flux being 55% higher than the peak flux obtained by considering either diffusion or advection independently.

In many practical situations involving clay liners, the hydraulic conductivity will be less than 10^{-9} m/s and the hydraulic gradient less than 0.2. These cases will involve Darcy velocities $v_a = 0.006\,\text{m/a}$ or less.

By performing calculations such as those for which the results are shown in Figure 1.21 for a range of velocities v_a, it is possible to establish when advection dominates over diffusion (and vice versa), while from equation 1.16 it is also possible to estimate the range of velocities over which diffusion dominates over mechanical dispersion (and vice versa) and the results are summarized in Figure 1.22. At velocities likely to be encountered with a well-constructed composite liner (geomembrane plus compacted clay) diffusion is likely to dominate. For velocities often encountered with outward gradients and a good compacted clay liner ($k \sim 1 \times 10^{-10}\,\text{m/s}$) diffusion and advection may both be of significance but not mechanical dispersion. On the other hand, in sand and gravel aquifers, advection will dominate over diffusion and dispersion may dominate over diffusion.

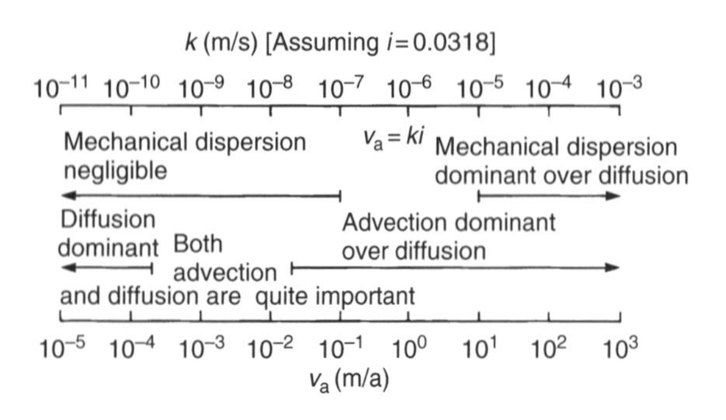

Figure 1.22 Range of Darcy velocities over which diffusion or mechanical dispersion may control the coefficient of hydrodynamic dispersion and the range of velocities over which diffusion or advection may control the magnitude of the exit flux for a 1.2 m thick liner (modified from Rowe, 1987).

1.4 Complicating factors

1.4.1 Unsaturated soils

The advective–diffusive movement of contaminants through unsaturated soils is somewhat more complicated than through saturated soils. The partial differential equation governing 1D movement is given by

$$\frac{\partial}{\partial t}(\theta c) = \frac{\partial}{\partial z}\left(\theta D \frac{\partial c}{\partial z}\right) - \frac{\partial}{\partial z}(v_a c) - \theta \lambda c \quad (1.18)$$

where θ is the volumetric water content (equal to the porosity for a saturated soil) [–]; λ is a term which takes account of sorption and biological, chemical and radioactive decay [T^{-1}]; and all other terms are as previously defined.

This equation bears a marked similarity to equation 1.14b, but this similarity may be deceptive. For an unsaturated soil the volumetric water content, θ, the coefficient of hydrodynamic dispersion, D, and the Darcy velocity may vary in both space and time.

The movement of contaminant through unsaturated soils is a very complex phenomenon as demonstrated by a number of laboratory and field studies (e.g., de Smedt, 1981; Gerhardt, 1984 and others). The simplest case is that in which there is negligible advective transport through the unsaturated soils. This situation can only arise when the net infiltration is negligible (e.g., below an intact composite liner). Under these circumstances, the migration of contaminant in solution will be very slow since the migration will be purely by diffusion and it has been shown (e.g., Klute and Letey, 1958; Porter *et al.*, 1960; Badv and Rowe, 1996; Rowe and Badv, 1996a,b) that the effective diffusion coefficient in unsaturated soils may (at least in some circumstances) be substantially lower than in similar saturated soils.

In humid climates the unsaturated soil will usually be hydraulically active and advective transport (which may vary with time) must be considered. As noted above, the diffusion coefficient is dependent on the volumetric water content and hence will vary both spatially and temporally in a hydraulically active region. The advective transport will depend, in part, on the hydraulic conductivity of the soil. This tends to increase with the volumetric water content of the soil up to a maximum value for a saturated soil (e.g., Gardner, 1958; Van Genuchten, 1978; Fredlund *et al.*, 2001). Thus, the hydraulic conductivity of an unsaturated soil will be far more sensitive to point-to-point variations in grain size distribution than that of saturated soils and this alone makes the determination of representative hydraulic conductivities substantially more difficult. Additional uncertainty arises from the effects of seasonal variations in infiltration and assumptions concerning the expected long-term weather pattern which may influence calculated contaminant transport through unsaturated soils.

Various investigators have questioned the direct application of equation 1.18 for unsaturated soils (e.g., Gaudet *et al.*, 1977; de Smedt, 1981). The problem tends to be manifest as an apparent dispersion well in excess of what would be expected for a saturated soil. In an attempt to explain this phenomenon various researchers (e.g., Rao *et al.*, 1974; Gaudet *et al.*, 1977;

de Smedt, 1981) have proposed multiple water phase models which involve advective–dispersive transport through the mobile water (typically in the smaller saturated pores) together with diffusion of contaminant into (or from) the immobile water (generally in the larger unsaturated pores). This approach appears to give reasonable agreement with experiments, although the parameters used are generally selected by matching the experimental and theoretical behaviours. de Smedt (1981) has shown that a reasonable fit to his experimental data could also be obtained using equation 1.18 and an effective dispersion coefficient D given by

$$D = \frac{\theta_m}{\theta} D_m + \frac{\theta_{im}^2 v^2}{\theta\theta_m \beta} \tag{1.19}$$

where D_m is the dispersion coefficient in the mobile water [L^2T^{-1}]; θ_{im}, θ_m, θ are the volumetric water contents in the immobile and mobile phases, and the bulk soil, respectively [–]; v is the seepage velocity (v_a/θ) [LT^{-1}]; and β is a coefficient for solute transport between phases [T^{-1}].

An inspection of equation 1.19 indicates the difficulties of using this approach in practice since the parameters β, v, θ_m, θ_{im}, θ may be expected to vary both temporally and spatially. Furthermore, the parameter β must be determined by curve fitting laboratory results for a particular situation.

It may be concluded that even though significant progress has been made concerning the prediction of contaminant transport through unsaturated soils, this is still a formidable undertaking. Diffusion and dispersion of dissolved contaminants in unsaturated barrier systems is discussed further in Section 8.3.5. Diffusion of gases is discussed in Sections 8.11 and 12.7.

1.4.2 Fractured porous media

When contaminants move through fractured porous media (e.g., fractured clay, shale, sandstone) the movement of contaminants along the fractures is typically by advection and dispersion. However, as contaminants move along these fractures, there is usually a difference between the concentration of a given species in the fracture at a given point and the concentration in the pore water of the adjacent intact material. This concentration difference (gradient) gives rise to diffusion of contaminants between the fluid in the fractures and the pore water in the matrix of the adjacent material (the matrix pore water). For example, if the concentration of contaminant in the fractures is higher than that in the matrix pore water then contaminant will diffuse into the matrix thereby reducing the concentration in the fracture. This phenomenon is referred to as matrix diffusion, and can have a significant effect on the movement of contaminants through fractured media as discussed in Chapter 11.

The 1D movement of contaminants in a fractured system consisting of a set of parallel fractures (more complex cases are considered in Chapter 7) can be written as

$$n_f \frac{\partial c_f}{\partial t} = n_f D \frac{\partial^2 c_f}{\partial z^2} - v_a \frac{\partial c_f}{\partial z} - \Delta k_f \frac{\partial c_f}{\partial t} - n_f \lambda c_f - q \tag{1.20}$$

where n_f is the fracture porosity [–]; c_f is the concentration in a fracture at depth z and time t [ML^{-3}]; D is the coefficient of hydrodynamic dispersion in the fractures [L^2T^{-1}]; v_a is the Darcy velocity [LT^{-1}]; Δ is the surface area of the fracture per unit volume of the fractured medium [L^{-1}]; K_f is the fracture distribution coefficient for sorption onto the fracture surface (defined by Freeze and Cherry, 1979) as the mass of solute adsorbed per unit area of surface divided by the concentration of solute in solution [L]; λ is the first-order decay constant [T^{-1}] and q is the rate at which contaminant is being transported into the matrix (per unit volume) by diffusion from the fractures [$ML^{-3}T^{-1}$].

The rate of contaminant transport into the matrix depends on the porosity of the intact material, the diffusion coefficient of the matrix material and fracture spacing. It is evaluated for 1D, 2D and 3D conditions in Chapter 7 and the effect of varying these parameters is examined in Chapter 11.

1.4.3 Consolidation and contaminant transport

Transient flow may be expected in clay liners due to consolidation of clay under the weight of the waste in the early stages of a landfill operation and due to a build-up in leachate head with either failure of the leachate collection system or termination of its operation later in the life of the landfill. Smith (2000) has developed equations that account for the effects of consolidation on the transport of chemicals through compacted clayey liners. Analytical solutions are presented for a quasi-steady-state problem (with time-dependent porosity due to consolidation). The results of this analysis suggest that consideration of consolidation has only a small effect on concentration distribution and mass flux. Smith's conclusions are consistent with the findings of Rowe and Nadarajah (1995) that the change in flow due to consolidation is within the typical zone of uncertainty regarding hydraulic conductivity and in any event the change in contaminant transport and hence impact on a receptor aquifer was not significant for the cases examined. It should, however, be noted that consolidation may shorten the time for early arrival of contaminants, especially organic contaminants, that can diffuse through the geomembrane component of composite liners underlain by a leak detection system.

Rowe and Nadarajah (1995) also examined the effect of transient flow during the development of a leachate mound and concluded that, to sufficient accuracy, this could be modelled as a series of steady-state flows in the transport analysis without the need for a full coupled consolidation analysis.

1.4.4 Thermal gradients

Recently there has been a growing interest in investigating the effects of thermal gradients on contaminant transport (e.g., Thomas and Missoum, 1999). This may be of significance with respect to radioactive waste containment but, to date, has not been seen as requiring attention in the modelling of most waste disposal facilities where the temperature on the base of the landfill remains relatively low and consistent. The liner temperature in modern landfills with an operating leachate collection system and no significant leachate mounding will depend on location but is typically in the range of 10–30 °C based on published data (e.g., around 15 °C in Toronto, Canada (Barone *et al.*, 2000) and 18–23 °C in Pennsylvania, 20–30 °C in Florida and 10–30 °C in California, USA (Koerner *et al.*, 1996a)). If a significant leachate mound develops then the temperature on the liner can increase significantly (Barone *et al.*, 2000 – see Section 2.6) and this can increase both advective and diffusive contaminant transport in a clay liner below the landfill (Rowe, 1998a, 2001). In the event that there is a composite liner there is also the potential for desiccation cracking of the clay below the geomembrane.

Under non-isothermal conditions, the temperature gradient and capillary pressure gradient are two driving forces for moisture transport in clay liner systems. When a temperature gradient is applied on an unsaturated landfill liner system, it causes changes in water and air pressures in the medium. Liquid water moves from higher capillary pressure towards lower capillary pressure and vapour water moves from higher temperature area towards lower temperature area due to vapour diffusion. In the unsaturated system, air moves from higher air pressure area to lower air pressure area. Air flow can increase or decrease vapour transport due to advection. The combined effect of liquid water, vapour water and air flow is a redistribution of water in the liner system; this redistribution has the potential to cause desiccation of the clay liner in the area of higher temperature.

The potential for desiccation of clay liners below geomembranes in composite landfill liners has been examined by Döll (1996, 1997), Heibrock (1997), Southen and Rowe (2002) and Zhou and Rowe (2003). To date there is no evidence that this is in fact a major problem however it is the subject of ongoing studies.

1.5 Modelling the finite mass of contaminant

In many practical situations, the mass of contaminant within a landfill (in the field) or a source reservoir (in the laboratory) is limited and mass will be reduced as contaminant is transported into the soil. In terms of the analogy shown in Figures 1.14 and 1.15, the number of people in the airport holding lounge (after disembarkation from their plane) is limited. As people move out of the lounge, the number of people remaining in the lounge drops and hence the concentration (i.e., number of people per unit area). The only way that the concentration can increase again is if another plane arrives and more people (mass) are added to the holding lounge. While this may happen in an airport holding lounge it does not normally happen in landfills after the landfill is closed.

Experience has shown that the concentration of potential contaminants generally increases during operation of the disposal facility, reaches a peak after closure and then declines. The increase in concentration may be related to:

1. the physical processes of leaching of contaminant from solid waste as water infiltrates through the waste; and/or
2. chemical and biological processes which generate the chemical species of interest from the synthesis, or breakdown, of existing chemical species in the waste (e.g., due to biological action).

Likewise, the decrease in concentration with time can be related to:

1. the physical process of removal of contaminant, in the form of leachate, from the landfill; and/or
2. chemical and biochemical processes which result in precipitation and/or the synthesis or breakdown of the chemical species of interest into other chemical forms.

In the design of barrier systems it is generally not practical to model the details of the leaching processes or of any associated chemical or biological processes. However, reasonable engineering approximations can be made which will allow the designer to obtain some insight into the potential impact of the finite mass of contaminant. Thus, for the purposes of performing design calculations, it is often conservative to assume that:

1. the concentration of a contaminant of interest reaches the peak concentration, c_0, instantaneously; and
2. all of the mass of this contaminant species, m_{TC}, is in solution at the time that the peak concentration occurs.

The mass of contaminant available for transport into the soil can be represented in terms of the peak concentration, c_0, and the reference height of leachate, H_r, or the equivalent height of leachate, H_f. In the case of laboratory diffusion tests H_r and H_f are identical and correspond to the actual height of source fluid (leachate) directly above the soil. In the case of the landfill H_r may be defined for each contaminant species of interest and corresponds to the volume of fluid (per unit area of landfill) that, at a concentration c_0, would contain the total mass, m_0, of that contaminant species which could be released either for transport or collection. It does not include contaminant that is, and is expected always to be, in a solid immobile form or contaminant that is released in the gas phase. The equivalent height of leachate, H_f, has a similar definition except that it only corresponds to that portion of the mass that is available for transport into the hydrogeologic system. Thus the essential difference between H_r and H_f is that H_r includes the mass collected by the leachate collection system while H_f excludes this mass.

Considering conservation of mass within the source solution, one can write:

$$\begin{bmatrix}\text{Mass of contaminant}\\ \text{within source}\\ \text{at time } t\end{bmatrix} = \begin{bmatrix}\text{Initial mass of}\\ \text{contaminant}\\ \text{within source}\end{bmatrix} - \begin{bmatrix}\text{Mass of}\\ \text{contaminant}\\ \text{transported}\\ \text{into the soil}\end{bmatrix} - \begin{bmatrix}\text{Mass of contaminant}\\ \text{lost due to first-}\\ \text{order decay processes}\end{bmatrix}$$

$$m_t = m_{Tc} - m - m_{dc} \tag{1.21}$$

Substituting equation for the mass of contaminant transported into the soil gives

$$m_t = m_{Tc} - A_0 \int_0^t \left(nvc(\tau) - nD\frac{\partial c(\tau)}{\partial Z} \right) d\tau - m_{dc} \tag{1.22}$$

where $m_t = A_0 H_r c(t)$ is the mass of contaminant in source at time t [M]; $m_{Tc} = A_0 H_r c_0$ is the initial mass of contaminant in the landfill [M]; $m_{dc} = A_0 H_r \int_0^t \lambda_T c(\tau)\, d\tau$ is the mass lost due to first-order decay processes [M]; $c(t)$ is the concentration in the landfill at time t [ML^{-3}]; A_0 is the area of landfill through which contaminant can pass into the soil [L^2]; $\lambda_T = \Gamma_R + \Gamma_{BT} + \Gamma_{ST}$ is the first-order decay constant [T^{-1}]; Γ_R is the decay constant for radioactive decay within the source [T^{-1}]; Γ_{BT} is the decay constant for biological decay within the source [T^{-1}]; Γ_{ST} is the volume of leachate withdrawn from unit volume of the source during unit time (e.g., from the leachate collection system) [T^{-1}]; and H_r is the reference height of leachate [L].

Thus, if f_T is the flux entering the surface of the deposit and c_T is the concentration at the surface, then

$$c_T(t) = c_0 - \frac{1}{H_r}\int_0^t f_T(\tau)d\tau - \int_0^t \lambda_T c_T(\tau)d\tau \tag{1.23a}$$

If $\lambda_T = 0$ and $H_r \rightarrow \infty$, equation 1.23a leads to the boundary condition $c_t(t) = c_0$.

Equation 1.23a explicitly models both the full mass of a given contaminant in the landfill ($m_{Tc} = A_0 H_r c_0$) and removal of contaminant from the landfill (e.g., by a leachate collection system) in terms of Γ_{ST}. As discussed in more detail in Chapter 10, an alternative approach is to reduce the total mass to the amount available for transport into the groundwater ($m_0 = A_0 H_f c_0$) (i.e., excluding that portion of the mass which will be collected by a primary leachate collection system). For this case the finite mass boundary condition can be written in terms of the equivalent height of leachate, H_f, as indicated in Figure 1.23. This gives rise to the boundary condition

$$c_T(t) = c_0 - \frac{1}{H_f}\int_0^t f_T(\tau)d\tau \tag{1.23b}$$

where H_f is the equivalent height of leachate [L].

For the case where $\lambda_T = 0$ (no mass removed from landfill except by migration into the underlying barrier system or groundwater) $H_r = H_f$ and equations 1.23a and 1.23b are identical.

1.6 Modelling a thin permeable layer as a boundary condition

In many situations (e.g., see Figures 1.2–1.8) it may be desirable to be able to model advective–diffusive transport through a relatively thin clayey barrier and along an aquifer of thickness

$$\begin{bmatrix}\text{Mass of contaminant} \\ \text{in the landfill at time } t \\ m_t = A_0 H_f c_T(t)\end{bmatrix} = \begin{bmatrix}\text{Initial Mass} \\ \text{of contaminant} \\ m_0 = A_0 H_f c_0\end{bmatrix} - \begin{bmatrix}\text{Mass which has} \\ \text{passed into the} \\ \text{soil up to time } t \\ A_0 = \int_0^t f_T(c,\tau)d\tau\end{bmatrix}$$

$$\therefore m_t = m_0 - A_0 \int_0^t f_T(c,\tau)d\tau$$

and

$$A_0 H_f\, c_T(t) = A_0 H_f c_0 - A_0 \int_0^t f_T(c,\tau)d\tau$$

where A_0 is the area of landfill through which contaminant can pass into the soil;
H_f is the equivalent height of leachate;
$f_T(c,\tau)$ is the surface flux (mass per unit area per unit time) passing into the soil.

Thus dividing by the equivalent volume of leachate ($A_0 H_f$) gives

$$c_T(t) = c_0 - \frac{1}{H_f}\int_0^t f_T(c,\tau)d\tau$$

Figure 1.23 Finite mass boundary condition to simulate the landfill on top of the liner.

h and porosity n_b (see Figure 1.24). If there is (say) advective transport through the barrier and into the aquifer then, strictly speaking, continuity requires that the velocity in the aquifer should vary with horizontal position below the barrier. However, in many practical situations where interest is focused on impact at the down-gradient edge of the landfill or a boundary compliance point close to the edge of the landfill the aquifer velocity, v_b, may be assumed to be uniform at a value equal to that at the down-gradient edge of the landfill. If it is also assumed that the concentration in the aquifer is uniform across its thickness, then consideration of

$$\begin{bmatrix}\text{Mass in base beneath} \\ \text{the landfill at the time } t\end{bmatrix} = \begin{bmatrix}\text{Mass transported into} \\ \text{the base up to time } t\end{bmatrix} - \begin{bmatrix}\text{Mass transported out from beneath} \\ \text{the base up to time } t\end{bmatrix}$$

$$\therefore m(t) = \int_0^t WLf_b(c,\tau)d\tau - \int_0^t Wh\, v_b c_b d\tau$$

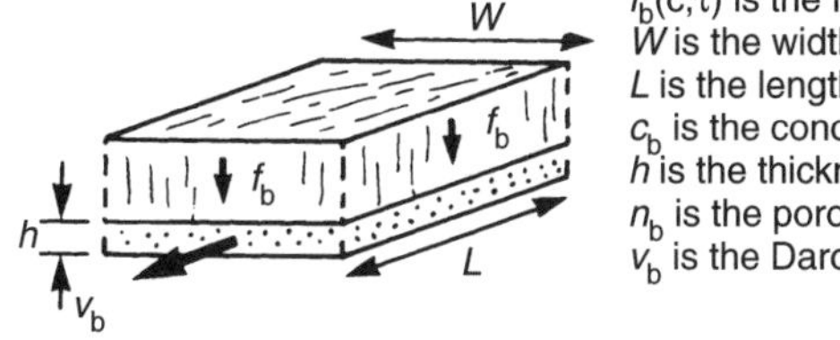

$f_b(c,\tau)$ is the flux into the base ($z = H$) at time τ;
W is the width of the landfill;
L is the length of the landfill;
c_b is the concentration in the base at time τ ($= m(\tau)/(WLhn_b)$);
h is the thickness of permeable layer;
n_b is the porosity of the permeable layer;
v_b is the Darcy velocity in permeable layer.

Dividing by the volume of fluid in the base ($WLhn_b$),

$$c_b(t) = \int_0^t \frac{f_b(c,\tau)}{n_b h} d\tau - \int_0^t \frac{v_b c_b}{n_b L} d\tau$$

Figure 1.24 Boundary conditions for modelling a thin aquifer beneath a liner.

conservation of mass within a small element of the aquifer between $(x, x + dx)$ gives:

$$
\begin{aligned}
n_b h \, dx \, c_b(x,t) = {} & dx \int_0^t f_b(x, c, \tau) \, d\tau \\
& -h \int_0^t [f_x(x + dx, c, \tau) \\
& -f_x(x, c, \tau)] \, d\tau \\
& - n_b h \, dx \int_0^t \lambda_b^* c_b(x, \tau) \, d\tau
\end{aligned}
\tag{1.24}
$$

which follows from the observation that the mass of contaminant in the element at a specific time t is equal to the total mass transported into the element from the clayey barrier minus the net mass transported out of the element in the x-direction and the mass loss due to first-order decay. Assuming that the mass flux, f_x, into the aquifer is governed by equation 1.8, dividing throughout by the pore volume, $n_b h \, dx$, and taking limits as dx tends to zero gives:

$$
\begin{aligned}
c_b(x,t) = {} & \int_0^t \left[\frac{f_b(x, c, \tau)}{n_b h} - \frac{v_b}{n_b} \frac{\partial c_b(x, \tau)}{\partial x} \right. \\
& \left. + D_H \frac{\partial^2 c_b(x, \tau)}{\partial x^2} \right] d\tau \\
& - \int_0^t \lambda_b^* c_b(x, \tau) \, d\tau
\end{aligned}
\tag{1.25}
$$

where n_b is the porosity of the aquifer [–]; h is the thickness of the aquifer [L]; v_b is the Darcy velocity in the aquifer (in the x-direction) [LT^{-1}]; D_H is the coefficient of hydrodynamic dispersion in the aquifer (x-direction) [L^2T^{-1}]; $c_b(x,\tau)$ is the concentration in the aquifer at position x at time τ [ML^{-3}]; $f_b(x,\tau)$ is the flux into the aquifer at position x at time τ [$ML^{-2}T^{-1}$]; λ_b^* is the first-order decay constant for the base and includes the effect of radioactive decay Γ_R and biological decay Γ_{Bb} but does not include Γ_{Sb}, the effect of "washing out" due to horizontal flow since this is explicitly considered by another term in the equation [T^{-1}].

For two-dimensional (2D) problems an aquifer can be modelled as a boundary condition by invoking equation 1.25. As will be discussed in Chapters 7 and 10, there are many practical situations where additional simplification is possible by considering 1D advective–diffusive transport down to the aquifer (e.g., vertically) combined with 1D advective (e.g., horizontal) transport out from beneath the landfill. Thus, for a landfill of width W and length L (where L is the dimension in the direction of groundwater flow in the underlying aquifer), as shown in Figure 1.24, the equation for concentration in the aquifer (equation 1.25) reduces to:

$$
c_b(t) = \int_0^t \left(\frac{f_b(c_b, \tau)}{n_b h} - \lambda_b c_b(\tau) \right) d\tau \tag{1.26}
$$

where $\lambda_b = \lambda_b^* + \Gamma_{sb} = \Gamma_R + \Gamma_{Bb} + v_b / n_b L$.

1.7 Hand solutions to some simple problems

Analytic solutions can be obtained for simplified cases typically involving a single homogeneous layer (barrier) subject to simplified boundary conditions. Numerous solutions have been reported in the literature (e.g., Lapidus and Amundson, 1952; Ogata and Banks, 1961; Lindstrom *et al.*, 1967; Ogata, 1970; Selim and Mansell, 1976; Rowe and Booker, 1985a; Booker and Rowe, 1987; and others). Although these analytic solutions have sometimes been expressed in graphical form suitable for hand computation (e.g., Ogata, 1970; Booker and Rowe, 1987), the evaluation of the analytic solution generally requires the use of a computer. The primary uses of analytic solutions are to:

1. perform quick sensitivity studies and preliminary design calculations and
2. check the results of more sophisticated analyses.

1.7.1 Ogata–Banks equation

Probably the best known analytic solution is that for the concentration c at time t and depth z beneath the surface of a barrier which is assumed to be infinitely deep and subject to a constant surface concentration c_0 (Ogata, 1970):

$$c(z, 0) = 0, \quad z > 0 \tag{1.27a}$$

$$c(0, t) = c_0, \quad t \geq 0 \tag{1.27b}$$

$$c(\infty, t) = 0, \quad t \geq 0 \tag{1.27c}$$

This solution can be written in the form

$$\frac{c}{c_0} = \frac{1}{2}\left\{\operatorname{erfc}\left(\frac{z - vt}{2\sqrt{Dt}}\right) + \exp\left(\frac{vz}{D}\right)\operatorname{erfc}\left(\frac{z + vt}{2\sqrt{Dt}}\right)\right\} \tag{1.28}$$

where $\operatorname{erfc}(x) = 1 - \operatorname{erf}(x)$. For contaminant species which are retarded due to sorption and have a retardation coefficient $R = 1 + \rho K_d/n$, equation 1.28 may be used by scaling the time of interest τ such that $t = \tau/R$. Thus when one assumes a constant source concentration, sorption simply slows the rate of advance of the contaminant plume.

Equation 1.28 may be readily programmed or may be solved graphically using Figure 1.25 as illustrated in Figures 1.26 and 1.27.

The arrows in Figure 1.25, located by the calculation in Figure 1.26, are for a non-retarded species with $v_a = 0.002\,\text{m/a}$ ($v_s = 0.005\,\text{m/a}$) and $D = 0.02\,\text{m}^2/\text{a}$, at a depth of 2 m the concentration would have increased to one-third of the source value (i.e., 500 mg/l in this case) after 80 years migration.

Figure 1.27 shows an example calculation assuming $v_a = 0.002\,\text{m/a}$, $D = 0.01\,\text{m}^2/\text{a}$, $\rho K_d = 1.2$. In particular, it is noted that a modest level of sorption as implied by $\rho K_d = 1.2$ gives rise to a retardation coefficient $R = 4$ and this reduced the effective time used in the calculation by a factor of 4 (i.e., from $\tau = 160$ years to $t = 40$ years). Hence, for example, the impact at $z = 2\,\text{m}$ after 160 years is 75 mg/l (i.e., 5% of the initial concentration).

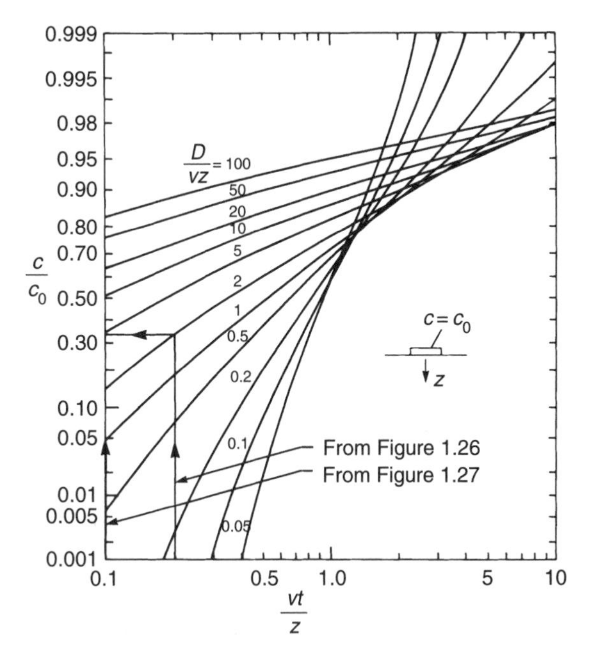

Figure 1.25 Ogata and Banks graphical solution to the advection–dispersion equation for an infinitely deep deposit with a constant surface concentration (modified from Ogata and Banks, 1961).

Assume:	Downward Darcy velocity	$v_a = 0.002\,\text{m/a}$
	Diffusion coefficient	$D = 0.01\,\text{m}^2/\text{a}$
	Porosity	$n = 0.4$
	No sorption	$\rho K_d = 0$
	Source concentration	$c_0 = 1500\,\text{mg/l}$
	Time of interest	$t = 80$ years
	Depth of interest	$z = 2\,\text{m}$

Deduce: $v = v_a / n = 0.002 / 0.4 = 0.005\,\text{m/a}$

$$\frac{D}{vz} = \frac{0.02}{0.005 \times 2} = 2$$

$$\frac{vt}{z} = \frac{0.005 \times 80}{2} = 0.2$$

From graphical solution: $c/c_0 = 0.33$

$\therefore\ c = 0.33 \times 1500 = 500\,\text{mg/l}$

Figure 1.26 Example problem A1 using the Ogata and Banks solution for a conservative contaminant: $H_f = \infty$.

Assume: Downward Darcy velocity $v_a = 0.002$ m/a
Diffusion coefficient $D = 0.01\ \mathrm{m^2/a}$
Porosity $n = 0.4$
Sorption $\rho K_d = 1.2$
Source concentration $c_0 = 1500$ mg/l
Time of interest $\tau = 160$ years
Depth of interest $z = 2$ m

Deduce: $v = v_a/n = 0.005$ m/a

$$R = 1 + \rho K_d/n = 1 + \frac{1.2}{0.4} = 4$$

$$\frac{D}{vz} = \frac{0.01}{0.005 \times 2} = 1$$

$$t = \frac{\tau}{R} = \frac{160}{4} = 40$$

$$\frac{vt}{z} = \frac{0.005 \times 40}{2} = 0.1$$

From graphical solution: $c/c_0 \approx 0.05$
$c = 0.05 \times 1500 = 75$ mg/l
(A more rigorous analysis gives $c = 61$ mg/l)

Figure 1.27 Example problem B1 using the Ogata and Banks solution for a reactive contaminant: $H_f = \infty$.

It should be noted that in the example shown in Figure 1.27 the value of $D = 0.01\ \mathrm{m^2/a}$ is half that used in Figure 1.25. If the same value of D had been used (i.e., $D = 0.02\ \mathrm{m^2/a}$) this would give $D/vz = 2$ and hence $c/c_0 = 0.15$ or $c \simeq 225$ mg/l.

1.7.2 Booker–Rowe equation

In Section 1.5 it was indicated that the finite mass of contaminant within a landfill could be represented in terms of an equivalent height of leachate H_f. An analytic solution for 1D migration in an infinitely deep deposit where the source concentration varies with time (as mass is transported into the barrier) has been given by Booker and Rowe (1987) and the concentration at any depth z and time t can be written as:

$$c(z,t) = \frac{c_0 \exp(ab - b^2 t)(bf(b,t) - df(d,t))}{b - d} \tag{1.29a}$$

where

$$f(b,t) = \exp(ab + b^2 t)\mathrm{erfc}\left(\frac{a}{2\sqrt{t}} + b\sqrt{t}\right) \tag{1.29b}$$

$$f(d,t) = \exp(ad + d^2 t)\mathrm{erfc}\left(\frac{a}{2\sqrt{t}} + d\sqrt{t}\right) \tag{1.29c}$$

$$a = z\left[\frac{n + \rho K_d}{nD}\right]^{1/2}$$

$$b = v\left[\frac{n}{4D(n + \rho K_d)}\right]^{1/2}$$

$$d = \frac{nD}{H_f}\left[\frac{n + \rho K_d}{nD}\right]^{1/2} - b$$

and all other terms are as previously defined. The function $f(p, t)$ (for $p = b$ or $p = d$) may be evaluated directly by computer or using a hand calculator by observing that:

$$f(p,t) = \exp\left(-\frac{a^2}{4t}\right)\phi(x)$$

where $x = (p\sqrt{t} + 0.5\,a/\sqrt{t})$ and the function $\phi(x)$ is given in Figure 1.28. With some algebra, equation 1.29 reduces to equation 1.28 for $H_f \to \infty$ (i.e., $d = -b$).

Figures 1.29 and 1.30 illustrate the use of the graphical method. The cases analyzed are the same as those examined in Figures 1.26 and 1.27, respectively, except that the equivalent height of leachate, H_f, is taken to be 1 m in Figures 1.29 and 1.30 compared to $H_f = \infty$ in Figures 1.26 and 1.27. Comparing the results obtained for Figures 1.26 and 1.29 (or Figures 1.27 and 1.30) it is seen

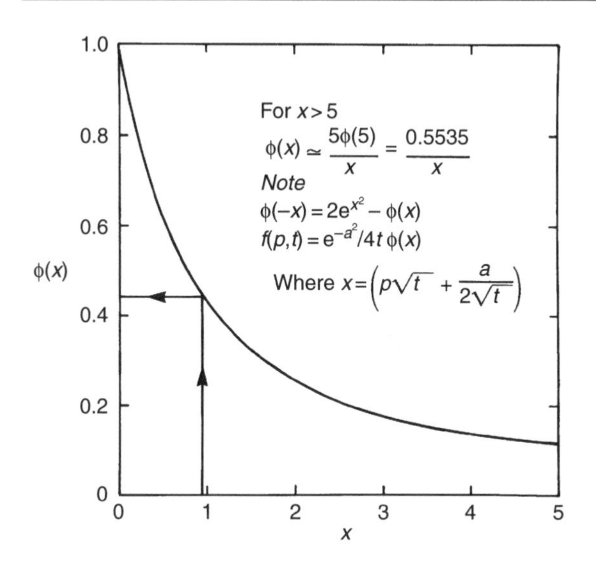

Figure 1.28 Booker and Rowe graphical solution to the advection–dispersion equation for an infinitely deep deposit with a finite mass of contaminant in the source (after Booker and Rowe, 1987; reproduced with permission of John Wiley & Sons Ltd).

that when one considers the finite mass of contaminant, the concentration at the point and time

Deduce: $v = 0.002/0.4 = 0.005$ m/a

$$a = z\left(\frac{n + \rho K_d}{nD}\right)^{1/2} = 2\left(\frac{0.4 + 0}{0.4 \times 0.02}\right)^{1/2} = 14.1$$

$$b = v\left(\frac{n}{4D(n + \rho K_d)}\right)^{1/2} = 0.005\left(\frac{0.4}{4 \times 0.02(0.4 + 0)}\right)^{1/2} = 0.0177$$

$$d = \frac{nD}{H_f}\left(\frac{n + \rho K_d}{nD}\right)^{1/2} - b = \frac{0.4 \times 0.002}{1}\left(\frac{0.4 + 0}{0.4 \times 0.02}\right)^{1/2} - 0.0177$$

$$= 0.039$$

$f(b,t) = \exp(-a^2/4t)\ \phi(bt^{1/2} + 0.5a/t^{1/2})$
$= 0.537\ \phi(0.946)$

From graphical solution: $\phi(0.946) = 0.44$
$\therefore f(b,t) = 0.537 \times 0.44 = 0.236$

$f(d,t) = \exp(-a^2/4t)\ \phi(dt^{1/2} + 0.5a/t^{1/2})$
$= 0.537\ \phi(1.137)$

From graphical solution: $\phi(1.137) = 0.39$
$\therefore f(d,t) = 0.537 \times 0.39 = 0.209$

$c = c_0 \exp(ab - b^2t)\ [bf(b,t) - df(d,t)]/(b - d)$
$= 1500 \exp(0.22)\ [0.0177 \times 0.236 - 0.039 \times 0.209]/(0.0177 - 0.039)$
$= 350$ mg/l

(cf. $c = 500$ mg/l if $H_f = \infty$)

Figure 1.29 Example problem A2 using the Booker and Rowe solution for a conservative contaminant: $H_f = 1$ m.

Deduce: $v = 0.002/0.4 = 0.005$ m/a

$$a = z\left(\frac{n + \rho K_d}{nD}\right)^{1/2} = 2\left(\frac{0.4 + 1.2}{0.4 \times 0.01}\right)^{1/2} = 40$$

$$b = v\left(\frac{n}{4D(n + \rho K_d)}\right)^{1/2} = 0.005\left(\frac{0.4}{4 \times 0.01(0.4 + 1.2)}\right)^{1/2} = 0.0125$$

$$d = \frac{nD}{H_f}\left(\frac{n + \rho K_d}{nD}\right)^{1/2} - b = \frac{0.4 \times 0.001}{1}\left(\frac{0.4 + 1.2}{0.4 \times 0.02}\right)^{1/2} - 0.0125$$

$$= 0.675$$

$f(b,t) = \exp(-a^2/4t)\ \phi(bt^{1/2} + 0.5a/t^{1/2})$
$= 0.082\ \phi(1.74)$

From graphical solution: $\phi(1.74) = 0.28$
$\therefore f(b,t) = 0.082 \times 0.28 = 0.023$

$f(d,t) = \exp(-a^2/4t)\ \phi(dt^{1/2} + 0.5a/t^{1/2})$
$= 0.082\ \phi(2.43)$

From graphical solution: $\phi(2.43) \approx 0.2$
$\therefore f(d,t) = 0.082 \times 0.2 = 0.0164$

$c = c_0 \exp(ab - b^2t)\ [bf(b,t) - df(d,t)]/(b - d)$
$= 1500 \times 0.024$
$= 36$ mg/l

(cf. $c = 60$ mg/l if $H = \infty$)

Figure 1.30 Example problem B2 using the Booker and Rowe solution for a reactive contaminant: $H_f = 1$ m.

of interest is substantially reduced. This matter will be discussed in greater detail in Chapter 10.

1.8 Design considerations

Attitudes regarding the design of municipal waste landfills and other disposal facilities for industrial and mining waste have changed significantly over the last couple of decades. It is not so long ago that waste was simply dumped in old rock quarries or gravel pits and it was assumed that the impact would be reduced to acceptable levels due to dilution of the leachate or chemical waste by the groundwater. In response to the environmental problems that this approach has caused, many governments are now regulating the design of waste disposal facilities. In the US, for example, the Congress has mandated regulations relating to the minimum design for waste disposal sites which involves a single composite liner and a landfill cover no more permeable than the liner as the standard design. Alternative designs may be permitted in approved states. In general, interest is only focused on a 30-year period after closure of the landfill. While this approach is much better than uncontrolled disposal of waste, it does not necessarily ensure that "good" designs are implemented and that a site which meets EPA requirements will not be an environmental hazard for the future.

In the Province of Ontario, Canada, the proponent of a disposal site is required to demonstrate that the proposed facility will not have any impact on the present or future "reasonable use" of groundwater at the site boundary (MoEE, 1993a), and the default designs in the Ontario Landfill Regulations (MoE, 1998) were developed on this basis. In essence, this recognizes that over the long term it is almost impossible to guarantee that there will be no impact on groundwater due to a waste disposal facility. While it is impossible to prove that any facility will never cause an environmental problem, it is possible to design a facility which, based on what we know today, should not impact on the reasonable use of groundwater and hence on the environment. Similarly, it is also possible to identify designs which, based on what we know today, have a reasonable likelihood of contaminating groundwater and the environment beyond the site. It may be that it will be tens or even hundreds of years before there is a serious impact; however, a long time to impact does not justify a design in the context of modern attitudes towards protecting the environment not only for ourselves but also for future generations. To meet the objective of providing long-term environmental protection, a barrier systems design must be evaluated in the context of a contaminant impact assessment (see Section 1.9) that considers both the contaminating lifespan of a landfill (Section 1.9) and the service life of the engineered components of the system (Section 1.10).

Modern containment systems can be quite complex and involve many components. For example, Figure 1.31 shows schematic of waste disposal facility with a double composite liner and a wide variety of applications for geosynthetics. Here the primary leachate collection system (discussed in detail in Chapter 2) on the base of the landfill consists of perforated leachate high-density polyethylene (HDPE) collection pipes (geopipes) in a gravel drainage blanket. This is separated from the waste by a geotextile filter and from the underlying primary liner by a geotextile protection layer. On the side slopes, a geonet drainage layer is used. The primary liner consists of an HDPE geomembrane (Chapter 13) and a geosynthetic clay liner (GCL; Chapter 12). A geonet is used as a leak detection/secondary leachate collection system between the primary and secondary liners. The secondary composite liner involves an HDPE geomembrane over a compacted clay liner (Chapter 3). The schematic shows the use of geosynthetic reinforcement to allow steepening of the side slopes. Each component of this system has its own design considerations and then there is the matter of the interactions between the components.

As important as base liners are another aspect of barriers is the cover (cap) placed over the waste. Figure 1.31 shows one possible configuration of a low-permeability cover (intended to limit percolation of fluid into the waste), involving extensive use of geosynthetics for erosion control, reinforcement, drainage, and a composite liner. Alternatively, covers may be designed with the objective of encouraging percolation into the waste (to accelerate stabilization of the waste). Other covers are designed with the primary objective of limiting gas (e.g., oxygen) advective–diffusive migration into the waste. Cover design is discussed in Chapter 14.

Under design conditions prior to termination of operation of a leachate collection system and/or failure of any geomembrane liner, the primary transport mechanism through modern liners is usually chemical diffusion. In particular, certain organic contaminants can diffuse through geomembranes, GCLs and compacted clay liners, and sufficient provision must be made for attenuation of these contaminants (e.g., by sorption or precipitation) before they reach any underlying aquifer.

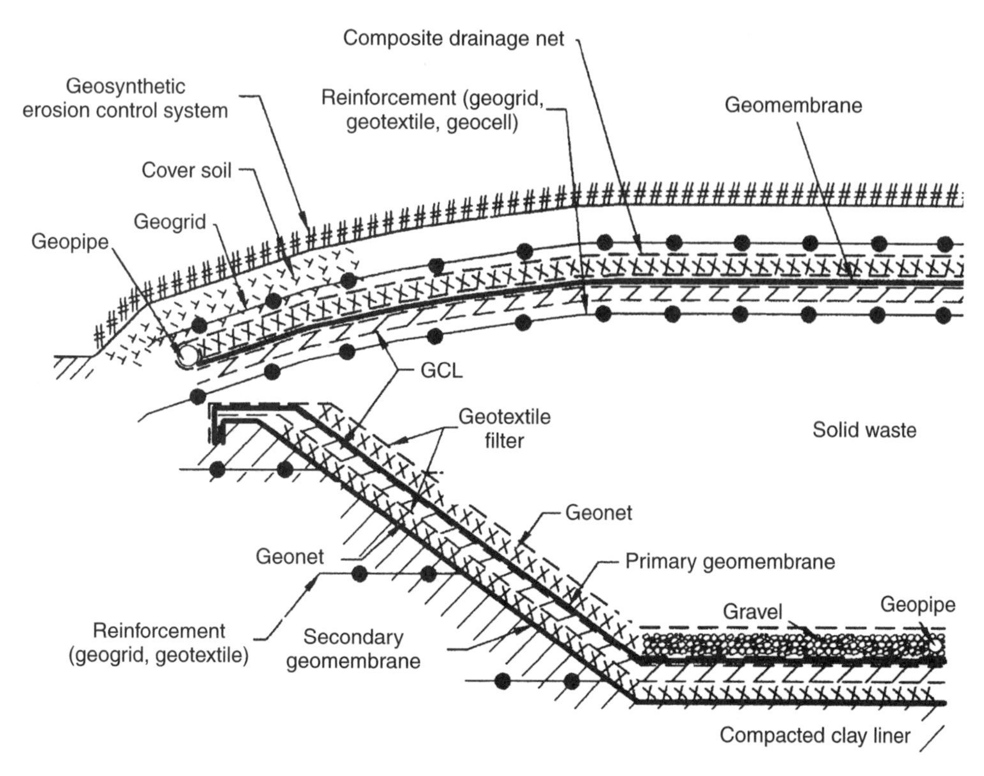

Figure 1.31 Multiple uses of geosynthetics in landfill design (modified from Zornberg and Christopher, 1999).

Leachate collection systems can experience clogging that can result in a substantial leachate mound on the base of a landfill. French drains and sand drainage blankets are particularly prone to clogging. The effect of clogging can be minimized and the service life of collection systems can be extended by appropriate design. Techniques for estimating the service life of granular drainage blankets in leachate collection systems are discussed in Chapter 2.

Leachate mounding appears to give rise to an increase in temperature on the underlying liner system. This has the potential to increase advective–diffusive contaminant transport and decrease the service life of some engineered component of the barrier systems (Chapters 12 and 13).

Leakage through geomembranes may be more than conventionally expected due to holes and wrinkles in the geomembranes. Data relating to the number of holes and hole size are presented in Chapter 13, and equations that may be used to estimate leakage are discussed in Chapter 5.

Diffusion through compacted clay liners, geosynthetic clay liners and geomembranes is discussed and typical parameters are given in Chapters 8, 12 and 13, respectively.

The service lives of compacted clay liners, geosynthetic clay liners and geomembranes are important considerations and are discussed in Chapters 3, 12 and 13.

While the focus of the design of barrier systems is on geoenvironmental issues, it is important not to overlook the geotechnical issues (see Chapter 15). Although there have been numerous landfills successfully constructed, there have also been a number of geotechnical failures that have included:

1. slides of the leachate collection layer;
2. sliding of waste and liner along a failure plane associated with liner construction;

3. slides associated with fluid pressures in landfills (e.g., due to leachate recirculation);
4. general shear failures associated with expansion of existing landfills;
5. general shear failures due to inadequate geotechnical stability assessment;
6. basal fracturing due to excessive water and gas pressures arising from an underlying aquifer.

Issues that need consideration include the geosynthetic (GS)–clay interface properties, the water content of clay near GS–CCL interface, the potential for a decrease in design interface strength during construction (e.g., due to rain during placement of GM), the selection of appropriate strength parameters (e.g., peak strength may only be mobilized over portion on failure surface), excess pore pressures developed in waste (e.g., due to recirculation of leachate or co-disposal of liquids), the effects of excavation at toe of exiting waste pile on stability, the risks associated with placing of waste above approved contours without checking stability, the selection of an appropriate waste density for stability calculations (e.g., neglecting the increase in density that occurs as water content increases has contributed to failures) and the effects of excavation on basal stability.

1.9 Impact assessment

The design of a barrier system is often intimately related to the environmental impact assessment of the proposed landfill. Environmental impact assessments are generally driven by regulatory requirements and as a consequence typical "acceptable" barrier systems can be expected to vary regionally as a result of variations in both hydrogeologic conditions and regulatory requirements. The fundamental question underlying most impact assessments is whether the proposed landfill will have no more than a negligible effect on groundwater quality at the site boundary. However, this perfectly reasonable question raises two subsidiary questions – what is a "negligible effect" and over what period of time must the effect be negligible? The answer to the latter question has very important implications since it is intimately related to the design life of the engineering features of the facility. The design of a barrier system for a landfill that is only required to have "negligible" impact for a 30-year post-closure period is likely to be different from the design for a 100-year period, which, in turn, may be quite different from that required to have negligible effect on groundwater quality in perpetuity.

Typically, environmental regulations fall into one of the following categories:

1. essentially no regulation;
2. prescriptive regulations which specify minimum requirements such as "two liners of which at least one is a synthetic liner";
3. regulations requiring "no impact" or "negligible impact" for a prescribed period of time (e.g., 30 years or 100 years post-closure); and
4. regulations requiring negligible impact in perpetuity.

The legal implications of different regulatory systems have been discussed by Estrin and Rowe (1995, 1997).

1.9.1 Non-existent regulations

The situation where there is no regulation provides considerable latitude to the landfill proponent and designer in terms of the barrier system adopted. It also provides little assurance that the environment will be protected unless the design is subjected to a rigorous pre-construction review.

1.9.2 Prescriptive regulations

Prescriptive regulations are simple. They are typically based on the perception of the regulators as to what constitutes a safe design and will implicitly have negligible effects on groundwater quality. Unfortunately, prescriptive regulations may not recognize that potential impact is not only related to the details of the barrier but may also be highly dependent on many other factors including (but not limited to) the local hydrogeologic conditions, the size of the landfill (both in areal extent and thickness of waste), the infiltration into the landfill and the detailed design of the leachate collection system. Prescriptive regulations may be easy to administer but they create a situation where for one landfill the design may be overly conservative while for another landfill the prescriptive design may provide no assurance that the long-term impact will be negligible.

1.9.3 No impact versus negligible impact

Regulations which require no impact are emotionally desirable but involve practical difficulties. Typically, some form of model is required to demonstrate that there will be no impact and a monitoring program is required to assess the performance of the facility. In terms of modelling, the no impact requirement is often impossible to achieve since for most designs the process of molecular diffusion will result in some contaminant migration from the landfill even for a 30 m thick clay deposit or a composite liner system composed of a geomembrane and compacted clay liner. The requirement of no impact is also impossible to enforce since, at the very least, the impact that can actually occur is controlled by the detection limits used in the chemical analysis of the groundwater. Thus no impact in practice becomes a negligible impact condition where negligible is related to sampling procedure and the analytic detection limits which vary depending on the chemicals being considered and the analytical technique used.

In short, no impact regulations may place unrealistic restrictions on the design of facilities since, if properly modelled the vast majority of facilities would not meet this requirement. Furthermore, this approach creates an arbitrary de facto limit on allowable potential impact which is directly related to field sampling and laboratory techniques and not to considerations of health and safety.

A more meaningful approach is to require that the facility has negligible effect on groundwater where negligible is quantified based on considerations of background chemistry, the chemical species and the potential aesthetic and health-related implications of an increase in concentration in the groundwater. An example of this type of approach is contained in the Ontario Ministry of the Environment (MoE) guidelines associated with its "Reasonable Use Policy" (MoEE, 1993a; MoE, 1998) which provides a means of quantifying a negligible effect in a rational and consistent manner.

1.9.4 Regulatory period

The regulated time period over which a landfill is required to have no or negligible effects on groundwater quality or on a neighbour's reasonable use of groundwater drawn from near the property boundary has a significant effect on the design. For example, one can readily design a barrier (and buffers) such that there is negligible impact at the site boundary for a 30-year period simply by designing the system such that the travel time for contaminants to reach the boundary (at detectable levels) is greater than 30 years (or 100 years or whatever arbitrary limit is set). However, this does not mean that there will not be a significant environmental impact. It may simply mean that the impact of that facility is being passed on to future generations. As discussed in Chapter 16, one can readily encounter situations where the proposed design is not expected to have more than a negligible impact for 100 years but can be predicted to have a significant and unacceptable impact after 150–200 years.

In the design of barriers, it is important to recognize that the clay and geosynthetic/clay barrier being commonly used today will, if properly constructed, greatly retard the migration of contaminants from a waste disposal facility but unless carefully designed may still ultimately result in significant long-term contamination of groundwater. The likelihood of this occurring increases with the trend to larger and larger landfills.

1.9.5 Service life and contaminating lifespan

A second factor intimately related to the regulated period of no or negligible impact is the service life of the facility. If properly designed, specified and constructed, the service life of many of the key components of a landfill barrier system, such as leachate collection systems and geomembrane liners, is likely to exceed 30 years. Under these circumstances, it is a relatively straightforward matter to design a landfill to meet requirements for no or negligible impact for a 30-year post-closure period. Monitoring and contingency measures should, of course, be established to detect and rectify any unpredictable failure that might occur during this period. However, when one is required to ensure negligible impact for periods exceeding 30 years, careful consideration must be given to the effect of degradation of the engineering components of the system.

The most stringent regulations with respect to environmental assessment require that the landfill has negligible effects on groundwater quality (typically at the site boundary) in perpetuity. For example, the Ontario Ministry of Environment and Energy's (MoEE, 1993a) "Reasonable Use Policy" places no time constraint on the period during which the landfill is to have a negligible effect on a neighbour's reasonable use of groundwater.

The MoEE's Engineered Facilities Policy (MoEE, 1993b) addresses the service life of the engineered facility and requires that the service life exceeds the contaminating lifespan, where the contaminating lifespan may be defined as the period of time during which the landfill will produce contaminants at levels that could have unacceptable impacts if they were discharged to the environment.

The contaminating lifespan will depend on the contaminant transport pathway, the leachate strength, the mass of waste (and, in particular, the thickness of waste) and the infiltration through the cover. The contaminating lifespan generally increases with increasing thickness of waste and decreasing infiltration into the landfill. The more hydrogeologically suitable the site (i.e., the greater the potential for natural attenuation of contaminants), the shorter the contaminating lifespan. For the limiting case of a site that has sufficient natural attenuation that no engineered barrier system or leachate collection system is required, the contaminating lifespan is zero. Chapter 16 provides a more detailed discussion of contaminating lifespan.

1.10 Choice of barrier system and service life considerations

The choice of barrier system should always be site-specific and should take account of:

1. the local hydrogeology;
2. the nature of the waste;
3. the size of the landfill (in both vertical and areal extent);
4. the climatic conditions;
5. the availability of suitable materials;
6. the potential impact of failure of the system on health and safety and the environment; and
7. the local social and regulatory systems.

It must be acknowledged that there is always more to learn and it is not possible to guarantee that a facility thought to be safe today will last, say, 500 years. On the other hand, professionals responsible for environmental management should, in the authors' opinion, design facilities to ensure that, based on what we know today and what we know that we do not know, a waste disposal facility will not cause an unacceptable environmental impact or risk to human health and safety, irrespective of when this might occur. This design philosophy recognizes that disposal sites will have a number of phases.

First, during the active life of the landfill, careful quality control and monitoring will be required to ensure that the engineered features are working as intended – and if not, problems should be rectified. This will involve careful inspection of existing clayey deposits, careful control and monitoring of compacted liners and careful control and monitoring of geosynthetic liners, drainage systems and leachate collection systems. Double liner systems (e.g., Figures 1.5, 1.6, 1.7, 1.12, 1.13 and 1.31) have an engineered backup which can contain much of the leachate in the event of a failure of the primary system. In other situations, the hydrogeologic setting must be selected such that it provides a backup in the event of a failure of the primary engineering. The facility should be designed such that it will not have any impact on the reasonable use of the groundwater or the environment during its operational life.

Second, it must be recognized that eventually some engineering components of the landfill may fail. In particular, the potential for clogging of the leachate collection system (Chapter 2) should be considered when performing environmental impact assessments. One cannot justifiably assume that because the pipes are clean that the leachate system is necessarily functioning correctly. Similarly, it is difficult to justify arguments that failure of the leachate collection system can be detected by monitoring the volume of leachate extracted. Monitoring of head in the granular component of the leachate collection system is one means of monitoring its performance. However, consideration should then be given to the potential impact of mounding which is not detected by these monitors and the potential impact should be quantified. Depending on the results of these analyses, additional levels of redundancy may be required as part of the engineered system (e.g., in some cases it may be necessary to go to a double liner with either secondary leachate collection capabilities or hydraulic control).

In addition to the consideration of clogging of the leachate collection system by a combination of biological and particulate clogging, reduction in transmissivity of geonet drainage systems due to time-dependent intrusion of soil should be considered (Section 2.4.7). Similarly, when dealing with geomembranes, time-dependent degradation of the geomembrane should also be considered (Chapter 13). Thus the barrier systems should be designed such that when allowance is made for the uncertainty concerning the service life of some components of the system, the system itself has a service life that exceeds the contaminating lifespan of the landfill. The material presented in subsequent chapters provides the basis for developing designs which meet this objective.

1.11 Summary

This chapter has provided an introduction to some of the design considerations associated with a number of different barrier systems. This has included a discussion which has dealt with clay and geomembrane liners, leachate collection and leak detection systems, HCLs and hydraulic traps.

The contaminant transport processes of advection, diffusion and dispersion have been described and it has been demonstrated that for low-permeability unfractured clayey barriers, diffusion may be a significant and, in many cases, the dominant transport mechanism.

The roles of sorption, biodegradation, partial saturation and matrix diffusion as retardation mechanisms have also been discussed briefly. Based on this introduction to concepts, the governing differential equations for contaminant transport have been developed for simple cases. The finite mass of contaminant has been discussed and simple boundary conditions for a landfill (or other source) having a finite mass of contaminant and for liners underlain by a thin aquifer have been presented and the solution of the governing equations for contaminant transport through a clayey barrier has been illustrated by means of a number of simple examples.

A number of the factors to be considered in environmental impact assessment of a number of barrier systems have been discussed. Barriers should not be designed in isolation. Careful consideration must be given to the site hydrogeology, geotechnical conditions and climate as well as the size of the footprint and capacity of the landfill. Depending on these conditions, landfills which will not significantly affect groundwater quality may range from very simple designs involving natural clayey barriers to highly engineered systems involving multiple liners and leachate collection systems. In general, the larger the landfill, the longer the contaminant lifespan and the more elaborate the barrier system required to provide adequate environmental protection.

A barrier that is perfectly adequate for a given landfill at one site may be totally inadequate for a larger landfill at the same site or for the same landfill at a second site. In particular, it should be emphasized that recommendations such as those by the US EPA represent minimum technological requirements for US conditions. While these documents provide a very useful resource, their recommendations should not be adopted without careful evaluation on a site-specific basis; ultimately the ethical responsibility and liability rest with the design engineer even if regulatory guidelines are followed in many legal systems (Estrin and Rowe, 1995, 1997).

Both natural and synthetic materials have advantages and disadvantages. A good design will recognize this and make appropriate use of both types of material. For example, if designed appropriately, giving due consideration to the function and service life of the various components in the context of the contaminating lifespan of the landfill, geosynthetics can provide a means of considerably enhancing the performance of these facilities. However, appropriate design is only the first step. It remains to construct the facility. Although it is beyond the scope of this book, it should be emphasized that Construction Quality Control and Assurance (CQC/CQA) are critical for both the natural and synthetic components of the engineered system. The CQC/CQA procedure should be such as to ensure that the designer's intentions are met (see Rowe and Giroud, 1994).

It is relatively straightforward to design landfills to meet environmental requirements for a 30-year post-closure period. The longer the period of environmental protection required, the more difficult is the design. The design of landfills which, given what we know today, can be reasonably expected not to cause unacceptable environmental impact at any time in the future often requires the balancing of a number of conflicting criteria. The challenge is to effectively use the engineered components of the barrier system such that eventual failure of key components, like the leachate collection system or a geomembrane liner, combined with the relatively long contaminant travel times through suitable clayey liners does not create environmental problems for future generations.

Chapter two

Leachate characteristics and collection

2.1 Introduction

Barrier systems are used to minimize the migration of contaminated fluids. Typically these fluids take the form of leachate associated with the percolation of water through a mass of waste (e.g., municipal solid waste leachate (MSWL)) or liquid hydrocarbons, associated with either spills or disposal, in aqueous systems. Knowledge of the composition of the fluids to be "contained" is important for both assessing the potential chemical interaction with the components of the barrier system (see Chapters 4, 12 and 13) and calculating the likely impacts on the environment (Chapter 16). This chapter will examine some typical characteristics of leachate and how they change with time, and will discuss the leachable mass of contaminants. It will also discuss factors affecting the long-term performance of leachate collection systems and leachate mounding.

Modern landfills will typically incorporate a primary leachate collection system which is intended to control the leachate head (i.e., the height of leachate) on any underlying liner and reduce the potential for advective migration of contaminant. Field observations suggested that collection systems experience a build-up of biofilm, chemical precipitates and small (e.g., silt and sand) particles that are deposited in pipes and the granular material (e.g., sand or gravel) used to drain and collect the leachate generated by a landfill (Brune *et al.*, 1991; Rowe *et al.*, [illegible]e, 1998a,b). This build-up progressively reduces the hydraulic conductivity of the drainage layer and hence its ability to drain the leachate and is called "clogging". However, these same processes that cause "clogging" also represent a form of "leachate treatment" serving to reduce both the organic and inorganic composition of the leachate by a combination of biodegradation and chemical precipitation. Hence the leachate collected at the end of a leachate collection system is not the same as the leachate that entered the system or the leachate that could potentially migrate through the barrier system at locations well away from the collection point. Similar processes can also occur at other leachate collection points, including leachate wells constructed in landfills to reduce leachate mounding (e.g., when other leachate collection is either not present or not effective) and purge wells used to control a leachate plume in the groundwater.

Leachate collection systems will begin to experience an accumulation of organic and inorganic material shortly after landfilling operations begin. The rate of accumulation will depend on a number of factors including the leachate characteristics, the leachate flow rate and the design of the system. Initially the accumulation will be undetected or unrecognized (e.g., the changes in the chemical composition of the collected leachate due to processes occurring in the collection system are not readily distinguished from the changes occurring due to biologically induced "ageing" of the waste). However when the decrease in hydraulic conductivity reaches the point where the system no longer satisfies the primary design function of controlling the leachate head on the underlying liners, it may be regarded as having "clogged" and to have reached its "service life" even though it may continue to collect leachate for a long period of time.

Early leachate collection systems (involving toe or perimeter drains only, see Figure 2.1) were useful for controlling leachate seeps through the landfill cover, but were ineffective in controlling the leachate mound within the landfill, and hence ineffective in controlling contaminant escape to underlying aquifers. The next generation of leachate collection systems involved drains (typically granular material placed around perforated leachate collection pipes as shown in Figure 2.2) at a spacing of between 50 and 200 m (Rowe, 1998b). Rowe (1992) cautioned against these designs because the confluence of flow near the collection pipe gives rise to high mass loading per unit time and hence increases the rate of clogging of the geotextile and the drainage gravel, assuming that the pipe clogging is mitigated by regular cleaning (Rowe, 1998b). Field studies of collection pipes wrapped with geotextiles have shown a drop of hydraulic conductivity of between two and five orders of magnitude (Koerner *et al.*, 1993). As clogging occurs there will be increased resistance of flow into the collection pipe and this will eventually result in a leachate mound developing in the waste between the drains (see Figure 2.2).

The third stage in the evolution of leachate collection systems (depicted in Figure 2.3) was the introduction of a continuous granular drainage blanket with perforated leachate collection pipes at regular spacing and a suitably selected geotextile filter above the primary drainage layer. As discussed in Section 2.4.4, the use of geotextiles as a filter above the drainage gravel has been reported (e.g., Fleming *et al.*, 1999) to result in substantially less clogging than that observed in areas with no geotextile filter. Geotextiles used in this configuration will experience some clogging; however even if

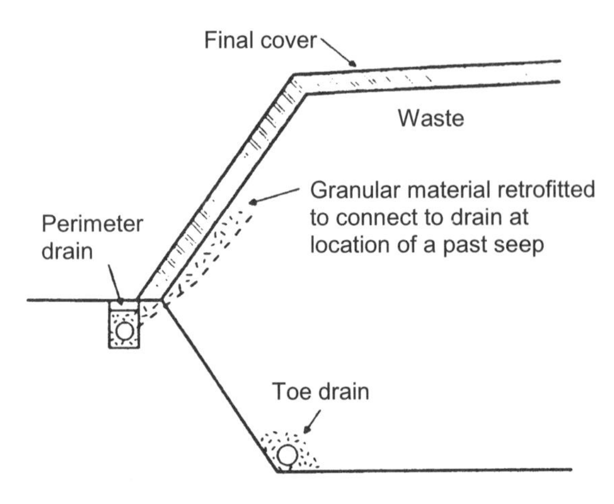

Figure 2.1 Schematic showing perimeter and toe drains at a landfill (not to scale; modified from Rowe, 1999).

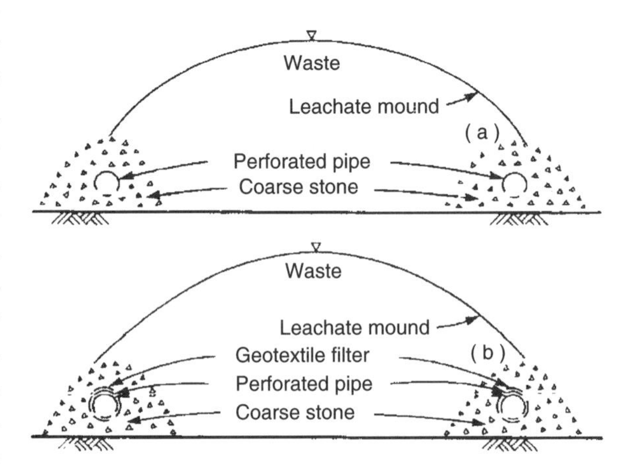

Figure 2.2 Schematic showing examples of poor leachate collection system designs: (a) problematic and (b) even worse. The "coarse stone" around the pipe is typically gravel or crushed rock. Schematic shows the leachate mound developed once there is excessive clogging of the geotextile filter and/or the "coarse stone" and/or the pipe. Note: There would also be a mound to the left and to the right of the section shown and the mound would only be symmetric as a special case; generally, it would not be symmetric due to factors such as variability of clogging, hydraulic conductivity of waste, etc. (modified from Rowe, 1992).

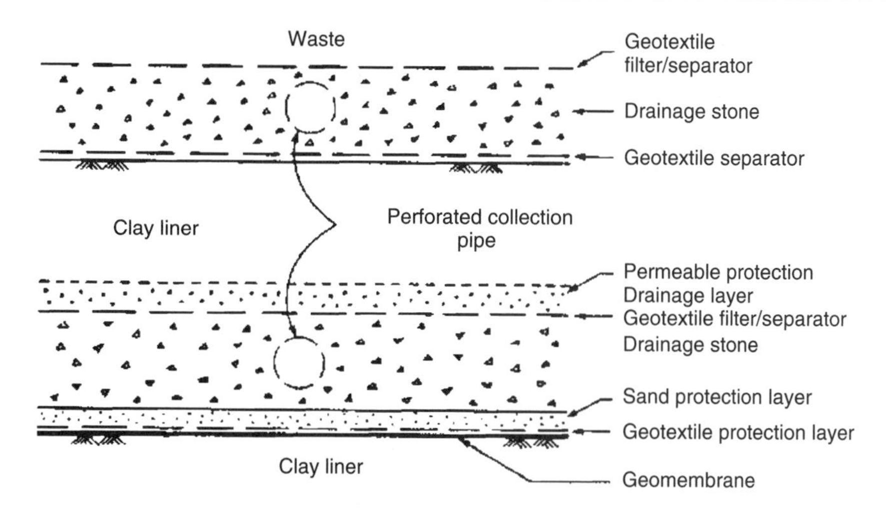

Figure 2.3 Schematic shows examples of blanket leachate collection system designs including a geotextile filter layer and uniform (>25 mm nominal particle diameter) drainage gravel (e.g., crushed stone or river gravel; modified from Rowe, 1992).

a perched leachate mound developed, this would have no effect on the underlying liner (Rowe, 1992); however special care is required in the design of these systems if leachate recirculation is envisaged as a part of leachate management plan. The design of leachate collection systems to prolong the service life will be discussed in this chapter.

Leachate collection systems typically include pipes and manholes. The structural performance of the pipes depends on: (a) the earth's pressure acting on the pipe (which is influenced by burial depth, overburden density and lateral confinement), (b) the backfill configuration (i.e., whether the pipe is placed within a trench, mound or blanket), (c) backfill materials, (d) properties of the pipe and (e) service conditions (e.g., temperature and chemical exposure). Design and construction issues for the pipes are discussed in Section 15.4. The design of manholes is discussed in Section 15.5.

2.2 Liquid hydrocarbons

The very complex subject of hydrocarbon liquids defies succinct summarization for purposes of a book such as this. Nevertheless, a useful summary of the classes of organic compounds and their properties has been prepared by Mitchell and Madsen (1987). The components considered by them are shown in Figure 2.4 and their properties are summarized in Table 2.1.

For purposes of this book, two classes of liquid hydrocarbons will be considered:

1. Organics which are essentially insoluble in water; this includes DNAPLs and light non-aqueous phase liquids (LNAPLs).
2. Organics which are soluble in water.

Since the solubility decreases rapidly with increasing size of a soluble organic molecule, consideration will be focused on small molecules consisting of up to about eight carbon atoms.

The solubility of organic liquids is generally controlled by their polarity. Non-polar or weakly polar substances dissolve in non-polar or weakly polar solvents. Highly polar compounds dissolve in highly polar solvents including water. Fortunately, the polarity or dipole moment (in debyes) is closely proportional to the dielectric constant, ε, which figures so significantly in the Gouy Chapman theory of potential (to be discussed in Chapter 3).

Type of compound		Functional group	Specific example	
Saturated hydrocarbons	alkanes	$\equiv C - C \equiv$	$H_3C - CH_3$	Ethane
Unsaturated hydrocarbons	(alkenes)	$= C = C =$	$H_2C = CH_2$	Ethylene
	(alkynes)	$- C \equiv C -$	$HC \equiv CH$	Acetylene
	(polyolefins)	$= C = C - C = C =$	$H_2C = CH - CH = CH_2$	Butadiene
Aromatic hydrocarbons		(benzene ring)	C_6H_6 (structural formula)	Benzene
Alcohols		$- OH$	$CH_3CH_2 - OH$	Ethyl alcohol
Phenols		$- OH$	$C_6H_5 - CH(OH)CH_3$	1-Phenyl-ethanol
Ethers		$- O -$	$CH_3CH_2 - O - CH_2CH_3$	Ethyl ether
Aldehydes		$H-C=O$	$CH_3 - C(=O) - H$	Acetaldehyde
Ketones		$>C = O$	$CH_3 - C(=O) - CH_3$	Acetone
Organic acids (carbolic acids)		$- C(=O)OH$	$CH_3 - C(=O) - OH$	Acetic acid
			$CH_3 - CH_2 - C(=O) - OH$	Propionic acid
Organic bases (amines)		$- NH_2$	$CH_3 - NH_2$	Methylamine
Halogenated hydrocarbons		$- Cl$	$H - CH(Cl) - Cl$	Dichloromethane
		$- F$	$C_6H_5 - Cl$	Chlorobenzene
		$- Br$	(structural formula)	Tetrachlorodibenzo-*p*-dioxin

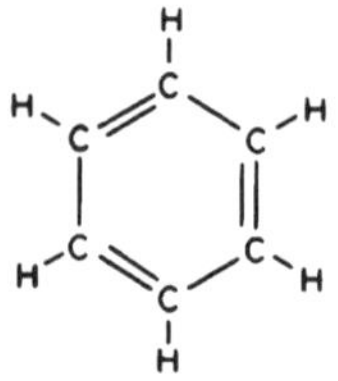

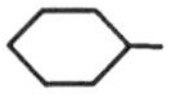

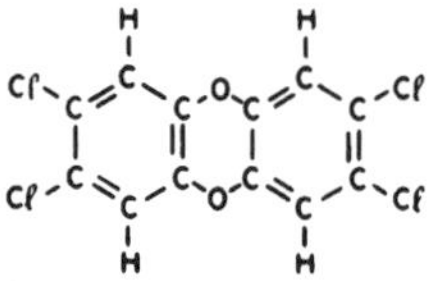

Figure 2.4 Classes of organic compounds (modified from Mitchell and Madsen, 1987).

Table 2.1 Properties of organic chemicals used in hydraulic conductivity testing of clays (after Mitchell and Madsen, 1987; reproduced with permission from American Society of Civil Engineers)

Class of compound	*Compound*	*Formula*	*Solubility in water* (g/l)	*Dielectric constant*	*Dipole moment (debye)*	*Density* (g/cm^3)
Hydrocarbons and related compounds	Heptane	C_7H_{16}	<0.3	1.9	0	0.684
	Cyclohexane	C_6H_{12}	<0.3	2.0	0	0.779
	Benzene	C_6H_6	0.7	2.3	0	0.879
	Xylene (dimethyl benzene)	C_8H_{10}	<0.3	2.27 (*para*)	0	0.880
				2.37 (*meta*)		
				2.57 (*ortho*)	0.62	
	Tetrachloromethane (carbontetrachloride)	CCl_4	0.8	2.2	0	1.594
	Trichloroethylene (TCE)	C_2HCl_3	1	3.4	0	1.464
	Nitrobenzene	$C_6H_5NO_2$	2	35.7	4.22	1.204
Alcohols and phenols	Methanol	CH_3OH	∞	33.6	1.70	0.791
	Ethanol	C_2H_5OH	∞	25.0	1.69	0.789
	Ethyleneglycol	$C_2H_6O_2$	∞	37.7	2.28	1.119
	Phenol	C_6H_5OH	86	13.1	1.45	1.072
Ethers	1,4-Dioxane	$C_4H_8O_2$	∞	2.2	0	1.034
Aldehydes and ketones	Acetone	C_3H_6O	∞	21.5	2.9	0.79
Organic acids	Acetic acid	$C_2H_4O_2$	∞	6.15	1.74	1.049
Organic bases	Aniline	$C_6H_5NH_2$	36	6.89	1.55	1.02
Mixed chemicals	Xylene	C_8H_{10}	∞ in acetone	2.27–2.57	0–0.62	0.880
	Acetone	C_3H_6O		21.5	2.9	0.79
	Sodium acetate	CH_3COONa				
	Glycerol	$C_3H_8O_2$	∞	42.5		1.261
	Acetic acid	$C_2H_4O_2$	∞	6.15	1.74	1.049
	Salicylic acid	$C_7H_6CO_3$	Slightly			1.443

Methanol, CH_3OH, is soluble in water because the hydrogen in a methanol molecule interchanges easily with hydrogen in water molecules producing nearly infinite miscibility. As shown in Table 2.2, the solubility decreases rapidly for alcohols consisting of chains longer than three carbons.

In water-wet clay barriers, solubility plays a critical role in the interrelationships between the barrier and a retained organic liquid. Several terms are used to describe solubility as follows:

1. hydrophilic is used to describe water-loving or soluble organics;
2. hydrophobic is used to describe water-hating or insoluble organics;
3. lyophilic is a general term describing mutual attraction (solubility) between two liquids; and
4. lyophobic is a general term describing mutual repulsion between two liquids.

When large volumes of contaminant are available, the source concentration may be controlled by the solubility limit, and the equivalent height of leachate may be quite large.

Finally, a few words about the dielectric constant, ε. Liquids with a high dielectric constant (water at 80, methanol at 34, etc.) dissolve ionic compounds not only because they are polar and efficiently hydrate the dissociated species but also because they have insulating properties. In

Table 2.2 Solubility of alcohols in water (adapted from Morrison and Boyd, 1983)

Alcohol	*Solubility* (g/100 g H_2O)
Methanol (CH_3OH)	∞
Ethanol (CH_3CH_2OH)	∞
Propanol ($CH_3CH_2CH_2OH$)	∞
Butanol ($CH_3CH_2CH_2CH_2OH$)	7.9
Pentanol ($CH_3CH_2CH_2CH_2CH_2OH$)	2.3
Hexanol ($CH_3CH_2CH_2CH_2CH_2CH_2OH$)	0.6
Heptanol $CH_3CH_2CH_2CH_2CH_2CH_2CH_2OH$)	0.2
Octanol ($CH_3CH_2CH_2CH_2CH_2CH_2CH_2CH_2OH$)	0.05

the case of NaCl in water, the force of attraction between the hydrated Na^+ and Cl^- atoms is greatly reduced by the high-dielectric solvent. This applies equally in the double layer (discussed in Chapter 3) where the presence of water reduces the force of attraction between the negative clay particle and positive cations. If organic liquids of lower dielectric constant enter the double layer they effect an increase in the force of attraction and contract the double layer.

For purposes of this book, organic liquids are subdivided into two groups, viz.:

1. insoluble (lyophobic) DNAPLs and LNAPLs which normally have dielectric constant values of 2–4 and
2. soluble (lyophilic) compounds which normally have dielectric constant values of about 10 or greater.

The use of dielectric constant is very convenient since it plays a prominent role in double-layer theory. The special case of low-dielectric liquids which are soluble in water (e.g., dioxane) will be considered in Chapter 4.

2.3 Landfill leachate

2.3.1 Characteristics

(a) MSWL

The composition of average municipal solid waste (MSW) (picked up at the curbside) may vary from country to country and will depend on the cultural background of the generating community. An example of the composition in North America is given in Table 2.3 (Ham *et al.*, 1979). Percolation of rainwater through solid waste leaches out soluble salts and biodegraded organic products to form a foul-smelling, grey coloured leachate. Fine-grained soils in the waste or that used as daily cover may also be incorporated as suspended solids in the leachate.

Bacterial growth starts immediately under oxidizing conditions generating temperatures which have been recorded to be in excess of 60 °C in some wastes (Collins, 1993). Since optimum growth takes place at C:N:P ratios of about 100:5:1, it is probably food wastes which fuel early biological reactions. Low average nitrogen levels in bulk waste (only 0.5%) and rapid depletion of oxygen soon render the system anaerobic, cooler and far less reactive. Internal temperatures in landfills of 30–40 °C are common and encompass the 35 °C temperature considered to be optimal for methanogenesis (Zehnder, 1978).

MSW leachate is a complex liquid which changes in characteristics as one passes from the early acetogenic phase of young leachate to the methanogenic phase of older leachate. Typically, the acetogenic phase is characterized by high organics with BOD_5/COD ratios greater than 0.4 (Ehrig, 1989). Because of the time period over which landfills may be constructed, the leachate that is collected may be a mixture of young and old leachate from different parts of the landfill. Leachate characteristics may also

Table 2.3 Municipal solid waste refuse composition

Component	Per cent of all refuse by weight	Moisture per cent by weight[b]	Analysis (per cent dry weight)[a]						
			Volatile matter	Carbon	Hydrogen	Oxygen	Nitrogen	Sulphur	Non-combustibles
Rubbish, 64%									
Paper	42.0	10.2	84.6	43.4	5.8	44.3	0.3	0.20	6.0
Wood	2.4	20.0	84.9	50.5	6.0	42.4	0.2	0.05	1.0
Grass	4.0	65.0	–	43.3	6.0	41.7	2.2	0.05	6.8
Brush	1.5	40.0	–	42.5	5.9	41.2	2.0	0.05	8.3
Greens	1.5	62.0	70.3	40.3	5.6	39.0	2.0	0.05	13.0
Leaves	5.0	50.0	–	40.5	6.0	45.1	0.2	0.05	8.2
Leather	0.3	10.0	76.2	60.0	8.0	11.5	10.0	0.40	10.1
Rubber	0.6	1.2	85.0	77.7	10.4	–	–	2.0	10.0
Plastic	0.7	2.0	–	60.0	7.2	22.6	–	–	10.2
Oils, paints	0.8	0.0	–	66.9	9.7	5.2	2.0	–	16.3
Linoleum	0.1	2.1	65.8	48.1	5.3	18.7	0.1	0.40	27.4
Rags	0.6	10.0	93.6	55.0	6.6	31.2	4.6	0.13	2.5
Street sweepings	3.0	20.0	67.4	34.7	4.8	35.2	0.1	0.20	25.0
Dirt	1.0	3.2	21.2	20.6	2.6	4.0	0.5	0.01	72.3
Unclassified	0.5	4.0	–	16.6	2.5	18.4	0.05	0.05	62.5
Food wastes, 12%									
Garbage	10.0	72.0	53.3	45.0	6.4	28.2	3.3	0.52	16.0
Fats	2.0	0.0	–	76.7	12.1	11.2	0.0	0.00	0.0
Non-combustibles, 24%									
Metals	8.0	3.0	0.5	0.8	0.04	0.2	–	–	99.0
Glass and ceramics	6.0	2.0	0.4	0.6	0.03	0.1	–	–	99.3
Ashes	10.0	10.0	3.0	28.0	0.5	0.8	–	0.5	70.2
All refuse	100.0	20.7	–	28.0	3.5	22.4	0.33	0.16	24.9

Sources: Ham *et al.* (1979) and Table 2.3 in EPA (1983).
[a] Analysis of the respective components.
[b] Moisture content of the respective component in the waste.

change in response to the development sequence of the landfill (Armstrong and Rowe, 1999; Demirekler *et al.*, 1999). For example, the placement of new waste over waste that has been decomposing for several years may result in lower strength (in terms of COD) leachate reaching the collection system than if the same waste were placed over relatively recently placed waste (Ham and Bookter, 1982).

Much of the organic phase in waste is extremely biodegradable resulting in high initial BOD values. Chian (1977) reports original total organic carbon values of ~17 g/l (~1.7%) in a 2-month-old Illinois leachate with a fatty acid composition as indicated in Table 2.4. These acids are very important since they render leachate weakly acidic and mobilize heavy metals. Table 2.5 is presented to illustrate the nature of these low-molecular-weight aliphatic acids. The remaining organics appear to be predominantly humic-carbohydrates, carboxyl compounds, carbonyl compounds, etc. (Chian, 1977).

With these comments in mind, Table 2.6 is presented to illustrate the composition of three US, three Italian and two German leachates, and Table 2.7 summarizes results for a number of Canadian leachates.

Experience would indicate that the concentration of key contaminants in leachate reaches a peak value after some point in time and then

Table 2.4 Fatty acid composition of 2-month-old leachate with total organic carbon content of 17 g/l (~1.7%) (adapted from Chian, 1977)

Acid	*Concentration* (g/l)
Acetic acid	1.75
Propionic acid	0.51
Isobutyric acid	0.31
Butyric acid	3.08
Isovaleric acid	0.72
Valeric acid	0.56
Hexanoic acid	1.49

decreases with subsequent time. Even for conservative species there is a sound theoretical basis for a decrease in concentration with time due to dilution as new leachate is generated and other leachate is removed by the leachate collection system (Rowe, 1991a; Ehrig and Scheelhaase, 1993). For non-conservative species, this decrease can be accelerated by both chemical and biological breakdown. To illustrate the basic trends, Figure 2.5 shows an empirical curve for the change in chloride concentration based on the age of MSW landfills in North America. Figure 2.6 shows the observed decay phenol in a landfill based on a number of years of monitoring data.

Based on a study of 50 MSW landfills in Germany, Kruempelbeck and Ehrig (1999) published mean values of leachate constituents in year ranges (1–5, 6–10, 11–20 and 21–30) as shown in Table 2.8. In general the peak average concentration of organic compounds was reached in years 1–5. However chloride, sulfate, the various nitrogen–oxygen compounds (NO_4, NO_3, NO_2) and some metals (Zn, Ni, Cr, As) peak at later times.

An examination of Tables 2.6–2.8 shows considerable variation in the concentration of chloride in landfill leachate; however, based on available data, a chloride concentration of between 1,500 and 2,500 mg/l would appear to be a reasonably representative source concentration for impact calculations for the MSW leachate. This is consistent with an average concentration for chloride of 2,100 mg/l that has been reported by Ehrig and Scheelhaase (1993) based on European data. In areas where there may be less leachate dilution, higher values may be more appropriate. Similarly, if the landfill is used to dispose road salt, sewage sludge or industrial salty wastes then higher values may be appropriate.

There is some question concerning the potential effect of recycling on leachate quality.

Table 2.5 Short-chain carboxylic acids

Acetic acid (CH_3COOH)

```
    H
    |   //O
 H-C-C
    |   \O-H
    H
```

Propionic acid (CH_3CH_2COOH) (carboxylic acid C_3)

```
    H  H
    |  |   //O
 H-C-C-C
    |  |   \O-H
    H  H
```

Butyric acid ($CH_3CH_2CH_2COOH$) (carboxylic acid C_4)

```
    H  H  H
    |  |  |   //O
 H-C-C-C-C
    |  |  |   \O-H
    H  H  H
```

Valeric acid ($CH_3CH_2CH_2CH_2COOH$) (carboxylic acid C_5)
Hexanoic acid ($CH_3(CH_2)_4COOH$) (carboxylic acid C_6)
Isobutyric acid ($(CH_3)_2CH\ COOH$) (2-methylpropionic acid)

Collins (1991) reported that the removal of paper, glass, metals, plastics and textiles (about 36% of the refuse stream) had no significant effect on chloride concentrations but significantly reduced the concentration and mass of iron and calcium collected in leachate. The removal of an additional 32% of the waste as organic material to be composted led to additional reductions in concentrations of iron and calcium and did not significantly affect chloride. This is beneficial with respect to extending the longevity of leachate collection systems (see Section 2.4.4).

Tables 2.7 and 2.9 provide data concerning organic compounds detected at a number of landfills. In Table 2.7, Ontario landfills 2, 5 and 7 are considered to be reasonably representatives of MSW. Some industrial waste is expected at the other landfills listed in Tables 2.7 and 2.9. It should be noted that the concentrations of organic compounds are low. The most commonly reported xenobiotic organic compounds (XOCs) are the monoaromatic hydrocarbons (BTEX – benzene, toluene, ethylbenzene and xylenes) and halogenated hydrocarbons (e.g., dichloromethane, trichloroethylene). However, as reported by Kjeldsen *et al.* (2002), leachates may also contain pesticides (e.g., Atrazin [0.12 μg/l], Bentazone [0.3–4 μg/l], Glyphosat [1.7–27 μg/l], Lindane [0.03–0.95 μg/l], MCPP or mecoprop [0.38–150 μg/l], MCPA [0.2–9.1 μg/l], 4-CPP [15–19 μg/l]) at low concentrations with 2-(2-methyl-4-chlorophenoxy) propionic acid (MCPP) having the highest concentration and being most frequently observed. Benzene and naphthalene sulfonates, which include some of the surfactants used in laundry detergents,

Table 2.6 Composition of MSW landfill leachates (concentrations in mg/l)

	USA[a]			*Italy*[a]			*Germany*[a]	
	1	*2*	*3*	*4*[b]	*5*[b]	*6*[c]	*7*[d]	*8*[e]
BOD_5	–	13,400	–	3,000	10,400	2,125	57–2,700	400–45,900
COD	42,000	18,100	1,340	38,520	28,060	7,750	1,450–6,340	1,630–63,700
Total solids	36,250	12,500	–	–	–	–	–	–
Total volatile fatty acids	–	930	333	1,574	435	–	–	–
Organic nitrogen as N	–	107	–	60	554	125	–	–
Ammonia nitrogen as N	950	117	862	1,293	1,203	1,040	620–2,080	620–3,500
pH	6.2	5.1	6.9	6.0	6.3	8.5	7.2–7.9	5.7–8.1
Total alkalinity as $CaCO_3$	8,965	2,480	–	5,125	4,250	8,250	–	–
Chemicals and metals								
Calcium	2,300	1,250	354	175	–	–	70–290	130–4,000
Chloride	2,260	180	–	2,231	1,868	3,650	1,490–3,550	1,490–21,700
Iron	1,185	185	4	47	330	–	8–79	8–870
Magnesium	410	260	233	1,469	827	–	100–270	100–840
Manganese	58	18	0.04	42	27	–	0.2–4.0	0.3–28
Phosphate	82	1.3	–	–	–	2.3	–	–
Potassium	1,890	500	–	1,200	1,200	–	–	–
Sodium	1,375	160	748	1,400	1,300	–	–	–
Sulphate	1,280	–	<0.01	1,600	1,860	219	1–115	1–121
Zinc	67	–	18.8	7	5	–	–	–

[a] Sources: 1, Wigh (1979); 2, Breland (1972); 3, Griffin and Shimp (1978); 4–6, Cancelli and Cazzuffi (1987); 7 and 8, Brune *et al.* (1991).
[b] "Young" landfill.
[c] "Old" landfill.
[d] Weak concentration.
[e] Strong concentration.

Table 2.7 MSW leachate from Canada

	Ontario								Alberta
	1	*2*	*3*	*4*	*5*	*6*	*7*	*8*	*9*
BOD_5[a]	3,500	2,330–12,500	450–12,500	–	1,850–3,875	23,500	30	>5,000	–
COD[a]	4,700	53–21,200	28–21,800	11,000–82,000	–	10,600	–	–	–
TOC[a]	1,650	–	6,190	–	–	3,060	–	5,710	1–10500
DOC[a]	–	18–7,090	10–7,360	270–25,000	–	4,600	20	–	2,610–29,500
TDS[a]	6,100	–	–	–	–	–	–	–	–
Ammonia[a] (as N)	–	–	355	–	–	715	–	295	–
TKN[a]	–	–	370	–	–	750	30	365	11–508
pH	6.6	6.2	6.7	–	–	7.0	7.0	6.6	5.3–7.1
Total alkalinity (as $CaCO_3$)[a]	2,600	–	6,330	–	–	5,490	770	6,720	1,280–7,620
Aluminium[a]	1	–	–	–	–	–	–	–	0.06–18.2
Boron[a]	5	–	–	–	–	–	–	–	1–24.9
Cadmium[a]	–	0.006	0.8–52	–	–	0.035	0.005	0.001	0.003
Calcium[a]	540	60–2,500	74–1,700	60–4,580	–	535	480	1,740	157–2,910
Chloride[a]	–	(750)	(505)	(1,915)	(1,400)	–	–	–	–
	670	30–3,800	28–3,130	30–12,000	730–2,230	1,270	240	3,400	795–18,000
Chromium[a]	0.15	0.06	0.28	–	0.09	0.05	0.02	0.001	0.006–0.75
Copper[a]	–	0.04	0.036	–	0.1	–	–	–	0.01
Iron[a]	80	4–540	0.4–560	6–1,670	45–160	270	2	520	1–2,650
Lead[a]	–	0.03	0.07	–	0.22	<0.05	–	0.014	0.04–1.7
Magnesium[a]	240	10–465	45–560	21–1,750	–	190	240	640	270–630
Manganese[a]	7	8	6	3	2	–	–	–	0.001–1.9
Mercury[a]	–	0.06	0.2	–	–	–	–	–	0.0006
Nickel[a]	0.15	0.22	0.50	–	0.39	–	–	–	0.03–28.6
Phosphate[a]	<0.8	0.8	–	–	–	–	–	–	–
Potassium[a]	240	227	37	18–1,680	–	800	40	545	20–1,660
Sodium[a]	–	(1,095)	(580)	(3,140)	–	–	–	–	–
	430	90–10,400	14–3,650	210–16,000	–	1,330	210	3,110	284–8,420
Sulphate[a]	–	(103)	(74)	(290)	–	–	–	–	–
	380	1–490	2–4,200	23–2,530	–	510	1,715	25	58–2,490
Zinc[a]	2.4	(1.5)	(2.2)	–	(3.8)	1.4	0.3	20	0.005–0.4
Phenol[b]	3,900	(2,750)	(970)	(1,130)	(1,280)	5,750	16	64,000	–
1,1-Dichloroethane[b]	60	–	14	–	–	–	–	–	ND–120
1,2-Dichloroethane[b]	6	–	18	–	–	–	–	–	–
Benzene[b]	7	–	18	–	–	7	–	–	26–750
Dichloromethane[b]	–	–	1	3,500	–	130	–	–	9–2,000
Ethylbenzene[b]	4	–	80	–	–	30	–	–	40–180
Toluene[b]	80	(90)	–	–	–	270	1.6	–	58–1,400
Trichloroethylene[b]	–	90	–	–	–	–	–	–	–
m,p-Xylene[b]	–	120	–	–	–	50	–	–	80–900
o-Xylene[b]	–	70	–	–	–	30	–	–	–

Numbers in brackets represent geometric mean of data; ND, not detected.
[a] In mg/l or ppm.
[b] In μg/l or ppb.

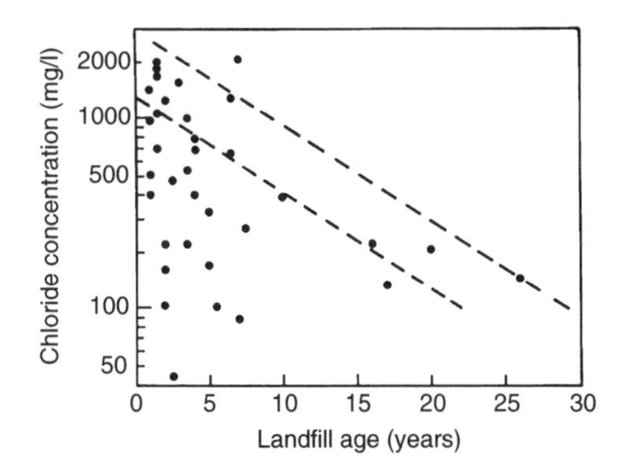

Figure 2.5 Chloride concentrations observed in landfills of different age. Note: No correlation has been made with size, climate or initial source concentration (based on data from Lu *et al.*, 1985).

were also reported at concentrations ranging from 2.5 to 1,188 µg/l. Finally it has been reported that the fuel additive methyl-tert-butyl-ether (MTBE) has been detected in eight Swedish landfills at concentrations up to 35 µg/l.

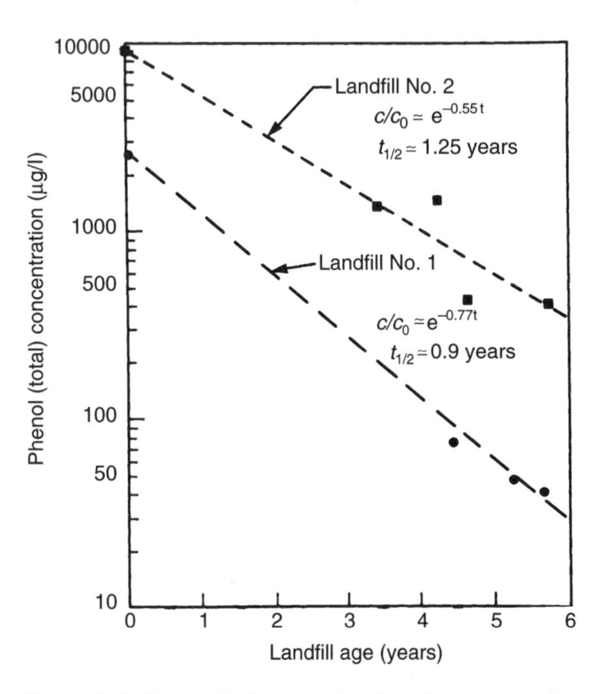

Figure 2.6 Observed decrease in phenol concentration with time at two landfills. Note: There was no significant decay in chloride concentration during this time period. These landfills do not have a leachate collection system.

Examinations of the toxicity of MSW leachate have been summarized by Kjeldsen *et al.* (2002). Several bioassay studies have concluded that ammonia was the primary cause of acute toxicity of MSW landfill leachate. However other studies have indicated that pH, alkalinity, conductivity and the concentrations of chloride, copper or zinc may also have an important effect in aquatic toxicity studies using bioassays. Due to the dominant effect of these factors, the contribution of XOCs to MSW leachate toxicity is largely unknown.

The concentrations of various potential contaminants existing in waste (e.g., see Tables 2.6–2.9) may be compared with typical drinking water standards as listed in Table 2.10.

(b) Comparison of leachates from different types of landfill

Bonaparte *et al.* (2002) summarized data from a number of different types of landfill (MSW; hazardous waste, HW; MSW ash, and construction and demolition waste, C&DW) for a number of post-1990 US landfill and the findings are presented in Table 2.11. This is a relatively small data set, especially for HW and C&DW, and it should be recognized that this is also still very early in the contaminating life of all of these landfills. However, despite the wide variation in concentrations for a given type of landfill, the data do highlight some of the potential similarities and differences.

The MSW landfill leachates were mineralized, biologically active liquids with relatively low concentrations of VOCs and heavy metals. The data are also representative of relatively "young" leachate with a ratio of average BOD/COD of 0.5 and a slightly acidic pH (average 6.8). The heavy metals and VOCs encountered at the highest concentrations were chromium, nickel, dichloromethane and toluene while the concentrations of cadmium, benzene, 1,2-dichloroethane, trichloroethylene and vinyl chloride were higher than US EPA maximum contaminant levels (MCLs) for drinking water.

Table 2.8 Long-term behaviour of leachate from 50 MSW landfills in Germany (adapted from Kruempelbeck and Ehrig, 1999)

Parameter	*Mean value 1–5 years*	*Range 1–5 years*	*Mean value 6–10 years*	*Range 6–10 years*	*Mean value 11–20 years*	*Range 11–20 years*	*Mean value 21–30 years*	*Range 21–30 years*
pH	7.3	5.4–8.7	7.5	6.1–8.7	7.6	6.4–8.9	7.7	1,602–1,609
Specific electrical conductivity	9,280	545–23,300	12,160	691–51,400	10,610	809–40,300	1,290	1,602–1,09,000
COD	3,810	46–22,700	3,255	22–22,500	1,830	28–29,150	1,225	76–6,997
BOD_5	2,285	10–16,000	1,210	6.3–64,880	465	7–25,800	290	5.7–1,100
BOD_5/COD	0.6	–	0.37	–	0.25	–	0.24	–
TOC	1,235	20–7,725	845	17–4,930	520	10–2,600	475	35–1,120
NH_4-N	405	0.4–7,000	600	1–2,360	555	1–2,870	445	12.4–1,571
NO_3-N	3.6	0.024–26	7.6	0.032–160	11.7	0.023–200	9.2	0.113–64
NO_2-N	0.064	0.003–0.3	0.63	0.003–11.7	0.54	0.003–9.1	0.84	0.010–7.18
AOH	2,765	452–7,500	1,930	58–6,200	1,505	20–5,300	1,130	130–5,600
Chloride	1,300	13–11,950	1,600	52–8,700	1,380	19–6,700	1,025	18–2,880
Sulphate	98	2–400	146	1.1–1,810	92.5	2.2–556	83	5.2–490
Sodium	815	66–2,200	930	45–1,950	830	13–2,300	645	50–1,500
Potassium	1,220	98–2,200	910	110–1,850	695	100–1,900	595	25–1,268
Magnesium	290	15–612	205	32–1,167	145	50–593	115	43–221
Calcium	375	46–2,290	230	44–430	170	12–604	155	71–863
Manganese	3.9	0.055–43	2.5	0.02–95	1.1	0.05–38.4	0.91	0.21–5.7
Iron	50	0.08–550	29.5	0.06–1,383	16.5	0.13–825	12.5	0.4–189
Lead	0.156	0.005–0.92	0.056	0.005–0.317	0.67	0.005–1.3	0.034	0.005–0.19
Zinc	1.1	0.02–24	1.5	0.016–125	0.53	0.01–43.5	0.538	0.05–9
Cadmium	0.011	0.0002–0.05	0.0058	0.0002–0.1925	0.0039	0.00013–0.07	0.0028	0.0002–0.018
Nickel	0.199	0.007–1.4	0.249	0.012–10.6	0.135	0.003–1.93	0.115	0.0036–0.348
Copper	0.711	0.003–40	0.115	0.002–3.3	0.062	0.0025–1.03	0.036	0.004–0.27
Chromium	0.156	0.013–0.48	0.224	0.008–2.57	0.164	0.006–1.16	0.177	0.005–1.62
Arsenic	0.015	0.003–0.03	0.021	0.002–0.097	0.042	0.001–0.37	0.036	0.0026–0.182

All parameters are in mg/l except pH (unit less) and specific electrical conductivity (μs/cm).

The hazardous waste leachates were alkaline ($7.9 < pH < 9.4$) but, nevertheless, the concentrations of heavy metals were several times to several orders of magnitude higher than for MSW leachate, with arsenic, nickel, 1,2-dichloroethane and vinyl chloride representing the heavy metals and VOCs with the highest concentrations. Chloride is at a high concentration with an average more than an order of magnitude greater for HW (7.758 mg/l) than for the MSW (463 mg/l).

The MSW ash landfill leachates were near neutral (average $pH = 7.06$ and $6.54 < pH < 7.44$). VOC concentrations were low as they also were for arsenic. Concentrations for chromium, cadmium and lead were similar to or greater than that for MSW, although still relatively low with only the average for cadmium exceeding the MCL. However, the TDS and chloride concentrations were very high with the average concentrations for the ash fills (24,493 and 10,426 mg/l, respectively) being more than an order of magnitude higher than for the conventional MSW leachate (2,758 and 463 mg/l, respectively).

Data for C&DW leachate are very limited; however, it is of note that the COD, BOD, TDS, chloride, heavy metal and VOC concentrations were all well within the MSW leachate

Table 2.9 Liquid hydrocarbon contaminants found in leachate-contaminated groundwater at six Canadian landfill sites reported by Barker *et al.* (1987) and based on a review of published data by Kjeldsen *et al.* (2002) (in μg/l)

	Landfill sites						
Contaminant	*Borden*	*Woolwich*	*North Bay*	*New Borden*	*Upper Ottawa Street*	*Tricil*	*Kjeldsen* et al. *published range*
Aliphatic, aromatic and carboxylic acids	20	>10,000	>300	–	>1,000	–	–
Carbon tetrachloride	<1	5	<1	9	p	ND	4–9
Chloroform	<1	20	<1	25	5	ND	–
Trichloroethylene	1	37	2	750	p	ND	0.05–750
Trichloroethane	ND	7	<1	90	20	8.44	0.01–3,910
Tetrachloroethylene	ND	2	<1	<1	<1	ND	0.01–250
Acetone	<1	–	6	–	6	–	6–4,400
Tetrahydrofuran	p	–	9	–	200	–	9–430
1,4-Dioxane	ND	–	<1	–	p	–	–
Benzene	3	70	51	50	60	7,920	0.2–1,630
Toluene	1	7,500	60	1,400	2,600	9,520	1–12,300
Xylenes	<1	700	140	500	3,500	–	0.8–3,500
Ethylbenzene	<1	1,100	64	120	700	3,320	0.2–2,330
Tetramethylbenzene	ND	10	250	70	450	–	0.3–250
Chlorobenzene	ND	ND	105	ND	110	–	0.1–110
Dichlorobenzenes	<1	ND	13	ND	5	–	0.1–32
Naphthalene	ND	50	15	260	p	2,350	0.1–260
Phenols	–	1,100	10	p	p	–	0.6–1,200
Benzothiozoles	<1	30	10	ND	p	–	–
PAHs	–	–	ND	–	ND	–	–
Phthalates	<1	p	110	p	p	–	0.1–14,000

Note: Many of these landfills are known to have accepted industrial waste; p = detected but concentration not estimated; ND, not detected; –, not determined/reported.

range. Based on this limited data set it is not possible to distinguish the C&DW leachate from MSW leachate.

Bonaparte *et al.* also compared pre- and post-1990 MSW landfill leachate and concluded that there was no statistically significant difference in the concentrations of heavy metals or VOCs between old (pre-1990) and newer (post-1990) MSW landfill leachates at the 90% confidence level, although average concentrations were generally lower in the post-1990 landfills. Likewise the average concentrations for newer (post-1990) HW landfills were within the range observed for older (pre-1990) HW landfills, although average concentrations were again generally lower in the post-1990 landfills.

2.3.2 Variation in concentration with time

It is well recognized that the characteristics of landfill leachate change with time. However, when leachate characteristics are based on chemical analyses of leachate after it has passed through the leachate collection system, it is not clear whether the changes that occur in the leachate with time are due to the changes in the landfill, the changes that occur once it gets into the leachate collection system or both. This can best be illustrated with reference to data from a large Ontario landfill as described below.

The landfill received MSW between 1984 and 2002. The variations in key leachate parameters with time during the operation of the landfill

Table 2.10 Some drinking water standards

Constituent	*Recommended concentration limit*[a] (mg/l)
Inorganic	
Total dissolved solids	500
Chloride	250
Copper	1
Hydrogen sulphide	0.05
Iron	0.3
Manganese	0.05
Sodium	200
Sulphate	250
Zinc	5
Organic	
Acetone	1
Phenols	3.5
Toluene	0.024
	Maximum permissible concentration (μg/l)
Inorganic	
Antimony	6
Arsenic	25
Barium	1,000
Boron	5,000
Cadmium	5
Chromium (Cr^{VI})	50
Selenium	10
Lead	10
Mercury	1
Nitrate as nitrogen	10,000
Silver	50
Organic	
Cyanide	100
DDT	30
Endrine	0.2
Lindane	4
Methoxychlor	100
Toxaphene	2
2,4-D	100
2,4,5-TP silvex	10
Synthetic detergents	500
cis-1,2-Dichloroethane	70
trans-1,2-Dichloroethane	100
1,2-Dichloroethane	5
1,4-Dichlorobenzene	5
Dichloromethane (methylene chloride)	50 (Ontario)
Dichloromethane (methylene chloride)	5 (EPA)
Tetrachloroethylene	30
Trichloroethylene	50
Trichloromethane (chloroform)	100
Xylenes	300
	Maximum acceptable concentration (Bq/l)
Radionuclides and radioactivity	
Radium 226	0.6
Strontium 90	5
Gross beta activity	1
Gross alpha activity	0.1
Bacteriological	
Escherichia coli (*E. coli*) or fecal coliform	Not detected per 100 ml

Sources (and order of priority in selecting value quoted where differences exist): Ontario Ministry of the Environment (2001); Health Canada (2002); US EPA (2002).
[a]Recommended concentration limits for these constituents are mainly to provide acceptable aesthetic and taste characteristics.

(based on analyses of leachate taken from the end of the leachate collection system (LCS)) are shown in Figures 2.7–2.10. The chloride concentration increased significantly within the first 7 years of operation (until 1991). Due to continued construction and operation at the landfill site, there was more variability in the chloride (and calcium) concentrations between 1991 and 1995; however it appears that chloride concentrations continued to increase until 2001 and to have stabilized at about 4,000 mg/l in 2001 and 2002. The calcium concentration reached a peak value in 1992 and has subsequently shown significant variability. It is evident that the calcium concentration at the main collection manhole (i.e., at the "end of the pipe") after flowing through the collection system decreased until it appears to have stabilized in 2001 and 2002. In 2002 the annual average calcium concentration of the collected leachate had dropped to 66 mg/l. It is not known whether this is because of a decrease in the concentration in the leachate as it enters the collection system and/or because calcium is being removed from the leachate as it moves through the collection system. Despite considerable variability in individual values, the annual average values of pH and COD were reasonably stable until 1993. Since then, the pH has increased about 1–1.4 pH units whereas the COD concentrations appear to have decreased from a high in 1992. This suggests that the soluble or "stable"

Table 2.11 Comparison of post-1990 leachate characteristics from different types of US landfills (modified from Bonaparte *et al.*, 2002; additional unpublished data added)

	Waste type (number of landfills)											
	MSW (13–26)			*HW (2–4)*			*MSW Ash (2–7)*			*C&DW (1–2)*		
Parameter	*Average*	*Minimum*	*Maximum*	*Average*	*Minimum*	*Maximum*	*Average*	*Minimum*	*Maximum*	*Average*	*Minimum*	*Maximum*
pH[a]	6.86	5.90	8.5	8.17	7.55	9.36	7.06	6.54	7.44	6.43	6.43	6.43
Specific conductance[b]	3,693	597	13,548	22,096	12,302	39,598	22,083	10,732	43,383	4,815	4,815	4,815
TDS[c]	3,133	480	11,000	–	–	–	24,493	6,067	46,733	3,553	2,880	4,225
COD[c]	1,939	<10	6,800	–	–	–	1,670	304	5,607	2,414	1,139	3,688
BOD[c]	976	<2	4,700	–	–	–	55	15	84	1,126	1,126	1,126
TOC[c]	527	24	2,609	1,623	7	3,239	54	30	109	839	443	1,235
Alkalinity[c]	1,536	203	5,800	–	–	–	1,942	99	5,010	2,450	2,450	2,450
Chloride[c]	660	5	5,600	7,758	3,783	11,734	14,941	2,940	33,000	681	671	690
Sulphate[c]	205	<7	1,376	2,985	704	5,267	881	85	3,430	255	48	463
Calcium[c]	398	66	1,994	–	–	–	2,718	96	8,170	292	203	382
Magnesium[c]	108	10	360	–	–	–	281	113	420	202	202	202
Sodium[c]	282	3	2,900	5,243	2,514	7,972	1,828	684	5,060	304	284	324
Arsenic[d]	23	<2	236	26,710	30	79,912	9	5	17	15	15	15
Cadmium[d]	<7	<1	<20	<119	<5	<233	<12	<2	49	<3	<1	<5
Chromium[d]	38	3	90	124	22	226	<30	<1	84	39	39	39
Lead[d]	15	1	50	109	24	249	23	3	74	7	3	10
Nickel[d]	82	10	220	738	285	1,190	<40	<24	48	<56	<56	<56
Benzene[d]	<19	<2	<100	<131	<7	370	<3	<1	<5	17	17	17
1,1-Dichloroethane[d]	66	<2	260	123	<14	<371	<12	<1	<33	92	92	92
1,2-Dichloroethane[d]	<16	<1	<100	<382	5	<1124	<3	<1	<5	3	3	3
cis-1,2-Dichloro-ethylene[d]	<57	<1	436	–	–	–	<2	<1	<3	–	–	–
trans-1,2-Dichloro-ethylene[d]	<18	<1	<110	<79	<14	<143	<3	<1	<5	–	–	–
Ethylbenzene[d]	35	<1	118	<133	<5	<512	<4	<2	<7	66	66	66
Dichloromethane[d]	334	<1	4,150	161	4	<447	<3	<1	<6	417	417	417
1,1,1-Trichloroethane[d]	<55	<1	270	<99	8	<347	<7	<1	<16	51	51	51
Trichloroethylene[d]	<24	<1	100	<76	33	<146	<3	<1	<5	<11	<11	<11
Toluene[d]	228	<1	740	<173	<9	616	<10	<1	<25	613	613	613
Vinyl chloride[d]	<34	<3	<300	<1475	<10	<4405	<5	<1	<10	8	8	8
Xylenes[d]	83	<5	220	14	9	18	<2	<1	<3	210	210	210

[a] In pH units.
[b] In μmho/cm.
[c] In mg/l.
[d] In μg/l.

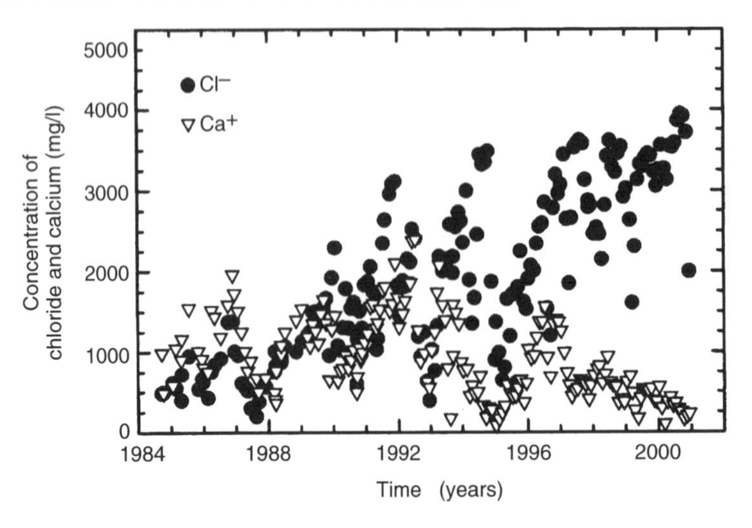

Figure 2.7 Variation in "end-of-the-pipe" chloride (Cl) and calcium (Ca) concentrations with time for the period 1984–2001 (modified from Rowe and VanGulck, 2001).

inorganic load (represented by chloride) in the leachate at the end of the collection system has increased or remained stable while the insoluble or "unstable" fraction of the inorganic loading (calcium) has undergone similar variability but had decreased relative to the "stable" inorganic load. There has also been a decrease in the "end of pipe" organic loading. Again it is unknown as to whether this reflects true leachate behaviour or treatment in the system (or a combination of both).

These leachate data suggest that there is a shift in the "end of pipe" landfill behaviour, with the "stable" inorganic load still increasing while the "unstable" inorganic load is decreasing. This trend is evident in Figure 2.10 which shows a

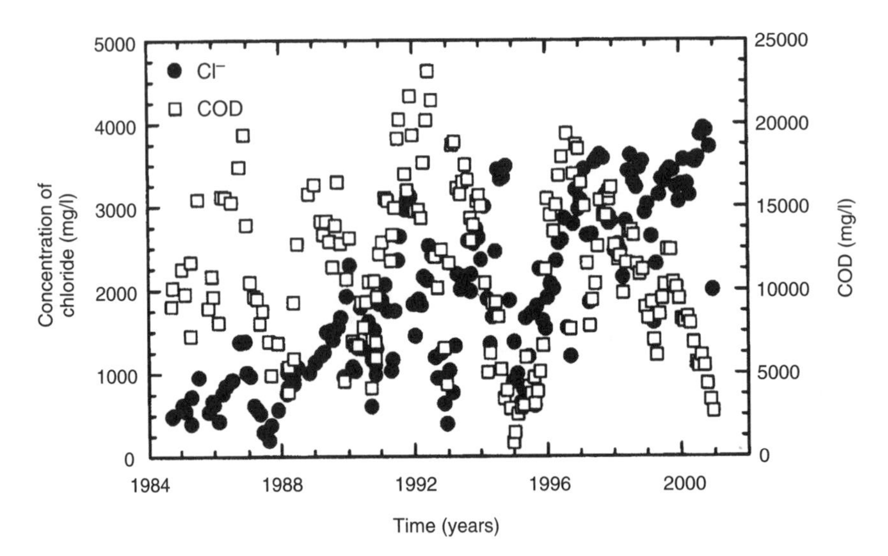

Figure 2.8 Variation in "end-of-the-pipe" chloride (Cl) and COD concentrations with time for the period 1984–2001 (modified from Rowe and VanGulck, 2001).

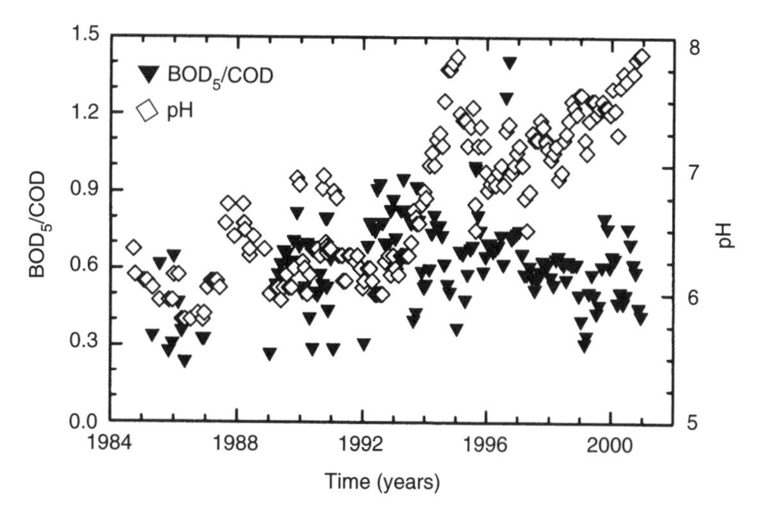

Figure 2.9 Variation in "end-of-the-pipe" BOD_5 to COD ratio and pH with time for the period 1984–2001 (modified from Rowe and VanGulck, 2001).

decrease in leachate organic load represented as COD (dissolved calcium followed a similar trend) normalized by the chloride concentration. It is noteworthy that the ratio of BOD_5/COD remained at about 0.5 suggesting "young leachate" until 2001. This provides some indication that, at least up to 2001, the leachate has been "treated" as it passed through the leachate collection system and does not represent the characteristics of leachate entering the collection system from the waste. Additional evidence is provided by physical observations of clogging of the main header line leading to the main manhole. In the spring of 2001, the main header was occluded with clog material to the extent that a pipe observation camera could not enter into the

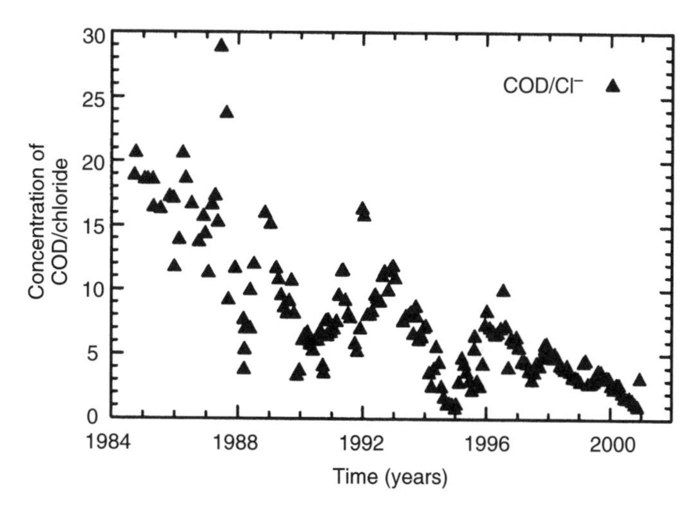

Figure 2.10 Ratio of organic load (COD) to "stable" inorganic load (Cl) in "end-of-the-pipe" leachate with time for the period 1984–2001 (modified from Rowe and VanGulck, 2001).

pipe. Thus, it is likely that some of the decrease in COD and Ca^{2+} and increase in pH in the recent data points in Figures 2.7–2.10 can be attributed to consumption and deposition in the collection system. This is the same leachate collection system examined by Fleming *et al.* (1999) as discussed in greater detail in Section 2.4.4. In 2001 and 2002 the ratio of BOD/COD has dropped and in 2002 the average ratio was about 0.1 suggesting "old" leachate corresponding to the stable methanogenic phase.

2.3.3 Stability in laboratory tests

The dissolved organics in raw MSWL are highly biodegradable so that biochemical alteration continues in the sample bottles brought back to the laboratory for testing. Bacteria growth results in a rapid rise in pH, reduction in E_h (oxygen depletion as in reduction of SO_4^{2-} to S^{2-}) and production of CO^{2-}. Since most leachates seem to have a high concentration of Ca^{2+}, flocs of calcite ($CaCO_3$) appear within a couple of days of laboratory storage if stored at room temperature (20 °C) and Ca^{2+} levels in the leachate experience a rapid reduction. Similarly, Fe^{2+} which may be abundant in acidic leachate declines rapidly as it forms an amorphous black slime of FeS_2. Typical curves showing these phenomena are presented in Figures 2.11 and 2.12.

Although these reactions are greatly inhibited by storage at 4 °C, this temperature is not suitable for long-duration hydraulic conductivity testing. This lack of stability may cause chemical control and interpretation problems in laboratory tests. For example, Ca and Fe concentrations and pH in the influent permeant must be continuously monitored along with the effluent chemistry during hydraulic conductivity testing for clay/leachate compatibility assessment as discussed in Chapter 4. Also, the potential for similar decreases should also be considered during diffusion testing. A review of these problems has been published by Quigley *et al.* (1990b).

2.3.4 Leachable mass of contaminants

The contaminating potential of a landfill depends on both the concentration of contaminants in the leachate and the leachable mass of contaminant in the landfill. One of the early compositional analyses of waste is given in Table 2.12 which, for example, indicates that chloride represents 0.097% (say 0.1%) of the total dry mass of the waste examined by Hughes *et al.* (1971). Belevi and Baccini (1989) estimated the element fraction of waste from an MSW landfill in Switzerland (in mg/kg MSW) as summarized in Table 2.13. Here chloride represents 0.1–0.15% of the mass of waste (1,000–1,500 mg/kg). A summary of the leachable mass of a number of key contaminants is given in Table 2.14.

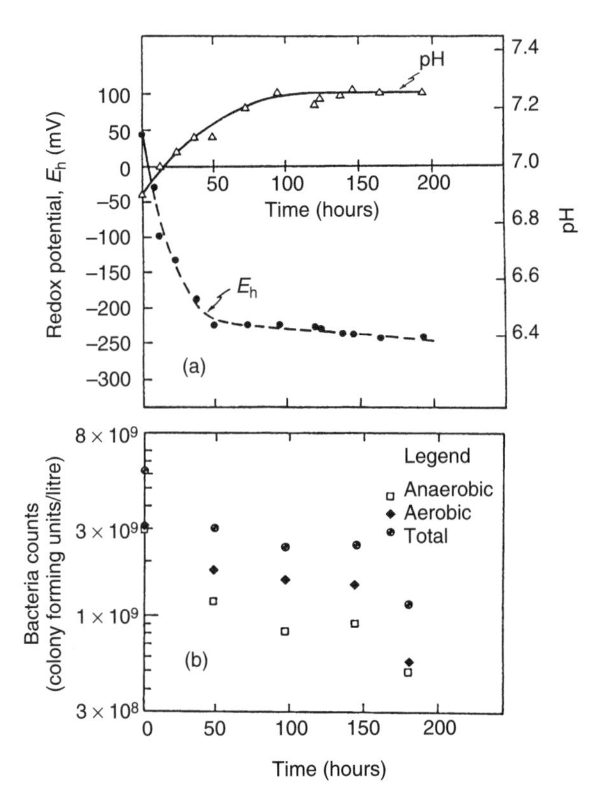

Figure 2.11 Changes in pH, E_h and bacterial population during 8-day experiment on the stability of MSW leachate from London, Canada (modified from Quigley *et al.*, 1990b).

Table 2.12 Municipal solid waste refuse composition (adapted from Hughes *et al.*, 1971)

Component	*Proportion (relative to dry weight of refuse or per cent by weights)*
Crude fibre	38.3%
Moisture content	18.2%
Ash	20.2%
Free carbon	0.57%
Nitrogen	
Free	0.02 mg/g
Organic	1.23 mg/g
Water solubles	
Sodium	2.33 mg/g
Chloride	0.97 mg/g
Sulphate	2.19 mg/g
COD	42.29 mg/g
Phosphate	0.15 mg/g
Hardness	10.12 mg/$CaCO_3$/g
Major metals	
Aluminium, iron, silicon	>5.00% (by spectrographic analysis)
Minor metals	
Calcium, magnesium, potassium	1.0–5.0% (by spectrographic analysis)

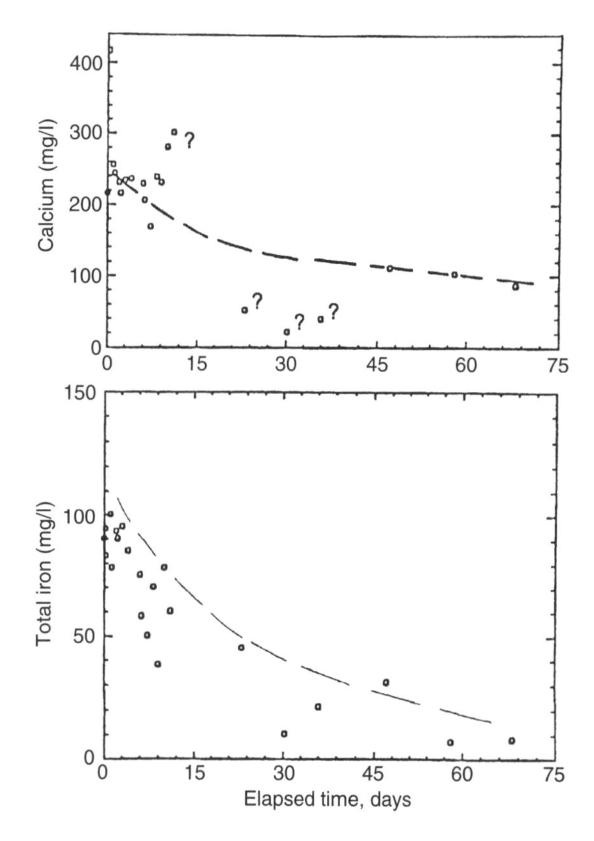

Figure 2.12 Concentration versus time curves for calcium and iron in a typical raw MSW leachate stored in sealed jars at 21 °C.

The mass of selected trace organics and hydrocarbons present in waste from Hamburg, Germany (deduced by Deipser and Stegmann, 1994) are given in Table 2.15. The highest concentrations of trace organics were measured during the acid phase (i.e., in a period of low gas production). The results showed that 28% of the organic chloride content and 29% of the BTEX content were emitted via leachate during the acid phase. With the start of the methane phase the content of trace organics in the leachate was considerably reduced, possibly related to the increase of pH in the leachate and higher gas production rates. Deipser and Stegmann (1994) suggest that lower mass loadings of readily volatile contaminants in the leachate could be obtained by reaching the methane phase in the landfill at earlier times. However, this could lead to greater increases of these components in the gaseous phase.

2.3.5 Contaminating lifespan

The contaminating lifespan of a landfill may be defined as the period of time during which the landfill will produce contaminants at levels that could have unacceptable impact if they were discharged into the surrounding environment. An upper estimate of the contaminating lifespan with respect to different contaminants can be calculated based on the leachate concentrations, leachable mass and regulatory levels permitted on release from the landfill (without treatment or attenuation). Thus, for example, Table 2.13 gives the estimated time, t_{FS}, required before leachate could be directly released as deduced by Belevi and Baccini (1989). This suggests that

Table 2.13 Leachable mass based on Belevi and Baccini (1989)

Element	*Mass proportion* (mg/kg)	c_0 (mg/l)	c_e (mg/l)	t_{FS} *(years)*
C_{org}	2,100–7,100	750	20	500–1,700
N	200–310	1,200	5[a]	55–80
F	ND	0.65	1	< 10
P	5–33	6.8	0.4	100–700
S	ND	2.7	30	< 10
Cl	1,000–1,500	1,300	100	100–150
Fe	20–39	8.0	10	< 10
Cu	1.0–6.7	0.1	0.1	< 10
Zn	14–98	0.6	0.6	< 10
Pb	0.1–2.5	0.07	0.5	< 10
Cd	0.007–0.024	0.002	0.05	< 10

c_0, initial concentration of leachate at the end of the reactor phase; c_e, Swiss wastewater quality standard (1975); t_{FS}, the estimated time for leachate to reach allowable discharge quality; ND, not determined; assumed leach rate/mass of waste = 0.02 l/kg MSW/year.

Table 2.14 Comparison of leachable mass of contaminants (mg contaminant/kg dry mass of waste)

	Mass as a proportion of total (dry) mass of waste (mg/kg) – *assumed dry mass*			
Contaminant	*Belevi and Baccini (1989)*	*Kruse (1994); reported by Kruempelbeck and Ehrig (1999)*	*Heyer* et al. *(1997)*	*Kruempelbeck and Ehrig (1999)*
COD	–	3,000	25,000–40,000	2,500–11,200
TKN and NH_4–N	–	2,200	2,000–4,000	1,400–3,400
AOH	–	9	–	2–22
Chloride	1,000–1,500	2,500	2,500–4,000	900–3,800
Cadmium	0.007–0.024	–	–	0.0016–0.0062
Lead	0.1–2.5	–	–	–
Zinc	–	–	–	0.2–16

Table 2.15 Mass of organics in German waste based on Deipser and Stegmann (1994)

Component	*Concentration* (mg/kg)	*Maximum leachate concentration* (μg/l)
Dichloromethane[a]	0.01–2.68	400
Vinylchloride[b]	ND–0.4	2,500
Trichloroethene[a]	ND–0.59	400
1,1,1-Trichloroethane[a]	0.01–3.65	26,500
Benzene[b]	0.02–0.68	470
Toluene[b]	0.1–2.08	90
Ethylbenzene[b]	ND–0.39	90
o-Xylene[b]	ND–1.21	50
m,p-Xylene[b]	ND–2.93	110

ND, not detected.
[a] Detection limit = 0.005–2 mg/m^3.
[b] Detection limit = 0.1–0.001 mg/l.

the contaminating lifespan with respect to organic carbon is greater than that for metals and other elements (even Cl). However, other studies as reported in Table 2.16 tell a somewhat different story with N controlling the contaminating lifespan. It is acknowledged that NH_4-H is probably the parameter for which it is most difficult to perform these calculations because the concentration curves often do not give a clear indication of the rate of decrease in concentration with time (even after 30 years). Based on this, it could be argued that NH_4-H will control the contaminating lifespan of the landfill rather than COD or chloride. Even leaving N aside, these data suggest that it is likely to take centuries for COD and chloride to reduce to levels where they could be discharged without attenuation or treatment.

To date, estimates of contaminating lifespan of landfill have been based on projections using data collected in the first four phases of a landfill lifetime (i.e., the aerobic, acid, initial methanogenic and stable methanogenic phases); however, Kjeldsen *et al.* (2002) have hypothesized that in the future landfills will experience four subsequent phases (i.e., methane oxidation, air intrusion, carbon dioxide and soil–air) before reaching their final "stable" state. While little is known about how the landfill will behave in these latter phases, Kjeldsen *et al.* indicate that there are a number of factors that will affect metal mobility as the refuse goes from an anoxic to aerobic state including pH, E_h, functional groups on humic acid and the sorptive capacity of refuse. Limited laboratory studies have resulted in conflicting findings suggesting both increased metal mobilization (Martensson *et al.*, 1999) and no observed metal mobilization (Flyhammer, 1997; Revens *et al.*, 1999); more research and long-term laboratory testing is needed in this area since it will be a very long time before relevant field data can be collected.

In Germany, concern regarding the long-term implications of organic matter in waste disposed in landfills has lead to a move to incinerate waste to reduce the carbon content entering landfills. However, there have been questions raised in the literature about this type of landfilling and the effects that the consequent high metal concentrations remaining in the waste after incineration will have on the long-term effects on the environment due to the consequent "ash fills" and these issues have not been adequately investigated. As discussed in Section 2.6, ash fills may also generate high temperatures (Klein *et al.*, 2001) that could significantly impact on the service life of composite liners.

2.3.6 Half-lives of volatile organic compounds

Even though volatile organic compounds (VOCs) generally only occur at low concentrations in modern MSW leachate (see Section 2.3.1), they

Table 2.16 Calculated potential contaminant lifespan of German landfills

	Threshold value (mg/l)	*Kruse (1994) (years)*	*Heyer* et al. *(1997) (years)*	*Kruempelbeck and Ehrig (1999) (years)*
COD	200	280	80–360	65–320
N (TKN, NH_4-N)	70	815	120–450	Decades to centuries
Chloride	100[a]	210	90–250	25–130
AOH	0.5	–	30–210	40–100
Heavy metals	0.1–2	–	–	<10

[a] Swiss wastewater regulation.

are still of concern because of the extremely low allowable concentrations in drinking water and the fact that they have the potential to diffuse through both clay and plastic (geomembrane) liners (see Chapters 8, 12, 13 and 16). Thus while barrier systems can slow the migration of these contaminants, the primary factors controlling the impact on groundwater quality will be the decease in the leachable mass of contaminant in the landfill due to emissions in landfill gas and biodegradation of the compounds in the waste, as well as due to a decrease in concentration as they migrate through the barrier system due to biodegradation. The decrease in concentration with time as these contaminants migrate through soil is often expressed in terms of a half-life. A summary of published half-lives of some chlorinated solvents and hydrocarbon (Tables 2.17 and 2.18, respectively) shows considerable variation in the half-life of each compound. This variation is typical of environmental and biologically mediated systems and is a function of, but not limited to, the effects of redox condition, type of medium, temperature and leachate characteristics (e.g., see Christensen *et al.*, 2001). Due to differences in environmental conditions, different half-lives may be anticipated in leachate and leachate collection systems, the loose sediment at the base of a landfill (or in the geotextile protection layer over a geomembrane), the clay forming part of the landfill barrier and in the granular material of an underlying aquifer. It may also be anticipated that the presence of a geomembrane liner could affect the degradation rates of VOCs in the underlying strata since it will act as a selective barrier (see Chapter 13), readily permitting the diffusion of VOCs but controlling the escape of primary substrates such as volatile fatty acids to minimal levels, thereby reducing the rate of degradation of the VOCs.

The rates of degradation have most extensively been examined in aquifers where there may be a significant variation in redox conditions as the leachate plume migrates away from the landfill. The effect of redox conditions (methanogenic, sulphate reducing, iron reducing and nitrate reducing) has been reviewed by Christensen *et al.* (1994, 2000). In an attempt to identify degradation rates for each redox environment, Tables 2.17 and 2.18 separate, where known, the reported half-lives for each of the primary anaerobic environments. Since not all studies reported either a specific redox condition or had more than one redox condition occurring in the test, the data from these cases are listed as either "unspecified redox" or "mixed redox" as appropriate. All half-lives reported are from anaerobic systems.

Since the redox conditions, medium type and leachate strength may all affect the half-lives of contaminants and there is still a paucity of data related to half-lives both within the landfill itself and in the underlying barrier system, conservative estimates of half-lives will need to be used for landfill approval and design.

2.3.7 Recirculation

There is growing interest in the use of leachate recirculation as a potential means of accelerating decomposition processes and hence waste stabilization. Doing so offers potential environmental and economic advantages in terms of reducing contaminating lifespan, reducing operating and monitoring periods, accelerating settlement, allowing more efficient gas collection and utilization, and providing greater post-closure land use flexibility. This section will summarize the basic concepts behind waste stabilization, the operational methods proposed to enhance this process, what is known about the effectiveness of recirculation and the potential concerns.

(a) Basic concepts

There are a number of very different philosophies regarding the design of cover systems to control leachate generation, and each has implications for the contaminating lifespan of the

Table 2.17 Summary of chlorinated solvent half-lives (years)

Compound	*Iron reduction*	*Sulphate reduction*	*Methanogenesis*	*Mixed redox*	*Anaerobic oxidation iron reduction*	*Unspecified redox*	*Reference*
DCE							
Number of rates	8.0	13.0	3.0	1.0			a
Mean	1.0	23.8	24.8	0.37			a
Maximum	2.7		105.0				a
TCE							
Number of rates	11.0	7.0	10.0	2.0			a
Mean	1.8	5.8	7.6	0.74			a
Maximum	5.8	12.0	57.0				a
Number of rates						47.0	b
Mean						0.76	b
Maximum						13.6	b
Laboratory study				8.0			c
VC							
Number of rates	2.0		3.0		7.0		a
Mean	137.0		121.0		22.2		a
Maximum					63.2		a
Number of rates						19.0	b
Mean						0.24	b
Maximum						5.7	b
DCM							
Number of rates						1.0	a
Rate						0.30	a
Leachate						1.5–5	d

a Reported by Bedient *et al.* (1999).
b Reported by Wiedemeier *et al.* (1999).
c Thornton *et al.* (2000).
d Rowe (1995).

landfill. The low-infiltration (dry) philosophy is generally employed in US Subtitle D landfills and seeks to construct a low-permeability cover as rapidly as possible following waste placement (the *dry-tomb* approach). This philosophy results in low leachate generation with correspondingly low treatment costs, but has the disadvantage of extending the contaminating lifespan. Thus from a short-term economic standpoint, this approach is advantageous to the landfill owner in regions with regulations that limit their responsibility for long-term environmental protection (e.g., to 30 years post-closure). However, from an environmental protection perspective, this may not be desirable.

In contrast to the *dry-tomb* approach, other jurisdictions normally require a final cover system that allows a minimum (e.g., of 0.15 m/a in Ontario, Canada) infiltration into the waste mass (unless otherwise justified). This *moderate infiltration* approach generates larger quantities of leachate than would otherwise be obtained but is intended to help accelerate the stabilization of the waste mass and reduce the contaminating lifespan of the landfill. This approach allows for a shorter aftercare period than the *dry-tomb* approach, although this period may still range from many decades to centuries depending on the size of the landfill. Such a strategy is thus better suited to regulations requiring environmental protection in perpetuity.

It might be hypothesized that increasing the rate of leachate production beyond that arising

Table 2.18 Summary of hydrocarbon half-lives (years)

Compound	*Nitrate reduction*	*Iron reduction*	*Sulphate reduction*	*Methanogenesis*	*Mixed redox*	*Unspecified redox*	*Reference*
Benzene							
Number of rates	41.0	20.0	16.0	15.0	25		a
Mean	4.4	4.5	4.0	5.3.0	4.8		a
Maximum	46.9	17.9	25.8	40.6	45.8		a
Number of rates						41	b
Mean						0.53	b
Maximum						No degradation	b
Grindsted							c, d
Landfill plume					0.29–4.5		
North Bay							e
Landfill plume					1.8–4.9		
Toluene							
Number of rates	49.0	13.0	14.0	24.0	17		a
Mean	242.0	6.1	32.7	19.5	159		a
Maximum	2,270	23.7	110.0	97.9	2,530		a
Number of rates						46	b
Mean						0.03	b
Maximum						1.9	b
Vejen Landfill							c, f
Plume					0.03–0.05		
Grindsted							c, d
Landfill plume					0.17–0.96		
North Bay							e
Landfill plume					0.3–2.1		
Landfill						1.8–12.3	f
Laboratory Study					6.4		g
Laboratory Study					0.04–2.71		h
Ethylbenzene							
Number of rates	37.0	7.0	8.0	12.0	17		a
Mean	142.0	1.8	1.1	5.4	5.1		a
Median	8.3	0.79	0.29	0.55	1.1		a
Maximum	3,180	9.0	3.8	28.4	41.1		a
Ethylbenzene							
Number of rates						37	b
Mean						0.13	b
Maximum						3.2	b
Vejen Landfill							c, f
Plume					0.07–0.57		
Grindsted							c, d
Landfill plume					0.21–1.7		
North Bay							e
Landfill plume					0.9–6.4		
Leachate						0.5–1.4	f
m-Xylene							
Number of rates	41.0	8.0	7.0	12.0	16		a
Mean	46.7	5.3	42.5	1.0	2.2		a
Median	8.9	1.3	29.5	0.53	1.1		a
Maximum	258.0	19.5	168.0	54.8	13.2		a
Number of rates						33	b
Mean						0.08	b
Maximum						1.6	b

p-Xylene							
Number of rates	21.0	8.0	4.0	10.0	18		a
Mean	35.7	5.2	5.7	9.5	3.32		a
Median	4.2	0.97	4.7	1.3	1.05		a
Maximum	231.0	19.5	11.6	42.7	16.3		a
Number of rates						34	b
Mean						0.05	b
Maximum						2.3	b
m,p-Xylene							
Vejen Landfill							c, f
Plume					0.03–0.04		
Grindsted							c, d
Landfill plume					0.11–0.64		
North Bay							e
Landfill plume					0.5–1.4		
Leachate						0.6–1.4	f
o-Xylene							
Number of rates	38.0	8.0	6.0	12.0	16		a
Mean	6.2	1.6	14.1	13.8	4.6		a
Median	2.4	0.95	5.5	0.55	1.6		a
Maximum	35.8	8.4	44.2	112.0	30.0		a
Number of rates						26	b
Mean						0.14	b
Maximum						2.2	b
Vejen Landfill							c, f
Plume					0.06–0.09		
Grindsted							c, d
Landfill plume					0.08–0.23		
North Bay							e
Landfill plume					0.5–9.5		
Leachate						0.6–1.6	f

a Reported by Bedient *et al.* (1999).
b Reported by Wiedemeier *et al.* (1999).
c Based on data reported by Christensen *et al.* (1993).
d Based on data reported by Rugge *et al.* (1995).
e Reported by Rowe (1994) based on data reported by Barker *et al.* (1987).
f Lyngkilde and Christensen (1992).
g Thornton *et al.* (2000).
h Bright *et al.* (2000).

from percolation of precipitation through a suitable cover (as achieved in the *moderate infiltration* approach) by the recirculation of leachate could have additional benefits in terms of further accelerating waste stabilization and decreasing the contaminating lifespan. The process of waste stabilization involves a complex interaction between hydrologic, hydraulic, physical, chemical and biological processes (Al-Yousfi and Pohland, 1998). The breakdown of organic matter in the waste is dominated by microbial activity which influences both the composition of leachate and the generation of landfill gas (Warith and Sharma, 1998). The degradation of waste involves a number of distinct phases (Pohland, 1980; Christensen and Kjeldsen, 1989; Reinhart and Al-Yousfi, 1996). In the first phase, aerobic decomposition, readily degradable organics are utilized by aerobic bacteria with the subsequent production of carbon dioxide. In the second phase, as oxygen is consumed and anaerobic conditions prevail, facultative acetogenic bacteria become active, hydrolysing and fermenting more complex organic matter into organic acids, alcohols, hydrogen and carbon dioxide, resulting in a decrease in pH. Over time, oxygen is further depleted and methanogenic bacteria become active and in the third phase the organic acids produced during the second phase are consumed and methane gas production increases, eventually reaching a steady state. The consequent conversion of organic matter into inorganic material and gas results in a stabilized waste mass and leachate with a greatly reduced organic content.

Aerobic decomposition is typically quite rapid, with completion in less than 2 years under optimum conditions (Baker and Eith, 2000). Anaerobic decomposition is a much slower process, taking place over years to decades, depending on conditions. Although waste stabilization will occur eventually in all landfills, it is the goal of bioreactor landfill design and operation to create and maintain optimum conditions within the waste such that degradation proceeds as rapidly as possible. The bioreactor landfill has been defined by the Solid Waste Association of North America as:

> a sanitary landfill operated for the purpose of transforming and stabilising the readily and moderately decomposable organic waste constituents within five to ten years following closure by purposeful control to enhance microbiological processes. The bioreactor landfill significantly increases the extent of waste decomposition, conversion rates and process effectiveness over what would otherwise occur within the landfill.

(b) Operational techniques

A wide variety of operational techniques have been used to accelerate waste decomposition. Although diverse in their details, these techniques may be broadly categorized into three groups as presented below (Baker and Eith, 2000).

1. *Anaerobic bioreactors*: This operational strategy seeks to increase anaerobic degradation of waste by increasing the moisture content of the waste. Although waste is typically placed at a moisture content of 10–20%, studies (e.g., Stegmann, 1983; Christensen and Kjeldsen, 1989; Barlaz *et al.*, 1990) have shown that anaerobic decomposition is optimized at moisture contents at or above field capacity, typically 45–65%. Thus the addition of moisture, either through leachate recirculation or the addition of other liquids, is an essential component of this operational strategy. Other factors such as pH and nutrient supply also play a key role in the rate of anaerobic decomposition and may have to be controlled through the addition of buffers or nutrient-rich material such as wastewater sludge if the process is to be optimized (Warith and Sharma, 1998).
2. *Aerobic bioreactors*: The operation of a landfill as an aerobic bioreactor seeks to accelerate waste decomposition by optimizing conditions for aerobic bacteria. Due to the greater efficiency of aerobic respiration, waste degradation may proceed more rapidly under this strategy (Baker and Eith, 2000). Oxygen must be added to the system through the injection of air or oxygen into the waste mass. As is the case for anaerobic decomposition, aerobic processes proceed more rapidly at elevated water contents. Moisture must thus be added through leachate recirculation or some other method. Aerobic respiration may generate significant heat, requiring abundant water and increasing the risk of fire, and may also lead to rapid clogging of leachate collection systems (Brune *et al.*, 1991). The higher level of operations management and higher moisture requirements tend to make aerobic bioreactors more expensive to implement than alternative techniques (Baker and Eith, 2000).
3. *Hybrid bioreactors*: The combination of the beneficial effects of aerobic and anaerobic degradation is referred to as a hybrid bioreactor. Typically, the waste will be treated

aerobically for a short period of time in order to rapidly pass the waste mass through the acetogenic phase. This is followed by an operational strategy similar to that of an anaerobic bioreactor. A typical operation would see waste placed in relatively thin layers at the landfill base, followed by a period of aerobic degradation and then full waste placement. This allows a layer of pre-treated waste to form at the base, which in turn is capable of attenuating the acidic leachate formed in the bulk of the waste mass. Thus, the hybrid bioreactor seeks to capitalize on the efficiency of aerobic decomposition and the relative simplicity of the anaerobic operation; it shares the same potential risks (Baker and Eith, 2000).

Details regarding any particular operational strategy will depend on site-specific conditions; however, in all instances the addition of moisture in the form of recirculated leachate or from some other source will be involved. Reinhart (1996) presents a summary of moisture introduction techniques practised at eight full-scale facilities. These methods include pre-wetting of waste (e.g., Doedens and Cord-Landwehr, 1989), leachate spraying (e.g., Robinson and Maris, 1985), surface ponds, vertical injection wells and horizontal subsurface introduction. Alternative methods, such as the injection of steam, have been proposed (e.g., Renaud, 2001), although their effectiveness has yet to be verified. These methods have varying degrees of efficiency and differ with respect to leachate recirculation capacity and compatibility with active and closed phases of landfill operation. Current practice is primarily based on the use of horizontal or vertical systems, or a combination of the two (Reinhart, 1996).

(c) Field studies

Several full-scale bioreactor studies have been conducted (e.g., Reinhart, 1996; Deusto *et al.*, 1998; Reinhart and Townsend, 1998; Yuen *et al.*, 1999; Yazdani *et al.*, 2000) and it has been reported that, in 1996, 130 North American landfills were recirculating leachate to varying degrees although only 10 were attempting to bring the landfill into full bioreactor status (Reinhart *et al.*, 2000). Unfortunately, due to the long period of time over which such studies are conducted, there is a paucity of data relating to the full-scale performance of leachate recirculation systems; the key, albeit limited, findings are summarized below.

The COD concentration half-life has been reported (Reinhart and Al-Yousfi, 1996) to be reduced from an average of approximately 10 years for conventional landfills to 0.78 and 1.05 years in bioreactor trials. Yazdani *et al.* (2000) have also reported a significant reduction in BOD, COD and inorganics such as iron, manganese and calcium in the leachate as a result of recirculation as well as a doubling of gas production and accelerated settlement (16% versus 3% in comparable times for bioreactor and conventional cells, respectively).

(d) Concerns

The limited available data discussed in Section 2.3.7(c) suggest that leachate recirculation has potential benefits. However, there are also potential drawbacks and limitations to the practical effectiveness of leachate recirculation that have been identified in full-scale applications. These include:

1. *Limited effectiveness*: The heterogeneity of waste may lead to significant variations in the moisture content throughout the waste body and may reduce the effectiveness of leachate recirculation operations (Yuen *et al.*, 1999). For example, a lithium tracer introduced at the Lower Spen Landfill was detected in the leachate collection system 40 min later (Blakey *et al.*, 1997). The presence of preferential flow paths may lead to large portions of the waste mass not experiencing the

beneficial effects of increased moisture content, and thus not degrading at an optimal rate.

2. *Inorganic contaminants*: A number of publications (e.g., Reinhart and Al-Yousfi, 1996; Walker *et al.*, 1997; Warith *et al.*, 1999; Jones-Lee and Lee, 2000) have noted that operation of a landfill as a leachate recirculating bioreactor does not decrease the concentrations of inorganic contaminants such as salts and heavy metals (indeed concentrations may potentially increase with recirculation). Thus, although the organic loading may decrease, other pollutants may remain in the landfill at significant concentrations. In order to decrease the concentration of these chemicals to acceptable levels, significant flushing of the waste must take place. Some authors (e.g., Walker *et al.*, 1997; Jones-Lee and Lee, 2000) have advocated a two-stage process whereby the waste mass is flushed with clean water to remove inorganics and heavy metals following the removal of organic matter through methanogenesis. The effectiveness of such systems must be examined in light of waste heterogeneity and the presence of preferential flow paths noted above.
3. *Reduced service lives*: The enhanced biological activity brought about by leachate recirculation may have an adverse effect on the service lives of engineered components of the lining system. Clogging of the leachate collection system may be expected to occur at a greater rate with recirculation than for conventional operation (see Section 2.4.4). Additionally, the operation of a landfill as a bioreactor generally results in increased temperatures which may reduce the service life of geomembrane liners as discussed in Chapter 13. These negative effects may be mitigated to some extent by the reduced contaminating lifespan that may be achieved with leachate recirculation, provided that the accelerated settlement is not used to allow placement of more waste and extend the operating life of the facility. More research is required to assess the implications of recirculation on the service life of the components of the engineered barrier system.
4. *Extended contaminant lifespan*: The potential effects of inorganic contaminant loading noted in (2) above is compounded if the accelerated waste settlement achievable with leachate recirculation landfills is used to increase landfill capacity and hence place more waste. Since leachate recirculation is unlikely to have a significant effect on reducing many inorganic contaminants with the landfill, the placement of additional waste following settlement will serve to increase the total inorganic contaminant load and thus extend the contaminating lifespan of the landfill. This factor must be carefully considered when assessing the feasibility of leachate recirculation, especially when the potential reduction in service life of liner system components is considered.
5. *Optimal conditions*: Recirculation of leachate is not, in and of itself, sufficient to achieve optimum conditions for bioreactor landfills (Phaneuf, 2000). There is some evidence that degradation can be enhanced by adding wastewater sludge, buffers and nutrient sources to the waste (e.g., Viste, 1997; Warith *et al.*, 1999; Yazdani *et al.*, 2000); however, considerable care would be required when adding sludge since it can also accelerate clogging of leachate collection systems (Rowe, 1998b).
6. *Stability*: Excess porewater pressures associated with leachate recirculation has led to instability in at least two cases. Additionally, increased densification and the placement of additional waste following settlement may lead to loading in excess of that for which lining systems and side slopes were designed if the landfill was not originally designed for recirculation. Monitoring of landfill conditions such as moisture

content, leachate head and density should be carried out during leachate recirculation operations.

7. *Leachate seeps*: Due to heterogeneity and anisotropy of waste (see Section 15.2.1), the addition of leachate during recirculation may give rise to leachate seeps created by lateral flow of leachate above relatively low permeability layers (e.g., intermediate cover soils).
8. *Costs*: Several studies have attempted to assess the economic implications of recirculation (e.g., Annex, 1996; Gambelin and Cochrane, 1998; Shaw and Knight, 2000). Bioreactor landfills generally realize economic benefits from enhanced and more rapid gas production, recovered landfill space and reduced post-closure care, while these factors are offset by increased capital and operational costs (Reinhart *et al.*, 2000). Unfortunately, these economic studies are inconclusive, with some favouring recirculation and others favouring conventional operations, depending on the parameters adopted. It is evident that the cost-effectiveness of leachate recirculation must be estimated on a case-by-case basis.

Although bioreactor technology shows promise for improving the performance of landfills, the benefits cannot be assessed in isolation from other aspects of the overall waste management strategy. Increasing levels of composting of organic materials is likely to reduce the effectiveness of biological decomposition processes within landfills. Furthermore, the current practice of disposing household waste in plastic bags limits the effectiveness of moisture addition (Jones-Lee and Lee, 2000). As indicated by Yuen *et al.* (1999), the incorporation of bioreactor technology to an integrated waste management system must be done in a manner that considers the implications of all aspects of the system.

2.4 Collection, clogging and mounding

2.4.1 Function of the collection system

The movement of contaminant through a barrier will, as discussed in Chapter 1, depend on both the advective (Darcy) velocity and diffusion. The Darcy velocity will, in turn, depend on the hydraulic conductivity of the barrier and the hydraulic gradient.

The hydraulic gradient will often depend on the engineering of the landfill since landfill construction will frequently change the hydraulic conditions at a site. For example, modern landfills are commonly designed to have a leachate collection system (see Figures 2.1–2.3) that may serve several functions. First, by lowering the height of leachate mounding, leachate seeps (and consequent contamination of surface waters) can be minimized. Second, by reducing the leachate head acting on the base of the landfill, the hydraulic gradient through the underlying barrier and the Darcy velocity out of the landfill can be reduced to acceptable levels in many cases. Third, by removing contaminant from the landfill, the mass of contaminant available for transport will be reduced.

Of the functions listed above, the second and third are of greatest significance to the impact of the landfill on groundwater quality. Various methods of estimating the height of the leachate mound (and hence the head within the landfill) have been proposed. Once the height of leachate mounding has been calculated, it is a relatively straightforward calculation to estimate the average hydraulic gradient and Darcy velocity through the barrier. As demonstrated in the following section, it is then possible to estimate the impact of the leachate collection system on the mass of contaminant available for transport into the barrier. The following paragraphs will discuss a number of methods available for estimating the height of leachate mounding.

2.4.2 Leachate mounding

If the barrier beneath the waste is flat and of low permeability compared to the waste, then the height, h, of the mound between two drains separated by a distance (see Figure 2.13a) may be estimated from an equation given by Harr (1962) which (on simplification) can be written as

$$h = [\Omega(l - x)x]^{0.5} \quad (2.1a)$$

$$\Omega = \frac{q}{k} \quad (2.1b)$$

where h is measured relative to the head at the leachate collection drains [L], l is the spacing between drains [L], x is the distance from one of these points [L], q is the portion of the steady-state infiltration that is being collected by the drains [LT^{-1}] and k is the hydraulic conductivity of the material between the drains [LT^{-1}]. At the midpoint between drawdown points, this equation reduces to

$$h_{max} = 0.5l\sqrt{\Omega} \quad (2.2)$$

where h_{max} is the maximum height of mounding above the barrier. Based on the variation in h with position x, it can be shown that the average value of h is given by

$$h_{avg} = 0.785h_{max} = 0.393l\sqrt{\Omega} \quad (2.3)$$

For the situation shown in Figure 2.13b, the maximum leachate head is often calculated using Moore's (1983) equation; however, as indicated by Giroud and Houlihan (1995), Moore's equation may result in a significant underestimate of h_{max} (by a factor of 2, or more) for slopes of more than a few per cent and/or values of $\Omega = (q/k) < 0.01$ and hence should not be used. They proposed an alternative, more accurate equation for the maximum head, h_{max}:

$$h_{max} = 0.25\,lj(\sqrt{\tan^2\alpha + 4\Omega} - \tan\alpha) \quad (2.4a)$$

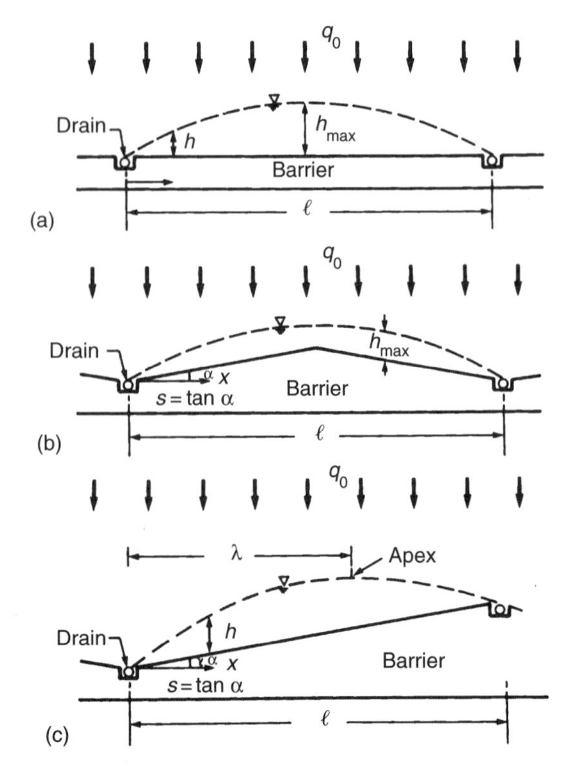

Figure 2.13 Schematic showing different collection systems (after Rowe, 1988; reproduced with permission of the *Canadian Geotechnical Journal*).

where

$$j = 1 - 0.12\exp\{-[\log_{10}(1.6\Gamma)^{0.625}]^2\} \quad (2.4b)$$

and

$$\Gamma = \frac{\Omega}{\tan^2\alpha} \quad (2.4c)$$

Equation 2.4 reduces to Equation 2.1 for $\alpha = 0$, and is easy to use (it can be further simplified by taking $j = 1$ for an error of 12% or less).

The average head, h_{avg}, acting on the liner varies and is given by

$$h_{avg} = \Lambda h_{max} \quad (2.5)$$

where typical values of Λ are given in Table 2.19.

Table 2.19 Relationship between average and maximum head on a liner below a drainage layer (adapted from Giroud and Houlihan, 1995)

Γ	Λ
0	0.5
0.05	0.56
0.1	0.6
0.15	0.63
0.2	0.67
0.35	0.7
0.6	0.75
0.8	0.79
1.0	0.80
1.2	0.83
1.5	0.85
3.0	0.87
10	0.85
30	0.82
50	0.81
1000	0.79
∞	0.785

Assuming that there is a granular drainage blanket, one can readily show that for a value of $k \geq 10^{-5}$ m/s, $q < 0.2$ m/a ($q < 6 \times 10^{-9}$ m/s) and drains at a spacing of 20 m, the maximum steady-state head is likely less than a typical drainage layer thickness of 0.3 m. For a 1% slope the spacing between drains as shown in Figure 2.13b could be 30 m and still maintain a maximum head of less than 0.3 m. As a consequence, these drainage layers are often specified to have a hydraulic conductivity greater than 10^{-5} m/s. Unfortunately, while this may provide the required drainage immediately after construction, a reduction in hydraulic conductivity due to biological, chemical or particulate clogging may quickly result in an excessive leachate mound as discussed in the next subsection. When there is potential for clogging, special care is required to ensure adequate long-term drainage capacity.

McBean *et al.* (1982) considered a different sloping collection system as shown in Figure 2.13c. This case is somewhat more complicated than the previous two cases and one cannot write a simple explicit equation for the height, h, of leachate above the barrier. However, assuming zero pressure head within the drains, the height, h, and the distance at which it occurs can be related by the equation

$$x = \lambda(1 - A \exp B) \tag{2.6a}$$

where

$$A = \frac{\sqrt{\Omega}}{\left(\sqrt{\dfrac{h^2}{(\lambda - x)^2} - \dfrac{sh}{(\lambda - x)} + \Omega}\right)} \tag{2.6b}$$

$$B = \frac{s}{\sqrt{4\Omega + s^2}}\left[\tan^{-1}\left(\frac{-s}{\sqrt{4\Omega + s^2}}\right) - \tan^{-1}\left(\frac{\dfrac{2h}{\lambda - x} - s}{\sqrt{4\Omega + s^2}}\right)\right] \tag{2.6c}$$

The leachate mounding between drains can be readily calculated by determining the location x for an assumed height of mounding h using a successive substitutions algorithm. However, to do so, one must first know the value of λ. This can be done by a process of trial and error:

1. estimate λ;
2. calculate the leachate mound to the right of the lower drain (see Figure 2.13c) over a distance λ;
3. calculate the leachate mound to the left of the upper drain over a distance $(-\lambda)$ (noting that both positive and negative slopes $s = \tan \alpha$ are permitted);
4. if the height h at the location λ (calculated in steps 2 and 3) does not agree, then revise the estimate of λ and repeat the procedure until the calculated mound is continuous between the two adjacent drains.

Equations 2.1–2.6 have been developed assuming that the infiltration q is equal to the flow passing into the collection points. When a portion of the infiltration water moves down through the barrier, the actual flow to the collection points q is equal to the difference between the infiltration q_0 and the flow into the barrier q_i (i.e., $q = q_0 - q_i$). These equations provide an approximate estimate of the head above the barrier.

More elaborate equations for estimating the height of leachate mounding have been proposed (e.g., see Wong, 1977; Demetrocopoulos *et al.*, 1984). These equations have been developed to explicitly consider leakage through the barrier and in this sense are more realistic than equations 2.1–2.6. However, to make the problem tractable, a number of other assumptions have been made. For example, Wong assumes that the leachate collection system is above the water table; the effects of groundwater flow beneath the barrier are negligible; and the leachate instantly saturates a rectilinear volume above the liner and retains this shape while draining towards the collection drain and through the barrier. The adoption of these more elaborate equations is only likely to improve the estimate of leachate mounding if these assumptions are reasonably applicable for a given situation. In many cases, they are not. If the hydraulic conductivity of the barrier is significantly lower than that of the overlying waste and drainage layers, then it may be appropriate to use the simple equations given above (equations 2.1–2.6), recognizing the simplifications involved. If the hydraulic conductivity of the "barrier" is of a similar order to that of the material between the pipes (e.g., the landfill is constructed directly on a silty sand or fractured/weathered till), then the simplified equations are not valid and it may be necessary to use the finite element technique to model the entire flow system (i.e., waste, barrier and underlying hydrostratigraphy).

Mounding with waste below the drains

A special case of mounding where the drains are located above the bottom of the landfill and flow can occur both above and below the level of the pipe can be encountered when considering inducing drawdown of the leachate mound by inserting pipes or trenches into the waste body at some distance, D, above a relatively impermeable base of the landfill. This situation also arises if the leachate mound is being controlled by perimeter drains hydraulically connected to the waste body at or near ground surface along the edges of a cell (e.g., see Figure 2.1). Under these circumstances the maximum leachate mound can be estimated from the Hooghoudt equations which can be written as

$$h_{max} = 0.5l\left(\sqrt{\left(\frac{2d}{l}\right)^2 + \Omega} - d\right) \qquad (2.7)$$

where h_{max} is the maximum height of the mound above the level of the drains and d (which is a function of l, D and the radius of the drain, r_0) represents the equivalent depth of waste below the drains contributing to flow to the drains and is given in Tables 2.20a and 2.20b.

2.4.3 Examples of mounding calculations

To focus discussion on design considerations, it is useful to begin with a number of examples.

(a) French drains

Figure 2.14a shows a schematic of a relatively flat landfill base with French drains (in this case pipe surrounded by granular material) at a spacing of 30 m. Based on equations 2.1–2.3, it is possible to estimate both the maximum and average leachate heads acting on the barrier. Assuming that the infiltration through the cover to the waste, q_0, is 0.15 m/a and conservatively neglecting flow through the barrier (i.e., taking $q = q_0$), the heads calculated for two different assumptions concerning the hydraulic conductivity of waste, k_w, are:

(a) Assume $k = k_w = 10^{-6}$ m/s $=31.5$ m/a, then:

maximum head (using equation 2.2): $h_{max} = 0.5 \times 30(0.15/31.5)^{0.5} = 1.03$ m;
average head (using equation 2.3): $h_{avg} = 0.785h_{max} = 0.81$ m.

Table 2.20a Values for equivalent depth, *d*, for Hooghoudt equation (equation 2.7) for typical drain radius $r_0 = 0.1$ m; *D* and *L* are in m

D \ *L*	*5*	*10*	*15*	*20*	*25*	*30*	*35*	*40*	*45*	*50*
0.5	0.47	0.49	0.49	0.49	0.5	0.5	0.5	0.5	0.5	0.5
1.0	0.67	0.80	0.86	0.89	0.91	0.93	0.96	0.96	0.96	0.96
1.5	0.70	0.97	1.11	1.19	1.25	1.28	1.31	1.34	1.35	1.36
2.0		1.08	1.28	1.41	1.5	1.57	1.62	1.66	1.70	1.72
2.5		1.13	1.38	1.57	1.69	1.79	1.87	1.94	1.99	2.02
3.0			1.45	1.67	1.83	1.97	2.08	2.16	2.23	2.29
4.0			1.52	1.81	2.02	2.22	2.37	2.51	2.62	2.71
5.0				1.88	2.15	2.38	2.58	2.75	2.89	3.02
6.0					2.20	2.48	2.70	2.92	3.00	3.26
7.0						2.54	2.81	3.03	3.24	3.43
8.0						2.57	2.83	3.13	3.35	3.56
9.0							2.89	3.18	3.43	3.66
10.0								3.23	3.48	3.74
∞	0.71	1.14	1.53	1.89	2.24	2.58	2.91	3.24	3.56	3.88

(b) Assume $k = k_w = 10^{-7}$ m/s ≈ 3.15 m/a, then:

maximum head (using equation 2.2): $h_{max} = 3.27$ m;
average head (using equation 2.3): $h_{avg} = 2.57$ m.

These values together with a number of other cases are summarized in Table 2.21.

It is evident from these calculations that substantial leachate mounds can develop on the base of the landfill when French drains are used to control the mounding. The height of the

Table 2.20b Values for equivalent depth, *d*, for Hooghoudt equation (equation 2.7) for typical drain radius $r_0 = 0.1$ m; *D* and *L* are in m

D \ *L*	*75*	*100*	*150*	*200*	*250*
2	1.80	1.85	1.90	1.92	1.94
3	2.49	2.60	2.72	2.79	2.83
4	3.04	3.24	3.46	3.58	3.66
5	3.40	3.78	4.12	4.31	4.43
6	3.85	4.23	4.70	4.97	5.15
7	4.14	4.62	5.22	5.57	5.81
8	4.38	4.95	5.68	6.13	6.43
9	4.57	5.23	6.09	6.63	7.00
10	4.74	5.47	6.45	7.09	7.53
12.5	5.02	5.92	7.20	8.06	8.68
15	5.20	6.25	7.77	8.84	9.64
17.5	5.30	6.44	8.20	9.47	10.4
20		6.60	8.54	9.97	11.1
25		6.79	8.99	10.7	12.1
30			9.27	11.3	12.9
40			9.44	11.8	13.8
50				12.1	14.3
∞	5.38	6.82	9.55	12.2	14.7

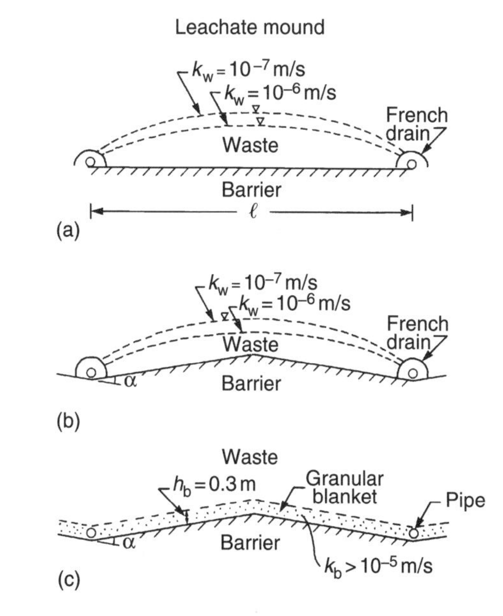

Figure 2.14 Leachate mounding for three cases (not to scale).

Table 2.21 Calculated maximum head acting on the base of a landfill for a number of assumed conditions; infiltration: 0.15 m/a

Spacing of pipes (m)	*Hydraulic conductivity*, k, *between pipes* (m/s)	*Slope of base*, s = tan α *(see Figure 2.13)*	*Maximum head*, h_{max} (m)
2.5	10^{-4}	0	0.01
	10^{-5}	0	0.03
	10^{-6}	0	0.09
	10^{-7}	0	0.27
10	10^{-4}	0	0.03
	10^{-5}	0	0.11
	10^{-6}	0	0.34
	10^{-7}	0	1.09
30	10^{-4}	0	0.10
	10^{-5}	0	0.33
	10^{-6}	0	1.03
	10^{-7}	0	3.27
30	10^{-4}	0.01	0.05
	10^{-5}	0.01	0.26
	10^{-6}	0.01	0.96
	10^{-7}	0.01	3.20
30	10^{-4}	0.02	0.03
	10^{-5}	0.02	0.21
	10^{-6}	0.02	0.9
	10^{-7}	0.02	3.12
50	10^{-4}	0.02	0.05
	10^{-5}	0.02	0.35
	10^{-6}	0.02	1.49
100	10^{-4}	0.02	0.1
	10^{-5}	0.02	0.69
	10^{-6}	0.02	2.98

mound can be reduced either by decreasing the pipe spacing or increasing the slope of the barrier between the drains (thereby encouraging more flow to the drains). A few calculations will quickly indicate that reducing the pipe spacing alone is not an effective means of reducing the maximum head on the liner to a nominal (say 0.3 m) value. For example, even with a hydraulic conductivity of the waste of $k_w = 10^{-6}$ m/s, a spacing of less than 8.7 m would be required. To allow for the potential reduction in the hydraulic conductivity to $k_w = 10^{-7}$ m/s (e.g., due to consolidation of the waste) the spacing of drains required to maintain a maximum head of 0.3 m would be about 2.75 m.

By increasing the slope between drains, as shown in Figure 2.14b, the head on the liner can be reduced as summarized in Table 2.21. Although the slope clearly has some effect, it only becomes significant in terms of substantially reducing the leachate mound if the hydraulic conductivity of the waste is of the order of 10^{-5} m/s or higher. While this may sometimes occur, it cannot be relied upon.

(b) Granular blanket drains

An alternative to using French drains to reduce leachate mounding is to use a blanket drain as shown in Figure 2.14c. When blanket drains have been used, they have often been constructed

Table 2.22 Summary of observations from exhumation of collection systems in North America

Waste type;[a,b] *age; leachate*	*Collection system design*[c]	*Key observations*
1. MSW and LI ~4 years COD = 14,800 mg/l BOD_5 = 10,000 mg/l pH = 6.3 Performance: • Adequate at time of exhumation	• Blanket underdrain: waste over 50 mm relatively uniform stone; 200 mm SDR 11 HDPE pipe; 8 mm holes • Pipe never cleaned • No geotextile between waste and stone	• 30–60% loss of void space in upper stone • 50–100% loss of void space near pipe • Permeability of stone decreased from $\sim10^{-1}$ to $\sim10^{-4}$ m/s • All lower holes in pipe blocked; majority of upper holes blocked • Large clog growth inside pipe
2. MSW and LI As for 1 above	• Blanket underdrain: waste over geotextile over 50 mm stone (rest as above) (GT: W; $M_A = 180\,g/m^2$, AOS = 0.475 mm, $t_{GT} = 0.6$ mm, $\psi = 0.04\,s^{-1}$)	Substantially less clogging than observed in 1 above where there was no GT • 0–20% loss of void space in upper stone below geotextile
3. MSW and LI; LR ~10 years COD = 31,000 mg/l BOD_5 = 27,000 mg/l pH ~ 6.9 Performance: • No flow in LCS • High leachate mound	• Toe drain only • Trench with 600 mm of crushed stone (6–30 mm) around geotextile wrapped 100 mm SDR 41 perforated PVC pipe (GT: HBNW; $M_A = 150\,g/m^2$, AOS = 0.15 mm; $t_{GT} = 0.30$ mm, $\psi = 1.1\,s^{-1}$)	• Flow reduction noted after 1 year • Pipe crushed (likely due to construction equipment) • Substantial reduction in void space and cementing of stone. k reduced from 2.5×10^{-1} to 1.2×10^{-4} m/s • Sand (SW; AASHTO 10) layer above GM was clogged and leachate drained on top (not through this layer). k reduced from 4×10^{-4} to 2×10^{-7} m/s • Excessive clogging of GT (see Table 2.24)
4. MSW and LI; LR 6 years COD = 10,000 mg/l BOD_5 = 7,500 mg/l pH ~ 7.5 Performance: • Drain functioning adequately	• Perimeter drain to control leachate seeps geotextile wrapped trench with 6–18 mm gravel and 100 mm SDR 30 HDPE perforated pipe (GT: W, $M_A = 170\,g/m^2$; POA = 7%, AOS = 0.25 mm; $t_{GT} = 0.41$ mm; $\psi = 0.9\,s^{-1}$)	• Only small reduction in k of gravel from 5.3×10^{-1} to 2.8×10^{-1} m/s • Marginal clogging of GT (see Table 2.24)
5. ISS (included slurried fines 70% finer than 150 μm) 0.5 years COD = 3,000 mg/l BOD_5 = 1,000 mg/l pH = 9.9 Performance: • No flow in LCS	• Blanket underdrain: waste over protection sand (0.075–4 mm) over geotextile (AOS = 0.19 mm) over pea gravel (1–20 mm) drainage layer; 100 mm diameter geotextile wrapped HDPE perforated pipe; 12 mm diameter holes (GT: NPNW, $M_A = 330\,g/m^2$, AOS = 0.19 mm; $t_{GT} = 2.7$ mm, $\psi = 1.8\,s^{-1}$)	• High leachate mound • Upper geotextile functioning (see Table 2.24, Case 5a) • Pea gravel relatively clean • Geotextile wrapping around perforated pipe excessively clogged (see Table 2.24, Case 5b) • Once geotextile sock removed, leachate flowed freely • Heavy geotextile sock clogging at location of perforations in pipe

[a] MSW, municipal solid waste; LI, light industrial; ISS, industrial solids and sludge; LR, leachate recirculation.
[b] References: Cases 1 and 2 – Rowe *et al.* (1995a), Fleming *et al.* (1999); Cases 3, 4 and 5 – Koerner and Koerner (1995b).
[c] GT, geotextile; W, woven; HBNW, heat bonded, nonwoven; NPNW, needle-punched, nonwoven; SDR, Standard dimension ratio.

from sand with hydraulic conductivities of 10^{-4}–10^{-5} m/s; although, it is now recognized that sand is highly susceptible to clogging (see Section 2.4.4) and hence uniform coarse gravel with an initial hydraulic conductivity of the order 10^{-2} m/s (or higher) is now preferred. For a pipe spacing of 30 m and a hydraulic conductivity of 10^{-5} m/s, a maximum leachate head of less than 0.3 m can be readily obtained, provided that there is a small slope (about 0.4%) on the base. With a 2% slope, the spacing between pipes can be increased to 43 m and still maintain a maximum head of less than 0.3 m on the base. With a coarse gravel drainage medium (k about 10^{-2} m/s), the head can be maintained below the nominal 0.3 m design value for a spacing of less than 870 m. In all cases consideration needs to be given to how the leachate level will rise as clogging occurs and the hydraulic conductivity of the drainage medium drops, as discussed in the following sections.

(c) Clogged granular blanket drain

While the foregoing works well in theory, experience has indicated that sand blankets placed directly beneath waste have a tendency to clog as discussed in Section 2.4.4. This clogging may be a result of a combination of particulate clogging, clogging due to chemical precipitation and clogging due to biofilm growth.

It must be recognized that "clogging" of a drainage layer is not synonymous with it becoming impermeable. On the contrary, a clogged sand blanket may still be substantially more permeable than, say, an underlying clay liner. "Clogging" of a drainage layer becomes significant when the hydraulic conductivity of the blanket drops to, or below, the hydraulic conductivity of the overlying waste. For example, consider a leachate collection system similar to that shown in Figure 2.14c where a 0.3 m thick granular blanket is constructed over a compacted clay liner. Assume the following:

1. design maximum head on liner: <0.3 m;
2. drain spacing: $l = 30$ m;
3. base slope: $s = 0.01$ (1%);
4. hydraulic conductivity of waste: $k_w = 10^{-6}$ m/s;
5. initial hydraulic conductivity of blanket: $k_s = 10^{-5}$ m/s $= 315$ m/a; and
6. infiltration through waste to collection system: $q_0 = 0.15$ m/a.

Design conditions

Under operating conditions the maximum height of leachate mounding is given by equation 2.4 to be 0.26 m (see Table 2.21). Since this is less than the thickness of the 0.3 m thick sand blanket, the design requirement of a maximum head of less than 0.3 m is met.

Moderate clogging

The sand/granular blanket may be regarded to have clogged once its horizontal hydraulic conductivity drops to a value similar to that of the waste. Once this occurs, leachate migration to the drains will no longer be preferential to the drainage blanket, although a significant proportion of the leachate may still be being transmitted horizontally to the drains through the "clogged" sand blanket.

In this example, assume k_s (moderately clogged) $= k_w = 10^{-6}$ m/s. This gives a maximum mound of 0.96 m. Thus the head on the base of the liner exceeds the design value by a factor of 3 despite the fact that at the location of maximum mound, approximately one-third of the lateral flow is still going through the (moderately clogged) sand blanket and near the drains almost all the lateral flow is still in the sand blanket.

Severe clogging

Severe clogging of a sand/granular blanket may be regarded to have occurred if the horizontal hydraulic conductivity drops to more than an order of magnitude less than that in the waste. Once this occurs, the majority of lateral flow towards the drains will occur in the waste. For example, suppose the hydraulic conductivity of the blanket drops to 10^{-7} m/s. Under these circumstances, the blanket effectively becomes part of the "barrier" and mounding will occur above the blanket.

It should be noted that while this level of clogging substantially increased the height of mounding and hence hydraulic gradient across the barrier, it does not significantly affect the hydraulic conductivity of the barrier because the flow will still be controlled by the clay liner (and/or geomembrane if one is present) whose vertical hydraulic conductivity is likely to be one to three orders of magnitude lower than that of even a severely clogged granular blanket. If the hydraulic conductivity of the liner is of the order of 10^{-10} m/s (or smaller) then even with mounding as calculated above, diffusion is still a major transport mechanism. For example, assuming a 1.2 m thick clay liner with a hydraulic conductivity of 10^{-10} m/s and underlain by a second unsaturated drainage layer, it can be shown that the Darcy velocity through the clogged sand and liner is approximately 0.006 m/a (the means of performing this calculation is discussed in Chapter 5). In this context, the reader may wish to refer again to the discussion of Figure 1.22 in Section 1.3.7. Under these circumstances, the sand blanket does not contribute to the hydraulic performance of the barrier, but it will become part of the "diffusion barrier" and the diffusion profile may be expected to begin at the interface between the more permeable waste (where lateral flow dominates) and the less permeable (clogged) granular blanket. An example of this is given with respect to the observed field diffusion profile at the Keele Valley Landfill (KVL) discussed in Section 9.3.2.

2.4.4 Clogging of leachate collection systems

Field examples of clogging of leachate collection systems can be found in a number of existing landfills including Toronto's large Brock West Landfill where a 20 m high leachate mound built up during the first 11 years of operation (Rowe, 1998b). Here, the primary leachate collection system consists of leachate collection pipes at a spacing ranging between 50 m (newer portions of the landfill) and 200 m (older portions of landfill). The collection system is reported to involve "French drains" with a perforated pipe in pea gravel (5–10 mm diameter) pipe bedding. No geotextile was used. The landfill had accepted a significant quantity of sewage sludge in addition to MSW.

Another example of clogging of the drainage material around leachate collection pipes in the drain around a landfill has been reported by McBean *et al.* (1993), who noted that extensive clogging had resulted in excessive leachate mounding and leachate seeps. This design also involved French drains at wide spacing.

A study of German landfills by Brune *et al.* (1991) provided insights into a number of issues related to clogging. One of these landfills had a high leachate strength (COD: 50,000–80,000 mg/l; Ca^{2+}: 3,500 mg/l) associated with rapid disposal of waste experienced heavy clogging (incrustation) upon visual inspection of the leachate collection pipes. A second site with lower leachate strength (COD: 1,000 mg/l; Ca^{2+}: 130 mg/l) associated with slow placement of pre-composted waste showed practically no clogging in the leachate collection pipes at the time of inspection. Comparing these two cases, it is evident that high concentrations of organic and inorganic substances have an important role in the incrustation (clogging) of a leachate collection system.

The KVL provides two different examples of clogging. First, there is a sand "protection" layer over the liner. Field exhumations (Reades *et al.*, 1989) have shown that this sand blanket has become clogged and does not contribute to the hydraulic performance of the collection system, but rather has become part of the "diffusion barrier" with the diffusion profile beginning at the interface between the more permeable waste (where lateral flow dominates) and the less permeable (clogged) sand blanket (see Section 9.3.2). Koerner and Koerner (1995b) reported similar clogging of a sand protection layer (Case 3, Table 2.22) where after 10 years the hydraulic conductivity dropped three orders of magnitude from 4×10^{-4} to 2×10^{-7} m/s and leachate was flowing through the waste rather than the sand.

Exhumation of portions of the leachate collection system at the KVL after 4 years of operation (see Case 1, Table 2.22 and Fleming *et al.*, 1999) indicated three orders of magnitude drop in the hydraulic conductivity of the relatively uniform 50 mm gravel near the leachate collection pipe (although the hydraulic conductivity was still sufficient to transmit leachate). As discussed in more detail in Section 2.4.4(b), clogging was observed to be substantially less (Case 2, Table 2.22) in areas where a geotextile was used between the waste and the drainage gravel.

Koerner and Koerner (1994, 1995) have described exhumations of three leachate collection systems (see Table 2.22, Cases 3, 4 and 5) and found excessive clogging of the geotextile in two cases where the geotextile was wrapped either around the perforated pipe or around the granular material in a drainage trench.

The mechanisms associated with the clogging observed in these cases and the practical implications are discussed in the following subsections.

(a) The clogging process

Field studies and long-term laboratory tests have shown that the clogging of the drainage and filter materials is the result of an accumulation of material within the pore space of the drainage medium resulting in a decrease in porosity and hydraulic conductivity. The material which clogged the drainage systems (Figures 2.15 and 2.16) consists of a network of aggregates of bacteria and deposits of inorganic material (Brune *et al.*, 1991; Rowe *et al.*, 1995a; Fleming *et al.*, 1999). This process is initiated in the bacterial cells and the slime fibrils which they excrete. Precipitation then centres around these seeds. Chemical analyses (Table 2.23) indicate that the precipitate predominantly consists of calcium, iron, magnesium and manganese combined with carbonate and sulphur.

As indicated by Ramke and Brune (1990), the precipitation and clogging of drainage material in landfills is primarily caused by two processes:

- Iron-reducing bacteria solubilize Fe(III) by reducing it to Fe(II) while sulphate-reducing bacteria reduce sulphate to sulphide. This bioreduction causes the region immediately around the bacteria to become more alkaline and leads to the precipitation of sulphur as an insoluble metal sulphide.
- Calcium is mobilized in the waste as a result of fermentative organisms producing organic acids which lower the pH of the leachate and hence mobilize calcium (and other metals). As in the case of sulphide, it would appear that calcium carbonate then precipitates around the bacteria as discussed above.

Thus, the clogging of drainage systems is the result of a mobilization process involving fermentative bacteria together with iron- and manganese-reducing bacteria, followed by precipitation processes involving primarily methane- and sulphate-reducing bacteria. These latter processes are identical to the precipitation processes described in Section 2.3.3.

The exhumation of a portion of the relatively uniform 50 mm gravel drainage blanket at the KVL after 4 years of operation found that clogging was considerably greater in the lower (saturated) zone of the drainage blanket (with 50–80% of the voids being filled with clog material) than in the upper (unsaturated) zone (where 30–60% of the voids were filled with clog material). This implies that clogging is likely to be far greater when the drainage material is saturated with leachate than when it is unsaturated and is consistent with the findings from large-scale laboratory (mesocosm) test results (Fleming *et al.*, 1999).

Also after 4 years of operation, the 8 mm diameter perforations in the exhumed leachate

Figure 2.15 Clog sample removed from inside the leachate collection pipe after 4 years of operation.

collection pipe (which had not been cleaned) were mostly blocked by clog material, especially the lower rows of holes. Within the perforated pipes, significant accumulation of solid material had occurred in the form of large, loosely cemented pieces. The clog samples removed from the inside pipes were substantially larger than the opening size of the perforated piping (Figure 2.15) and had formed inside the pipe due to precipitation of minerals from the leachate (Table 2.23). Likewise at the modern Halton Landfill, which began accepting waste in 1992 (Rowe *et al.*, 2000c) and where the pipes have been cleaned annually, significant clogging (requiring special effort to clean) was observed in the leachate collection pipes after 9 years of operation, necessitating an increase in cleaning frequency to 2–3 times per annum. It is significant that the waste disposed in the landfill is the residual after an intensive effort to divert waste that can be recycled and composted from the landfill, and hence this leachate can be regarded as leachate generated from a modern waste stream where all reasonable efforts have been taken to minimize the waste going to the landfill.

(b) Role of geotextile filter–separator layer

A comparison of clogging where the waste was in direct contact with the underlying drainage blanket with that where a woven, slit-film, geotextile filter had been placed between the waste and the underlying drainage blanket at the KVL (Fleming

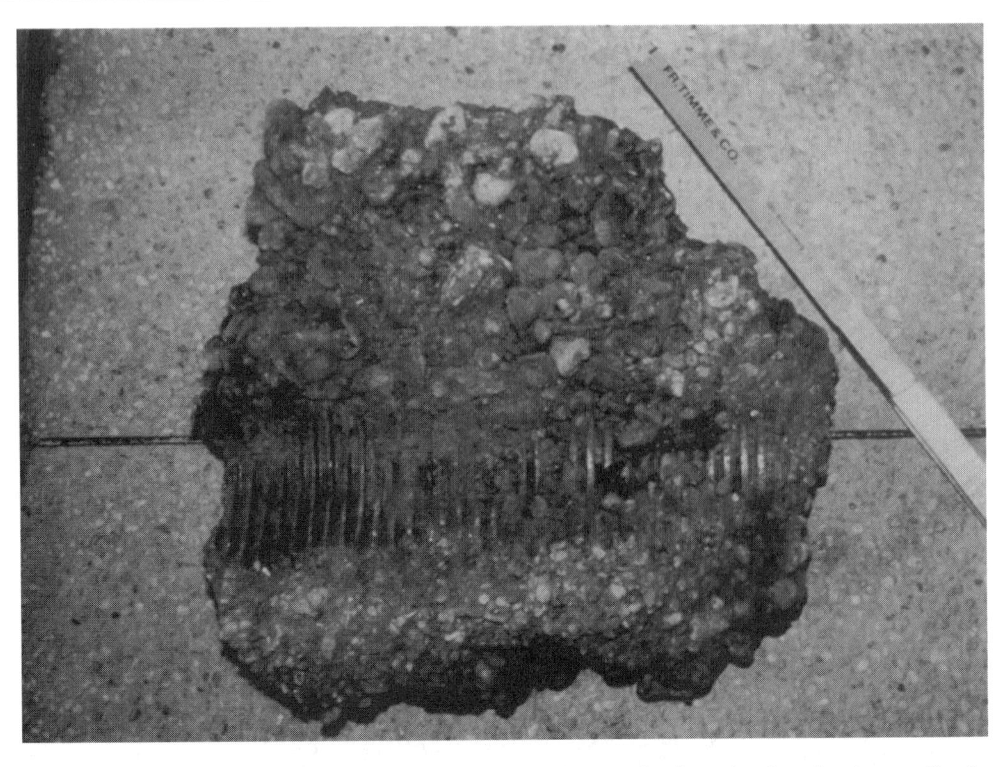

Figure 2.16 Clogging of granular material from around a leachate collection pipe in a leachate collection system (Photo courtesy of Prof. H.J. Collins).

et al., 1999) showed a significant difference in the upper unsaturated portion of the collection system (i.e., the portion influenced by the presence of the geotextile filter). Where the geotextile filter had been used, between 0 and 20% of the pore space was filled with clog material whereas there was a 30–60% loss of void space in a comparable area where no geotextile was used. This highlighted the advantages of having a suitable separator/filter between the waste and any underlying coarse gravel drainage blanket.

Rowe *et al.* (1995a), Fleming (1999) and McIsaac *et al.* (2000) described the findings of a mesocosm study designed to examine, under otherwise similar conditions, the performance of different combinations of materials used to construct leachate collection systems. The mesocosms were full-scale collection system designs that were fed KVL leachate at a vertical and horizontal flow rate representative of the field at a location near the collection pipe and run under anaerobic conditions.

The mesocosm collection system designs generally consisted of a waste layer overlying a coarse gravel drainage layer. Selected mesocosms also had a filter/separator layer between the waste and drainage gravel. The various separator layers included a nonwoven needle-punched polypropylene geotextile and a slit-film woven geotextile. A comparison of the drainable porosity of the drainage gravel after 3.75 years showed that the woven geotextile was less effective in reducing clogging of the drainage layer than the nonwoven geotextile but that both were much better than the case where no filter was used. Mesocosms with a filter had about 20–40% higher porosity in the drainage layer than those without a filter (for similar drainage media and flow conditions).

Table 2.23 Comparison of clog composition (% of total dry mass except for Ca/CO_3 ratio) in field and column studies

	Ca	*CO_3*	*Si*	*Mg*	*Fe*	*Sum*	*Ca/CO_3*
Brune *et al.* (1991)							
German landfill (field study)	21	34	16	1	8	80	0.62
Fleming *et al.* (1999)							
Toronto landfill (field study)	20	30	21	5	2	78	0.67
Rowe *et al.* (2000b)							
Particle size series							
4 mm	27	49	4	1	3	84	0.55
6 mm	24	50	3	1	4	82	0.48
15 mm	26	52	2	<1	3	84	0.50
Rowe *et al.* (2000a)							
Mass loading series							
0.51 $m^3/m^2/d$	24	50	3	1	4	82	0.54
1.02 $m^3/m^2/d$	27	58	3	1	4	93	0.47
2.04 $m^3/m^2/d$	27	49	3	<1	4	84	0.55
Armstrong (1998)							
Temperature							
21 °C	30	47	2	<1	2	82	0.64
27 °C	24	50	3	1	3	81	0.48

Tests conducted by Brune *et al.* (1991) using a nonwoven geotextile filter above the drainage material also indicated that the geotextile substantially reduces clogging of the underlying drainage material but with the formation of a cake of clogged waste in the upper portion of the geotextile and above the geotextile (see Figure 2.17). The beneficial role of the geotextile was most evident for the coarse gravel (16–32 mm) which remained uncemented with only a very light biofilm around the gravel.

(c) Clogging of geotextile filters

The geotextile filters discussed in the previous section provided protection to the underlying gravel drainage layer but at the same time did experience some clogging. This was most evident for the high-strength leachate columns examined by Brune *et al.* (1991; see Figure 2.17). An indication of the degree of clogging of geotextile filters that can occur due to accumulations of leachate fines and microbial growth may be obtained from studies by Cancelli and Cazzuffi (1987), Koerner and Koerner (1989, 1990), Cazzuffi *et al.* (1991), Brune *et al.* (1991), Fourie *et al.* (1994), and Colmanetti and Palmeira (2002). The study by Koerner and Koerner (1989) indicates a rough correlation between the TS/BOD levels of the leachate and the degree of geotextile clogging. The study by Brune *et al.* (1991) indicated only modest clogging of a geotextile for leachate 7 given in Table 2.6 but considerable clogging for leachate 8 in Table 2.6.

The studies noted above show that the magnitude of the decrease in hydraulic conductivity k_n (and permittivity, ψ) will depend on the geotextile (e.g., openness of the pore structure), the flow rate and the concentration of the leachate. Koerner and Koerner (1995b) provided data from a series of tests where mild leachate was passed through a number of different geotextiles yielding a decrease in hydraulic conductivity of between about two and five orders of magnitude at the highest flow rate examined. The worst hydraulic conductivity observed for nonwoven needle-punched geotextile was similar to or higher than that observed in the field (6×10^{-8} m/s in Table 2.25 versus 4×10^{-8} m/s in Table 2.24) under the most extreme conditions. Generally, the level of decrease was similar (four orders of magnitude) to that in the field under severe conditions. The implications of these findings with respect to design are discussed in Section 2.4.6.

(d) Interactions between leachate and drainage gravel

Brune *et al.* (1991) hypothesized that carbonate drainage gravel were unsuitable for use in the drainage layer because they could dissolve and contribute to subsequent calcite crystallization on alkaline biofilms covering drainage gravel and potentially add to the clogging process. The use of carbonate aggregate as a drainage material for landfill site is discouraged by the United States Environmental Protection Agency (Daniel and Koerner, 1993a) and forbidden in some states (Niemann and Hatheway, 1997).

Figure 2.17 Clogging of a geotextile in laboratory tests performed by Brune *et al.* (1991) (Photo courtesy of Prof. H.J. Collins).

Niemann and Hatheway (1997) performed a laboratory study of the dissolution of carbonate as leachate with a pH of 6–6.5 passed over a limestone aggregate. A negligible weight loss occurred and they concluded that significant dissolution of these types of carbonate material would not occur in landfills. Manning and Robinson (1999) and Jefferis and Bath (1999) assessed the potential for carbonate dissolution by considering the leachate chemistry. Both studies concluded that calcium carbonate was supersaturated in leachate if the landfill was in the methanogenic stage and that it is unlikely that dissolution would occur under these conditions.

Jefferis and Bath (1999) questioned whether calcium carbonate was supersaturated during the acetogenic phase of the landfill. However, Owen and Manning's (1997) United Kingdom leachate study showed supersaturation with respect to calcite, aragonite and dolomite, and undersaturation with respect to gypsum, and they suggested that carbonate drainage gravel would not dissolve in the presence of such leachates. Additionally, the United Kingdom leachate data suggest that carbonate saturation is reached very early in landfilling (<1 year). This may be partly the result of bicarbonate derived from dissolution of cover-soil carbonates and other carbonates present in the waste (Bennett *et al.*, 2000), but mostly from the mineralization of organic carbon by microbes (Brune *et al.*, 1991; Owen and Manning, 1997; Bennett, 1998).

Table 2.24 Summary of hydraulic conductivity (k_n) and permittivity (ψ) changes for geotextile exhumed from field application in landfills (based on Koerner *et al.* (1993) but as modified by Koerner, personal communication)

Case	Leachate COD (mg/l)	Leachate TS (mg/l)	Geotextile	M_A (g/m^2)	AOS (mm)	t_{GT} (mm)	Initial k_n (m/s)	Final k_n (m/s)	Initial ψ (s^{-1})	Final ψ (s^{-1})
3[a]	31,000	28,000	HBNW	150	0.15	0.38	4.2×10^{-4}	3.1×10^{-8}	1.1	8.2×10^{-5}
4[a]	10,000	3,000	W	170	0.25	0.41	3.7×10^{-4}	1.4×10^{-4}	0.9	3.3×10^{-1}
5a[a]	3,000	12,000	NPNW	220	0.21	2.7	4.9×10^{-3}	8.5×10^{-5}	1.8	3.1×10^{-2}
5b	3,000	12,000	NPNW	220	0.21	2.7	4.9×10^{-3}	4.4×10^{-8}	1.8	1.6×10^{-5}
6a[b]	24,000	9,000	NPNW	176	0.21	2.2	2.3×10^{-3}	3.7×10^{-5}	1.1	1.7×10^{-2}
6b	24,000	9,000	NPNW	176	0.21	2.2	2.3×10^{-3}	1.6×10^{-7}	1.1	7.3×10^{-5}
6c	24,000	9,000	NPNW	176	0.21	2.2	2.3×10^{-3}	7.5×10^{-7}	1.1	3.4×10^{-4}

[a] Refer to Table 2.21.
[b] Geotextile around gas collection wells at depth of 3, 7.5 and 15 m.

Table 2.25 Decrease in hydraulic conductivity of geotextiles permeated with leachate (average COD: 3000–4000 mg/l; BOD: 2000–2500 mg/l; TSS: 300–600 mg/l) at a rate of 2×10^{-5} m/s (620 m^3/a/m^2) (adapted from Koerner *et al.*, 1994)

Type of filter	Unit mass, M_A (g/m^2)	POA (%)	AOS (mm)	Thickness, t_{GT} (mm)	Initial k_n (m/s)	Initial permittivity, ψ (s^{-1})	Equilibrium k_n (m/s)	Flow to equilibrium (m^3/m^2)
Uniform sand					4×10^{-3}		2×10^{-6}	119
Well-graded sand					6×10^{-4}		4×10^{-7}	170
W: Monofilament	200	32		0.7	3.4×10^{-3}	4.8	2.5×10^{-6}	51
W: Multifilament	270	14		0.8	1.9×10^{-3}	2.4	7×10^{-6}	51
W: Slit film	200	7		0.4	1.6×10^{-4}	0.4	8×10^{-8}	43
W: Monofilament	250	10		0.6	6.4×10^{-4}	1.0	5×10^{-8}	76
Special NW/W	740		0.3	6.3	1.5×10^{-2}	2.4	3.5×10^{-6}	102
NPNW	130		0.21	1.1	2.3×10^{-3}	2.1	6×10^{-8}	76
NPNW	270		0.18	2.4	3.6×10^{-3}	1.5	1×10^{-7}	93
NPNW	540		0.15	4.7	2.4×10^{-3}	0.5	2×10^{-7}	85
HBNW	120		0.165	0.4	2.4×10^{-4}	0.6	4×10^{-8}	76
NPNW	220		0.12	2.0	3.2×10^{-3}	1.6	1.5×10^{-7}	68

Bennett *et al.* (2000) evaluated the suitability of dolomitic limestone in landfill leachate drainage systems. Similar to the United Kingdom leachate study (Owen and Manning, 1997), the leachate samples collected from southern Ontario MSW landfills showed that dolomite and other carbonates are thermodynamically stable, whereas gypsum is not. However, to further address the concern of acetogenic leachate and the dissolution of dolomitic limestone, drainage gravel collected from the KVL-LCS by Fleming *et al.* (1999) was examined. Incipient dissolution is commonly characterized by the formation of small pits, which can be detected at high magnification of the crystal surfaces (Herman and White, 1985). Bennett *et al.* (2000) showed that the surface of the dolomitic limestone exhumed by Fleming *et al.* from the KVL-LCS was devoid of such pits. Furthermore, the dolomitic limestone retrieved from a laboratory mesocosm study showed that after being exposed to leachate for 4 years, a coating of calcite crystallized around the gravel. The primary porosity was filled with calcite (within the gravel to about 100 µm below the surface). A sharp grain boundary between the original gravel material and secondary calcite cement suggested that little, if any, dissolution occurred prior to or during formation of calcite. The authors concluded that despite the abundant presence of microbes, there is little evidence that microbial membrane effects caused significant dissolution of the dolomitic limestone. Additionally, they found significant detrital fines (dolomite, quartz, feldspar) within the secondary calcite formed around the gravel that added to the clogging of the collection system.

(e) Size of drainage material

Research has demonstrated a clear relationship between the grain size distribution and clogging. For example, Brune *et al.* (1991) examined samples exhumed from clogged leachate collection systems (Figure 2.16) and from laboratory studies. They demonstrated that a graded sandy gravel with grain size in the range 1–32 mm and fine gravel (2–4 mm size) experience severe clogging in laboratory column tests conducted using high-strength leachate under anaerobic conditions (see landfill 8 listed in Table 2.6). Coarser material (medium gravel: 8–16 mm size) experienced clogging of the smaller pores but the larger pores remained open at the end of the experiment (16 months). Coarse gravel (16–32 mm) in direct contact with waste was covered with a thick film and locally clumped together; however, the large pores remained unclogged throughout the (16 months) test and maintained its hydraulic conductivity.

A comparison of the performance of mesocosms (Rowe *et al.*, 1995a; Fleming, 1999) with different-sized drainage gravel showed that the 19 mm diameter gravel clogged much faster than the 38 mm gravel, as a result of a lower pore volume and increased surface area of the smaller gravel.

Rowe *et al.* (2000a) reported experiments where a number of columns filled with different-sized particles (nominal bead diameters of 4, 6 and 15 mm) were permeated with landfill leachate under anaerobic conditions at the same flow rate and temperature. These tests showed that the rate of clogging was much quicker for the smaller diameter particles compared to larger diameter particles. It was also shown that the rate of clogging occurring in the columns was generally consistent with what was observed for an actual landfill with a French drain collection system design.

(f) Effect of mass loading

Koerner *et al.* (1994) suggested that the hydraulic conductivity of a geotextile decreased with increasing leachate flow rate. More generally, Rowe *et al.* (2000b) have shown that clogging is directly related to the leachate mass loading in terms of the mass of COD (chemical oxygen demand), calcium and suspended solids per unit area perpendicular to leachate flow. The leachate mass loading is a function of (a) the concentration of volatile fatty acids (represented in terms of COD) and inorganic constituents (especially calcium and, to a much lesser extent, iron) in the leachate; (b) the flow rate per unit area and (c) elapsed time. Thus when leachate that is generated over a large area of the landfill is directed through a filter surrounding isolated drainage material (e.g., French drains in Figure 2.2a) or, even worse, a filter wrapped around the pipe itself (Figure 2.2b), there is a confluence of flow across the filter and the confluence factor is approximately equal to the area of the landfill from which leachate is directed to the drain divided by the area of the filter material through which the leachate must flow.

Since clogging depends on flow rate per unit area, confluence of flow increases the mass loading per unit time and hence increases the rate of clogging of the geotextile and the drainage gravel (it is assumed here that clogging of pipes is mitigated by regular cleaning – see Section 2.4.4(i)). However, even more important is the effect that clogging has when it does occur. When the resistance to flow into the drainage pipe is increased, a leachate mound will develop in the waste between the drains (see Figure 2.2). This mound will maintain some flow to the pipe but will also increase flow out through the underlying liner. Based on this, Rowe (1992) cautioned against designs such as that shown in Figure 2.2. The field investigation discussed previously in Section 2.4.4(c) (see Table 2.22) confirms the soundness of this recommendation since the geotextiles used to wrap drains or pipes showed a drop in hydraulic conductivity of between two and five orders of magnitude (to as low as 4×10^{-8} m/s: Case 5b, Table 2.24).

(g) Relationship between COD consumption and clogging

Fleming (1999) showed that the drainage layer of the mesocosms could be modelled as an anaerobic fixed film reactor using techniques found in wastewater applications. Additionally, a key link was found between the reduction in COD concentrations due to microbial consumption, resulting in an increase in pH of the bulk leachate solution and subsequent decrease in calcium in solution. Here the increase in pH is driven by the microbial conversion of volatile acids to CO_2 (which also leads to an increase in carbonates in the leachate – see Rittmann *et al.*, 1996). This increase in pH and carbonates allows, or accelerates, the chemical precipitation of $CaCO_3$. Fleming (1999) showed that by decreasing the temperature, the rate of COD consumption decreased relative to higher temperature but given enough time similar treatment occurred. Similar results were obtained by VanGulck (1998) using flasks packed with marbles (gravel size material) at 10, 20 and 27 °C. Additionally, a similar effect was noticed by varying the surface area available for microbial attachment. Tests with a smaller surface area took a longer time to remove COD compared to larger surface area tests but similar overall treatment occurred given enough time.

Column experiments (Rowe *et al.*, 2000a,b, 2002b) have provided the empirical correlations and parameters needed for predictive clogging models: specifically the calcium carbonate yield coefficient, bulk densities of the clog material, clog chemical composition and a relationship between porosity of the medium and hydraulic conductivity. The issues related to carbonate yield have been examined by Rittmann *et al.* (2003) and VanGulck *et al.* (2003). The column experiments gave relatively similar yield coefficients, densities and clog composition. Additionally, organic and inorganic (in terms of COD and Ca^{2+}) removal in the leachate over the entire bed volume of the column and relation to amount of clogging (drainable porosity) has been used to calibrate predictive clogging models (e.g., Cooke *et al.*, 2001).

(h) Use of tyre shreds to replace drainage gravel

Tyre shreds are increasingly being considered as a replacement for coarse gravel in the blanket drain of leachate collection systems for MSW landfills. To provide data that could be used to assess the extent to which tyre shreds will clog and how the service life might compare with that of more conventional gravel, McIsaac and Rowe (2003) conducted a series of laboratory experiments in which four columns of drainage material were permeated with KVL leachate. Two columns contained tyre shred denoted as "P", one involved a tyre shred denoted as "G" and the fourth (control) contained 38 mm gravel. The G shreds were cleanly cut and had an average size of 100 × 50 × 10 mm. The P shreds had a torn appearance, a wider size distribution than the G shreds and an average size of 125 × 40 × 10 mm. The compressed (at 150 kPa) thickness of tyre shreds was similar to that of the gravel. The average initial drainable porosity of the G shreds was 0.22 and that for the duplicate P shreds was 0.25. The corresponding value for the gravel column was 0.45. The initial hydraulic conductivities of the G and P shreds (under the applied pressure of 150 kPa) and the gravel drainage material were found to be about 0.007, 0.02 and 0.8 m/s, respectively.

The gravel column outperformed the rubber shred columns by maintaining a hydraulic conductivity of greater than 10^{-5} m/s for a substantially longer period of time than that of the tyre shreds. Based on this empirical evidence, it was inferred that the service life for a given thickness of leachate collection system with 38 mm gravel will be 3 times longer than that of a similar thickness of compressed tyre shreds. This is primarily due to the difference in void structure in the rubber shred and gravel drainage material and its effect on the exposed surface area available for clog growth, the resulting pore connectivity through the drainage materials, the pore opening sizes, and the growth and development of organic and inorganic clog materials. In particular, the compressibility of the shreds allowed the void space between the shreds to decrease as the load applied to the material increased resulting in very tight shred-to-shred contacts, low initial void volumes and small constrictions between larger open void space. There were many bottleneck pores in the shred material, with the constrictions between the rubber shreds being much smaller than that for the gravel which had a far more uniform and open void structure than the tire shreds.

The bulk (wet) densities of the clog material ranged from 1.50 to 1.75 g/cm^3 and the total calcium fraction was approximately 29% of the total clog material. It was found that metals (predominantly iron but also including zinc and aluminium) leached from both the P and G shreds exposed to typical MSW leachate. However, these metals precipitated into the clog material and were not detected at elevated levels in the effluent leachate.

Based on calculations using the approach described in Section 2.5, the service lives of leachate collection systems composed of rubber tyre shreds were significantly less (by about a factor of 2, other things being equal) than those composed of 38 mm gravel. In these simplified calculations, this was primarily due to the lower initial porosity in the rubber tyre shred drainage medium. This lower porosity could be potentially offset by increasing the thickness of the drainage layer and modifying the design of the leachate collection system. Initial estimates suggest that a similar service life could be achieved by increasing the compressed drainage layer thickness by a factor of about 3 if tyre shreds similar to those tested were to be used.

Within the context of the factors examined by McIsaac and Rowe (2003), recognizing that a greater compressed thickness of material will be required to achieve the same service life and assuming appropriate design to address these issues, there appears to be scope for the use of tyre shreds as a partial replacement for conventional granular drainage material in leachate collection systems (e.g., on side slopes and in non-critical areas). However, given the superior performance of uniform coarse gravel, the use of gravel would appear appropriate in critical areas where severe clogging could result in unacceptable performance (e.g., adjacent to leachate collection pipes, leachate sumps, etc.). More field verification of the performance of tyre shreds in leachate collection systems is required.

(i) Cleaning of leachate collection pipes

The development of clog material begins with the formation of a soft biofilm (sometimes called bioslime). This is followed by the precipitation of inorganic material which, initially, is loosely contained in the biofilm but subsequently coalesces into a solid material (sometimes called biorock). Once hardened, the clog material is very difficult to remove from pipes and even harder to remove from pipe perforations. In the early stages, the formation of the clog material may be relatively slow. However, once the biofilm has become established, clog formation can be quite rapid. Thus an appropriate inspection and cleaning program is essential for ensuring the long-term performance of the leachate collection pipes.

While in its soft stage, the clog material is relatively easy to remove by high-pressure water jetting. Once the clog development moves into its steady-state phase, solid clog material becomes very difficult to remove by jetting. For landfills that have biorock-clogged pipes, chain flailing has sometimes worked. For chain flail-resistant deposits, dissolution using a weak acid has also been attempted, although significant build-ups of clog material are still very difficult to remove and the dissolution process also has the potential to generate a hazardous waste leachate due to the mobilization of heavy metals that have precipitated into the clog material. The best approach is regular inspection and cleaning to disrupt the development of the biofilm in the pipe perforation and on the walls and bottom of the pipe before it hardens into solid clog material.

The required frequency of cleaning will vary from landfill to landfill and even for a given landfill will likely change with time. At least annual inspection and/or cleaning is typically appropriate. However, after the conditions become conducive to clog formation more frequent cleaning may be necessary. For example, as noted earlier, pipes in the leachate collection system at the KVL which had never been cleaned and were collecting leachate from older portions of the landfill experienced significant clogging only 4 years after they were placed. At the Halton Landfill, where the pipes had been cleaned annually, significant clogging (requiring special effort to clean) was observed in the leachate collection pipes after 9 years of operation, necessitating an increase in cleaning frequency to 2–3 times per annum.

None of the cleaning methods can be expected to do more than clean the pipe and, if they are large enough to be effectively cleaned, its perforations. The material around the pipe cannot be cleaned and hence it is essential that it be selected such that it has a relatively long service life and that the system be designed to minimize clogging of the blanket drain (Section 2.4.5).

2.4.5 Minimizing clogging of blanket drains

Clogging of granular blanket drains may be the result of a combination of particulate clogging, clogging due to chemical precipitation and biofilm growth (Section 2.4.4). Thus, the likelihood of clogging occurring can be minimized by: (a) maximizing the flow velocity in the drain; (b) maximizing the void size; (c) minimizing the surface area available for biofilm growth and (d) minimizing the mass loading per unit area adjacent to the collection pipes.

By maximizing the flow velocity, one reduces the residence time for leachate in the collection system, thereby reducing the amount of sedimentation of particulates and chemical precipitation that can occur. The flow velocity can be increased either by increasing the gradient (e.g., by increasing the slope of the landfill base) and/or by increasing the hydraulic conductivity of the granular layer. Ideally both the slope and hydraulic conductivity will be as large as practicable.

Maximizing the void size within the drainage blanket tends to increase the initial hydraulic conductivity and reduces the likelihood of the voids becoming blocked.

Biofilm growth is related to the surface area available on the particles forming the drainage layer. Since for a sphere (or approximately for near-spherical particles) the surface area is proportional to the square of a particle's diameter and the volume is proportional to the cube of a particle's diameter, it is evident that the surface area can be minimized by increasing the diameter of particles used to construct the leachate collection system.

Some specifications require that the hydraulic conductivity of a granular drainage blanket be greater than about 1×10^{-5} m/s. There are many granular materials that would meet this specification and some are far more susceptible to clogging than others. Based on the foregoing argument, it is evident that the potential for clogging is greatest for well-graded sands (which will have a hydraulic conductivity close to 10^{-5} m/s, small voids and a large available surface area), will be less for uniform fine gravel and will be minimum for uniformly graded coarse gravel (e.g., 35–50 mm nominal size). This would lead one to specify uniformly graded coarse gravel as the granular blanket; however, if uniform gravel is placed in direct contact with the waste, one can expect the migration of particles (e.g., sand and silt size particles) of the waste into the gravel. This in turn would reduce the effective void space and the hydraulic conductivity and would increase the surface area available for biofilm growth. This leaves the options of (1) placing a select waste above the coarse granular blanket and/or (2) having a multicomponent collection system involving a filter which will minimize the migration of particles into the granular blanket. If the gravel is to be placed on top of a compacted clay liner, it may also be desirable to install a geotextile separator between the gravel and the clay to minimize the spoiling of gravel and liner. Both geotextile applications are shown in Figure 2.3.

Rowe (1992) recommended the use of a suitably selected geotextile as a granular filter above a blanket gravel underdrain layer (Figure 2.3) and the findings of both Fleming *et al.* (1999: Cases 1 and 2, Table 2.22) and Koerner and Koerner (1995: Case 5a, Table 2.22) provide some evidence that a suitable blanket geotextile will function adequately. These investigations demonstrated (Cases 2, 4 and 5, Table 2.22) that the geotextile provided good protection to the granular drainage layer which experienced relatively little clogging compared to that observed when there was no geotextile filter between the waste and drainage medium (Cases 1 and 3, Table 2.22). Geotextiles used as shown in Figure 2.3 will experience some clogging (see Section 2.4.4(c)); however, even if a perched leachate mound developed above the geotextile, there would be no effect in terms of contaminant transport through the underlying liner because the leachate can readily drain to the collection pipes once it breaks through the geotextile. The level of leachate perching above the geotextile will depend on the leachate generated per unit area of the landfill and the hydraulic conductivity of the geotextile filter, as discussed in Section 2.4.6.

A granular layer (e.g., sand) can also be used as a "protection layer" either in conjunction with a geotextile (see Figure 2.3b) or as part of a graded granular filter. This layer will experience clogging in the same way as a geotextile (see McIsaac *et al.*, 2000) and will also provide some leachate treatment. More research is required to establish the cost-effectiveness of multilayered systems for leachate pre-treatment within the landfill itself (see Fleming and Rowe, 2000).

The mass loading per unit area adjacent to collection pipes can be reduced by decreasing pipe spacing and/or increasing the drainage layer thickness near the leachate collection pipes. It should be noted that pipes in trenches pose a particular challenge in this respect since the flow localizes in the trench above the pipe and clogging in the trench will likely control the system performance. It is best to avoid placing leachate collection pipes in trenches. When this cannot be avoided, the mass loading will need to be controlled either by adjusting the width of the trench and/or pipe spacing.

Particular care is required in the design of collection systems where leachate recirculation is part of the leachate management plan since recirculation has the potential to increase the mass loading on the leachate collection system.

2.4.6 Geotextiles as filters and separators in leachate collection systems

The studies of geotextile interaction with leachate (Section 2.4.4(c)) generally indicate that the more open the geotextile, the less is the biologically induced clogging. This would suggest that if geotextiles are to be used in locations where clogging would have a significant impact on the performance of the system then the use of an open geotextile would be desirable. Koerner and Koerner (1995b) recommended that for "mild" leachate, either a woven or nonwoven geotextile could be used in contact with the waste, provided that it had the properties that met the requirements given in Table 2.26. Giroud (1996a) recommends the use of monofilament-woven geotextiles with a minimum filtration opening size (AOS) of 0.5 mm and a minimum relative open area (POA) of 15%, with a preference for a POA greater than 30%. The rationale for Giroud's recommendations arises from the observations that (a) the specific surface area for monofilament-woven geotextile is much smaller than that for nonwoven geotextiles and this decreases the surface area for biofilm growth; (b) the woven filter allows more effective and rapid movement of fine material (i.e., material not intended to be retained) and leachate through the filter and (c) due to their compressibility, the filtration characteristics of a nonwoven geotextile vary with applied pressure and the critical filtration characteristics should be assessed under design pressures. There is some evidence to suggest that geotextiles selected in accordance with Giroud's (1996) recommendations are likely to experience less clogging and reduction in hydraulic conductivity with time (e.g., see Table 2.25) than geotextiles that simply meet the requirements of Koerner and Koerner (1995b; Table 2.26).

It is important to recognize that Giroud's (1996a) recommendations are based on the premise that one wishes to minimize clogging of the filter. This may indeed be the case for some design situations (e.g., if one has no choice but to use a design such as is shown in Figure 2.2). However, when a continuous filter/separator layer across the site is underlain by a continuous drainage blanket the situation is somewhat different. While excessive clogging is undesirable even under these circumstances, provided that leachate can continue to pass through the filter, some biological clogging of the filter is actually desirable since the biofilm associated with clogging also provides leachate treatment and, in so doing, (a) decreases the potential for clogging at more critical zones (e.g., near collection pipes) and (b) reduces the level of leachate treatment required after removal of leachate from the landfill. Clogging of the continuous filter above a continuous drainage blanket will result in some perching of leachate over the geotextile. This is not problematic with respect to contaminant impact on underlying groundwater aquifers since the leachate is separated from the barrier by the drainage layer and hence the perched mounding does not represent mounding on the barrier. Clearly, as

Table 2.26 Koerner and Koerner (1995b) recommended minimum values for geotextile filters for use with mild leachate and select waste over the geotextile (no hard or coarse material; for coarse or hard material over GT, the strength requirements may need to be increased)

Property	*Woven monofilament geotextiles*	*Nonwoven needle-punched geotextiles*
Mass, M_A (g/m^2)	200	270
Per cent open area (POA) (%)	10	–
Apparent opening size (AOS) (mm)	–	0.21
Grab strength (N)	1400	900
Trapezoidal tear (N)	350	350
Puncture strength (N)	350	350
Burst strength (kPa)	1300	1700

the perched mounding increases so too does the gradient across the clogged geotextile until the gradient is sufficient to provide transmission of infiltration to the drainage layer.

One consideration regarding clogging of the continuous filter is whether the hydraulic conductivity might drop to such a point that it would cause so much perching of leachate that would have negative effects such as side seeps. However, based on published data (e.g., see Tables 2.24 and 2.25), it appears unlikely that the hydraulic conductivity of the geotextile selected in accordance with either Giroud's (1996a) or Koerner and Koerner's (1995b) recommendations would be below 4×10^{-8} m/s (1.3 m/a) for normal conditions and more likely it would be of the order of 1×10^{-7} m/s or higher. As shown below, if the geotextile was used in a blanket drain (e.g., see Figure 2.3), there would be negligible perched leachate on the geotextile for typical rates of leachate generation (i.e., less than 3×10^{-8} m/s or $1\,m^3/a/m^2$). Thus, while recognizing that geotextiles will clog, based on the available data it appears that an appropriately selected geotextile (e.g., conforming with Koerner and Koerner's recommendation in Table 2.26) used to protect gravel in a blanket drain will improve the performance and the service life of the drainage stone and not cause excessive perched leachate mounding. As noted in Section 2.4.4(b), there is some evidence to suggest that a needle-punched nonwoven geotextile will perform better in this regard than a woven geotextile (although both are superior to the situation where there is no filter). Concern regarding potential perched mounding on top of a continuous filter can also be addressed in terms of design details such as that adopted for the Halton landfill (Rowe *et al.*, 2000c) where alternative paths were provided for any perched leachate to enter the manholes. Selection of an appropriate geotextile requires consideration of construction survivability as discussed in Section 15.3.2.

To illustrate the implications arising from the use of a filter above a continuous gravel drainage blanket, consider the cross-section shown

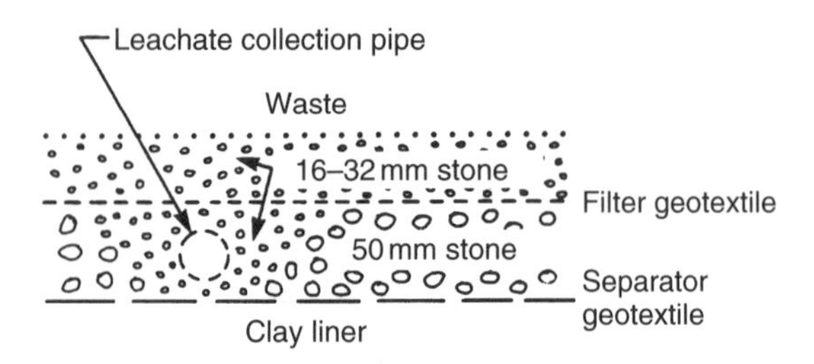

Figure 2.18 Schematic of a two-layered leachate collection system. A geotextile filter is used to separate the upper and lower granular layers.

in Figure 2.18 applied to a geometry shown in Figure 2.14c for the following parameters:

- design maximum head on liner: <0.3 m;
- drain spacing: $l = 30$ m;
- base slope: $s = 0.01$;
- hydraulic conductivity of waste: $k_w = 10^{-6}$ m/s;
- hydraulic conductivity of blanket: $k_s = 10^{-2}$ m/s;
- infiltration through landfill to collection layer: $q_0 = 0.15$ m/a; and
- initial hydraulic conductivity of geotextile above drainage layer: $k = 10^{-3}$ m/s.

Design conditions

Under operating conditions under a unit gradient ($i = 1$) the upper gravel layer and geotextile can transmit a vertical flow of

$$q = ki = 10^{-3}\,\text{m/s} = 31{,}540\,\text{m/a}$$

which well exceeds the infiltration of 0.15 m/a and hence there should be no leachate perching above the drain. Once in the drain the maximum height of leachate mounding is calculated by equation 2.4:

$$h_{max} = 0.0006\,\text{m}$$

is negligible and the maximum height of mounding would be controlled by irregularities in the surface of the liner and ponding which may occur at and below the invert of the collection pipe.

Moderate clogging

Assume that the hydraulic conductivity of the geotextile and overlying clogged granular material drops to a value similar to that of the waste (viz. $k = 10^{-6}$ m/s) under a unit gradient ($i = 1$) across the clogged geotextile; the geotextile can transmit a vertical flow (to the blanket drain) of

$$q = ki = 10^{-6}\,\text{m/s} = 31\,\text{m/a}$$

which still greatly exceeds the infiltration of 0.15 m/a and hence there should be no long-term leachate perching above the drain. The maximum height of leachate mounding in the drain would be the same as calculated above.

Severe clogging

If clogging resulted in the lowest value of hydraulic conductivity (3×10^{-8} m/s = 0.95 m/a) obtained from samples exhumed from landfills (Table 2.24), there would be no long-term leachate perching (short-term perching may occur during periods of high infiltration) and the flow of 0.15 m/a could pass through the clogged geotextile. If the hydraulic conductivity of the geotextile dropped to the lowest value (10^{-9} m/s) found by Cazzuffi *et al.* (1991) under extreme laboratory conditions, then the clogged geotextile could transmit a vertical flow of

$$q = ki = 10^{-9} i\,\text{m/s} = 0.031 i\,\text{m/a}$$

and so, in order to transmit the average annual infiltration of 0.15 m/a, a gradient

$$i = \frac{0.15}{0.031} = 4.8$$

would be required. This would be achieved with perched mounding of about 0.48 m above the geotextile assuming a 0.1 m thick clogged zone above and including the geotextile. Once the leachate reaches the granular drainage layer, it would readily drain. Furthermore, even if the hydraulic conductivity of the drainage layer was also reduced by three orders of magnitude, the design criteria of less than 0.3 m head on the liner would still be met.

The foregoing example serves to illustrate the potentially beneficial role of combining a coarse granular drainage blanket with a geotextile filter in the design of a leachate collection system. The extreme case of complete clogging of the geotextile can be addressed by installing the 16–32 mm thick layer of gravel over the geotextile as indicated in Figure 2.18. This zone should be hydraulically connected to the drains and provides a means of minimizing perched mounding of leachate.

2.4.7 Geosynthetic drainage layers

Geosynthetic drainage layers are typically net materials (geonets) that are structured to include open channels for the in-plane conveyance of relatively large quantities of fluids. The use of these geosynthetic drainage layers in caps and covers for leachate collection on side slopes and leak detection below a composite liner at waste disposal facilities is becoming quite common. For this application, the geosynthetic drainage layer is often used in conjunction with an upstream geotextile filter and downstream geomembrane barrier to form a composite system. For example, Figure 1.31 shows a geonet used in a composite drainage net in the cover for leachate collection on a bottom liner side slope and as a leak detection system below the primary liner. The use of geonets on the base of the landfill as the primary means of providing leachate collection on the landfill base is not encouraged because of the potential for clogging. A key design consideration for geonet drainage layers is their *in-situ* flow capacity and the leachate residence time within the drainage layer as discussed below.

(a) Flow capacity and compressibility

The flow regime within the geosynthetic drains is usually turbulent and so flow calculations based on the assumption of an intrinsic material permeability or transmissivity are not possible since flow is not linearly proportional to the hydraulic gradient. Thus the design must be based on direct comparisons of the required and actual material flow rates. In addition to the hydraulic gradient, several other factors have been identified which will influence the flow behaviour and capacity of the geosynthetic drainage (Williams *et al.*, 1984; Koerner *et al.*, 1986; Lundell and Menoff, 1989; Fannin *et al.*, 1998). These include compression due to overburden loads, intrusion, compression-induced creep, thickness (or number of layers), boundary conditions, the temperature and viscosity of the flowing fluids and the directional drainage characteristics. An additional consideration is environmental stress cracking and there are geonets on the market with both good and poor environmental stress crack resistance (Siebken and Cunningham, 1999).

In order to select suitable geosynthetic drainage material to meet a design flow, it is necessary to have results from laboratory simulation tests on candidate materials using the actual loads and boundary conditions of the field situation. The ASTM D-4716 test protocol provides guidance for these simulation tests. The test program should be conducted to provide relations between the flow rate, the hydraulic gradient and the applied pressure. The compressive strength of the drainage layer must be determined to be adequate under expected overburden loads. Consideration must also be given to compressive creep behaviour.

The level of confidence in the long-term behaviour of these systems decreases with the length of time that the system is required to be operative. In cover applications the drainage layer can be excavated and repaired as needed. When buried to depths where replacement is not feasible, careful consideration should be given to the contaminating lifespan of the landfill. Generally, the larger the landfill thickness and the lower the infiltration, the greater will be the concern regarding long-term performance. The longer the contaminating lifespan, the greater is the need for backup systems in the event that the geonet fails.

(b) Particulate clogging of geonets

To avoid particulate clogging it is necessary to: (a) ensure that there is a suitable filter between the geonet and adjacent soil or waste and (b) develop an appropriate design and construction plan that will minimize the accumulation of fines.

The design of filters is discussed in detail by Giroud (1996a) and Giroud and Bonaparte (2001). However, it is noted here that even a relatively small proportion of fines (0.5–1%) in a sand protective layer separated from an underlying geonet drainage layer by a geotextile filter has the potential to cause significant particulate clogging of the underlying geonet (e.g., see Giroud and Soderman, 2000). The same amount of fines would have a negligible effect on the performance of a 0.3 m thick coarse gravel drainage layer.

(c) Reinforcement function of adjacent geosynthetics

The use of geonets for detection and leachate collection imposes a reinforcement function on the adjacent materials (usually geotextiles or geomembranes), which must span openings on the surface of the mat. Tests by Hwu *et al.* (1990) indicate that a flow rate reduction of one order of magnitude is possible due to geotextile intrusion into the apertures of geonets. Modest decreases in intrusion (and increase in flow) were observed for geotextiles stiffened by resin treating, burnishing and scrim reinforcing, and a large decrease in intrusion results with the use of a composite fabric (a needle-punched nonwoven over a woven slit-film geotextile). Some drainage layers have been designed with limited opening size adjacent to the geotextile or with a three-layer system to minimize the effect of geotextile/soil intrusion. In some cases there may be advantages in using a geocomposite drainage layer with the geotextile heat-bonded to the geonet, although care is required to ensure that the permittivity (and hence permeability) of the geotextile is not significantly decreased by this process. Particular care is needed when the geonet is overlain by a GCL (e.g., in a leak detection system below a composite liner) due to the potential for intrusion of bentonite into the geonet. Here the choice of the geotextile between the bentonite core and the geonet may be critical as discussed in the context of internal erosion in Section 12.5.6.

(d) Selection and acceptability

Published data show wide variations in flow rate of various geosynthetic drainage products; some have flow capacities of the same order as 0.3 m of gravel while others are similar to the flow capacity of 0.3 m of sand. The wide variations in flow rate illustrate the importance of laboratory simulation of the actual field situation for selection purposes. It should be noted that some products may be placed in multiple layers to increase the flow capacity. Test results (Koerner and Hwu, 1989) show that the flow rate increases from the use of additional layers cannot be reliably estimated from tests with a single layer but must also be determined from simulation testing. An alternative to the use of multiple layers is to use thicker geonets, however, the greater the three dimensionality of the grid, the greater the need to carefully examine the time-dependent response of the system. The design should carefully consider the type and orientation of the geonet, and the location and nature of the geonet seams (recognizing that geonet transmissivity may be considerably reduced when overlapping is used, due to intrusion of one geonet into the other, as shown by Zagorski and Wayne, 1990).

2.5 Estimating the service life of leachate collection blankets

A clogging model (BIOCLOG) has been developed for the purpose of predicting the rate of clogging of leachate collection systems (Cooke *et al.*, 1999, 2001). The model utilizes transient anaerobic, fixed-film wastewater treatment processes combined with geotechnical engineering concepts within a time marching algorithm. This algorithm is used to model the evolution of the influent and effluent organic and calcium concentrations, biofilm thickness, inert biofilm plus mineral film thickness and porosity at any position or time. Using published biofilm growth kinetic parameters for the methanogenesis of acetate and the acetogenesis of propionate and butyrate, and an experimentally determined calcium carbonate yield coefficient, a good fit was obtained for predicted effluent organic and calcium concentrations and porosity profiles for laboratory column tests conducted under saturated conditions (Cooke *et al.*, 2001; VanGulck, 2003). Since powerful models such as these require a high level of sophistication to use them, as yet they are not ready for use in standard design. This creates the need for a simpler approximate method for estimating service life as descried below.

A practical engineering technique for estimating the service life of LCSs has been developed by Rowe and Fleming (1998) assuming that relatively uniform coarse gravel material is used for the drainage blanket. This technique provides a means of estimating the service life of LCSs in MSW landfills and can be used in engineering practice for comparing the potential performance of different proposed designs. The model uses the finding that $CaCO_3$ is the dominant fraction in the clog formation and uses calcium as surrogate for other clog fractions (i.e., magnesium, iron, silica) as well as the leachate calcium concentration to calculate the reduction in porosity within a drainage aggregate. The model assumes that all calcium entering the drainage layer gets deposited in the granular layer. For example, they give a time to clog (t_c) as

$$t_c = \frac{(1 + 2a/L)B\rho_c f_{Ca} v_f}{3q_0 c_{L2}} - \frac{(c_{L1} - c_{L2})(T1 + T2)}{2c_{L2}} \tag{2.8}$$

where a [m] is the length of the zone where clogging is likely to develop through the full thickness of the blanket drain (Rowe and Fleming (1998) used $a = 5$ m in their example calculations, but this may need to be selected on a site-specific basis); L [m] is the length of the maximum flow path to the leachate collection pipe being considered; B [m] is the thickness of the drainage blanket near the leachate collection pipes that, when clogged, would effectively prevent leachate from percolating into the perforated pipe; ρ_c [Mg/m^3] is the density of clog material; f_{Ca} [–] is the mass fraction of calcium in the clog material; v_f [–] is the porosity reduction required to cause clogging; n_0 is the initial porosity [–]; q_0 [m^3/m^2/a] is the average vertical infiltration into the drainage blanket drain; c_{L1} [Mg/m^3] is the average calcium concentration in the leachate entering the collection system for time $t \leq T1$ [a]. After time $T1$, the concentration of calcium is assumed to reduce linearly to a steady value c_{L2} [Mg/m^3] at time $T2$.

Predictions of the time to clog a blanket drainage system with a 50 mm uniform gravel and a cubic packing arrangement are presented in Table 2.27 for various pipe spacings, drainage layer thickness and infiltration into the collection system. In Table 2.27, the time to clog was defined as being the time for the hydraulic conductivity to reduce to 10^{-6} m/s. Calculations were performed for the case of (a) a constant calcium concentration with time and (b) a variable concentration of calcium in the leachate with time. The results indicate that with a smaller pipe spacing, larger drainage layer thickness and lower infiltration into the collection system, the time to clog increases. It should be noted that the concentrations of calcium are typical for infiltrations over the range examined here. Much lower infiltration (e.g., associated with very low permeability covers) may result in different concentrations and hence different service lives. Likewise recirculation of leachate may also result in different concentrations, greater mass loading and faster clogging. The parameters used in Table 2.27 are illustrative and must be selected for each particular case depending on the nature of the waste and leachate, the rate of leachate generation and the mode of operation of the landfill.

It should be noted that this simplified technique is not suitable for use with sands or well-graded materials; it may significantly overestimate the time to clog for these cases.

2.6 Leachate mounding and liner temperature

Barone *et al.* (2000) reported a direct correlation between an increase in leachate mounding and an increase in liner temperature at the base of the KVL. Figure 2.19 shows that temperatures were relatively stable during the early operation of both the older (Stage 1) and newer (Stage 3) portions of the landfill. However, when leachate mounding began to become significant in the older portion of the landfill (Stage 1, where the primary leachate collection system consisted of lateral French drains at spacings of about 65 m) temperature increased to between 30 and 40 °C when the leachate head increased to between 4.5 and 6.5 m. In the newer (<8 years old) area (Stage 3, where there is a 0.3 m thick continuous drainage blanket of 50 mm gravel and the leachate head was less than 0.3 m despite some clogging as discussed in Section 2.4.4), the liner temperature was only 10–16 °C. The relationship increasing leachate head and temperature was attributed to two factors: (1) a faster rate of anaerobic degradation of organic matter and hence a faster rate of heat production with increasing moisture content of the refuse and (2) a slower rate of heat loss from a saturated refuse layer due to its lower thermal conductivity compared to unsaturated waste. By compiling data from a number of sources they also showed a correlation between leachate mound in different landfills and the liner temperature in these landfills (Figure 2.20). As will be discussed in later chapters, this increase in temperature can cause an increase in both advective and diffusive transport and reduces the service life of geomembranes used for landfill liners.

Table 2.27 Time to clog drainage gravel using a distributed clog formation model (adapted from Rowe and VanGulck, 2001)

L (m)	B (m)	q_0 (m/a)	a^b(m)	*Service life* [a] *(years)* Constant concentration with time [c]	*Variable concentration with time* [d]
100	0.5	0.2	5	69	355
100	0.5	0.15	5	92	545
100	0.3	0.2	5	42	126
100	0.3	0.15	5	55	240
75	0.5	0.2	5	71	372
75	0.5	0.15	5	95	568
75	0.3	0.2	5	43	136
75	0.3	0.15	5	57	254
50	0.5	0.2	5	76	407
50	0.5	0.15	5	101	615
50	0.3	0.2	5	45	157
50	0.3	0.15	5	61	282
25	0.5	0.2	5	88	511
25	0.5	0.15	5	118	753
25	0.3	0.2	5	53	219
25	0.3	0.15	5	71	365

Analysis completed using a fraction of calcium of 26% and a bulk density of 1.5 Mg/cm^3 for the clog material with initial porosity of 0.48.

[a] $v_f = 0.32$ is the decrease in porosity that corresponds to a hydraulic conductivity of 10^{-6} m/s as calculated from the empirically derived porosity versus hydraulic conductivity relationship where v_f represents porosity filled with clog material. Note that the concentration of calcium is that entering the LCS and not what is being collected (which will be severely depleted due to losses caused by clogging).

[b] *a* is the distance from the pipe where there is uniform clog formation.

[c] Calcium concentration constant with time at 1650 mg/l.

[d] Calcium concentration at 1650 mg/l for 10 years, then reduced linearly to 200 mg/l at 50 years, and then remains constant at this concentration.

Monitoring of liner temperature at a landfill in Pennsylvania where up to 50 m of waste was placed above a leachate collection and composite liner system (Koerner, 2001) indicated that the temperature at the composite liner was constant at 20 °C for about 5 years; however, it subsequently rose to 35 °C and then fluctuated. Thus it appears that temperatures of 30–35 °C are not unusual in MSW landfills (Yoshida and Rowe, 2003).

The move to incinerate waste reduces the volume and organic matter in the waste but studies (Klein *et al.*, 2001) have shown that temperatures in ash landfills can reach 90 °C due to exothermic reactions (heat of hydration). The temperature at the liner in one of these landfills peaked at 45.9 °C 17 months after the start of landfilling and is subsequently decreasing at 0.6 °C per month. Even relatively short exposure to high temperatures can have a significant impact on the service life of composite liners and needs to be considered in design.

2.7 Summary

Leachate may change with time both as the waste ages and as it passes through the leachate collection system to the point of removal. Thus the leachate that is collected is not the same as the leachate that entered into the system. Changes include a reduction in both organic (in terms of COD) and inorganic constituents as a result of the growth of a

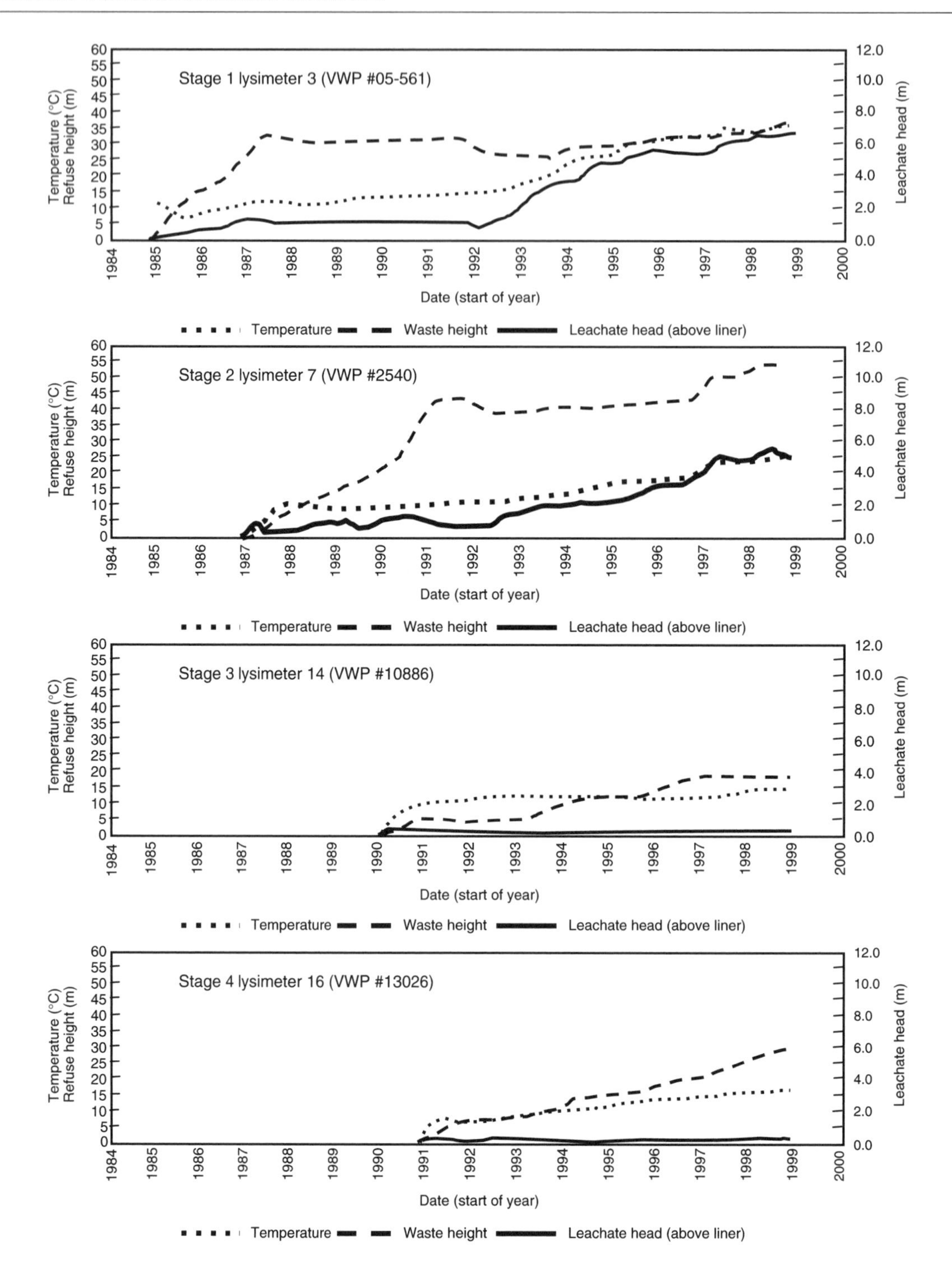

Figure 2.19 Variation in leachate mounding and liner temperature with time for the KVL: Stage 1 has French drains at 65 m spacing; Stage 3 has a continuous blanket drain (modified from Barone *et al.*, 2000).

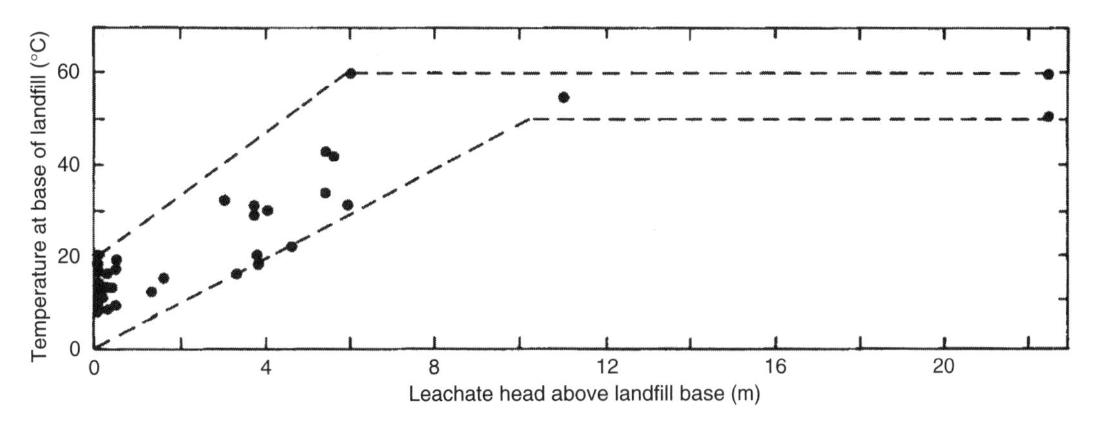

Figure 2.20 Variation in temperature at the landfill base with leachate head for a number of landfills (modified from Barone *et al.*, 2000).

biofilm in the collection system and biologically induced precipitation of inorganic matter (e.g., $CaCO_3$) combined with physical capture of particulates in the leachate by the biofilm. Poorly designed collection systems have experienced problems in less than a decade. The evidence would suggest that well-designed systems may have service lives ranging from many decades to centuries. Field and laboratory studies have shown that the rate of clogging is related to the leachate strength and mass loading, grain size and grain size distribution, initial pore volume and the use of a filter/separator between the waste and the drainage layer. For design purposes it is advantageous to use a drainage material with a large initial void volume thus (1) maximizing the amount of clog that can accumulate in the pores before the hydraulic conductivity of the clog medium is reduced to a point where leachate mounding occurs and (2) reducing the initial specific surface area to which active biofilm can attach and grow, thereby reducing the amount of treatment and chemical precipitation of minerals relative to a large surface area material (smaller gravel or sand). Also important is the placement of drainage layer during construction. Efforts should be made to minimize the amount of fines initially present in the gravel and the use of construction practices that will minimize the generation of fines by rock crushing and shearing.

A filter/separator used between the continuous drainage blanket and the waste can reduce the physical ingress of waste and soil into the drainage material (due to compaction and settlement of waste) and also reduce the mobilization of sand and silt particles into the drainage layer by the leachate. When used directly below the waste, some of the studies have shown that a nonwoven geotextile was more effective than a woven geotextile in reducing the amount of clogging in the underlying drainage system. However, the geotextile filter may also be used in conjunction with a granular layer above the geotextile. The purpose of this upper granular layer is to (a) minimize perched leachate mounding when clogging of the geotextile does occur (best achieved by using gravel for this layer combined with a design that includes a means for leachate in this layer to enter the manholes) and (b) provide protection to the geotextile to minimize damage during waste placement. Sand above the geotextile would serve to provide similar protection to gravel, better filtration and more leachate pre-treatment than gravel, but would likely be less effective in controlling a perched leachate mound.

Techniques for estimating the height of leachate mounding and the service life of leachate collection systems have been described. The importance of inspecting and regularly cleaning leachate collection pipes has been highlighted. Finally it is noted that an increase in leachate mounding on a liner can result in significant increase in the liner temperature which may have implications for both the rate of contaminant transport and the service life of liners (see Section 1.4.3 and Chapter 13).

Chapter three

Clayey barriers: compaction, hydraulic conductivity and clay mineralogy

3.1 Introduction

Soil barriers, containing enough clay minerals to provide low permeability (hydraulic conductivity), are used extensively to prevent the rapid advective migration of various leachates from waste disposal sites. The clayey barriers vary from geosynthetic clay liners (GCLs) ($\sim$1 cm thick, see Chapter 12), to thin ($\sim$3 cm thick) bentonite liners, to (0.6–3 m thick) compacted clayey liners, to natural undisturbed clayey barriers up to 30 or 40 m thick.

Undisturbed clayey barriers have a long record of performance with respect to the containment of chemical species often found in a waste disposal site. For example, the 10,000-year salt profile through a freshwater clay deposit overlying a naturally occurring high-salt bedrock is discussed in Chapter 9. The study of this migration profile provides excellent data on the long-term behaviour of clayey barriers. Chapter 9 also discusses a 20-year long investigation of migration of contaminants from a municipal waste disposal site which indicates long-term performance consistent with what one would expect based on short-term laboratory and field data. However, good performance of clay as a barrier to municipal, industrial and hazardous wastes cannot be assumed *a priori*. The hydraulic performance will depend on a number of important factors to be discussed in this chapter. These include the method of placement and compaction of the clay and the clay mineralogy. The potential interaction between clay minerals and leachate and its effect on hydraulic conductivity will be examined in Chapter 4.

3.1.1 Undisturbed clayey deposits

The hydraulic conductivity of undisturbed clayey deposits will depend on the mineralogy, the manner of deposition and the stress history of the deposits. The *in-situ* hydraulic conductivity of these deposits may be assessed by a number of means including

- laboratory tests such as triaxial and fixed wall hydraulic conductivity apparatus;
- rising head, falling head and constant head tests conducted in piezometers in the field (slug tests); and
- pumping tests conducted on aquifers underlying (or overlying) the undisturbed clayey deposit.

These techniques for assessing the hydraulic conductivity of undisturbed clay are discussed in Section 3.2. Attention will then be focussed on clay mineralogy and potential changes in the hydraulic conductivity of clayey soils due to clay–leachate interaction (Sections 3.4 and 3.5).

3.1.2 Compacted clayey liners

As with undisturbed clayey deposits, the hydraulic conductivity of compacted clayey liners will also depend on the mineralogy, the manner of placement and the stress history of the liner. The primary difference between natural and compacted barriers is that man has some control over the manner of placement and the stress history for compacted clay liners. The performance of a clay liner will depend on both the choice of material (i.e., the geologic origin, grain size distribution and mineralogy) and the manner in which this material is broken up and recompacted. The methods of assessing hydraulic conductivity for compacted clay liners include:

- laboratory tests on liner samples by triaxial and fixed wall hydraulic conductivity apparatus;
- large ring infiltrometers;
- large-scale field lysimeters; and
- falling head tests in short boreholes into liners, etc.

These techniques are discussed in Section 3.2.

3.1.3 Liner specifications

Although the detailed requirements for natural and compacted clayey barriers vary greatly from one jurisdiction to another, the following criteria seem to commonly apply.

1. The liner shall have a hydraulic conductivity, k, of 10^{-9} m/s or less and be free of natural or compaction-induced fractures. From a chemical flux point of view 10^{-10} m/s is preferable since then diffusion often becomes the dominant migration mechanism.
2. Since a low hydraulic conductivity is normally associated with the presence of clay minerals, a minimum of 15–20% of the soil with a particle size smaller than 2 μm (< 2 μm), a plasticity index, I_p, greater than 7%, an activity of 0.3 or greater, and less than 50% gravel (i.e., $D_{50} < 4.75$ mm) may be specified. A minimum cation exchange capacity (CEC) of about 10 meq/100 g of soil may also be specified. Special care is needed with soils of low plasticity ($7\% < I_p < 15\%$) to ensure that the desired hydraulic conductivity can be achieved. Care is also required for high-plasticity clay ($I_p > 30\%$) since these can (a) form hard clods that are difficult to break up when constructing the liner and (b) be particularly susceptible to shrinkage and cracking if allowed to dry after compaction.
3. The clayey barrier shall be compatible with the leachates to be retained. In other words, the hydraulic conductivity, k, shall not increase significantly on exposure to leachate similar to that to be contained by the barrier. Alternatively, the hydraulic conductivity obtained from compatibility tests run to chemical equilibrium may be used as the design hydraulic conductivity.
4. The minimum thickness of a compacted clayey liner for a municipal waste facility will generally be about 0.9–1 m; however, this is highly variable from jurisdiction to jurisdiction and in some cases is reduced to 0.6 m if the clayey liner is used together with a geomembrane to form a composite liner.
5. The minimum thickness of a clayey liner for an industrial/toxic facility shall generally be 3–4 m although some jurisdictions require up to 15 m or multiple composite liner systems such as those discussed in Chapter 1.
6. For compacted clay liners, compaction at a water content of 2–4% above (wet) of Standard Proctor optimum water content (see Section 3.3.1) is often specified subjected to

it being demonstrated that the target hydraulic conductivity (typically 10^{-9} m/s) can be achieved in laboratory tests conducted on samples in this moisture content range. Daniel (1998) recommends a laboratory program that involves (a) compaction of the soil over a range of compactive energy (this allows one to establish the line of optimums see Section 3.3.1); (b) performing hydraulic conductivity tests on these compacted samples; (c) defining a water content range over which acceptable hydraulic conductivity is achieved; (d) limit acceptable dry density/water contentrelationship to ensure adequate shear strength (and possibly to limit susceptibility to desiccation cracking). This would result in an "acceptable zone" as discussed in Section 3.3.1

7. The suitability of a compacted clay liner specification should be verified by the construction of a test pad prior to construction of the landfill liner (e.g., see Rowe *et al.*, 1993).

It should be emphasized that while the foregoing represent common characteristics of clayey liners, the design of a liner is site-specific and its characteristics must be selected such that the barrier system (of which the liner is usually only part) will provide adequate control of contaminant migration. This can only be assessed by appropriate site-specific contaminant impact modelling of the landfill-barrier system. The reader is referred to Chapters 10, 11 and 16 for more details.

Compaction–hydraulic conductivity (permeability) relationships are discussed in Section 3.3 and clay mineralogy and clay colloid chemistry are discussed in Sections 3.4 and 3.5. Clay leachate compatibility and its effect on hydraulic conductivity are discussed in Chapter 4.

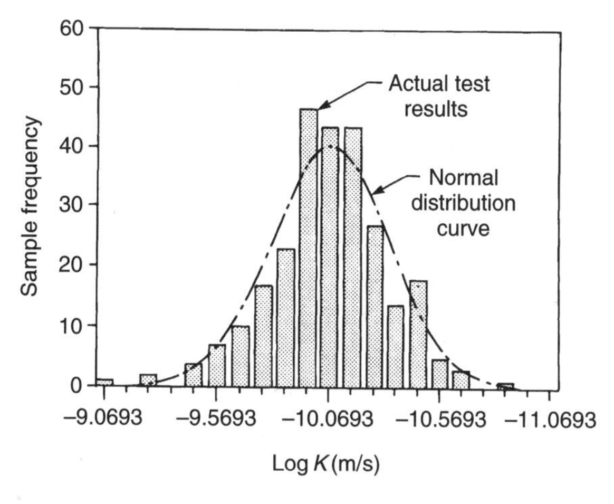

Figure 3.1 Normal distribution of hydraulic conductivity test values obtained on samples of compacted clayey silt from a high-quality landfill barrier (after King *et al.*, 1993; reproduced with permission of the *Canadian Geotechnical Journal*).

3.2 Methods of assessing hydraulic conductivity

3.2.1 Laboratory tests

Laboratory hydraulic conductivity or k-testing involves application of a total head drop across a soil specimen and calculation of k from Darcy's law. The many methods available are discussed in most standard texts and in specialty volumes devoted to clay barriers like *Pollution Technology Review* No. 178 (Noyes Data, 1990).

The objective of this section is not to review all the test procedures but to highlight problems that develop with those most commonly used, namely fixed wall and flexible wall testing.

(a) Fixed wall testing

This form of testing normally involves application of either a constant or falling head across a soil sample compacted into a fixed wall cell. Since a good seal is required between the soil and the cell walls, application of a static vertical effective stress is highly recommended, so lateral yield guarantees a seal. Such a system is often called a consolidation permeameter. Special variations involving constant flow rates, q, have been successfully used by Fernandez and Quigley (1991) (see Chapter 4).

Although small gradients close to field values are preferred, high gradients give correct values for hydraulic conductivity, k, provided samples are saturated and seepage stresses do not cause consolidation and thus reduction in void ratio and k-values.

If set up properly, fixed wall tests can generally provide a good simulation of field conditions and should provide reliable k-test results. However, problems do develop, as itemized below:

1. Tests at zero vertical static stress are very prone to failure by sidewall leakage, especially for soils compacted at a water content less than the Standard Proctor optimum water content (referred to as being compacted "dry-of-optimum") or for fractured natural soils.
2. Very high vertical stresses may be required for stiff fractured soils to obtain a side seal.
3. Most soils compacted at a water content about 2% greater than the Standard Proctor optimum water content (referred to as being compacted "wet-of-optimum") will have a saturation of about 95%. Since most fixed wall devices do not have back-pressure systems, testing is done on unsaturated specimens. This may or may not be bad since the field soil would also be in this condition initially.
4. Clay–leachate compatibility tests on soils that shrink normally result in abrupt failure by sidewall leakage. Thus, fixed wall compatibility k-testing must be done at various stress levels such that the relationship between leachate interaction (shrinkage) and applied stress can be evaluated.

(b) Flexible wall testing

Triaxial or flexible wall testing involves wrapping the soil specimen in a membrane and running the test in a triaxial cell at the appropriate cell pressure. A major benefit is that the degree of saturation may be increased by applying a high back-pressure to dissolve air in the voids. Normally full saturation produces the highest possible hydraulic conductivity, k, values.

Although triaxial k-testing tends to be favoured by industry (using gradients less than 50), it does have both advantages and disadvantages as follows:

1. Cell pressure limitations prevent the use of high gradients so that even after several weeks of flow, rarely has more than 10% of the total pore volume (PV) of a sample been replaced. Thus, compatibility testing requiring replacement of several PVs is not normally feasible in a flexible wall device even if bladders are part of the system.
2. If a soil shrinks in a compatibility test, the membrane contracts with the specimen preventing any abrupt increases in k which is easily observed in a fixed wall device.
3. Very low stress environments are much easier to simulate in a flexible wall permeameter.
4. Flexible wall testing is probably the only option for rock specimens and hard fractured clay specimens.

In summary, the authors consider that both types of testing are viable and in some cases necessary for the same project.

If the test soils are soft enough, k-values calculated from consolidation tests provide a very useful third check on the magnitude of the results of fixed and flexible wall testing. However, k-values from consolidation tests are usually not as reliable as k-values from fixed and flexible wall testing.

(c) Stress levels

Of more importance than the actual test method employed are the stress levels at which the soils are tested. For example, Shelby tube sampling of a barrier clay, subsequent cutting and trimming, etc., all cause disturbance, stress release, gas release, fracture separation, etc. Testing at very low effective stresses may yield hydraulic conductivity, k, values 10–100 times higher than those run at a final field stress of say 100–150 kPa. A major consideration in developing proper protocol is establishing the appropriate confining stress. In a comparative study presented by King *et al.* (1993), flexible wall k-tests run at a confining stress of 145 kPa yielded k-values ranging from 1.2×10^{-10} to 1.2×10^{-11} m/s and averaging 8.5×10^{-11} m/s (see Figure 3.1). The values were quite similar to field values of k deduced from lysimeter data after field loading to similar stress levels.

3.2.2 Borehole/piezometer field tests

Piezometers installed in natural clay, till, silt, sand or gravel deposits are frequently used to assess the hydraulic conductivity of the deposit. The properties often play an important role in modelling the potential impact of a disposal facility. Figure 3.2 shows an idealized hydrostratigraphy into which a series of piezometers have been placed. Typically, the piezometer installation will involve a screen and/or a sand pack (of diameter D) in the intake portion of the piezometer (e.g., see Figures 3.2 and 3.3) and a tube (of diameter d) for controlling flow or monitoring water levels (see Figure 3.3). Two basic types of test can be performed using these piezometers, viz.

1. variable head tests (sometimes called falling head, rising head, slug, or Hvorslev tests) in which the change in water level with time is monitored following some perturbation of the water level in the piezometer;
2. constant head tests in which the head in the piezometer is maintained at some specified level (relative to static conditions) and the flow is monitored.

In principle, either test can be performed on all types of deposits; however, in practice variable head tests are commonly used for low-permeability units (e.g., silts, clays, tills) and constant head tests are commonly used for highly permeable deposits (e.g., sands, gravels). The reason for this is related to the ease of measurement. For low-permeability deposits, it is relatively easy to monitor the gradual rise or fall in water level with time after some perturbation

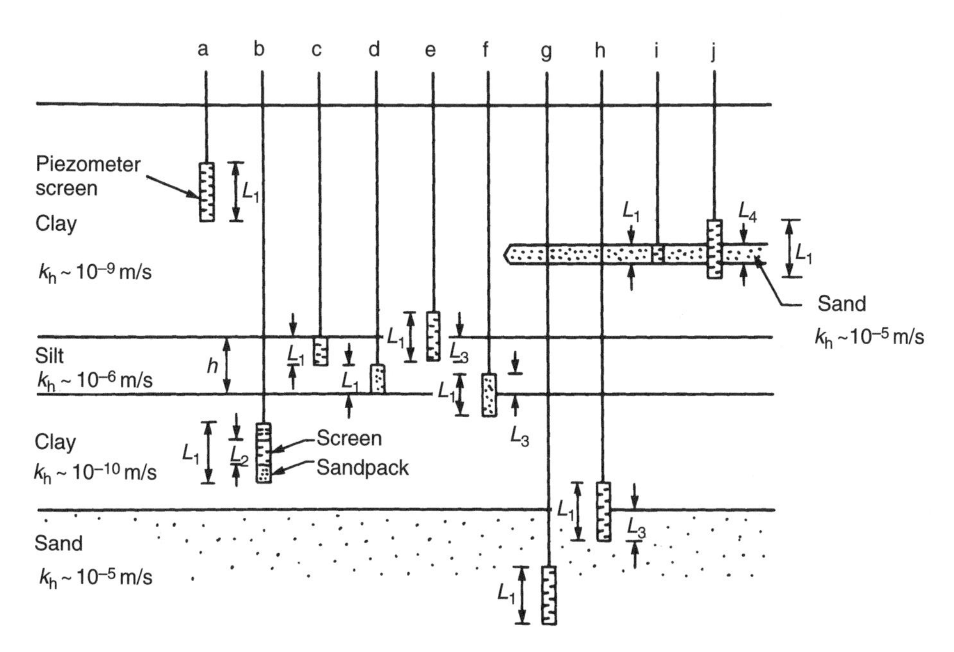

Figure 3.2 Schematic diagram of various piezometer configurations that have been used in performing "falling head", "rising head" or "constant head" tests to estimate the hydraulic conductivity of different strata (not to scale).

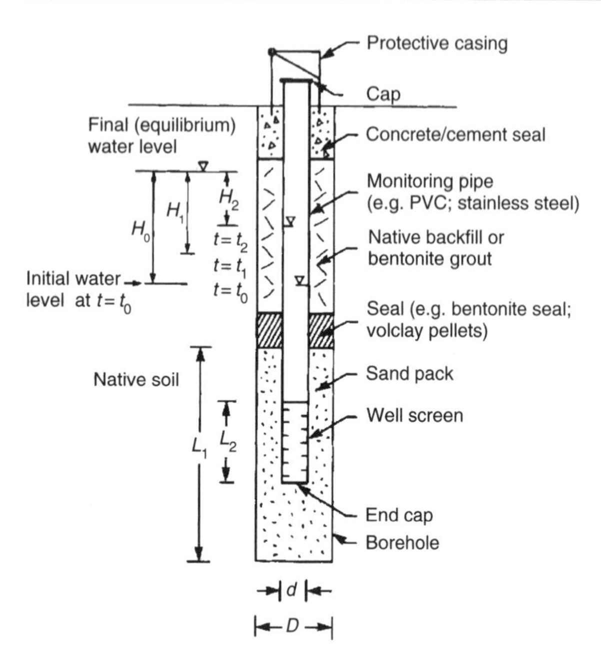

Figure 3.3 Schematic diagram of a piezometer installation.

in water level but it is somewhat more difficult and/or expensive to accurately monitor the flow corresponding to a constant head difference between the piezometer and the equilibrium water level for these soils. For permeable deposits ($k > 10^{-6}$ m/s), the change in head in the piezometer resulting from some perturbation may be so quick that it is difficult and/or expensive to accurately record. In these cases, however, it is usually quite easy to measure the flow corresponding to a constant head difference.

The following sections will discuss the interpretation of results from these types of test, both for simple cases such as that shown for piezometers "a", "b", "c", "d", "g" and "i" in Figure 3.2 and for more complex situations such as shown for piezometers "e", "f", "h" and "j". In all cases, piezometer tests such as those discussed in this section:

1. primarily give an estimate of horizontal hydraulic conductivity;
2. provide an estimate of the hydraulic conductivity within a relatively small region of soil close to the piezometer.

For large-scale estimates of horizontal hydraulic conductivity in aquifers, or vertical hydraulic conductivities in aquitards adjacent to aquifers, a pumping test (as discussed in Section 3.2.3) may be required.

(a) Variable head tests

Variable head tests are performed by perturbing the water level in the piezometer and then monitoring the change in piezometer water level (head) with time. This perturbation may be generated by:

1. raising the water level above the equilibrium value. For example, by putting a metal slug down the hole which, by displacing the water, causes a water level rise. With time, this water will flow from the piezometer into the surrounding soil and the water level in the piezometer will fall and so this is a falling head test;
2. removing water from the piezometer by pumping or bailing and then monitoring the increase in water level to the static level, with time, as water flows into the piezometer from the surrounding soil. This is sometimes called a bail down or rising head test.

The analysis of both forms of the variable head tests is similar and is based on the classic paper by Hvorslev (1951); for this reason, variable head tests are sometimes generically referred to as Hvorslev tests. They are also sometimes generically referred to as slug tests.

Piezometers in a thick uniform soil (piezometers "a", "b" and "g"; Figure 3.2)

For this case, the horizontal hydraulic conductivity is approximately given by

$$k_{\mathrm{h}} = \frac{d^2 \ln[mL/D + \sqrt{1 + (mL/D)^2}]}{8L(t_2 - t_1)} \ln\left(\frac{H_1}{H_2}\right) \tag{3.1a}$$

where k_h is the horizontal hydraulic conductivity [LT^{-1}], d the diameter of monitoring tube (see Figure 3.3) [L], $m = \sqrt{k_h/k_v}$ ($m = 1$ for isotropic soil $k_h = k_v$) [–], D the diameter of piezometer intake (see Figure 3.3) [L], H_1 the piezometer head at time $t = t_1$ (relative to static water level) [L], H_2 the piezometer head at time $t = t_2$ (relative to static water level) [L] and L the intake length of the piezometer [L]; the selection of this length will be discussed below for three situations "a", "b" and "g" shown in Figure 3.2.

For long piezometers such that $mL/D > 4$, equation 3.1a reduces to

$$k_h = \frac{d^2 \ln(2mL/D)}{8L(t_2 - t_1)} \ln\left(\frac{H_1}{H_2}\right) \tag{3.1b}$$

The horizontal hydraulic conductivity can also be calculated based on the basic time lag T. This time lag can be established graphically by plotting the piezometric head H (on a log scale) versus the time t (on a linear scale) as shown in Figure 3.4. The basic time lag T_0 is the time corresponding to $H = 0.37H_0$ (i.e., $\ln(H_0/H) = 1$) where H_0 is the difference in maximum head from the final static level (i.e., it is the perturbation head at $t = 0$; see Figure 3.3). As noted by Hvorslev, the advantage of plotting this diagram is that it reveals irregularities in the data caused by volume changes or stress adjustment time lag and permits easy advance adjustment of the results of the test.

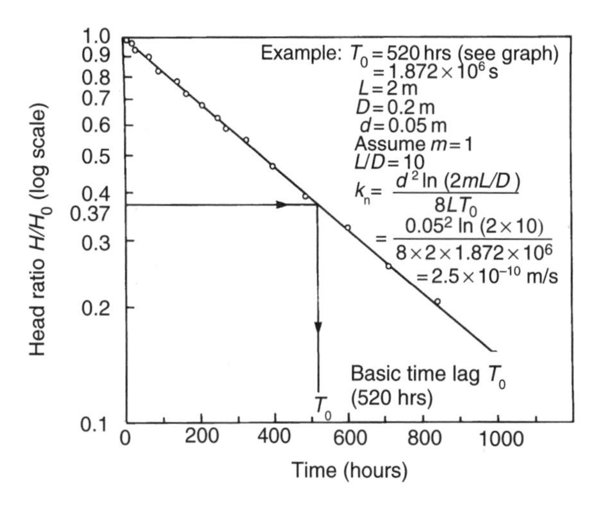

Figure 3.4 Determination of Hvorslev's basic time lag and example calculation of hydraulic conductivity for piezometer "b" shown in Figure 3.2.

Once the basic time lag has been evaluated, the hydraulic conductivity can be calculated from

$$k_h = \frac{d^2 \ln[mL/D + \sqrt{1 + (mL/D)^2}]}{8LT_0} \tag{3.2a}$$

which, for $mL/D > 4$, reduces to

$$k_h = \frac{d^2 \ln(2mL/D)}{8LT_0} \tag{3.2b}$$

and all terms are as previously defined.

When the final equilibrium water level is not known, the basic time lag, T_0, may be estimated prior to full stabilization by observing successive changes in piezometer level for equal time intervals. Suppose that the change in head over some time interval Δt since the beginning of the test is Δh_1 and that over the subsequent time interval Δt the *change* in head is Δh_2 then

$$T_0 = \frac{\Delta t}{\ln(\Delta h_1/\Delta h_2)} \tag{3.3}$$

The evaluation of the piezometer intake length L warrants some discussion. Consider piezometers "a", "b" and "g" shown in Figure 3.2.

Piezometers "a" and "g" involve a screen length which is the entire length of the borehole below the seal. In this case, the intake length is equal to the screen length ($L = L_1$); this is the case irrespective of whether the soil in which the piezometer is placed is a uniform clay, silt or sand.

Piezometer "b" involves a screen length L_2 and a total sandpack length L_1. If the sandpack is substantially more permeable than the surrounding soil (e.g., the clay in Figure 3.2), then the hydraulic conductivity is best estimated by taking the intake length L equal to the entire length L_1 (i.e., $L = L_1$) which includes the sandpack.

Piezometers at the boundary of a confined aquifer (piezometers "c", "d", "e", "f" and "h"; Figure 3.2)
Consider the situation indicated by piezometers "c" and "d" which monitor a portion of a more permeable layer (e.g., silt) overlain (or underlain) by a much lower permeability layer (e.g., clay). Provided that the penetration length L_1 is small compared to the thickness of the layer being monitored ($L_1 \ll h$), then the hydraulic conductivity of the monitored layer may be estimated from

$$k_h = \frac{d^2 \ln[2mL/D + \sqrt{1 + (2mL/D)^2}]}{8L(t_2 - t_1)} \ln\left(\frac{H_1}{H_2}\right) \tag{3.4a}$$

which, for $mL/D > 2$, reduces to

$$k_h = \frac{d^2 \ln(4mL/D)}{8L(t_2 - t_1)} \ln\left(\frac{H_1}{H_2}\right) \tag{3.4b}$$

where all the variables are as previously defined.

In terms of the basic time lag T_0, the hydraulic conductivity is given by

$$k_h = \frac{d^2 \ln[2mL/D + \sqrt{1 + (2mL/D)^2}]}{8LT_0} \tag{3.5a}$$

which, for $mL/D > 2$, reduces to

$$k_h = \frac{d^2 \ln(4mL/D)}{8LT_0} \tag{3.5b}$$

Provided that the sandpack is substantially more permeable than the monitored layer by an order of magnitude or more, the length L is equal to the full length of the screen/sandpack in the monitored layer ($L = L_1$).

Consider the situation shown for piezometers "e", "f" and "h" where the screen/sandpack is of length L_1 but only part of the length (L_3) penetrates the monitored layer (i.e., $L_3 < L_1$). If the hydraulic conductivity of the aquitard is more than an order of magnitude lower than that of the monitored layer, then the length, L, may be reasonably estimated as being equal to the portion of the intake interval actually in the more permeable layer ($L = L_3$).

Factors affecting piezometer performance
The foregoing equations (equations 3.1a–3.5b) may be used to estimate hydraulic conductivity from the observed response of piezometers to a perturbation in piezometer water level. However, it should be recognized that these equations are approximate and that the response of the piezometer may be affected by:

1. leakage through the piezometer seal;
2. clogging of the intake;
3. removal of fine-grained particles from the surrounding soil;
4. storage in the piezometer casing; and
5. accumulation of gases (air bubbles) near the intake.

(b) Constant head tests

There are a number of advantages to the use of the constant head test in preference to the more commonly used variable head tests discussed above. These include:

1. constant head tests tend to be less affected by storage effects in the sandpack and are unaffected by storage within the piezometer casing;
2. for permeable deposits, the constant head test is conducted over a much longer time period than variable head tests and hence can test a much larger volume of soil around the piezometer/well;
3. for permeable deposits, this test is easier to perform than the variable head test and involves less potential error.

The constant head test may involve either adding or removing water from the piezometer as required to maintain the water level to some value H_c relative to the equilibrium (static) water level and then observing the flow required to maintain this constant water level.

Piezometer in a thick uniform soil (piezometers "a", "b" and "g"; Figure 3.2)
The most commonly used expression for assessing the hydraulic conductivity k_h in a constant head test is that developed by Hvorslev (1951), viz.

$$k_h = \frac{q \ln[mL/2R + \sqrt{1 + (mL/2R)^2}]}{2\pi L H_c} \qquad (3.6a)$$

where q is the steady-state flow required to maintain a constant head H_c [L^3T^{-1}], H_c the change in water level (relative to the static/equilibrium water level) [L], R the radius of the piezometer/well as discussed below [L], L the length of the intake as discussed below [L] and $m = \sqrt{k_h/k_v}$.

For $mL/R > 8$, this reduces to

$$k_h = \frac{q \ln[mL/R]}{2\pi L H_c} \qquad (3.6b)$$

If the test is performed in a unit whose hydraulic conductivity is substantially less than that of the sandpack (e.g., piezometers "a" and "b"), then the length $L = L_1$ and the radius $R = D/2$ (Figure 3.3). If the piezometer is installed in a granular unit with a hydraulic conductivity similar to that of the sandpack (e.g., if the sandpack is really material from the same unit), then the length L is best approximated by the screen length (i.e., $L = L_2$) and the radius is the radius of the well/screen, viz. $R = d/2$ (Figure 3.3).

Although not commonly used, constant head tests can be reliably performed even in low-permeability materials. In the past, problems with performing tests in clay have been related to remolding of the soil (e.g., by borehole construction) and clogging of the permeameter during installation. Selfboring permeameters have now been developed which minimize these effects (e.g., see Tavenas *et al.*, 1983, 1986). These permeameters tend to be shorter than conventional piezometers and the approximation involved in Hvorslev's equation needs to be carefully evaluated. Accordingly, various investigators have also refined Hvorslev's equation (e.g., Brand and Premchitt, 1982; Randolph and Booker, 1982; Tavenas *et al.*, 1986, 1990). Based on these studies, a simplified (approximate) equation can be developed (assuming isotropy; $m \simeq 1$) for $2 \le L/D \le 10$:

$$k_h = \frac{q}{(5D + 2.1L)H_c} \qquad (3.7)$$

where q is the steady-state flow [L^3T^{-1}], H_c the change in water level giving rise to the flow q [L], D the diameter of the intake [L] and L the length of the intake [L].

Equation 3.7 will give lower estimates of hydraulic conductivity than equations 3.6a and 3.6b. There are data (e.g., Tavenas *et al.*, 1986) to support the lower estimates based on equation 3.7.

Piezometer at boundary of a confined aquifer (piezometers "c", "d", "e", "f" and "h"; Figure 3.2)
Based on Hvorslev (1951), the hydraulic conductivity k_h (m/s) can be estimated from a constant head test using

$$k_h = \frac{q \ln[mL/R + \sqrt{1 + (mL/R)^2}]}{2\pi L H_c} \qquad (3.8a)$$

and for $mL/R > 8$, this reduces to

$$k_h = \frac{q \ln[2mL/R]}{2\pi L H_c} \qquad (3.8b)$$

where all terms are as defined in the discussion of equations 3.6a and 3.6b, including the choice of L and R.

Piezometer screened across a confined aquifer (piezometers "i" and "j"; Figure 3.2)
Again, based on Hvorslev (1951), the hydraulic conductivity k_h (m/s) for a constant head test can be estimated from

$$k_h = \frac{q \ln(R_0/R)}{2\pi L H_c} \quad (3.9)$$

where R_0 is the effective radius to the source of supply (i.e., the estimated zone of influence of the test) [L], R the radius of the screen/well [L], q the steady-state flow [L^3T^{-1}], H_c the change in water level (head) giving rise to the flow q [L] and L the length of the intake [L]. For piezometer "i", $L = L_1$. For piezometer "j" in Figure 3.3 the length of the screen exceeds the thickness of the deposit, L_4 (i.e., where $L_1 > L_4$) and so the length L is taken as the thickness of the permeable deposit, $L = L_4$.

Factors affecting interpretation
Hvorslev's equations only consider the hydraulics of the permeable layer being monitored. They ignore effects such as leakage from adjacent aquitards and boundary conditions such as that shown in Figure 3.2 where piezometers "i" and "j" are monitoring a lens of finite extent. These effects may make it difficult to obtain a true steady-state flow q. Judgement is required to assess the reasonableness of using Hvorslev's equations for a given field situation. Where serious doubt exists, more elaborate methods of analysis may be adopted as discussed in the following section.

3.2.3 Pumping tests

Much has been written on the subject of pumping tests and space does not permit a detailed review here. Rather, this section is intended to

1. summarize some commonly adopted methods for estimating the transmissivity and storativity of an aquifer using a pumping test; and
2. discuss methods of estimating the bulk hydraulic conductivity of an adjacent aquitard based on its response to the pumping of an adjacent aquifer.

Typically, the performance of a pumping test involves construction of a pumping well in the aquifer to be pumped. The well is then pumped at a constant flow rate and the drawdown of the water level is observed with time and is plotted on semi-log paper (drawdown versus log(time)). The primary differences between this test and the constant head test discussed in Section 3.2.2 are:

1. the pumping test is a transient test (i.e., steady-state conditions are not required for interpretation); and
2. the pumping well is usually larger than the piezometer used in a constant head test.

In many cases, piezometers will be installed at different distances from the pumping well and the response of these piezometers to the pumping test can be used to assess the hydraulic characteristics of the aquifer. In some cases, monitors are also placed in an adjacent aquitard as shown in Figure 3.5. The response of these piezometers may be used to estimate aquitard properties.

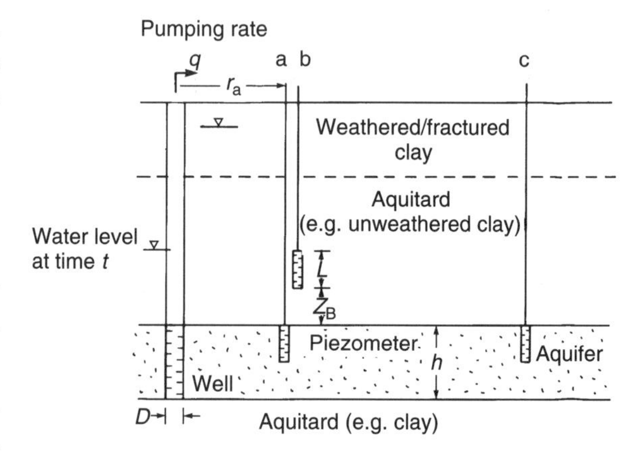

Figure 3.5 Schematic diagram of a pumping test (not to scale).

(a) Cooper and Jacob's (1946) interpretation of aquifer properties

Assuming:

1. a uniform aquifer with the same hydraulic conductivity in all directions;
2. uniform thickness h of the aquifer and infinite extent;
3. a fully penetrating pumping well (e.g., see Figure 3.5);
4. a well with 100% efficiency;
5. laminar flow in the well and aquifer;
6. all water removed comes from aquifer storage (i.e., no recharge or leakage); and
7. the potentiometric surface has no significant slope.

The drawdown response of a well (e.g., see Figure 3.6) can be used to estimate the transmissivity T (i.e., hydraulic conductivity k times aquifer thickness h; $T = kh$) and hence the hydraulic conductivity from the equation

$$T = \frac{0.183q}{\Delta s} \tag{3.10}$$

where T is the transmissivity [L^2T^{-1}], q the flow [L^3T^{-1}] and Δs the slope of the drawdown graph (expressed as the change in drawdown per log cycle time) [L].

The hydraulic conductivity is then given by

$$k_h = \frac{T}{h} \tag{3.11}$$

where k_h is the horizontal hydraulic conductivity [LT^{-1}], T the transmissivity (from equation 3.10) [L^2T^{-1}] and h the thickness of the aquifer [L].

An example calculation is shown in Figure 3.6. The drawdown curve shown in Figure 3.6 exhibits a significant change in slope at about 2,000 min. This suggests that some boundary effect such as a change in aquifer transmissivity occurs at some distance away from the pumping well.

The storativity (S) of the aquifer can also be calculated from the time/drawdown plot of a piezometer by extrapolating the drawdown curve back until it intercepts the zero drawdown

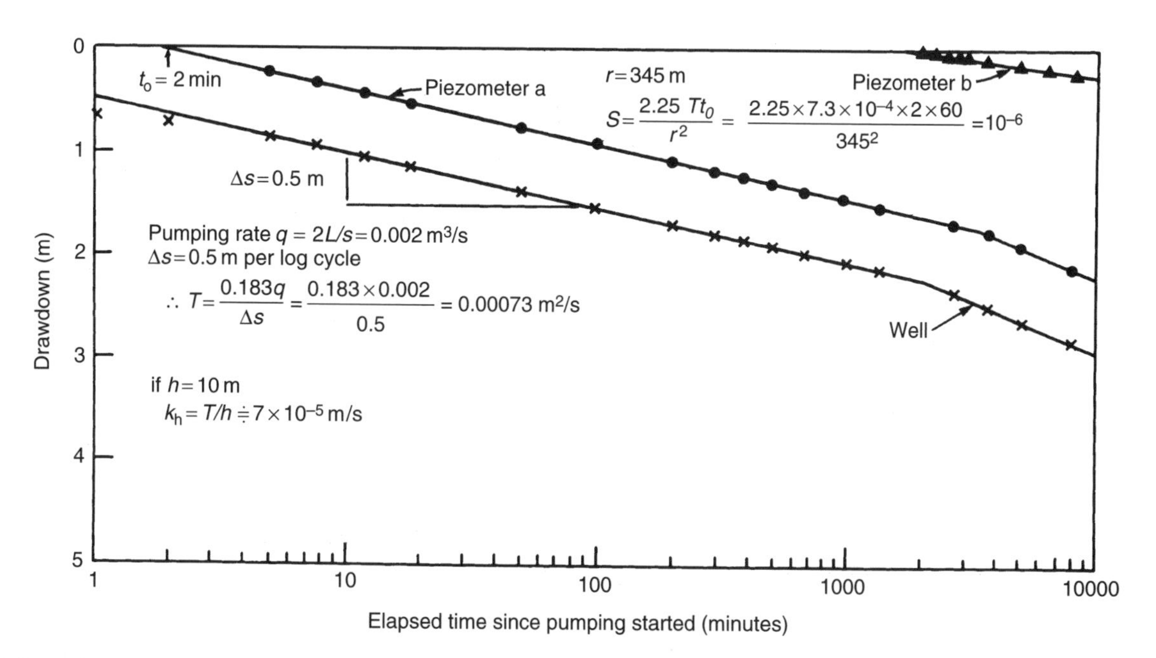

Figure 3.6 Example drawdown versus elapsed time plot for a pumping test.

axis; the corresponding time t_0 is noted and the storativity calculated as follows:

$$S = \frac{2.25Tt_0}{r^2} \tag{3.12}$$

where S is the storativity [–], T the transmissivity [L^2T^{-1}], t_0 the zero drawdown intercept [T] and r the distance from the pumped well to the piezometer [L]. An example calculation is shown in Figure 3.6.

The method of interpretation presented above assumes no leakage into the aquifer from above or below. In reality, there is likely to be leakage but, as noted by Neuman and Witherspoon (1972), the errors associated with the use of equations 3.11 and 3.12 which neglect leakage will be small if the data are collected close to the pumping well. The errors increase with the distance of the monitoring point from the piezometer. As a general rule, early drawdown data are less affected by leakage than later time data and hence, provided the data are reliable, most emphasis should be given to early time data.

If leakage is considered important, techniques such as that proposed by Hantush (1956, 1960) can be used for estimating transmissivity and storage coefficient.

(b) Neuman and Witherspoon's (1972) interpretation of aquitard hydraulic conductivity properties

Neuman and Witherspoon (1972) proposed a simple method of estimating the bulk hydraulic conductivity of an aquitard by monitoring the response of the aquitard to a pumping test conducted on an adjacent aquifer. For example, consider the schematic diagram shown in Figure 3.5. Suppose that a piezometer "a" is installed in an aquifer at some distance r from the pumping well. Suppose that a second piezometer "b" is installed in the aquitard at some distance z_B above the aquifer and that the length of the piezometer is L. The drawdown of the aquifer, s, and of the aquitard, s', in piezometers "a" and "b" can be monitored with time. From the early time response of piezometer "a", the aquifer transmissivity T and storativity S may be calculated using equations 3.11 and 3.12, respectively. At some time t after the aquitard piezometer begins to respond (e.g., see piezometer "b" in Figure 3.6), the ratio (s'/s) of the drawdown of the aquitard piezometer, s', to that of a nearby aquifer piezometer, s, can then be calculated. Also, a dimensionless time factor t_D can be calculated for this time

$$t_D = \frac{Tt}{Sr^2} \tag{3.13}$$

where T is the transmissivity of the aquifer [L^2T^{-1}], S the storativity [–], r the distance from the pumping well to piezometer "a" [L] and t the time of interest [T].

Using t_D and s'/s, Figure 3.7 can be used to obtain a second dimensionless time factor t'_D, where

$$t'_D = \frac{k_V t}{S'_s z^2} \tag{3.14}$$

where k_v is the bulk vertical hydraulic conductivity [LT^{-1}], t the time of interest [T], S'_s the specific storage ($S_s = m_v\gamma_w$) [L^{-1}] and z the distance from the aquifer to the monitoring point [L].

Knowing t'_D, t, z and S'_s (e.g., from a consolidation test), the hydraulic conductivity can be calculated from equation 3.14. Neuman and Witherspoon's charts were developed assuming that the monitoring piezometer is a point. In practice, the monitor length L is usually significant compared with the distance z_B from the aquifer to the bottom of the piezometer; this complicates the interpretation as illustrated in the following example.

Example

Consider the data presented in Figure 3.6. Based on the response of piezometer "a"

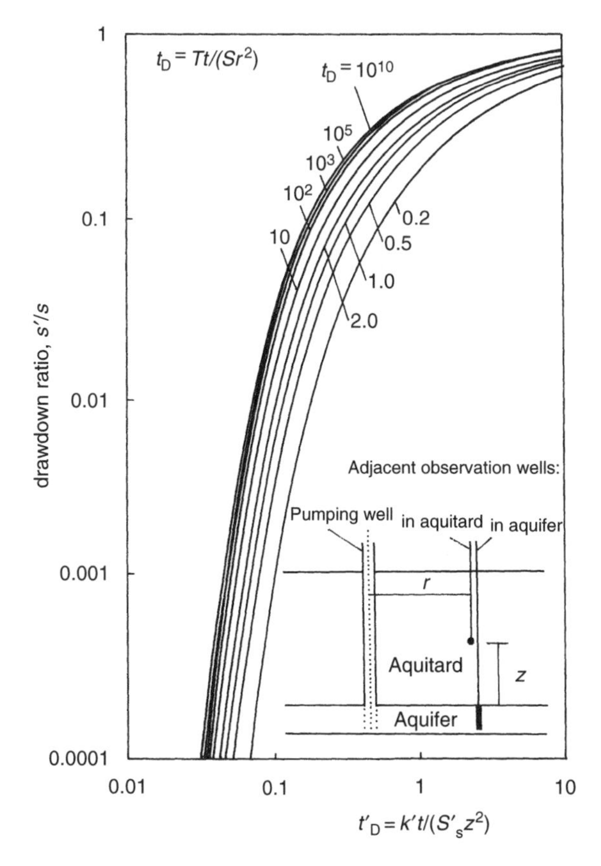

Figure 3.7 Variation of s'/s with t'_D for semi-infinite aquitard (after Neuman and Witherspoon, 1972; reproduced with permission of *American Geophysical Union*).

$$T = \frac{0.183 \times 0.002}{0.5} = 0.00073\,\mathrm{m^2/s}$$

$$S = \frac{2.25Tt_0}{r^2} = \frac{2.25 \times 0.00073 \times 2 \times 60}{345^2}$$

$$= 1.66 \times 10^{-6}$$

At about 2,000 min, piezometer "b" begins to respond. At $t = 2{,}500$ min the drawdowns are:

$s = 1.65\,\mathrm{m}$ (for piezometer "a")

$s' = 0.05\,\mathrm{m}$ (for piezometer "b")

Therefore the dimensionless time factor

$$t_D = \frac{Tt}{Sr^2} = \frac{0.00073(2{,}500 \times 60)}{1.66 \times 10^{-4} \times 345^2} = 555$$

and the drawdown ratio $s'/s = 0.05/1.65 = 0.03$.

From Neuman and Witherspoon's charts (see Figure 3.7) for $t_D = 555$, $s'/s = 0.03$:

$$t'_D \simeq 0.11$$

Assuming that the specific storage $S'_s = 10^{-3}\,\mathrm{m^{-1}}$ (e.g., as obtained from consolidation tests on the aquitard; $S'_s = \gamma_w m_v$ where γ_w is the unit weight of water and m_v the coefficient at volume change), the bulk vertical hydraulic k_v conductivity may be estimated as

$$k_v \simeq \frac{S'_s z^2 t'_D}{t} = \frac{10^{-3} \times z^2 \times 0.11}{2{,}500 \times 60}$$

$$= 7.3 \times 10^{-10} z^2\,\mathrm{m/s}$$

The major problem is selecting the distance z. Suppose the piezometer is installed such that $z_B = 0.9\,\mathrm{m}$ and $L = 1.6\,\mathrm{m}$. The value of z would be a minimum for $z = z_B = 0.9\,\mathrm{m}$ giving $k_v = 5.9 \times 10^{-9}\,\mathrm{m/s}$ whereas for the average $z = z_B + L/2 = 1.7\,\mathrm{m}$, $k_v = 2 \times 10^{-9}\,\mathrm{m/s}$. In this case, the choice of z makes a threefold difference to the estimated hydraulic conductivity. Furthermore, Neuman and Witherspoon (1972) (a) used diffusion theory which fails to consider the effect of stress changes in the aquitard due to pumping and (b) did not consider time lag in the piezometer. It may be concluded that Neuman and Witherspoon's method is most applicable for small-length piezometers with a small time lag (i.e., where a pressure transducer is used to monitor pressure changes in the piezometer).

Rowe and Nadarajah (1993) have examined the effect of stress change in the aquifer and the finite size of the piezometer as discussed in the following section.

(c) Rowe and Nadarajah's (1993) interpretation of aquitard hydraulic conductivity

Using Biot theory (Biot, 1941), Rowe and Nadarajah (1993) examined the effect of the finite size of piezometers, the proximity of the piezometer to the aquifer, time lag and effective stress changes within the aquitard on the interpretation of pumping test results. They developed a number of simple correction factors β_1, β_2 for the effect of time lag (β_1), and size and proximity of the piezometer to the aquifer (β_2) which can be used in association with the conventional Neuman and Witherspoon (1972) chart given in Figure 3.7.

The correction factors β_1 and β_2 can be expressed in terms of a dimensionless parameter λ, viz.

$$\lambda = 1.5\, l S_s' \left(\frac{1-\nu}{1+\nu}\right) \frac{k_h}{k_v} \frac{\pi D^2}{A} \qquad (3.15)$$

where l is the length of the piezometer screen plus the length of any sand/gravel packed between the screen and the aquifer being pumped [L], D the diameter of the borehole [L], A the cross-sectional area of riser pipe used for monitoring the change in water level in the piezometer ($0.25\pi d^2$) [L^2], ν the drained Poisson's ratio of the aquitard soil [–] (typically 0.3–0.35 for silt, 0.3–0.4 for soft clay and 0.2–0.3 for stiff clay), S_s' the specific storage, where $S_s' = m_v \gamma_w$ [L^{-1}], m_v the coefficient of volume change [$M^{-1}LT^2$], γ_w the unit weight of water [$ML^{-2}T^{-2}$], k_h the expected horizontal hydraulic conductivity of aquitard [LT^{-1}], k_v the expected vertical hydraulic conductivity of aquitard [LT^{-1}], z_B the thickness of aquitard from aquifer being pumped to the bottom of the borehole [L] and $z = z_B + 0.5l$.

Figure 3.8a shows the relationship between the correction factor β_1 and the parameter λ for a number of values of s'/s which is the ratio of the drawdown in the aquitard (s') to the drawdown in the aquifer (s). Curves are shown for different values of z/D.

The horizontal hydraulic conductivity k_h can be estimated from a falling head test in the aquitard piezometer as described in Section 3.2.2.

Figure 3.8b shows the relationship between the correction factor β_2 and parameter λ for a number of values of s'/s. In this case, curves are shown for different values of z/l.

The effects of time lag and piezometer location can then be incorporated into the assessment of hydraulic conductivity, k, as follows:

$$k_V = \frac{t_D' S_s' z^2}{t} \frac{\beta_2^2}{\beta_1} \qquad (3.16)$$

where t_D' is the time factor determined from Neuman and Witherspoon's chart (Figure 3.7) [–], β_1 the correction factor for time lag determined from Rowe and Nadarajah's chart (Figure 3.8a) [–], β_2 the correction factor for piezometer location determined from Rowe and Nadarajah's chart (Figure 3.8b) [–] and the other parameters are as defined for equation 3.15.

The time lag of a piezometer can also be affected by entrapped air. The potential effect of entrapped air increases with the size of the sand/gravel pack. Thus, the size of the sand/gravel pack should be kept as small as practical.

Changes in atmospheric pressure can give rise to apparent changes in head in the piezometers. The atmospheric pressure should be monitored during the test and this should be used in the interpretation of the field data.

The specific storage may be estimated from the results of 1D consolidation tests. However, these test results may overestimate S_s' due to:

1. sample disturbance;
2. entrapped air; and
3. larger stress/strain changes than expected in the field.

In some cases, the specific storage can be bracketed by the values obtained in compression and expansion over the relevant stress range.

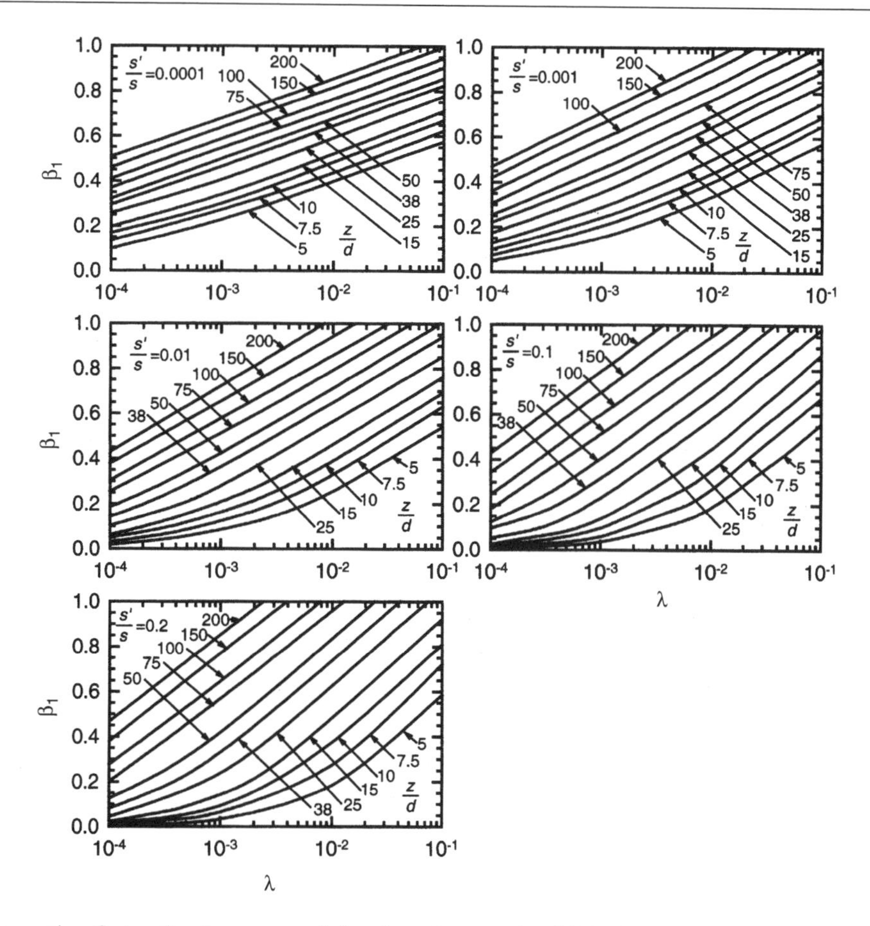

Figure 3.8a Correction factor for time versus λ for drawdown ratios (s'/s) of 0.0001, 0.001, 0.01, 0.1 and 0.2 (after Rowe and Nadarajah, 1993; reproduced with permission of the *Canadian Geotechnical Journal*).

The effect of uncertainty regarding the specific storage can be assessed by evaluating the hydraulic conductivity for the upper and lower values over the reasonable range of uncertainty.

Example

Figure 3.9 shows the results from a pumping test conducted to help define the hydraulic conductivity of an aquitard. The drawdowns were monitored using observation wells which consist of a 100-mm diameter flush-joint PVC pipe, installed with a 0.31-m machine slot screen in a 0.23-m diameter borehole with a gravel pack extending 0.3 m above the top level of the screen (no gravel pack was placed below the bottom of the screen).

From Figure 3.9 when $t = 400\,\text{min}$, $s' = 0.029\,\text{m}$ and $s = 3.66\,\text{m}$, so the drawdown ratio becomes

$$\frac{s'}{s} = \frac{0.029}{3.66} = 0.0079$$

The transmissivity $T = 0.0184\,\text{m}^2/\text{s}$ and the storativity $S = 1.1 \times 10^{-4}$ were obtained based on a conventional analysis of the aquifer response using initial straight-line portion of the drawdown curve and Jacob's method. For a monitoring nest located 22 m from the pumping well:

$$t_{\text{D}} = \frac{Tt}{Sr^2} = \frac{0.0184(400 \times 60)}{1.1 \times 10^{-4} \times 22^2} = 8.3 \times 10^3$$

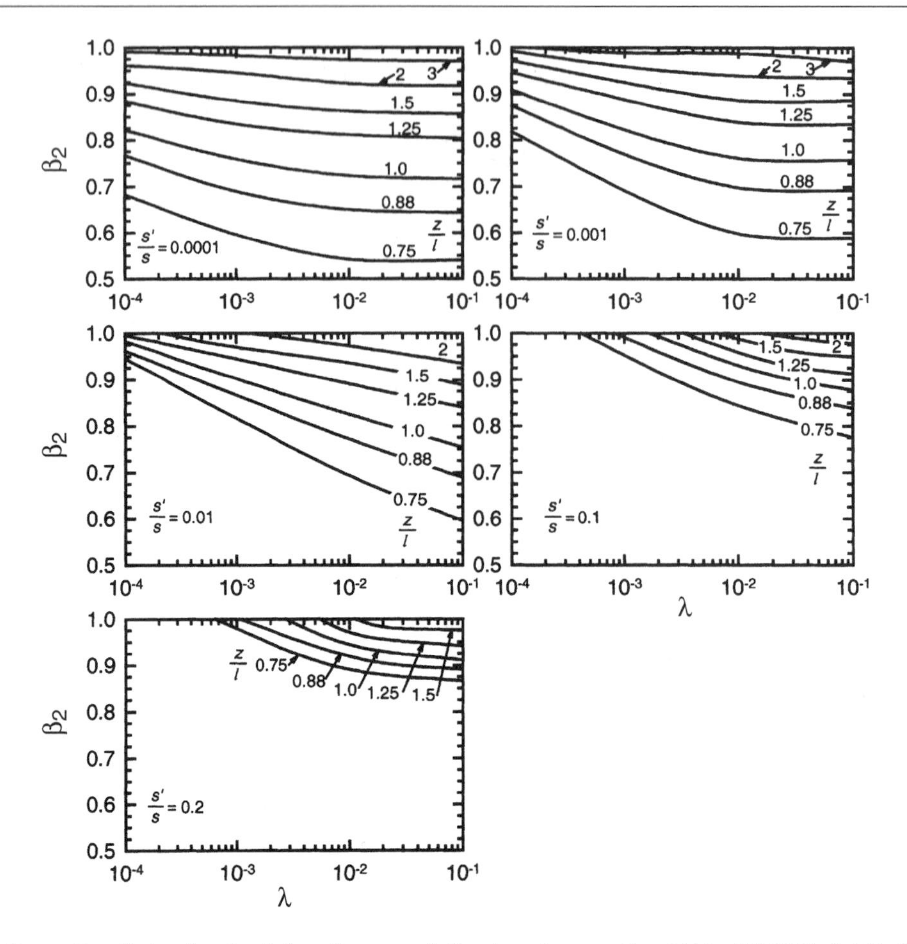

Figure 3.8b Correction factor for depth/length versus λ for drawdown ratios (s'/s) of 0.0001, 0.001, 0.01 and 0.1 (after Rowe and Nadarajah, 1993; reproduced with permission of the *Canadian Geotechnical Journal*).

For the *in-situ* stress range, an average specific storage of the aquitard obtained from consolidation tests is taken to be $7.9 \times 10^{-4}\,\mathrm{m}^{-1}$. Assuming $\nu = 0.3$ and $k_h/k_v = 1$, λ can be calculated as

$$\lambda = 1.5lS'_s \frac{(1-\nu)}{(1+\nu)} \frac{\pi D^2}{A}$$

$$\therefore \lambda = 1.5 \times 0.31(7.9 \times 10^{-4}) \frac{(1-0.3)}{(1+0.3)}$$

$$\times \frac{\pi(0.23)^2}{\pi(0.05)^2} = 0.004$$

$$z = 3.2\,\mathrm{m}, \qquad D = 0.23\,\mathrm{m}, \qquad l = 0.31\,\mathrm{m}$$

$$\therefore \frac{z}{D} = \frac{3.2}{0.23} = 13.9, \qquad \frac{z}{l} = \frac{3.2}{0.31} \approx 10.3$$

From Figure 3.8a the time correction factor, β_1, is 0.33 and from Figure 3.8b the depth/length correction factor, β_2, is 1 since $z/l > 4$. So the gross correction factor is

$$\frac{\beta_2^2}{\beta_1} = \frac{(1.0)^2}{0.33} = 3$$

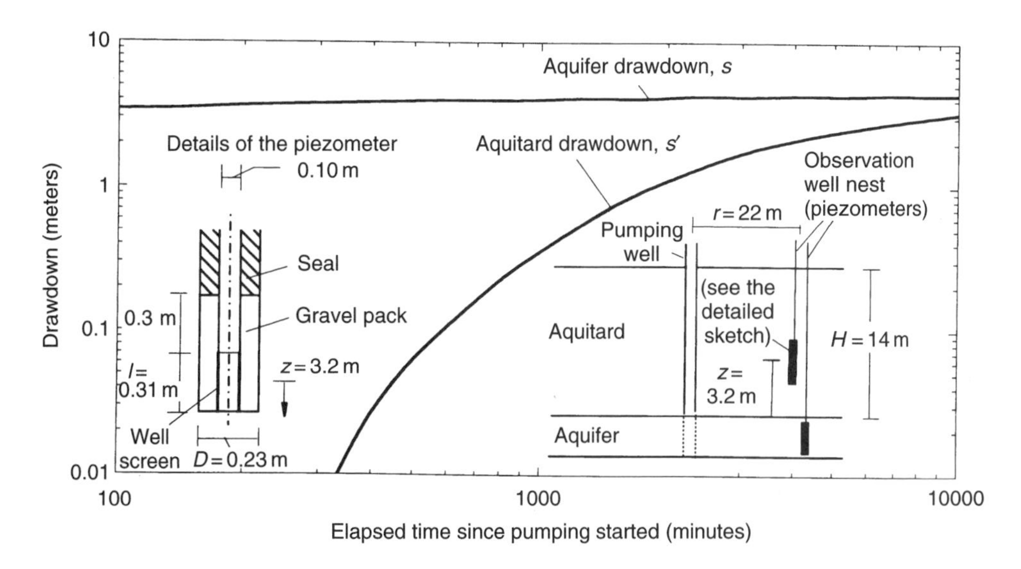

Figure 3.9 Drawdown versus elapsed time in a pumping test – example (after Rowe and Nadarajah, 1993; reproduced with permission of the *Canadian Geotechnical Journal*).

Thus, in this particular case the errors that would arise from neglecting the time lag are such that the hydraulic conductivity obtained from the normal application of the ratio method would underestimate the bulk hydraulic conductivity by a factor of 3.

From Figure 3.7, $t'_D = 0.075$ for $s'/s = 0.0079$ and $t_D \simeq 8.3 \times 10^3$. The hydraulic conductivity of the aquitard is therefore given by

$$k_v = \frac{t'_D S'_s z^2}{t} \frac{\beta_2^2}{\beta_1}$$

$$= \frac{0.075(7.9 \times 10^{-4}) \times 3.2^2}{400 \times 60} \times 3$$

$$= 7.6 \times 10^{-8}\,\text{m/s}$$

(d) Estimating hydraulic conductivity for complex aquitard systems

As discussed in the preceding sections, one common means of estimating the bulk vertical hydraulic conductivity is to perform a pumping test on the underlying aquifer and then monitor the pore pressure response in the overlying aquitard. The techniques discussed so far were developed for a uniform soil layer above (or below) the aquifer (e.g., Neuman and Witherspoon, 1972; Rowe and Nadarajah, 1993). These techniques can be readily applied to situations where the layer adjacent to the aquifer is the primary barrier to contaminant migration. However, in many practical situations, the clayey deposit is layered (e.g., see Figure 3.10) and the primary barrier (Layer 3 in Figure 3.10) may be separated from the aquifer by one or more layers of clayey material with different hydraulic conductivities (and coefficient of consolidation) from that of the primary layer. Under these circumstances, the techniques of Neuman and Witherspoon (1972) and Rowe and Nadarajah (1993) cannot be used to provide a suitable estimate of the hydraulic properties of the aquitard. Usher and Cherry (1988) have modified a two-layer heat flow solution to allow the analysis of a two-layer aquitard. However, even this approach is not adequate in some practical situations.

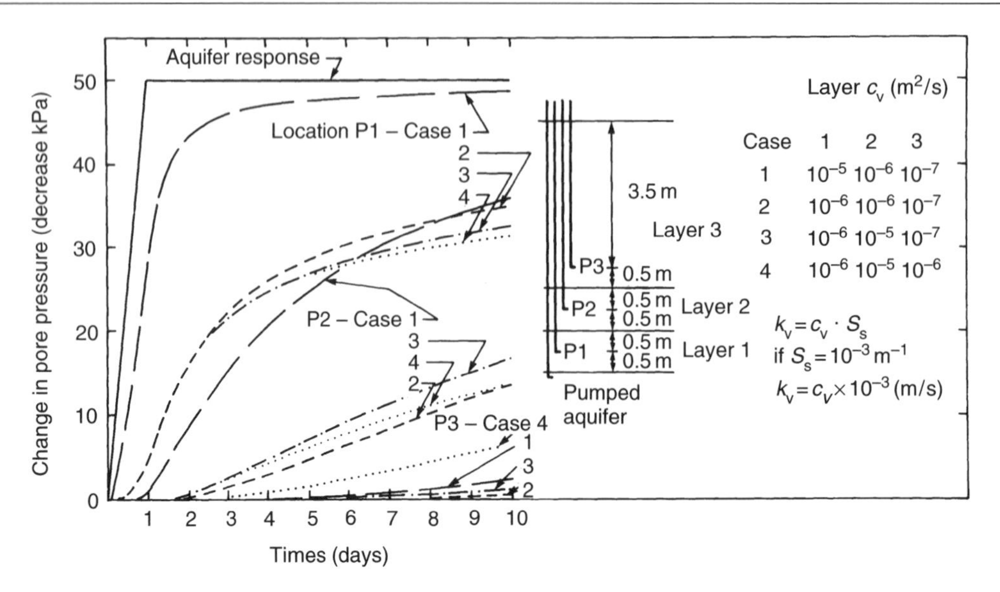

Figure 3.10 Response of a three-layered aquitard to a pumping test (after Rowe, 1991d; reproduced with permission, Balkema).

Finite layer techniques are ideally suited to the analysis of multilayer problems such as that indicated in Figure 3.10. For example, the finite layer contaminant transport model proposed by Rowe and Booker (1986, 1987) and described in Chapter 7 can be readily used to analyse any number of layers and since the solution is analytically based, large contrasts in hydraulic properties between layers do not cause any numerical error or numerical problems. Recognizing that in the absence of advection and sorption, the equations governing contaminant diffusion are the same as those governing pore pressure diffusion (where contaminant concentration is analogous to excess pore pressure), one can model the aquifer loading and pore pressure response using the theory presented in Chapter 7.

To illustrate the application of this approach, the hydrogeologic conditions shown in Figure 3.10 were modelled for a 10-day pumping test. The aquifer was drawn down approximately 5 m beneath the monitoring points P1, P2 and P3 over a 1-day period. The pumping rate was then adjusted to maintain a 5-m drawdown for the remaining 9 days. Figure 3.10 shows the response of a piezometer located in each of the three layers as calculated using program POLLUTE (Rowe and Booker, 2004), for four different combinations of hydraulic properties. An inspection of the relative response of piezometers P1, P2 and P3 for the four cases clearly shows the sensitivity of the analysis (and expected system response) to changes in hydraulic properties. Using small-volume piezometer sampling points and pressure transducers to monitor the response, the effects evident in Figure 3.10 could be readily detected and by adjusting the hydraulic diffusivity (i.e., coefficient of consolidation, c_v) for each layer to obtain a fit to the observed response, the hydraulic properties of the layered system can be readily estimated from $k_v = c_v S'_s$.

3.2.4 Ring infiltrometers

Large-diameter (0.5–2 m in diameter) ring infiltrometers have been recommended (Day and Daniel, 1985) for field use to establish the hydraulic conductivity, k, values of liners prior to waste disposal. In some government jurisdictions they are mandatory. Such testing is expensive, time-consuming and fraught with so many problems that mandatory use is most unfortunate.

A good discussion of large infiltrometers was presented by Daniel (1989) and the schemes presented in Figure 3.11. These devices do a good job of measuring fairly high k-values on inferior liners ($k > 10^{-8}$ m/s) which are poorly compacted, slickensided, fractured and/or fissured by desiccation or frost, etc. Laboratory tests on such barrier soils tend to be conducted on intact interfissure clods or after closure of fractures by application of triaxial cell pressures. Therefore, as presented in Table 3.1, the resulting ratio k_{field}/k_{lab} can be as high as 2,000–100,000 in the extreme case.

Use of ring infiltrometers (be they single, double, open or sealed) on high-quality barriers ($k < 10^{-9}$ m/s) may involve many weeks of testing and for this reason alone may be quite impractical, especially open systems where evaporation losses may far exceed infiltration

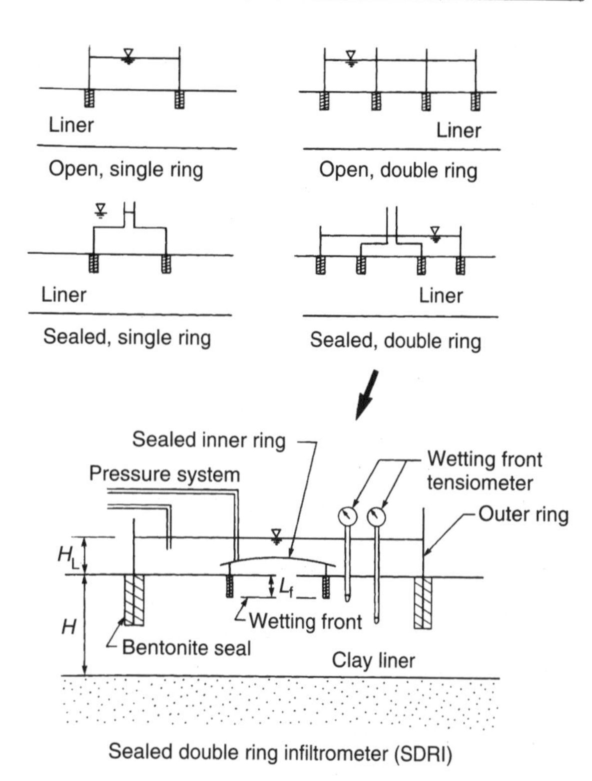

Figure 3.11 Open and sealed, single- and double-ring infiltrometers (modified from Daniel and Trautwein, 1986; Daniel, 1989).

Table 3.1 Hydraulic conductivity of clay liners

1. Daniel (1984)

Project	k_{field}/k_{lab}
Central Texas	25–100,000
Northern Texas	10–2,000
Southern Mexico	25–200
Northern Mexico	5–200

2. Bracci *et al.* (1991)
   $k_{field} = 10^{-6}–10^{-8}$ m/s
   $k_{lab} = 10^{-8}–10^{-11}$ m/s

3. Three Wisconsin landfill liners (specified $k = 10^{-9}$ m/s) (Gordon *et al.*, 1989)
   "Marathon" k (lysimeter) $= (0.5–2) \times 10^{-10}$ m/s
   "Portage" k (lysimeter) $= (1–9) \times 10^{-11}$ m/s
   "Sauk" k (lysimeter) $= (0.6–4) \times 10^{-10}$ m/s

4. Keele Valley liner (Specified $k = 10^{-10}$ m/s) (King *et al.*, 1993)
   1984 k (lysimeter) $\approx 5 \times 10^{-10}$ m/s (stabilizing)
   1988 k (lysimeter) $\approx 5 \times 10^{-11}$ m/s (stabilized)

into the barrier. Another major problem is determining the length of the flow path, L, in calculations of gradient, $i = \Delta h_t / L$. Normally this is taken as the distance from the water reservoir to the wetting front, and suction head at this front is ignored in the estimation of Δh_t. Special potentiometers or post-test excavation are required to estimate the wetting front location.

Inactive clayey barriers compacted ~2% wet-of-standard Proctor optimum have very positive self-healing characteristics during consolidation as the overlying waste is placed. Thus, for these clays, a combination of flexible wall laboratory testing and field lysimeter monitoring probably represents adequate control. Again, stringent construction quality control and construction quality assurance are required during barrier compaction to obtain a low hydraulic conductivity liner. In addition, a suitable post-placement protocol is required to prevent drying or swelling since this can be critical to final performance at the compacted clay liner.

3.2.5 Lysimeters

Lysimeters are large-scale seepage collection devices installed within or below clayey liners in the field to measure bulk hydraulic conductivity. A typical lysimeter might be 15 × 15 m in horizontal area and lined on the bottom and sides with an HDPE geomembrane (see Figure 3.12). The lysimeter would be filled with sand, possibly filtered upwards to prevent migration of clay from the liner. However, care is required in selecting the size of the lysimeter and the soil used to fill the lysimeter, since in an unsaturated state the hydraulic conductivity of sands may be considerably less than that of the near-saturated clay liner. Under these circumstances water may go around, rather than into, the lysimeter if it is too small (Barone *et al.*, 1999). Assuming that this issue has been adequately addressed, water passing through the liner is collected in the lysimeter and transferred through drainage tubes to a collection or sampling point at the margin of the liner. Using Darcy's law, $Q = kiA$, the hydraulic conductivity, k, may be calculated if flow, Q, gradient, i, and area, A, can be determined.

Like most field instrumentation, measurement is complicated by problems such as those discussed by King *et al.* (1993). Difficulties include the following:

1. intermittent flow, Q, from a lysimeter resulting from barometric fluctuations so that long-term averaging is required;

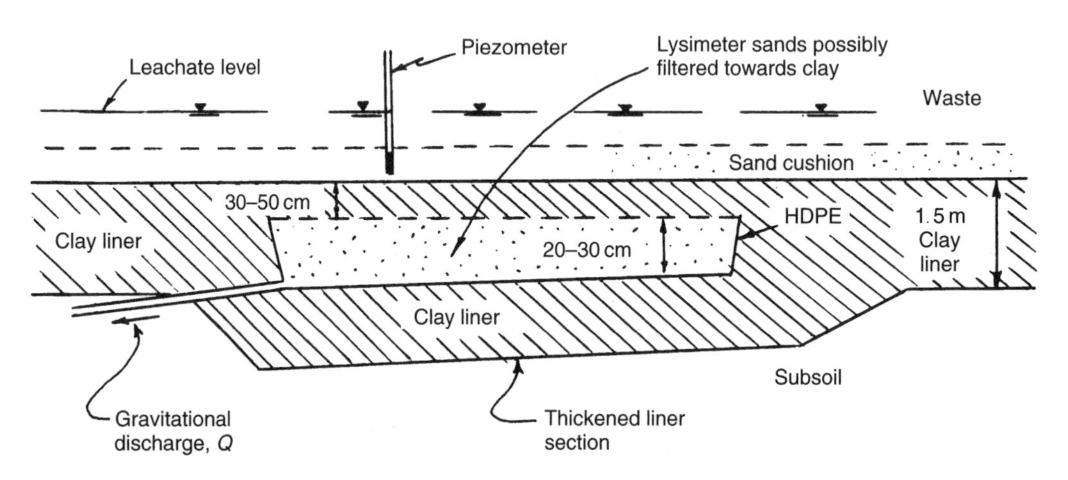

Figure 3.12 Schematic diagram of a lysimeter installation.

2. the true area of the lysimeter and the true thickness of the overlying clay barrier (typically 0.3–0.5 m) are subject to construction errors;
3. the true pressure heads on the liner above a lysimeter require installation of some kind of strategically located piezometer for determination of i;
4. possible suction effects created by negative pore pressures associated with capillary rise within the lysimeter sands to the bottom surface of the clay may influence i;
5. the competing effects of decreasing void ratio and increasing saturation with time as waste is placed in the landfill above the lysimeter make the hydraulic conductivity k time-dependent;
6. the effects of double-layer expansion as Na^+ is adsorbed and the potential effects of mineralogical changes due to potassium or ammonium fixation may influence k with time.

As discussed by King *et al.* (1993), it is probably good policy to collect liner samples from an adjacent dummy lysimeter for laboratory k-testing. The lysimeter k-test results presented by King *et al.* for the Keele Valley clayey silt till liner showed a significant (one order of magnitude) decrease in hydraulic conductivity over 3 years as the lysimeter was loaded with waste. This particular liner was compacted $\sim$2% wet-of-standard Proctor optimum and was amenable to self-healing on load application above $\sim$100 kPa.

Chemical analyses on the lysimeter discharge Q may also be effectively used to monitor the time rate of arrival of effluent chemicals and to check any advection–diffusion modelling done prior to construction.

3.3 Compacted clay liners

3.3.1 Compaction and hydraulic conductivity

The compaction characteristics of soils are well understood (Lambe, 1958, 1960; Mitchell *et al.*, 1965). As shown in Figure 3.13b, the dry unit weight of a soil (γ_d) (and hence the dry density, ρ_d) varies with the molding water content (ω) at which it is compacted. By definition, the maximum dry density is obtained at what is referred to as the optimum water content (ω_{opt}). Both maximum dry density and optimum water content are not unique soil parameters but depend on the energy imparted to the sample during compaction. The maximum dry density increases while the optimum water content decreases with increasing compactive effort as shown in Figure 3.14b. Generally speaking, liner compaction specifications are based on standard Proctor laboratory compaction data (ASTM D698). The modified Proctor laboratory (ASTM D1557) test imparts greater compaction effort to the soil. The line of optimums is used to quantify the influence of compactive effort on a soil and corresponds to a line connecting the maximum dry density and optimum water content for different compactive efforts.

Figure 3.13a shows the corresponding relationship between the hydraulic conductivity, k, and molding water content. The hydraulic conductivity normally decreases from high values when compacted at water contents less than the optimum water content for the soil (this is referred to as being dry-of-optimum) to minimum values at water contents typically 2–4% above the optimum water content (this is referred to as being wet-of-optimum). In Figure 3.13a, the decrease amounts to about two orders of magnitude from 10^{-8} m/s at optimum to 10^{-10} m/s for kneading compaction 4% wet-of-optimum for this particular soil.

These decreases in hydraulic conductivity are normally attributed to the macrostructure of the compacted soil and are related to the presence of clods and interclod macropores in the soil. Clods are macroparticles (i.e., agglomerates of smaller particles) that arise from capillary effects in fine-grained soils. Since the size of a clod is very large relative to the individual soil particles, the presence of clods within a matrix of a fine-grained

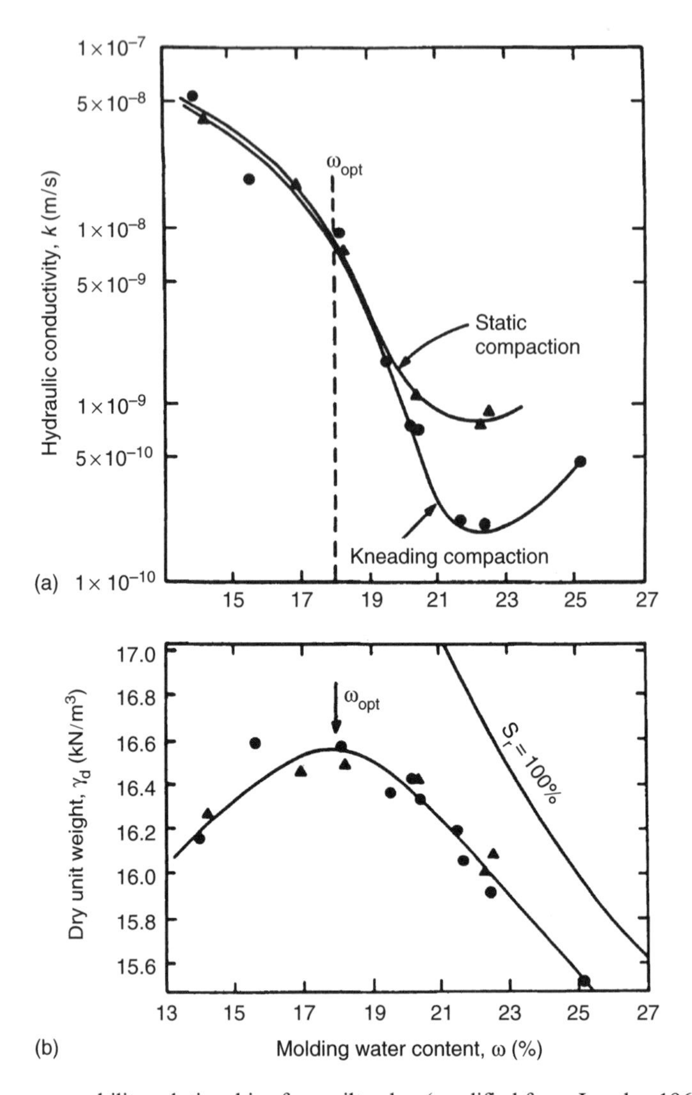

Figure 3.13 Compaction–permeability relationships for a silty clay (modified from Lambe, 1960; Mitchell *et al.*, 1965).

soil will therefore introduce voids in between clods. These are often referred to as interclod macropores. Preferential flow can then occur in the interclod macropores (rather than the low-permeability clods) resulting in an increase in hydraulic conductivity of the compacted soil. In this case, the hydraulic conductivity is then a function of the number and interconnectivity of the macropores. Thus for a given fine-grained soil, the objective of attaining low hydraulic conductivity is attained by the reduction of interclod macropores. This is normally achieved by reducing the size of clods by adding water to reduce the effects of capillary, and/or by imparting greater energy during compaction.

To illustrate the importance of soil macrostructure, consider the dry density–water content relationship plotted in Figure 3.15 obtained from standard Proctor compaction tests on a silty clay (samples obtained from the

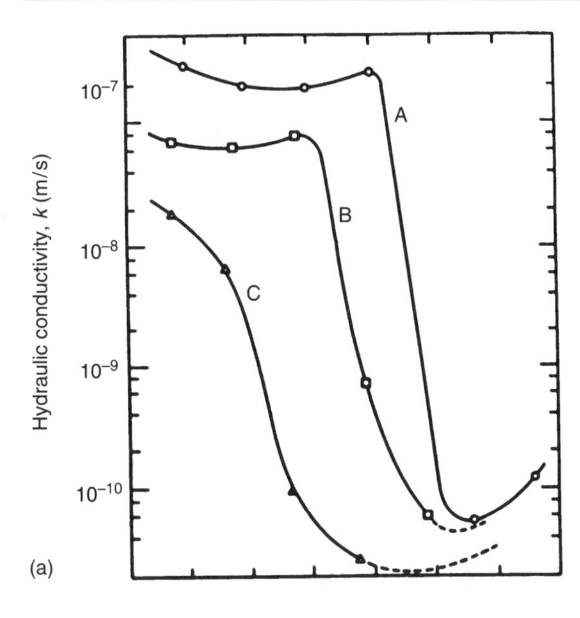

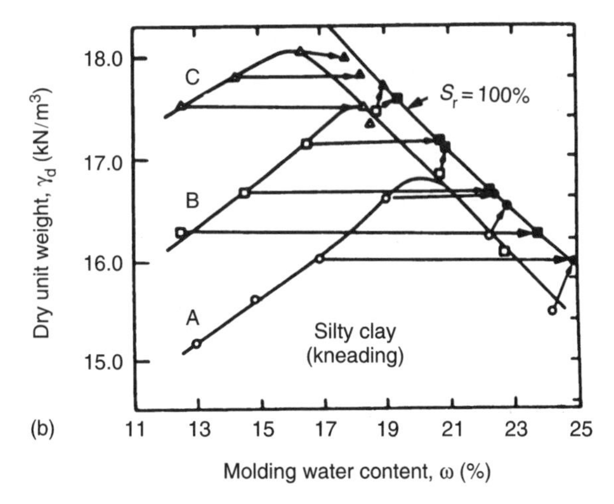

Figure 3.14 Compaction–permeability relationships for low-energy (A) to high-energy (C) compaction (modified from Mitchell *et al.*, 1965).

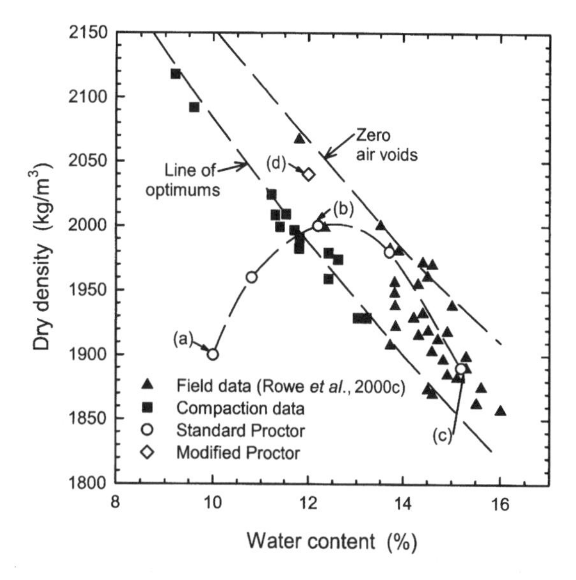

Figure 3.15 Dry density–water content relationship for the Halton landfill liner material and field compaction data.

Halton landfill as described in Section 3.3.4). Photographs of the clay when removed from the compaction mold are shown in Figure 3.16. When compacted dry-of-optimum, standard Proctor compaction is insufficient to break-up the clods. For example, one large clod (approximately 1.5 cm in diameter) and the surrounding macropores are visible in Figure 3.16a (many more clods are visible if the sample is sliced down the middle). With the same compactive effort, addition of water reduces the size of the interclod macropores as shown in Figure 3.16b; however, these pores still appear to be largely interconnected when compacted near standard Proctor optimum. At water contents wet-of-optimum, the clay is easier to remold and thereby break-up interclod pores. Figure 3.16c shows the clay when compacted 4% wet-of-standard Proctor optimum. Although macropores are visible in this photograph, these pores are essentially isolated from one another. Flow is forced to occur through the compacted clay resulting in low hydraulic conductivity. In the absence of interconnected macropores, the hydraulic conductivity of the clay is largely a function of the clay mineralogy, void ratio, degree of saturation and type of permeant.

Since the lowest values of k are obtained usually with kneading compaction (see Figure 3.13a), it is common to make a great effort in the field to repetitively knead the wet-of-optimum soil with many passes by pad-foot or

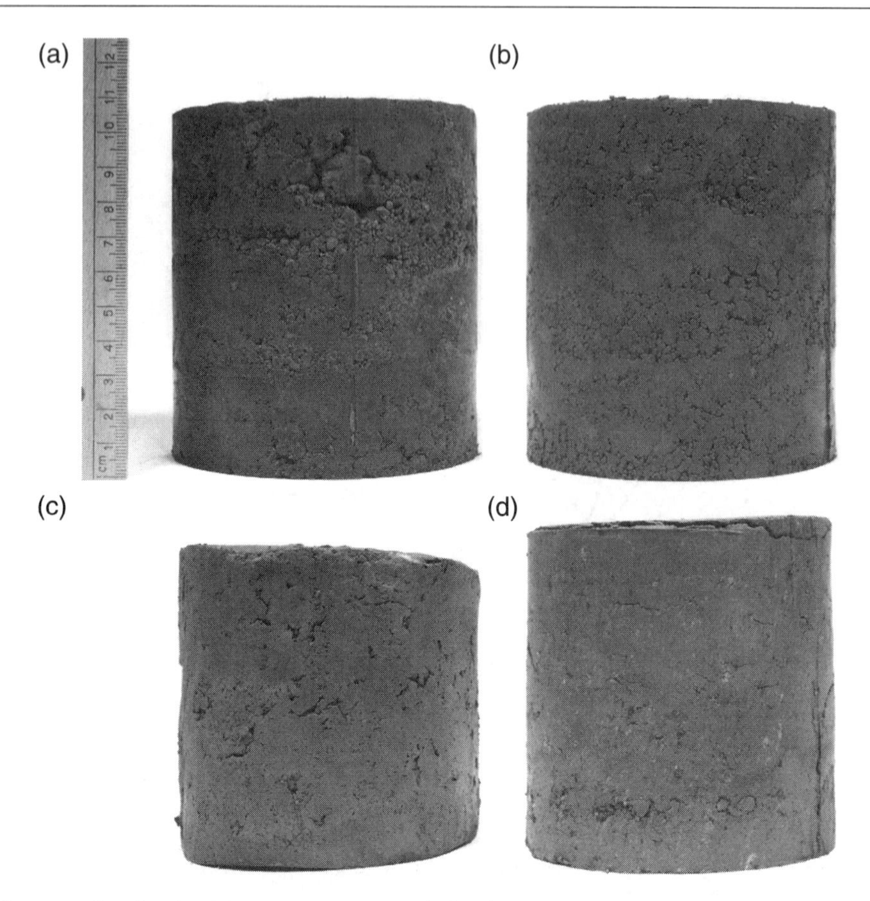

Figure 3.16 Photographs showing clay macrostructure following compaction test at different water contents: (a) standard Proctor, $\omega = 10.4\%$; (b) standard Proctor, $\omega = 12.2\%$; (c) standard Proctor, $\omega = 15.2\%$; and (d) modified Proctor, $\omega = 11.7\%$. (NB Points (a),(b),(c) and (d) shown on Figure 3.15 correspond to samples above).

wedge-foot rollers that remold and break-up the soil clods and interclod macropores.

As shown in Figure 3.14a, somewhat lower hydraulic conductivity can be achieved by laboratory compaction using more energy (i.e., heavier equipment) at lower water contents. This occurs as the increased energy is able to break-up more soil clods. For example, comparison of Figures 3.16b and d shows that the modified Proctor compacted sample (d) appears to have less connectivity between interclod macropores than sample (b) at the same molding water content. However, care is required when specifying field compaction at water contents drier than standard Proctor optimum. If too dry, the soil clods may become hard and difficult to compact, resulting in interclod macropores. Also, the soil dry-of-optimum is brittle, increasing the likelihood of compaction-induced fractures along the sides of the compaction feet. This combination of interclod flow channels and fracture flaws makes representative sampling of drier liners for control *k*-testing more difficult because of the extra scatter in the *k*-test results.

This type of scatter is illustrated in Figure 3.17 which shows that a silty clay soil compacted by kneading to a unit weight of $17\,kN/m^3$ at moisture contents between 19.0 and 20.3% can exhibit laboratory-measured *k*-values varying between 10^{-7} and $10^{-10}\,m/s$. Obviously, a 1,000-fold

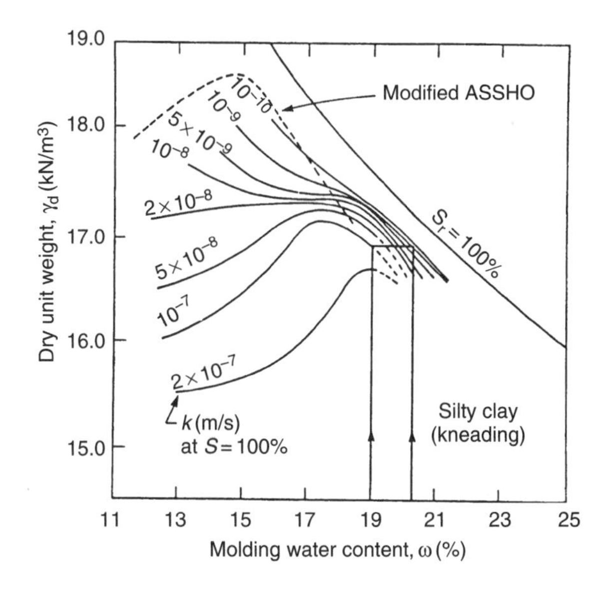

Figure 3.17 Fracture-induced variability in laboratory-measured hydraulic conductivity versus dry unit weight and moulding water content (modified from Mitchell *et al.*, 1965).

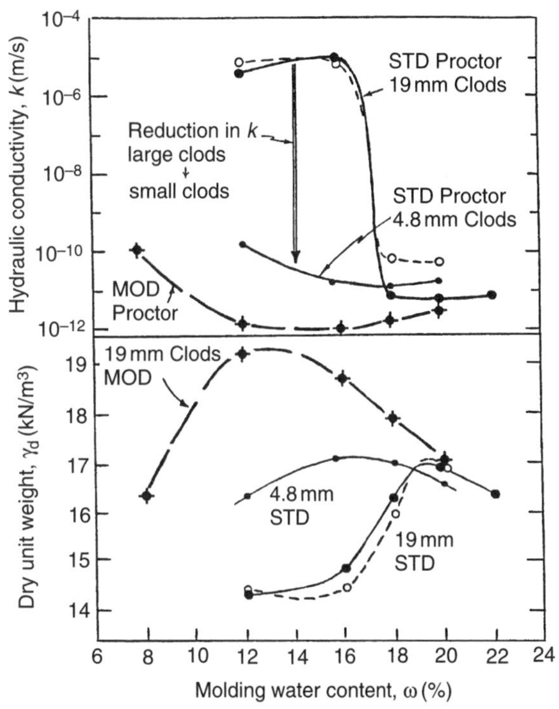

Figure 3.18 Density–permeability relationships for lab compaction of a soil of variable initial clod size (modified from Benson and Daniel, 1990; Elsbury *et al.*, 1990).

range in k caused by lack of a proper laboratory compaction protocol could completely sabotage a clay/leachate compatibility testing program.

Papers emphasizing clod destruction have been published by Benson and Daniel (1990) and Elsbury *et al.* (1990). Figure 3.18, adapted from these authors, illustrates how reduction of clod size from 19 down to 4.8 mm before standard Proctor compaction produces a 4–5 orders of magnitude reduction in laboratory measured k-values. Such extreme variations in k imply open voids between hard clods that must be destroyed by field compaction procedures such as pre-crushing, extended wetting, heavier equipment, extra passes, etc.

To provide sufficient plasticity to break-up clods and develop a suitable clay fabric in the liner, the soil should be compacted at a water content near, or above, the plastic limit (PL). Since the PL is often close to, or slightly above, the standard Proctor optimum water content, ω_{opt} (Leroueil *et al.*, 1992a,b), standard Proctor optimum is generally a lower limit for the water content in order to obtain a good liner. Increasing water content, ω, above standard Proctor optimum, ω_{opt}, substantially increases the probability of achieving a hydraulic conductivity of 1×10^{-9} m/s or lower (Benson *et al.*, 1994). Benson *et al.* (1994) reported that the average compacted water content of eleven compacted clay liners with $1 \times 10^{-10} < k < 1 \times 10^{-11}$ m/s was 2% greater than the PL (i.e., $\omega \geq PL + 2\%$) while the average for thirteen liners with $1 \times 10^{-9} < k < 1 \times 10^{-10}$ m/s was PL + 1%. The average for four liners with $k > 1 \times 10^{-9}$ m/s was PL − 7.5%. The relationship between optimum water content, ω_{opt}, and PL will vary from soil to soil and care is required in the use of any empirical relationship.

Typically the initial degree of saturation, S_r, of well-compacted clay liners will exceed 90% although some low-hydraulic conductivity liners

($k < 1 \times 10^{-9}$ m/s) have had initial degrees of saturation as low as 83% (Benson *et al.*, 1994; Trast and Benson, 1995). Liners with $S_r < 83\%$ usually have $k > 1 \times 10^{-9}$ m/s. Thus there is a strong correlation between achieving $k < 1 \times 10^{-9}$ m/s and a combination of dry density and water content that plot on or above the line of optimums (see Figure 3.19 and Benson and Boutwell, 1992). Daniel (1998) suggested that for an acceptable liner, a minimum of 70–80% of the water content–dry density (ω, ρ_d) points should plot above the line of optimums (Figure 3.19).

While the compacted dry density, ρ_d, is important in terms of where the liner material lies relative to the line of optimums, as discussed above, there is no significant correlation between the field hydraulic conductivity and the per cent compaction. Thus, specifying ρ_d to be 95% of that obtained by standard Proctor compaction (or other arbitrary percentage) does not, in and of itself, provide any assurance of a low hydraulic conductivity liner.

Although the clay should be compacted such that the water content and dry density points (ω, ρ_d) generally plot above the line of optimums, if the water content is too high there may be problems with trafficability and obtaining

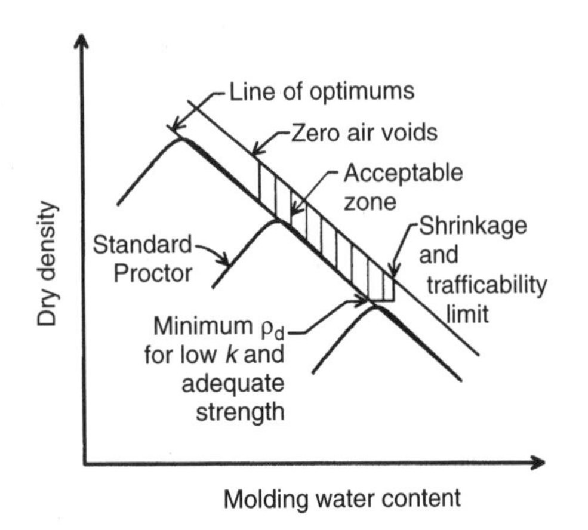

Figure 3.19 Method of defining acceptable zone of water content based on compaction, hydraulic conductivity and shear strength considerations (modified from Benson *et al.*, 1999).

adequate compaction with heavy equipment (Daniel, 1998). There is also a potential problem due to insufficient shear strength to ensure stability. Leroueil *et al.* (1992a) proposed an empirical relationship for estimating the undrained shear strength s_u of clays that plot above the A-line and have a value of $(\omega - \omega_{opt})/I_p > -0.06$, viz.

$$s_u \text{ (kPa)} = 140 \exp\left[\frac{-5.8(\omega - \omega_{opt})}{I_p}\right] \quad (3.17)$$

This equation may be useful for estimating the shear strength to be used in assessing both stability and trafficability; however, like all empirical equations, it must be used with caution and appropriate laboratory tests should be performed to obtain the actual shear strength for important projects or where there are significant cost or safety implications if the shear strength is not correctly evaluated. This may be especially important for ensuring the stability of composite liners.

While it is generally desirable to compact wet-of-standard Proctor using extra passes to achieve extra kneading and destruction of soil clods and the contacts between them, a few risks may still remain. As shown in Figure 3.20, all such soils shrink on drying. The kneaded soil demonstrates the greatest axial (vertical) shrinkage, probably because of the horizontal preferred orientation of the clay platelets. Since horizontal shrinkage would cause immediate cracking rather than shrinkage, it is important that the clay liner not be allowed to dry out since this can rapidly induce vertical desiccation cracking.

Many clay barriers are covered with a drainage layer of sand and/or a geosynthetic and kept moist by watering as needed. This permits time for quality control testing for the hydraulic conductivity of the clay liner which may require a month or so before waste is applied. With good field control, the liner should stay damp and the risk of shrinkage can be minimized. For example,

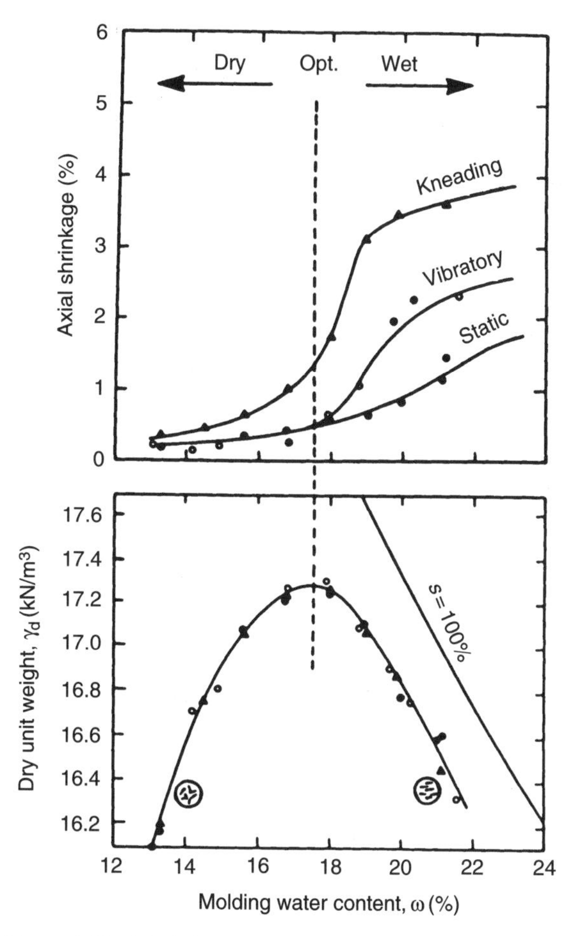

Figure 3.20 Axial shrinkage as a function of compaction method and molding water content (modified from Mitchell *et al.*, 1965; see also Lambe, 1960).

three exhumations of liners in Ontario showed no evidence of vertical desiccation cracking. This is in part attributable to wet-of-optimum compaction which produced horizontal clay platelet parallelism and one-directional shrinkage (vertical) rather than horizontal shrinkage and cracking. The liner also needs to be protected from frost damage to avoid a potential increase in hydraulic conductivity with time of several orders of magnitude. The most effective means of protecting landfill liners against frost is to schedule liner construction such that an adequate thickness of waste or, when that is not possible, other protective material is placed over the liner before the onset of freezing conditions.

Another factor to be considered is thixotropy. Figure 3.21, also adapted from Mitchell *et al.* (1965), shows increases in hydraulic conductivity k for certain soils if allowed to "age" after compaction. This figure is not a simple one since all soils are noted to be at a $\gamma_d = 17\,kN/m^3$. For the range of water contents and saturations shown, a wide variety of compaction energies must have been employed. Thixotropy as an influence on k was also mentioned specifically by Dunn and Mitchell (1984).

Two of the test soils show increases in k by factors of 2–3 as a result of ageing at constant volume. This "thixotropic" increase in k probably results from re-orientation of the clay platelets into a more flocculated or aggregated state with larger interped pores and less tortuosity. Such reorientation would certainly be more likely to occur wet-of-standard Proctor optimum and under conditions of low confining stress as in a new liner awaiting waste after protocol testing. Normally active soils (smectites) are considered more thixotropic than inactive

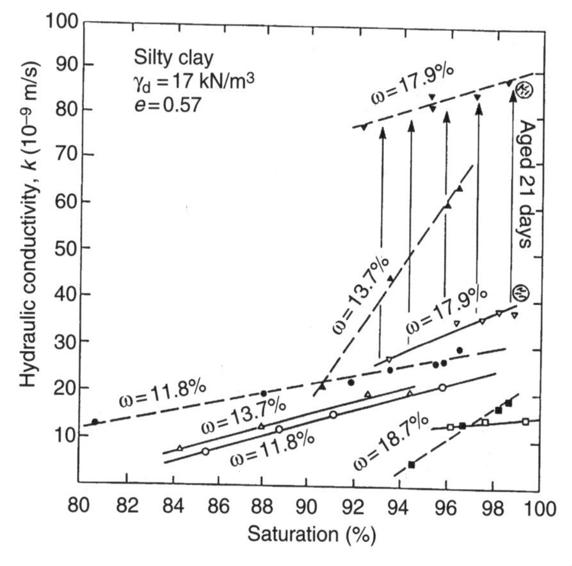

Figure 3.21 Examples of thixotropically induced increases in k caused by ageing (modified from Mitchell *et al.*, 1965).

soils. While the increases are not large, they are certainly significant enough to greatly complicate interpretation of long-term clay/leachate compatibility test results obtained on soils compacted wet-of-optimum.

Thixotropic strength gain is very common in geotechnical practice wherever very soft or active clays are involved. Examples are unwanted strength gain in liquid limit tests and increased remolded strength measurement for calculation of sensitivity. It is not commonly considered in the context of *k*-testing. The authors have noted that freshly compacted, inactive clayey soils from southern Ontario, Canada, exhibit an apparent preconsolidation pressure of ~100 kPa, that may be somewhat "age-dependent" or thixotropically controlled. This set-up in soil structure tends to resist compression early in the loading history of a liner but will be eliminated once field stresses exceed ~100 kPa.

3.3.2 Construction of compacted clay liners

Compacted clay liners constructed using soil that meets the requirements outlined in Section 3.1.3 can be expected to have low hydraulic conductivity provided that there is appropriate (a) water content control, (b) break-up of clods of soil and homogenization of non-uniform soils, (c) lift thickness, (d) method of compaction and equipment used and (e) protection of the liner from desiccation or frost action after construction (see Section 3.3.1).

The liner material must be broken up to minimize clod size and for dry, hard, clay-rich soil it is essential that these clods be wetted to the point that they can be readily remolded by the compaction equipment as discussed in Section 3.3.1. Daniel and Koerner (1995) recommend a maximum particle size of between 25 and 50 mm; however, success has been achieved in constructing low hydraulic condition liners with stones up to 100 mm in size provided that the liner is of a thickness exceeding 0.9 m (Rowe *et al.*, 2000c). Large stones may be a problem with thin liners (< 0.6 m) due to the presence of voids adjacent to these stones that develop during compaction (see Rowe *et al.*, 1993). Oversize particles need to be removed using a mechanized stone picker or by hand.

Each new lift should be placed on a rough surface and it may be necessary to scarify the surface using an industrial disk before placing the soil for the new lift. The loose thickness of lifts used to construct compacted clay liners generally should not exceed 230 mm so that the compacted thickness does not exceed about 150 mm. The sheeps-foot/pad-foot compactor should have feet that are long enough (i.e., >150 mm) to fully penetrate the lift being compacted and extend into the underlying lift. This serves to break-up interfaces between the lifts (see Rowe *et al.*, 1993). The weight of the compactor should be greater than or equal to 180 kN. Typically at least five passes are required to achieve the desired compaction.

When constructing the liner over an existing engineered component of the system (e.g., when constructing the primary liner in a double-liner system), it may be necessary to compact a sacrificial layer using a lightweight smooth drum roller to avoid damaging the underlying system (e.g., the secondary leachate collection/leak detection layer). The light smooth drum roller will generally not be sufficient to achieve the hydraulic conductivity requirement and hence cannot be considered as contributing to the effective thickness of the clay liner (see Rowe *et al.*, 1993). A smooth drum roller is also needed to obtain a smooth surface on the final lift after it has been compacted with the sheeps-foot/pad-foot compactor.

3.3.3 Construction quality control and assurance

While a liner will be generally required to meet at specified hydraulic conductivity (typically $<1 \times 10^{-9}$ m/s), in order to achieve this performance criterial the liner specifications need to define the soil characteristics, and the compacted water content and density that must be achieved (see Sections 3.1.3 and 3.3.1). An experienced engineer/technician can usually assess whether the water content is in the correct range by eye and feel (as noted in Section 3.3.1, the desired water content is often close to the PL). A soil that looks and feels dry will usually give poor performance if compacted into the liner. Although testing is essential to confirm that the specified water content and density requirements are met, there is no substitute for having an experienced engineer/technician overseeing the liner construction on a full-time basis.

Quality control procedures typically involve checking the soil meets the specifications in terms of Atterberg limits, percentage of clay and gravel, and that the compacted clay liner meets the specifications in terms of water content, dry density and hydraulic conductivity. Recommended testing methods and minimum testing frequency are given in Table 3.2.

As discussed in Section 3.2.4, sealed double-ring infiltrometers (SDRIs) may be used to establish the field hydraulic conductivity. However, flexible wall hydraulic conductivity tests (see Section 3.2.1) performed on Shelby tube samples taken from the liner can provide a very good indication of the hydraulic conductivity of the liner provided that (a) there is no visible evidence of macrostructure (e.g., desiccation cracks), (b) adequate water content control has been achieved (see Sections 3.3.1 and 3.3.2) and (c) the liner has been compacted to a degree of saturation of 87% or higher.

Where Shelby tube samples have been taken, the liner in this area must be stripped to below the depth of the Shelby tube penetration over a width greater than the width of the compactor and the liner must be reconstructed according to the original procedure.

To confirm the suitability of the proposed liner specification, acquaint the contractor with potential problems, and calibrate the quality control and assurance procedures, a test pad should be constructed prior to construction of the actual liner. Test pads will typically be about 10–15 m in width and 15–30 m in length.

The inherent variability of soil and the challenges of large-scale construction are such that there will be times when the parameters

Table 3.2 Recommended minimum testing frequencies (adapted from Daniel and Koerner, 1995)

		Frequency (one test per)		
Parameter	*Method (ASTM)*	*Borrow pit*	*In loose lift*	*After compaction*
Water content (field)	D-3017	–	–	750 m^2/lift
Water content	D-2216	2,000 m^3*	800 m^3	750 m^2/lift
Density (field)	D-2922	–	–	750 m^2/lift
Density	D-1556/2167	–	–	15,000 m^2/lift
Atterberg limits	D-4318	5,000 m^3*	800 m^3	750 m^2/lift
Per cent clay	D-422	5,000 m^3*	800 m^3	750 m^2/lift
Per cent gravel	D-422	5,000 m^3 *	800 m^3	750 m^2/lift
Compaction curve	D-698	5,000 m^3*	4,000 m^3	–
Hydraulic conductivity	D-5084	10,000 m^3*	–	2,000 m^2 of finished liner

* Or each change in material type.

evaluated in accordance with Table 3.2 do not meet the specifications. Daniel and Koerner (1995) recommend that up to 3% of water content and dry density results be allowed to fall outside the specified range in any one lift or area provided that no water content is less than 2% below or more than 3% above the specified values and no dry density is less than 80 kg/m^3 below the required value. Some specifications also require that no water content be below standard Proctor optimum. If the liner is found to be out of specifications, additional tests will likely be required to establish the extent of the problem area requiring repair. In situations where the water content is within range but the dry density is too low, the problem can usually be rectified by more passes of the compactor.

3.3.4 Example case – the Halton landfill liner

The design and construction of the Halton landfill as a hydraulic containment landfill (see Sections 1.2.1 and 16.3.4) has been described in detail by Rowe *et al.* (2000c). This design required the construction of a primary compacted clay liner over a hydraulic control layer. The specifications for the liner, and the geotextile separator layers above and below the liner, were developed based on a detailed trial liner investigation (see Rowe *et al.*, 1993).

The soil used as liner material had a clay content typically between 21 and 38% and plasticity index between 9 and 15%. The clay minerals consisted predominately of illite (35–37%) and the CEC ranged between 14 and 18 meq/100 g. All cobbles, stones and clods greater than 100 mm in size were to be removed. There was some variability in the borrow material and the standard Proctor optimum water content ranged between 10.5 and 12.4% while the maximum dry density ranged between 1,930 and 2,025 kg/m^3. The compaction water content was specified to be between 2 and 4% above standard Proctor optimum, subject to the additional restriction that it should not be less than 12% or more than 16%. The compacted dry density was specified to exceed 95% of standard Proctor optimum. Compaction was specified using a tamping foot compactor with a minimum mass of 15,000 kg and feet between 150 and 230 mm in length except for the first lift (i.e., the lift directly over the hydraulic control layer) which was a sacrificial lift compacted with a smooth drum finishing roller. A lift was considered acceptable if all the following requirements were met: (a) the average water content was between 1 and 4% wet-of-standard Proctor optimum water content, ω_{opt}; (b) no water content was less than ω_{opt}; (c) 90% of all water contents, ω, were within the range $12\% \leq w \leq 16\%$; (d) the average dry density exceeded 95% of standard Proctor maximum dry density, ρ_d; and (e) the minimum dry density exceeding 90% of ρ_d.

The line of optimums (Figure 3.15), obtained from tests performed at several levels of compaction energy, corresponded to a degree of saturation of 86%. Cell 1 was constructed in three stages between 1992 and 1995. The mean and standard deviation of the water content over this period were 14.5 and 1.1%, respectively. For Cell 2 (west), constructed in 1998, the mean and standard deviation of the water content were 14.1 and 0.8%, respectively.

Typical liner data from the 1992, 1995 and 1998 liner constructions are plotted in Figure 3.15 (due to the large amount of available data, 704 data points, only 12 randomly selected points are plotted for each year for clarity of presentation). The data typically fell on or above the line of optimums with the degree of saturation ranging between 86 and 100%.

Since the site was visually inspected for macrostructures and the field water content and density data corresponded to a degree of saturation of 86% or greater and generally fell on or above the line of optimums, the hydraulic conductivity was evaluated based on flexible wall permeameter tests on Shelby tube samples. The data from the 66 tests conducted for Cell 1 (6.5 tests/ha) were approximately log-normally distributed with a geometric mean of 1×10^{-10} m/s and a standard deviation ($\log_{10}$ unit) of 0.1. The maximum measured hydraulic conductivity was 1.6×10^{-10} m/s. The liner 26 hydraulic conductivity tests (4.7 tests/ha) conducted for Cell 2 (west) gave similar results to those for Cell 1, with a maximum hydraulic conductivity of 1.4×10^{-10} m/s, a geometric mean of 0.98×10^{-10} m/s and standard deviation ($\log_{10} k$) of 0.11. Thus the liner constructed at the Halton landfill between 1992 and 1998 appears to be similar to that obtained at the KVL (King *et al.*, 1993).

A key aspect of the liner construction protocol was the full-time presence of a soil technician, under the direction of a qualified geotechnical engineer, who was responsible for (a) evaluating soil suitability and water content, and (b) ensuring the liner was kept moist and inspecting for any desiccation cracking.

3.3.5 Field behaviour of compacted liners

The reliability of clayey liners has been a contentious issue. A number of failures related to very poor control have led to suspicion and prescriptive regulations involving the use of geomembrane liners along with clay. Thus it is appropriate to say a few words here about field performance, although this is discussed in more detail in Chapter 9.

Daniel (1984) published a comparative study of four liners in the southern United States and demonstrated the presence of field k-values 5–100,000 times greater than laboratory values (Table 3.1). This led to the recommended use of large-diameter, double-ring infiltrometers to measure field k before placement of waste. Daniel noted that post-construction desiccation cracking may have played a role in the failure of the four liners. A similar comparison implying ratio of field to laboratory k-values of between 2 and 1,000 times was published by Bracci *et al.* (1991).

A series of problems related to under-compaction of two prototype clay liners (Day and Daniel, 1985) and field k-testing at low effective stresses led to a long sequence of discussions in ASCE regarding the question of proper control and protocol.

Articles by Gordon *et al.* (1989) and King *et al.* (1993) have demonstrated marked reductions in field permeability a few years after construction as measured by lysimeter installations. Their k-test results are also summarized in Table 3.1 and show field k-values generally varying from 5×10^{-10} to 5×10^{-11} m/s. These data are discussed again in Chapter 9.

It would appear that the inconsistency between the k-test results for various barriers may be related to as many as four factors. First, it would appear that the high field values obtained in some cases may have been related to under compaction by equipment that was too light. Second, cracking by desiccation on dry hot days and fissures from ice lenses in winter appear to be major problems. When cracking occurs it will obviously result in a liner having much higher hydraulic conductivity than is desired and the cracked material should be excavated and recompacted or replaced. Third, undamaged liners may become damaged during installation of liner k-test systems such as ring infiltrometers or shallow wells and in these cases the results are not representative. Fourth, field tests are often conducted at stress levels which are not representative of later loaded field conditions or the corresponding laboratory test conditions. For example, the data presented by King *et al.* (1993), as summarized in Table 3.1, show a significant decrease in hydraulic conductivity with time which appears to be related to consolidation under the applied stress caused by the overlying waste and cover materials.

In summary, it would appear that for well-designed and constructed clay liners, there is a good correlation between laboratory and field values under true stress conditions when field lysimeter data are compared with laboratory data obtained with significant applied stresses. However, laboratory tests may not detect macrostructures such as large uncompacted clods or desiccation cracking. This must be controlled by good field supervision. Field tests such as the ring infiltrometer may detect faulty construction but typically only in the upper portion of the liner. The results obtained from this test represent low-stress k-values not characteristic of higher stress field conditions.

3.3.6 Service life of compacted clay liners

A properly designed and constructed compacted clayey liner is likely to maintain a hydraulic conductivity less than or equal to the design value for thousands of years provided that (a) appropriate consideration has been given to any potential for increase in hydraulic conductivity when exposed to chemical permeants (as discussed in Chapter 4) and (b) the clay liner is not permitted to desiccate following placement. Desiccation refers to shrinkage cracks that arise related to a decrease in water content of the clay, which would increase the bulk hydraulic conductivity of the liner.

Desiccation can occur within hours after liner placement if the clay is not kept moist. The extent of desiccation cracking will depend on the mineralogy of the clay, the compaction water content, and the degree and length of exposure to drying conditions. Rowe *et al.* (1994) reported on an experiment at the KVL where a compacted clay liner (20% clay size of predominately illite and chlorite) had cracks form to a depth of 120 mm over a 5-week period (mostly in June). The potential for such desiccation can be minimized by adopting appropriate construction procedures (e.g., the timely placement of cover material). For composite liners consisting of a geomembrane overlying a compacted clay liner, visual inspection of the clay for cracking prior to placement of the geomembrane is important. Desiccated portions of the clay liner must be removed and replaced. Simply rewetting the desiccated liner will not close up cracks and, furthermore, the additional construction moisture may reduce the interface shear strength between the clay liner and overlying geomembrane and may lead to slope failure (Bonaparte *et al.*, 2002).

Special consideration is required to prevent dessication of a compacted clay liner beneath a geomembrane prior to placement of cover materials. The temperature of a black geomembrane exposed to the sun may be as large as 80 °C (Felon *et al.*, 1992) and temperatures of 60–70 °C are not unusual. Cycles of daytime heating followed by night-time cooling of an exposed geomembrane may cause moisture to evaporate from the underlying clay liner into any airspace between the geomembrane and clay liner (e.g., see Figure 13.12 for potential sources of gaps between a geomembrane and clay liner). This may be of particular concern on side slopes where there can be condensation and movement of water downslope during the cooler portions of the thermal cycle. For example, Basnett and Bruner (1993) reported on such a case where a compacted clay liner beneath a geomembrane on a side slope was observed to desiccate through its entire 0.3-m depth following 3 years of exposure. Rowe *et al.* (1998, 2003a) also reported severe desiccation cracking of a compacted clay liner below a geomembrane in a leachate lagoon exposed to temperature cycles for 14 years. In this case, cracking was restricted to the side slopes above the leachate level. Provided the geomembrane is covered relatively quickly, the available evidence (Corser *et al.*, 1992; Bowders *et al.*, 1997) tends to suggest that the depth of desiccation cracking should be limited to the upper portion of the clay liner. However, the presence of cracks even in the upper portion of the liner could still be significant (especially on the base of the landfill) if these cracks were beneath a hole in the geomembrane since the cracked clay liner would more readily allow movement of leachate through the hole and along the crack, and thus should be considered in assessment of leakage through composite liners (Section 13.4). Clearly there is a need to cover the geomembrane with the protection layer and the leachate collection layer as soon as possible after placement of the geomembrane to minimize the potential for desiccation cracking.

There is still much unknown about the potential for desiccation of a clay liner due a sustained thermal gradients induced by the landfill. It may be postulated that moisture could move from regions of higher temperature in the landfill (Section 2.6) to regions of lower temperature towards the groundwater. In this case, the risk of desiccation of a clay liner likely depends on the properties of the clay liner and the subgrade, overburden pressures, temperature gradient across the liner and the depth to groundwater (Holzlöhner, 1989, 1995). This highlights the need to control leachate mounding to keep the temperatures as low as practicable at the level of the liner.

3.3.7 Soil–bentonite liners

When suitable natural soils are not available for use in a compacted clay liner, bentonite may be added to a non-cohesive soil (e.g., silty sand) to achieve a liner with the required hydraulic conductivity, although GCLs (see Chapter 12) often represent a more convenient and cost-effective alternative to soil bentonite liners. The key considerations in the design of these mixed-soil liners are (a) the grain size distribution of the base soil, (b) the amount of bentonite and (c) the mineralogy of the bentonite. Typically the bentonite used is a sodium bentonite, calcium bentonite or sodium-activated calcium bentonite although a wide range of modified bentonites (e.g., organobentonites) are also available. Goldman *et al.* (1990) recommend that only powdered bentonite be used since it mixes more uniformly with the base soil.

The use of between 4 and 10% bentonite in the soil mixture may give hydraulic conductivities of 10^{-9} to 10^{-11} m/s. However, it is essential to establish the optimum proportion of bentonite and water content on a site-specific basis by performing laboratory hydraulic conductivity tests on compacted soil mixes with different amounts of bentonite at different water contents. Chapuis (1990) has described a simple technique for estimating the hydraulic conductivity and has provided suggestions regarding the associated laboratory study. As part of the liner evaluation, it is important to assess the potential for internal erosion of the bentonite (i.e., transport of the bentonite out of the liner by water flow) by examining the hydraulic conductivities under the maximum gradient that could occur (e.g., with the development of a leachate mound).

Good performance of soil–bentonite liners is highly dependent on: (a) maintaining a homogeneous mixture of the base soil and bentonite, and avoiding segregation prior to and during placement; (b) control of compaction density and water content; and (c) lift thickness. If the soil–bentonite liner is to be used as a landfill liner, it is recommended that, like other compacted clay liners, the soil–bentonite liner be compacted in a number of layers and that the construction specifications be verified by the construction and evaluation of a test pad prior to liner construction.

A critical consideration in the selection and design of soil–bentonite liners is the potential for clay-leachate interaction, particularly when Na or Na-modified bentonite is used. Despite the finding that, in the laboratory, sand–bentonite specimens appear to be more resistant to free-thaw cycles than conventional compacted clay (Wong and Haug, 1991), it is still prudent to protect the sand–bentonite liners against desiccation and frost. Testing for construction quality control should not be less than that given in Table 3.2.

3.3.8 Summary comments concerning compaction and hydraulic conductivity

1. Field compaction should normally be wet-of-standard Proctor optimum and probably close to the Plastic limit (PL). This should produce a clayey barrier soft enough to self-heal on stress application.
2. A minimum of 70–80% of the water content–dry density (ω, ρ_d) points should plot above the line of optimums (Figure 3.19).
3. If the water content is too high there may be problems with trafficability, obtaining adequate compaction with heavy equipment, or shrinkage and cracking on drying.
4. Compaction should normally be performed using a sheeps-foot, pad-foot or club-foot roller so that the interclod voids are destroyed. The compactor should have feet that are long enough to fully penetrate the lift being compacted and extend into the underlying lift. The weight of the compactor should be greater than or equal to 180 kN. Typically at least five passes are required to achieve the desired compaction.
5. Most borrow pits contain stratified materials, so careful control may be required to obtain a uniform field distribution of minimum hydraulic conductivity (k) values.
6. Even homogeneous clays are often variably weathered so that successive lifts have different ω_{opt}, $\gamma_d(\max)$ and k-values.
7. Quality control procedures (Table 3.2) should involve checking that the borrow material meets the specifications in terms of Atterberg limits, and percentage of clay and gravel, and that the compacted liner meets the specifications in terms of water content, dry density and hydraulic conductivity.
8. Liners should be immediately covered and kept moist enough to prevent shrinkage until covered by waste. Similarly, smectitic liner components must not be allowed to free swell.
9. Desiccated liner sections should be removed, pulverized, rewetted and recompacted.
10. Liner sections must be protected from frost action and ice lensing, and if damaged should be replaced.
11. Wet-of-optimum compaction that produces a minimum k-value produces a clay wet enough to experience chemical shrinkage on exposure to certain leachates. Clay/leachate compatibility testing is therefore necessary for soils for which no data exist or whose clay mineralogy indicates reactive mineral constituents.
12. A major benefit of wet-of-optimum compaction is that the preconsolidation pressure of inactive clayey soils seems to be 100–150 kPa. Thus, if the amount of effective stress applied by the mounded waste exceeds about 200 kPa, the wet clay liner will consolidate. The resultant decrease in void ratio should enable the wet clay to self-heal even in the presence of deleterious chemicals as indicated by the lysimeter k-values of King *et al.* in Table 3.1 which is discussed further in Chapter 9.
13. High swell Na^+ bentonites used in sand–bentonite mixes require special care. Like other compacted clay liners, they should be compacted in a number of layers and the construction specifications should be verified by the construction and evaluation of a test pad prior to liner construction. The potential for clay and leachate interaction, and internal erosion require very careful examination.
14. GCLs are discussed in Chapter 12.

3.4 Clay mineralogy

3.4.1 Introduction

Only a brief overview of clay mineralogy and colloid chemistry are provided in this text, with emphasis being placed on those factors which are responsible for critical interactions with leachate. Good texts on clay mineralogy include: Grim (1953). Van Olphen (1977) and Mineralogical Society (1980). Good applied texts with useful abbreviated discussions include: Grim (1962), Yong and Warkentin (1975), Lambe and Whitman (1979) and Mitchell (1993).

3.4.2 Abbreviated classification of the clay minerals

Most clay minerals are sheet silicates (phyllosilicates) and a useful abbreviated classification adapted from Bailey (1980) is presented in Table 3.3. The single most important characteristic of the clay minerals is that they carry a net negative charge which is expressed as the number of deficient electrons/unit cell of 10 oxygens with respect to charge neutrality as emphasized in the table. If there are no fixed cations between the sheets, the negative charge deficiency correlates directly with the CEC which is the measure of the number of positive cations in milliequivalents to neutralize 100 g of clay. Since the layer combinations and both the size and the location of the charge deficiency influence all aspects of engineering behaviour including the CEC, it is important to understand these phenomena, at least conceptually as presented in the following sections.

3.4.3 Structural components of clay minerals

The two most important elementary structural elements of the clay minerals are the silicon tetrahedron and the aluminium or magnesium octahedron. The charges present on a given clay mineral are located in either or both of these units due to isomorphous substitution of cations for one another.

(a) The silicon (and aluminium) tetrahedron

The silicon tetrahedron, illustrated in Figure 3.22, is comprised of four oxygens (O^{2-}) surrounding

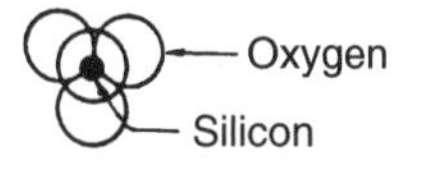

Space available in center ≈ 0.50 Å
Si^{4+} ionic radius = 0.42 Å (usually present)
Al^{3+} ionic radius = 0.51 Å (often present)

Figure 3.22 Illustration of a tetrahedron (note 10 Å = 1.0 nm).

Table 3.3 Abbreviated classification of the clay minerals[a]

Layer type	*Negative[b] charge per unit cell of 10 oxygens*	*Mineral group*	*Selected mineral species*
Amorphous	N/A	Allophane[c]	
1:1	0–0.015	Kaolin[c] Serpentine	Kaolinite, dickite, halloysite, amesite
2:1	0	Talc/pyrophyllite	Talc, pyrophyllite
	0.2–0.6	Smectite[c]	Montmorillonite, beidellite, saponite, hectorite, sauconite, etc. (bentonite is a genetic name)
	0.6–0.9	Vermiculite[c]	Vermiculites, often weathering products
	0.6–1.0	Mica/illite[c]	Muscovite, biotite, phlogopite, illite, etc.
	0.6–1.0	Chlorite[c]	Chamosite, etc.

[a] Roughly adapted from Bailey (1980).
[b] Number of deficient electrons/unit cell of 10 oxygens.
[c] The important soil minerals and the name normally used in geotechnical practice.

a central Si^{4+} which fits snugly into a central space ~0.50 Å (0.050 nm) in diameter. The Si—O bond is the strongest in the silicate mineral system and is the basis of silicone technology. The net charge on an individual tetrahedron is −4 electrons (−4*e*).

During formation of tetrahedra, probably in the high-temperature/pressure atmosphere of a molten rock magma, aluminium (Al^{3+}) may substitute for silicon, Si^{4+}. The net charge on an aluminium tetrahedron is −5*e* compared to −4*e* for silicon.

(b) The aluminium and magnesium octahedron

The individual octahedron illustrated in Figure 3.23 consists of six hydroxyls (OH^-) surrounding one Al^{3+} or Mg^{2+}, both of which are frequently present in clay minerals.

The charge on an individual Al octahedron is −3*e* compared to −4*e* for a Mg octahedron. Many other elements fit conveniently into the 0.65 Å space between the hydroxyls, and the basis of detailed clay mineral classification is often the dominant cationic species in the octahedral positions.

(c) Tetrahedral sheets

The silica tetrahedra are linked together by sharing oxygens in a wide variety of nets, chains and sheets to form many of the minerals in the silicate system. Except for a few species, the majority of the clay minerals are sheet silicates or phyllosilicates.

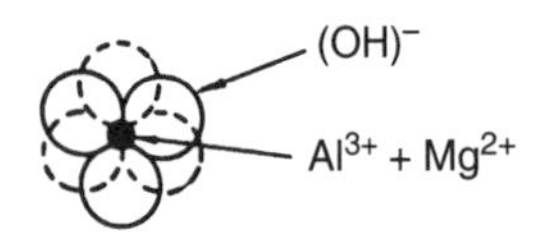

Figure 3.23 Illustration of an octahedron (note 10 Å = 1.0 nm).

The structure of the silica sheet is illustrated in both cross-section and plan in Figure 3.24, complete with a repeating unit cell of dimensions a_0 and b_0. The individual tetrahedra are tied together by sharing oxygens in such a way that the top of the sheet consists of a network of hexagonal holes surrounded by oxygens. These holes are about the same size as the potassium cation, K^+, which occupies them in the structure of mica and illite.

If all the tetrahedral positions are filled with Si^{4+}, the tetrahedral sheet is neutral. If Al^{3+} is present, the sheet becomes negative.

Both potassium, K^+, and ammonium, NH_4^+, are abundant in domestic waste leachate. Both tend to adsorb onto certain high-charge clay minerals, fixing themselves into the holes in the surface of the tetrahedral sheets. If this causes the *c*-axis of the mineral to contract and/or

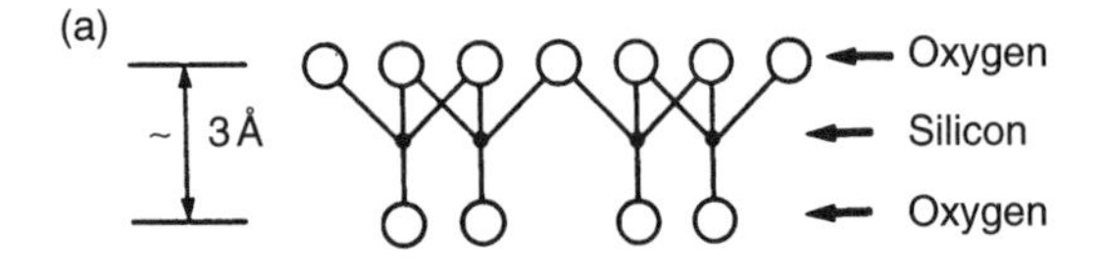

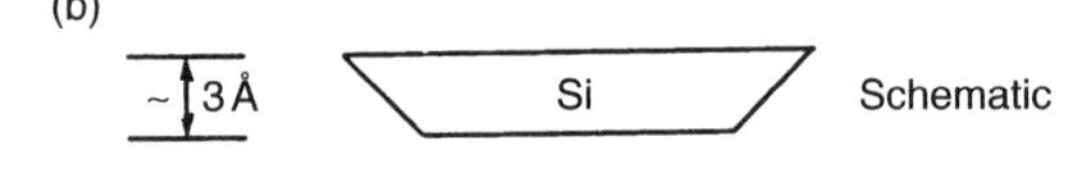

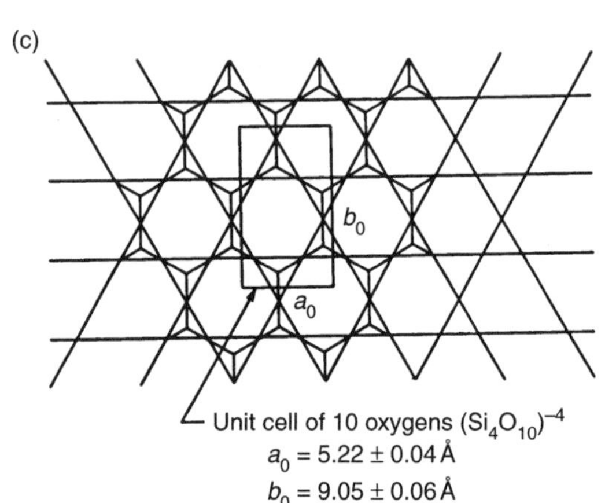

Figure 3.24 Structure of the tetrahedral sheet: (a and b) cross-sections; (c) top view showing hexagonal holes (note 10 Å = 1.0 nm).

causes the CEC to decrease, a potential incompatibility problem may exist.

(d) Octahedral sheets

The individual octahedra are linked together by sharing hydroxyls to form sheets as shown in Figure 3.25. If only Al^{3+} is present, two-thirds of the positions are filled to form a neutral, dioctahedral sheet known as gibbsite. Gibbsite, a mineral in its own right, is a common product of tropical weathering and is found in bauxite deposits.

If only magnesium, Mg^{2+}, is present, all the octahedral positions are filled and a trioctahedral brucite sheet is formed. Brucite is also a mineral in its own right, which is mined for magnesium.

The sheets are normally represented by rectangles labelled G (for gibbsite) and B (for brucite), as shown in Figure 3.25c.

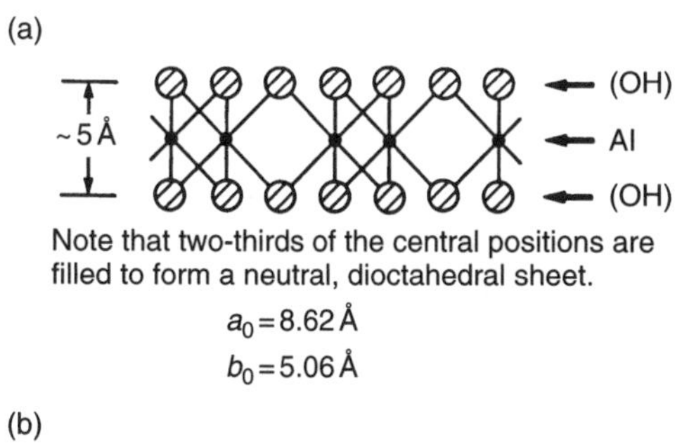

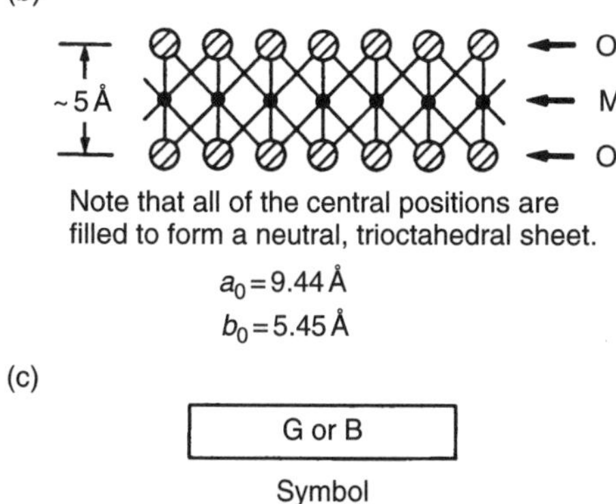

Figure 3.25 Structure of the octahedral sheet: (a) gibbsite sheet; (b) brucite sheet; (c) schematic view (note 10 Å = 1.0 nm).

Isomorphous substitution of divalent cations such as Fe^{2+} or Mg^{2+} for Al^{3+} in the octahedral gibbsite sheet will render it negative. Alternatively, if Mg^{2+} or Fe^{2+} fill some of the empty holes in a gibbsite sheet it could become positive.

(e) Two-layer and three-layer units

When clay minerals form during weathering, the silica sheets grow simultaneously with the octahedral sheets to form two-layer and three-layer units illustrated in Figure 3.26.

Under slightly acidic tropical conditions (pH 5–6), aluminium is retained and silica is gradually depleted from the soil so that two-layer units or 1:1 clays tend to form. As shown in Figure 3.26a, these clays consist of one silica

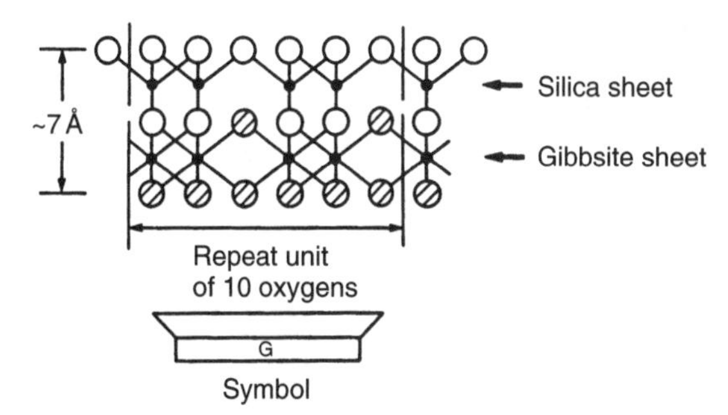

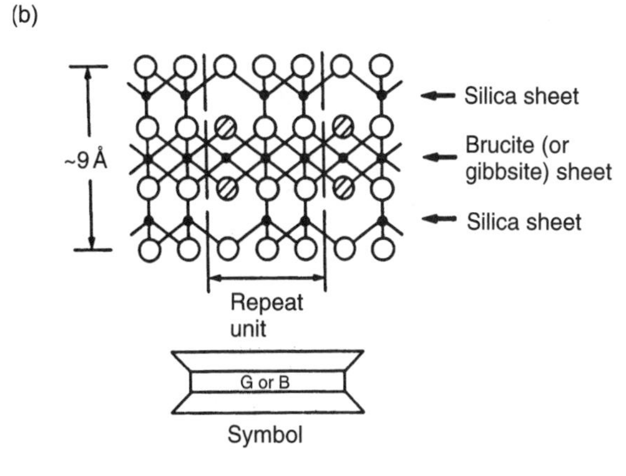

Figure 3.26 Structure of the principal phyllosilicate units: (a) 1:1 layer clays; (b) 2:1 layer clays (note 10 Å = 1.0 nm).

sheet merged with one octahedral sheet. The kaolins are typical of these clays.

Under conditions of incomplete leaching (arid, flooded, etc.) three-layer units or 2:1 layer clays tend to form. As shown in Figure 3.26b, these clays consist of two silica sheets sandwiching one octahedral sheet. The smectites and illites are typical of these clays.

The clay minerals form very small crystals rarely exceeding 2 or 3 µm in diameter. The reason for this is distortion of the layer structure because of the poor match between the unit cell dimensions (a_0 and b_0) of the tetrahedral and octahedral sheets (compare the a_0 and b_0 values in Figures 3.24 and 3.25). Halloysite, a form of tubular kaolin, is the most spectacular example of distortion since it forms in curved sheets with the gibbsite sheet inside the tetrahedral sheet (Figure 3.27). Montmorillonite forms very small curved particles since there is no interlayer bonding between the 2:1 layer units whereas muscovite may form large, table-sized slabs due to a very strong, interlayer K^+ bond which resists the distortion.

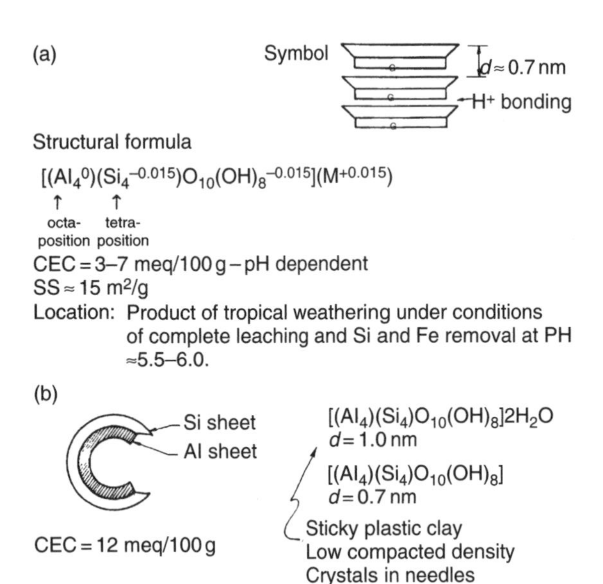

Figure 3.27 The 1:1 layer clays: (a) kaolinite and (b) halloysite.

The formulae of the clay minerals are all conveniently written in terms of a unit cell of 10 oxygens. For clarity, repeat units comprising 10 oxygens are shown schematically for both the 1:1 and 2:1 clays in cross-section in Figure 3.26. A count of the elemental species present in the repeat unit actually yields the correct formula for the phyllosilicate part of the clay minerals shown.

3.4.4 Important clay minerals found in soil

The clay mineral species normally encountered in engineering practice are the kaolins, illites, smectites, vermiculites and chlorites. Emphasis is placed on the importance of the CEC which is the measured abundance of exchangeable cations required to be adsorbed onto the negative clay platelets to render them neutral (measured in milliequivalents of cations/100 g of clay). These five mineral types are now described in some detail.

(a) Kaolinite (and halloysite)

A sketch showing the structure of kaolinite is presented in Figure 3.27a along with its structural formula. This clay mineral consists of stacked 1:1 layer units which have a *c*-axis lattice spacing, *d*, of 0.7 nm (7 Å).

As shown by the structural formula, there is very little charge on this mineral. The formula shows neutral octahedral layers and very slightly charged tetrahedral layers, probably by isomorphous substitution of Si^{4+} by Al^{3+}. As a result, very few cations are adsorbed onto this clay and the CEC is very low at ~5 meq/100 g.

The bonding between the sheets is a fairly strong hydrogen ion bond. The H^+ is supplied by the $(OH)^-$ forming the gibbsite sheet and tends to co-bind with O^{2-} on the surface of the tetrahedral sheet. The strength of this bond enables kaolinite to form fairly large crystals so that kaolinite has a fairly low surface area (specific surface) of only 15 m^2/g. The specific surface of a mineral is its measured surface area per unit mass or volume as desired.

Although the double layers around kaolinite may be fairly thick, this mineral is generally very inactive, has little adsorptive capacity and a hydraulic conductivity k that is rarely less than 10^{-8} m/s unless very pure and fine-grained. Fairly rapid advective transport and little retarding capacity are characteristics. Kaolinite, therefore, is not an ideal clay mineral for a liner even though it is relatively immune to damage when exposed to many chemicals.

The clay mineral halloysite is actually a tubular form of kaolinite as shown in Figure 3.26b. This tropical mineral crystallizes with water between the sheets. Since the unit cell of the Al sheet is slightly smaller than that of the Si sheet, the sheets curl with the gibbsite sheet on the inside. When dried, the interlayer water is irreversibly lost and d changes from 1.0 nm to a permanent 0.7 nm similar to kaolinite.

This mineral is difficult to compact because of its tubular structure so that k-values tend to be high and erratic (see Lambe and Martin, 1955, for a discussion of the properties of hallyosite).

(b) Illite

A sketch showing the structure of illite is presented in Figure 3.28a. Illite is really a form of mica and much soil illite appears to be fine-grained muscovite and occasionally biotite derived by mechanical weathering from igneous and metamorphic rocks (glacial grinding).

This mineral consists of stacks of 2:1 layer units held together by a very strong potassium bond to form a c-axis repeat unit, $d = 1.0$ nm (10 Å). The high strength of this bond is related to two main factors: (1) a very high negative charge within the tetrahedral sheets (-0.6 to -1.0 e/unit cell of 10 oxygens) and (2) the holes in the tetrahedral sheets into which the K^+ is partially counter-sunk. As shown by the structural formula, the cause of this negative charge is isomorphous substitution of one in four Si^{4+} by Al^{3+}.

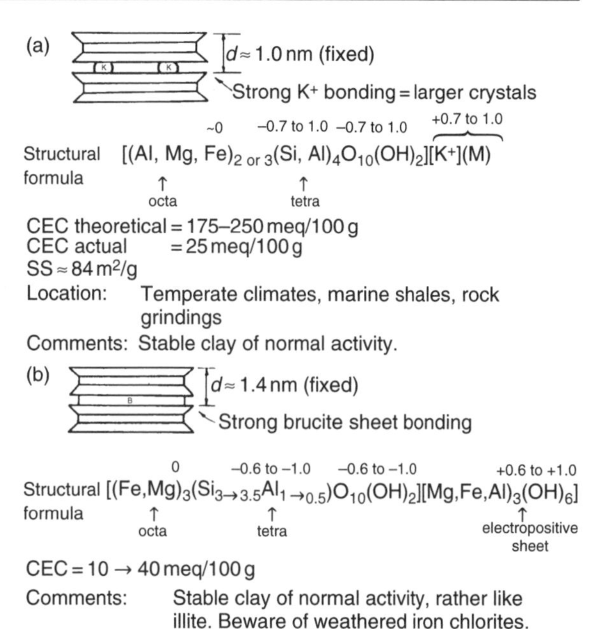

Figure 3.28 The non-swelling, 2:1 layer clay minerals: (a) illite (hydrous or soil mica) and (b) chlorite.

Most of the large negative charge on illite is satisfied by fixed potassium shown in square brackets in the structural formula. A small unbalanced charge corresponding to a CEC of 25 meq/100 g remains and these cations are what make illite a clay mineral of normal activity.

Illite is efficiently compacted or consolidated to form clayey soils having a hydraulic conductivity, k, of 10^{-9} to 10^{-11} m/s depending on the void ratio. Also, the CEC of 25 meq/100 g is adequate to permit abundant adsorption of undesirable species such as heavy metals. Finally, there is no interlayer, c-axis expansion or contraction possible, so illite is often considered to be one of the most desirable clay minerals for use in engineered clay liners for MSW.

In temperate areas of acid weathering, the interlayer K^+ is sometimes leached out of illite enabling c-axis expansion and formation of degraded illite or soil vermiculite. Such degradation would also occur if illite was leached by acid mine or tailings waters with a pH of about 2. Further discussion of *c*-axis expansion/contraction problems are presented later.

(c) Chlorite

Chlorite is a 2:1 layer clay bonded together by positively charged octahedral sheets having Mg, Al or Fe in the central position (Figure 3.28b). The resulting mineral has a *d* value of 1.4 nm and the bond is normally strong enough to prevent *c*-axis swelling.

In many respects, chlorite has properties like illite and is often favoured as an effective, non-reactive barrier clay for MSW.

Some iron chlorites, however, are highly susceptible to oxidation weathering. Fe^{2+} in the octahedral position oxidizes to Fe^{3+}, reducing the charge deficiency enough to permit swelling and formation of vermiculites or smectites. These latter two minerals can be very troublesome and are discussed next.

(d) Smectites

Smectite is a group name for the 2:1 layer *swelling* clays which have a range in charge deficiency from −0.2 to −0.6 *e*/unit cell (Table 3.3). Normally, this charge is present in the octahedral layer and is too low to fix cations and bind the layers together. This mineral, therefore, experiences interlayer, *c*-axis swelling from about 1.7 nm to over 10 nm depending on the chemical state of the clay. These features are illustrated in Figure 3.29.

It is important to realize that a smectite with Ca^{2+} cations surrounding and neutralizing its negative charge will have a *d* value of ~1.8 nm and actually consists volumetrically of 0.9 nm of water for each 0.9 nm of solid (i.e., a void ratio of unity for the solid phase!). When the double-layer water and free pore water are added in, it is not hard to understand why smectites have such high void ratios.

Smectites consist of very tiny particles which produce a specific surface of up to $800\,m^2/g$.

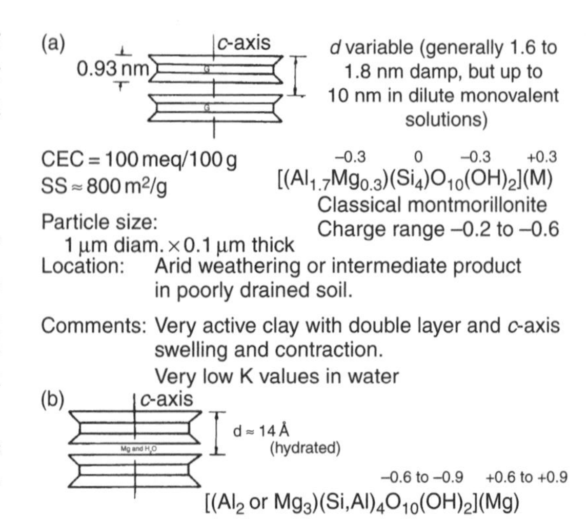

Figure 3.29 The swelling, 2:1 layer clays: (a) smectites (montmorillonite) and (b) vermiculite.

Thus, smectites may have very low hydraulic conductivities ($k \simeq 10^{-11}$–10^{-15} m/s) even at high void ratios. For this reason they (bentonites) are marketed for use as additives to sand to make clayey liners (or in combination with geotextiles to form a GCLs). Great care must be taken, however, to ensure that they are compatible with the leachate to be retained, since they may be prone to chemically induced *c*-axis contraction, double-layer shrinkage and cracking. High-charge smectites (−0.6 *e*/unit cell, Table 3.3) require particular caution since they fix K^+ making them even more prone to shrinkage and reduction in CEC.

(e) Vermiculites

Vermiculite is a common soil additive used by gardeners as a modifier. A fine-grained equivalent occurs in many soils where it can be troublesome if abundant enough.

The illustration for vermiculite in Figure 3.29b shows a high-charge, 2:1 layer clay. The charge of −0.6 to −0.9 *e*/unit cell is satisfied by hydrated interlayer Mg^{2+}. Vermiculite rarely expands beyond 1.4 nm but if it comes into contact with K^+ it promptly fixes it and collapses to 1.0 nm illite. This *c*-axis contraction causes shrinkage of the soil phase and an abrupt decrease in CEC. This converts an active, adsorbent clay into a much less active illite. If sufficient quantities of vermiculite are present, shrinkage and possibly cracking seems probable in the absence of high effective stresses.

Smectites and vermiculites "overlap" at a charge deficiency of −0.6 (Table 3.3) and it is very difficult to decide which mineral is present and how it will behave in a barrier contacted by leachate. The amount of *c*-axis contraction to be expected can be assessed by X-ray diffraction analysis, and examples are presented later.

(f) Summary

A summary tabulation showing the values of CEC and specific surface of the clay minerals is presented in Table 3.4. Generally speaking, inert clay minerals of modest CEC such as chlorite and illite probably produce the most reliable clayey liners.

Vermiculites tend to fix K^+ and NH_4^+ causing reductions in CEC and thus require compatibility testing if present in significant amounts.

Table 3.4 Specific surface and CEC of common clay minerals (adapted from Yong and Warkentin, 1975)

Mineral	*Surface area* (m^2/g)	*CEC* (meq/100 g)
Montmorillonite	800	~100
Clay mica (illite)	80	~25
Chlorite	80	~25
Vermiculite	250±	~150
Kaolinite	15	~5

The kaolins are the most inert but suffer from very low CEC values and problems in obtaining low values of hydraulic conductivity.

Smectites may in the end be the most difficult clay mineral to control (especially in the Na^+ state) due to possible *c*-axis contraction caused by cation exchange and changes in electrolyte concentration. Another factor to be considered is that smectite barriers tend to be very thin (in some cases, as little as a few centimetres), and consequently a fairly large diffusion flux may be expected to pass through them unless they are used in conjunction with another soil which will act as a diffusion barrier (see Chapter 16).

Most natural soils contain two or three different clay minerals often mixed with quartz, feldspar and carbonates. To make matters even more complex, some clay minerals are interstratified with one another. For example, illite/smectite, illite/vermiculite and illite/chlorite are quite common in southern Ontario, Canada.

Quantitative mineralogical analyses, complete with CEC measurements and an assessment of the adsorbed cation regime, are now normally run as part of each site appraisal. Assessment of diffusion coefficients and occasionally clay/leachate compatibility assessment by hydraulic conductivity testing may also be required. Chapters 4 and 8 deal with these topics.

3.5 Clay colloid chemistry

3.5.1 The nature of the clay–water micelle

A single, tabular-shaped clay particle is shown in Figure 3.30. The negatively charged particle is surrounded in nature by a swarm of hydrated cations which balance the charge. The particle and its surrounding double layer of water together comprise a "micelle".

Even though most clay particles are far larger than the typical colloidal size of 1 μm to 1 nm, they behave like colloids because they carry a negative charge. Their electrical properties,

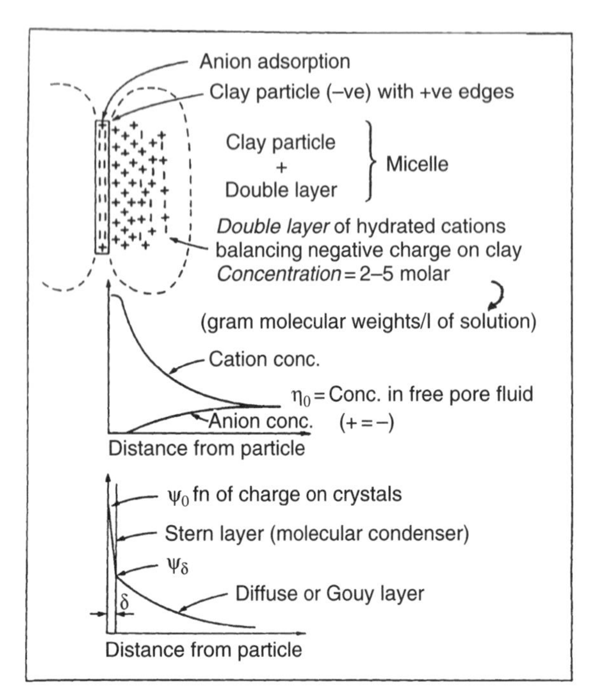

Figure 3.30 Nature of the clay–water micelle including the Stern layer concept.

therefore, have a far greater effect on their physical properties than do gravitational factors, and under the right chemical conditions, clays (especially smectites) will form a stable suspension.

The concentration of hydrated cations is highest next to the charged particle, decreasing with distance to the same value as the free pore water. Similarly, the concentration of anions is virtually zero next to the clay particle (except the edges) and increases to that of the free pore water with distance. Cation concentrations of 2–5 molar (gram molecular weights/l) are typical of the double layer.

The electric potential illustrated in Figure 3.30 incorporates the Stern layer concept in which the proximity of the first layer of firmly attached cations results in an abrupt drop in potential through the Stern layer as shown. The distance δ corresponds to the centreline of the first layer of cations and ψ_δ corresponds to the reduced potential which actually controls the size of the double layer. ψ_δ is somewhat variable since an increase in concentration will add more cations to the Stern layer reducing ψ_δ. Also, changes in the cation species can greatly alter δ.

Perhaps the most important factor in clay/leachate interaction is expansion and contraction of the double layers. As illustrated in Figure 3.31, a large double-layer contraction at constant void ratio, sometimes referred to as flocculation, creates a large increase in free void space. This may cause increases in hydraulic conductivity and possibly in the diffusion coefficient as well. Conversely, a chemical change that peptizes, disperses or expands the double layers may eliminate most of the free pore space and reduce hydraulic conductivity.

For most conceptual needs in engineering practice, the Gouy–Chapman theory of potential adequately explains double-layer behaviour. The Gouy–Chapman equation is presented in

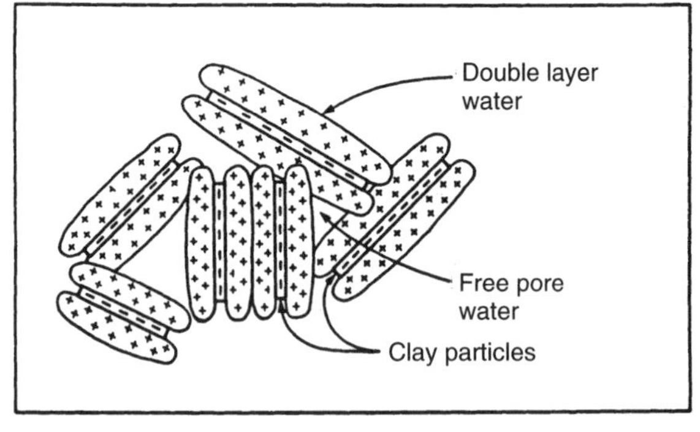

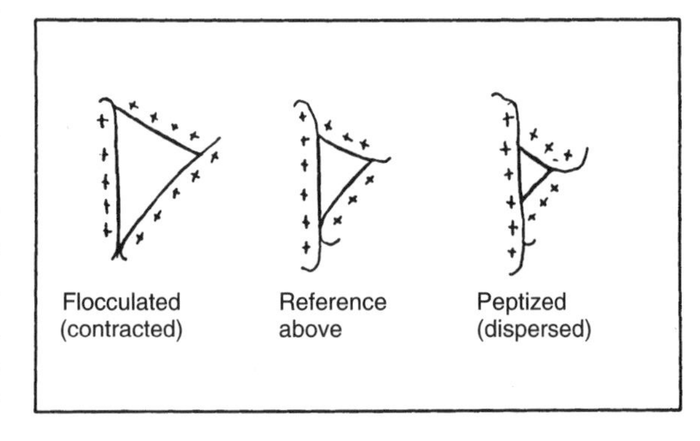

Figure 3.31 Nature of a soil–water system showing effects of "flocculation" or "dispersion" on free pore space at constant void ratio.

Figure 3.32 along with a sketch of the double-layer potential. The centre of gravity of the area under the potential curve occurs at a value of $x = 1/\kappa$ and many discussions now refer the $1/\kappa$ as the double-layer thickness (Mitchell, 1993).

The main messages to be derived from the equation for potential are as follows:

1. If the pore water concentration is increased, the double layers contract.
2. If the cations are changed from monovalent to divalent or trivalent, the double layers contract (e.g., from Na^+ to Ca^{2+}).
3. If the dielectric constant, ε, is reduced, the double layers contract (e.g., from 80 for water to 35 for ethanol).
4. If the temperature is increased, the dielectric constant ε decreases and the double layers contract (this is not indicated by the equation).
5. All four of the above factors will cause the hydraulic conductivity to increase (sometimes dramatically) if the void ratio remains constant.

Two plots of potential versus distance from the clay particle are presented in Figure 3.33 to

$$\psi = \psi_0 \exp(-\kappa x)$$

$$\kappa = \sqrt{\frac{4\pi e^2 \Sigma n_{i0} z_i^2}{\varepsilon k T}}$$

n_{i0} = conc. of ions (i) in bulk suspension
z_i = valence of ions
k = Boltzmann's constant (1.38×10^{-16} erg/°K)
T = Absolute temperature (°K)
$kT = 0.4 \times 10^{-13}$ erg at room temperature
ε = dielectric constant of pore fluid
e = elementary charge = 4.77×10^{-10} esu = 16.0×10^{-20} Coulomb

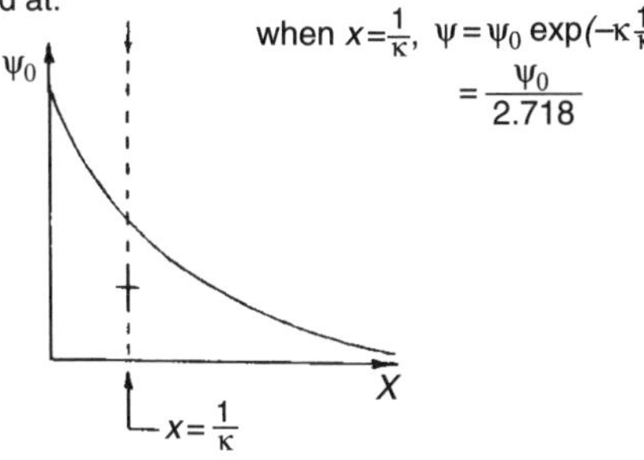

Figure 3.32 The Gouy–Chapman theory of potential.

(a)

Ion conc. (moles/l) (ψ_0 const)	Double-layer thickness	
	Monovalent	Divalent
0.01	100 nm	50 nm
1.0	10 nm	5 nm
100	1 nm	0.5 nm

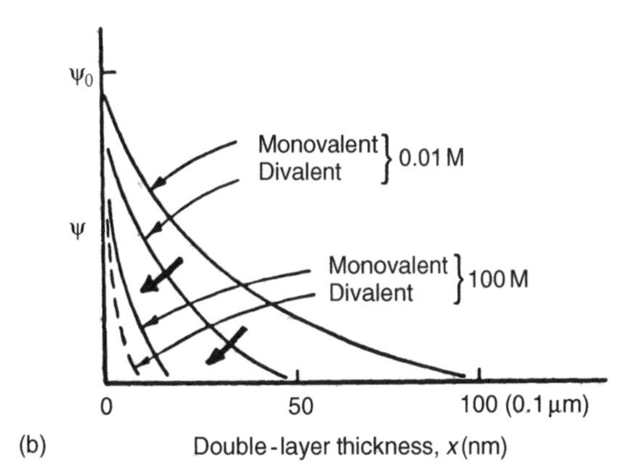

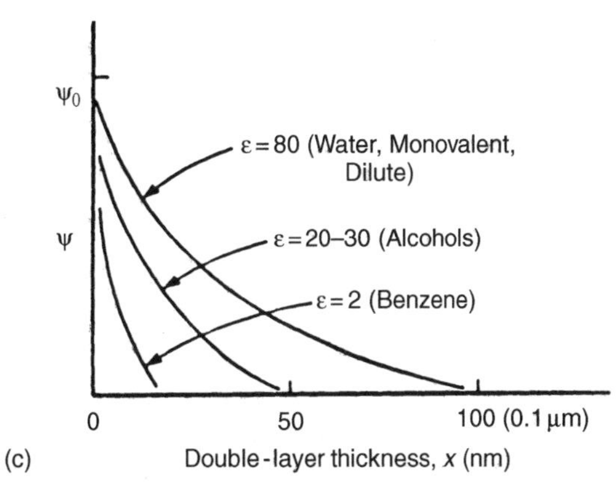

Figure 3.33 Double-layer thickness variations: (a) data from Van Olphen, 1977; (b) concentration and valency effects; (c) dielectric effects.

illustrate concentration, valency and dielectric effects. The upper plot (Figure 3.33b) demonstrates how monovalent to divalent cation exchange can reduce the double-layer thickness from 100 to 50 nm at dilute concentrations (0.01 M). At *very* high concentrations, double-layer thicknesses of only 0.5–1.0 nm are indicated (Van Olphen, 1977, p. 35).

The second plot for changes in dielectric constant (Figure 3.33c) are equally dramatic and will form the basis of the discussion of clay/liquid hydrocarbon compatibility in Chapter 4.

3.5.2 Adsorbed cation regimes and cation exchange

Cations normally found in nature are summarized as follows:

1. Ca^{2+} and Mg^{2+} dominate most systems;
2. Na^+ and K^+ are common in marine soils and certain volcanic ash deposits;
3. Al and Fe hydroxide cations having "partial valences" of $\sim 0.5^+$ in low pH podsolic environments (pH < 4);
4. H^+ or H_3O^+ common in acid soils;
5. Metal cations are rare and rapidly adsorbed or precipitated, except in low pH environments.

The replacing power of the normally occurring cations can be arranged in the following sequence (Yong and Warkentin, 1975):

$$Al^{3+} >> Ca^{2+} > Mg^{2+} >> NH_4^+ > K^+ > H^+ > Na^+ > Li^+$$

The Gapon exchange equation can be used to calculate roughly the relative abundance of the various cations on a clay but it is normally more useful to do an NH_4Cl exchange or a combined Ag–thiourea/KCl exchange and actually measure the adsorbed cations.

Since Na^+ is the most common and troublesome monovalent cation found on natural soils, measurement of the sodium adsorption ratio (SAR) is useful and often required during geotechnical assessment:

$$\mathrm{SAR} = \frac{Na^+}{\sqrt{0.5(Ca^{2+} + Mg^{2+})}} \quad (\mathrm{meq/L})^{1/2}$$

Analyses are usually run on soils wetted to their saturation water content (i.e., the water content close to the liquid limit at which a shine appears on the wet soil) and all concentrations are in meq/l.

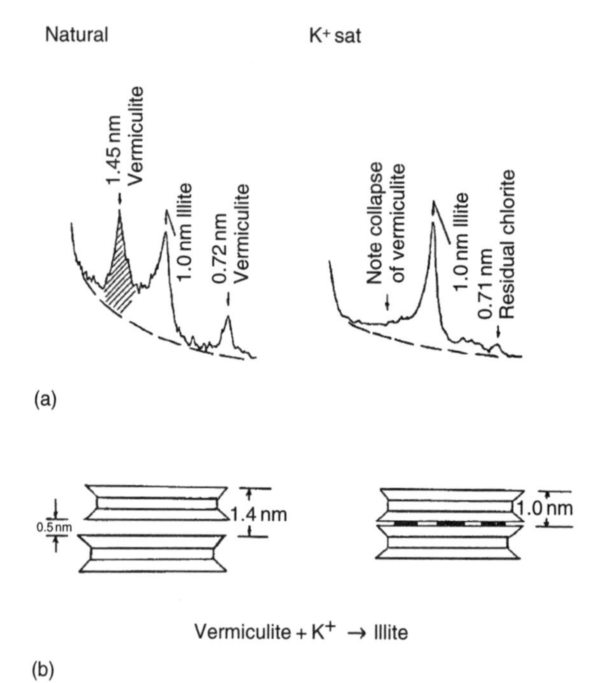

Figure 3.34 Potassium fixation by soil vermiculites and high-charge smectites.

The SAR (which is actually a form of Gapon's equation) is particularly useful because it often correlates with the amount of exchangeable sodium (ESP) (see Sherard *et al.*, 1972; Mitchell 1993). Dispersed, low-salt soils containing abundant adsorbed Na^+ may initially have low hydraulic conductivities but they may also be susceptible to chemical shrinkage and cracking if chemically altered, especially at low confining stresses.

(a) Cation exchange

The levels of Na^+, K^+ and NH_4^+ in MSW leachate are sufficiently high that they effectively exchange some of the Ca^{2+} and Mg^{2+} present on natural clays during advection and diffusion. Since it takes two Na^+ to exchange one Ca^{2+}, this reaction should expand the double layers and decrease k.

The role of the similar-sized cations, K^+ and NH_4^+, is quite different. These two species tend to fix onto vermiculites and high-charge smectites and are highly retarded. Worse than this, however, they may contract the c-axis spacing causing a reduction in the mineral or solids volume which should increase hydraulic conductivity, k. If they do fix into the holes in the tetrahedral sheets they also may reduce the CEC. This may contract the double layers owing to charge reduction, causing large increases in free pore space at constant void ratio. In turn, this may cause increase in k and possibly even liner cracking. These phenomena were discussed previously in Sections 3.4 and 3.5 and are illustrated by the X-ray traces and sketches shown in Figure 3.34.

Chapter Four

Clay/leachate compatibility by measurement of hydraulic conductivity

4.1 Introduction

Clay/leachate compatibility is now well recognized as an important consideration in the design of clay liners for landfill sites. This chapter discusses clay/leachate compatibility testing with examples cited with respect to clays from southern Ontario, Canada. The issue is subsequently discussed with respect to liners GCLs in Section 12.5.1.

Major concerns regarding potential pollution problems related to contamination by toxic liquid hydrocarbons (both soluble and insoluble) exist and so a major portion of this chapter will be devoted to organic liquids. It is also necessary to consider the potential interaction between municipal waste leachate (MSW) and proposed clayey liners for modern landfills (e.g., see MoE, 1998). This subject will be considered first.

The hydraulic conductivity equipment employed for the work presented in this chapter is a constant flow rate system described by Fernandez and Quigley (1985 and 1991). This equipment has proven to be highly versatile and useful for assessing compatibility of both conventional compacted clay and GCLs (Petrov and Rowe, 1997; Petrov *et al.*, 1997a; see Section 12.5.1). The system (Figure 4.1) generates a constant flow, q, through the test specimens and the induced total head drop across the specimen is used to calculate the hydraulic conductivity, k, using Darcy's law. This procedure, extensively described by Olsen (1966), is particularly useful for the volatile, toxic liquid hydrocarbons which are sealed in the reservoir cylinders (syringes). The single circuit shown for one permeameter is actually expanded to eight circuits with the reservoir syringes mounted in a triaxial frame. The upward displacement generates constant flow that can be varied from 6×10^{-6} to 1×10^{-2} ml/s. A dial gauge mounted on the compression frame allows continuous monitoring of the flow rates. The pressure head at the inlet end of the specimen is measured using a pressure transducer and the outlet end is kept at atmospheric pressure. The total head drop is essentially equal to the pressure head drop for high-pressure heads, the elevation head correction becoming more important at lower inlet pressures.

The permeameter cell, illustrated in Figure 4.2, consists of either an aluminium or stainless

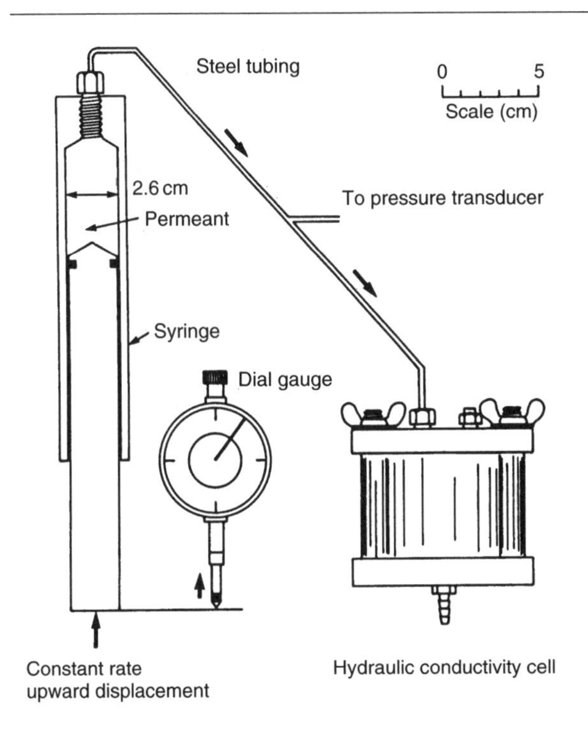

Figure 4.1 Schematic diagram of constant flow permeameter (after Fernandez and Quigley, 1985; reproduced with permission of the *Canadian Geotechnical Journal*).

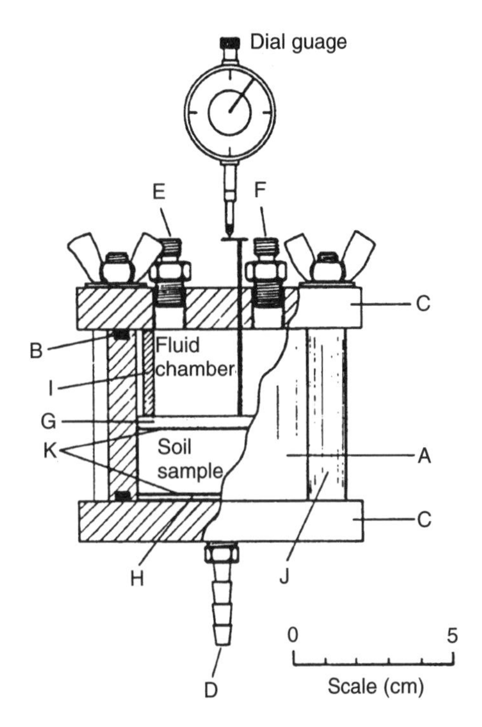

Figure 4.2 Schematic diagram of one-permeameter cell (modified from Fernandez and Quigley, 1985).

steel cylinder (A). Both ends of the cylinder are machined to contain viton O-rings (B) sealing the contact between the cylinder and the aluminium or steel plates (C). The fluid outlet (D) allows collection of the effluent permeant for constant flow assessment or chemical analysis. The two fittings in the upper plate are the fluid inlet (E) and a valve for escape of air during filling of the fluid chamber (F). Port F can also be used as the pressure transducer mount. A brass porous disk (G) approximately 3 mm in thick and a polyethylene filter (H) 1.5 mm in thick are placed on top and below the soil sample, respectively. I is either a rigid spacer on the porous disk to prevent swelling at zero static effective stress ($\sigma'_v \approx 0$) or a set of springs capable of applying static effective stress, σ'_v, up to 320 kPa before and during permeation. The assembled cell is held together by four-threaded and sleeved rods (J) fixed to the lower plate. Filter paper (K) is normally placed between the soil specimens and the filter disks (G and H). Finally, a settlement-measuring rod extends through the cap to the top of the sample and, by means of another dial gauge, chemically induced consolidation may be measured. This apparatus might be regarded as an elaborate constant-flow-rate consolidation permeameter.

The test soils to be described consist of brown and grey clays from Sarnia, Ontario. A brief description of the mineralogy of samples from 0.3 and 11 m in depth is presented in Table 4.1. The grey soils at depth, which are parent materials to the brown-weathered surface soils, contain more carbonate, more chlorite and very little (<1%) smectite. Near the soil surface, smectite constituted about 15% of the soil. This smectite

Table 4.1 Silty clay composition (Sarnia)

	Brown (0.3 m)	*Gray* (11.0 m)
Quartz and feldspar	~20%	~17%
Carbonates	~10%	~35%
Illite	~50%	~25%
Chlorite	~8%	~22%
Smectite	~15%	~1%
CEC	~25 meq/100 g	10 meq/100 g
<2 μm	~60%	~42%

is derived from the iron-rich chlorite by oxidation weathering.

These soils provide a range of potential responses to leachate since, at depth, they are fairly inactive yet good liner clays and at surface they contain enough smectite to demonstrate the high-activity effects of montmorillonite.

The effluent chemistry is carefully monitored during hydraulic conductivity (k) testing for clay/leachate compatibility assessment. Normally the criterion to stop a test is not a constant value for k, but chemical equilibrium between the clay and the influent leachate. Once the effluent concentrations, c, reach influent concentrations, c_0, this equilibrium is normally considered to have been reached.

Typical c/c_0 curves are presented in Figure 4.3 to illustrate a wide spectrum of possible interactions between soil (either undisturbed or compacted) and the chemical species used as a test permeant. These curves will be considered individually.

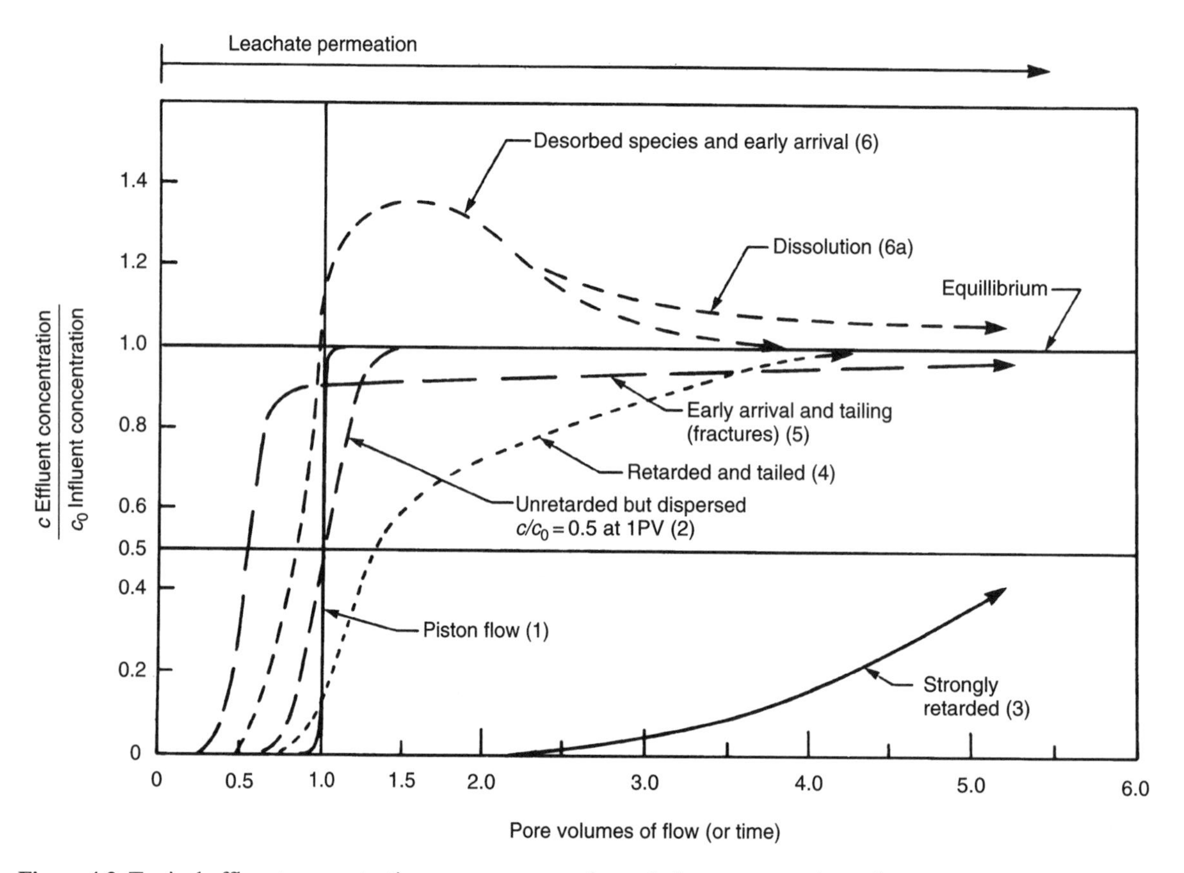

Figure 4.3 Typical effluent concentration curves expressed as relative concentration, c/c_0.

1. Uniform, undispersed advective movement (piston flow) would arrive as a 100% concentration front (i.e., $c/c_0 = 1$) at exactly one pore volume (PV) of flow. At this point, all fluid previously in the soil would be displaced by the permeant.
2. Dispersion tends to spread the front so that some chemicals arrive before 1 PV and c/c_0 does not reach unity until after 1 PV. Curve (2) therefore has a $c/c_0 = 0.5$ at 1 PV and is unretarded but somewhat dispersed.
3. Curve (3) represents a strongly retarded species which is totally retained by the soil for 2 PV and then slowly approaches equilibrium at more than 7 PV.
4. Curve (4) is described as retarded and tailed yet still demonstrates some arrival before 1.0 PV. This type of behaviour normally indicates channel or fracture flow of a highly reactive species with equilibrium obtained by diffusion into adjacent soil peds.
5. In comparison, curve (5) represents early arrival and tailing of a non-reactive species, c/c_0 of 0.5 arriving at only half a PV. Again, the only explanation is rapid channel flow with equilibrium reached by a long process of tailing caused by diffusion from the fractures (or macropores) into the adjacent micropores (see also Sections 1.4.2 and 7.6 and Chapter 11 for more information on this phenomenon which is referred to as matrix diffusion). It is conceivable that with an even higher flow rate, c/c_0 would approach unity and one would never know if equilibrium was reached.
6. Curve (6) is a typical desorption curve, often reflecting calcium Ca^{2+} and magnesium Mg^{2+} displaced by sodium Na^+ and potassium K^+ in a permeating leachate. Occasionally the Ca^{2+} curve never returns to unity, in which case long-term dissolution of carbonate or gypsum might be inferred. It is often important not to run such tests too long in case loss of solids causes a large increase in laboratory hydraulic conductivity (breakthrough) that could never happen in a field situation.

With the above summary of typical effluent curves, it is now appropriate to look at some actual clay/leachate compatibility tests.

4.2 Soil–MSW leachate compatibility

Most inactive soils whose clay minerals consist of illites and chlorites are relatively insensitive to leachate from MSW. In fact, the hydraulic conductivity will normally decrease (Griffin *et al.*, 1976) probably because of Na^+ adsorption, double-layer expansion and possibly bacterial clogging.

If soils contain significant amounts of swelling clay (here defined as vermiculite, montmorillonite, or interlayered illite-smectite; see Section 3.4), *c*-axis contraction or expansion is another complicating factor that can, respectively, increase or decrease hydraulic conductivity. For example, K^+ fixation by vermiculite causes a 28% decrease in crystal volume plus a contraction of the double-layer thickness because the charge deficiency, and hence the cation exchange capacity (CEC), is reduced equivalent to the amount of K^+ fixed. Such phenomena should contract the soil peds, open further the voids or fractures between them, and thus cause increases in the hydraulic conductivity. If a high effective stress is present on the clay liner, chemically induced consolidation should help compensate for any chemically induced increase in the size and frequency of the macropores. An illustration of these phenomena was presented in Figures 3.31 and 3.34.

4.2.1 Hydraulic conductivity

Hydraulic conductivity tests run on three samples of Sarnia clay from depths of 0.3 and 11 m are presented in Figure 4.4. These results will be used to demonstrate both the effect of clay

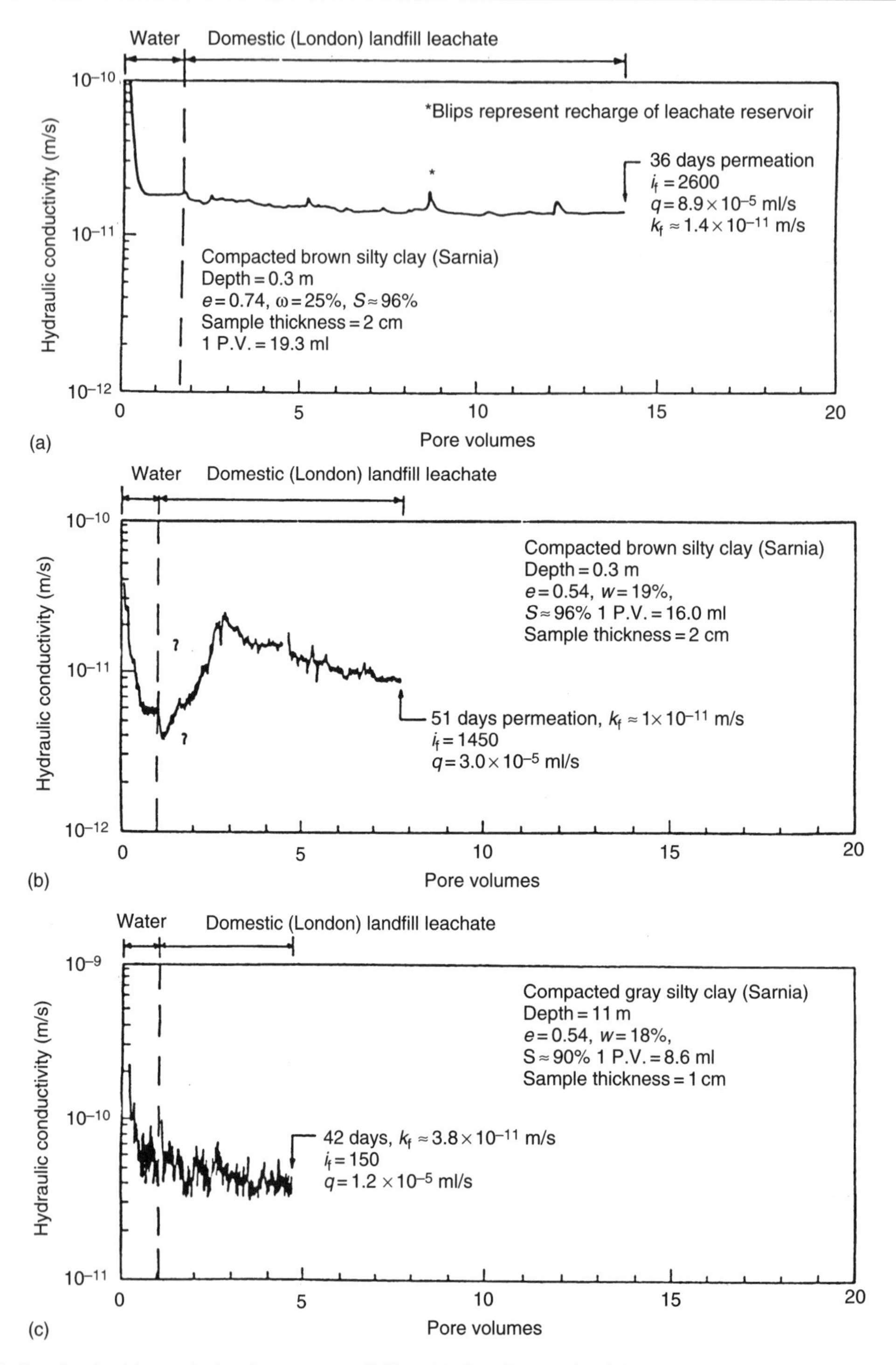

Figure 4.4 Sarnia clay/domestic leachate compatibility. Hydraulic conductivity versus PVs: (a) brown, compacted, $e = 0.74$, $\omega = 25\%$; (b) brown, compacted, $e = 0.54$, $\omega = 19\%$; (c) grey, compacted, $e = 0.54$, $\omega = 18\%$.

mineralogy and the effect of compaction at a water content too close to standard Proctor optimum which may result in a dense, strong liner but can also give rise to microfractures in the clay which will influence both the hydraulic conductivity and chemical characteristics of the movement of leachate through the barrier. The tests were run for the most conservative case of zero vertical static stress on the soil samples. The gradients developed by the constant flow rate test system are dependent on hydraulic conductivity, and for the three tests shown in Figure 4.4 were 2,600, 1,450 and 150 for the two brown and one grey clay, respectively. The corresponding values of the maximum seepage stresses, J_{max}, at the base of the test specimens, were approximately 510, 284 and 15 kPa, respectively.

The following observations can be made:

1. Domestic waste leachate causes a slight reduction in hydraulic conductivity for both the brown and grey test clays (Figure 4.4a and c).
2. The grey clay has a higher hydraulic conductivity than the brown clay (3.8×10^{-11} m/s compared with 1.4×10^{-11} m/s), even though it is at a lower void ratio ($e = 0.54$ compared with $e = 0.74$).
3. The greater smectite content of the brown clay is in part responsible for its lower k value. Also, as will be shown later, the denser, grey clay (sample a) contains compaction-induced fractures that allow channel flow.
4. The brown clay compacted to $e = 0.54$ (Figure 4.4b) produced an erratic hydraulic conductivity versus PVs curve that suggests rapid clogging of fractures followed by stabilization at a reasonable hydraulic conductivity of 1×10^{-11} m/s.

4.2.2 Chemical controls

Companion curves for the concentration ratio (c/c_0) are shown in Figure 4.5 for the three hydraulic conductivities versus PV curves shown in Figure 4.4. The upper set of curves (Figure 4.5a) for the looser sample compacted to a void ratio of 0.74 indicates homogeneous flow through equal-sized voids and arrival of the 50% chloride front ($c/c_0 = 0.5$) at close to 1 PV of leachate flow.

The set of curves in Figure 4.5b for the denser brown clay compacted to a void ratio of 0.54 demonstrates early arrival and tailing, both indicative of channel flow through compaction-induced fractures. This is quite certain since the presence of more clay at $e = 0.54$ should cause greater K^+ retardation, not less!

The final set of curves in Figure 4.5c for the grey soil represents a problem from a chemical point of view even though the hydraulic conductivity test results in Figure 4.4c appear to be quite reliable. The presence of fractures in this sample has allowed early arrival of all chemical species, at least at the gradient of 150 employed.

Considering Figures 4.5a and b for the brown clay only, the following aspects merit detailed comment:

1. Both sets of curves show Na^+ and K^+ retardation due to adsorption onto the clays and resultant displacement of Ca^{2+} and Mg^{2+} to form a hardness halo effect which arrives early.
2. Conservative chloride reaches $c/c_0 = 0.5$ at $\sim$0.8 PV of leachate flow for the looser soil (Figure 4.5a) compared with 0.1 PV for the dense soil (Figure 4.5b), indicating almost homogeneous flow through the former and channel (fracture) flow through the latter.
3. In the absence of macropores, the denser soil should retard more K^+ than the looser soil, whereas the opposite appears to occur. The higher (early) K^+ values indicate channel flow along compaction-induced fractures.
4. For sodium, c/c_0 reaches 0.5 at 1.4 PV of leachate flow, indicating the expected retardation (by adsorption) for the looser soil (Figure 4.5a). For the denser soil, c/c_0 reaches 0.5 at only 0.8 PV, again indicating early

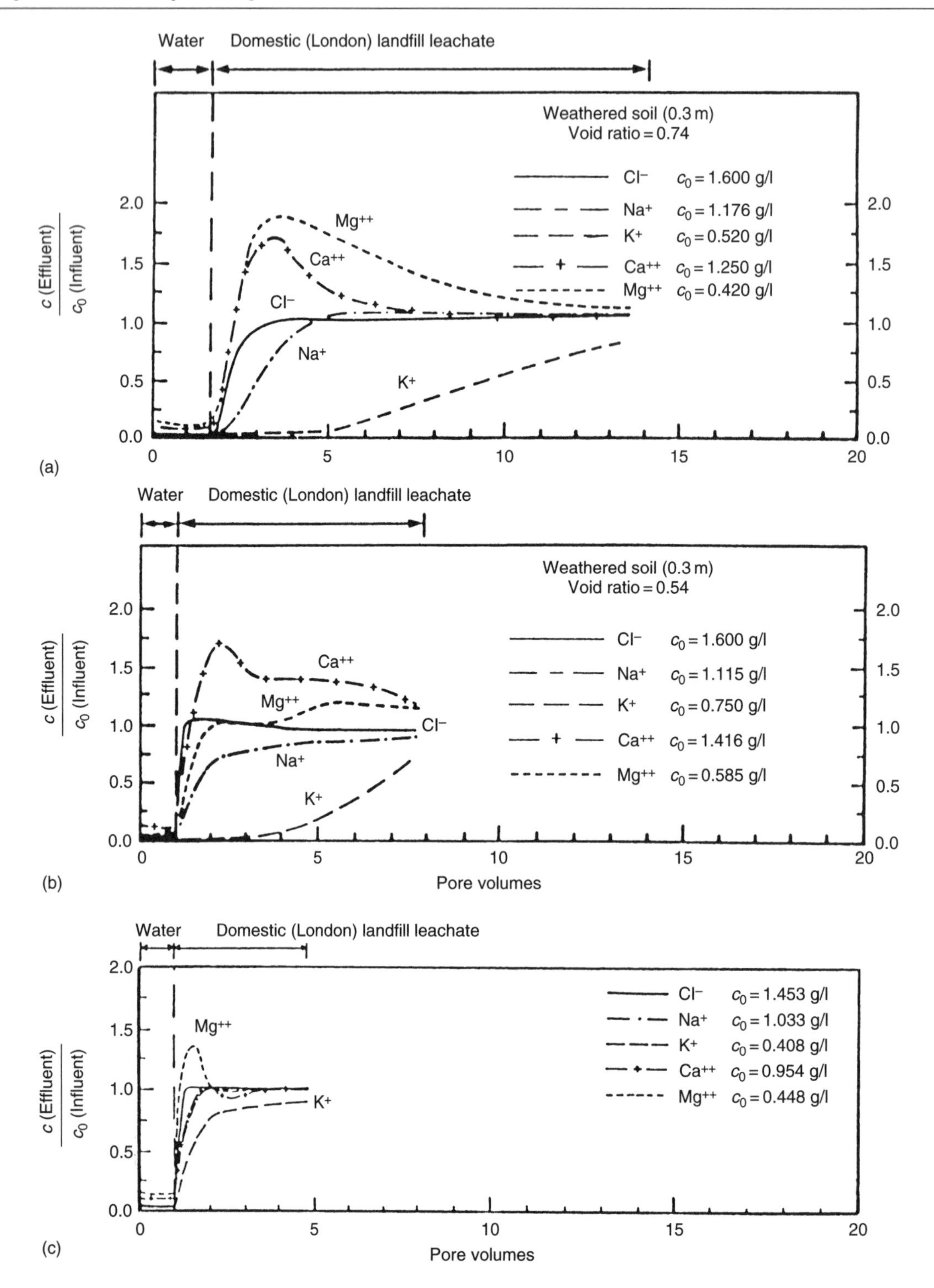

Figure 4.5 Effluent chemistry (c/c_0 curves) for the three test clays shown in Figure 4.4.

arrival due to channel flow through this sample. This occurs in spite of the much lower gradient on the denser sample compared with the looser sample ($i_f = 1{,}450$ compared with 2,600).

5. The abrupt flattening of the sodium curve (Figure 4.5b) and the long slope towards $c/c_0 = 1$ (tailing) indicates fracture or macropore flow with gradual diffusion of Na^+ into the pore fluid of the adjacent soil peds and ultimately cation exchange onto the clay and desorption of Ca^{2+} and Mg^{2+}.

The c/c_0 curves in Figure 4.5c, while technically a problem (as noted previously), do yield some useful information.

1. First, Mg^{2+} appears to be the major exchange cation since it yields a desorption peak rather than Ca^{2+} observed for the brown soils.
2. Some mineral component is also retarding K^+ in the grey clay, possibly an interlayered species.
3. Finally, the test should be repeated at a lower gradient than the 150 used so that retardation and equilibrium may be properly observed.

(a) The importance of K^+ retardation

The extent of K^+ adsorption from domestic leachate by these soils increases with increasing amounts of vermiculitic smectite in the specific test specimens. Since the smectite increases from about 1% in the grey clays to about 15% near surface as a result of oxidation weathering of chlorite, much more retardation should occur with passage of leachate through near-surface soils.

Many borrow pits used for construction of clayey barriers demonstrate weathering changes similar to the Sarnia site. The mineralogy of a clayey soil barrier thus may vary significantly from one stage of construction to the next as the borrow depth varies.

The significance of this weathering phenomenon is illustrated in Figure 4.6, which shows c/c_0 curves for K^+ generated by MSW leachate influent passing through clayey samples taken from 0.3, 1.7 and 11.0 m depths. The three samples were essentially free of fractures and macropores, and thus the various degrees of K^+ retardation reflect increasing vermiculite/smectite towards surface.

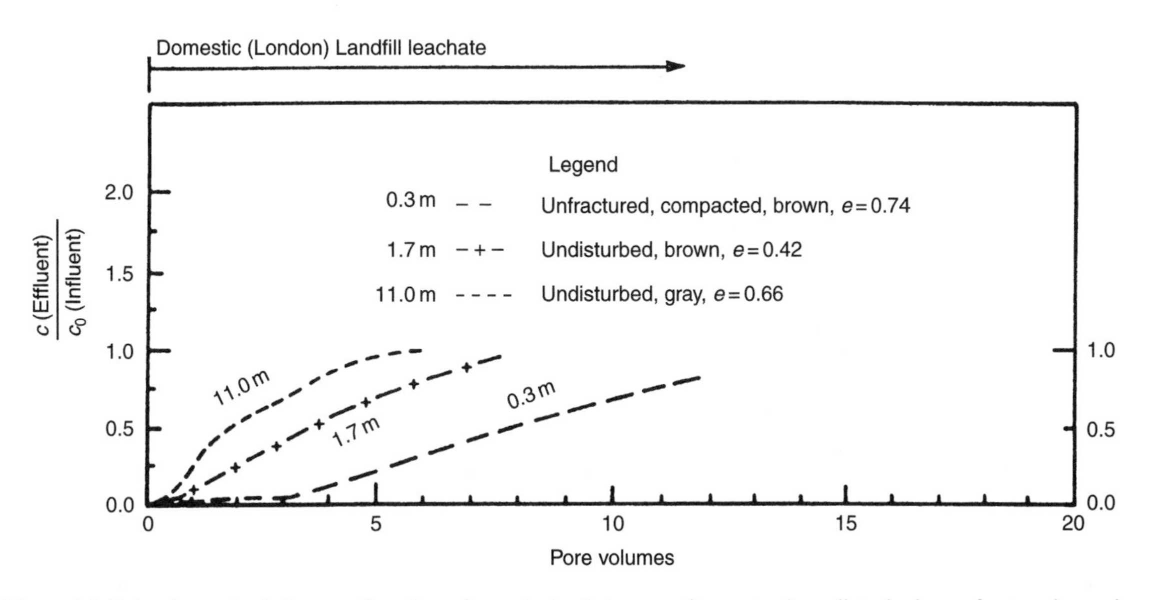

Figure 4.6 Potassium retardation as a function of sample depth (or smectite content), undisturbed or unfractured samples.

The c/c_0 curves for K^+ are also useful to indicate the role of fracture flow. Figure 4.7 compares curves for fractured (dashed) and unfractured (solid) samples from 0.3 and 11.0 m depths. The early arrival of K^+ evident for both soils with a void ratio $e = 0.54$ occurs as a result of fracture flow. For the two unfractured soils, from 0.3 and 11.0 m, passage of ~15 and ~6 PV of leachate was required for the effluent and influent liquids to contain equal amounts of K^+. If flow through the soil was homogeneously through equal pore sizes, it could be reasonably assumed that equilibrium had been reached, at least with respect to soluble chemical species. Testing times for these two tests (0.3 and 11.0 m) were 5 and 7 weeks at gradients of 2,600 and 2,000, respectively.

For the two fractured samples, equilibrium might be difficult to establish because K^+ would have to diffuse from the flow channels into the pore fluid in the adjacent soil peds (by matrix diffusion as discussed in Section 1.4.2) and then sorb onto the clay minerals before equilibrium is reached. Arrival of c/c_0 values equal to 1 after a long period of tailing would probably indicate equilibrium. If, however, the fractures opened further owing to soil shrinkage, the hydraulic conductivity would increase and c/c_0 would equal unity after a short testing time without equilibrium being reached.

Like K^+, NH_4^+ is also abundant in raw domestic waste leachate and it will also fix onto vermiculitic minerals in borrow soils.

(b) Post-testing assessment

Post-testing assessment is quite an elaborate process often requiring:

1. A pore water squeeze and chemical analysis to ensure constant pore fluid composition at the end of testing equal to the effluent and to compare with pre-testing conditions.
2. For thick specimens, pore water compositions at the top, middle and bottom might

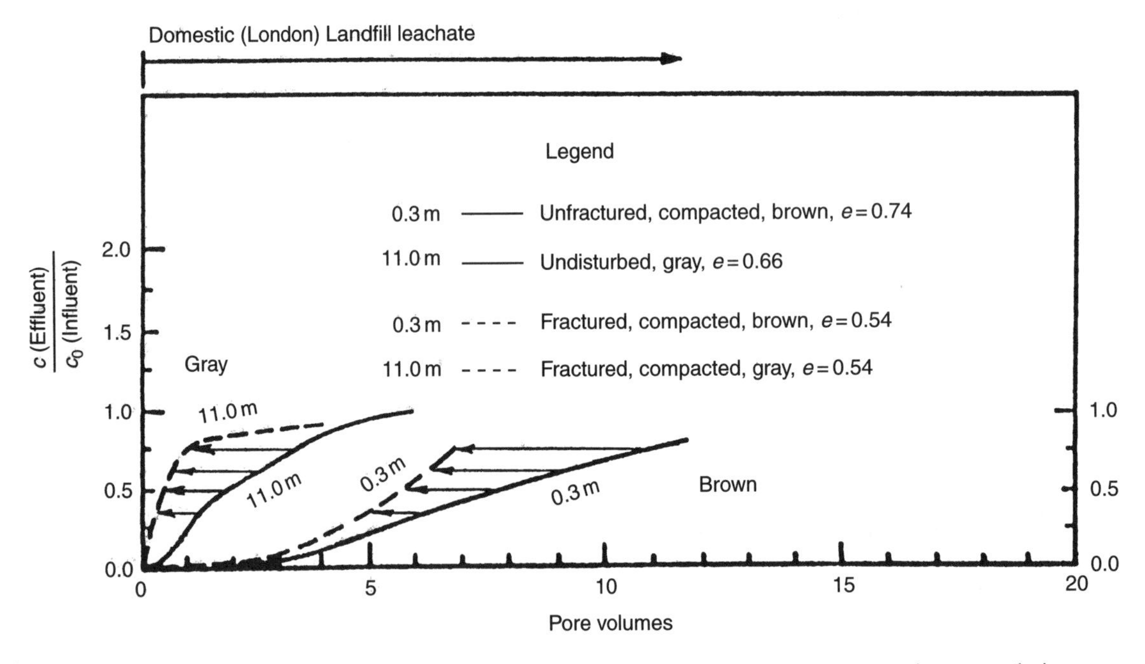

Figure 4.7 Influence of compaction-induced fractures and interped macropores on potassium retardation curves (comp = kneading compaction).

be required to demonstrate identical composition throughout.

3. Assessment of the adsorbed cation regime and CEC before and after testing. This is particularly important if significant K^+ retardation has been observed and *c*-axis contraction of vermiculite is suspected. A drop in CEC would certainly signal the possibility of increased *k*-values unless consolidation has occurred.
4. A comparison of X-ray diffraction traces for the test soil before and after permeation, again looking for *c*-axis changes.
5. A very careful inspection of shrinkage cracks which might have developed near the top of the samples complete with a photographic record.

For the Sarnia example soil described to this point, a series of X-ray traces were obtained as illustrated in Figure 4.8. Considering the grey unweathered clay first (bottom traces on (a) and (b)), it can be seen that the clay minerals are dominated by illite and iron chlorite. Both KCl treatment and air-drying have relatively little effect on the X-ray traces, indicating an absence of smectite/vermiculite, and a soil probably relatively immune to changes caused by MSW leachate, which the *k*-tests indicated to be the case.

The weathered brown surface clay in the air-dry state (dashed trace, Figure 4.8b) shows significant collapse of smectite due to 0.5 N KCl treatment if air-dried (solid trace). The 1.4 nm peak is greatly reduced and the 1.0 nm peak greatly strengthened. In the water-wet state (Figure 4.8a), however, little collapse occurs except in the interlayered illite/smectite phase (I/S) indicated by the high background between 1.0 and 1.4 nm that develops after KCl treatment.

As noted in the figure, water-wet and leachate-wet samples may have to be X-rayed to properly define the behaviour of a clayey barrier which will normally remain wet in the field. For example, in Ontario, Canada, many of the apparent vermiculites in the brown-weathered surface soils do not collapse on exposure to MSW leachate but stay expanded and perform satisfactorily as compacted landfill barriers.

In other cases, higher charge vermiculites may well collapse in the wet state possibly causing shrinkage cracks and loss of barrier integrity.

4.2.3 Clay–leachate compatibility testing costs

Hydraulic conductivity testing performed as an assessment tool in clay–leachate compatibility studies (when performed with appropriate chemical controls) is expensive. Once experience has been gained with local soils, it is probable that a mineralogical report including an X-ray assessment of the effects of K^+ saturation by both KCl and MSW leachate will suffice. A few hydraulic conductivity tests using leachate as permeant are always a useful confirmation of compatibility.

4.2.4 Weathered Leda clay

The high moisture content and often sensitive Champlain Sea clays of eastern Canada are generally not suitable for compaction unless they have been desiccated as is commonly the case in stiff near-surface crusts. The desiccated soils are also weathered brown and original soil chlorites may be weathered to vermiculite. A study of such a weathered Leda clay from Ottawa, Canada, found that this soil was susceptible to K^+ fixation when contacted by MSW leachate. After permeation with 9 PVs of MSW leachate, the X-ray traces on pressure-oriented post-test samples altered as illustrated in Figure 4.9. The natural clay contained abundant vermiculite/ smectite and interlayered illite/vermiculite that produced a CEC value of 32 meq/100 g. The traces for the wet, post-testing sample show complete collapse of the vermiculite to illite and a drop in CEC to 17 meq/100 g.

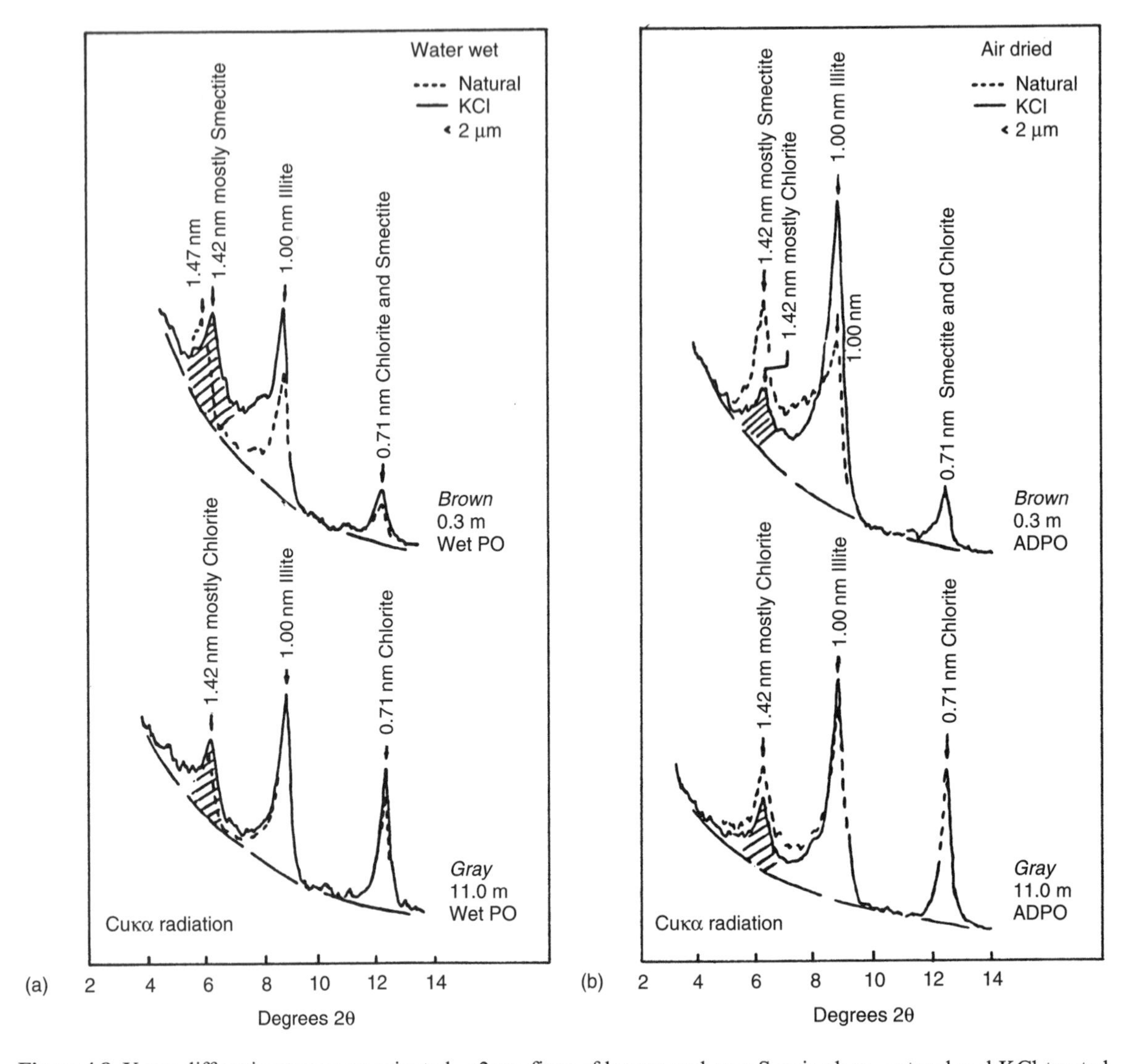

Figure 4.8 X-ray diffraction traces on oriented, <2 μm fines of brown and grey Sarnia clays, natural and KCl-treated.

The sample was compacted for hydraulic conductivity testing at its natural water content of 34.5% and subjected to a static effective stress of 40 kPa. During leachate permeation, the soil creep-consolidated throughout the test period as illustrated in Figure 4.10b. One might expect a steady decrease in hydraulic conductivity as the void ratio decreased but this did not happen (Figure 4.10a). Potassium fixation and double-layer contraction appear to have compensated for the void ratio decrease.

This example shows that if compaction is carried out on a soil that is wet enough (e.g., 2% wet of standard Proctor optimum for the clays discussed here), consolidation in the presence of effective vertical stresses may compensate for deleterious chemical and mineralogical alterations.

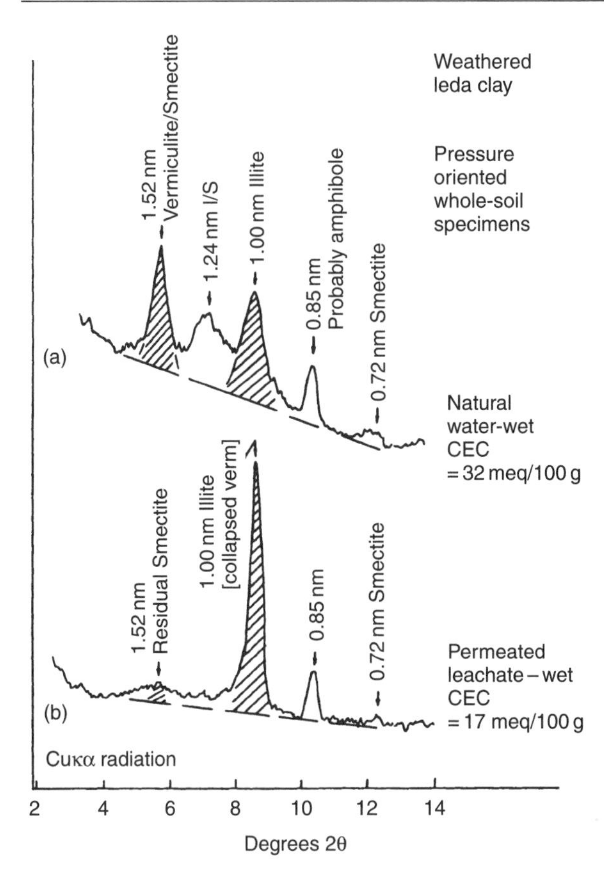

Figure 4.9 X-ray traces obtained on pressure-oriented bulk Leda clay: (a) natural water-wet and (b) permeated with 9 PV of MSW leachate.

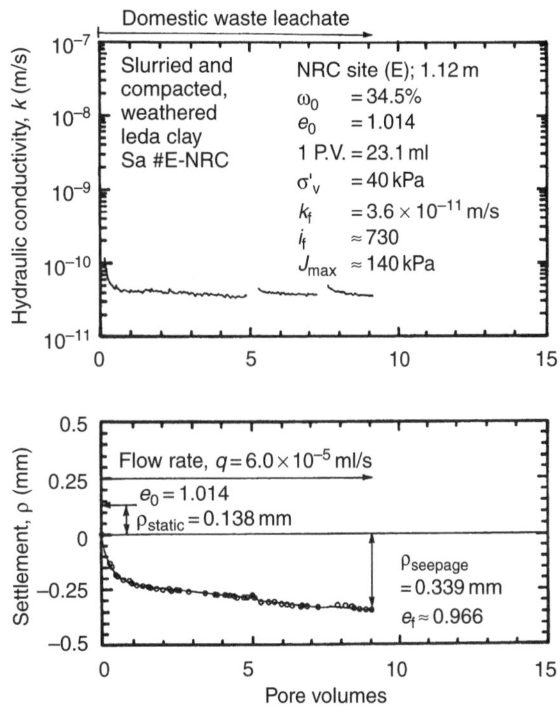

Figure 4.10 Hydraulic conductivity and settlement versus PVs; NRC site (E); 1.12 m depth, slurried and compacted. Permeant: MSW leachate.

4.3 Compatibility of clays with liquid hydrocarbon permeants

The following discussion of hydrocarbon liquids is presented under six major headings in an attempt to organize and rationalize a very complex physicochemical problem. Much has been written on this subject and this section simply seeks to illustrate a number of key points. Selected references are noted in the text.

4.3.1 Dry clay–organic mixtures

The only way to obtain complete clay mineral interaction with many liquid hydrocarbons is to mix dry clay with the test liquid under consideration. In Figure 4.11, a plot of hydraulic conductivity for smectitic Sarnia soil versus void ratio is presented for a variety of liquid organics having dielectric constants varying from ~80 (polar water) to ~25 (ethanol) to ~2 (aromatics). At a void ratio of 0.8, for example, an increase in hydraulic conductivity from 3×10^{-11} m/s for water ($\varepsilon = 80$) to 1×10^{-6} m/s for simple aromatics ($\varepsilon = 2$) is observed.

These trends are well known and were especially well described by Mesri and Olson (1971) on dry clay–organic mixtures subjected to oedometer testing (Figure 4.12). Smectite (CEC = 100) is particularly susceptible to organics demonstrating up to 1,000,000-fold increases in hydraulic conductivity as the dielectric constant of the molding fluid decreases from 80 to 2. Illite (CEC = 25) is much less susceptible

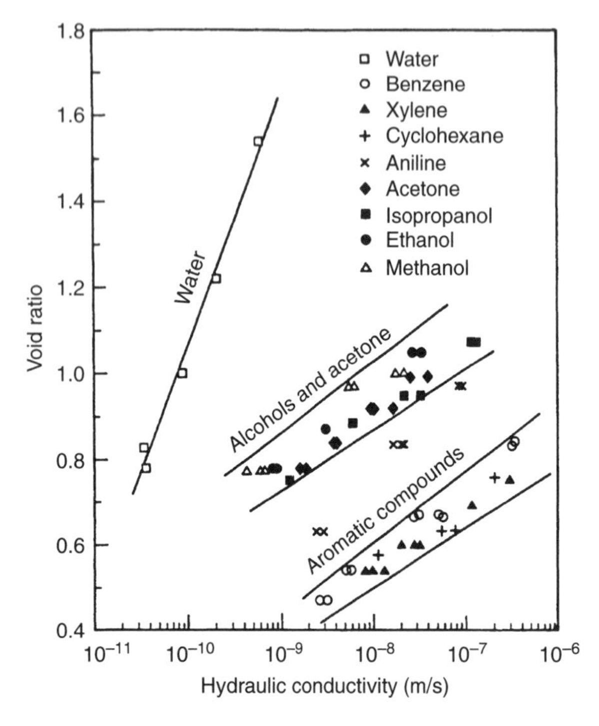

Figure 4.11 Directly measured hydraulic conductivity versus void ratio, all samples molded and permeated with the fluid indicated (modified from Fernandez and Quigley, 1985).

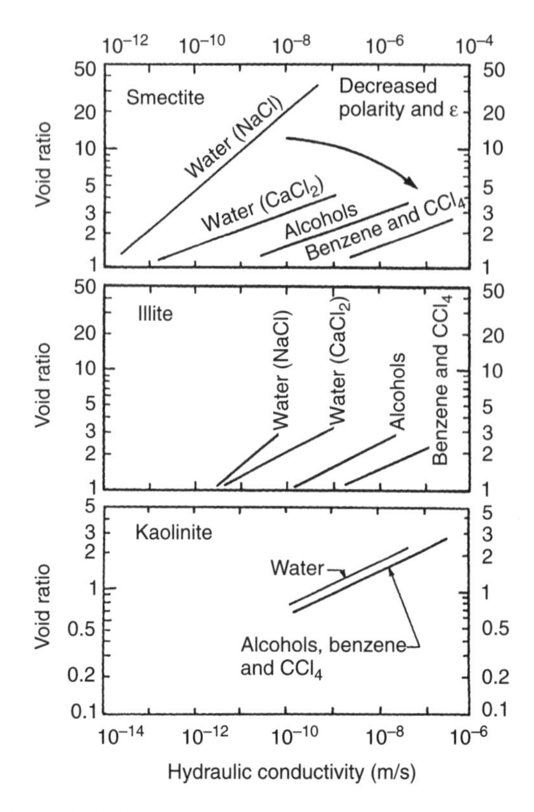

Figure 4.12 Hydraulic conductivity from consolidation tests versus void ratio for smectite, illite and kaolinite (modified from Mesri and Olson, 1971).

demonstrating increases in hydraulic conductivity of up to 1000 times. Finally, kaolin (CEC = 5) seems relatively inert to large changes. Similar work dates back to the 1950s (e.g., Michaels and Lin, 1954). Since dry clays flocculate in the presence of liquid organics, due to their low ε values (see Figure 3.33c), soil fabric or structure is directly responsible for the range in observed hydraulic conductivities. To illustrate this, scanning electron photomicrographs taken of vertical fracture surfaces after freeze-drying are presented in Figure 4.13. Much larger interfloc pores occur in the benzene-molded clay (Figure 4.13d) compared with that observed for ethanol-molded or water-molded clays (Figure 4.13a–c). When visually inspected, the benzene-flocculated clay actually looks like a loose assortment of sand particles.

A further correlation with dielectric constant is presented in Figure 4.14 (Acar and Seals, 1984). Two plots of liquid limit versus dielectric constant of the molding fluid are presented for smectite and kaolinite. Again it appears that smectites are markedly influenced, whereas the kaolins are fairly inert.

4.3.2 Compacted, water-wet clays and insoluble organics

Flow through clayey soils is generally governed by Darcy's law which can be written as:

$$q = kiA \qquad (4.1)$$

which relates the flow q [L^3T^{-1}] through a cross-sectional area of soil A [L^2] to the hydraulic

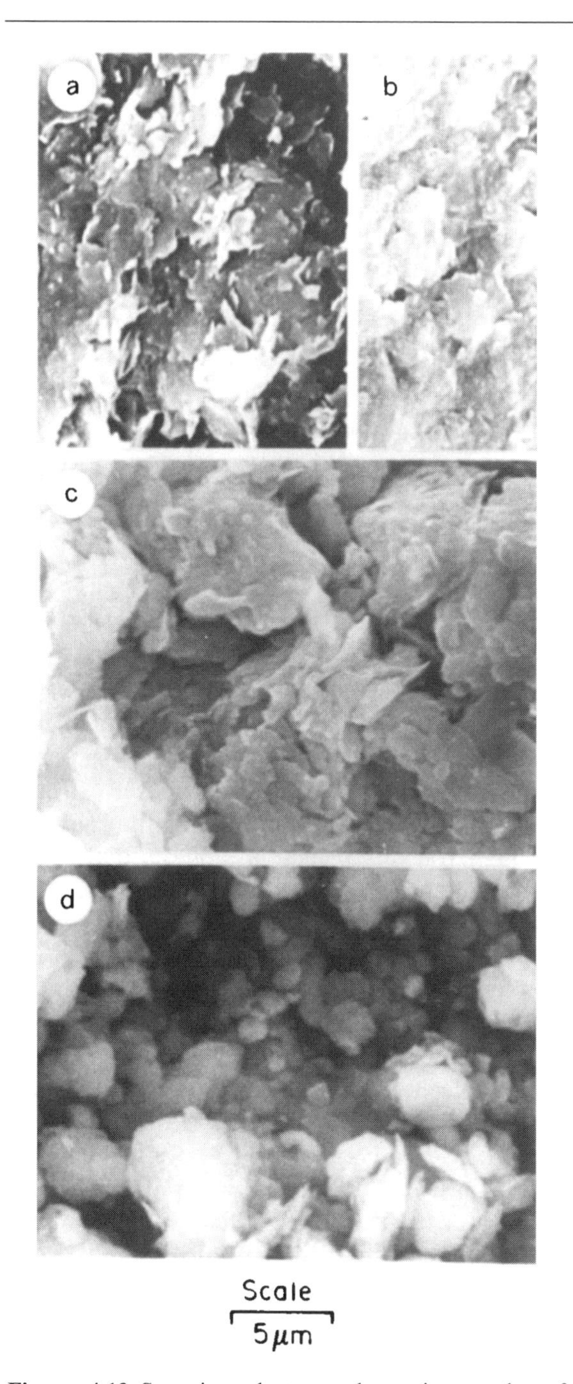

Figure 4.13 Scanning electron photomicrographs of vertical surface test samples: (a) fractured, water-molded clay; (b) cut and smeared water-molded; (c) alcohol-molded; (d) benzene-molded (after Fernandez and Quigley, 1985; reproduced with permission of the *Canadian Geotechnical Journal*).

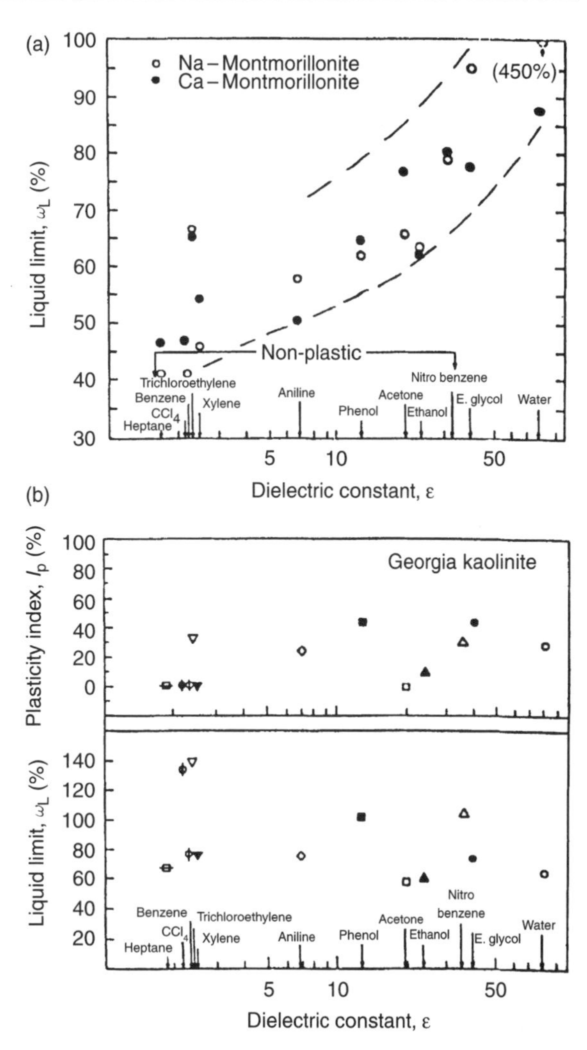

Figure 4.14 Liquid limit versus dielectric constant for a variety of molding fluids: (a) smectite; (b) kaolinite (after Acar and Seals, 1984; reproduced with permission, *Hazardous Waste*).

gradient i [–] and hydraulic conductivity k [LT^{-1}]. A plot of q versus i for water-saturated clays yields a series of straight lines having a slope proportional to the hydraulic conductivity, k, as illustrated in Figure 4.15. Inactive kaolins and relatively inactive illites and chlorites compacted dry-of-optimum with compaction-induced fractures have relatively high k-values and thus yield a high q at low gradients. Illites and chlorites

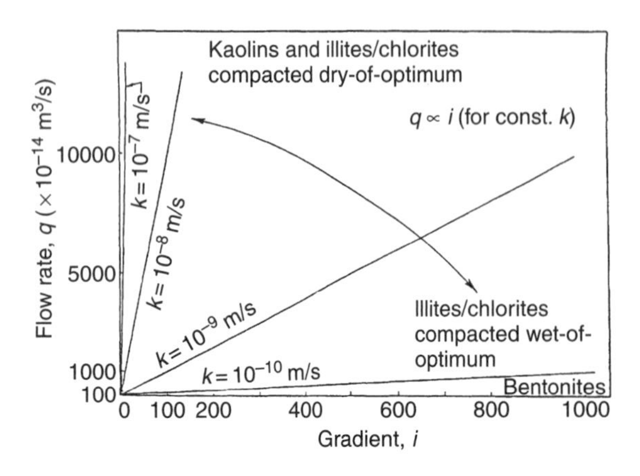

Figure 4.15 Darcy's law plotted as linear q versus i lines with the slope of each line representing a different k-value (unit cross-sectional area A assumed).

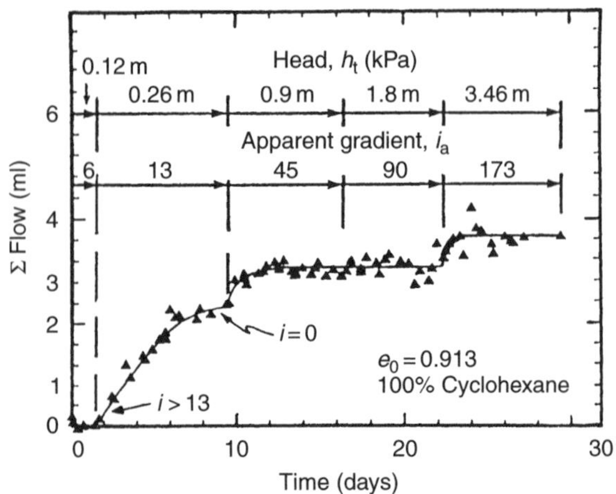

Figure 4.16a Constant head tests for insoluble cyclohexane penetration into water-compacted clay (all data before breakthrough).

compacted wet-of-optimum without the fractures have a much lower k-value and thus yield low q and high gradients. High-quality Na^+ bentonites have such low k-values that low flows are generated even at extraordinary gradients. This is one factor that makes their study very difficult.

The most important aspect of Figure 4.15 with respect to the current discussion is the linearity of the q versus i lines provided no changes in chemistry, void ratio, etc., occur during testing of the water-saturated clays.

If an insoluble lyophobic organic liquid (i.e., a non-aqueous phase liquid or NAPL) is placed on top of a water-wet clay, the interparticle surface tension of the highly structured water creates an extremely strong membrane which resists penetration and flow. For this reason, DNAPLs perch on top of clayey layers in field situations.

If constant flow rate hydraulic conductivity equipment is used to study NAPL flow through water-wet clays, an abrupt experimental nonlinearity of the q versus i plots is produced and this can be used to infer breakthrough pressures and gradients.

To start this discussion, Figure 4.16a and b are presented to illustrate the resistance to lyophobic cyclohexane flow caused by the surface tension of water at the face of an unfractured, water-compacted clay. At a constant head of 0.26 m above a 20-mm thick clay specimen, total penetration amounted to 2.4 ml increasing in two steps to 3.7 ml at 3.46 m head without breakthrough for this soil at a void ratio of 0.913 (Figure 4.16a). Although the initial static effective stress on the clay specimens was zero, the non-penetrating liquid head would operate like an effective stress in this case.

The sketch in Figure 4.16b illustrates schematically this increasing penetration which has almost reached breakthrough. At this point, the penetration amounts to ~17% of the total PV of 21.9 ml in this compacted fine-grained sample. The presence of flaws such as cracks and fractures in a soil would probably enable much lower breakthrough values.

Because of the lyophobic behaviour of cyclohexane and water, the constant flow rate hydraulic conductivity test system can also be used to measure approximate values of breakthrough pressure or gradient. This is illustrated in Figure 4.17 which combines both constant head and constant flow tests to illustrate the breakthrough point. Breakthrough is inferred

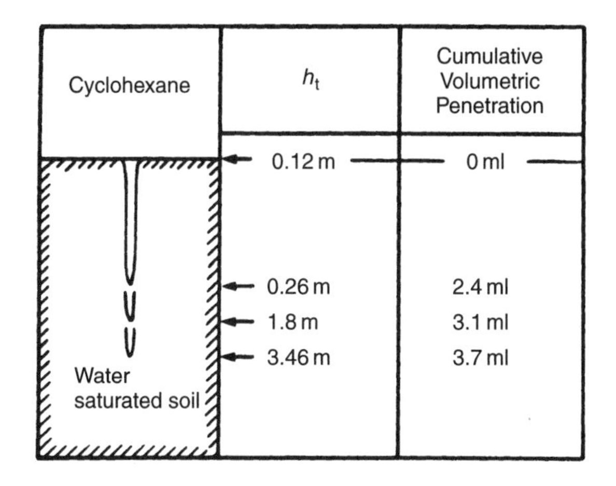

Figure 4.16b Schematic diagram illustrating volumetric penetration of cyclohexane without breakthrough at three values of constant total head (1 PV = 21.9 ml).

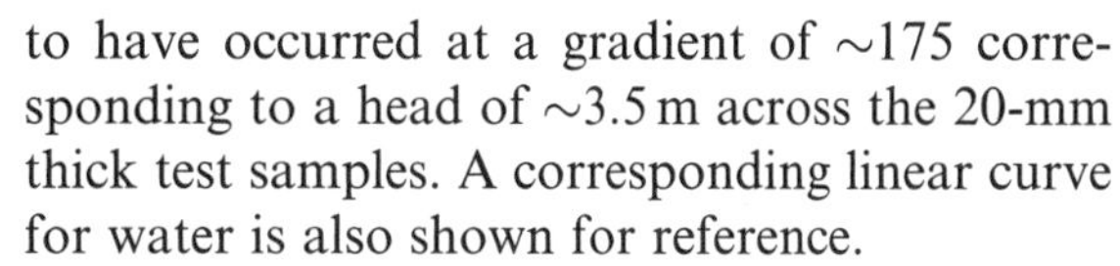

to have occurred at a gradient of ~175 corresponding to a head of ~3.5 m across the 20-mm thick test samples. A corresponding linear curve for water is also shown for reference.

Once breakthrough occurs, the insoluble cyclohexane is forced along macropore channels or microfissures which expand and contract (in this relatively loose sample) in proportion to the head. This causes the measured hydraulic conductivity to be gradient-dependent as shown by the curved plot of flow rate q versus hydraulic gradient. A corresponding linear curve of q versus i for water is also shown for reference.

Provided the soil has no fractures, the breakthrough head or gradient should increase with decreasing void ratio. This is demonstrated in Figure 4.18 for cyclohexane at two lower soil

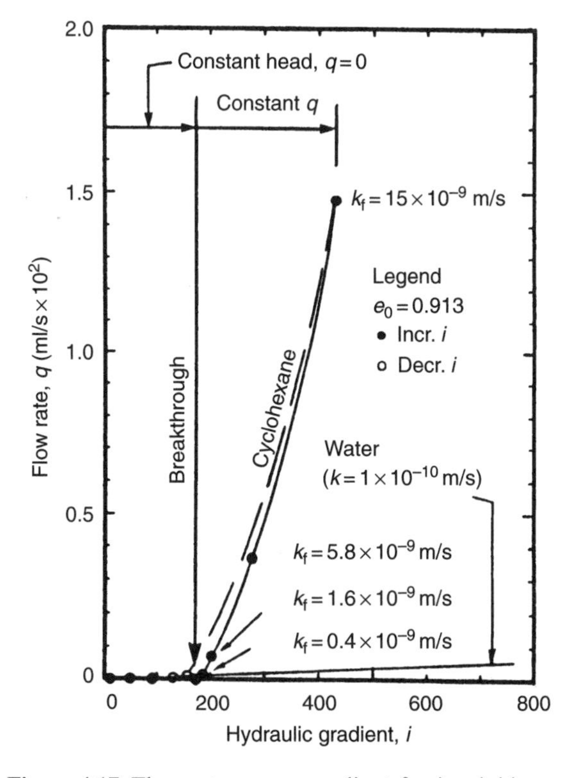

Figure 4.17 Flow rate versus gradient for insoluble cyclohexane flow through water-compacted brown silty clay. (Note that five constant head and four constant flow rate values are plotted.)

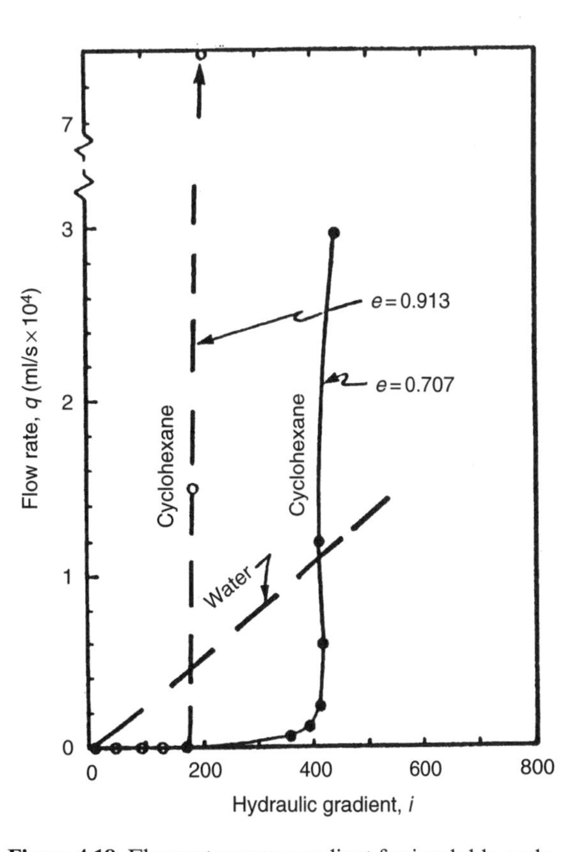

Figure 4.18 Flow rate versus gradient for insoluble cyclohexane flow through water-compacted silty clay. (Note increase in breakthrough gradient at lower void ratio.)

void ratios of 0.707 and 0.580. In these lower void ratio tests, the breakthrough gradients were ~410 and ~920, respectively.

A three-point summary plot of breakthrough gradient versus void ratio is presented in Figure 4.19. The breakthrough gradient (based on constant flow rate tests) increases with decreasing void ratio on a hyperbolic curve that plots as a straight line on a semi-log plot. Since the capillary strength of a water film increases with decreasing pore size, it is inferred that this is a major factor in defining breakthrough heads or gradients in the case of lyophobic liquids.

Finally, it is noted that cyclohexane and other insoluble liquid hydrocarbons rarely occupy more than 6–10% of the soil pore space at the end of testing, a factor indicating macropore or channel flow. Also, previous studies and reviews by Acar *et al.* (1985), Fernandez and Quigley (1985), Foreman and Daniel (1986) Mitchell and Madsen (1987), and all indicate that forced permeation of insoluble, lyophobic organics does not increase hydraulic conductivity beyond that of water. This is illustrated in Figure 4.20 where almost 2 PV of cyclohexane flow left only 7% cyclohexane in the pore fluid and yielded the same hydraulic conductivity as did water.

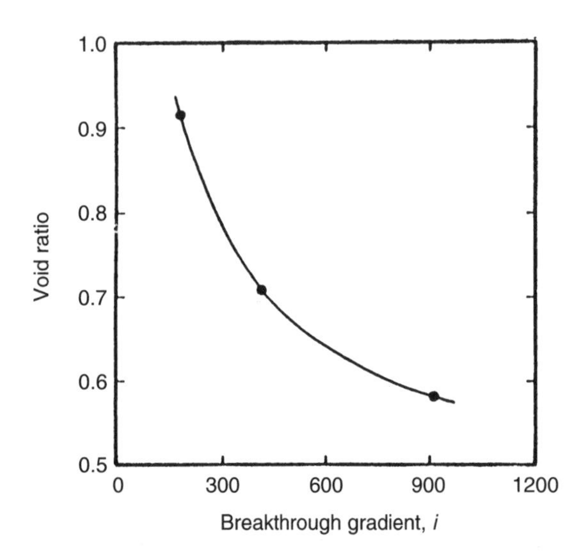

Figure 4.19 Breakthrough gradient versus void ratio.

In Figure 4.17, however, it was clearly shown that the constant flow rate system used measures an apparent hydraulic conductivity which may be highly gradient or head-dependent. If enough different flow rates are employed to establish gradients both below and above the breakthrough gradient, the resulting curves may be extrapolated back to a zero flow condition thereby identifying a gradient which is believed to be close to the breakthrough gradient measured by constant head testing.

Quigley and Fernandez (1992) have shown that an increasing static effective stress on a soil also caused a large increase in breakthrough gradient, possibly related to the greater difficulty in expanding the flow channels associated with fingering once the water meniscus is penetrated.

The principles just described have been known for a long time and were effectively used by Lambe (1956) to store oil in a water-compacted clayey reservoir in Venezuela.

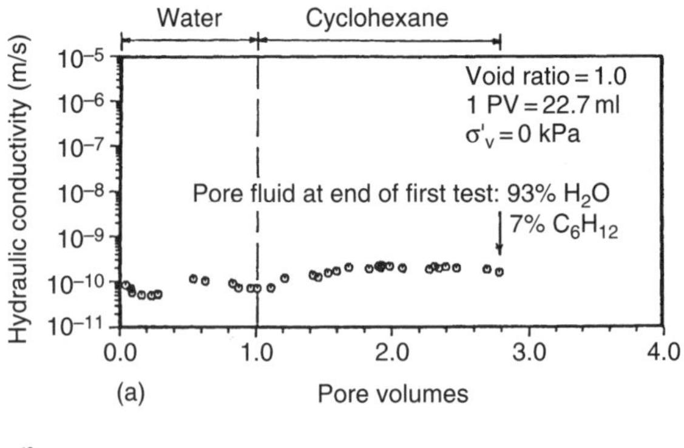

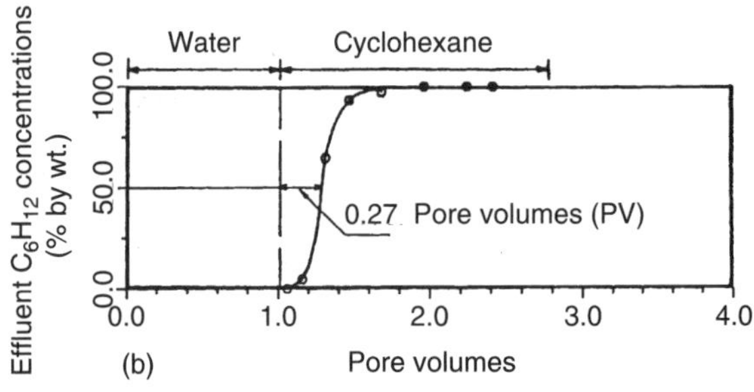

Figure 4.20 Permeation of cyclohexane through water-saturated clayey soil: (a) hydraulic conductivity; (b) relative concentration versus PVs.

4.3.3 The role of surfactants

A surfactant is a surface-active agent which reduces the interfacial surface tension of liquids and should greatly reduce the strength of the water menisci of a water-compacted clay in contact with insoluble liquid hydrocarbons.

The presence of 5% surfactants (such as soap) in the cyclohexane permeant discussed in the previous section has the important effect illustrated in Figure 4.21. The constant head testing conditions are exactly the same as those used to produce Figure 4.16a for pure cyclohexane (replotted in Figure 4.21 as a solid curve). Surface tension in water at the cyclohexane–soil interface was apparently completely destroyed and there appears to have been little resistance to cyclohexane penetration and flow even at low gradients. Also, the cumulative flow versus time plot is only slightly non-linear.

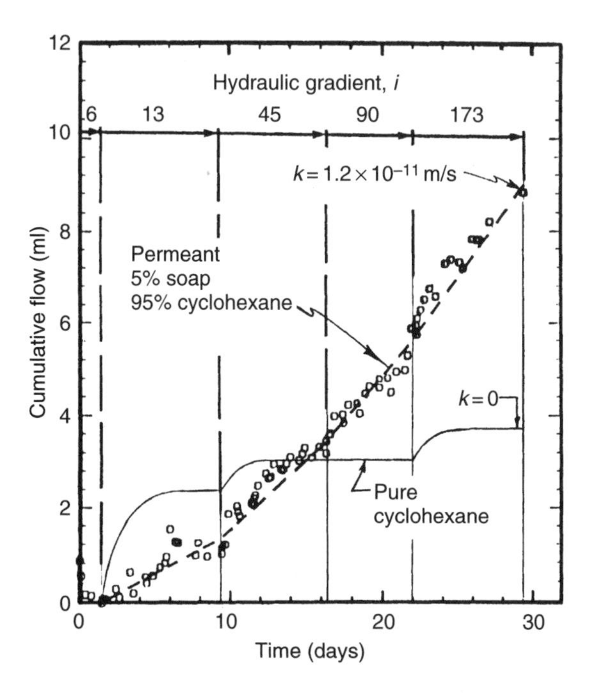

Figure 4.21 Constant head tests for insoluble cyclohexane penetration into water-compacted clay in the presence of 5% surfactant (soap) (after Quigley and Fernandez, 1989; reproduced with permission of Swets & Zeitlinger).

Comparison of the cumulative flow at 29 days at a gradient $i = 173$ shows non-breakthrough penetration of 3.7 ml for pure cyclohexane compared with a total flow volume of ~9 ml for the 5% soap–95% cyclohexane mixture. It is important to note that a hydraulic conductivity of zero ($k = 0$) was obtained for the pure cyclohexane system since there was no continuing flow, whereas a hydraulic conductivity $k = 1.2 \times 10^{-11}$ m/s was obtained for the cyclohexane/surfactant system in which there was slow continuous flow through the sample. Again, since many types of surfactants are probably present in domestic and industrial leachates, it is suggested that one should not rely on the high breakthrough pressures of a good clay barrier to retain insoluble hydrocarbon liquids.

4.3.4 Water-soluble ethanol and dioxane mixed with MSW leachate

Mixing water-soluble liquid hydrocarbons with water or MSW leachate produces some important changes in their physical and electrochemical properties as shown in Figure 4.22. Both ethanol and dioxane, when mixed with water, produce 2–3-fold increases in kinematic viscosity at concentrations of 50–60% (Figure 4.22a). The same two organics cause the dielectric constant of the mixture to decrease as their percentage increases (Figure 4.22b).

The effect of these two factors on the hydraulic conductivity, k, of a water-compacted clay is shown in Figure 4.23a for ethanol–leachate mixtures used as permeant. At all concentrations up to ~70% ethanol in MSW leachate, the hydraulic conductivity decreases in a manner which mirrors the increases in viscosity shown in Figure 4.22a. Above 70%, ethanol apparently enters the double layers around the clay particles in sufficient quantity to cause double-layer contraction in accordance with Gouy–Chapman theory (see Figure 3.33c). Since the tests shown in Figure 4.23a are for zero static stresses (only

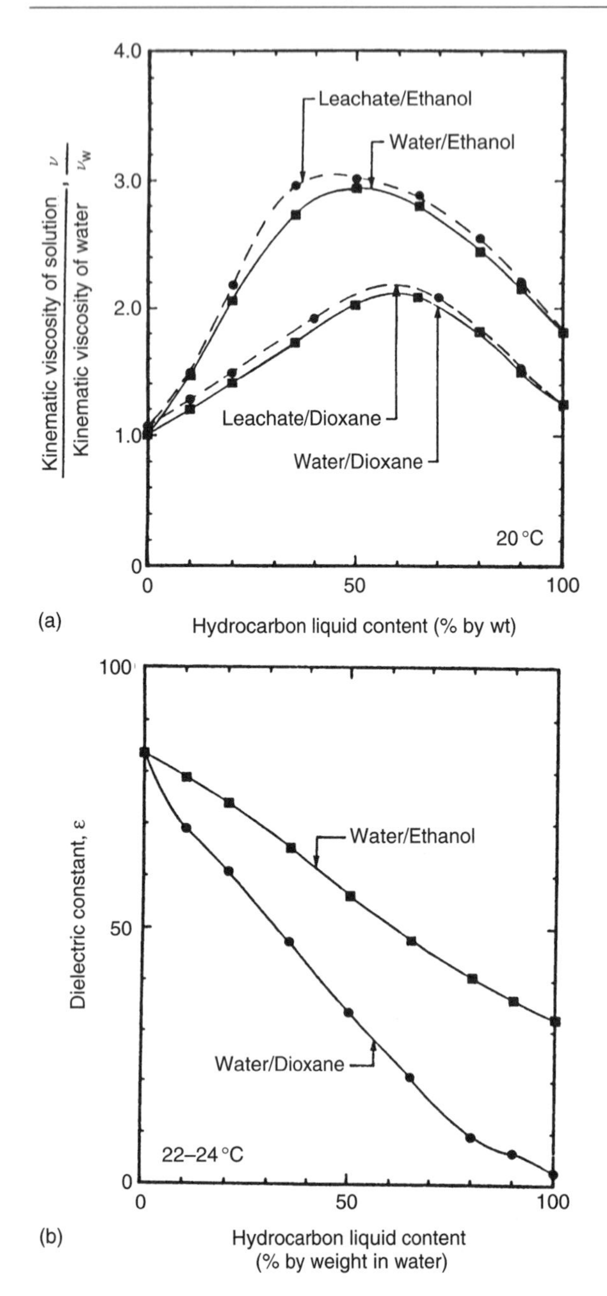

Figure 4.22 Properties of aqueous test mixtures of water-soluble liquid hydrocarbons: (a) relative kinematic viscosity; (b) dielectric constant (modified from Fernandez and Quigley, 1988a).

seepage stresses influence the sample) large increases in free pore space develop at a constant void ratio (see Figure 3.31) resulting in 100-fold increases in hydraulic conductivity, k. This work was discussed in considerable detail by Fernandez and Quigley (1988a).

The intrinsic permeability, K, is defined by

$$K = \frac{k\mu}{\gamma} \tag{4.2}$$

where K is the intrinsic permeability [L^2], k the hydraulic conductivity [LT^{-1}], μ the dynamic viscosity [$ML^{-1}T^{-1}$] and γ the unit weight of permeant [$ML^{-2}T^{-2}$].

Figure 4.23b shows a plot of intrinsic permeability versus the percentage of ethanol. By considering the intrinsic permeability, one removes the complicating effects of viscosity and density. This plot shows nearly constant K at $\sim 6 \times 10^{-18}\,m^2$ up to about 50% ethanol at which point K increases to about 100-fold for 100% ethanol. This suggests that ethanol does not significantly access the double layers at concentrations below 50 or 60%. This is to be expected since the double-layer cations are hydrated with more polar water ($\varepsilon = 80$) and should reject the less polar ethanol ($\varepsilon = 32$).

Similar plots are presented in Figure 4.24 for water-soluble dioxane which has a dielectric constant, ε, of only 2.3. Dioxane is non-polar yet completely soluble in water owing to the presence of two oxygens replacing carbon in the benzene ring (see Figure 4.24a). The oxygen apparently associates readily with the H-O system of polar water. The plots for k and K are very similar to those for ethanol. Decreases in k for less than 75% dioxane relate to the viscosity increases and exclusion of dioxane from the double layers. Once the concentration reaches 75%, however, high increases are observed in both k and K as the double layers contract.

The data shown in Figures 4.23 and 4.24 were obtained on samples with a static applied vertical effective stress of zero ($\sigma'_v \approx 0\,kPa$). Two important data points for a static effective stress

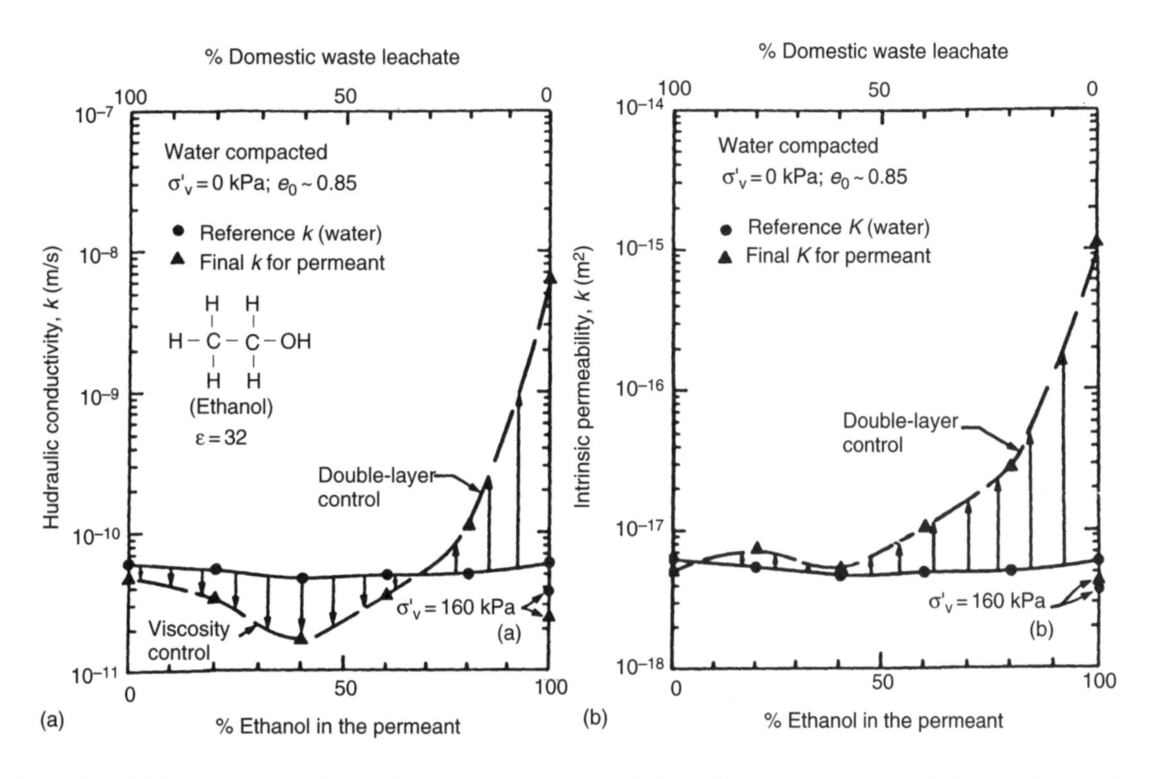

Figure 4.23 Hydraulic conductivity (a) and intrinsic permeability (b) of water-compacted brown Sarnia clay permeated with leachate/ethanol mixtures (modified from Fernandez and Quigley, 1988a).

of 160 kPa applied to the samples prior to pure hydrocarbon permeation are also shown. These two points indicate a decrease in hydraulic conductivity k for ethanol and only a slight increase for dioxane. Chemically induced consolidation in the presence of stresses was shown to greatly inhibit the large increases in k observed at low stress levels (Fernandez and Quigley, 1988a,b).

By way of summary, the intrinsic permeability ratio K_f/K_w is plotted in Figure 4.25 for water-compacted brown smectitic Sarnia clay permeated with mixtures of both ethanol-leachate and dioxane-leachate. Since the effects of viscosity and density have been removed, the values of K_f/K_w reflect the effects of dielectric constant ε on the double layers and hence on both intrinsic permeability K and hydraulic conductivity k.

For dioxane concentrations greater than 70%, intrinsic permeability K increases rapidly reaching K values 1,000 times greater than water for pure dioxane. Remarkably, no effects on K are observed below 70%. Although dioxane is completely soluble in water, because of oxygen replacing two carbons in the benzene ring, it is also non-polar. This means that cations present in the double layers around the clay particles must retain their affinity for polar water and reject dioxane, preventing double-layer collapse until concentrations in the range of 70% are reached.

In the case of ethanol, which is more polar than dioxane, the trends are somewhat different. Small increases in K are apparent at concentrations possibly as low as 25%, suggesting partial entry into the double layers.

Chemical confirmation of these hypotheses is quite difficult since the double layers occupy only a small portion of the pore space in these high void ratio soils. Also, salt dumping

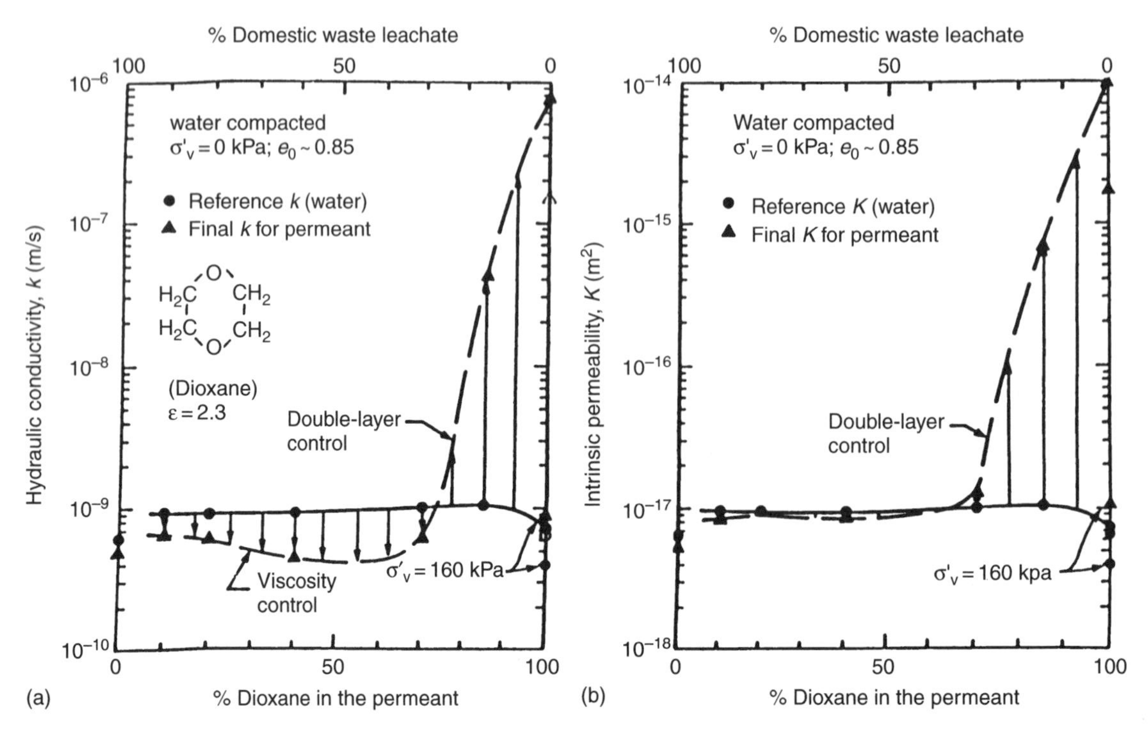

Figure 4.24 Hydraulic conductivity (a) and intrinsic permeability (b) of water-compacted brown Sarnia clay permeated with leachate/dioxane mixtures (modified from Fernandez and Quigley, 1988a).

(precipitation) by addition of these organics may also create large increases in salinity or free salt crystals.

4.3.5 The role of effective stress

One of the most significant conclusions of the review of hydraulic conductivity by Mitchell and Madsen (1987) was: "Permeation with pure hydrocarbon leads to a decrease in hydraulic conductivity when flexible wall permeameters are used." The reason for this is that the triaxial cell pressures consolidate the test specimens if any chemical reactions occur, effectively masking any increases in hydraulic conductivity k. For this reason, the above authors recommended the use of "consolidometer permeameters".

In Figures 4.23–4.25, data points were plotted for hydraulic conductivity tests run with pure ethanol and dioxane at an applied effective stress $\sigma'_v = 160\,\text{kPa}$ to compare with those run at $\sigma'_v = 0\,\text{kPa}$. In each plot, the presence of the stresses eliminated the increases in k-value for ethanol and reduced them to small increases for dioxane.

The effective stress work of Fernandez and Quigley (1991) is presented in Figures 4.26 and 4.27 to illustrate the importance of stress level during k-testing. For ethanol (Figure 4.26a), each water-compacted specimen was consolidated to the appropriate stress level and the hydraulic conductivity, k, was measured for water (solid circles on Curve 1). Pure ethanol was then passed through each sample and k measured at equilibrium (open circles on Curve 2). At $\sigma'_v = 0$, hydraulic conductivity k increased by up to 100 times as noted in Figure 4.23, whereas at $\sigma'_v = 20\,\text{kPa}$ there was no change and at all higher stresses up to 160 kPa, k actually decreased.

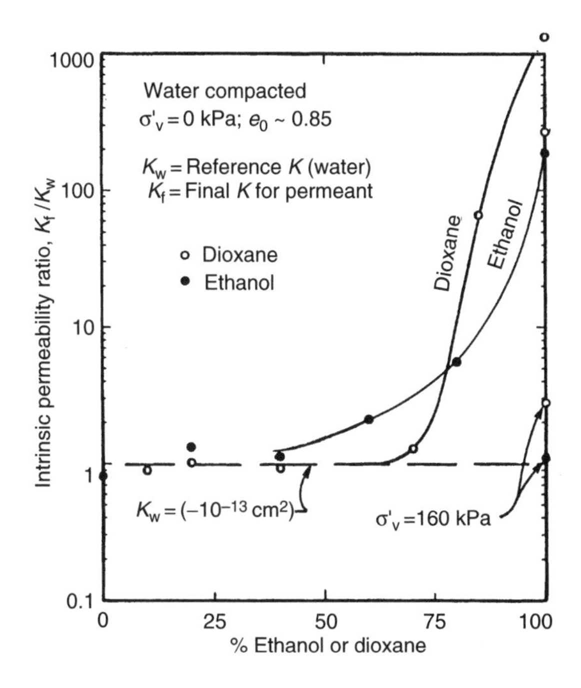

Figure 4.25 Ratio of final to initial values of intrinsic permeability, K, for water-compacted clay permeated with mixtures of ethanol/leachate and dioxane/leachate (after Quigley and Fernandez, 1989; reproduced with permission of Swets and Zeitlinger).

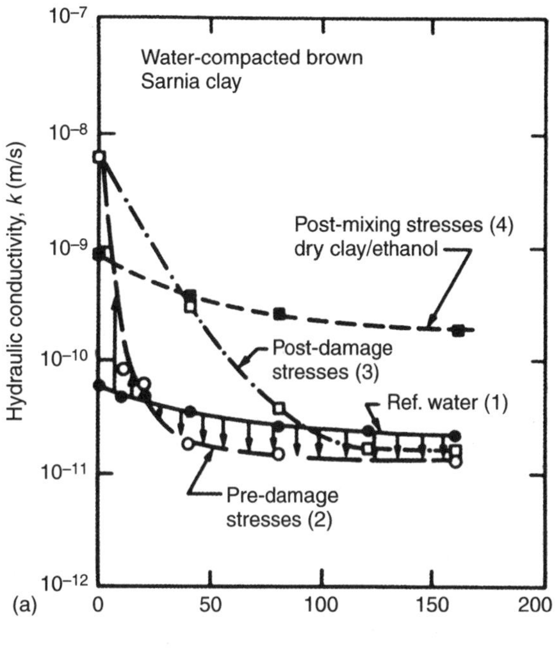

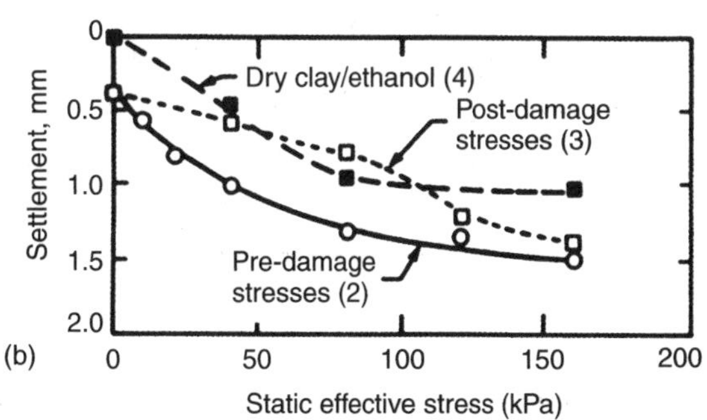

Figure 4.26 Hydraulic conductivity and settlement for pure ethanol permeation: (a) k versus stress level for four stress scenarios described in the text; (b) settlement versus stress level (modified from Fernandez and Quigley, 1991).

The third curve was obtained from tests on water-wet soil damaged by pure ethanol permeation at $\sigma'_v = 0$ kPa (open squares). Stress was then applied in an attempt to heal the damaged barrier. The figure shows that up to 100 kPa of static vertical stress is required, but that healing can be effected in the case of pure ethanol damage.

The fourth curve in Figure 4.26 (solid squares) represents soil molded and compacted with ethanol, then permeated with ethanol at the indicated stress level. These samples were flocculated as discussed in Section 4.3.1, and the strength and structure derived from this pretreatment apparently resisted efficient soil consolidation and recovery of low k-values. In spite of the settlement shown in Figure 4.26b, the flow channels apparently remained open.

For dioxane (Figure 4.27) the trends are quite different. Permeation by pure dioxane resulted in large increases in hydraulic conductivity, k, that seemed inversely proportional to the stress level on the water-compacted sample, and even at 160 kPa there was a small increase (open circles on Curve 2).

Curve #3 in Figure 4.27 represents post-damage application of stress and shows only minor healing even at 160 kPa. This, in spite of significant consolidation which apparently, was not accompanied by flow channel closure.

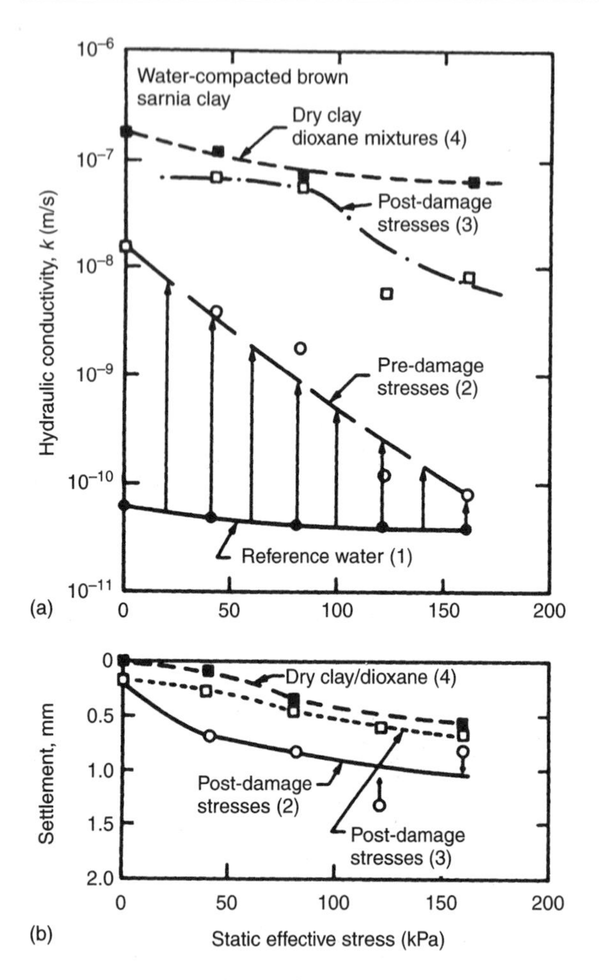

Figure 4.27 Hydraulic conductivity and settlement for pure dioxane permeation: (a) k versus stress level for four stress scenarios described in the text; (b) settlement versus stress level (modified from Fernandez and Quigley, 1991).

The fourth curve in this figure represents dioxane-moulded and compacted specimens which were then consolidated at the σ'_v shown and subsequently permeated with dioxane. Again the highly flocculated structure of this pretreated soil enabled it to resist consolidation and closure of the flow channels so the k-values remained very high even at $\sigma'_v = 160\,\text{kPa}$.

On the basis of this work, it is suggested that water-compacted clayey barriers which are heavily stressed before a spill of water-soluble organics may self-heal during exposure by consolidation. If the spill occurs on an unloaded or unstressed clay barrier, however, large increases in hydraulic conductivity k may occur with little chance of subsequent healing by stress application.

4.3.6 The role of association liquids

An association liquid (for the purposes of this book) is one that is mutually soluble in two immiscible lyophobic liquids and which if added produces a single-phase liquid over a certain concentration range. In complex mixtures of waste containing hundreds of compounds, it is probable that such liquids and processes exist.

The effects of these complex mixtures on the integrity of clay barriers have received little study. A single, complex plot is presented in Figure 4.28 to illustrate the probable effects. In this figure ethanol is used as the association liquid and cyclohexane as the insoluble LNAPL (light non-aqueous phase liquid).

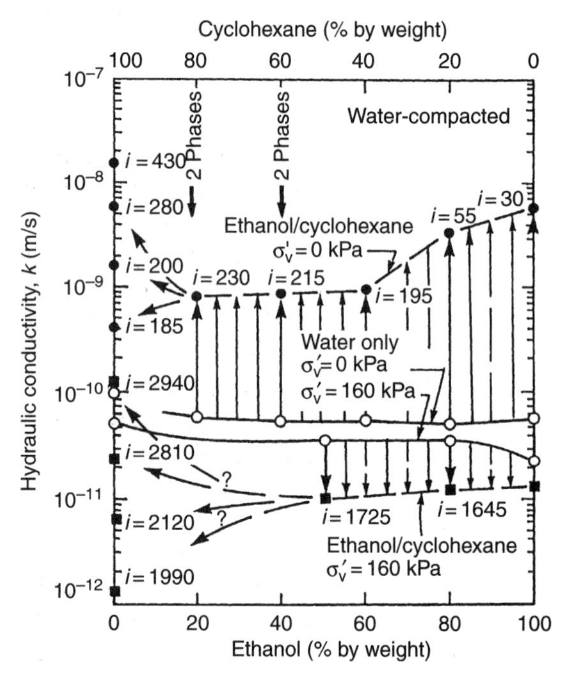

Figure 4.28 Influence of mixed organics on the hydraulic conductivity of water-compacted clay at $\sigma'_v = 0$ and 160 kPa (modified from Quigley and Fernandez, 1991).

Considering zero effective stresses first: if pure cyclohexane is passed through the sample, the hydraulic conductivity, k-value, is close to that of water but is gradient-dependent as illustrated by the four solid circles on the 100% cyclohexane line. If mixtures of alcohol and cyclohexane are passed through, k increases by a factor of 10–100 depending on the ratio of cyclohexane to ethanol. In all cases the hydraulic conductivity k was greater than for water. This feature is quite opposite to the trends for alcohol–MSW leachate mixtures shown in Figure 4.23 until ethanol concentrations exceed 60–70%.

Considering $\sigma'_v = 160$ kPa: at 160 kPa static vertical effective stress, all samples yielded k-values below those for water except the pure cyclohexane at one high gradient of 2,940 kPa. Even at the slowest flow rate, the low hydraulic conductivity of the clay ($\sim 10^{-11}$ m/s) gives rise to high values of gradient i from 1,645 to 2,810.

In summary, the curves in Figure 4.28 indicate that the application of an effective stress, $\sigma'_v = 160$ kPa prevents the increases in hydraulic conductivity k observed at zero effective stress $\sigma'_v = 0$ kPa for cyclohexane–ethanol mixtures. The gradients indicated next to each point are generally highly variable and often very high. In addition, two of the mixtures split into two phases making interpretation difficult. Figure 4.28, therefore, well illustrates the complex problems associated with the study of mixed organic liquids containing association liquids and surfactants.

4.4 Summary

Engineered clayey liners are often used as barriers for municipal waste leachate. Based on the discussion presented in this chapter, the following concluding comments may be made:

1. MSW leachate will normally cause a slight decrease in the hydraulic conductivity of inactive clays with divalent cations on the exchange site.
2. High concentrations of K^+ and NH_4^+ cause c-axis contraction of high-charge smectites and vermiculites resulting in decreases in CEC, double-layer contraction and increases in the free pore space at constant void ratio. The clay mineralogy should carefully identify the clay types and their collapse characteristics.
3. The amount of vermiculite/smectite necessary to cause significant damage remains to be established especially in relation to the stress level on the barrier clays.
4. Three field studies of hydraulic conductivity below landfills are presented in Chapter 9.

The effects of liquid hydrocarbons on the integrity of clayey liners range from negligible to large. Although there is the need for further research, the following conclusions may be drawn from the work presented herein.

1. Water-soluble liquid hydrocarbons may cause large increases in hydraulic conductivity at concentrations above 70% in low-stress environments. In higher stress environments, the increases may be largely eliminated by chemical consolidation. Contraction of the double layers related to low values of dielectric constant appears to be the cause.
2. Water-soluble liquid hydrocarbons present at concentrations below 70% often induce decreases in hydraulic conductivity due to increases in viscosity of the aqueous mixtures.
3. Water-compacted clayey barriers are resistant to penetration by insoluble organic liquids due to surface tension effects. Even after breakthrough, the hydraulic conductivity, k, remains low due to lyophobicity effects and flow restricted to macropores. The presence of fractures, however, creates higher velocity flow channels and much higher hydraulic conductivity values.
4. Surfactants and mutually soluble liquid organics destroy surface tension effects permitting easy entry of the insolubles suggesting that little reliance should be placed on the presence of breakthrough pressures in the field.

5. Dry clay–organic mixtures are a useful guide to behaviour if carefully interpreted with respect to soil fabric and pore size. The latter play an adverse role not necessarily encountered in a water-compacted clay liner system.
6. The presence of high effective stresses on a water-wet clayey liner is beneficial since consolidation occurs and compensates for chemical damage. Since post-damage application of stresses is much less efficient in healing a barrier, it is suggested that a suitable HDPE geomembrane on top of a clay barrier would represent a good design. The geomembrane would protect the clay against organic contact long enough for establishment of high effective stresses on the clayey portion of the composite system.
7. Hydration of a dry clay using uncontaminated water results in a substantial improvement in performance compared to clays that are brought into contact with contaminant (e.g., organic chemicals) before full hydration by uncontaminated water.

Chapter five

Flow modelling

5.1 Introduction

The design of barrier systems typically requires some associated contaminant impact assessment. As discussed in Chapters 1 and 2, this should involve consideration of both advective and diffusive transports. The calculation of the advective components requires that an assessment be made of the flow in the barrier system. In some cases, this may be isolated from the general groundwater system (e.g., in an arid region where there is a thick unsaturated zone above the aquifer and negligible on-site recharge to the aquifer either before or after landfill construction). However, in most cases, the construction of a landfill will involve some interaction between the engineered system and the local hydrogeology. Thus the objective of this chapter is to summarize a number of approaches for calculating flows in barrier systems.

In Section 5.2, we consider 1D flow systems where simple hand calculations can be used to estimate flows. These simple calculations are then used to provide background for the interpretation of field data.

In some cases, the flow system may be too complicated to use 1D hand calculations. For these situations, it is usual to use an analytical model (e.g., Rowe and Nadarajah, 1996a) or perform numerical (finite element) analyses. Section 5.3 provides the theoretical basis for 2D and 3D flow modelling. Sections 5.4, 5.5 and 5.6 provide background regarding finite difference, finite element and boundary element techniques, respectively, for solving the governing differential equations (given in Section 5.3) subject to appropriate boundary conditions. Finally Section 5.7 examines the advective flow through holes in either a geomembrane alone or a geomembrane that is part of a composite liner system and hence is underlain by a clay liner (e.g., compacted clay as discussed in Chapter 3 or geosynthetic clay as discussed in Chapter 12).

5.2 One-dimensional flow models

5.2.1 Basic concepts

Flow modelling is based on Darcy's law:

$$v_a = ki = -k\frac{d\phi}{dz} \tag{5.1a}$$

$$q = v_a A = kiA \tag{5.1b}$$

$$q = ki \quad \text{for } A = 1 \tag{5.1c}$$

where v_a is the Darcy velocity (flux) [LT^{-1}], k the hydraulic conductivity at the location where flow is being evaluated [LT^{-1}], $i = -d\phi/dz$ the hydraulic gradient at the same location [–], q the flow at the location of interest [L^3T^{-1}] and, when $A = 1$, q is flow per unit area, and A the cross-sectional area through which flow is occurring at the location of interest [L^2].

Hydraulic gradient [–] may be defined as a change in (total) head $\Delta\phi$ [L] over a given distance, H [L].

$$i = \frac{\Delta\phi}{H} \tag{5.1d}$$

Total head is the sum of two significant components – the pressure head h_p and the position head z. To illustrate this, consider the situation shown in Figure 5.1 where a 0.3-m deep pond of water is above a 1.2-m thick clayey liner of

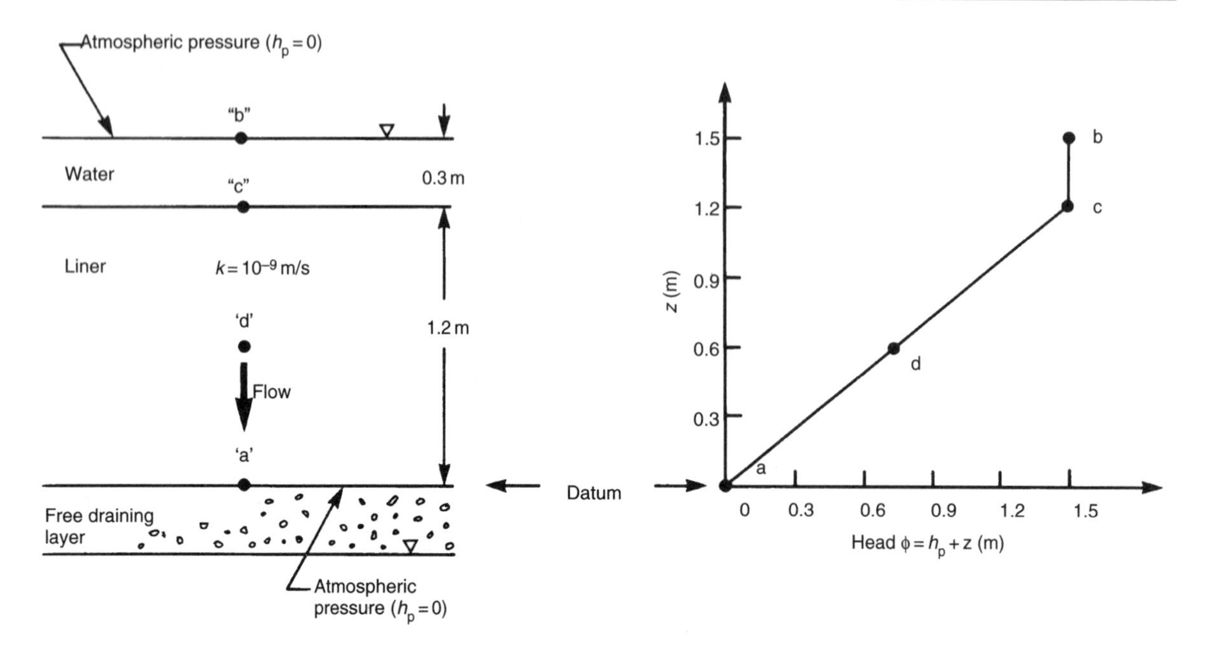

Figure 5.1 Simple flow model – head distribution through a liner from a pond when the liner is underlain by a free draining layer.

low-permeability soil. For simplicity, it is assumed that the clayey liner is underlain by a free draining layer and that atmospheric pressure exists at the bottom of this liner. This layer might correspond to a coarse gravel secondary leachate collection system which is kept pumped, as shown in Figures 1.5 and 1.12.

Total head is always measured relative to some datum. Since flow only depends on the change in total head (over a given distance), the choice of datum is not important. For purposes of illustration, in Figure 5.1, the datum is taken to be at the bottom of the liner. If the position head is defined as the vertical distance, z, above the datum, then the total head ϕ at any point of interest is given by

$$\phi = h_p + z \tag{5.2}$$

where ϕ is the total head [L], h_p the pressure head [L] ($= u/\gamma_w$, where u is the pore water pressure [$ML^{-1}T^{-2}$] and γ_w the unit weight of water [$ML^{-2}T^{-2}$]); and z the position head [L].

For example, in Figure 5.1, the total head at point "a" at the bottom of the liner is equal to the pressure head which is zero for atmospheric pressure ($h_p = 0$) plus the position head (which is also zero since the datum is at point "a")

$$\therefore \phi_a(\text{at "a"}) = h_p + z = 0 + 0 = 0 \tag{5.3a}$$

At point "b", the pressure head $h_p = 0$ (atmospheric conditions above the water) and the position head is 1.5 m.

$$\therefore \phi_b(\text{at "b"}) = h_p + z = 0 + 1.5 = 1.5\,\text{m} \tag{5.3b}$$

At the top of the liner (point "c"), the pressure head h_p is equal to the height of water above this point, $h_p = u/\gamma_w = 0.3$ m and the pressure head is $z = 1.2$ m.

$$\therefore \phi_c(\text{at "c"}) = 0.3 + 1.2 = 1.5\,\text{m} \tag{5.3c}$$

The gradient across the clayey liner is then given by

$$i = \frac{\Delta\phi}{H} = \frac{\phi_c - \phi_a}{H} = \frac{1.5 - 0}{1.2} = 1.25 \qquad (5.4)$$

and the flow across the liner (per unit area) is

$$q = ki = 1\times10^{-9}\times1.25 = 1.25\times10^{-9}\,\text{m/s} = 0.0394\,\text{m/a} \qquad (5.5)$$

Consider a point "d" at the middle of the liner. The flow between points "c" and "d" is

$$q_1 = ki = k\left(\frac{\phi_c - \phi_d}{0.5H}\right) \qquad (5.6a)$$

Similarly, the flow between points "d" and "a" is

$$q_2 = ki = k\left(\frac{\phi_d - \phi_a}{0.5H}\right) \qquad (5.6b)$$

Now for 1D flow conditions, the flow between points "c" and "d" must be the same as the flow between points "d" and "a", since there is nowhere else for the water to go for 1D flow conditions.

Thus, continuity of flow requires that

$$q_1 = q_2$$

and from equations 5.6a and 5.6b,

$$k\left(\frac{\phi_c - \phi_d}{0.5H}\right) = k\left(\frac{\phi_d - \phi_a}{0.5H}\right) \qquad (5.6c)$$

For a uniform layer the hydraulic conductivity k is a constant throughout and it follows (from equation 5.6c) that

$$\phi_c - \phi_d = \phi_d - \phi_a \qquad (5.6d)$$

hence

$$\phi_d = \frac{\phi_a + \phi_c}{2} = \frac{0 + 1.5}{2} = 0.75 \qquad (5.6e)$$

This implies that the change in total head is linear across a uniform layer (as shown graphically in Figure 5.1).

5.2.2 Multilayered systems

Consider the 1D flow situation, shown in Figure 5.2, in which one liner of thickness $H_1 = 0.3$ m and hydraulic conductivity $k_1 = 2 \times 10^{-9}$ m/s overlies a second liner of thickness $H_2 = 1.2$ m and hydraulic conductivity $k_2 = 1 \times 10^{-9}$ m/s assuming that the water table is at the surface of liner 1, and that liner 2 is underlain by a drain such that the pressure head at the bottom of liner 2 is zero. Taking account of datum at the bottom of liner 2, the total head, ϕ_a, at the bottom of liner 2 ($h_p = 0$, $z = 0$) is given by

$$\phi_a = h_p + z = 0\,\text{m} \qquad (5.7a)$$

and is shown as point "a" in Figure 5.2. The total head, ϕ_b, at the top of liner 1 ($h_p = 0$, $z = 1.5$ m) is given by

$$\phi_b = h_p + z = 1.5\,\text{m} \qquad (5.7b)$$

and is shown as point "b" in Figure 5.2.

In order to plot the head distribution, it is necessary to calculate the total head ϕ_c at point "c" which is at the boundary of liners 1 and 2. This can be obtained from consideration of continuity of flow per unit area as follows:

$$\text{Flow in layer 1:} \quad q_1 = k_1 i_1 = k_1\left(\frac{\phi_b - \phi_c}{H_1}\right) \qquad (5.8a)$$

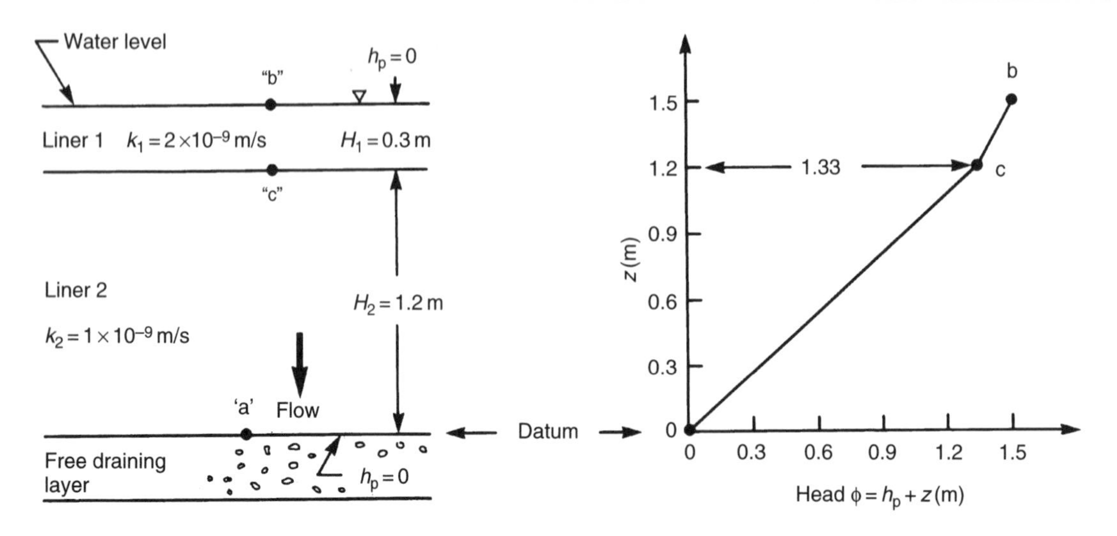

Figure 5.2 Simple flow model – head distribution for a two-layered liner system.

Flow in layer 2:
$$q_2 = k_2 i_2 = k_2\left(\frac{\phi_c - \phi_a}{H_2}\right) \tag{5.8b}$$

Continuity of flow requires that $q_1 = q_2$, thus

$$k_1 i_1 = k_2 i_2 \tag{5.9a}$$

or

$$k_1\left(\frac{\phi_b - \phi_c}{H_1}\right) = k_2\left(\frac{\phi_c - \phi_a}{H_2}\right) \tag{5.9b}$$

where k_1, k_2, ϕ_a, ϕ_b, H_1, H_2 are known and ϕ_c is to be determined. Rearranging equation 5.9b gives

$$\phi_c = \frac{1}{(k_1/H_1 + k_2/H_2)}\left(\frac{k_1}{H_1}\phi_b + \frac{k_2}{H_2}\phi_a\right)$$

and substituting $\phi_a = 0$, $\phi_b = 1.5$, $H_1 = 0.3$, $H_2 = 1.2$, $k_1 = 2 \times 10^{-9}$ m/s, $k_2 = 1 \times 10^{-9}$ m/s,

$$\phi_c = \frac{1}{(2\times10^{-9}/0.3 + 1\times10^{-9}/1.2)}\left(\frac{2\times10^{-9}}{0.3}\times 1.5 + 0\right)$$

Multiplying the numerator and denominator by 10^9, we get

$$\phi_c = \frac{1}{(2/0.3 + 1/1.2)}\left(\frac{2\times1.5}{0.3}\right) = \frac{10}{7.5} = 1.33$$

The flow can then be calculated from either equation 5.8a or 5.8b:

$$\begin{aligned} q_1 &= 2\times10^{-9}\times\left(\frac{1.5-1.33}{0.3}\right) \\ &= 1.11\times10^{-9}\,\text{m/s} = 0.035\,\text{m/a} \end{aligned}$$

$$\begin{aligned} q_2 &= 1\times10^{-9}\times\left(\frac{1.33-0}{1.2}\right) \\ &= 1.11\times10^{-9}\,\text{m/s} = 0.035\,\text{m/a} \end{aligned}$$

Alternatively, one can show that the equivalent hydraulic conductivity of the system is given by the harmonic mean $\overline{k}$ where

$$\frac{\Sigma H_i}{\bar{k}} = \Sigma \frac{H_i}{k_i} \tag{5.10}$$

where the summation is performed over all layers. In this case,

$$\frac{H_1 + H_2}{\bar{k}} = \frac{H_1}{k_1} + \frac{H_2}{k_2}$$

$$\therefore \quad \frac{1.5}{\bar{k}} = \frac{0.3}{2 \times 10^{-9}} + \frac{1.2}{1 \times 10^{-9}}$$

$$\therefore \quad \bar{k} = 1.11 \times 10^{-9}\,\text{m/s}$$

The flow can then be calculated from

$$\bar{q} = \bar{k}\,\bar{i} \tag{5.11}$$

where $\bar{i}$ is the hydraulic gradient across the entire system. In this case,

$$\bar{i} = \frac{\phi_b - \phi_a}{H_1 + H_2} = \frac{1.5 - 0}{0.3 + 1.2} = 1$$

$$\therefore \; \bar{q} = \bar{k}\,\bar{i} = 1.11 \times 10^{-9} \times 1\,\text{m/s} = 0.035\,\text{m/a}$$

Knowing

$$\bar{q} = q_1 = k_1 \left(\frac{\phi_b - \phi_c}{H_1} \right) \tag{5.12}$$

one could then calculate ϕ_c by rearranging terms to give

$$\phi_c = \phi_b - \frac{H_1}{k_1}\bar{q}$$

$$= 1.5 - \frac{0.3}{2 \times 10^{-9}} \times 1.11 \times 10^{-9} = 1.33\,\text{m}$$

This latter approach (equations 5.10–5.12) often represents the most convenient method of calculating a head distribution.

It also follows from equation 5.9a that

$$\frac{i_1}{i_2} = \frac{k_2}{k_1} \tag{5.13}$$

Thus, the change of slope in the total head distribution is directly related to the change in hydraulic conductivity between layers.

5.2.3 Layered systems with high hydraulic conductivity contrasts

Figure 5.3 shows a layer system identical to that shown in Figure 5.2 except that in this case it is assumed that the upper 0.3 m is waste with an hydraulic conductivity of 1×10^{-6} m/s. Based on equation 5.13 (which is based on continuity of flow)

$$\frac{i_1}{i_2} = \frac{(\phi_b - \phi_c)/H_1}{(\phi_c - \phi_a)/H_2} = \frac{k_2}{k_1} = \frac{10^{-9}}{10^{-6}} = 10^{-3} \tag{5.14}$$

Since $\phi_b = 1.5$ m and $\phi_a = 0$ as already demonstrated in the previous section, and $H_1 = 0.3$ m, $H_2 = 1.2$ m, it follows that

$$\frac{(\phi_b - \phi_c)/0.3}{(\phi_c - \phi_a)/1.2} = 10^{-3}$$

$$\therefore \quad \frac{\phi_b - \phi_c}{\phi_c - \phi_a} = 0.25 \times 10^{-3}$$

$$\therefore \quad \phi_b - \phi_c = 0.25 \times 10^{-3}(\phi_c - \phi_a)$$

$$\therefore \quad 1.00025\phi_c = (\phi_b + 0.25 \times 10^{-3}\phi_a) = 1.5$$

$$\therefore \quad \phi_c = 1.499625 \simeq 1.5$$

which implies that the total head change across the waste is negligible. This is because the waste is much more permeable than the clay. The gradient across the liner is given by

$$i \simeq \frac{1.5}{1.2} = 1.25$$

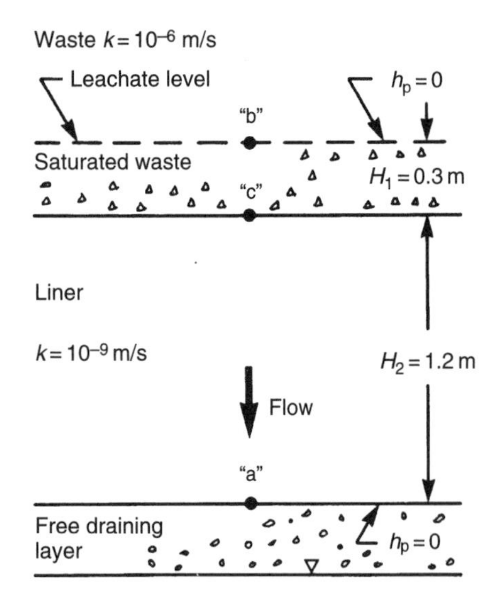

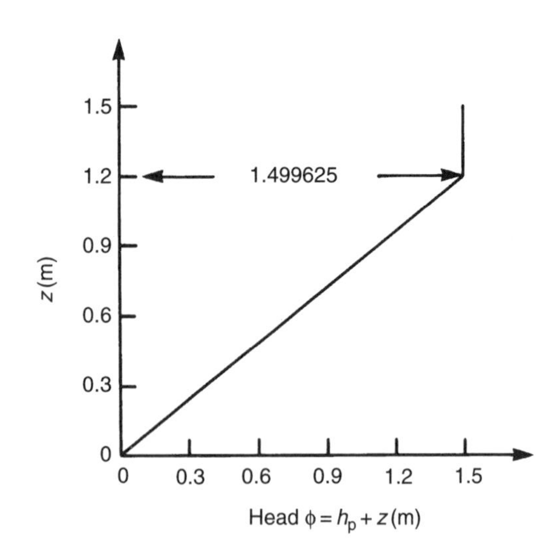

Figure 5.3 Simple flow model – head distribution for a two-layered system with a high hydraulic conductivity contrast between layers.

which is essentially the same as that calculated in Section 5.2.1,

$$q = ki \simeq 1 \times 10^{-9} \times 1.25 = 1.25 \times 10^{-9}\,\text{m/s} \simeq 0.0394\,\text{m/a}$$

Thus when there are high hydraulic conductivity contrasts, the flow is controlled by the least-permeable layer for 1D conditions.

5.2.4 Interpretation of data

The change in hydraulic gradient which may be associated with significant changes in hydraulic conductivity between different layers, as illustrated in the previous section, should be considered when interpreting data from field installation of piezometers. For example, consider the hydrostratigraphy shown in Figure 5.4. The head distribution from the upper fine sand aquifer to the bottom gravel layer is shown together with a number of piezometers and their corresponding water levels (relative to some datum).

The important point to be made is that when calculating vertical gradients for systems such as that shown in Figure 5.4, it is essential to consider the hydrostratigraphy and not simply calculate gradient as the difference in water level in piezometers divided by the distance between piezometers. The calculation of gradients in the various units will be discussed below.

Fine sand unit

Piezometers "a" and "b" record essentially the same water level (50 masl – meters above sea level) to the accuracy of measurement and hence one would calculate essentially zero vertical gradient in this unit. However, this does not mean that there is zero vertical flow – only that the gradient and flow are small.

Clay unit

Piezometer "b" is partly screened across the fine sand and partly screened across the clay unit. However, because of the large hydraulic conductivity contrast, piezometer "b" is recording the head in the fine sand rather than the clay and for purposes of calculating

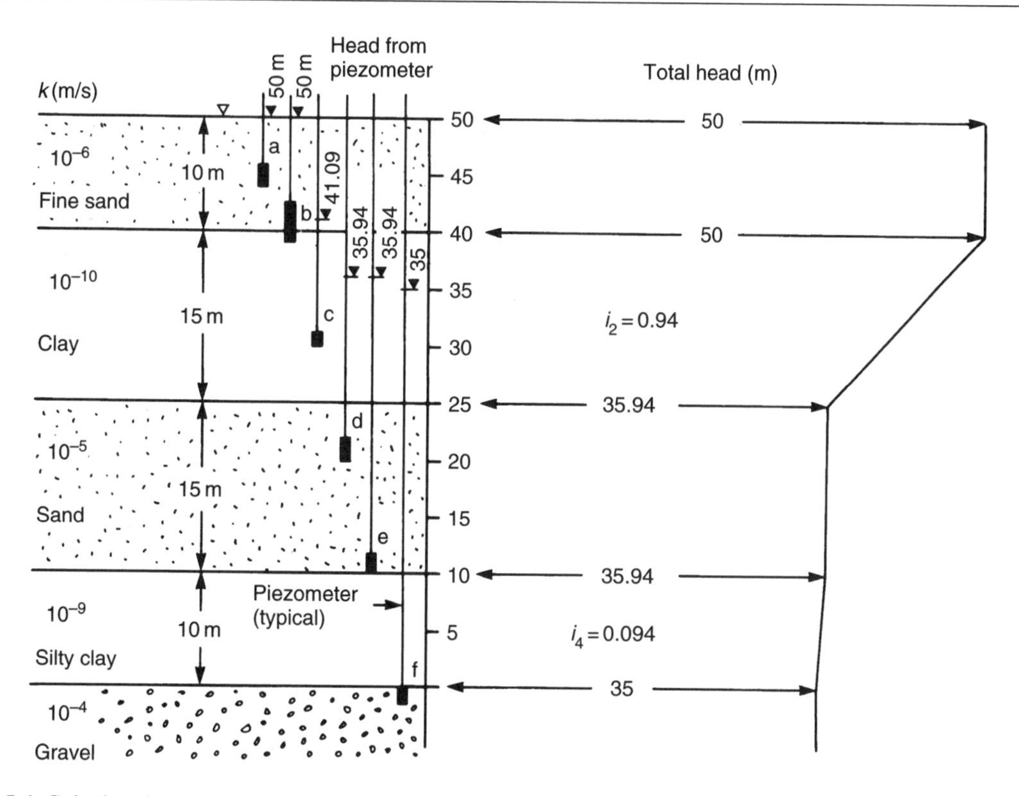

Figure 5.4 Calculated and observed head distribution through a multilayered system.

a gradient, the head for piezometer "b" may be regarded as corresponding to the top of the clay unit.

Piezometer "c" is located 9–10 m below the top of the clay (having a 1-m long sand pack). The total head in piezometer "c" is 41.09 masl. Since this piezometer is embedded completely in the clay, the head is considered to correspond to that at the centroid of the sand pack.

$$\therefore \; i_{\text{b-c}} \simeq \frac{50 - 41.09}{9.5} = 0.937 \simeq 0.94$$

where the distance of 9.5 m is measured from the centroid of piezometer "c" to the top of the clay unit.

A second estimate of the gradient across the clay unit can be obtained using piezometers "c" and "d" but recognizing that, because of the hydraulic conductivity contrast, the head of $\simeq$35.94 masl recorded by piezometer "d" is essentially the same as at the bottom of the clay unit.

$$i_{\text{c-d}} \simeq \frac{41.09 - 35.94}{5.5} = 0.937 \simeq 0.94$$

where 5.5 m is the distance between the centroid of piezometer "c" and the bottom of the clay unit.

A third estimate of gradient across the clay unit can be obtained using piezometers "b" and "d", recognizing that these record the head at the top and bottom of the clay unit, respectively. Thus

$$i_{\text{b-d}} = \frac{50 - 35.94}{15} = 0.937 \simeq 0.94$$

where the 15-m distance is the thickness of the clay unit.

The agreement between the three calculations of gradient across the clay unit indicates that the unit has a uniform vertical hydraulic conductivity.

It should be noted that the gradient that one would calculate is based on the distance between the centroid of piezometers "b" and "d", viz.

$$\frac{50 - 35.94}{20} = 0.70$$

When used in conjunction with conventional interpretations, this gives a misleading indication of the vertical flow in the system which is being controlled by the clay layer.

The problems of gradient interpretation can be minimized by placing piezometers close to boundaries between materials having significantly different hydraulic conductivities. For example, as shown for piezometers "e" and "f" in Figure 5.4. Based on the head measured in these piezometers, the gradient across the silty clay layer is fairly readily estimated

$$i = \frac{35.94 - 35}{10} = 0.094$$

based on the 10-m thickness of the silty clay layer.

When dealing with deposits of wide lateral extent there are often situations where there is a gradual change in grain size with depth. This will generally give rise to a gradual change in hydraulic conductivity and, in situations where there are significant differences in vertical head, this will be manifested by a change in gradient. For example, Figure 5.5 shows a number of piezometers located in a silt to silty clay layer where there is a change from lower clay content at the top to a high clay content near the bottom. The water levels and their corresponding total head distributions are shown in Figure 5.5. This data alone indicates a significant change in hydraulic conductivity with depth with the bottom 5 m being responsible for 60% of the head loss. The change in gradient for a 1D system such as this indicates the relative changes in hydraulic conductivity. If data is available to allow an estimate of hydraulic conductivity to be made for one layer, then an estimate can also be made for the entire deposit. For example, suppose that the result of a pumping test (e.g., see Section 3.2.3) conducted on the lower aquifer and monitored at piezometers "d" and "c" indicates a bulk (vertical) hydraulic conductivity of the soil between "c" and "d" of 1.6×10^{-10} m/s.

The gradient between each pair of piezometers can be readily calculated as

$$i_{c-d} = \frac{29.09 - 20}{5} = 1.82$$

(where the 5-m distance is from the centroid of the sand pack of piezometer "c" and the bottom of the clay layer for the reasons discussed earlier).

$$i_{b-c} = \frac{34.19 - 29.09}{10} = 0.51$$

$$i_{a-b} = \frac{35 - 34.19}{10} = 0.08$$

For essentially 1D (vertical) flow through this silt–silty clay layer, the ratio of gradient is equal to the inverse ratio of bulk hydraulic conductivity. Thus from equation 5.13,

$$\frac{i_{c-d}}{i_{b-c}} = \frac{k_{b-c}}{k_{c-d}}$$

$$\therefore \frac{1.82}{0.51} = \frac{k_{b-c}}{1.6 \times 10^{-10}}$$

$$\therefore k_{b-c} \simeq 5.7 \times 10^{-10}\ \text{m/s}$$

and similarly,

$$\frac{i_{c-d}}{i_{a-b}} = \frac{k_{a-b}}{k_{c-d}}$$

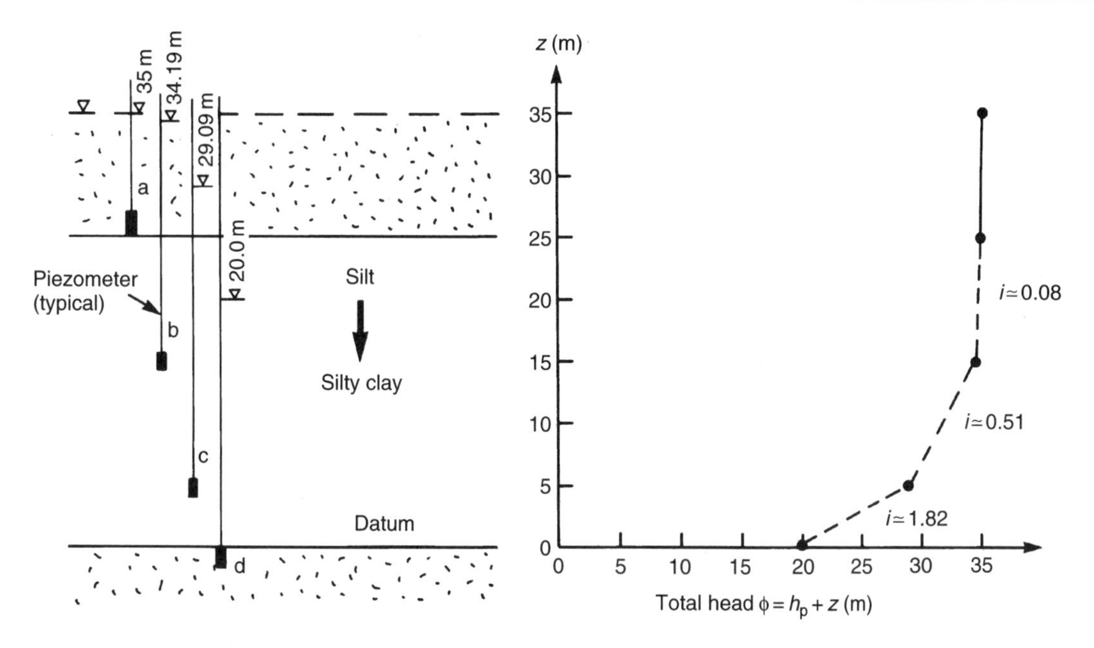

Figure 5.5 Head distribution through a gradually varying silt–silty clay layer and the estimation of hydraulic conductivity distribution based on water level data.

$$\therefore \frac{1.82}{0.08} = \frac{k_{a-b}}{1.6 \times 10^{-10}}$$

$$\therefore k_{a-b} \simeq 3.6 \times 10^{-9}\,\text{m/s}$$

The total vertical flow in this system can be estimated from the hydraulic conductivities and gradients evaluated between any pair of piezometers. Thus

$$q = k_{c-d} \times i_{c-d} = 1.6 \times 10^{-10} \times 1.82$$
$$\simeq 2.9 \times 10^{-10}\,\text{m/s} \equiv 0.009\,\text{m/a}$$

This is a small but potentially significant flow with regard to contaminant transport through the silt–silty clay system. Note that it could be quite misleading to use a gradient determined between one pair of piezometers (e.g., a–d) and a hydraulic conductivity between a second pair of piezometers (e.g., c–d). For example, this would give an estimate of flow

$$q = \frac{15}{25} \times 1.6 \times 10^{-10} = 9.6 \times 10^{-11}\,\text{m/s}$$
$$\equiv 0.003\,\text{m/a}$$

which is quite incorrect. If the gradient across the entire unit is to be used, it must be used with the appropriate harmonic mean hydraulic conductivity, $\bar{k}$, which can be calculated using the values of k_{a-b}, k_{b-c}, k_{c-d} estimated above, viz.

$$\frac{H}{\bar{k}} = \frac{H_{a-b}}{k_{a-b}} + \frac{H_{b-c}}{k_{b-c}} + \frac{H_{c-d}}{k_{c-d}}$$

$$\therefore \frac{25}{\bar{k}} = \frac{10}{3.6 \times 10^{-9}} + \frac{10}{5.7 \times 10^{-10}} + \frac{5}{1.6 \times 10^{-10}}$$

$$\therefore \bar{k} = 4.85 \times 10^{-10}\,\text{m/s}$$

$$\therefore\ q = \bar{k}\bar{i} = 4.85 \times 10^{-10} \times \frac{15}{25}$$
$$= 2.9 \times 10^{-10}\,\text{m/s}$$
$$\equiv 0.009\,\text{m/a}$$

5.2.5 Head distributions in some simple barrier systems

Figure 5.6 shows a barrier system consisting of a primary leachate collection system, a primary clay liner with a hydraulic conductivity $k_1 \simeq 3 \times 10^{-10}$ m/s underlain by a leak detection/secondary leachate collection layer which, it is assumed, is kept free draining without significant build-up of head in this layer. This is then underlain by a 1.5-m thick natural clay (secondary liner) ($k_2 = 10^{-9}$ m/s) and an aquifer. It is assumed that leachate mounding in the primary collection system is 0.3 m above the primary liner and that the water level (potentiometric surface) in the aquifer is 1.0 m above the top of the aquifer.

The vertical flow in the two liner systems can be readily calculated using the concepts presented in the previous sections. Since both the primary and secondary leachate collection systems are far more permeable than the liner, the gradient across the primary liner is given by

$$i_1 = \frac{\phi_a - \phi_b}{z_a - z_b} = \frac{3.3 - 1.8}{3.0 - 1.8} = 1.25$$

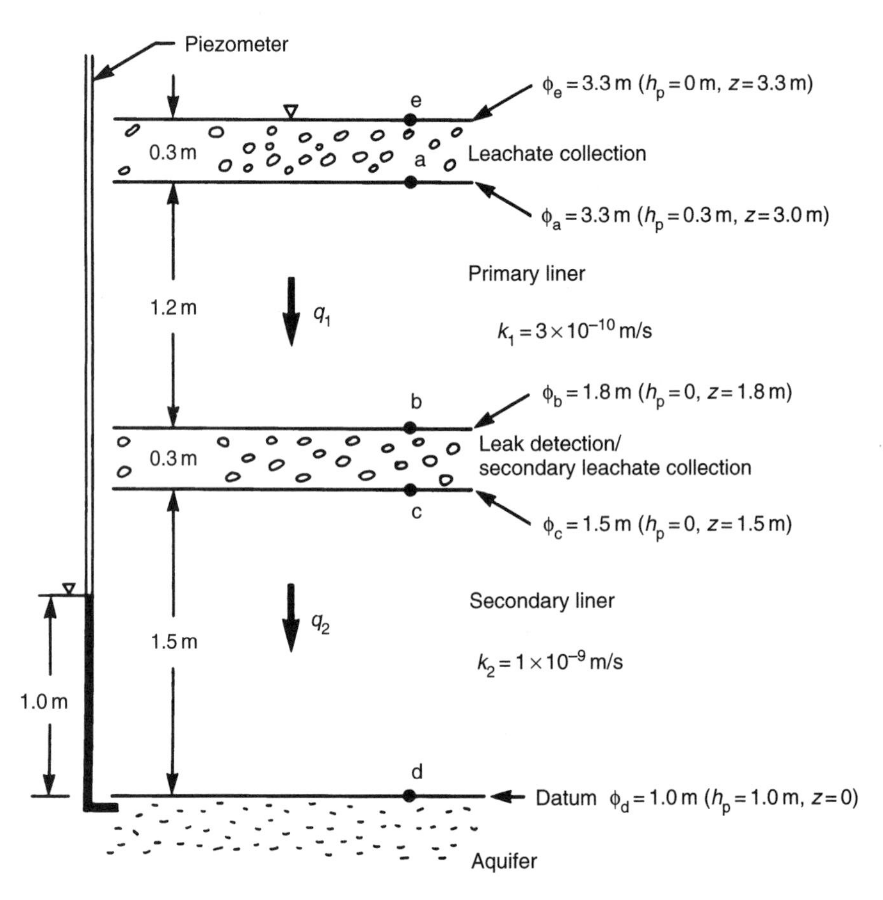

Figure 5.6 Head distribution and schematic description for a multilayered system involving a leak detection secondary leachate collection layer.

$$q_1 = k_1 i_1 = 3 \times 10^{-10} \times 1.25$$
$$= 3.75 \times 10^{-10}\,\text{m/s} \equiv 0.012\,\text{m/a}$$

The vertical flow through the secondary liner can be calculated as follows:

$$i_2 = \frac{\phi_c - \phi_d}{z_c - z_d} = \frac{1.5 - 1.0}{1.5 - 0} = 0.33$$

$$q_2 = k_2 i_2 = 3.3 \times 10^{-10}\,\text{m/s} \equiv 0.0105\,\text{m/a}$$

Since $q_2 < q_1$, some of the flow will be collected by the secondary leachate collector; however, this system has the potential to allow the majority of flow passing through the primary liner to also pass through the secondary liner and hence the effectiveness of the "leak detection/secondary leachate collection" for the system must be questioned.

If the head in the aquifer was higher (e.g., if $\phi_d \simeq 1.5\,\text{m}$) or if the secondary liner was less permeable, either because of lower hydraulic conductivity clay or the use of a geomembrane to form a composite liner, then the secondary collection system would be more effective.

Figure 5.7 shows a system with similar physical characteristics to that in Figure 5.6 but with a quite different hydraulic head distribution arising from a much higher head in the aquifer and not pumping the second granular layer. Under these conditions, the head at points "b" and "c" will depend on the hydraulic conductivity contrast between the primary and secondary liners. Since it is assumed that no fluid is added (or removed) from the hydraulic control layer and since this layer is far more permeable than either the primary or secondary liner, continuity of flow requires that $q_1 = k_1 i_1 = q_2 = k_2 i_2$.

$$\therefore \frac{i_1}{i_2} = \frac{k_2}{k_1} \tag{5.15a}$$

(as already established in equation 5.13).

Thus, taking $k_1 = 3 \times 10^{-10}\,\text{m/s}$ and $k_2 = 1 \times 10^{-9}\,\text{m/s}$ gives

$$\frac{i_1}{i_2} = \frac{1 \times 10^{-9}}{3 \times 10^{-10}} = 3.33 \tag{5.15b}$$

Now

$$i_1 = \frac{\phi_a - \phi_b}{1.2} \tag{5.16a}$$

$$i_2 = \frac{\phi_c - \phi_d}{1.5} \tag{5.16b}$$

where

$$\phi_b \simeq \phi_c \tag{5.16c}$$

then from equations 5.15a, 5.15b and 5.16a–c

$$\frac{\phi_a - \phi_b}{\phi_b - \phi_d} = 3.33 \times \frac{1.2}{1.5} = 2.66$$

$$\therefore \phi_b = \frac{\phi_a + 2.66\phi_d}{3.66} = \frac{3.3 + 2.66 \times 4.2}{3.66}$$
$$= 3.95\,\text{m}$$

$$\therefore q_1 = k_1 i_1 = 3 \times 10^{-10} \times \frac{3.3 - 3.95}{1.2}$$
$$= -1.64 \times 10^{-10}\,\text{m/s} \equiv -0.005\,\text{m/a}$$

$$q_2 = k_2 i_2 = 1 \times 10^{-9} \times \frac{3.95 - 4.2}{1.5}$$
$$= -1.64 \times 10^{-10}\,\text{m/s} \equiv -0.005\,\text{m/a}$$

where the negative sign indicates upward flow from the aquifer to the leachate collection system. This provides a hydraulic trap as discussed in Chapter 1.

The preceding calculation assumes that the head in the aquifer is not affected by the upward flow. This is only likely to be true if it

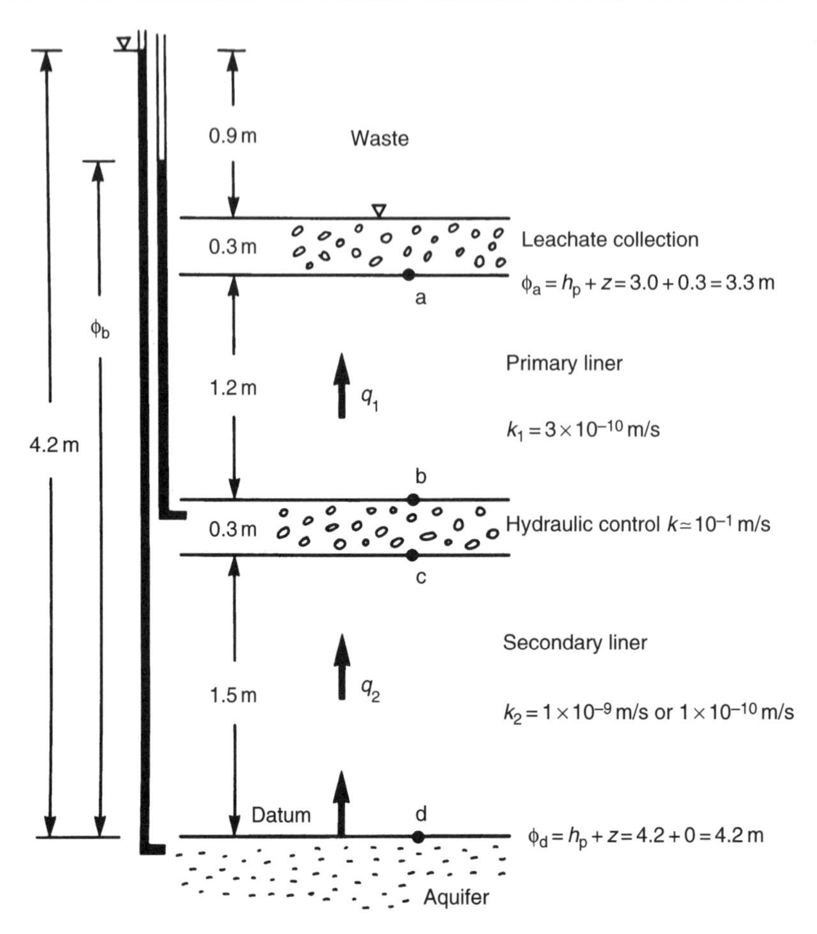

Figure 5.7 Head distribution and schematic description for a multilayered system with a hydraulic controlled layer between two liners.

is a highly transmissive aquifer which did not rely on recharge from the overburden beneath the waste site for a significant proportion of its flow. This will be discussed in more detail in Chapter 16.

The effect of the hydraulic conductivity contrast between the layers can be illustrated by taking $k_2 = 10^{-10}$ m/s and repeating the foregoing calculations. Thus

$$\frac{i_1}{i_2} = \frac{10^{-10}}{3 \times 10^{-10}} = 0.33$$

$$\frac{\phi_a - \phi_b}{\phi_b - \phi_d} = 0.33 \times \frac{1.2}{1.5} = 0.26$$

$$\therefore \ \phi_b = \frac{\phi_a + 0.26 \times 4.2}{1.26} = 3.49 \text{ m}$$

$$\therefore \ q_1 = k_1 i_1 = 3 \times 10^{-10} \times \frac{3.3 - 3.49}{1.2}$$
$$= -4.7 \times 10^{-11} \text{ m/s} \equiv -0.0015 \text{ m/a}$$

$$q_2 = k_2 i_2 = 1 \times 10^{-10} \times \frac{3.49 - 4.2}{1.5}$$
$$= -4.7 \times 10^{-11} \text{ m/s} \equiv -0.0015 \text{ m/a}$$

Thus the lower hydraulic conductivity of the secondary liner relative to the previous example reduces flow through the entire system. This will reduce the effectiveness of the hydraulic trap as will be discussed in subsequent chapters.

In these two examples, the hydraulic control layer was used in a passive mode wherein no water was added or subtracted from this layer. By active operation of this layer the head, $\phi_b(\cong\phi_c)$, can be controlled and either increased or decreased relative to the leachate head, ϕ_a, and potentiometric surface in the aquifer, ϕ_d.

5.3 Analysis of 2D and 3D flow

While the simple 1D calculations are adequate for some important practical applications, there are situations where the flow regime is more complex and it is necessary to use more elaborate analysis techniques to model it adequately, although one can often do useful checks using simple 1D hand calculations.

5.3.1 Governing equations

Suppose that the occurrence of some event, such as the construction of a landfill, the installation of cut-off wall or the introduction of a well or drain, disturbs the equilibrium of the pore water in the soil. At first there will be a period during which the pore water pressure and the stress distribution will change, but ultimately the pore water pressure and the stress state will reach a new equilibrium and steady-state flow condition will be established. It is usually convenient to analyse the steady-state flow in terms of the total head ϕ, defined by equation 5.2. For 2D and 3D cases, the form of Darcy's Law becomes

$$v_x = k_{xx} i_x, \qquad v_y = k_{yy} i_y, \qquad v_z = k_{zz} i_z \tag{5.17}$$

where $\boldsymbol{v} = [v_x, v_y, v_z]^T$ is the Darcy velocity vector and $\boldsymbol{i} = (i_x, i_y, i_z)^T$ is the hydraulic gradient vector, with

$$i_x = -\frac{\partial\phi}{\partial x}, \qquad i_y = -\frac{\partial\phi}{\partial y}, \qquad i_z = -\frac{\partial\phi}{\partial z} \tag{5.18}$$

and where the negative sign has been introduced in recognition that water flows from high head to low head and so $\partial\phi/\partial x$ will be negative for flow in the positive x-direction, and similarly for $\partial\phi/\partial y$ and $\partial\phi/\partial z$ in the y- and z-directions.

Since a steady state has been reached there will be no change in the effective stress state and thus an element of soil will undergo no deformation and in particular will undergo no change in volume. If the pore fluid can be considered as incompressible, then continuity of flow within an element will be satisfied provided that

$$\frac{\partial v_x}{\partial x} + \frac{\partial v_y}{\partial y} + \frac{\partial v_z}{\partial z} = 0 \tag{5.19}$$

This is called the continuity equation.

It follows from Darcy's Law (equation 5.17) that the total head satisfies

$$\frac{\partial}{\partial x}\left(k_{xx}\frac{\partial\phi}{\partial x}\right) + \frac{\partial}{\partial y}\left(k_{yy}\frac{\partial\phi}{\partial y}\right) + \frac{\partial}{\partial z}\left(k_{zz}\frac{\partial\phi}{\partial z}\right) = 0 \tag{5.20a}$$

If the soil is homogeneous and isotropic, this reduces to Laplace's equation

$$\nabla^2\phi = \frac{\partial^2\phi}{\partial x^2} + \frac{\partial^2\phi}{\partial y^2} + \frac{\partial^2\phi}{\partial z^2} = 0 \tag{5.20b}$$

For 2D plane flow (in the x, z-plane), this becomes

$$\frac{\partial^2\phi}{\partial x^2} + \frac{\partial^2\phi}{\partial z^2} = 0 \tag{5.20c}$$

5.3.2 Boundary conditions

Figure 5.8 illustrates the flow from a leachate pond through a clay cut-off wall, overlaying an impermeable stratum and into a drain. This example serves to illustrate four common types of boundary conditions:

1. On AB the water pressure is given by $u = \gamma_w h$ and thus for the datum shown in Figure 5.8 $\phi = h + z = H$ at any arbitrary point along AB. Thus the line AB is a line of constant head or equipotential. This is a special case of what is often called a Dirichlet or first-type boundary condition where the head is specified along a boundary.
2. The line AD is an impermeable surface. There can be no flow across such a surface and thus the direction of flow must be parallel to AD. The line AD is called an impermeable surface or a flow line. This is a special case of the Neumann or second-type boundary condition where the flux (flow) normal to the boundary is specified; in this case, the flow across the boundary is zero.
3. On CD the pore water pressure is equal to atmospheric pressure (usually taken as zero) and since $\phi = +z$, the head varies linearly with distance above the datum. Such a line is called a line of constant pressure.
4. When seepage occurs in the clay cut-off wall illustrated in Figure 5.8, the entire soil mass does not remain saturated. The material above BC becomes unsaturated and ultimately the water drains from there to a boundary called the phreatic surface which represents the boundary between unsaturated and saturated materials. The phreatic line BC is both a line of constant pore water pressure, which is equal to atmospheric pressure and taken as zero, and a flow line. The position of this phreatic line is not known initially and is determined by an iterative algorithm.

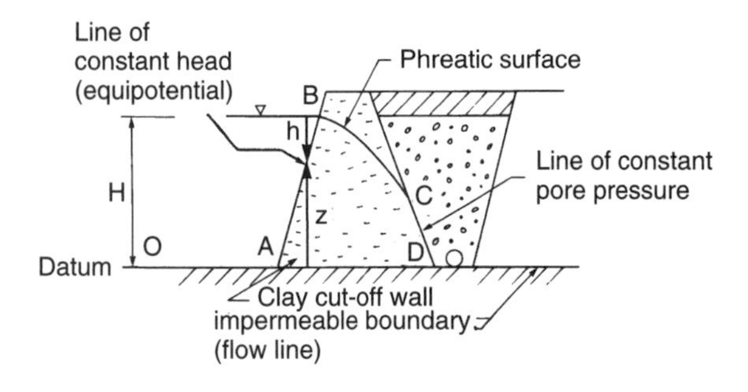

Figure 5.8 Boundary conditions for steady-state seepage.

5.3.3 Continuity conditions

Natural soil deposits and man-made barriers (e.g., below a waste facility) often consist of a number of distinct layers as illustrated in Figure 5.9. At the interface of two such layers, the pore water pressure just above the interface must equal the pore water pressure just below the interface and thus

$$\phi_A = \phi_B \tag{5.21a}$$

Also, there can be no net flow into the interface and thus

$$v_{nA} = v_{nB} \tag{5.21b}$$

Any solution of equations 5.20a–5.20c must satisfy these continuity conditions.

5.4 A finite difference approximation

The determination of the variation of head in any particular case depends upon the solution

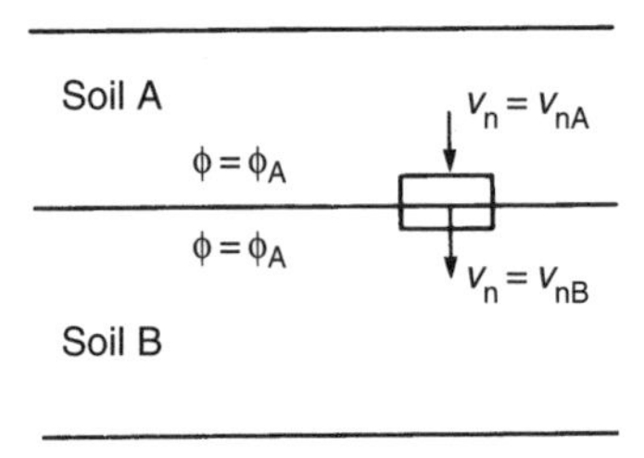

Figure 5.9 Continuity conditions.

of the governing differential equation, subject to boundary conditions of the type discussed in Section 5.3.2 and continuity equations of the type given by equations 5.21a and 5.21b.

The explicit determination of the distribution of head is only feasible for very simple flows and for more complex cases it is necessary to employ numerical techniques. The earliest approach to this was the finite difference approach. Figure 5.10a shows the section of a cut-off wall where it is assumed that the sand layer is far more permeable than the waste and the underlying silt to clayey silt soil. The head in the sand is assumed to be approximately uniform on each side of the cut-off wall but with a lower head in the sand to the right of the wall. In the finite difference method, there is no attempt to determine the distribution of head at all points; rather, the heads are determined at a finite number of points on a grid as shown in Figure 5.10b.

The governing differential equation, in this case

$$\frac{\partial^2 \phi}{\partial x^2} + \frac{\partial^2 \phi}{\partial z^2} = 0 \qquad (5.22)$$

is then approximated in terms of the values of head at these grid points, thus at the point P, equation 5.22 is approximated by

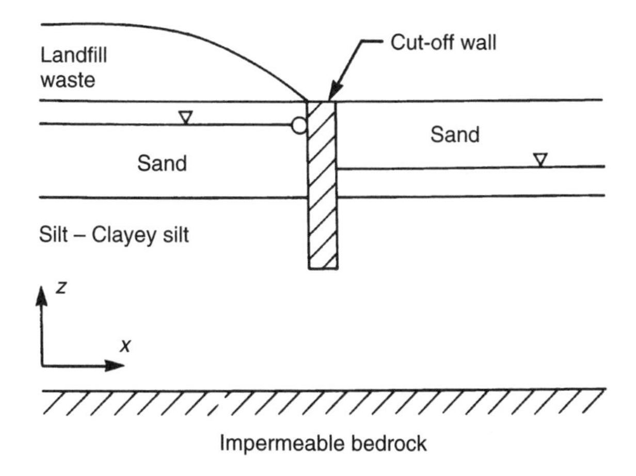

Figure 5.10a Schematic description of cut-off wall.

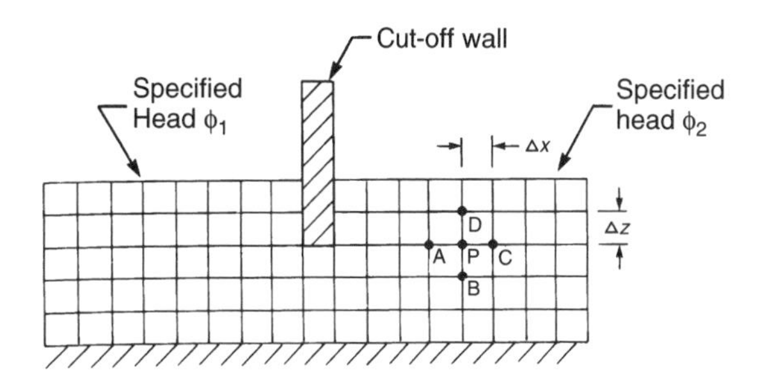

Figure 5.10b Finite difference grid.

$$\frac{\phi_C - 2\phi_P + \phi_A}{\Delta x^2} + \frac{\phi_D - 2\phi_P + \phi_B}{\Delta z^2} = 0 \qquad (5.23)$$

It is possible to incorporate boundary conditions and to take into account continuity conditions at the junction of different soils, although this is difficult for anything other than the simplest rectangular configurations.

Equation 5.23 together with the boundary conditions and continuity conditions lead to a set of linear equations which can be solved to obtain the head at each grid point. The finer the finite difference mesh, the more accurate the approximation to the governing differential equation.

5.5 Application of the finite element method to the analysis of plane flow

As noted above, there are some difficulties in applying the finite difference method to seepage analyses involving complex geometries and non-homogeneous soils. These difficulties are largely overcome by the finite element method.

The solution of any seepage problem involves satisfying

1. the continuity equation (equation 5.19);
2. the constitutive behaviour (Darcy's Law), equation 5.17;
3. the boundary conditions; and

4. the continuity conditions at the interface of soils having different properties.

Consider the landfill situation shown in Figure 5.11 where there is leachate mounding between a set of drains along the base of the landfill and in addition there is a known location of the water table outside the landfill which means that the head is defined along the boundary S_ϕ (this includes boundary conditions 1 and 4 as discussed in Section 5.3.2). At the upgradient and downgradient edge of the aquifer (along boundary S_ψ), either the head may be specified (i.e., boundary condition 1, Section 5.3.2) or the flow may be specified (i.e., a Neumann or second-type boundary condition). Along the boundary S_q the flow is specified to be zero (i.e., boundary condition 2, Section 5.3.2). Thus the problem is defined along the surface $S = S_\phi + S_\psi + S_q$. Within the volume, V, inside this boundary surface there are three materials with three different sets of properties – the liner, the natural aquitard and the aquifer. Thus we need to solve equations 5.17 and 5.19 within the region V, subject to the boundary conditions along S, while requiring that continuity conditions are satisfied at the material boundaries within the volume V.

The finite element method has a totally different philosophy to the finite difference method. In the finite element method, the continuity equation, Darcy's Law and the boundary conditions are combined to derive an equation of "virtual work". The finite element method then seeks to approximately satisfy the virtual work equation:

$$\text{Internal “virtual work”} = \text{Applied “virtual work”} \qquad (5.24)$$

This method of analysis has been applied extensively to the analysis of a wide variety of physical phenomena (Zienkiewicz, 1977). The procedure is straightforward and is outlined below.

First, the body is divided into a number of elements as shown schematically in Figure 5.12. Each element has a number of nodes, e.g., at the corners of the elements shown in Figure 5.12. The finite element method seeks to determine the value of head at these nodes and does so by a systematic approximation of the equation of virtual work. The quantities to be determined at the heads at the nodal points 1, 2, ..., viz. ϕ_1, ϕ_2, ... The essential idea is to approximate the continuous variation of head ϕ within an element in terms of nodal heads.

Similarly, the variation of the continuous "virtual head" $\delta\phi$ is approximated in terms of

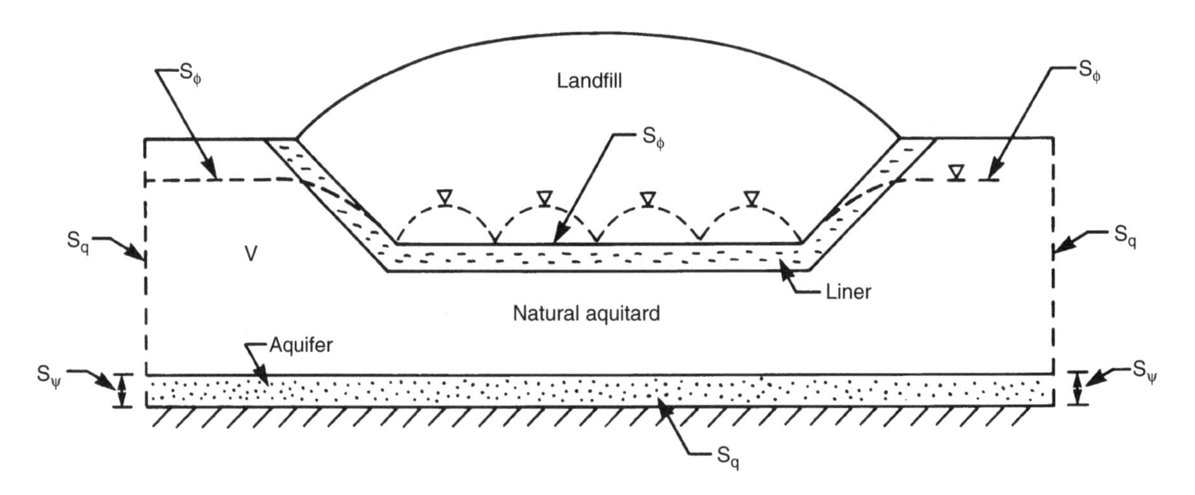

Figure 5.11 Simple seepage problem.

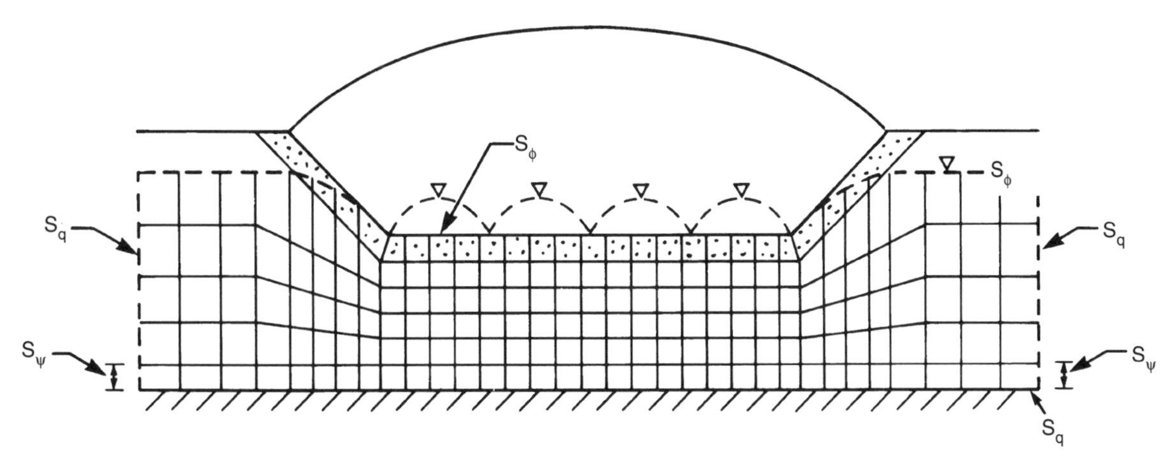

Figure 5.12 Division of body into elements.

the values of "virtual" nodal head $\delta\phi_1$, $\delta\phi_2, \ldots$. It is perhaps worth remarking that this process of approximation is essentially a local one so that the variation of head within a given element is expressed in terms of the heads at the nodes of that element. This approximation technique leads to approximations of the form:

$$\text{“Internal virtual work”} = \int_V \boldsymbol{\delta i}^{\mathrm{T}} \boldsymbol{v}\, \mathrm{d}V \simeq \delta \boldsymbol{a}^{\mathrm{T}} \boldsymbol{K} \boldsymbol{a} \qquad (5.25)$$

$$\text{“Applied virtual work”} = \int_S \boldsymbol{q} \boldsymbol{\delta\phi}\, \mathrm{d}S \simeq \delta \boldsymbol{a}^{\mathrm{T}} \boldsymbol{r} \qquad (5.26)$$

In these equations, $\boldsymbol{\delta\phi}$ is a virtual change in head which is zero at any point where the head is specified but is otherwise arbitrary, $\boldsymbol{\delta i} = [\delta i_x, \delta i_y, \delta i_z] = \partial\delta\phi/\partial x, \partial\delta\phi/\partial y, \partial\delta\phi/\partial z$ is the vector of virtual hydraulic gradients, v the vector of Darcy velocities, q the component of the Darcy velocity in the direction of the positive outward normal to the surface S, $\boldsymbol{a} = (\phi_1, \phi_2, \ldots)^{\mathrm{T}}$ the vector of nodal heads, $\delta\boldsymbol{a} = (\delta\phi_1, \delta\phi_2, \ldots)^{\mathrm{T}}$ the vector of "virtual" nodal heads, $\boldsymbol{K}$ the "flow stiffness matrix" which is known and can be calculated from the elements' configurations and properties and $\boldsymbol{r}$ the vector of applied nodal flows which can be calculated from the specified boundary flows on S_q and, if appropriate, S_ψ.

If the approximations, equations 5.25 and 5.26, are substituted into the equation of "virtual work" (equation 5.24), it follows that it will be approximately satisfied if

$$\delta\boldsymbol{a}^{\mathrm{T}}(\boldsymbol{K}\boldsymbol{a} - \boldsymbol{r}) = \boldsymbol{0} \qquad (5.27)$$

This equation is true for any variation of "virtual head", $\delta\phi$, that is for arbitrary values of δa and so equation 5.27 implies that

$$\boldsymbol{K}\boldsymbol{a} = \boldsymbol{r} \qquad (5.28)$$

Equation 5.28 is a set of linear equations which can be solved to determine the nodal heads $\boldsymbol{a}$.

5.6 Boundary element methods

The finite element method provides a powerful method of analysis for the determination of flow. It can deal with complex geometric configurations, a number of different material types and complex flow behaviour. However, particularly if it is desired to model 3D behaviour, it can involve the solution of a very large set of linear equations. This disadvantage can be overcome if the soil being considered is homogeneous by using the boundary element method (Brebbia and Dominguez, 1989).

The boundary element method considers a slight generalization of the flow equation

$$k\nabla^2\phi = f \qquad (5.29)$$

where f represents the volume of water removed from unit volume of soil in unit time.

Suppose that an arbitrary function ϕ^* satisfies certain conditions of differentiability such that

$$\int_V \left(v_x \frac{\partial \phi^*}{\partial x} + v_y \frac{\partial \phi^*}{\partial y} + v_z \frac{\partial \phi^*}{\partial z} \right) dV = \int_V f\phi^* \, dV + \int_S q\phi^* \, dS \qquad (5.30)$$

If, instead, the function ϕ^* satisfies the equation

$$k\nabla^2\phi^* = f^*$$

then it follows by interchanging the roles of ϕ and ϕ^* that

$$\int_V \left(v_x^* \frac{\partial \phi}{\partial x} + v_y^* \frac{\partial \phi}{\partial y} + v_z^* \frac{\partial \phi}{\partial z} \right) dV = \int_V f^*\phi \, dV + \int_S q^*\phi \, dS \qquad (5.31)$$

where $(v_x^*, v_y^*, v_z^*) = -k(\partial\phi^*/\partial x, \partial\phi^*/\partial y, \partial\phi^*/\partial z)$ are the Darcy velocities generated by the head ϕ^* and q^* is the normal component of this Darcy velocity at the boundary.

Finally, if ϕ^* is chosen to be the solution corresponding to a point source at position $\boldsymbol{r}_0$ in an unbounded medium, and if f is zero so that there is no mechanism removing water from the soil, then substitution of Darcy's Law and subtraction of equations 5.30 and 5.31 lead to the boundary integral equation:

$$\varepsilon\phi(\boldsymbol{r}_0) = \int_V (q\phi^* - q^*\phi) \, dS \qquad (5.32)$$

where $\varepsilon = 1$ if $\boldsymbol{r}_0$ is within V, $\varepsilon = 1/2$ if r_0 is on a smooth portion of S and $\varepsilon = 0$ if $\boldsymbol{r}_0$ is outside V.

Equation 5.32 represents the head at any point and in particular at a point on the boundary in terms of the value of head, ϕ, and the value of outward flow, q, on the boundary of the body.

The boundary element method then seeks to approximate the actual field behaviour in terms of the values of ϕ and q at a certain number of nodes on the surface of the body. A systematic application of this approach (see Grouch and Starfield, 1983; Brebbia and Dominguez, 1989) leads to the approximating set of equations

$$\boldsymbol{H}\,\boldsymbol{U} = \boldsymbol{G}\,\boldsymbol{Q} \qquad (5.33)$$

where $\boldsymbol{U} = (\phi_1, \phi_2, \ldots)$ is the vector of modal boundary heads, $Q = (q_1, q_2 \ldots)$ the vector of nodal flows and H, G are known square matrices.

At every boundary point, j, either the head ϕ_j is specified or the outward normal velocity q_j is specified so that equation 5.33 can be used to determine all unknown heads and all unknown flows by solving a set of linear equations. Once this has been done, the head and flow on the boundary are completely defined and so equation 5.32 can be used to determine the distribution of head throughout the body.

5.7 Calculating flow through holes in geomembranes

Advective flow is impeded by an intact geomembrane liner. However, liquid can pass through holes in a geomembrane. These holes can arise during construction or from gravel punctures or cracking when subject to overburden pressures (see Section 13.4.1). In this section, equations to estimate the flow (or leakage) through holes in geomembranes are presented for several cases depending on nature of the material above and below the geomembrane, and the contact conditions between the geomembrane and the underlying material. The implications of leakage rates for common composite liners with a geomembrane and low-permeability material (either compacted clay or a geosynthetic clay liner) are examined in Sections 13.4.3 and 13.4.4.

5.7.1 Circular hole in a geomembrane

The "maximum" possible leakage through a circular hole in a geomembrane occurs when the hydraulic resistance of the material below the geomembrane is zero. For this case with a circular hole of radius r_0 in a geomembrane and height of liquid h_w on top of the geomembrane (Figure 5.13a) leakage is given by Bernoulli's equation:

$$Q = \pi C_B r_0^2 \sqrt{2gh_w} \tag{5.34}$$

where g is the acceleration due to gravity (m/s^2) and C_B is a dimensionless coefficient related to the shape of the edges of the hole with $C_B = 0.6$ for sharp edges (Giroud and Bonaparte, 1989a).

5.7.2 Circular hole in a geomembrane on an infinite layer with perfect contact

The equation derived by Forchheimer (1930) may be used to calculate leakage through a circular hole with radius r_0 in a geomembrane resting on an infinitely deep layer with isotropic hydraulic conductivity k_L, viz.

$$Q = 4r_0 k_L h_d \tag{5.35}$$

where h_d is the head loss across the geomembrane and the underlying layer, normally taken to be equal to the leachate head, h_w, above the geomembrane (Figure 5.13b).

This equation is only applicable for the idealized conditions of no lateral flow along the interface between the geomembrane and the underlying soil (i.e., perfect contact). As discussed in Section 13.4.2, the actual contact conditions between the geomembrane and underlying

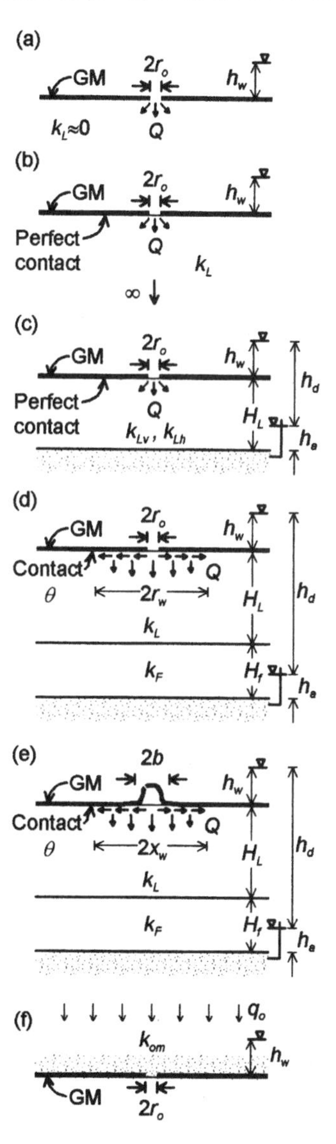

Figure 5.13 Geometry for leakage equations: (a) geomembrane alone; (b) geomembrane in perfect contact with semi-infinite deposit; (c) geomembrane in perfect contact with finite layer; (d) geomembrane in non-perfect contact with layered materials; (e) geomembrane wrinkle with holes (f) capacity of fluid to drain into hole.

material can be highly variable and lead to substantial lateral flow along the interface. Assuming perfect contact will underestimate the leakage if any lateral flow occurs. Hence Forchheimer's equation provides an estimate of the "minimum" possible leakage through the geomembrane.

5.7.3 Circular hole in geomembrane with perfect contact – general solution

Rowe and Booker (2000) derived a general 3D solution for leakage through a circular hole in a geomembrane that is in perfect contact with a layer of thickness H_L. The layer has vertical and horizontal hydraulic conductivity k_{Lv} and k_{Lh}, respectively, and is underlain by a more permeable layer with pressure head h_a, as shown in Figure 5.13c.

For the case with isotropic hydraulic conductivity k_{Lv}, leakage is given by

$$Q = \left(4 + F\frac{r_0}{H_L}\right) r_0 k_{Lv} h_d \tag{5.36}$$

where

$$F = 2.455 + 0.685 \tanh\left(0.6 \ln\left(\frac{r_0}{H_L}\right)\right) \tag{5.37}$$

For a very deep layer (i.e., $r_0/H_L \rightarrow 0$), leakage is equal to that of Forchheimer's equation while for a very shallow layer (i.e., $r_0/H_L \rightarrow \infty$) the solution becomes one-dimensional and

$$Q = \pi \frac{r_0}{H_L} r_0 k_{Lv} h_d \tag{5.38}$$

Flow for all cases between $0 \leq r_0/H_L \leq \infty$ is plotted in Figure 5.14. It can be seen that Forchheimer's equation is only valid (within 5%) for holes with $r_0/H_L < 0.1$. For example, a large hole (as defined by Giroud and Bonaparte, 1989a) in a geomembrane overlying a thin geosynthetic clay liner ($H_L = 10$ mm) corresponds to $r_0/H_L \cong 0.56$, and in this case Forchheimer's equation would underestimate the flow by 30%. The results in Figure 5.14 also show that the solution approaches the one-dimensional case when $r_0/H_L > 20$.

Anisotropy in the layer beneath the geomembrane will increase leakage in proportion to $(k_{Lh}/k_{Lv})^{1/2}$. As $r_0/H_L \rightarrow \infty$, the effect would be negligible and k_{Lv} would control leakage. Rowe and Booker (2000) found that leakage in an anisotropic layer can be obtained from

$$Q_a = \frac{Q}{r_0 k_{Lv} h_d}\left(1 + f\sqrt{\frac{k_{Lh}}{k_{Lv}}}\right) \tag{5.39}$$

where

$$f = 0.5\left(1 - \sqrt{\frac{k_{Lv}}{k_{Lh}}}\right)\left(1 - \tanh\left(\gamma + 0.6 \ln\left(\frac{r_0}{H_L}\right)\right)\right) \tag{5.40}$$

and

$$\gamma = -0.167 - 0.0073\sqrt{\frac{k_{Lh}}{k_{Lv}}} \tag{5.41}$$

Equations 5.37, 5.40 and 5.41 were obtained by curve-fitting to the results obtained from the analytical solution and are applicable for most practical cases of $1 \leq k_{Lh}/k_{Lv} \leq 100$.

5.7.4 Circular hole in a geomembrane with imperfect contact – general 1D solution

The solution presented in the previous section assumed perfect contact conditions between the geomembrane and the underlying soil. Rowe (1998a) derived a solution for the case where lateral flow is permitted along the interface between the geomembrane and underlying

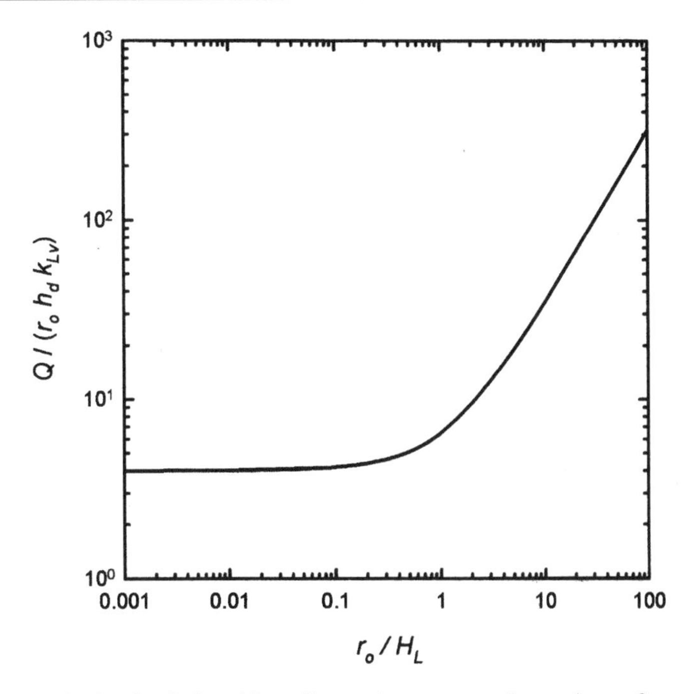

Figure 5.14 Leakage through circular hole with radius r_0 in a geomembrane in perfect contact with underlying material of thickness H_L and hydraulic conductivity k_{Lv} (modified from Rowe and Booker, 2000).

material with transmissivity θ. The geometry of the problem is shown in Figure 15.3d. The geomembrane rests on a low-permeability layer of thickness H_L and isotropic hydraulic conductivity k_L (e.g., a compacted clay liner, CCL; or geosynthetic clay liner, GCL) which in turn rests on some more permeable (but not highly permeable) foundation layer with thickness H_f and isotropic hydraulic conductivity k_f (e.g., natural soil deposit or engineered attenuation layer) which rests on a highly permeable layer in which there is no significant matric suction. The leachate is mounded to a depth h_w on the liner and taking a datum at the top of the underlying aquifer, the head in the aquifer is h_a. It is assumed that the geomembrane is not in perfect contact with the clay (e.g., from undulations in the geomembrane or underlying soil creating gaps) that can be quantified by transmissivity θ.

The solution considers: (a) lateral flow along the interface between the geomembrane and the underlying soil radially outward from the hole to wetted radius r_w and (b) 1D (i.e., vertical) flow through soil beneath the wetted radius. The wetted radius and thereby the leakage can be calculated for a given change in total head, underlying soil conditions and interface transmissivity using the procedure outlined below.

Provided that the hydraulic conductivity of the liner is low relative to the other soil layers, the flow through the liner and foundation may be approximated as vertical and is given by

$$Q = k_s i_s A \tag{5.42a}$$

where k_s is the harmonic mean hydraulic conductivity of the liner and foundation layer, i_s the average gradient through the liner and foundation layers, and

$$\frac{H_L + H_f}{k_s} = \frac{H_L}{k_L} + \frac{H_f}{k_f} \tag{5.42b}$$

$$i_s = \frac{H_f + H_L + h_w - h_a}{H_f + H_L} = 1 + \frac{h_w - h_a}{H_f + H_L} \tag{5.42c}$$

The head distribution beneath the geomembrane and acting on the low-permeability liner is given by the solution to the equation (Giroud and Bonaparte, 1989b):

$$\frac{d^2h}{dr^2} + \frac{1}{r}\frac{dh}{dr} - \alpha^2 h = \alpha^2 C \tag{5.43a}$$

where

$$\alpha^2 = \frac{k_s}{(H_L + H_f)\theta} \tag{5.43b}$$

$$C = H_L + H_f - h_a \tag{5.43c}$$

Equations 5.43a–5.43c are solved subject to the boundary conditions:

$$h = h_w \quad \text{when } r = r_0 \tag{5.44a}$$

$$h = 0 \quad \text{when } r = r_w \tag{5.44b}$$

and r_w is the wetted radius, and $r = 0$ corresponds to the centre of the circular hole.

An analytical solution to equations 5.43a–5.43c, 5.44a and 5.44b and for the head h at position r can be obtained by extending the work of Jayawickrama *et al.* (1988):

$$h = (H_L + H_f + h_w - h_a)\Omega_0 + (H_L + H_f - h_a)(\Omega_1 - 1) \tag{5.45a}$$

where

$$\Omega_0 = \frac{K_0(\alpha r)I_0(\alpha r_w) - K_0(\alpha r_w)I_0(\alpha r)}{K_0(\alpha r_0)I_0(\alpha r_w) - K_0(\alpha r_w)I_0(\alpha r_0)} \tag{5.45b}$$

$$\Omega_1 = \frac{K_0(\alpha r)I_0(\alpha r_0) - K_0(\alpha r_0)I_0(\alpha r)}{K_0(\alpha r_w)I_0(\alpha r_0) - K_0(\alpha r_0)I_0(\alpha r_w)} \tag{5.45c}$$

where K_0 and I_0 are modified Bessel functions of order zero. Wetted radius r_w can be evaluated by finding the head at $r = r_w$ that satisfies

$$\frac{dh}{dr} = 0 \tag{5.46}$$

where

$$\frac{dh}{dr} = (H_L + H_f + h_w - h_a)\Lambda_1 + (H_L + H_f - h_a)\Lambda_2 \tag{5.47a}$$

and

$$\Lambda_1 = \frac{-\alpha K_1(\alpha r_w)I_0(\alpha r_w) - \alpha K_0(\alpha r_w)I_1(\alpha r_w)}{K_0(\alpha r_0)I_0(\alpha r_w) - K_0(\alpha r_w)I_0(\alpha r_0)} \tag{5.47b}$$

$$\Lambda_2 = \frac{-\alpha K_1(\alpha r_w)I_0(\alpha r_0) - \alpha K_0(\alpha r_0)I_1(\alpha r_w)}{K_0(\alpha r_w)I_0(\alpha r_0) - K_0(\alpha r_0)I_0(\alpha r_w)} \tag{5.47c}$$

where K_1 and I_1 are modified Bessel functions of order 1. Equations 5.47a–5.47c can be evaluated to find the value of R for which equation 5.46 is satisfied.

The flow, Q, through the hole and the low-permeability soil within the zone defined by the wetted radius is given by

$$Q = \pi k_s \left[r_0^2 i_s + 2 i_s \Delta_1 + 2 i_s \Delta_2 - \frac{2h_w}{H_L + H_f}\Delta_2 \right] \tag{5.48a}$$

where

$$\Delta_1 = \frac{-[r_w\lambda_1(r_0, r_w)K_1(\alpha r_w) + r_w\lambda_2(r_0, r_w)I_1(\alpha r_w)]}{\alpha} + \frac{r_0\lambda_1(r_0, r_w)K_1(\alpha r_0)}{\alpha} + \frac{r_0\lambda_2(r_0, r_w)I_1(\alpha r_0)}{\alpha} \quad (5.48b)$$

and

$$\Delta_2 = \frac{[-r_w\lambda_1(r_w, r_0)K_1(\alpha r_w) - r_w\lambda_2(r_w, r_0)I_1(\alpha r_w)}{\alpha} \frac{+ r_0\lambda_1(r_w, r_0)K_1(\alpha r_0) + r_0\lambda_2(r_w, r_0)I_1(\alpha r_0)]}{\alpha} \quad (5.48c)$$

$$\lambda_1(X, Y) = \frac{I_0(\alpha Y)}{K_0(\alpha X)I_0(\alpha Y) - K_0(\alpha Y)I_0(\alpha X)} \quad (5.48d)$$

$$\lambda_2(X, Y) = \frac{K_0(\alpha Y)}{K_0(\alpha X)I_0(\alpha Y) - K_0(\alpha Y)I_0(\alpha X)} \quad (5.48e)$$

Leakage Q can be evaluated using equations 5.48a–5.48e for any combination of layer properties, hole size or head. Equations 5.47a–5.47c and 5.49 can be programmed using published code for Bessel functions (Press *et al.*, 1986). The implementation by Rowe and Lake (1997) was used to obtain the solutions for leakage through common composite liners presented in Sections 13.4.3(a) and 13.4.4(a).

The assumption of 1D flow through the soil beneath the geomembrane can be assessed using the numerical modelling of the same problem reported by Foose *et al.* (2001). Using a 3D finite difference model, they showed that there is negligible difference between the 1D approach and modelling 3D flow in the underlying soil, provided that the transmissivity is greater than $2 \times 10^{-10}\,m^2/s$ for a GM/CCL composite liner (with $k = 1 \times 10^{-9}$ m/s) and greater than $8 \times 10^{-13}\,m^2/s$ for a GM/GCL composite liner (with $k = 2 \times 10^{-10}$ m/s). The solution of Rowe (1998a) can therefore be used to estimate leakage through GM/CCL and GM/GCL composite liners for expected practical ranges of transmissivity (see Section 13.4.2).

The solution of Rowe (1998a) does underestimate the leakage for "perfect" contact between the GM and a GCL by a factor of 2. For the case of "perfect" contact, the semi-analytical solution given in Section 5.7.2 or the numerical solution of Foose *et al.* (2001) can be used, although it is likely that "perfect" contact rarely, if ever, exists in a landfill.

5.7.5 Hole in a geomembrane wrinkle and imperfect contact – general 1D solution

The previous section assumed that the hole occurred in an area away from a wrinkle in the geomembrane (see Section 13.2.4). However, there is no reason that the hole could not be at (or very close to) the location of the wrinkle. Since the gap beneath the wrinkle and the underlying soil provides essentially no resistance to lateral flow, the wrinkle will result in larger leakage.

Rowe (1998a) also presented a solution for the case where a hole coincides with a wrinkle in the geomembrane of length L and width $2b$ (see Figure 5.13e). The height of the wrinkle is such that the transmissivity beneath the wrinkle is much greater than that between the geomembrane and the soil beyond. The interface transmissivity is taken as θ where the geomembrane is in contact with the underlying soil. It is also assumed that $L >> b$ such that the effects of leakage at the ends of the wrinkle can be neglected.

This solution assumes unobstructed lateral flow to the width of the wrinkle and then lateral flow to wetted distance x_w. Vertical flow through the underlying soil is considered beneath the wetted distance over the length of the wrinkle. For this case, the head distribution beneath the geomembrane and acting on the low-permeability liner is given by

$$\frac{d^2h}{dx^2} - \alpha^2 h = \alpha^2 C \tag{5.49}$$

where α and C are as defined in equations 5.43b and 5.43c, and $x = 0$ occurs at the centre of the wrinkle. Equation 5.49 has the solution:

$$h = (h_w + C)\, e^{-\alpha(x_w - b)} - C \tag{5.50}$$

and the total flow is given by

$$Q = 2Lk_s\left[b + \frac{1}{\alpha}(1 - e^{-\alpha(x_w-b)})\right] \frac{(H_L + H_f + h_w - h_a)}{(H_L + H_f)} \tag{5.51}$$

where x_w is the wetted distance away from the centre of the wrinkle. Similar to calculating the wetted radius in Section 5.7.4, the wetted distance is found where $h = 0$ for at x_w and is given by

$$x_w = b - \ln\left(\frac{C}{h_w + C}\right) \times \left(\frac{(H_L + H_f)\theta}{k_s}\right)^{0.5} \tag{5.52a}$$

$$C = H_L + H_f - h_a \tag{5.52b}$$

Knowing the value of x_w, the leakage Q can be calculated from equation 5.51.

This approach assumes that wrinkles with holes are spaced far enough apart such that they do not interact. This is valid provided that the distance between wrinkles is greater than $2x_w$. If wrinkles with holes are close enough that they could possibly interact, then x_w should be taken as half the distance between the wrinkles with holes. Leakage through wrinkles with holes for common composite liners is examined in Sections 13.4.3(b) and 13.4.4(b).

5.7.6 Capacity of the fluid to drain to the hole

The leakage calculated using equations 5.48 and 5.51 cannot exceed (a) the flow from Bernoulli's

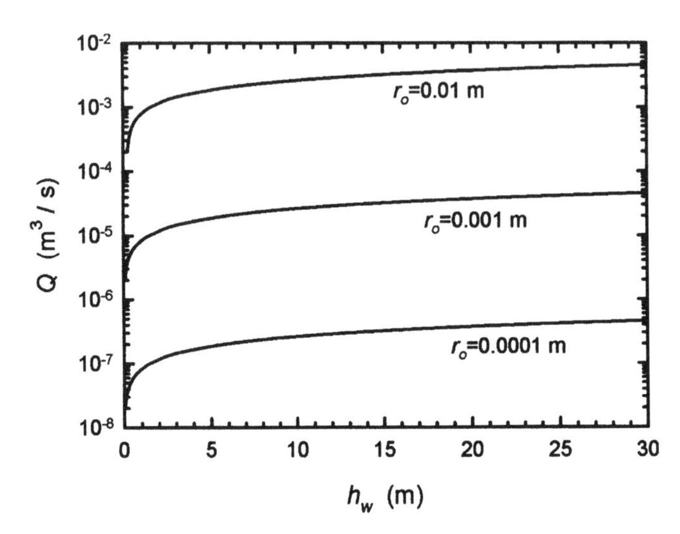

Figure 5.15 Capacity of fluid to drain to a circular hole with radius r_0 ($q_0 = 10^{-9}$ m/s and $k_{om} = 10^{-2}$ m/s).

equation if the permeability of the overlying layer is high enough or (b) the capacity of the fluid to drain to the hole. The latter case corresponds to the case examined by Giroud *et al.* (1997b) where the limiting value of Q is obtained (iteratively) from

$$h_w = \left\{ \frac{r_0^2 q_0}{2k_{om}} + \frac{Q}{2k_{om}\pi}\left[\ln\left(\frac{Q}{\pi r_0^2 q_0}\right) - 1\right] + \frac{1}{4g^2}\left(\frac{Q}{1.88r_0^2}\right)^4 \right\}^{0.5} \quad (5.53)$$

where q_0 is the liquid supply (e.g., permeation through the waste per unit area reaching the leachate collection system), k_{om} the hydraulic conductivity of the permeable leachate collection layer over the geomembrane (see Figure 5.13f) and all other terms are as previously defined. This expression reduces to the Bernoulli equation (equation 5.34) as k_{om} tends to infinity. Equation 5.53 is plotted in Figure 5.15 for the specific case of $q_0 = 10^{-9}$ m/s and $k_{om} = 10^{-2}$ m/s for a range of h_w and for three different values of r_0. As shown in Section 13.4, equation 5.53 may only limit leakage for very long wrinkles.

Chapter six

Chemical transfer by diffusion

6.1 Introduction

The basic concepts of pollutant migration by diffusion through clayey soils and geomembranes are presented in this chapter. Chapter 7 deals with modelling of advective–diffusive transport, Chapter 8 deals with laboratory evaluation of parameters, Chapter 9 describes observed field behaviour, Sections 12.6 and 13.5 address issues specifically related to GCLs and geomembranes, and Chapters 10, 11 and 16 examine the implications of diffusion with respect to landfill design and performance.

In field situations where there are very low hydraulic conductivities or very small hydraulic gradients, there will be little or no fluid flow. In these cases, the mass flux through a barrier may be controlled by chemical concentration gradients. This phenomenon was discussed in Chapter 1. Since clay minerals are surrounded by adsorbed cations in amounts of 5–150 milliequivalents/100 g of solid, the replacement of exchange cations, and in some cases anions, becomes an important factor in retardation of certain dissolved species. Therefore, retardation and the use of distribution coefficients are also discussed.

During hydraulic conductivity testing of compacted soils, diffusion into soil peds from adjacent compaction induced macropores and fractures causes tailed breakthrough curves of effluent concentration. This requires extended testing times and increased costs for clay/leachate compatibility testing, especially if K^+ and NH_4^+ fixation occur.

6.2 Free solution diffusion (D_0)

Although the modelling techniques for prediction of pollutant migration are presented later in Chapter 7, a brief review of the simple 1D transport equations at this point will serve to demonstrate the nature of diffusion and the component parts of an effective diffusion coefficient.

Maximum rates of migration by diffusion occur in bulk or free water at extreme dilution. The driving force for diffusion is a concentration gradient, or, more correctly, a chemical potential gradient as shown schematically in Figure 6.1. The equation for the mass flux, f, is Fick's first law as follows:

$$f = -D_0 \frac{dc}{dx} \tag{6.1}$$

where f is the mass flux produced by the concentration gradient dc/dx [$ML^{-2}\ T^{-1}$], D_0 the free

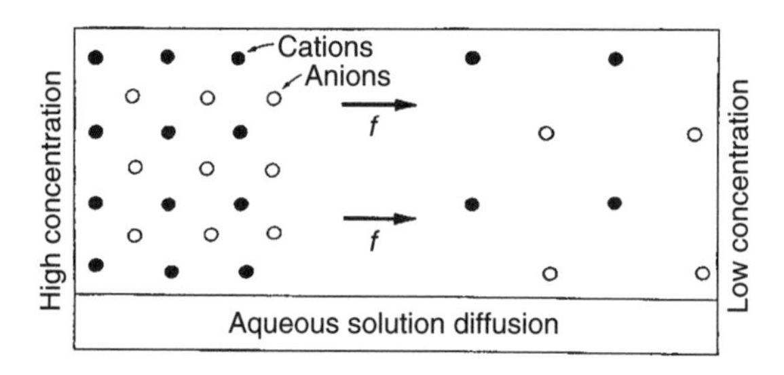

Figure 6.1 Schematic illustration of diffusion cations and anions from high to low concentration in water. Equilibrium would be reached when a uniform concentration develops.

solution diffusion coefficient [$L^2 T^{-1}$], c the solute concentration in the fluid at any point x [ML^{-3}] and x the distance from the high-concentration interface [L].

Typical values of D_0 (Lerman, 1979) are presented in Table 6.1 for several species of cations and anions which might be encountered in leachate. These values, determined at infinite dilution, would normally represent the maximum values measurable in any laboratory or field situation at the nominal temperature. It is important to note that requirements for electroneutrality generally demand that a cation/anion pair diffuse together, frequently slowing one or other of the migrating components.

Table 6.1 Tracer diffusion coefficients, D_0, of selected species at infinite dilution at 25 °C (adapted from Lerman, 1979; Cussler, 1997)

Species	$D_0(\times 10^{10} m^2/s)$	
Cation		
H^+	93.1	Note high D_0
Li^+	10.3	Monovalent
Na^{2+}	13.3	
K^+	19.6	
NH_4^+	19.8	
Mg^{2+}	7.05	Divalent
Ca^{2+}	7.93	
Mn^{2+}	6.88	
Fe^{2+}	7.19	
Cu^{2+}	7.33	
Zn^{2+}	7.15	
Cd^{2+}	7.17	
Pb^{2+}	9.45	
Anion		
OH^-	52.7	Note high D_0
Br^-	20.8	Monovalent
I^-	20.5	
Cl^-	20.3	
HS^-	17.3	
NO_2^-	19.1	
NO_3^-	19.0	
F^-	14.7	
HCO_3^-	11.8	
CH_3COO^-	10.9	
$CH_3CH_2COO^-$	9.5	
SO_4^{2-}	10.7	Divalent
CO_3^{2-}	9.5	
PO_4^{2-}	6.12	
CrO_4^{2-}	11.2	
$Fe(CN)_6^{3-}$	9.8	Trivalent

Even D_0 is a complex coefficient as shown by the Nernst equation (Lerman, 1979) which says

$$D_0 = \frac{RT}{F^2|z|}\lambda_i^0 \tag{6.2}$$

where R is the universal gas constant = 8.314 J/K/mol, F the Faraday constant = 96,500 C, z the valence, T the absolute temperature in Kelvin and λ_i^0 the limiting ionic conductance which is maximum at infinite dilution when the ions are most mobile and increases with increasing temperature (cm^2/Ω equivalent).

For strong electrolytes, ionization is complete over a wide range of concentrations and λ stays close to λ^0. For less easily dissociated species, λ decreases with increasing concentration as many ions diffuse in molecular form, thus decreasing the diffusion coefficient. The ionic mobility, u_i, which is similar in form to D_0, is related to λ_i^0 by the equation

$$u_i = \frac{N\,\lambda_i^0}{F^2|z_i|} \tag{6.3}$$

where N is Avogadro's number (the number of units in a mole = 6.022×10^{23}).

The effects of ionic potential, $|z|/r$, on values of D_0 calculated from equation 6.2 are shown in Figure 6.2. The plot shows a decrease in D_0 with increasing $|z|/r$ which indicates that the greater the degree of ion hydration, the lower is D_0. This is particularly well shown by the monovalent alkalis, Li, Na, K, Rb and Cs, the latter being the largest cations with the smallest hydrated radii (Mitchell, 1993). The atomic weight also increases significantly from Li^+ to Cs^+, another factor which could influence the rate of diffusion in a gravitational field.

The effects of temperature on D_0 are shown in Figure 6.3 for a number of species common to

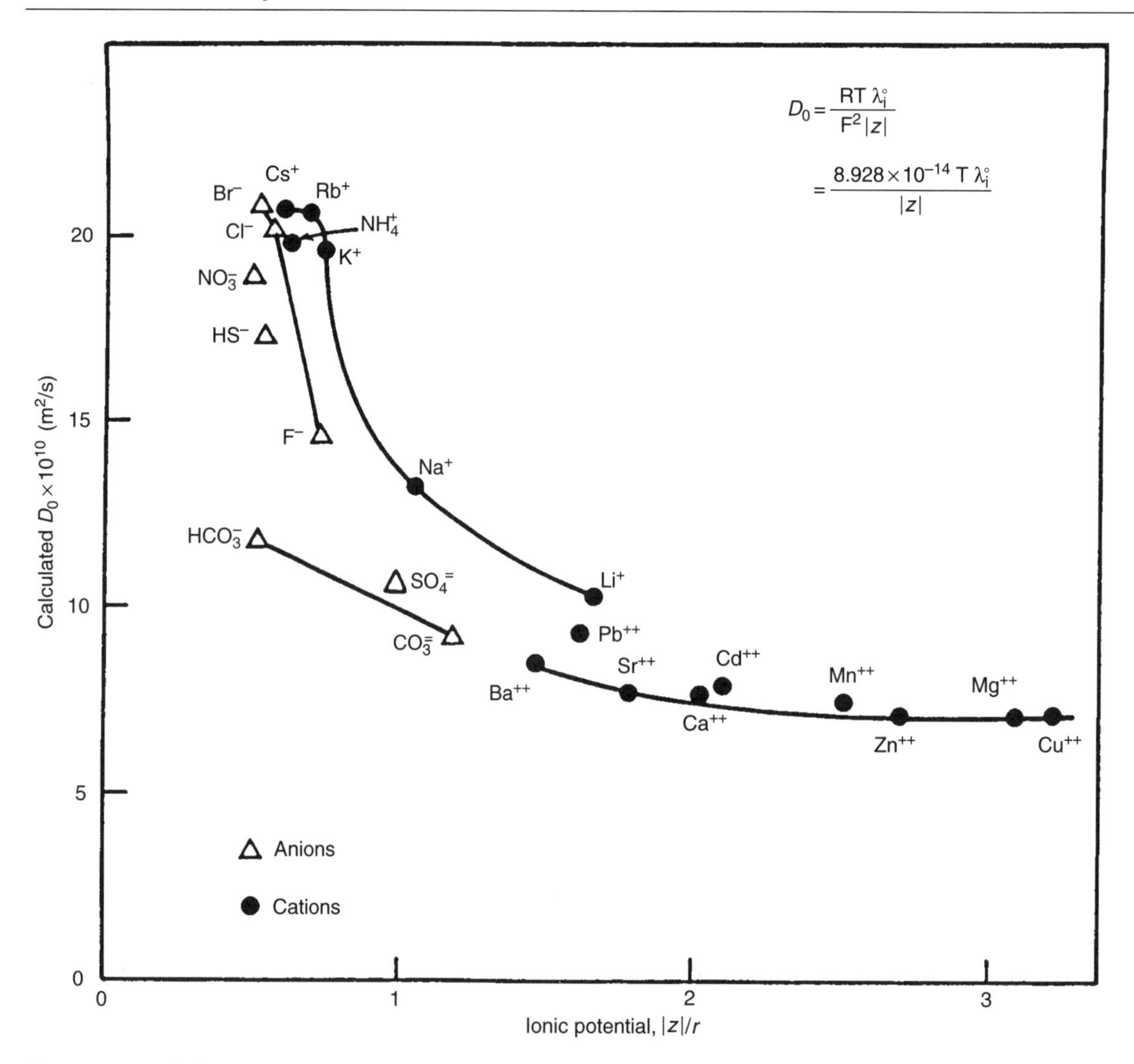

Figure 6.2 Relation between free solution diffusion coefficient, D_0, and ionic potential, $|z|/r$, at 25 °C ($|z|$ = valence, r = ionic radius) (after Quigley *et al.*, 1987b; reproduced with permission, ASCE).

municipal solid waste (MSW) leachate. Chloride, which is frequently considered as a conservative reference species, has a relatively high diffusion coefficient. Despite the similar high diffusion values for K^+ and NH_4^+, their rate of migration in the field may be substantially lower than implied by the diffusion coefficient above since both are highly retarded by clays in most field situations. If landfill temperatures reach and stay at 50 °C for several years, consideration should be given to the temperature in the liner and if this exceeds 20–25 °C then the diffusion coefficient D may be higher than typical laboratory values which are normally obtained either at 20–25 or at 10 °C.

In summary, the mobility of a species in free solution, as expressed by its D_0 value, is a complex function of its mass, radius, valence, concentration/

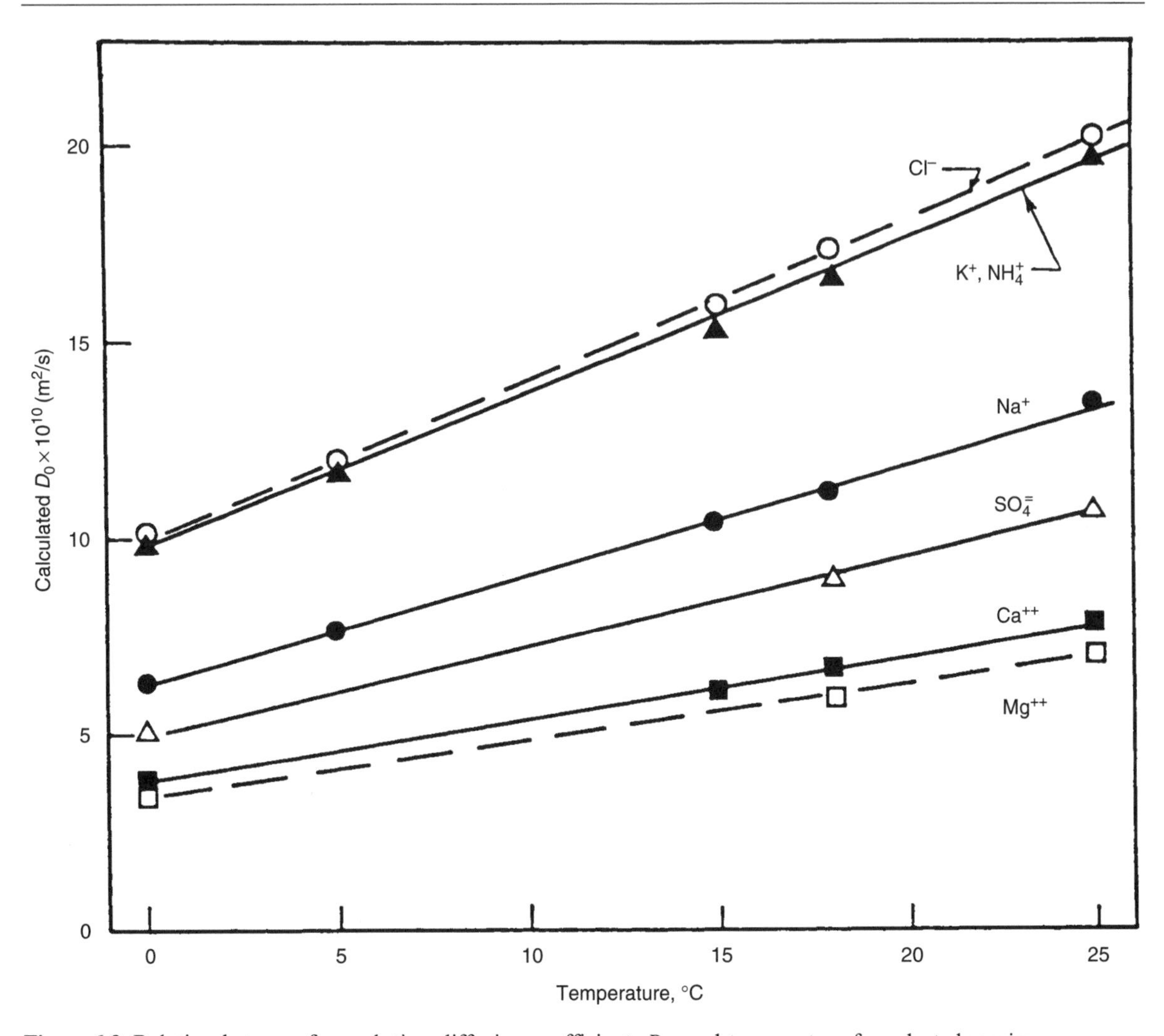

Figure 6.3 Relation between free solution diffusion coefficient, D_0, and temperature for selected species.

dissociation state, and the viscosity, dielectric constant and temperature of the diffusing medium.

6.3 Diffusion through soil

The presence of soil particles, particularly adsorptive clay minerals and organic matter complicates the diffusion process. Diffusion through a network of clay particles involves the diffusive movement of the species of interest in the pore water between the clay particles as illustrated in Figure 6.4.

The rate of chemical movement or mass flux through a soil may be slower than by diffusion in pure water for the following reasons:

1. tortuous flow path around particles (τ);
2. small fluid volume for flow (n or θ);
3. increased viscosity especially the double-layer portion of the pore water;
4. retardation of certain species by cation and anion exchange (retardation) by both clay minerals and organics;
5. biodegradation of diffusing organics;

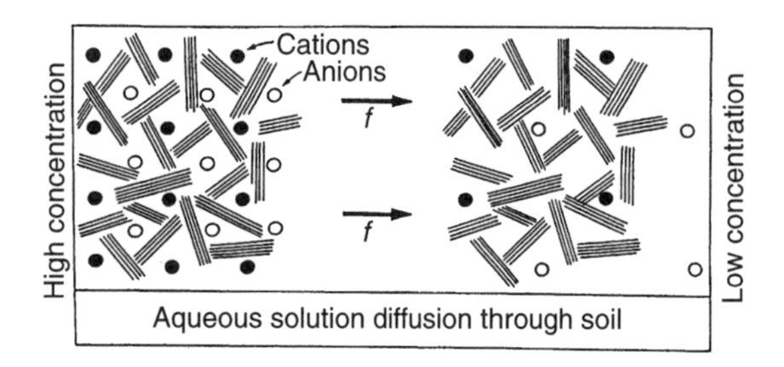

Figure 6.4 Schematic illustration of diffusing cations and anions from high to low concentration through a saturated clayey soil.

6. counter-osmotic flow;
7. electrical imbalance possibly by anion exclusion.

However, certain events in a soil may also accelerate movement. For example:

1. Decreased viscosity caused by any of the factors which cause double-layer contraction at constant void ratio.
2. Decreased viscosity caused by K^+ fixation by vermiculite which decreases the cation exchange capacity (CEC) and increases the free water pore space at constant void ratio.
3. Electrical imbalances such as accelerated anion migration that would electro-osmotically pull cations.
4. Electrical imbalances related to cation exchange and hardness halos that might pull chloride to help balance desorbed Ca^{2+} and Mg^{2+}.
5. Attainment of chemical equilibrium eliminating retardation of certain species.

The focus of this book is on a representation of diffusive transport that is of practical value in modelling contaminant transport through barrier systems. As illustrated in Chapter 9, this approach has provided good results in field application. However, it is important that in each case where one is considering to adopt a simplified approach, one recognize the nature of the simplifications and ask oneself whether they are acceptable. If they are, then one can proceed; if not then a more complex formulation is required. With this in mind, it is appropriate to look first at the process of steady-state diffusion, then at the process of transient diffusion.

6.4 Steady-state diffusion

Two useful examples of steady-state diffusion through a bentonitic soil were presented by Dutt and Low (1962) and Kemper and van Schaik (1966). In both papers, the concentration gradient was produced by a chloride salt across a very porous saturated montmorillonite barrier homoionized in the same cation as the diffusing salt. A sketch of the Dutt and Low experimental set-up is presented in Figure 6.5. The 1D equation for the mass flux used by the previous two authors was:

$$f = -D_p \frac{dc}{dx} = -D_p i_c \qquad (6.4a)$$

where D_p is the porous media diffusion coefficient $[L^2 T^{-1}]$ which is related to D_0 by the relationship $D_p = D_0 W_T$, and W_T is the complex tortuosity factor which equals unity for a pure solution and is normally less than unity in soils, and the concentration gradient $i_c = dc/dx$. In reality, many factors are lumped into W_T as discussed later. It is most important to note that the porosity, n (which equals the volumetric

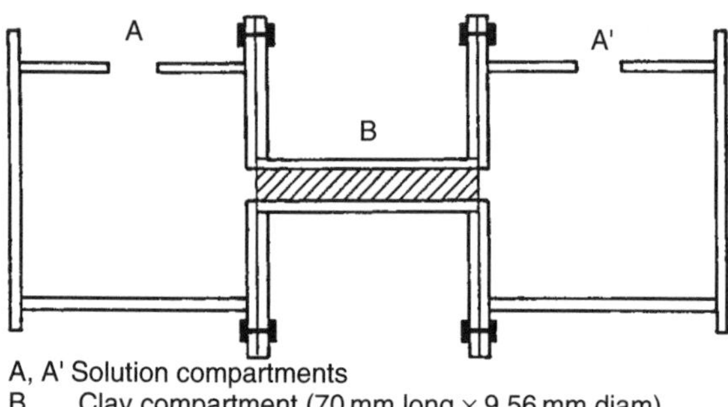

Figure 6.5 Schematic illustration of diffusion cell used by Dutt and Low (1962).

water content, θ, for a saturated soil), is included in D_p in equation 6.4a, possibly because the authors' slurries had high porosities of 0.85. In the case of clayey barriers and porosities close to 0.3, D and n should probably be clearly separated from one another. Thus, $D_p = \theta D_e = nD_e$ where D_e is the effective diffusion coefficient in the soil pores (the reader should be aware that the literature is not consistent in its definition of the term "effective diffusion coefficient" and in some literature the term D_p is called the effective diffusion coefficient – in this book it is called to porous media diffusion coefficient). Also, in practical work, τ is used as the symbol for tortuosity as subsequently defined in equation 6.7d.

The experimental procedure of the above authors, illustrated schematically in Figure 6.6, appears to be a deceptively simple way of measuring diffusion coefficients, just by measuring the mass flux, f, across the barrier. Equation 6.4a becomes

$$f = -D_p \frac{\Delta c}{\Delta x} \tag{6.4b}$$

where $\Delta c = c_2 - c_1$ and Δx is the barrier thickness (Figure 6.6).

An important observation from these two sets of experiments was a sharp drop in concentration gradient across both reservoir/soil interfaces, especially the input or high concentration side. Dutt and Low argued that the chloride part of the Na^+:Cl^- ion pair would tend to be rejected by the Na^+:clay$^-$ system, hence resisting

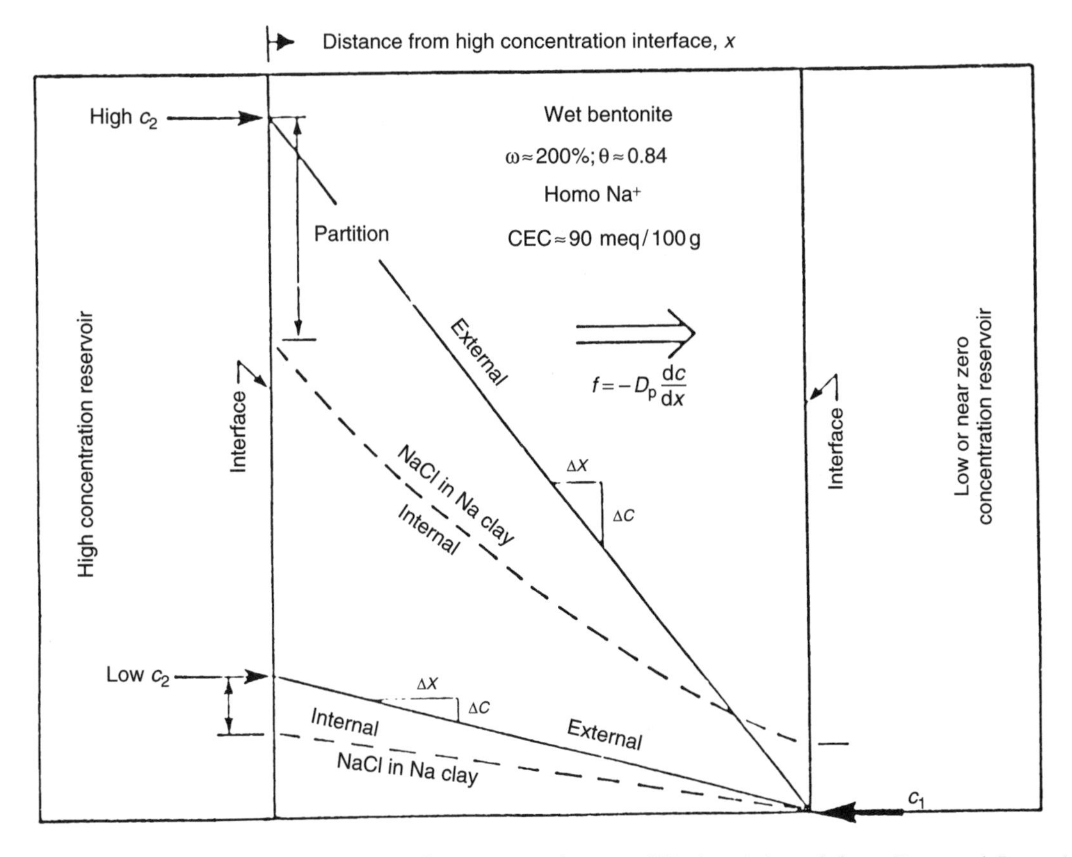

Figure 6.6 Chemical profiles and other gradients at steady-state diffusion (adapted from Dutt and Low, 1962; Kemper and van Schaik, 1966).

diffusion (anion exclusion) and effectively producing a very low diffusion coefficient and a large concentration gradient in a thin interfacial layer.

A similar sharp concentration drop across the waste/clay interface at the Confederation Road landfill site was believed to have been observed by Crooks and Quigley (1984). This has been attributed to plugging by heavy metal precipitation, Na^+ adsorption and possibly bacterial growth. The concentration profiles were modelled by Quigley and Rowe (1986) assuming a greatly reduced diffusion coefficient in a thin interfacial zone. Interface partitioning in the field remains unresolved, however, since it has not been observed at other sites. At Confederation Road, a mat of fine black organics may have shifted the effective interface up 25 or 30 cm thus confusing the observations.

Since the mass flux is constant at all points in a steady-state system, both external (across the barrier) and internal (within the clay) diffusion coefficients may be calculated and these are presented in Table 6.2 for the Dutt and Low data.

It is significant that the much larger internal D_p values were similar to those obtained by deuterium diffusion and were considered characteristic of the soil, whereas the lower external values seem more characteristic of the system.

Major curvature of the pore water concentration profiles at high concentrations of NaCl (Figure 6.6) was explained by Kemper and van Schaik (1966) as being caused by water movement to the low salt side (0.01 N) of their wet Na^+ bentonite (Figure 6.7) at the expense of the

Table 6.2 Average external and internal diffusion coefficients obtained by Dutt and Low (1962)

	D_p (m²/s)	
Salt/clay system	*External (across barrier)*	*Internal (within clay)*
LiCl/LiMont	$\sim 1.3 \times 10^{-10}$	$\sim 4.1 \times 10^{-10}$
NaCl/NaMont	$\sim 2.1 \times 10^{-10}$	$\sim 3.7 \times 10^{-10}$

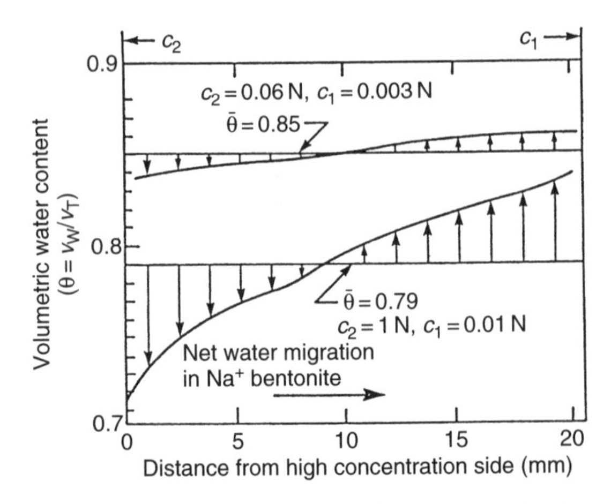

Figure 6.7 Water content profiles at steady-state diffusion for high NaCl contents and wet clays (adapted from Kemper and van Schaik, 1966).

high salt side (1.0 N). Other experiments with more dilute NaCl systems and with $CaCl_2$ yielded little water movement and nearly linear salt concentration profiles across the barrier. Such water migration might occur in slurry walls but seems unlikely for most natural or compacted clays unless soils are left completely unconfined.

Chemical concentration gradients, while in themselves easy to visualize, induce complex, coupled electrical (Section 6.4.1) and osmotic gradients (Section 6.4.2) that affect the net flow attributable to diffusion. In Table 6.3, transfer rates calculated by Kemper and van Schaik for only two of their pure salt systems are presented. This was possible because Kemper and van Schaik employed electrode screens at the ends of the soil to assess the relative speed of Na^+ and Cl^- migration by measuring charge build-up. A positive transfer rate indicates flow from the high-concentration to the low-concentration side. The concentration gradient produces positive cation and anion migration, the rate of anion migration being greater. This collection of surplus anions (Cl^- in this case) generates a negative potential at the low-concentration side of the clay. This negative potential in turn

Table 6.3 Components of cation and anion movement during steady-state diffusion through bentonite (adapted directly from Kemper and van Schaik, 1966)

Adsorbed cation *Salt solution*	*Na* *NaCl*	*Ca* *$CaCl_2$*
Conc. bulk solution		
High side	3.21 g/l (0.055 N)	5.6 g/l (0.05 N)
Low side	0.18 g/l (0.003 N)	0.3 g/l (0.003 N)
Average water content, θ	0.85	0.84
Transfer rates of cations (meq/m^2/s × 10^{12})		
By diffusion	+1.8	+0.84
By electrical flow	+4.1	+0.98
By bulk flow	3.8	-0.12
Net flow	, 2.1	-1.70
Transfer rates of anions (meq/m^2/s × 10^{12})		
By diffusion	+2.7	-2.14
By electrical flow	−0.2	-0.37
By bulk flow	4	-0.07
Net flow	+2.1	+1.70

attracts hydrated cations by positive-direction electro-osmotic flow and repels anions by negative-direction flow. The large, negative-direction bulk flow is osmotic flow through the clay as water from the low-concentration side seeks to enter the high-concentration side. The net mass flux is in the positive direction, and, while complex, all effects have resulted from the initial concentration gradient.

6.4.1 Modelling ionic diffusion

Rigorous modelling of the diffusion of ions requires the solution of the extended Nernst–Planck equations in which the flux of each species is given by

$$f_i = -D_i\left(\frac{dc_i}{dx} + \frac{c_i F}{RT} z_i \frac{dV}{dx} + c_i \frac{d(\ln \gamma_i)}{dx}\right) \qquad (6.5)$$

where f_i is the flux of ionic species i, D_i the diffusion coefficient, c_i the concentration of species i, z_i its valence, F the Faraday constant, R the ideal gas constant, T the temperature (in Kelvin), V the electrical potential and γ_i the chemical activity coefficient. Conservation of mass (Section 1.3.7) must then be invoked for each ionic species. To complete the system of equations, a relation is required to account for the electrical potential that is locally induced by the movement of all ions (Samson and Marchand, 1999). The relationship that should be adopted will depend on the nature of the problem being examined. In some cases, assumptions of electroneutrality or null current may be appropriate and can be used to calculate the potential, V, in the simplified Nernst–Planck equation (i.e., equation 6.5 without the chemical activity term). However, more generally, it may be necessary to use the Poisson equation (Samson and Marchand, 1999) to relate the electrical potential to the electrical charge in space. The Debye–Hückel model or the extended Debye–Hückel model can then be used to calculate the chemical activity coefficient for each ionic species, i (Samson and Marchand, 1999). A numerical (finite element) approach to solving this series of coupled equations proposed by Samson *et al.* (1999) could be used to model multi-ion diffusion in well-defined systems (e.g., simple laboratory experiments); however, the practical utility of the approach comes into question when wishing to model contaminants from a complex landfill leachate and a simpler, albeit approximate, approach is required as discussed in the remainder of this book.

6.4.2 Concentration-dependent diffusion

Diffusion coefficients are generally assumed to be independent of concentration; however, in reality they may be concentration-dependent for a number of reasons. For example, solutions of weak electrolytes (e.g., acetic acid that is often found in leachates) will contain cations, anions and molecules in local equilibrium. The diffusion coefficient for ions alone, assuming no sorption and electrical neutrality is maintained, is given by the harmonic mean of the diffusion coefficients of the individual ions in the ion pair. However, the diffusion of the ions and the molecules are interdependent and are linked by the association constant, $K = 10^{pK_a}$, with the average diffusion coefficient of the dimerizing solute changing from that of the monomer to that of the dimer at a concentration that is roughly the reciprocal of the association constant (Cussler, 1997). Fortunately, in many cases, the difference is not sufficiently large to be of practical significance in barrier design. For example, in free solution, the diffusion coefficient for acetic acid is reported (Cussler, 1997) to range between 18×10^{-10} and $19.5 \times 10^{-10}\,m^2/s$ depending on the degree of ionization (where the fully ionized value can be deduced from the diffusion coefficients given in Table 6.1 for H^+ and CH_3COO^-: $D_0 = 2/(1/93.1 + 1/10.9) \times 10^{-10} = 19.5 \times 10^{-10}\,m^2/s$).

The diffusion coefficient can also be concentration-dependent because of membrane effects as discussed in the following section.

6.4.3 Osmotic flux

Electrostatic repulsion associated with the diffuse double layer (see Section 3.5) of clays gives rise to a membrane action which restricts the movement of ions through the pores of clays (e.g., Cey *et al.*, 2001) and, by an analogous mechanism, in polymer gel membranes (Darwish *et al.*, 2004). This membrane action gives rise to the osmotic movement of water in a direction opposite to the concentration gradient. Clays are non-ideal membranes and the chemico-osmotic efficiency, ω, depends on factors such as the clay mineralogy of the soil, the void ratio and the contaminants being considered. For a given ionic contaminant, the chemico-osmotic efficiency generally increases with increasing amounts of high-activity clay (e.g., sodium montmorillonite) and lower void ratio (e.g., due to higher applied stress). The osmotic efficiency can vary with solute concentration (Shackelford *et al.*, 2001; Darwish *et al.*, 2004) and this may have some effect on the transient mass transfer of contaminants through clay liners.

Accounting for membrane behaviour, the mass flux of a contaminant through a clay (Mitchell, 1993) is given by

$$f = (1 - \omega)\, k i_h c + \omega k i_\Pi c - nDi_c \qquad (6.6)$$

where ω is the osmotic efficiency ($0 < \omega < 1$), k the hydraulic conductivity, c the contaminant concentration, n the porosity, D the diffusion coefficient and i_h, i_Π, and i_c are the hydraulic, chemico-osmotic and concentration gradients, respectively. In the absence of an applied hydraulic gradient (i.e., $i_h = 0$), the advective flux, given by the first term in equation 6.6, is zero. The second term in equation 6.6 represents the osmotic flux (Barbour and Fredlund, 1989) which is opposite to the direction of diffusion and hence will slow mass transport from the high-concentration to low-concentration side of the membrane. The osmotic efficiency, ω, can be calculated from Bresler (1973) while the osmotic pressure gradient, i_{Π}, can be calculated from Mitchell (1993). Examples, where osmotic flow has been examined, can be found in the literature and it may be important in the interpretation of diffusion tests conducted on membranes with high osmotic efficiency (e.g., Darwish *et al.*, 2004). The effect of osmotic flow is generally, and conservatively, neglected in the prediction of contaminant transport from waste disposal sites although there may be instances where taking account of membrane effects will improve the assessment of the suitability of a barrier for containing contaminants at low concentrations (e.g., Shackelford *et al.*, 2001).

6.4.4 Effective diffusion coefficient, D_e

The porous system diffusion coefficient, D_p, calculated from Fick's law for a system such as that shown in Figure 6.5 is a combination of a number of factors and might be written:

$$D_p = D_0 W_T = \theta D_e \tag{6.7a}$$

in which W_T, referred to as the complex tortuosity factor, has the following components:

$$W_T = \alpha_{ff}\gamma_e\left(\frac{L}{L_e}\right)^2\theta = \tau\theta \tag{6.7b}$$

and hence from equations 6.7a and 6.7b,

$$D_e = \alpha_{ff}\gamma_e\left(\frac{L}{L_e}\right)^2 D_0 = \tau D_0 \tag{6.7c}$$

$$\tau = \alpha_{ff}\gamma_e\left(\frac{L}{L_e}\right)^2 \tag{6.7d}$$

where α_{ff} is the decreased fluidity factor related to adsorbed double-layer water, γ_e the electrostatic interaction factor, $(L/L_e)^2$ the geometric tortuosity factor and θ the volumetric water content ($= n$, the porosity for a saturated system), and $\tau = \alpha_{ff}\gamma_e(L/L_e)^2$ is the tortuosity factor for the clay.

In texts such as Bear (1979), the use of the term "tortuosity" relates only to the geometric tortuosity $(L/L_e)^2$ and ignores the effects of α_{ff} and γ_e. While this may not be a bad approximation for granular materials, it is not correct to ignore α_{ff} and γ_e for clayey soils and as a result, the tortuosity factor for clayey soil, τ, is often less than the geometric tortuosity calculated by Bear for an ideal problem (i.e., $(L/L_e)^2 = 0.67$).

Using values of D_p calculated from their experiments ($(1.8-3.4) \times 10^{-10}\,\mathrm{m^2/s}$), Kemper and van Schaik calculated values for W_T which varied from about 0.07–0.3. Most of the variation in D_p was believed to be caused by changes in volumetric water content and this is shown in Figure 6.8 with θ expressed as the calculated number of water layers surrounding the bentonite particles.

An important study of non-steady-state diffusion offering further clarification of the nature of diffusion and its coupled electrical and hydraulic gradients was presented by Elrick *et al.* (1976) for Na^+ bentonite. In this study, a constant volume cell (Figure 6.9) prevented osmotic flow across

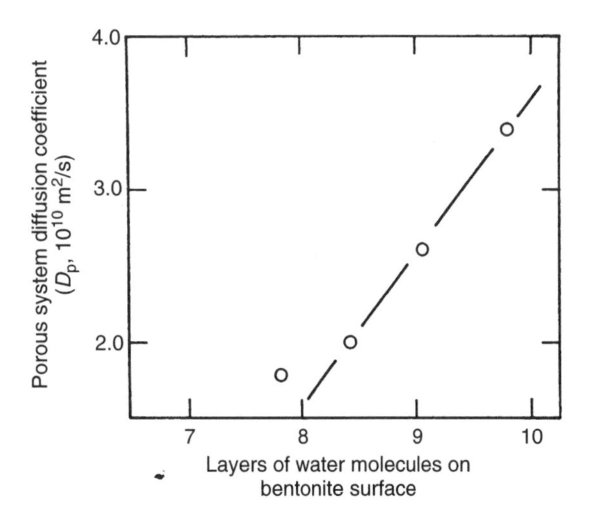

Figure 6.8 Porous system diffusion coefficient versus water content expressed as water thickness on clay particles (adapted from Kemper and van Schaik, 1966).

the barrier (from low to high concentration) and a large pressure head developed on the high-concentration side (Figure 6.10). The low-concentration side of the barrier developed a positive charge due to Na^+ diffusion and retarded Cl^- diffusion due to exclusion (opposite effect to the open system of Kemper and van Schaik, 1966).

Silver electrodes coated with silver chloride were used to short-circuit the system at 20 h (see Figure 6.10) enabling electron flow from the high-concentration side of their barrier to the low-concentration side without the necessity of Cl^- migration through the barrier. The electric potential was eliminated, the rate of Na^+ diffusion, f, increased greatly and the abundant water of hydration travelling with the Na^+ caused the hydraulic potential originally on the high side to reverse itself to the low-concentration side (Groenevelt, personal communication). In

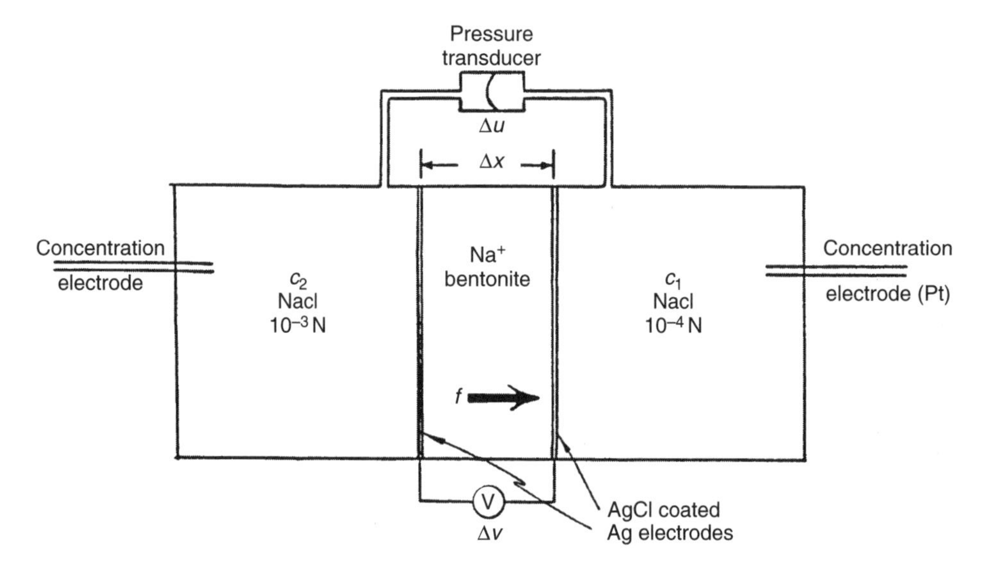

Figure 6.9 Schematic illustration of closed system diffusion cell used by Elrick *et al.* (1976) (osmosis not allowed).

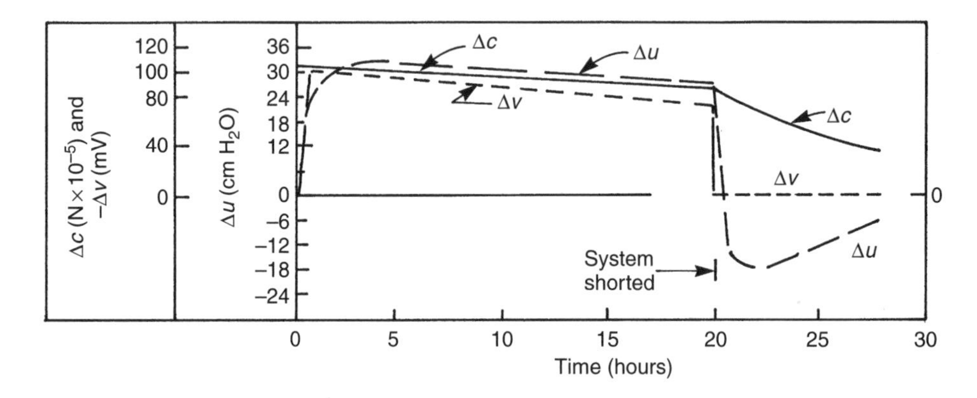

Figure 6.10 Gradients of concentration, pressure head and voltage versus time before and after electric shorting across the sample (adapted from Elrick *et al.*, 1976).

this particular experiment, eliminating the need for chloride migration by allowing direct electron flow increased the diffusion rate. It appears, therefore, that although Cl^- is normally treated as an unretarded species, in certain experimental set-ups exclusion may occur, especially at interfaces resulting in reduced diffusion coefficients of the migrating salt.

This \non-conservative nature of chloride migration appears in other experimental measuring systems and must be understood as it affects measurement of reliable diffusion coefficients for use in field situations.

In summary, the mass transport of ionic contaminant from a source, through a clay, to a receptor may involve a number of processes, including counter-osmotic flow and electrical imbalance as discussed above, and hence the effective diffusion coefficient, D_e, deduced by consideration of the mass transfer and application of Fick's law (i.e., from equations 6.5 and 6.7a) is an empirical parameter that incorporates the effect of these processes as manifest in the particular test under consideration. Thus this "effective diffusion coefficient" is not a true material property of the clay for a particular ion of interest but rather it can be expected to change depending on test conditions. For example, changing the chemical composition of the contaminant source or even the concentration of the ion of interest may influence the membrane behaviour and osmotic flow and hence the inferred effective diffusion coefficient based on simple application of Fick's laws. Fortunately, the simplified approach is adequate for most practical engineering applications, provided that this limitation is recognized and considered when selecting appropriate diffusion parameters.

6.5 Transient diffusion

In the previous homoionic, monomineralic systems, steady-state diffusion is reached fairly rapidly compared to a natural heterogeneous soil where extensive cation exchange, precipitation/dissolution and biodegradation may occur. Use is then made of Fick's second law which describes the time rate of change of concentration with distance, as follows:

$$\frac{\partial c}{\partial t} = D_R \frac{\partial^2 c}{\partial x^2} \tag{6.8}$$

where D_R is the retarded diffusion coefficient which may include the chemical/biological retardation factors just mentioned and which itself may be described as follows:

$$D_R = \frac{D_0 \tau}{R} = \frac{D_0 \tau}{1 + (\rho_d/\theta)K_d} = \frac{D_e}{1 + (\rho_d/\theta)K_d} \tag{6.9}$$

where ρ_d is the dry density of the soil $[ML^{-3}]$, θ the volumetric water content ($\theta = V_W/V_T$ [–], $\theta = n =$ porosity, V_V/V_T $[L^3/L^3]$ for 100% saturation), K_d the distribution coefficient $[L^3M^{-1}]$ which may reflect the degree of retardation by reversible ion exchange, but may also include partitioning of organics into organic matter in the soil and even precipitation, and D_e the effective diffusion coefficient $[L^2T^{-1}]$.

An analytical solution to equation 6.8 (Ogata, 1970) for an infinite layer with a constant surface concentration c_0 and zero advection is given by

$$\frac{c}{c_0} = \operatorname{erfc}\left[\frac{x}{2\sqrt{D_R t}}\right] = 1 - \operatorname{erf}\left[\frac{x}{2\sqrt{D_R t}}\right] \tag{6.10}$$

A series of relative concentration versus depth curves for chloride are plotted in Figure 6.11 along with a velocity plot of the 50% relative concentration front ($c/c_0 = 0.5$). The figure shows that the rate of diffusion is quite rapid for the first two or three years (150 → 50 mm/a) but decreases very rapidly with time due to the decrease in dc/dx as x increases. The initial speed of diffusion makes it possible to measure values of diffusion coefficient in laboratory tests lasting

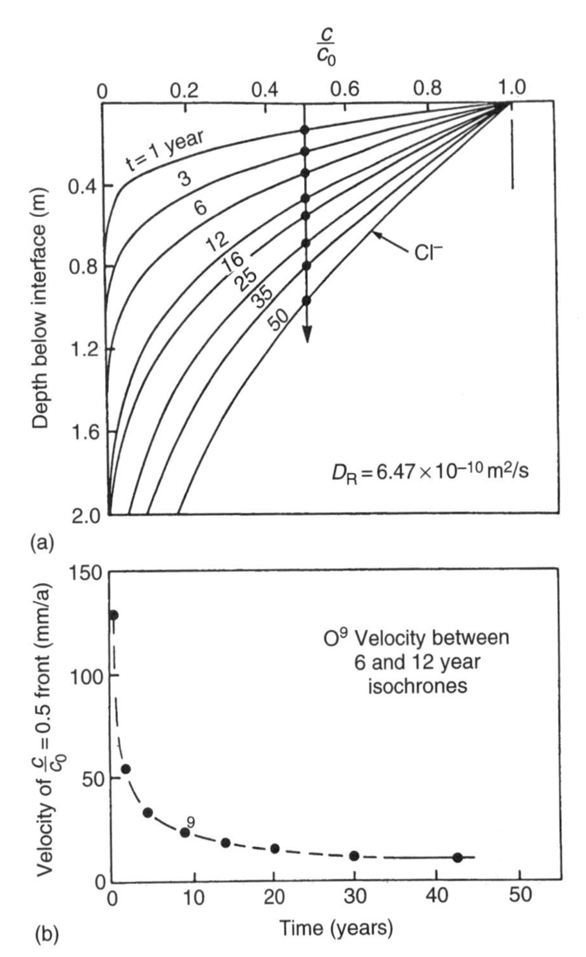

Figure 6.11 Time rate of change of migration by diffusion: (a) relative concentration–depth–time plots; (b) velocity of migration of $c/c_0 = 0.5$ front.

only a week or so. As will be illustrated later, these short-term values are quite close to long-term values obtained from field profiles.

If c_0 can be kept constant and interface partitioning is not a problem, experiments can be set up so that the concentration versus distance profile can be measured chemically or radioactively after a known period of elapsed time (e.g., Lai and Mortland, 1962; Gillham *et al.*, 1984). By assuming values for D_R, a calculated curve can be fitted to the experimental curve, and D_R selected. If a non-reactive species is also used (Cl^-, tritium, deuterium, for example), maximum values of $D_R = D_e$ are obtained and the tortuosity factor can be calculated from $D_e = D_0\tau$. Using this τ value, retardation factors and K_d values may be estimated for the other migrating species of interest.

If c_0 varies during the time of the experiment or during field exposure, more elaborate modelling methods are required to calculate D_e and K_d (as described in Chapter 8); however, both parameters can be estimated from one test on a single soil sample.

6.5.1 Sorption and the evaluation of K_d

(a) Empirical correlations

Values of K_d for organic compounds may be estimated from empirical correlations (e.g., Karickhoff *et al.*, 1979; Schwarzenback and Westall, 1981)

$$K_d = K_{oc} f_{oc} \tag{6.11}$$

where K_{oc} is the partition coefficient of a compound between organic carbon and water $[L^3M^{-1}]$ and f_{oc} [–] the weight fraction of organic carbon in the soil $[M_{organic}\ carbon/M_{solids}]$. Values of f_{oc} can be measured in the laboratory. An examination of f_{oc} data from a number of sites (Domenico and Schwartz, 1998) indicates that this parameter can be highly variable with values between 0.0001 and 0.0057 being reported. While K_{oc} values are available for many contaminants (Montgomery and Welkom, 1990), there is a reasonable correlation between $\log K_{oc}$ and the octanol–water partition coefficient, K_{ow}. Various regression equations between $\log K_{oc}$ and $\log K_{ow}$ have been published including that due to Karickhoff *et al.* (1979):

$$\log_{10} K_{oc} = -0.21 + \log_{10} K_{ow} \tag{6.12a}$$

and to Schwarzenbach and Westall (1981):

$$\log_{10} K_{oc} = 0.49 + 0.72 \log_{10} K_{ow} \tag{6.12b}$$

(b) Batch tests

Values of K_d may be obtained from batch tests (ASTM, D4646; Griffin *et al.*, 1985; Roy *et al.*, 1992; Cunha *et al.*, 1993) in which the soil is mixed with the leachate solution in question at different proportions or concentrations, until chemical equilibrium is established, and the loss of a species from the leachate to the soil is measured, usually by chemical analysis of the leachate.

The mass of sorbed contaminant per unit mass of soil, S [MM^{-1}], is then obtained through the difference between initial, c_0 [ML^{-3}], and equilibrium concentrations, c_e [ML^{-3}]:

$$S = \frac{(c_0 - c_e)V}{M} \tag{6.13}$$

where V [L^3] is the contaminant solution volume and M [M] the dry mass of soil.

The effect of the soil solution ratio (i.e., the mass of dry soil divided by the volume of solution) used in batch tests can influence the deduced sorption parameters (e.g., sorption may increase non-linearly with the amount of sorbent) due to the competition exerted by other ions released in solution when increasing amounts of soil are used (Roy *et al.*, 1992). Roy *et al.* (1992) noted that increasing the amount of solids can also alter the chemical characteristics of the system (e.g., pH, ionic strength), which, in turn, can affect the sorption. They concluded that the soil solution ratio values used in batch tests must be standardized or should be reported along with the sorption data. Similar observations and the fact that sorption non-linearity may increase as the soil solution ratio increases were reported by Puls *et al.* (1991), Manassero *et al.* (1998) and You *et al.* (1999).

It is important to evaluate the sorption capacity after equilibrium between the solid and liquid phases has been reached. This is typically considered to have occurred when there is no more than a 5% of variation in the concentration of the suspensions over a 24-h period (Roy *et al.*, 1992).

Either linear, a Langmuir or a Freundlich type of plot may be used as discussed in Chapter 1. The plot shown in Figure 6.12 is a log–log form of the Freundlich equation as follows:

$$\ln S = \ln K_f + \varepsilon \ln c_e \tag{6.14}$$

where S is the mass removed from solution per unit mass of soil solids, c_e the concentration of dissolved solute in the leachate at equilibrium, K_f a constant which is proportional to adsorption capacity and ε the slope of the line which reflects the intensity of adsorption with increasing c_e.

It is noted that while K_f and ε are constant, K_d is not and varies as a function of concentration. Thus the distribution coefficient K_d may be estimated at a given concentration level as discussed in Section 1.3.5.

The second and third lines drawn in Figure 6.12 (labelled soils 2 and 3) represent soils with greater adsorptive capacities than soil 1 and therefore higher K_d values for the species being measured. The soil 2 line is parallel to that for soil 1 indicating the same influence of liquid concentration on the amount adsorbed whereas soil 3 with a greater slope demonstrates a greater influence of liquid concentration.

Finally, adsorption isopleths of the nature shown in Figure 6.12 are only valid within certain soil and liquid concentration limits, so the mixes used have to be designed in order

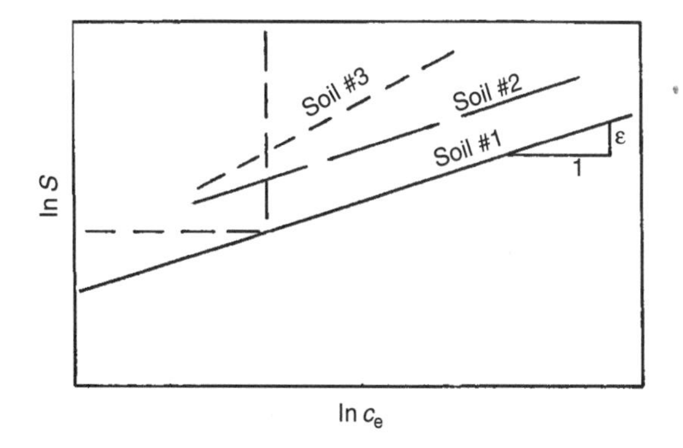

Figure 6.12 Linear log/log form of Freundlich adsorption equation.

to produce values representative of field conditions.

Adsorption isopleths are generally assumed to represent cation exchange but frequently incorporate the effects of fixation and precipitation in the presence of carbonate at high pH values (Griffin *et al.*, 1976). An example showing the effects of pH on removal of copper from a spiked leachate is shown in Figure 6.13. In this figure, the concentrations are plotted arithmetically as distinct from the log–log plots in Figure 6.12. The amount of copper removed from the solution is far greater for the carbonate-rich soil at pH 8.2 than for the carbonate-free soil at a slightly acidified leachate pH of 5.5. Since the amount removed is greater than the exchange capacity of the soil, precipitation as copper carbonate is probable. The apparent K_d values are therefore much greater for carbonate-rich clayey barriers with respect to heavy metal retardation.

In summary, batch tests, while useful, are usually considered as only a guide to K_d values. Non-linearity with increasing solution concentration, pH and temperature factors, precipitation, etc., all make it preferable to measure D and K_d from laboratory or field diffusion profiles.

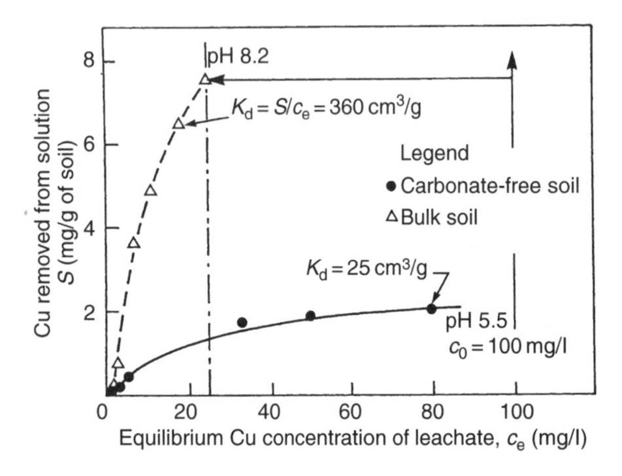

Figure 6.13 Mass of Cu removed from spiked landfill leachate at 22 °C versus equilibrium concentration (soil : solution = 0.25 g : 50 ml) (adapted from Yanful *et al.*, 1988b).

6.6 Use of laboratory and field profiles to obtain diffusion coefficient D_e

The use of laboratory and field profiles to estimate diffusion coefficient D_e and distribution coefficient K_d is deferred to Chapter 8, after the modelling required to interpret some of these profiles has been discussed. It is noted that laboratory measurement of diffusion coefficients is not easy; however, the values so obtained appear to be quite close to values back-figured from observed field behaviour and hence can be used with a considerable degree of confidence.

6.7 Diffusion during hydraulic conductivity testing for clay/leachate compatibility

Environmental approval often requires an assessment of clay/leachate compatibility – usually by means of hydraulic conductivity testing (Chapter 4). Such tests should be run until soil/leachate equilibrium is reached, or at least until effluent concentrations equal influent values (Daniel *et al.*, 1984; Bowders *et al.*, 1986). Fracture flow and diffusion of chemicals into adjacent soil peds greatly complicate chemical control as discussed in Section 4.1 and elaborated here.

A typical chemical control plot of relative concentration (effluent, c, divided by influent, c_0) versus pore volumes (PVs) is shown in Figure 6.14 for an MSW passing through a compacted, weathered, carbonate-rich, brown, southern Ontario soil containing about 8% smectite. This figure, which is similar to the plot presented in Figure 4.5, shows that even after passage of 12 PVs of leachate, potassium is still being retarded by the soil, probably by K^+ fixation on high charge smectites. This slow development of a condition of equilibrium may require testing times as long as 3 months on soil specimens as thin as 20 mm with high gradients (>500).

Effluent chloride concentration reaches 50% of the c_0 value at about 0.5 PVs indicating early

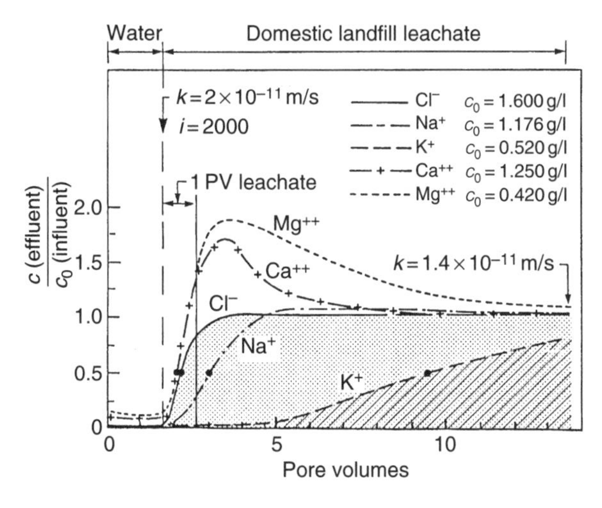

Figure 6.14 Typical curves of relative concentration of effluent obtained from high gradient hydraulic conductivity testing.

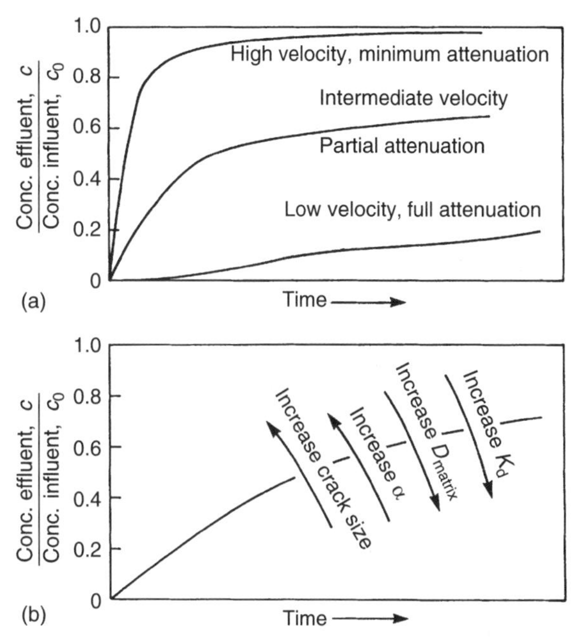

Figure 6.15 Factors affecting chemical attenuation during hydraulic conductivity testing through soils containing fractures or connected macropores: (a) fracture velocity effects; (b) controls at constant velocity (α = dispersivity) (adapted from Grisak and Pickens, 1980).

arrival due to channel flow. Sodium is mildly retarded ($c/c_0 = 0.5$ at 1.4 PV) and K^+ is severely retarded ($c/c_0 = 0.5$ at 8 PV). Adsorption of Na^+ and K^+ from the leachate onto the clay is compensated by Ca^{2+} and Mg^{2+} desorption resulting in early arrival ($c/c_0 = 0.5$ at 0.35 PV) and effluent concentrations far above influent values.

Interped flow channels or fractures which can occur between peds in compacted soils may significantly affect the measured values of hydraulic conductivity (Mitchell *et al.*, 1965). Under these circumstances, leachate flowing under a high hydraulic gradient moves primarily along the macropore channels and attenuation of contained dissolved species is primarily by diffusion into the adjacent soil peds.

The factors affecting attenuation or retardation were described mathematically by Grisak and Pickens (1980) and demonstrated in the laboratory for fractured till by Grisak *et al.* (1980). A summary plot of this work as it applies to clay/leachate compatibility testing in an advection-dominated system is shown in Figure 6.15. At low gradients which produce low velocity flow along the channels, all attenuating mechanisms function efficiently resulting in low c/c_0 values early in the hydraulic conductivity test (Figure 6.15a). At high gradients which produce high velocity flow along channels, minimum attenuation develops and early breakthrough occurs with high c/c_0 values. The actual components of such retardation are illustrated in Figure 6.15b. Both the increased diffusion coefficient of the matrix peds, D_{matrix}, and/or the increased distribution coefficient, K_d, increase the retardation. Both the increased fracture size and/or dispersivity, α, of the system accelerate channel flow causing early arrival in the effluent and inefficient clay/leachate interaction.

The important balance between the gradients employed and diffusion from the flow channels makes clay/leachate compatibility testing much more complicated than normally appreciated.

Using a method similar to that employed by Griffin *et al.* (1976), attenuation numbers have been

Table 6.4 Attenuation numbers for selected leachate species permeated through brown Sarnia soil at $e = 0.74$

Species	*Arrival of 50% relative concentration (number of pore volumes)*	*Attenuation number relative to chloride*
Cl^-	0.52	0.0
Na^+	1.36	5.6
K^+	7.89	56.1
Ca^{2+}	0.35	−12.0
Mg^{2+}	0.35	−31.3

calculated for the data shown in Figure 6.14. The chloride curve is used for the stippled reference area for input to compensate for channelling (dispersivity) factors and the attenuation number is defined as follows:

$$\text{ATN\#} = \frac{\text{stippled area } (Cl^-) - \text{hatched area (species)}}{\text{stippled area } (Cl^-)} \times 100 \quad (6.15)$$

with all curves extrapolated to equilibrium at $c/c_0 = 1$. The hatched area is shown for K^+. Table 6.4 contains both the calculated attenuation numbers and arrival time for the 50% relative concentrations expressed in PVs. The attenuation numbers reflect retardation in exactly the same sense as K_d values.

6.8 Diffusion through geomembranes

Geomembranes are commonly used in composite liners with compacted clay (Chapter 3) and/or GCLs (Chapter 12) with the primary function of impeding the advective flow of contaminants through the liner. Contaminants can migrate through geomembranes by (a) "leakage" through holes and defects in the geomembrane and (b) diffusion through the intact geomembrane. Sources of leaks and methods to estimate leakage through geomembranes are discussed in Chapter 13. In this section, attention is focussed on the diffusive movement of both water and contaminants through an intact geomembrane.

Geomembranes are not a conventional porous medium in which the pore size is large relative to both water and contaminant molecules. There is no "flow", in the conventional or Darcian sense, through an intact geomembrane. However, there can be movement of water and contaminant through the geomembrane due to molecular diffusion. The diffusive movement of a penetrant (contaminant or water) through a geomembrane where there are no defects such as pores, cracks or holes involves a cooperative rearrangement of the penetrant molecule and the surrounding polymer chain segments. The process requires a localization of energy to be available to allow a diffusive jump of the penetrant molecule in the polymer structure. The penetrant molecule and part of the polymer's molecular chain may share some common volume both before and after the jump. However, this jump will involve the breaking of some van der Waals forces or other interaction between the component molecules and polymer segments (Rogers, 1985). Thus the diffusive motion depends on the energy available and the relative mobilities of the penetrant molecules and polymer chains. This will depend on temperature and concentration, the size and shape of the penetrant, and the nature of the polymer.

6.8.1 Contaminant transport through an intact geomembrane

Migration of dilute aqueous solutions through an intact geomembrane involves three main stages (Haxo and Lahey, 1988; Park and Nibras, 1993; Prasad *et al.*, 1994): (i) adsorption at the interface between the medium containing the permeant and the inner surface of the geomembrane, (ii) diffusion of the permeant through the polymeric structure of the geomembrane and (iii) desorption at the interface between the outer surface of the geomembrane and the outer medium.

With respect to the first stage, there will be a relationship between the final equilibrium concentration in the geomembrane, c_g [ML^{-3}], and the equilibrium concentration in the adjacent fluid, c_f [ML^{-3}], when a geomembrane is in contact with a fluid (gas or liquid) for sufficient time (Figure 6.16). This relationship may often be described by the Nernst distribution function (Rogers, 1985) which may take a linear form (Henry's law):

$$c_g = S_{gf}\, c_f \tag{6.16}$$

where S_{gf} is variously called a solubility, partitioning or Henry's coefficient [–]. The term partitioning coefficient will be used here to denote S_{gf}. Alternatively, it may take a non-linear (e.g., Langmuir) form:

$$c_g = \frac{S_a b c_f}{1 + b c_f} \tag{6.17}$$

where S_a and b are experimentally determined constants. The coefficients S_{gf}, S_a and b may all vary with contaminant, phase of the fluid (gas or liquid) and temperature.

In the present context (see Rogers, 1985; Naylor, 1989), the term "sorption" is a generalized term for the removal of the penetrant molecules from the fluid and its dispersal on or in the polymer. This may involve numerous processes including adsorption, absorption, incorporation in microvoids, etc. The distribution of penetrant may change with concentration, temperature and time, and may change due to potential interaction with the polymer and swelling of the polymer matrix. Although much research has been conducted on sorption relating to plastics, there is limited recognition of its importance with respect to the design of landfills using geomembrane liners.

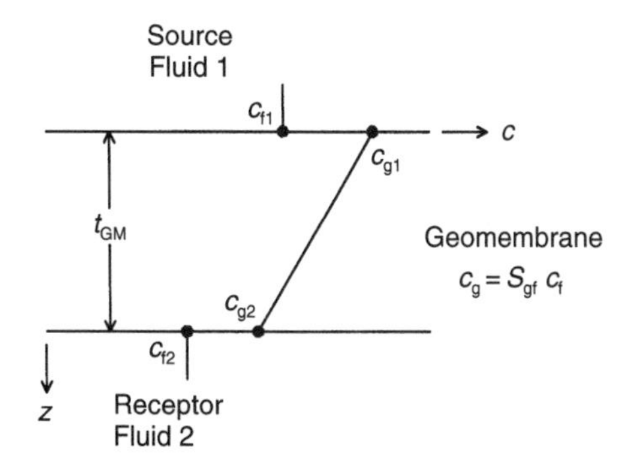

Figure 6.16 Concentration profile from diffusion across a geomembrane showing partitioning between the concentration in the solution and the concentration in the geomembrane.

For the simplest case, where the permeant does not interact with the polymer (often, but not always, the case for HDPE geomembranes) or at low concentrations (Rogers, 1985), Henry's law is obeyed and for this "Type I" sorption (Naylor, 1989) the relationship between the concentration in the fluid and solid is given by equation 6.16 where, in principle, S_{gf} is a constant for the given molecule, fluid, geomembrane and temperature of interest (as will become evident later, S_{gf} will depend on the chemical composition of the fluid).

Once sorbed into the geomembrane, the penetrant molecule can then diffuse within the polymer medium. The diffusion of penetrant molecules through a geomembrane can be modelled by Fick's law

$$f = -D_g \frac{\partial c_g}{\partial z} \tag{6.18}$$

in which f is the mass flux [$ML^{-2}T^{-1}$], D_g the diffusion coefficient in the geomembrane [L^2T^{-1}], c_g the concentration of the penetrant of interest in the geomembrane [ML^{-3}] and z the distance parallel to the direction of diffusion [L]. Consideration of conservation of mass then gives the following governing differential equation for transient diffusion:

$$\frac{\partial c_g}{\partial t} = D_g \frac{\partial^2 c_g}{\partial z^2} \tag{6.19}$$

which must be solved for the appropriate boundary and initial conditions.

The final stage of contaminant migration involves desorption from the geomembrane to the outside fluid. This surface partitioning is similar to the adsorption stage and can also be described using Henry's law:

$$c'_g = S'_{gf} c_f \qquad (6.20)$$

where S'_{gf} is the contaminant partitioning coefficient between the geomembrane and the outside fluid. The two partitioning coefficients S_{gf} and S'_{gf} may be assumed to be the same for dilute aqueous solutions.

Thus, substituting equation 6.16 into equation 6.19 gives the flux from a fluid on one side of a geomembrane to a similar fluid on the other side

$$f = -D_g \frac{\partial c_g}{\partial z} = -S_{gf} D_g \frac{\partial c_f}{\partial z} = -P_g \frac{\partial c_f}{\partial z} \qquad (6.21)$$

where P_g is given by

$$P_g = S_{gf} D_g \qquad (6.22)$$

and is often referred to in the polymer literature as the permeability or permeation coefficient of the geomembrane. This should not be confused with the old soil mechanics term "coefficient of permeability" (now called hydraulic conductivity) or the intrinsic permeability of a porous medium. It has nothing to do with Darcy's law or flow through the interconnected voids within porous media, rather P_g is a mass-transfer coefficient that accounts for partitioning and diffusion processes. Steady-state laboratory tests will allow the measurement of P_g, while transient laboratory tests may be used to obtain S_{gf} and D_g (see Chapter 8).

It is important to note that there is a difference between phase change at the fluid/geomembrane boundary and sorption that occurs within the geomembrane itself as contaminant migrates through the geomembrane. Because plastic sheets are often very thin, the two have been confused in the literature. While this may be of little importance for very thin membranes, it is potentially important for thicker geomembranes. Sorption may occur, e.g., when the contaminant absorbs to sorption sites in the geomembrane such as dispersed porous particles of high surface area (e.g., carbon black or silica gel). For this type of sorption, equation 6.19 is modified to

$$\frac{\partial c_g}{\partial t} = D_g \frac{\partial^2 c_g}{\partial z^2} - \frac{\partial S_g}{\partial t} \qquad (6.23)$$

where S_g may have a number of forms. For linear sorption

$$S_g = K_g c_g \qquad (6.24)$$

where S_g is the mass sorbed per unit volume [ML^{-3}], K_g the linear sorption coefficient [-] and c_g the dissolved concentration [ML^{-3}]. For Langmuir sorption,

$$S_g = \frac{\alpha \beta c_g}{1 + \beta c_g} \qquad (6.25)$$

where α is the limiting sorption mass per unit volume [ML^{-3}] and β is an experimentally derived coefficient [$M^{-1} L^3$]. This sorption is distinct from the partitioning defined by equation 6.17. By combining equation 6.25 with linear partitioning between phases (equation 6.16), one can simulate the case where some of the molecules dissolved in the geomembrane are free to diffuse down the concentration gradient while other molecules are bound at a fixed number of adsorption sites or voids within the polymer.

6.8.2 Factors affecting diffusion through a geomembrane

Partitioning, diffusion and permeability coefficients that control contaminant transport through an intact geomembrane depend on properties of the permeant and the polymer as well as temperature, each of which are now discussed in turn.

One of the most important factors is the similarity of the penetrant species relative to the polymer, which strongly influences partitioning, diffusion and thus permeation coefficients. Strongly polar penetrant molecules (e.g., water, alcohols) may have very low permeabilities through non-polar polyethylene whereas species that are more similar to polyethylene (e.g., hydrocarbons) have larger permeabilities. In general, the permeation coefficient increases in the following order: alcohols < acids < nitroderivatives < aldehydes < ketones < esters < ethers < aromatic hydrocarbons < halogenated hydrocarbons (August and Tatzky, 1984; Rowe *et al.*, 1996a).

Diffusion of inorganic species may be very slow. For example, August *et al.* (1992) found that there was negligible diffusion of heavy metal salts (Zn^{2+}, Ni^{2+}, Mn^{2+}, Cu^{2+}, Cd^{2+}, Pb^{2+}) from a concentrated (0.5 M) acid solution (pH 1–2) through HDPE over a 4-year test period. Additionally, Rowe *et al.* (1996a) measured only negligible diffusion of Na^+ and Cl^- through a geomembrane after 4 years and indeed at the time of writing these same tests have been ongoing for 10 years with concentration of Na^+ and Cl^- still at or below the detection limit of the receptor side of the geomembrane (at 22 °C). Thus the available data suggest that HDPE geomembranes should provide an excellent diffusive barrier for inorganic contaminants.

Sorption and thereby the partitioning coefficient are mainly controlled by contaminant characteristics (Müller *et al.*, 1998). The affinity of a chemical for geomembrane can be expressed using the *n*-octanol/water coefficient, expressed as $\log K_{ow}$, which quantifies the ability of a chemical to partition between water and *n*-octanol (i.e., organic matter) when they are in solution, and incorporates effects of solubility and polarity of the chemical. For organic contaminants, S_{gf} increases with increasing contaminant molecular weight and increasing hydrophobicity, i.e., increasing $\log K_{ow}$ (Sangam and Rowe, 2001).

The diffusion coefficient also tends to decrease with increasing penetrant weight and size (see Berens and Hopfenberg, 1982; Park and Nibras, 1993). The shape of the molecule is also important with linear, flexible and symmetrical molecules having higher mobility than rigid or non-symmetrical molecules (Salame, 1961; Berens and Hopfenberg, 1982; Saleem *et al.*, 1989). For example, the diffusion coefficient for *o*-xylene was found to be much lower than for *p*-xylene (Saleem *et al.*, 1989). This is attributed to the symmetrical nature of *p*-xylene (that has a dipole moment of 0) and the distorted shape of *o*-xylene with its two adjacent methyl groups (dipole moment $\simeq 0.62$ debye). Berens and Hopfenberg (1982) showed that the diffusion coefficient for *n*-alkane and other elongated or flattened molecules are higher, by a factor of up to 10^3, than the diffusion coefficients of spherical molecules of similar molecular weight.

For crystalline polymers, the crystalline areas act as "diffusion barriers" and diffusion appears to be primarily through the amorphous phase and hence one would expect that, as the crystallinity increases, the diffusion coefficients and solubility will decrease. For example, Park *et al.* (1995) reported that the permeability ($P_g = S_{gf}D_g$) of xylene in VLDPE was almost twice that through HDPE. Ashley (1985) has indicated that orientation can reduce permeability ($P_g = S_{gf}D_g$) of amorphous polymers by 10–15% and of crystalline polymers by over 50%.

The solubility and diffusion of a contaminant in a geomembrane will increase with increasing temperature. Over small temperature ranges, the Arrhenius relationship may be used to quantify the temperature dependence of solubility, diffusion and permeation coefficients (Naylor, 1989; Chainey, 1990):

$$S_{gf} = S_{gf0}\, e^{(-\Delta H_s/RT)} \tag{6.26a}$$

$$D_g = D_{g0}\, e^{(-E_d/RT)} \tag{6.26b}$$

$$P_g = P_{g0}\, e^{(-E_p/RT)} \tag{6.26c}$$

where ΔH_s is the heat of solution of the penetrant in the polymer, E_d and E_p are the activation energies of diffusion and permeation, and S_{gf0}, D_{g0} and P_{g0} are constants.

It has been reported (Rogers, 1985) that the diffusion coefficient, $D_g(c)$, may be a function of aqueous concentration, c [ML^{-3}], such that

$$D_g(C) = D_{c0}\, e^{(B_c C)} \tag{6.27}$$

where D_{co} is the diffusion coefficient as $c \to 0$ [L^2T^{-1}] and B_c an experimentally determined constant [$M^{-1}L^3$].

6.8.3 Simplified estimates of partition, diffusion and permeation coefficients

A description of the laboratory methods used to estimate partition, diffusion and permeation coefficients and a summary of these values reported in the literature are provided in Chapter 8. However, it is useful to have simple empirical relationships to provide initial estimates of the transport coefficients to assist with the selection and evaluation of geomembrane liners.

Sangam and Rowe (2001) compiled values of permeation coefficients for dilute organic contaminants through HDPE geomembranes and developed correlations with: (a) the *n*-octanol/water coefficient ($\log K_{ow}$) and (b) chemical molecular weight (M_w). Correlations of partition, diffusion and permeation coefficients with K_{ow} and M_w obtained by Sangam and Rowe (2001) are plotted in Figures 6.17 and 6.18. It can be noted that:

(a) As $\log K_{ow}$ increases (i.e., decreased polarity of the permeant) the contaminant has a high attraction to the geomembrane and thus a larger partitioning coefficient S_{gf}.

(b) S_{gf} increases with increasing molecular weight M_w with partitioning coefficients in an increasing order for: oxygenated $<$ chlorinated $<$ aromatic $<$ aliphatic organic compounds.

(c) For many organic contaminants the diffusion coefficient D_g typically lies in the range 10^{-12}–10^{-13} m^2/s. The decrease in D_g for large values of $\log K_{ow}$ may be attributed to larger molecule sizes for permeants with large $\log K_{ow}$.

(d) Correlation of D_g with $\log K_{ow}$ alone is poor indicating that properties of the polymer (e.g., crystallinity, branching, cross-linking) have a substantial effect on diffusion coefficient.

(e) Permeation coefficient P_g may vary by up to four orders of magnitude with variations in $\log K_{ow}$ and M_w.

The equations for the empirical correlations developed by Sangam and Rowe (2001) are provided in Table 6.5. These values may be used to provide initial estimates of permeation coefficients for dilute organic aqueous solution passing through HDPE geomembranes. Caution is required in using these correlations since large variations in partition coefficient can occur between different species and because the simple models do not account for the influence of polymer properties on diffusion coefficient. Also, permeation coefficients are temperature- and maybe concentration-dependent. Improved estimates for use with specific contaminants and geomembranes can be obtained from the laboratory tests described in Chapter 8.

6.8.4 Diffusion of water through HDPE geomembranes

Eloy-Giorni *et al.* (1996) performed a detailed study of water movement through intact geomembranes. Based on a series of experiments, they concluded that the concept of hydraulic conductivity (as defined by Darcy's law) is not appropriate for describing water transport (and

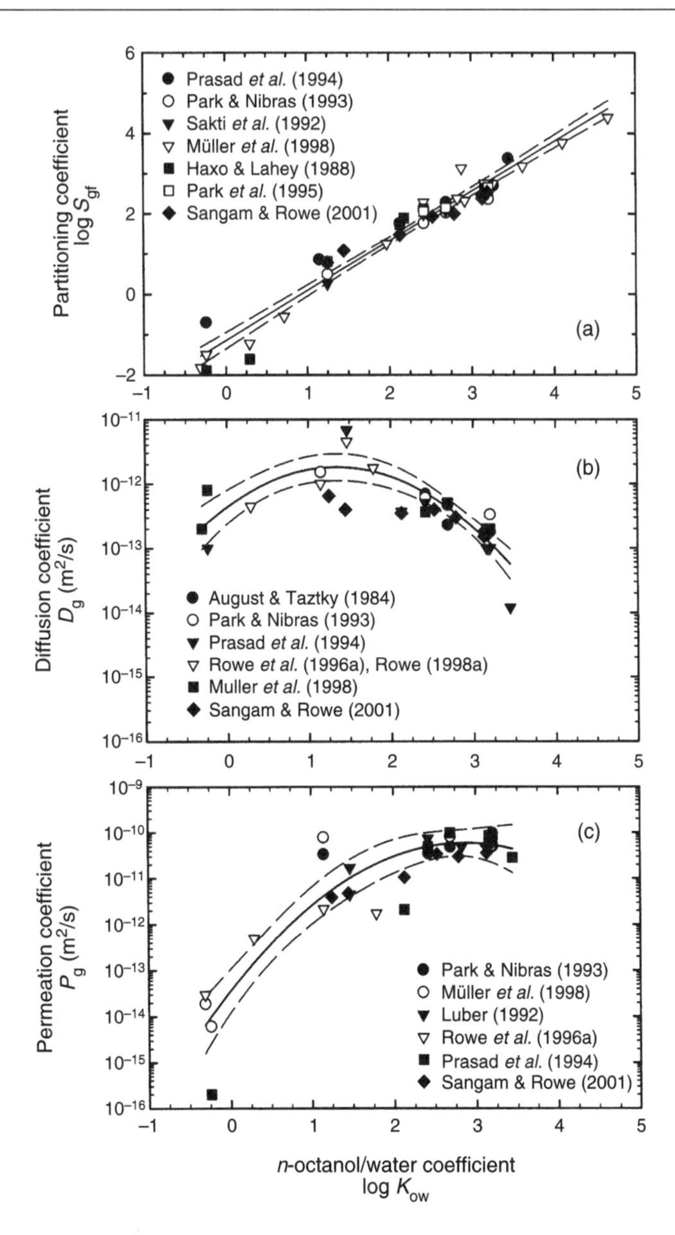

Figure 6.17 Variation in (a) partitioning; (b) diffusion; (c) permeation coefficients with *n*-octanol/water coefficient for aqueous organic solutions migrating through geomembranes (modified from Sangam and Rowe, 2001).

by inference the advective transport of contaminants) through hydrophobic geomembranes. In particular, they demonstrated that a difference in pressure head across a geomembrane of up to 200 m had no significant effect on the movement of water molecules across a 1.7-mm thick HDPE geomembrane ($\rho_g = 940\,kg/m^3$). Likewise, a head difference of up to 400 m had no significant effect on water movement across a 1- and 1.6-mm thick PVC geomembrane ($\rho_g = 1260\,kg/m^3$). In these experiments, the water flux (i.e., the flow of water across a geomembrane per unit

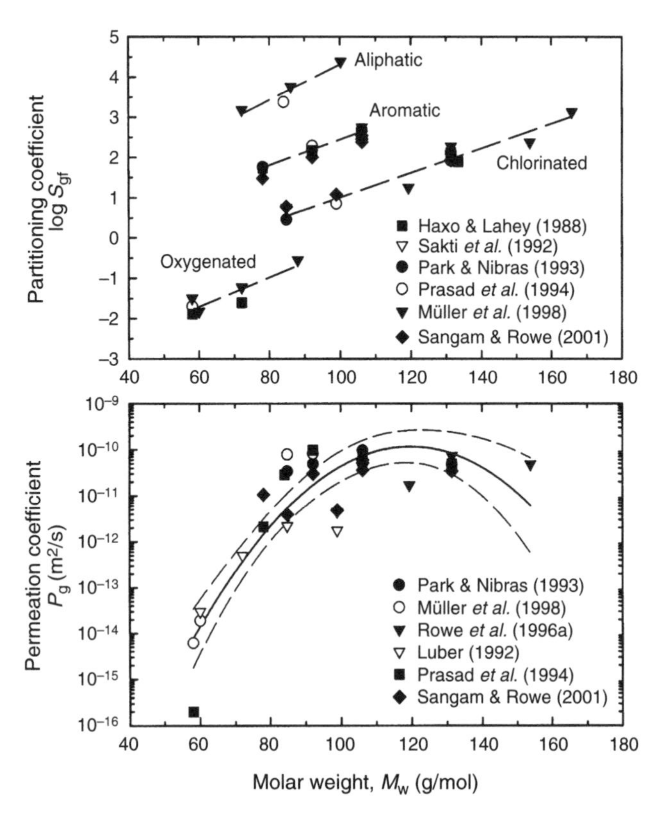

Figure 6.18 Variation in (a) partitioning; (b) permeation coefficients with molar weight for aqueous organic solutions migrating through geomembranes (modified from Sangam and Rowe, 2001).

area per unit time under conditions where there is a pressure head difference across the geomembrane) was less than or equal to the limit of accuracy of the apparatus ($10^{-7}\,m^3/m^2/day \leq 3 \times 10^{-5}\,m^3/m^2/a$) and hence was negligibly small.

These findings, which were verified in several ways, are at variance with earlier results used by Giroud and Bonaparte (1989) to establish permeation rates for geomembranes. Given the rigour of the recent experiments and the

Table 6.5 Empirical relationships between partition, diffusion and permeation coefficients for dilute organic solutions passing through HDPE geomembranes (adapted from Sangam and Rowe, 2001)

Method	*Parameter*	*Relationship*		r^2
n-Octanolwater	S_{gf}	$\log S_{gf} = -1.1523 + 1.2355 \log K_{ow}$		0.97
	D_g	$\log D_g = -12.3624 + 0.9205 \log K_{ow} - 0.3424(\log K_{ow})^2$		0.72
	P_g	$\log P_g = -13.4476 + 2.2437 \log K_{ow} - 0.3910(\log K_{ow})^2$		0.84
Molar weight	S_{gf}	Oxygenated	$\log S_{gf} = -3.8883 + 0.0363 M_w$	0.81
		Chlorinated	$\log S_{gf} = -2.0467 + 0.0305 M_w$	0.94
		Aromatic	$\log S_{gf} = -0.0776 + 0.0322 M_w$	0.95
		Aliphatic	$\log S_{gf} = -0.1107 + 0.0442 M_w$	0.91
	P_g	$\log P_g = -25.6933 + 0.2633 M_w - 1.099 \times 10^{-3} M_w^2$		0.81

sophisticated techniques used, there is a reasonable likelihood that the results of Eloy-Giorni *et al.* (1996) are more reliable than those used by Giroud and Bonaparte. If one accepts the more recent results, then the only significant mechanism for water migration through the geomembrane is diffusion.

One may expect negligible water diffusion across a geomembrane in a composite MSW landfill liner for two reasons. First, the permeability ($P_g = S_{gf} D_g$) of the geomembrane to water is extremely low. Based on the parameters given in Table 6.6 and equation 6.21, one can show that the water flux, f, through a 1.5-mm thick HDPE geomembrane would be $1.5 \times 10^{-13} \Delta c_f$ kg/m²/s (where Δc_f is in kg/m³ units) ($1.3 \times 10^{-4} \Delta c_f$ L/ha/day) where Δc_f is the difference in the "concentration" of water on the two sides of the geomembrane. Thus, even with water on one side and zero relative humidity on the other side, the water flux would be very small. Second, apart from the low permeability of the geomembrane to water, one would expect there to be very little diffusive flux since the concentration gradient Δc_f is likely to be very small. For compacted clay liners (CCLs), the soil water potential at the time of construction is typically −10 to −50 kPa and this corresponds to a relative humidity of about 99.5% (Daniel, personal communications). With water (leachate) on one side of a geomembrane and compacted clay on the other side, the difference in relative humidity (and hence Δc_f) is negligible. Thus, even with the presence of salts in the MSW leachate to influence the energy difference between the two sides of the geomembrane, the gradient in energy is so close to zero that, for all practical purposes, there will be no diffusion of water or water vapour across the geomembrane. This would appear to be the likely case for most of the service life of a geomembrane used as part of composite liner systems at the base of an MSW landfill. Thus, it would seem appropriate to revise the permeation rates suggested by Giroud and Bonaparte (1989) to be, to all practical purposes, zero for HDPE geomembrane liners.

Table 6.6 Solubility and diffusion coefficient for water in three types of geomembrane (adapted from Eloy-Giorni *et al.*, 1996)

Material	*Partition coefficient,* S_{gf} (–)	*Diffusion coefficient,* D_g (m²/s)	*Density,* ρ_g (kg/m³)	*Thickness,* t_{GM} (mm)
HDPE	8×10^{-4}	2.9×10^{-13}	940	1.7
PVC	7×10^{-2}	4.4×10^{-13}	1,260	1.0, 1.6
Bituminous	9×10^{-3}	8×10^{-13}	1,150	5.0

6.9 Summary

The complexities of diffusion and its importance both as a transport mechanism and as an attenuating mechanism have been briefly discussed in this chapter. A few conclusions may be made as follows:

1. The process of steady-state diffusion expressed by Fick's first law is complex even in homoionic, monomineralic systems due to coupled osmotic and electro-osmotic flow, and anion exclusion particularly at soil–liquid interfaces.
   (a) Fick's first law, $f = -nD_e \frac{dc}{dx}$.
   (b) For charge balance, the forward flow of cations should equal that of anions.
   (c) Concentration differences in open, steady-state diffusion systems create forward cation flow, forward cation electro-osmotic flow and reverse cation osmotic flow.
   (d) These same concentration differences create forward anion diffusion flow, reverse anion electro-osmotic flow and reverse anion osmotic flow.
2. The variability in terminology leads to confusion and one must carefully distinguish between different definitions of the diffusion coefficient and its component parts (e.g., D_p, D_R, D_e as discussed herein). In particular, since D_R incorporates adsorption it may increase as retardation equilibrium occurs.

3. The importance of diffusion in clay/leachate compatibility assessment by hydraulic conductivity testing was illustrated by a test which had not reached equilibrium after passage of even 12 PVs of domestic leachate.

   (a) K^+ fixation remains a critical assessment problem because it can result in clay mineral contraction, reduced CEC values and double-layer contraction, all potential causes of cracking.
   (b) The interrelationships of fracture-flow velocity and matrix diffusion control the effluent concentration versus time curves.
   (c) K^+ fixation (by diffusion) may create testing times of up to 3 months even at gradients as high as 500.

4. Contaminant transport through an intact geomembrane liner does not involve conventional flow through a porous medium, but rather involves partitioning of the permeant at the interface with the geomembrane and subsequent diffusion through the polymer.

   (a) Henry's law and Fick's first law may be used to quantify partitioning and diffusion processes in a geomembrane.
   (b) The permeation coefficient P_g (sometimes called permeability of the geomembrane) is the product of the partitioning coefficient, S_{gf} and diffusion coefficient, D_g, and depends on the properties of the permeant, the geomembrane and temperature.
   (c) Geomembranes act as diffusive barrier for inorganic contaminants.
   (d) Organic contaminants may diffuse through geomembranes; hence diffusive transport must be considered during assessments of contaminant impact through a geomembrane liner.
   (e) Negligible diffusion of water is expected through an intact (i.e., no holes or leaks) HDPE geomembrane in a barrier system.

Chapter seven

Contaminant transport modelling

7.1 Introduction

A contaminant transport model consists of the governing equations together with the boundary and initial conditions. Once the model has been formulated and the appropriate parameters have been determined, it remains to find a solution to the governing equations subject to the appropriate boundary and initial conditions. The most frequently used solution techniques can be subdivided into five broad categories, namely analytic, finite layer, boundary element, finite difference and finite element techniques.

The other techniques will be briefly discussed here and then attention will be directed at providing some additional analytical solutions and describing the finite layer formulation.

7.1.1 Finite layer techniques

The finite layer technique is applicable to situations where the hydrostratigraphy can be idealized as being horizontally layered with the soil properties being the same at any horizontal location within the layer. For these conditions, the governing equations can be considerably simplified by introducing a Laplace and Fourier transform (the latter only being required for 2D or 3D problems). The transformed equations can then be readily solved. This procedure parallels that adopted in the development of many analytic solutions. The difference between finite layer solutions and analytic solutions arises from the fact that in the finite layer approach, the solution is inverted numerically rather than analytically. As a consequence, it is possible to examine more complicated and realistic situations.

Finite layer techniques have been described by Rowe and Booker (1985a,b, 1987) and are available as computer programs for 1½D (POLLUTE: Rowe and Booker, 1983, 1994, 1997, 2004) and 2D (MIGRATE: Rowe and Booker, 1988b) conditions.

For situations involving a clayey liner overlying a drainage layer which can be pumped or overlying a thin natural aquifer (e.g., see Figures 1.2–1.8), a reasonable initial estimate of contaminant impact can be obtained in seconds using a personal computer and 1½D finite layer programs such as POLLUTE (e.g., Rowe and Booker, 2004). The designation of these programs as 1½ D is intended to indicate that they consider 1D transport down through aquitard layers and into the aquifer(s) or drainage layer(s) while approximately taking account of contaminant removal due to lateral flow or pumping of these permeable layers. These techniques will often provide a reasonable estimate of both the magnitude of the peak impact beneath the landfill and the time at which this occurs.

A more rigorous solution of this problem can be obtained using a full 2D analysis (e.g., program MIGRATE). A comparison of the two approaches has been given by Rowe and Booker (1985b) and will be discussed in Section 10.8. The 2D approach allows one to consider impact at points outside the landfill. The full 2D analysis can be readily performed on a personal computer but it does involve substantially more computation than the 1½D analysis. Before performing 2D analyses, a 1½D analysis should always be performed to estimate the likely magnitude of impact and its time of occurrence beneath the landfill; the 2D program can then run for appropriate times using the 1½D results as a reference. One of the advantages of finite layer techniques over conventional finite element methods is that it is unnecessary to determine solutions at times prior to the time period of interest.

Finite layer techniques have many of the advantages of analytic techniques. They are easy to use and the user need not be an expert in numerical analysis. They also require minimal input and only give results at the times and locations of interest. This technique is particularly well suited for performing sensitivity studies to identify the potential impact of uncertainty regarding the value of key design parameters. The technique is also well suited for performing checks on the results of numerical analyses using finite element or finite difference techniques.

Finite layer techniques are not appropriate for modelling situations where there is a complex geometry or flow pattern which cannot reasonably be idealized in terms of horizontal layers with uniform properties.

7.1.2 Boundary element techniques

The boundary element technique is suitable for solving the advection–dispersion equation (e.g., see Brebbia and Skerget, 1984). Its primary advantage over finite layer methods is the ability to model more complicated geometries. To date, boundary elements have not found wide applications in contaminant migration studies.

7.1.3 Finite difference and finite element techniques

There has been extensive research into the use of finite difference and finite element techniques for the analysis of contaminant migration through soils. These numerical techniques are likely to be used for:

(i) calculating the steady-state flow pattern within a hydrogeologic system, thereby defining the velocity field;
(ii) calculating the rate of migration of contaminants by solving the advection–dispersion equation (after using velocities determined from (i)).

The techniques for modelling steady-state flow are well established (e.g., see Frind, 1987) and numerous commercial software packages are available. Many of the packages will run on personal computers; however, there is a potential danger that inappropriate finite element mesh arrangements may be used simply to fit the problem onto a personal computer. It is essential that the suitability of any finite element mesh be checked either against a known solution or by comparison with results obtained by substantially refining the mesh.

Finite element techniques provide the opportunity of modelling problems with complex geometries, complicated flow patterns, heterogeneity and non-linearity. There is a wealth of literature dealing with different algorithms which have been proposed for solving the advection–dispersion equation. The sheer volume of literature is itself a warning that the use of these techniques is not as simple as might first appear. This is particularly so when dealing with problems where there are high advective velocities, low dispersivities and/or high contrast in dispersivity (e.g., see Allan, 1984; Yeh, 1984). While good results can be obtained (e.g., Frind and Hokkanen, 1987), particular care is required to ensure the selection of a suitable computational algorithm, finite element mesh and time step.

7.1.4 General comments

All five classes of techniques discussed above have a role to play in the design of barriers. However, of these techniques, the two most useful are finite layer and finite element methods. Finite layer techniques require some idealization of the problem; however, they also provide a means of quickly assessing the implications regarding different design scenarios and different assumed parameters. The cost of performing these analyses is relatively small. The finite element method provides the most general technique for solving the governing equations including complex geometry, velocity fields and non-linearity. However, use of the approach requires an experienced user as well as relatively large data preparation and computational cost.

7.2 Analytical solutions

In the preliminary investigation of contaminant migration, there is often insufficient information to warrant more than a simple idealization of the physical situation. In some cases, one may be able to formulate the problem in such a way that an analytic solution is possible. Analytical solutions are ideal for making quick preliminary calculations, are valuable in performing sensitivity studies and are useful in checking the results of more sophisticated analyses.

Numerous solutions have been reported in the literature (e.g., Lapidus and Amundson, 1952; Ogata and Banks, 1961; Lindstrom *et al.*, 1967; Ogata, 1970; Selim and Mansell, 1976; Rowe and Booker, 1985a; Booker and Rowe, 1987; and others). Although these analytic solutions have sometimes been expressed in graphical form suitable for hand calculations as discussed in Section 1.7 (e.g., Ogata, 1970; Booker and Rowe, 1987), it is often more convenient to use a computer. The computation involved in such calculations is usually not particularly sophisticated and is made easier because of the existence of efficient approximations for many mathematical functions which can often be evaluated on programmable calculators (Abramowitz and Stegun, 1964), and because of robust readily portable subroutines (e.g., Press *et al.*, 1986) which can be incorporated into relatively simple programs which can be run on a personal computer.

7.2.1 Solutions for deep deposits (pure diffusion)

In the absence of advection and first-order decay, the equation governing contaminant migration (equation 1.15) reduces to a simple diffusion or heat flow equation. Thus the extensive literature on heat flow (Carslaw and Jaegar, 1959) provides a rich source of analytic solutions. The following provides a number of additional solutions that may be found to be useful.

Figure 7.1 shows a transmissive layer ($-H/2 < z < H/2$) which is bounded by non-transmissive strata. A portion of the layer ($-a < x < +a$) has become contaminated. If the initial concentration in the contaminated zone is c_0 then it is not difficult to show that subsequently

$$c = \frac{c_0}{2}\left[\operatorname{erfc}\left(\frac{|x|+a}{2\sqrt{D_{\mathrm{R}}t}}\right) - \operatorname{erfc}\left(\frac{|x|-a}{2\sqrt{D_{\mathrm{R}}t}}\right)\right] \quad (7.1)$$

where $D_{\mathrm{R}} = nD/(n + \rho K_{\mathrm{d}})$.

The concentration may be evaluated either by utilizing tabulated values of the complementary error function, erfc (x), or by employing a polynomial approximation (Abramowitz and Stegun,

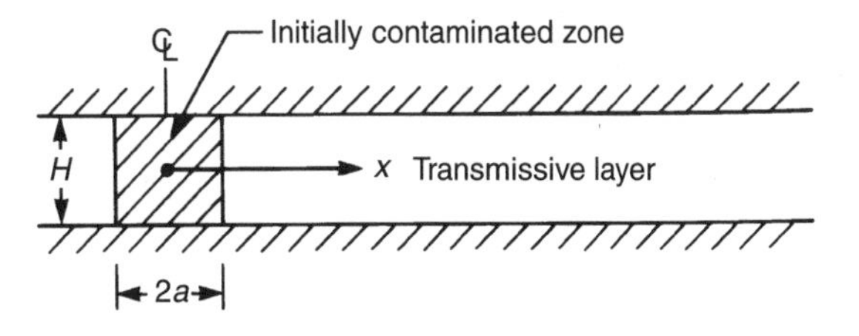

Figure 7.1 Contaminated zone in a transmissive layer.

1964) or by using the graphical values of the function given in Figure 1.28.

$$\phi(x) = e^{x^2} \operatorname{erfc}(x) \tag{7.2}$$

The variation of the concentration profile with time is shown in Figure 7.2 and as would be expected this figure shows a reduction of concentration within the initially contaminated zone as contaminant is transported by diffusion to the initially uncontaminated portion of the transmissive layer.

Figure 7.3 shows a deep deposit which contains a contaminated spherical zone ($r < a$). Initially, the contaminant concentration in the spherical zone is c_0; subsequently, analysis of the diffusion equation shows that the concentration is given by

$$c = \frac{c_0}{2}\left[\left(\operatorname{erfc}\left(\frac{r-a}{2\sqrt{D_R t}}\right) - \operatorname{erfc}\left(\frac{(r+a)}{2\sqrt{D_R t}}\right)\right) - 2\sqrt{\frac{D_R t}{\pi r^2}}\left(\exp\left[\frac{-(r-a)^2}{4D_R t}\right] - \exp\left[\frac{-(r+a)^2}{4D_R t}\right]\right)\right] \tag{7.3}$$

The process of diffusion from the initially contaminated zone into the initially uncontaminated zone is evident from Figure 7.4. The rate of diffusion in this case is far faster than for the one previously examined because of 3D effects.

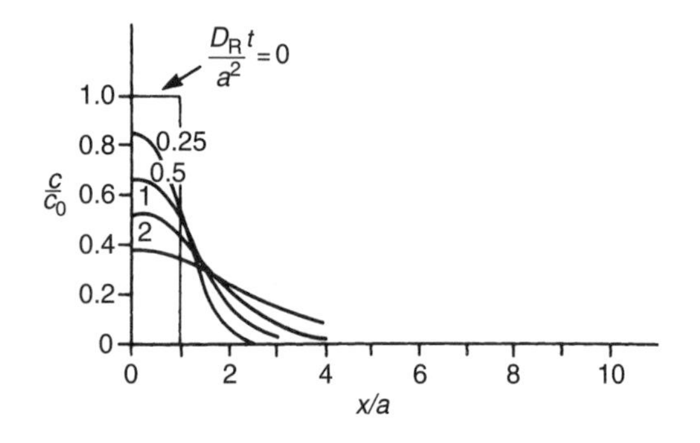

Figure 7.2 Variation of concentration with time in an initially contaminated transmissive layer.

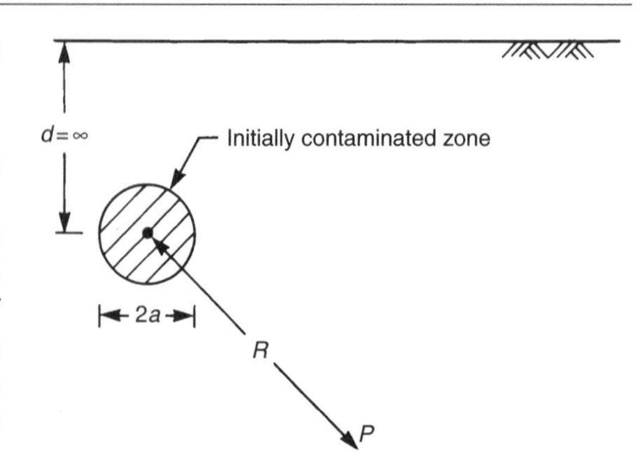

Figure 7.3 Spherical contaminated zone.

7.2.2 Solutions for deep deposits (advection and diffusion)

The solutions derived in the previous section can be modified to incorporate the effect of advection. The modification depends on the following observation. Suppose that

$$c = F(x, y, z, t) \tag{7.4a}$$

is a solution of the transport equation with no advection ($v_x = 0$, $v_y = 0$, $v_z = 0$) and no first-order (e.g., radioactive or biological) decay ($\lambda = 0$), then a solution to the equation of contaminant transport with advection and first-order decay is found to be

$$c = e^{-\Lambda t} F[x - v_{Rx}t, y - v_{Ry}t, z - v_{Rz}t, t] \tag{7.4b}$$

where

$$v_{Rx} = \frac{nv_x}{(n + \rho K_d)}$$

$$v_{Ry} = \frac{nv_y}{(n + \rho K_d)}$$

$$v_{Rz} = \frac{nv_z}{(n + \rho K_d)}$$

$$\Lambda = \frac{n\lambda}{(n + \rho K_d)}$$

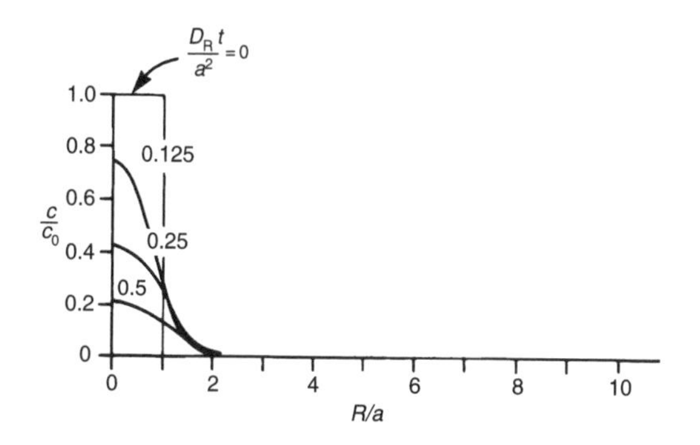

Figure 7.4 Variation of concentration with time due to an initially contaminated spherical zone.

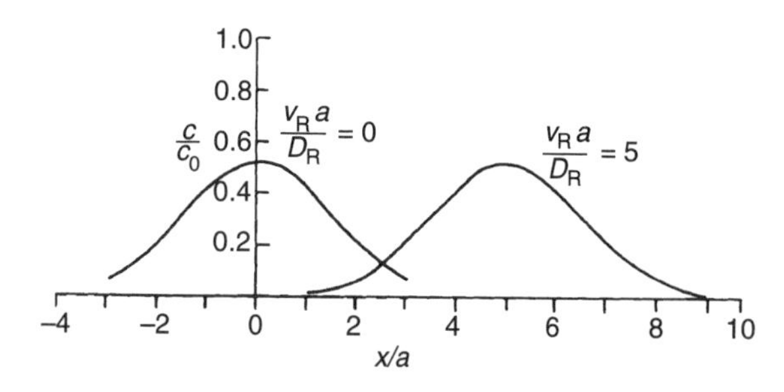

Figure 7.5 Concentration profile for different advective velocities when $D_R t/a^2 = 1$.

The solution given by equation 7.4b obeys identical initial conditions ($t = 0$) to the solution given by equation 7.4a. There is no guarantee that the boundary conditions of the problem will be satisfied; however, in problems involving deep deposits where the boundary conditions are especially simple, the exact solution is found to be given by equation 7.4b.

To illustrate the application of this method, return to the situation shown in Figure 7.1 and suppose that there is a groundwater velocity v in the transmissive layer. The distribution of contaminant is then found to be

$$c = \frac{1}{2} c_0 \, e^{-\Lambda t} [\operatorname{erfc}(X_m) - \operatorname{erfc}(X_p)] \tag{7.5}$$

where

$$X_m = \frac{|x - v_R t| - a}{2\sqrt{D_R t}}$$

$$X_p = \frac{|x - v_R t| + a}{2\sqrt{D_R t}}$$

A comparison of the distribution of contaminant in the transmissive layer for a particular case of advection [$v_R a/D_R = 5$] and no advection [$v_R a/D = 0$] is shown in Figure 7.5. As would be expected from equation 7.5, the contaminant profile when advection is present is identical to the profile when there is no advection present except that it has been translated downstream with a velocity v_R.

Similarly, it is possible to determine the solution for the initially contaminated spherical zone shown in Figure 7.3 when there is a uniform advective velocity v in the x-direction. It then follows from equations 7.3, 7.4a and 7.4b that

$$c = \frac{c_0}{2} e^{-\Lambda t} \left[(\operatorname{erfc}(R_m) - \operatorname{erfc}(R_p)) - 2\sqrt{\frac{D_R t}{\pi R^2}} (\exp[-R_m^2] - \exp[-R_p^2]) \right] \tag{7.6}$$

$$R = \sqrt{(x - v_R t)^2 + y^2 + z^2}$$

$$R_m = \frac{R - a}{2\sqrt{D_R t}}$$

$$R_p = \frac{R + a}{2\sqrt{D_R t}}$$

Again, it can be seen that the concentration profile in the advective case is identical in shape to the non-advective case but has been translated downgradient. The contaminant profiles, in the plane $y = 0$, along the horizontal (x) axis and vertical (z) axis are illustrated in Figure 7.6 for the particular advective

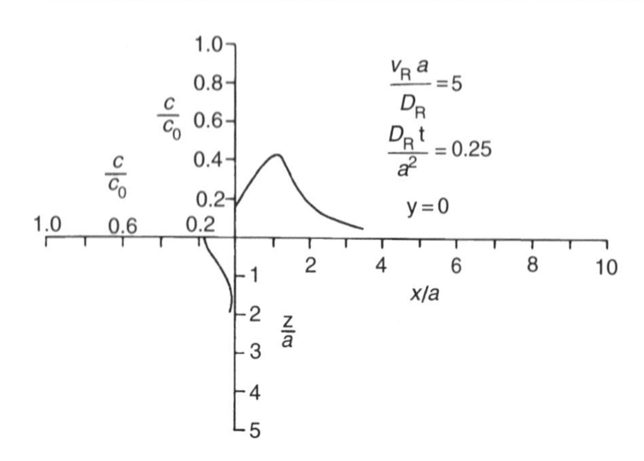

Figure 7.6 Concentration profile along horizontal and vertical axes.

velocity $v_R a/D_R = 5$ and at the particular time $D_R t/a^2 = 0.25$.

7.2.3 Solutions for surface repositories underlain by a deep deposit

In the two previous sections, the concentration distribution near a contaminated zone within a very deep deposit was examined. In this section, the contaminant distribution in the neighbourhood of a surface repository overlaying a deep deposit is examined.

The simplest situation to analyse is that of an extensive surface repository overlaying a deep layer. This is shown schematically in Figure 7.7.

The simplest case to consider is that in which there is no advective transport and no first-order decay, and where it can be assumed that the concentration in the repository remains constant. If it is assumed that there is no background concentration of the contaminant in the layer, then the distribution of contaminant is given by the well-known expression

$$c = c_0 \operatorname{erfc}\left(\frac{z}{2\sqrt{D_R t}}\right) \tag{7.7}$$

and is plotted in Figure 7.8.

Ogata and Banks (1961) extended this solution to take account of advection in the vertical direction and found that

$$c = \frac{c_0}{2}\left[\operatorname{erfc}\left(\frac{z - v_R t}{2\sqrt{D_R t}}\right) + \exp\left(\frac{v_R z}{D_R}\right) \operatorname{erfc}\left(\frac{z + v_R t}{2\sqrt{D_R t}}\right)\right] \tag{7.8}$$

This solution is shown graphically in Figure 1.25 and was discussed in Section 1.7.1.

Booker and Rowe (1987) pointed out that the assumption of constant surface concentration

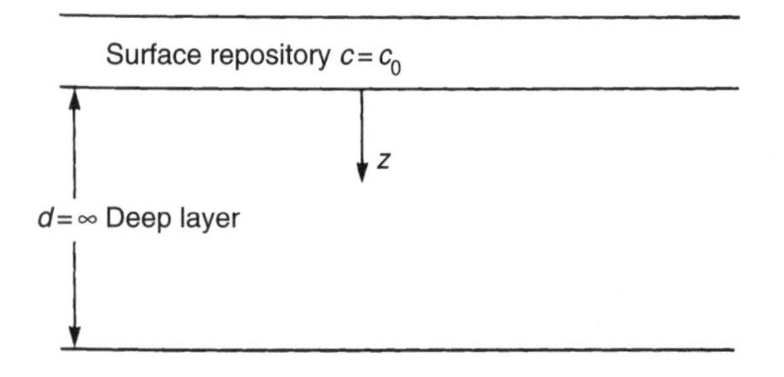

Figure 7.7 Extensive surface repository overlaying a deep layer.

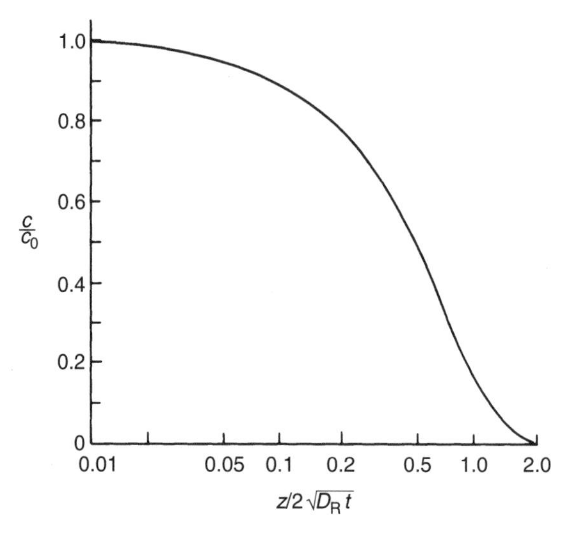

Figure 7.8 Variation of concentration for an extensive surface repository (no advection, constant surface concentration).

was generally unrealistic since most landfills contain a specific mass of contaminant which is not augmented after closure. As contaminant is transported into the underlying deposit, the concentration of contaminant in the landfill reduces. They introduced the concept of the equivalent height of leachate H_f (see Chapter 10 for a detailed discussion) and found that in the absence of first-order decay ($\lambda = 0$) for a landfill in which the initial concentration is c_0, the concentration distribution beneath the landfill is given by

$$c\ (z,t) = \frac{c_0 \exp(ab - b^2 t)(bf(b,t) - df(d,t))}{(b-d)} \tag{7.9}$$

where

$$f(b,t) = \exp(ab + b^2)\,\mathrm{erfc}\left(\frac{a}{2\sqrt{t}} + b\sqrt{t}\right)$$

$$f(d,t) = \exp(ad + d^2)\,\mathrm{erfc}\left(\frac{a}{2\sqrt{t}} + d\sqrt{t}\right)$$

and all other terms are as previously defined. The function $f(q,t)$ (for $q = b$ or d) may be evaluated directly by computer or using a hand calculator or by observing that

$$a = \frac{z}{\sqrt{D}}$$

$$b = \frac{v}{2\sqrt{D}}$$

$$d = \frac{(n + \rho K_d)\sqrt{D}}{H_f} - \frac{v}{2\sqrt{D}}$$

$$f(q,t) = \exp\left(\frac{-a^2}{4t}\right)\phi(X)$$

where

$$X = \frac{a}{2\sqrt{t}} + q\sqrt{t}$$

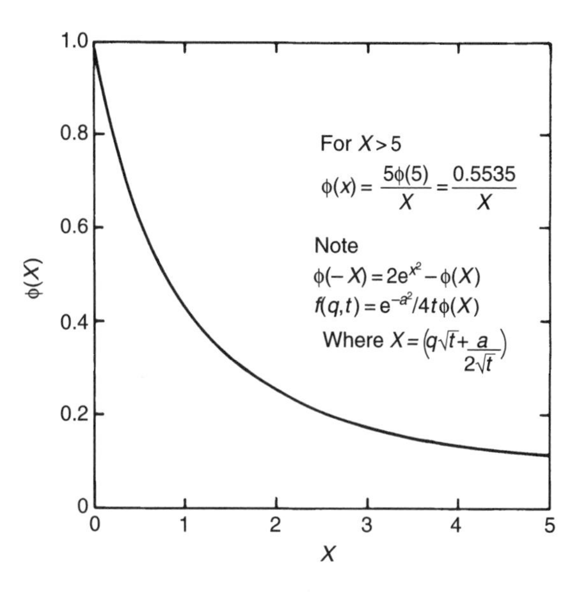

Figure 7.9 Booker and Rowe graphical solution to the advection–dispersion equation for an infinitely deep deposit with a finite mass of contaminant in the source (modified from Booker and Rowe, 1987).

and

$$\phi(X) = e^{X^2}\,\mathrm{erfc}(X)$$

The function $\phi(X)$ is presented graphically in Figure 7.9 and the operation of this approach was illustrated in Section 1.7.2.

7.2.4 Evaluation of analytic solutions

The analytic solutions discussed in this section can all be expressed in terms of the complementary error function which in turn is closely related to the error function, viz.

$$\mathrm{erfc}(x) = 1 - \mathrm{erf}(x)$$

where

$$\mathrm{erf}(x) = \frac{2}{\sqrt{\pi}}\int_0^x e^{-u^2}\,\mathrm{d}u$$

The error function is extensively tabulated (see Abramowitz and Stegun, 1964). The complementary error function can also be evaluated from Figure 7.9 using the relation

$$\mathrm{erfc}(x) = \mathrm{e}^{-x^2}\,\phi(x)$$

Often, it is more convenient to evaluate the solutions directly and this can be easily done by using a polynomial approximation for the error function (Abramowitz and Stegun, 1964). The evaluation can be performed on personal computers or, in many circumstances, on a programmable calculator.

7.3 Application of Laplace transforms to develop a finite layer solution for a single layer

7.3.1 Introduction

The Laplace transformation technique (Carslaw and Jaegar, 1948) provides a powerful method for the development of analytic solutions to the equations of contaminant transport. It will be shown in the following sections that it can also be used as a most effective method of numerical solution. To illustrate this, some problems involving 1D contaminant transport in the z-direction will be considered.

7.3.2 Governing equations

The equations governing advective–diffusive–dispersive transport have been discussed in Chapter 1. For ease of presentation they will be repeated here. The equation (equation 1.14a) of mass balance may be written as

$$-\frac{\partial f}{\partial z} = (n + \rho K_\mathrm{d})\frac{\partial c}{\partial t} + n\lambda c \qquad (7.10\mathrm{a})$$

If it is assumed that initially the ground is uncontaminated, then

$$c = 0 \quad \text{when } t = 0 \qquad (7.10\mathrm{b})$$

Finally, if it is assumed that contaminant migration is by advective–diffusive transport, then

$$f = nvc - nD\frac{\partial c}{\partial z} \qquad (7.10\mathrm{c})$$

In equations 7.10a–7.10c, c is the concentration at depth z and time t [ML^{-3}], f the mass flux in the z-direction [$ML^{-2}T^{-1}$], n the effective porosity of the soil [–], ρ the dry density of the soil [ML^{-3}], K_d the distribution coefficient [$M^{-1}L^3$], v the seepage velocity [LT^{-1}] and λ the first-order decay constant [T^{-1}].

$$\lambda = \Gamma_\mathrm{R} + \Gamma_\mathrm{B} + \Gamma_\mathrm{S}$$

where Γ_R (= ln 2/(radioactive half-life)) is the radioactive decay constant [T^{-1}], Γ_B (= ln 2/(biological decay half-life)) the biological decay constant at this location [T^{-1}] and Γ_S the flow removed per unit volume of soil in a permeable layer beneath a landfill (see Section 1.3.7) [T^{-1}].

Equations 7.10a and 7.10c may be combined to give the transport equation:

$$nD\frac{\partial^2 c}{\partial z^2} - nv\frac{\partial c}{\partial z} = (n + \rho K_\mathrm{d})\frac{\partial c}{\partial t} + n\lambda c \qquad (7.10\mathrm{d})$$

Equation 7.10d must then be solved subject to the initial condition (equation 7.10b) and any boundary conditions.

As mentioned in the introduction, a powerful way of solving these equations is by employing a Laplace transform:

$$\overline{\psi} = \int_0^\infty \mathrm{e}^{-st}\,\psi(t)\,\mathrm{d}t \qquad (7.11)$$

The variable s is called the Laplace transform parameter, the superior bar is often used to indicate that a Laplace transform has been applied. Some typical transforms are given in Table 7.1. More detailed lists of transforms are given in Carslaw and Jaegar (1948), Abramowitz and Stegun (1964) and a very comprehensive list in Erdelyi *et al.* (1954).

If equation 7.10d is transformed using a Laplace transform; it is found that after incorporating the initial condition (equation 7.10b) it reduces to the ordinary differential equation:

$$nD\frac{\partial^2 \bar{c}}{\partial z^2} - nv\frac{\partial \bar{c}}{\partial z} = nS\bar{c} \tag{7.12a}$$

where

$$S = \left(1 + \frac{\rho K_\mathrm{d}}{n}\right)s + \lambda \tag{7.12b}$$

An expression for transformed flux follows from equation 7.10c so that

$$\bar{f} = nv\bar{c} - nD\frac{\partial \bar{c}}{\partial z} \tag{7.12c}$$

Equations 7.12a–7.12c are solved, subject to any transformed boundary conditions, leading to the solution in the Laplace transform domain. The solution in the time domain is then found by inverting the Laplace transform. This procedure will be illustrated by a simple example in the following section.

Table 7.1

Typical transforms	
$\psi(t)$	$\bar{\psi}(s)$
1	$\frac{1}{s}$
e^{-bt}	$\frac{1}{s+b}$
$\frac{\partial \phi}{\partial t}$	$s\bar{\phi} - \phi(0)$
$\mathrm{erfc}\left(\frac{k}{2\sqrt{t}}\right)$	$\frac{\mathrm{e}^{-k\sqrt{s}}}{s} \quad (k>0)$
$\mathrm{e}^{a(k+ak)}\,\mathrm{erfc}\left(\frac{a\sqrt{t}+k}{2\sqrt{t}}\right)$	$\frac{\mathrm{e}^{-k\sqrt{s}}}{\sqrt{s}(a+\sqrt{s})} \quad (k>0)$

7.3.3 Solution to a simple problem

One of the simplest situations that can be envisaged is 1D diffusion in a single uniform layer of thickness H subject to the initial condition ($t = 0$) of zero concentration in the layer, viz.

$$c(z,0) = 0 \qquad 0 \le z \le H \tag{7.13a}$$

and the boundary condition of constant surface concentration at the top of the layer and zero concentration at depth H, viz.

$$c = c_0 \quad \text{when } z = 0, \quad t > 0 \tag{7.13b}$$

$$c = 0 \quad \text{when } z = H, \quad t > 0 \tag{7.13c}$$

The case for which $H = \infty$ corresponds to an infinitely deep layer.

Referring to Table 7.1, the boundary conditions (equations 7.13b and 7.13c) become

$$\bar{c} = \frac{c_0}{s} \quad \text{when } z = 0 \tag{7.14a}$$

$$\bar{c} = 0 \quad \text{when } z = H \tag{7.14b}$$

For the sake of simplicity, attention will be restricted to the case where there is no advection and no first order-decay ($v = 0$, $\lambda = 0$). The transformed transport, equation 7.12a, was developed in the previous section; it becomes

$$\frac{\partial^2 \bar{c}}{\partial z^2} = \frac{S}{D}\bar{c} \tag{7.15a}$$

which may be conveniently written as

$$\frac{\partial^2 \bar{c}}{\partial z^2} = \alpha^2 \bar{c} \tag{7.15b}$$

where

$$\alpha^2 = \frac{S}{D} = \frac{s}{D_R}$$

$$S = \left(1 + \frac{\rho K_d}{n}\right)s$$

$$D_R = \frac{D}{(1 + \rho K_d/n)}$$

Equations 7.15a and 7.15b have the general solution

$$\bar{c} = A\,e^{\alpha z} + B\,e^{-\alpha z} \tag{7.16}$$

The constants A and B can be found from the transformed boundary conditions (equations 7.14a and 7.14b) and thus the solution in transform space is found to be

$$\bar{c} = \frac{c_0}{s}\frac{e^{\alpha z}}{(1 - e^{2\alpha H})} + \frac{c_0}{s}\frac{e^{-\alpha z}}{(1 - e^{-2\alpha H})} \tag{7.17}$$

Equation 7.17 is the solution in transform space. It is now necessary to invert the transform and find the solution in physical space. This can be done numerically using a technique developed by Talbot (1979) and this will be the approach adopted for the more complex problem considered in the following sections. In this present case, it is relatively simple to obtain an analytic conversion of equation 7.17 for the case where the layer is infinitely deep ($H \simeq \infty$), in which case equation 7.17 reduces to

$$\bar{c} = \frac{c_0\,e^{-\alpha z}}{s} \tag{7.18a}$$

or

$$\bar{c} = c_0\frac{e^{-k\sqrt{s}}}{s} \tag{7.18b}$$

where

$$k = \frac{z}{\sqrt{D_R}}$$

It now follows immediately from Table 7.1 that

$$c = c_0\,\mathrm{erfc}\left(\frac{z}{2\sqrt{D_R t}}\right) \tag{7.18c}$$

thus providing the derivation of equation 7.7.

7.4 Contaminant transport into a single layer considering a landfill of finite mass and an underlying aquifer

Consider the situation shown schematically in Figure 7.10 of an extensive landfill overlaying a clay which in turn overlays a transmissive layer.

Since the landfill is extensive, transport will be predominantly vertical and thus the migration of contaminant will be governed by equation 7.10d in the physical domain and equation 7.12a in the Laplace transform domain. The general solution of equation 7.12a is

$$\bar{c} = A\,e^{\alpha z} + B\,e^{\beta z} \tag{7.19a}$$

where

$$\alpha = \frac{v}{2D} + \sqrt{\left(\frac{v^2}{4D^2}\right) + \frac{S}{D}}$$

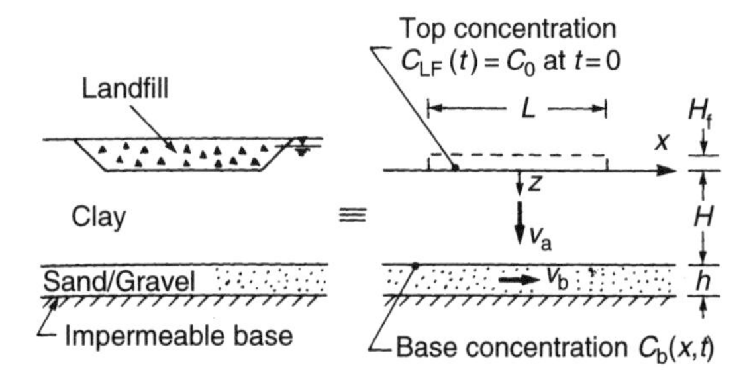

Figure 7.10 Problem description – single aquitard layer underlain by an aquifer.

$$\beta = \frac{v}{2D} - \sqrt{\left(\frac{v^2}{4D^2}\right) + \frac{S}{D}}$$

It is not difficult to show that the transformed flux (equation 7.12c) is then given by

$$\bar{f} = nD(\beta A\, e^{\alpha z} + \alpha B\, e^{\beta z}) \tag{7.19b}$$

It was shown in Section 1.5 that it was possible to take into account the fact that the landfill contains only a finite amount of contaminant and that (equation 1.23),

$$c_T(t) = c_0 - \frac{1}{H_r}\int_0^t f_T(\tau)\,d\tau - \int_0^t \lambda_T c(\tau)\,d\tau \tag{7.20 a}$$

which in Laplace transform space becomes

$$\bar{c}_T = \frac{c_0}{S_T} - \frac{\overline{f_T}}{S_T H_r} \tag{7.20 b}$$

where

$$S_T = s + \lambda_T = s + \Gamma_R + \Gamma_{BT} + \frac{q_c}{H_r}$$

In the above equation, c_T is the concentration of contaminant in the landfill [ML^{-3}], c_0 the initial concentration of contaminant in the landfill [ML^{-3}], f_T the flux entering the clay from the landfill [ML^{-2}T^{-1}], H_r the reference height of the leachate as briefly discussed in Section 1.5 and as will be discussed in greater detail in Section 10.2 [L], Γ_R (= ln 2/(radioactive half-life)) the radioactive decay constant [T^{-1}], Γ_{BT} (= ln 2/(biological half-life in the landfill)) the biological decay constant [T^{-1}] and q_c the volume of leachate collected per unit area of landfill [LT^{-1}].

It was shown in Section 1.6 that it was also possible to model the underlying aquifer and it is found that (equation 1.25b)

$$c_b(t) = \int_0^t \left[\frac{f_b(\tau)}{n_b h} - \lambda_b c_b(\tau)\right] d\tau \tag{7.21a}$$

or in Laplace transform space

$$\bar{f}_b = n_b h S_b \bar{c}_b \tag{7.21b}$$

where

$$S_b = s + \lambda_b = s + \Gamma_R + \Gamma_{Bb} + \frac{v_b}{n_b L}$$

In the above equation, n_b is the porosity of the aquifer [–], h the thickness of the aquifer [L], v_b the Darcy velocity in the aquifer [LT^{-1}], c_b the concentration in the aquifer [ML^{-3}], f_b the flux entering the aquifer from the clay [ML^{-2}T^{-1}], Γ_{Rb} the radioactive decay constant [T^{-1}], Γ_{Bb} the biological decay constant [T^{-1}] and L the length of the landfill parallel to the direction of flow in the aquifer [L].

Substitution of equations 7.19a and 7.19b into equations 7.20a, 7.20b, 7.21a and 7.21b lead to the set of equations

$$\left(1 + \frac{nD\beta}{H_r S_T}\right)A + \left(1 + \frac{nD\alpha}{H_r S_T}\right)B = \frac{c_0}{S_T}$$

$$e^{\alpha H}\left(1 - \frac{nD\beta}{h n_b S_b}\right)A + e^{\beta H}\left(1 - \frac{nD\alpha}{h n_b S_b}\right)B = 0 \tag{7.22}$$

These equations can easily be solved for the constants A and B, and this together with equation 7.19 complete the solution in transform space. The solution in the physical plane is found by using the algorithm developed by Talbot (1979).

7.5 Finite layer analysis

7.5.1 1D transport of contaminant from a landfill overlaying a layered deposit

The procedures developed in the previous section will now be extended to include the effects of layering of the soil beneath the landfill. The simplest case which can be considered is that of 1D contaminant transport in the clayey deposit, parallel to the z-direction as shown in Figure 7.11. It is assumed that this deposit is divided into a number of layers by node planes $z = z_0, z_1, \ldots, z_n$ and that each layer $(z_{k-1} < z < z_k)$ may be considered homogeneous. Thus the vertical flux per unit area per unit time at a point in layer k is given by equation 7.10c:

$$f_z = nv_z c - nD_{zz}\frac{\partial c}{\partial z} \tag{7.23a}$$

The equation of contaminant transport (equation 7.10d) may be written as

$$nD_{zz}\frac{\partial^2 c}{\partial z^2} - nv_z\frac{\partial c}{\partial z} = (n + \rho K_{\mathrm{d}})\frac{\partial c}{\partial t} - n\lambda c \tag{7.23b}$$

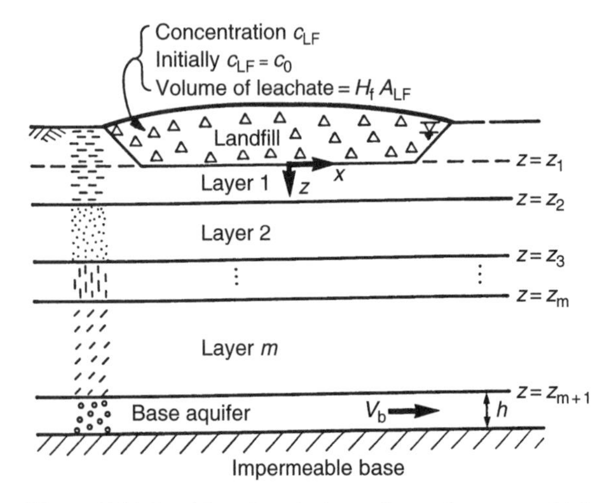

Figure 7.11 Problem description – layered system which may contain multiple aquitards and aquifers.

where the subscript z denotes that this parameter applies to the z-direction and the quantities n, v_z, D_{zz}, ρ, K_{d} and λ assume constant values appropriate for the layer k under consideration. As in the previous section, it will be assumed that there is no pre-existing distribution of contaminant within the deposit so that

$$c = 0 \quad \text{when } (z_{k-1} \le z \le z_k) \text{ at } t = 0$$

Equations 7.23a and 7.23b can be simplified by introducing the Laplace transform

$$(\bar{c}, \bar{f}_z) = \int_0^\infty (c, f_z)\,\mathrm{e}^{-st}\,\mathrm{d}t$$

yielding

$$\bar{f}_z = nv_z\bar{c} - nD_{zz}\frac{\partial \bar{c}}{\partial z} \tag{7.24a}$$

$$nD_{zz}\frac{\partial^2 \bar{c}}{\partial z^2} - nv_z\frac{\partial \bar{c}}{\partial z} = nS\bar{c} \tag{7.24b}$$

where

$$S = \left(1 + \frac{\rho K_{\mathrm{d}}}{n}\right)s + \lambda$$

$$\bar{c} = A\,\mathrm{e}^{\alpha z} + B\,\mathrm{e}^{\beta z} \tag{7.24c}$$

$$\bar{f}_z = nD_{zz}(\beta A\,\mathrm{e}^{\alpha z} + \alpha B\,\mathrm{e}^{\beta z}) \tag{7.24d}$$

Equations 7.24a and 7.24b have the solution where A and B are constants to be determined

$$\alpha = \frac{v_z}{2D_{zz}} + \sqrt{\frac{v_z^2}{4D_{zz}^2} + \frac{S}{D_{zz}}}$$
$$\beta = \frac{v_z}{2D_{zz}} - \sqrt{\frac{v_z^2}{4D_{zz}^2} + \frac{S}{D_{zz}}} \tag{7.24e}$$

If the constants A and B appearing in equations 7.24c and 7.24d are evaluated in terms of the concentrations at the nodal planes z_j and z_k (where $j = k - 1$), it is found that

$$\bar{c} = \bar{c}_j \left\{ \frac{e^{\alpha(z-z_k)} - e^{\beta(z-z_k)}}{e^{\alpha(z_j-z_k)} - e^{\beta(z_j-z_k)}} \right\} + \bar{c}_k \left\{ \frac{e^{\alpha(z-z_j)} - e^{\beta(z-z_j)}}{e^{\alpha(z_k-z_j)} - e^{\beta(z_k-z_j)}} \right\} \qquad (7.25a)$$

and thus

$$\frac{\bar{f}_z}{nD_{zz}} = \bar{c}_j \left\{ \frac{\beta\, e^{\alpha(z-z_k)} - \alpha\, e^{\beta(z-z_k)}}{e^{\alpha(z_j-z_k)} - e^{\beta(z_j-z_k)}} \right\} + \bar{c}_k \left\{ \frac{\beta\, e^{\alpha(z-z_j)} - \alpha\, e^{\beta(z-z_j)}}{e^{\alpha(z_k-z_j)} - e^{\beta(z_k-z_j)}} \right\} \qquad (7.25b)$$

It is now possible to evaluate the flux on each of the node planes, using equation 7.25b and it is found that

$$\begin{bmatrix} \bar{f}_{zj} \\ -\bar{f}_{zk} \end{bmatrix} = \begin{bmatrix} Q_k & R_k \\ S_k & T_k \end{bmatrix} \begin{bmatrix} \bar{c}_j \\ \bar{c}_k \end{bmatrix} \qquad (7.26)$$

where

$$Q_k = \frac{nD_{zz}(\beta\, e^{\mu\beta} - \alpha\, e^{\mu\alpha})}{e^{\mu\beta} - e^{\mu\alpha}}$$

$$R_k = -\frac{nD_{zz}(\beta - \alpha)}{e^{\mu\beta} - e^{\mu\alpha}}$$

$$S_k = -\frac{nD_{zz}(\beta - \alpha)\, e^{\mu(\beta+\alpha)}}{e^{\mu\beta} - e^{\mu\alpha}}$$

$$T_k = \frac{nD_{zz}(\beta\, e^{\mu\alpha} - \alpha\, e^{\mu\beta})}{e^{\mu\beta} - e^{\mu\alpha}}$$

and $\mu = z_k - z_{k-1}$.

Noting that both the flux f_z and concentrations c must be continuous, the layer matrices defined by equation 7.26 may be assembled for each layer in the deposit to give the following equation:

$$\begin{bmatrix} Q_1 & R_1 & & & & \\ S_1 & T_1+Q_2 & R_2 & & & \\ & S_2 & T_2+Q_3 & R_3 & & \\ & & & \cdot & & \\ & & & \cdot & & \\ & & & \cdot & & \\ & & & S_{n-1} & T_{n-1}+Q_n & R_n \\ & & & & S_n & T_n \end{bmatrix} \times \begin{bmatrix} \bar{c}_T \\ \bar{c}_1 \\ \bar{c}_2 \\ \cdot \\ \cdot \\ \cdot \\ \bar{c}_{n-1} \\ \bar{c}_b \end{bmatrix} = \begin{bmatrix} \bar{f}_T \\ 0 \\ 0 \\ \cdot \\ \cdot \\ \cdot \\ 0 \\ -\bar{f}_b \end{bmatrix} \qquad (7.27)$$

where $\bar{c}_T$ and $\bar{f}_T$ are the values of concentration and flux at the top of the deposit, and $\bar{c}_b$ and $\bar{f}_b$ are the values of concentration and flux at the base of the deposit.

We now wish to solve this equation subject to the appropriate boundary conditions. The boundary condition at the bottom of the layered system was examined in the previous section (equation 7.21) and it follows in identical fashion that

$$\bar{f}_b = Q_{n+1}\bar{c}_b \qquad (7.28a)$$

where $Q_{n+1} = hn_b S_b$.

The boundary condition at the top of the layered system was also considered in the previous section (equation 7.20b) and it follows that

$$\bar{f}_T = B_0 - T_0\bar{c}_T \qquad (7.28b)$$

where $B_0 = H_r c_0$ and $T_0 = H_r S_T$.

$$\begin{bmatrix} T_0+Q_1 & R_1 & & & & & \\ S_1 & T_1+Q_2 & R_2 & 0 & & & \\ 0 & S_2 & T_2+Q_3 & R_3 & & & \\ & & & \cdot & & & \\ & & & \cdot S_{n-1} & T_{n-1}+Q_n & R_n & \\ & & & \cdot & 0 & S_n & T_n+Q_{n+1} \end{bmatrix} \begin{bmatrix} \bar{c}_{\mathrm{T}} \\ \bar{c}_1 \\ \bar{c}_2 \\ \cdot \\ \bar{c}_{n-1} \\ \bar{c}_{\mathrm{b}} \end{bmatrix} = \begin{bmatrix} B_0 \\ 0 \\ 0 \\ \cdot \\ 0 \\ 0 \end{bmatrix} \tag{7.29}$$

Introducing the expressions for $\bar{f}_{\mathrm{T}}$ and $\bar{f}_{\mathrm{b}}$ into equations 7.28a and 7.28b then gives the complete set of equations: where c_j denotes the concentration in the layer plane $z = z_j$; Q_k, R_k, S_k and T_k are defined by equation 7.26; and Q_{n+1}, B_0 and T_0 are defined by equations 7.28a and 7.28b. Equation 7.29 can now be solved giving $\bar{c}_j$ at each node plane. The Laplace transform can then be inverted using a very efficient scheme proposed by Talbot (1979). Using this approach, an accuracy of order 10^{-6} and 10^{-10} can typically be achieved using 11 and 18 sample points respectively. For many practical problems, an accuracy of 10^{-6} is more than adequate. The theory described above has been coded in program POLLUTE (Rowe and Booker, 1983, 1994, 2004) and can be run on a personal computer.

7.5.2 Finite layer analysis in two dimensions

The procedure described in the previous section can be readily generalized for contaminant transport in both the x- and z-directions. Again, it is assumed that the deposit is divided into a number of layers with node planes $z = z_0, z_1, \ldots, z_n$ and that each layer k ($z_{k-1} < z < z_k$) may be considered as homogeneous. Thus the fluxes f_x and f_z transported in the x- and z-directions within layers k are given by

$$f_x = nv_x c - nD_{xx}\frac{\partial c}{\partial x} \tag{7.30a}$$

$$f_z = nv_z c - nD_{zz}\frac{\partial c}{\partial z} \tag{7.30b}$$

For 2D transport in the soil, the equation governing contaminant migration can be written as

$$(n+\rho K_{\mathrm{d}})\frac{\partial c}{\partial t} = nD_{xx}\frac{\partial^2 c}{\partial x^2} + nD_{zz}\frac{\partial^2 c}{\partial z^2} - nv_x\frac{\partial c}{\partial x} - nv_z\frac{\partial c}{\partial z} - n\lambda c \tag{7.31a}$$

where the quantities n, v_x, v_z, D_{xx}, D_{zz}, λ, ρ and K_{d} are physical constants appropriate to the layer k under consideration. For simplicity of presentation, it will be assumed that there is no initial background concentration in the deposit so that

$$c = 0 \quad \text{when } t = 0 \tag{7.31b}$$

Equations 7.30a, 7.30b, 7.31a and 7.31b can be simplified by the introduction of a Laplace transform:

$$(\bar{c}, \bar{f}_x, \bar{f}_z) = \int_0^{\infty} (c, f_x, f_z)\,\mathrm{e}^{-st}\,\mathrm{d}t$$

and a Fourier transform:

$$(\overline{C}, \overline{F}_x, \overline{F}_z) = \frac{1}{2\pi}\int_{-\infty}^{\infty} (\bar{c}, \bar{f}_x, \bar{f}_z)\,\mathrm{e}^{-i\xi x}\,\mathrm{d}x$$

$$nD_{zz}\frac{\partial^2\overline{C}}{\partial z^2} - nv_z\frac{\partial\overline{C}}{\partial z} = nS\overline{C} \quad (7.32a)$$

$$\overline{F}_x = nv_x\overline{C} - nD_{xx}i\xi\overline{C} \quad (7.32b)$$

$$\overline{F}_z = nv_zC - nD_{zz}\frac{\partial\overline{C}}{\partial z} \quad (7.32c)$$

These transforms reduce equations 7.30a, 7.30b, 7.31a and 7.31b to

$$S = \left(1 + \frac{\rho K}{n}\right)s + \xi^2 D_{xx} + i\xi v_x + \lambda \quad \text{and}$$

$$\lambda = \Gamma_R + \Gamma_B + \Gamma_s$$

The solution of equations 7.32a and 7.32c is

$$\overline{C} = A\,e^{\alpha z} + B\,e^{\beta z} \quad (7.33a)$$

$$\frac{\overline{F}_z}{nD_{zz}} = \beta A\,e^{\alpha z} + \alpha B\,e^{\beta z} \quad (7.33b)$$

where A and B are constants to be determined, and

$$\alpha = \frac{v_z}{2D_{zz}} + \sqrt{\frac{v_z^2}{4D_{zz}^2} + \frac{S}{D_{zz}}}$$

$$\beta = \frac{v_z}{2D_{zz}} - \sqrt{\frac{v_z^2}{4D_{zz}} + \frac{S}{D_{zz}}} \quad (7.34)$$

If the constants A and B are evaluated in terms of the nodal plane concentrations, then interpolation formulae similar to equation 7.25a are found, viz.

$$\overline{C} = \overline{C}_j\left\{\frac{e^{\alpha(z-z_k)} - e^{\beta(z-z_k)}}{e^{\alpha(z_j-z_k)} - e^{\beta(z_j-z_k)}}\right\} + \overline{C}_k\left\{\frac{e^{\alpha(z-z_j)} - e^{\beta(z-z_j)}}{e^{\alpha(z_k-z_j)} - e^{\beta(z_k-z_j)}}\right\} \quad (7.35)$$

This equation can be used to determine the node plane fluxes for the element and leads to the relationship

$$\begin{bmatrix} \overline{F}_{zj} \\ -\overline{F}_{zk} \end{bmatrix} = \begin{bmatrix} Q_k & R_k \\ S_k & T_k \end{bmatrix}\begin{bmatrix} \overline{C}_j \\ \overline{C}_k \end{bmatrix} \quad (7.36)$$

where Q_k, R_k, S_k and T_k are precisely as defined in equation 7.26 but where in this case α and β are defined by equation 7.34.

Next, consider the lower boundary. It will be assumed that the clay layer is underlain by a more permeable stratum of thickness h and porosity n_b. Assuming that the concentration is uniform across the thickness h but may vary with position x, the general equation (equation 1.25) governing the concentration in the base aquifer was developed in Section 1.6 and it becomes

$$c_b = \int_0^t \left[\frac{f_b}{hn_b} - \frac{v_b}{n_b}\frac{\partial c_b}{\partial x} + D_H\frac{\partial^2 c_b}{\partial x^2} - \lambda_b^* c_b\right] d\tau$$

where

$$\lambda_b^* = \Gamma_R + \Gamma_{Bb}$$

or on applying both Laplace and Fourier transforms,

$$\overline{F}_b = \Omega_b\overline{C}_b \quad (7.37)$$

where $\overline{C}_b$ and $\overline{F}_b$ are the transformed concentration and flux at the base and

$$\Omega_b = hn_b\left[s + \lambda_b^* + \frac{i\xi v_b}{n_b} + \xi^2 D_H\right]$$

It now follows from the continuity of normal flux at the node planes that, and $\overline{C}_T$ and $\overline{F}_T$, respectively, denote the transformed concentration and flux at the top.

$$\begin{bmatrix} Q_1 & R_1 & & & & & \\ S_1 & T_1+Q_2 & R_2 & & & & \\ & S_2 & T_2+Q_3 & R_3 & & & \\ & & & S_{n-1} & T_{n-1}+Q_n & R_n \\ & & & & S_n & T_n+\Omega_b \end{bmatrix} \begin{bmatrix} \overline{C}_T \\ \overline{C}_1 \\ \overline{C}_2 \\ \overline{C}_{n-1} \\ \overline{C}_b \end{bmatrix} = \begin{bmatrix} \overline{F}_T \\ 0 \\ 0 \\ 0 \\ 0 \end{bmatrix} \tag{7.38}$$

It remains to model the interaction between the layered soil and the landfill. It will be assumed that the distribution of contaminant within the landfill maintains the same spatial distribution, shown in Figure 7.12. This takes into account that the concentration of contaminant will be virtually constant within the landfill but there will be a transition zone at the edge of the landfill where the contaminant decreases from its maximum value to a value of zero just outside the landfill. It follows that

$$c_T = p(x)c_{LF} \tag{7.39}$$

where $p(x)$ represents the distribution shown in Figure 7.12 and c_{LF} is the concentration away from the landfill. When equation 7.39 is transformed it becomes

$$\overline{C}_T = \frac{2}{\pi}\frac{[\cos(\xi l/2) - \cos(\xi L/2)]}{\xi^2(L-l)}\overline{c}_{LF} \tag{7.40}$$

The equation governing mass balance in the landfill was discussed in Section 1.5 and it follows in similar fashion to the derivation of equation 1.23 that

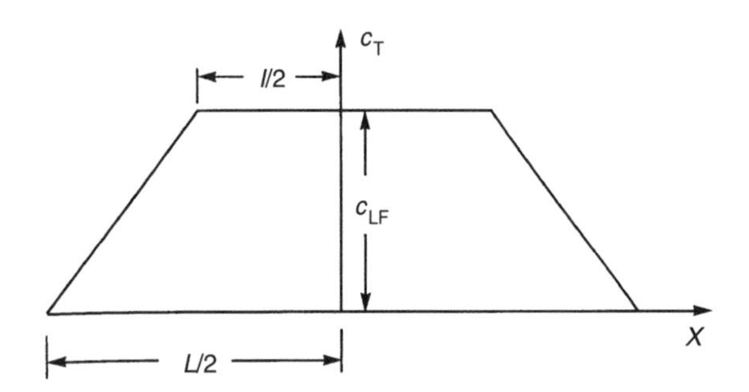

Figure 7.12 Assumed distribution of contaminant within a landfill.

$$c_{LF} = c_0 - \frac{1}{L_{av}H_r}\int_0^t \left(\int_{-L/2}^{L/2} \overline{f}_T(x,\tau)\,dx\right)d\tau - \lambda_T\int_0^t c_{LF}\,dt \tag{7.41a}$$

where $L_{av} = (L+l)/2$, whereupon, utilizing Fourier's inversion theorem,

$$\overline{c}_{LF} = \frac{c_0}{S_T} - \frac{2}{S_T H_r L_{av}}\int_{-\infty}^{\infty} \overline{F}_T \frac{\sin(\xi L/2)}{\xi}\,d\xi \tag{7.41b}$$

where $S_T = s + \lambda_T$

At this stage, the value of c_{LF} is as yet undetermined. The key to its evaluation is to realize that although $\overline{C}_T$ and $\overline{F}_T$ cannot be individually evaluated, their ratio can be. This is most easily done by arbitrarily setting $\overline{F}_T = 1$ solving equation 7.38 for all the transformed node plane concentrations and then evaluating

$$\overline{\chi} = \frac{L\overline{F}_T}{D_{zz}\overline{C}_T} \tag{7.42}$$

Once $\overline{\chi}$ has been determined, equations 7.40, 7.41b and 7.42 can be utilized to show that

$$\overline{c}_{LF} = \frac{c_0}{S_T + (D_{zz}/LH_r)\Omega_T} \tag{7.43}$$

where

$$\Omega_T = \frac{8}{\pi}\int_{-\infty}^{\infty} \frac{\overline{\chi}\sin(\xi L/2)[\cos(\xi l/2) - \cos(\xi L/2)]}{\xi^3(L^2 - l^2)}\,d\xi$$

Once $\overline{c}_{\mathrm{LF}}$ is known, the complete solution can be obtained from equation 7.38 and the Talbot inversion.

The theory described above for 2D conditions has been coded in program MIGRATE (Rowe and Booker, 1988b). The major computational effort involved is associated with the numerical inversion of the Laplace and Fourier transforms for the locations and times of interest. The Laplace transform can again be inverted using Talbot's algorithm. The Fourier transform can be efficiently inverted using 20-point Gauss quadrature. The width and number of integration subintervals which are needed to achieve a reasonable accuracy (say 0.1%) depend somewhat on the geometry and properties of the problem under consideration. These parameters can be determined from a few trial calculations for a representative point and time of interest. Similarly, it should be noted that numerical experiments are also required to determine an appropriate finite element mesh and time integration procedure if alternative finite element or finite difference codes are used.

7.5.3 1D transport with an initial contaminant distribution

In Section 7.5.1 it has been assumed that:

1. there was no initial distribution of contaminant within the soil; and
2. no environmental changes, such as a change in advective velocity, or a change in the general nature of conditions within the landfill or in the underlying aquifer, occur.

The analysis developed in Section 7.5.1 will now be extended to incorporate these effects. Attention will be focussed on the effect of initial concentrations since it will be shown that modelling of environmental change can always be reduced to a case involving initial concentrations.

The configuration to be considered will be shown schematically in Figure 7.10 with the exception that it will be assumed that there is an initial distribution c_{I} of contaminant throughout the deposit, so that

$$c = c_{\mathrm{I}} \quad \text{when } t = 0 \tag{7.44}$$

The equation governing contaminant transport within a particular layer $z_{k-1} < z < z_k$ will again be equation 7.23b. The Laplace transform of this equation is, after incorporation of the initial condition equation 7.44, found to be

$$nD_{zz}\frac{\partial^2 \overline{c}}{\partial z^2} - nv_z\frac{\partial \overline{c}}{\partial z} = n(S\overline{c} - \phi) \tag{7.45a}$$

where

$$S = \left(1 + \frac{\rho K_{\mathrm{d}}}{n}\right)s + \lambda \tag{7.45b}$$

and

$$\phi = c_{\mathrm{I}}\left(1 + \frac{\rho K_{\mathrm{d}}}{n}\right) \tag{7.45c}$$

and, as before, the parameters n, v_z, D_{zz}, λ, c_{I} refer to the particular layer under consideration.

In order to find the solution of equations 7.45a–7.45c, it will be assumed that ϕ may be approximated in the form

$$\phi = E\,\mathrm{e}^{\varepsilon z} \tag{7.46}$$

The quantities E and ε can be found from the condition that $\phi = \phi_j$ when $z = z_j$ $(j = k - 1)$ and $\phi = \phi_k$ when $z = z_k$ and thus

$$\varepsilon = \frac{\ln \phi_j - \ln \phi_k}{z_j - z_k}$$

$$\ln E = \frac{z_k \ln \phi_j - z_j \ln \phi_k}{z_k - z_j} \tag{7.47}$$

Equations 7.45a–c are now readily solved and it is found that the transformed flux and concentration are given by

$$\bar{c} = \frac{\phi}{G} + A\,e^{\alpha z} + B\,e^{\beta z}$$
$$\frac{\bar{f}_z}{nD} = \frac{(\alpha + \beta - \varepsilon)}{G}\phi + \beta A\,e^{\alpha z} + \alpha B\,e^{\beta z} \tag{7.48}$$

where $G = -D(\varepsilon - \alpha)(\varepsilon - \beta)$ and

$$\alpha = \frac{v_z}{2D_z} + \sqrt{\frac{v_z^2}{4D_z^2} + \frac{S}{D_{zz}}}$$

$$\beta = \frac{v_z}{2D_z} - \sqrt{\frac{v_z^2}{4D_{zz}^2} + \frac{S}{D_{zz}}}$$

are identical to the values given in equation 7.24e. The quantities A and B can now be found in terms of the nodal values c_j and c_k, it then follows

$$\begin{bmatrix} +\bar{f}_{zj} \\ -\bar{f}_{zk} \end{bmatrix} = nD \begin{bmatrix} +(\alpha + \beta - \varepsilon) & \frac{\phi_j}{G} \\ -(\alpha + \beta - \varepsilon) & \frac{\phi_k}{G} \end{bmatrix} + \begin{bmatrix} Q_k & R_k \\ S_k & T_k \end{bmatrix} \begin{bmatrix} \bar{c}_j & -\frac{\phi_j}{G} \\ \bar{c}_k & -\frac{\phi_k}{G} \end{bmatrix} \tag{7.49}$$

There are no changes to the boundary conditions at the upper and lower surface and so if the layer matrices, equation 7.49 can be assembled and boundary conditions added as before to give

$$\begin{bmatrix} T_0 + Q_1 & R_1 & & & & \\ S_1 & T_1 + Q_2 & R_2 & & & \\ \cdot & \cdot & \cdot & \cdot & \cdot & \\ & & \cdot & \cdot & & \\ & & S_{n-1} & T_{n-1} + Q_n & R_n & \\ & & & S_n & T_n + Q_{n+1} \end{bmatrix} \begin{bmatrix} \bar{c}_T \\ \bar{c}_1 \\ \cdot \\ \cdot \\ \bar{c}_{n+1} \\ \bar{c}_b \end{bmatrix} = \begin{bmatrix} B_0 + A_1 \\ B_1 + A_2 \\ \cdot \\ \cdot \\ B_{n-1} + A_n \\ B_n + A_{n+1} \end{bmatrix} \tag{7.50}$$

where

$$A_k = \frac{1}{G}[nD(\varepsilon - \alpha - \beta)\phi_j + Q_k\phi_j + R_k\phi_k]$$
$$B_k = \frac{1}{G}[-nD(\varepsilon - \alpha - \beta)\phi_k + S_k\phi_j + T_k\phi_k]$$
$$\text{for } k = 1, 2, \ldots, n \quad \text{and} \quad A_{n+1} = 0$$

Equation 7.50 can now be solved for the transformed node plane concentrations $\bar{c}_k$ and the solution in the physical plane found by Talbot (1979) inversion.

7.5.4 Finite layer analysis when there are environmental changes

Consider a layered system which undergoes some environmental change at time $t = t^*$. This environmental change may lead to a change in the governing differential equation such as might arise from a change in the advective flow in the deposit (e.g., due to the failure of a primary leachate collection system with time). It might lead to a change in boundary conditions such as that arising from the expansion of an old existing landfill by the addition of extra lifts of waste; or where the source concentration remained relatively constant for a period of time (e.g., due to solubility limits) and then decreases with future time (e.g., see Rowe and San, 1992).

In what follows it will be shown that these types of events can be analysed by the techniques developed above. This is facilitated by the introduction of the elapsed time

$$t' = t - t^* \tag{7.51}$$

There can be no abrupt jump in concentration in the deposit and thus

$$c = c(t^*) \quad \text{when } t' = 0 \tag{7.52}$$

The form of the governing differential equation in any layer of the deposit does not change and still has the form of equation 7.23b; however, the physical parameters may have changed and thus n, D_{zz}, v_z, ρ, K_d and λ may also have changed values to become n', D'_{zz}, v'_z, ρ', K'_d and λ', etc. Introduction of a Laplace transform with respect to elapsed time, viz.

$$\bar{c}' = \int_0^\infty c\,\mathrm{e}^{-s't'}\,\mathrm{d}t' \tag{7.53}$$

leads to the transformed equations

$$n'D'_{zz}\frac{\partial^2 \bar{c}'}{\partial z^2} - n'v'_z\frac{\partial \bar{c}'}{\partial z} = n'(S'\bar{c}' - \phi') \tag{7.54a}$$

where

$$S' = \left(1 + \frac{\rho' K'_d}{n'}\right)s' + \lambda' \tag{7.54b}$$

$$\phi = \frac{c(t^*)}{(1 + \rho' K'_d/n')} \tag{7.54c}$$

Analysis of equations 7.54a–7.54c in a specific layer will lead to a layer matrix similar in form to equation 7.49 but with any quantity Q replaced by the equivalent quantity Q'.

Consideration of conservation of mass in the landfill leads to

$$c_T = c'_0 - \frac{1}{H'_r}\int_{t^*}^{t} f_T(\tau)\,\mathrm{d}\tau - \int_{t^*}^{t} \lambda'_T c_T(\tau)\,\mathrm{d}\tau \tag{7.55}$$

where c'_0 represents the average concentration in the landfill just after $t = t^*$, λ'_T is the value of the first-order decay constant and H'_r the current reference height of leachate.

The Laplace transform of equation 7.55 is

$$\bar{f}'_T = B'_0 - T'_0\bar{c}'_T \tag{7.56}$$

where

$$B'_0 = H'_r c'_0, \qquad T'_0 = H_r S'_T = H_r(s' + \lambda'_T)$$

An examination of conservation of mass in the base aquifer leads to the equation

$$c_b = \int_{t^*}^{t} \frac{f_b}{n'_b h'} - \lambda'_b c_b\,\mathrm{d}t \tag{7.57}$$

where again the prime indicates the current value of the particular parameter.

The Laplace transform of equation 7.57 is

$$\bar{f}'_b = Q'_{n+1}\bar{c}'_b \tag{7.58}$$

where

$$Q'_{n+1} = h'n'_b S_b = h'n'_b(s + \lambda'_b)$$

It is now clear that the assembly procedure for layer matrices will lead to a set of equations similar to equation 7.29 where all quantities have been replaced by the corresponding appropriate primed quantity and thus the calculation can proceed.

Thus one can model environmental changes by starting with the initial conditions (c_I at $t = 0$) and parameters and then obtain the concentration field, c_{I1}, at time $t = t_1^*$ (the time when the first change in conditions occurs) by solving equation 7.29 (if $c_I = 0$) or equation 7.50 (if $c_I \neq 0$). This concentration field c_{I1} can then be used to evaluate concentrations at time $t > t_1^*$ by solving equation 7.50 using $c_I = c_{I1}$, $t' = t - t_1^*$ and the new boundary conditions and values of D'_{zz}, v'_{zz}, n', etc., for $t > t_1$. If there is another change in conditions at $t = t_2^*$, then the concentration field c_{I2} can be evaluated at time $t = t_2^*$ as described above for $t > t_1^*$ and then for $t > t_2$ one solves equation 7.50 using $c_I = c_{I2}$, $t' = t - t_2^*$ are the next boundary conditions and values of D'_{zz}, v'_{zz}, λ' etc., for $t > t_2^*$. This process can be repeated many times.

In the finite layer formulation presented in Sections 7.5.1 and 7.5.2, the number of layers was controlled by physical considerations and any subdivision of physical layers into sublayers has no effect on the accuracy of the solution. However, the theory, as described in this section, uses an exponential approximation (equation 7.46) to interpolate between the concentration at the top and bottom of each sublayer, k, at time t^* and hence to evaluate concentrations at $t > t^*$ to model variable environmental conditions. Since this is an interpolation function rather than the exact distribution, it follows that the number of sublayers may influence the results for $t > t^*$.

In general, the smaller the thickness of the sublayers or the more uniform the variation in concentration between the top and bottom of each sublayer, the smaller will be the error resulting from the interpolation. Thus it is prudent to do a sensitivity study of the effect of the number of sublayers when using this theory, as implemented in a computer program (e.g., POLLUTE). This is particularly important if the source concentration is changed significantly at time t^*. Under these circumstances, the most important thing is to have one thin layer near the boundary.

7.6 Contaminant migration in a regularly fractured medium

A major consideration in many landfill designs is the potential for contaminant migration in fractured clay or rock. For any fractured media which can be idealized as being regularly fractured (e.g., as shown in Figure 7.13) it is possible to develop a very simple solution as described below.

Consider an extensive landfill which is adjacent to fractured ground. It will be assumed that there are three possible sets of planar fractures.

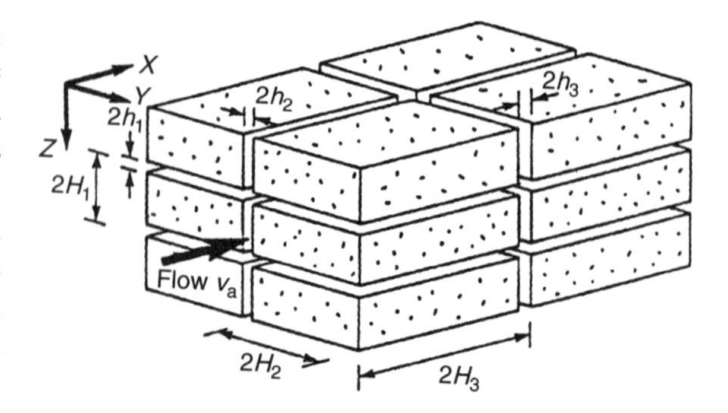

Figure 7.13 Definition of fracture geometry.

Suppose that 0_x, 0_y, 0_z are a set of Cartesian reference axes. Referring to Figure 7.13, it will be assumed that the primary set of fractures are distance $2H_1$ apart, of width $2h_1$, and parallel to the xy-plane, the secondary set of fractures is assumed to be spaced at an interval of $2H_2$, of width $2h_2$, and parallel to the xz-plane, the tertiary set of fractures is assumed to be spaced at an interval $2H_3$, of width $2h_3$, and parallel to the zy-plane. The ground adjacent to the landfill is thus assumed to consist of a series of rectangular blocks made up of a homogeneous matrix material separated by fractures having a width far smaller than the smallest dimension of the box.

It will be assumed that the interface of the landfill and the adjacent ground is the yz-plane ($x = 0$) and that the landfill is quite extensive so that the predominant mechanism for transport of the leachate will be by advective–dispersive transport along fracture sets 1 and 2 accompanied by 1D, 2D or 3D diffusion of leachate into the intact blocks.

7.6.1 Development of basic equations for a 1D, 2D or 3D system

The net flux with components F_x, F_y, F_z may be defined by

$$F_x = \frac{h_1 H_2 F_{1x} + h_2 H_1 F_{2x}}{H_1 H_2} = \frac{h_1}{H_1} F_{1x} + \frac{h_2}{H_2} F_{2x}$$

$$F_y = \frac{h_1 H_3 F_{1y} + h_3 H_1 F_{3y}}{H_1 H_3} = \frac{h_1}{H_1} F_{1y} + \frac{h_3}{H_3} F_{3y}$$

$$F_z = \frac{h_2 H_3 F_{2z} + h_3 H_2 F_{3z}}{H_2 H_3} = \frac{h_2}{H_2} F_{2z} + \frac{h_3}{H_3} F_{3z} \tag{7.59}$$

where F_{1x}, F_{1y} are the x and y components of the flux in the first set of fractures with similar definitions for F_{2x}, F_{2z} and F_{3y}, F_{3z}.

Then conservation of mass in the fracture system leads to the equation

$$-\left(\frac{\partial F_x}{\partial x} + \frac{\partial F_y}{\partial y} + \frac{\partial F_z}{\partial z}\right) = n_\mathrm{f}\left(\frac{\partial c_\mathrm{f}}{\partial t} + \lambda_\mathrm{f} c_\mathrm{f}\right) + \dot{g} + \dot{q} - \dot{r} \tag{7.60}$$

where c_f is the concentration of contaminant at a point in the fracture and

$$n_\mathrm{f} = \frac{h_1}{H_1} + \frac{h_2}{H_2} + \frac{h_3}{H_3}$$

where λ_f is the first-order decay constant of the solute ($\lambda_\mathrm{f} = \Gamma_\mathrm{R} + \Gamma_\mathrm{B} + \Gamma_\mathrm{s}$) which may result from the first-order radioactive decay Γ_R, biological degradation Γ_B or flow removed per unit volume Γ_s; $\dot{q}$ the rate at which the contaminant is being transported into the matrix per unit volume of matrix and fissures; $\dot{r}$ the rate at which the contaminant is being "injected" into the fissure system per unit volume of matrix and fissures.

Flow in the fissures is governed by

$$F_{1x} = v_{1x} c_\mathrm{f} - D_{1x} \frac{\partial c_\mathrm{f}}{\partial x}$$

$$F_{1y} = v_{1y} c_\mathrm{f} - D_{1y} \frac{\partial c_\mathrm{f}}{\partial y} \tag{7.61}$$

$$F_{2x} = v_{2x} c_\mathrm{f} - D_{2x} \frac{\partial c_\mathrm{f}}{\partial x}$$

$$F_{2z} = v_{2z} c_\mathrm{f} - D_{2z} \frac{\partial c_\mathrm{f}}{\partial z}$$

$$F_{3y} = v_{3y} c_\mathrm{f} - D_{3y} \frac{\partial c_\mathrm{f}}{\partial y}$$

$$F_{3z} = v_{3z} c_\mathrm{f} - D_{3z} \frac{\partial c_\mathrm{f}}{\partial z}$$

where the definitions of v_{1x}, v_{1y}, etc., and D_{1x}, D_{1y}, etc., parallel those of F_{1x}, F_{1y}, etc., and where v_{1x}, etc., represent the groundwater velocity and D_{1x}, etc., represent the coefficient of hydrodynamic dispersion in the referenced fracture set, and it is assumed that the fractures are open.

It follows that

$$F_x = v_\mathrm{ax} c_\mathrm{f} - D_\mathrm{ax} \frac{\partial c_\mathrm{f}}{\partial x}$$

$$F_y = v_\mathrm{ay} c_\mathrm{f} - D_\mathrm{ay} \frac{\partial c_\mathrm{f}}{\partial y} \tag{7.62}$$

$$F_z = v_\mathrm{az} c_\mathrm{f} - D_\mathrm{az} \frac{\partial c_\mathrm{f}}{\partial z}$$

where

$$v_\mathrm{ax} = v_{1x} \frac{h_1}{H_1} + v_{2x} \frac{h_2}{H_2}$$

$$D_\mathrm{ax} = D_{1x} \frac{h_1}{H_1} + D_{2x} \frac{h_2}{H_2}$$

with similar definitions for v_{ay}, D_{ay}, v_{az}, D_{az}.

For the special case where the groundwater velocity and coefficient of hydrodynamic dispersion is the same in both fracture sets (i.e., $v = v_{1x} = v_{2x}$, $D = D_{1x} = D_{2x}$), the above equations reduce to

$$v_{\mathrm{a}i} = n_\mathrm{f} v$$

$$D_{\mathrm{a}i} = n_\mathrm{f} D$$

where

$$n_\mathrm{f} = \frac{h_1}{H_1} + \frac{h_2}{H_2}$$

From equations 7.60 and 7.62 and the assumption that there are a homogeneous layer and constant groundwater velocity in the fracture system within this layer, it follows that

$$
\begin{aligned}
&D_{\mathrm{ax}}\frac{\partial^2 c_{\mathrm{f}}}{\partial x^2}+D_{\mathrm{ay}}\frac{\partial^2 c_{\mathrm{f}}}{\partial y^2}+D_{\mathrm{az}}\frac{\partial^2 c_{\mathrm{f}}}{\partial z^2}-v_{\mathrm{ax}}\frac{\partial c_{\mathrm{f}}}{\partial x}\\
&\quad-v_{\mathrm{ay}}\frac{\partial c_{\mathrm{f}}}{\partial y}-v_{\mathrm{az}}\frac{\partial c_{\mathrm{f}}}{\partial z}\\
&=(n_{\mathrm{f}}+\Delta K_{\mathrm{f}})\frac{\partial c_{\mathrm{f}}}{\partial t}+n_{\mathrm{f}}\lambda_{\mathrm{f}}c_{\mathrm{f}}+\dot{q}-\dot{r}
\end{aligned}
\qquad (7.63)
$$

where it has been assumed that the rate at which contaminant is being sorbed onto the fracture walls per unit volume is linear, viz.

$$\dot{g}=\Delta K_{\mathrm{f}}\frac{\partial c_{\mathrm{f}}}{\partial t},\quad \Delta=\left(\frac{1}{H_1}+\frac{1}{H_2}+\frac{1}{H_3}\right)$$

where K_{f} is the fracture distribution coefficient, defined by Freeze and Cherry (1979) as the mass of solute adsorbed per unit area of surface divided by the concentrations of solute in solution and Δ represents the surface area per unit volume. Equation 7.63 can be solved by means of a Laplace transform and it is found that

$$
\begin{aligned}
&D_{\mathrm{ax}}\frac{\partial^2 \bar{c}_{\mathrm{f}}}{\partial x^2}+D_{\mathrm{ay}}\frac{\partial^2 \bar{c}_{\mathrm{f}}}{\partial y^2}+D_{\mathrm{az}}\frac{\partial^2 \bar{c}_{\mathrm{f}}}{\partial z^2}-v_{\mathrm{ax}}\frac{\partial \bar{c}_{\mathrm{f}}}{\partial x}\\
&\quad-v_{\mathrm{ay}}\frac{\partial \bar{c}_{\mathrm{f}}}{\partial y}-v_{\mathrm{az}}\frac{\partial \bar{c}_{\mathrm{f}}}{\partial z}\\
&=[(n_{\mathrm{f}}+\Delta K_{\mathrm{f}}+\overline{\eta})s\\
&\quad+(n_{\mathrm{f}}\lambda_{\mathrm{f}}+\Lambda_{\mathrm{m}}\overline{\eta})]\bar{c}_{\mathrm{f}}-\bar{\dot{r}}
\end{aligned}
\qquad (7.64)
$$

where it is shown in Appendix C that the Laplace transform of $\dot{q}$ has the form

$$\bar{\dot{q}}=(s+\Lambda_{\mathrm{m}})\overline{\eta}\,\bar{c}_{\mathrm{f}} \qquad (7.65)$$

7.6.2 An infinite fractured medium

Consider now the case of 1D transport from a landfill, located at $x=0$, into an extensively fractured zone, $x>0$, which is initially uncontaminated. In this case, there is no injection of contaminant so that $r=0$. The process is governed by the equation

$$D_{\mathrm{ax}}\frac{\partial^2 \bar{c}_{\mathrm{f}}}{\partial x^2}-v_{\mathrm{ax}}\frac{\partial \bar{c}_{\mathrm{f}}}{\partial x}=S\,\bar{c}_{\mathrm{f}}$$

where $S=(n_{\mathrm{f}}+\Delta K_{\mathrm{f}})s+n_{\mathrm{f}}\lambda_{\mathrm{f}}+(s+\Lambda_{\mathrm{m}})\overline{\eta}$.

The above equation has the general solution

$$\bar{c}_{\mathrm{f}}=A\,\mathrm{e}^{\alpha x}+B\,\mathrm{e}^{\beta x}$$

where

$$
\begin{aligned}
\alpha&=\frac{v_{\mathrm{ax}}}{2D_{\mathrm{ax}}}+\sqrt{\frac{v_{\mathrm{ax}}^2}{4D_{\mathrm{ax}}}+\frac{S}{D_{\mathrm{ax}}}}\\
\beta&=\frac{v_{\mathrm{ax}}}{2D_{\mathrm{ax}}}-\sqrt{\frac{v_{\mathrm{ax}}^2}{4D_{\mathrm{ax}}}+\frac{S}{D_{\mathrm{ax}}}}
\end{aligned}
\qquad (7.66)
$$

Recalling that the solution must remain bounded as $x\to\infty$, and the concentration in the landfill is c_{LF}, it follows that the distribution of concentration is given by

$$\bar{c}=\bar{c}_{\mathrm{LF}}\,\mathrm{e}^{\beta x} \qquad (7.67a)$$

while the flux distribution is given by

$$\overline{F}_x=\bar{c}_{\mathrm{LF}}\,D_{\mathrm{ax}}\,\alpha\,\mathrm{e}^{\beta x} \qquad (7.67b)$$

The behaviour within the landfill has been examined previously (equations 1.23 and 7.20) where it was shown that

$$\bar{c}_{\mathrm{T}}=\frac{c_0}{S_{\mathrm{T}}}-\frac{\overline{f}_{\mathrm{T}}}{S_{\mathrm{T}}H_{\mathrm{r}}}$$

so that

$$\bar{c}_{\mathrm{LF}}=\frac{c_0}{(S_{\mathrm{T}}+D_{\mathrm{ax}}/H_{\mathrm{r}})} \qquad (7.68)$$

Equations 7.67a, 7.67b and 7.68 define the concentration plume (in the fractures) in transformed space. The concentration at any given time and location can then be readily obtained by numerically inverting the Laplace transform using the method proposed by Talbot (1979). This has been implemented in program POLLUTE.

7.6.3 Finite layer analysis of fissured material

The solution of problems involving contaminant transport through fissured material has been illustrated by a simple example. It is not difficult to extend the finite layer methods developed in Sections 7.5.1 and 7.5.2 to incorporate the behaviour of fissured material. For 1D transport, the finite layer equations reduce to equation 7.29, with the factor nD_{zz} appearing in equation 7.26b replaced by D_{az} and the values of α, β given by equation 7.24e replaced by those given by equation 7.66.

For 2D transport, the finite layer equations reduce to equation 7.38, 7.43 with the factor nD_{zz} appearing in equation 7.26b again replaced by D_{az} and the values of α, β given by equation 7.24a replaced by

$$\alpha = \frac{v_{ax}}{2D_{ax}} + \sqrt{\frac{v_{ax}^2}{4D_{ax}^2} + \frac{S}{D_{ax}}}$$
$$\beta = \frac{v_{ax}}{2D_{ax}} - \sqrt{\frac{v_{ax}^2}{4D_{ax}^2} + \frac{S}{D_{ax}}} \qquad (7.69)$$

with $S = (n_f + \Delta K_f)s + n_f\lambda_f + (s + \Lambda_m)\bar{\eta} + i\xi v_{ax} + \xi^2 D_{ax}$. The full details are given by Rowe and Booker (1990a).

A similar approach can also be used to consider fracture media for 2D contaminant transport (see Rowe and Booker, 1991a). Using the finite layer approach, one can consider both fractured and unfractured layers (e.g., see Rowe and Booker, 1991b).

7.7 Finite layer analysis of geomembranes and gas-filled layers where there is phase change

When modelling contaminant transport in a layered system such as that shown in Figure 1.12 there is potential for phase change between layers. For example, considering the migration of a volatile organic compound through the system shown in Figure 1.12 there would be an initial phase change from the aqueous state in the leachate to the dissolved state in the geomembrane and a change back to the aqueous phase in the underlying clay liner. This would be followed by a change from the aqueous phase in the clay liner to the gaseous phase in an unsaturated secondary leachate collection system, a change back to aqueous phase in the leachate above the secondary geomembrane, a change to the dissolved state in the geomembrane, and finally a last change back to the aqueous phase in the secondary clay liner. These phase changes can be described in terms of a dimensionless Henry's coefficient. Thus taking the concentration in the aqueous state to be c, the concentration in the dissolved state in a geomembrane, c_g, is given by

$$c_g = S_{gf}\, c \qquad (7.70)$$

where S_{gf} is variously called a solubility, partitioning or Henry's coefficient [–] as discussed in Section 6.8.1. Likewise the concentration in a gaseous phase, c_a, in an unsaturated layer is given by

$$c_a = K'_H\, c \qquad (7.71)$$

where K'_H is a dimensionless form of Henry's coefficient [–] as discussed in Section 8.11.2. The mathematics behind dealing with both cases is essentially identical and will be discussed below for the 1D case but can be readily generalized to 2D conditions.

The mass flux in the layer where there has been a phase change is given by

$$f = nv_z c_a - nD_{zz}\frac{\partial c_a}{\partial z} = K'_H\, nv_z c - K'_H\, nD_{zz}\frac{\partial c}{\partial z} \tag{7.72}$$

where n is the "porosity" through which transport is occurring ($n = \theta_a$, the effective air porosity for gas diffusion, while for diffusion in a solid geomembrane porosity has no meaning and drops out of the equations by taking $n = 1$), nv_z the Darcy flux through the layer, D_{zz} the diffusion coefficient for the layer (i.e., $D_{zz} = D_a$ the diffusion coefficient in air for gas diffusion, or $D_{zz} = D_g$ for diffusion through a geomembrane), K'_H a dimensionless form of Henry's coefficient (S_{gf} for a geomembrane), and all other terms are as previously defined. Taking the Laplace transform of equation 7.72 then gives

$$\bar{f} = K'_H\left[nv_z\bar{c} - nD_{zz}\frac{\partial \bar{c}}{\partial z}\right] \tag{7.73}$$

which is identical to equation 7.24a except for the multiplier K_H.

Contaminant transport in the layer is described by

$$K'_H\, nD_{zz}\frac{\partial^2 c}{\partial z^2} - K'_H\, nv_z\frac{\partial c}{\partial z} = K'_H(n+\rho K_d)\frac{\partial c}{\partial t} - K'_H n\lambda c \tag{7.74}$$

but since K_H cancels out this reduces to equation 7.23b and, upon taking the Laplace transform, becomes equation 7.24b which has a solution given by equation 7.24c with α and β defined by equation 7.24 e.

It follows from equations 7.73 and 7.24 d that the transformed flux is given by

$$\bar{f} = K'_H\, nD(\beta A\, e^{\alpha z} + \alpha B\, e^{\beta z}) \tag{7.75}$$

and thus, following the development given in Section 7.5,

$$\frac{\bar{f}_z}{K'_H\, nD_{zz}} = \bar{c}_j\left\{\frac{\beta\, e^{\alpha(z-z_k)} - \alpha\, e^{\beta(z-z_k)}}{e^{\alpha(z_j-z_k)} - e^{\beta(z_j-z_k)}}\right\} + \bar{c}_k\left\{\frac{\beta\, e^{\alpha(z-z_j)} - \alpha\, e^{\beta(z-z_j)}}{e^{\alpha(z_k-z_j)} - e^{\beta(z_k-z_j)}}\right\} \tag{7.76}$$

It is then possible to evaluate the flux on each of the node planes bounding the layer with a phase transformation and it is found that

$$\begin{bmatrix} \bar{f}_{zj} \\ -\bar{f}_{zk} \end{bmatrix} = \begin{bmatrix} K'_H\, Q_k & K'_H\, R_k \\ K'_H\, S_k & K'_H\, T_k \end{bmatrix}\begin{bmatrix} \bar{c}_j \\ \bar{c}_k \end{bmatrix} \tag{7.77}$$

where Q_k, R_k, S_k and T_k are as defined in equation 7.26.

Since the equations have been formulated in terms of the aqueous concentration at the layer boundaries, the layer matrices for the layers with a phase transformation (equation 7.77) can be assembled together with those for layers without a transformation (equation 7.26) to give a full set of equations analogous to equation 7.27. Boundary conditions can be applied as described in Section 7.5 except that if the top layer is a geomembrane then the top boundary condition follows from

$$\bar{f}_T = K'_H\, Q_k\,\bar{c}_T + K'_H\, R_k\,\bar{c}_1 = B_0 - T_0\bar{c}_T \tag{7.78a}$$

which, upon re-arrangement becomes

$$(T_0 + K'_H\, Q_k)\bar{c}_T + K'_H\, R_k\,\bar{c}_1 = B_0 \tag{7.78b}$$

where B_0 and T_0 are defined by equation 7.28 and $K_H = S_{gf}$ for the geomembrane. This theory has been implemented in program POLLUTE (Rowe and Booker, 1997, 2004). Relevant parameters are discussed in Sections 8.11.2 and 8.12.2, and its application will be illustrated in Chapter 16.

Chapter eight

Evaluation of diffusion and distribution coefficients

8.1 Introduction

The key transport mechanisms and the development of the governing equations have been discussed in Chapter 1 and the basic concepts of diffusion were examined in Chapter 6. However, in order to use theoretical approaches to predict the potential impact of a waste disposal site, it is necessary to estimate appropriate design parameters. These parameters should be obtained using samples of the proposed barrier material and using a leachate as similar as possible to that expected in the facility.

The key soil-related parameters identified in Chapter 1 were the advective velocity, diffusion coefficient, distribution/partitioning coefficient, effective porosity and dispersivity. The advective velocity will depend on the hydraulic gradient and hydraulic conductivity as already discussed. In this chapter, consideration will be focussed on diffusion and distribution/partitioning coefficients.

8.2 Obtaining diffusion and partitioning/distribution coefficients: basic concepts

As discussed in Chapter 1, the coefficient of hydrodynamic dispersion has two components: mechanical dispersion and molecular diffusion. In this chapter, attention will be focussed on the movement of contaminants through clayey soils at relatively low velocities and a technique for estimating both the diffusion coefficient and the distribution (partitioning) coefficient using a single test will be described. Attention will initially be directed at establishing parameters for clayey soils but it will then be demonstrated that a similar approach can be adopted for polymer gels (Section 8.3.5) and geomembranes (Section 8.12).

In the proposed test, an undisturbed sample of soil is placed in a column and the leachate of interest is placed above the soil. Contaminant is then permitted to migrate through the specimen under the prescribed head (which may be zero). The volume of leachate above the soil will normally be selected to be sufficiently small to allow a significant drop in concentration of contaminant within the source solution. Typically the height of leachate, H_f, in the column above the clay will range from 0.05 to 0.3 m. This drop in concentration with time should be monitored.

A number of possible boundary conditions at the base of the sample may be considered. If the test is to be conducted with advective transport through the specimen, then a porous collection plate can be placed beneath the sample and the effluent collected and monitored (Figure 8.1). The volume in the reservoir is maintained constant by the addition of background reference

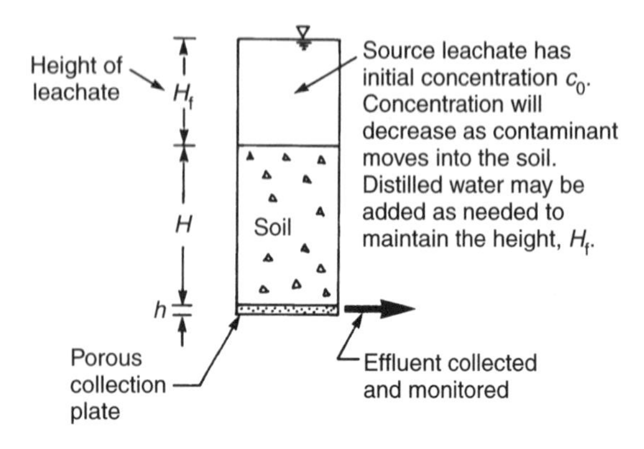

Figure 8.1 Schematic illustration of an advection–diffusion column test with a finite mass of contaminant source boundary condition (after Rowe *et al.*, 1988; reproduced with permission, *Canadian Geotechnical Journal*).

fluid (not leachate). If there is no advective flow, then two other base boundary conditions may be considered. First, the base could be an impermeable plate (see Figure 8.2a). Second, a closed collection chamber (reservoir), similar to that for the leachate but initially having only a background concentration of the contaminant of interest, may be used (see Figure 8.2b). Thus, as contaminant passes through the soil, it accumulates in this collection chamber and the concentration in this reservoir can be monitored.

Suppose that the volume of source solution (leachate) is equal to AH_f where A is the plan area of the column and H_f the height of the leachate in the column (e.g., see Figure 8.3). At any time t, the mass of any contaminant species of interest in the source solution is equal to the concentration $c_T(t)$ in the solution multiplied by the volume of solution (assuming here that the solution is stirred so that $c_T(t)$ is uniform throughout the solution). The principle of conservation of mass requires that at this time t, the mass of contaminant in the source solution is equal to the initial mass of the contaminant minus the mass which has been transported into the soil up to this time t. This can be written algebraically (as shown in Chapter 1) as

$$c_T(t) = c_0 - \frac{1}{H_f}\int_0^t f_T(\tau)\,d\tau \tag{8.1}$$

where $c_T(t)$ is the concentration in the source solution at time t [ML^{-3}], c_0 the initial concentration in the source solution ($t = 0$) [ML^{-3}], A the plan area of the column [L^2], H_f the height of leachate (i.e., the volume of leachate per unit area) [L] and $f_T(\tau)$ the mass flux of this contaminant into the soil at time τ [$ML^{-2}T^{-1}$].

Contaminant is allowed to migrate from the source chamber through the soil and, if present, into the collection chamber. If no additional contaminant is added to the source chamber, then the concentration of contaminant will decrease with time as mass of contaminant diffuses into the soil (see Figure 8.4). The rate of decrease can be controlled by the choice of the height of

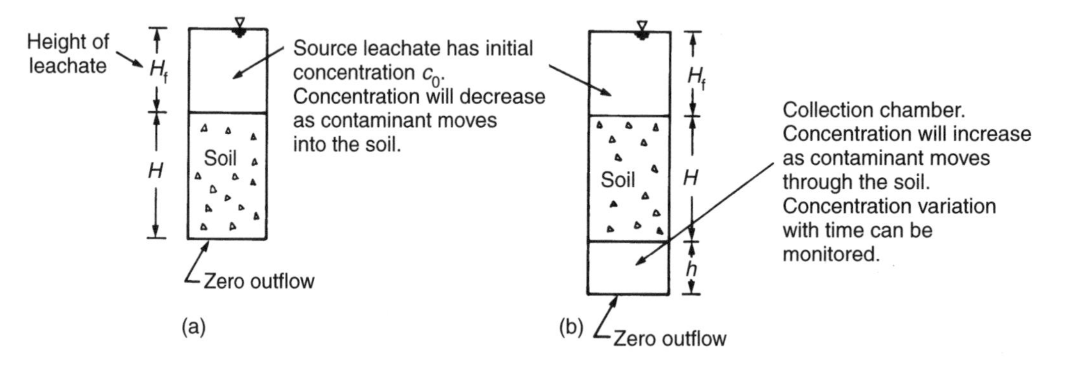

Figure 8.2 Schematic illustration of pure diffusion test with a finite mass of contaminant source boundary condition (after Rowe *et al.*, 1988; reproduced with permission, *Canadian Geotechnical Journal*).

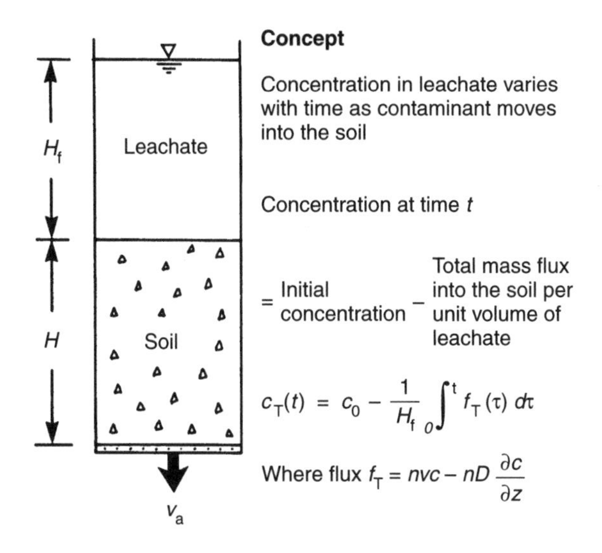

Figure 8.3 Schematic illustration showing how the concentration of contaminant in the source varies as contaminant is transported into the soil.

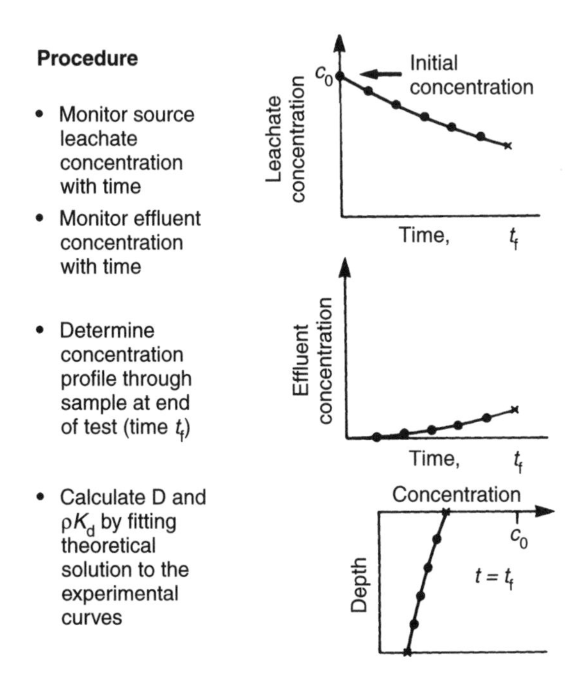

Figure 8.4 Experimental procedure used to determine the diffusion coefficient D and distribution coefficient K_d (modified from Rowe, 1988).

leachate, H_f. Conversely, as contaminant moves into the collection chamber (Figures 8.1 and 8.2b), the increase in mass gives rise to an increase in contaminant concentration in this reservoir (see Figure 8.4). The rate of decrease in concentration in the source and increase in the collection chamber should be monitored with time. At some time t_f, the test is terminated and the concentration profile through the soil sample may be obtained (see Figure 8.4). Assuming linear sorption, theoretical models can then be used to estimate the effectivity porosity, n, effective diffusion coefficient, D_e, and the product of dry density and distribution coefficient ρK_d. This theoretical analysis has been described in detail by Rowe and Booker (1985a, 1987) as outlined in Chapter 7 and has been implemented in the computer program POLLUTE (Rowe and Booker, 1994, 2004). This approach permits accurate calculation of concentration in only a few seconds on a personal computer and hence it is well suited for use in interpretation of the results of the column tests.

The sensitivity of this approach for the estimation of D and ρK_d can be illustrated by considering the diffusive migration of a contaminant through a 10-cm thick sample given a soil porosity of 0.4, a Darcy velocity ($v_a = nv$) of 0.0 and a height of leachate (volume of leachate divided by the plan area of specimen) $H_f = 0.05$ m, as indicated in the inset to Figure 8.5. Taking a typical diffusion coefficient D (e.g., for Na^+) of $0.015\,m^2/a$, Figure 8.5a shows the theoretical concentration profile at 0.1 years for values of $\rho K_d = 0$, 0.2 and 0.4. It can be seen that even this relatively small difference in sorption results in a measurable difference in the concentration profiles. Figure 8.5b shows the theoretical concentration profile at 0.1 years, for three combinations of parameters (D, ρK_d), viz. ($0.01\,m^2/a$, 0), ($0.015\,m^2/a$, 0.2) and ($0.02\,m^2/a$, 0.4). These three combinations of the parameters correspond to the same value of $D_R = 0.01\,m^2/a$ (see Section 6.5 for a discussion regarding D_R). If leachate concentration was held constant

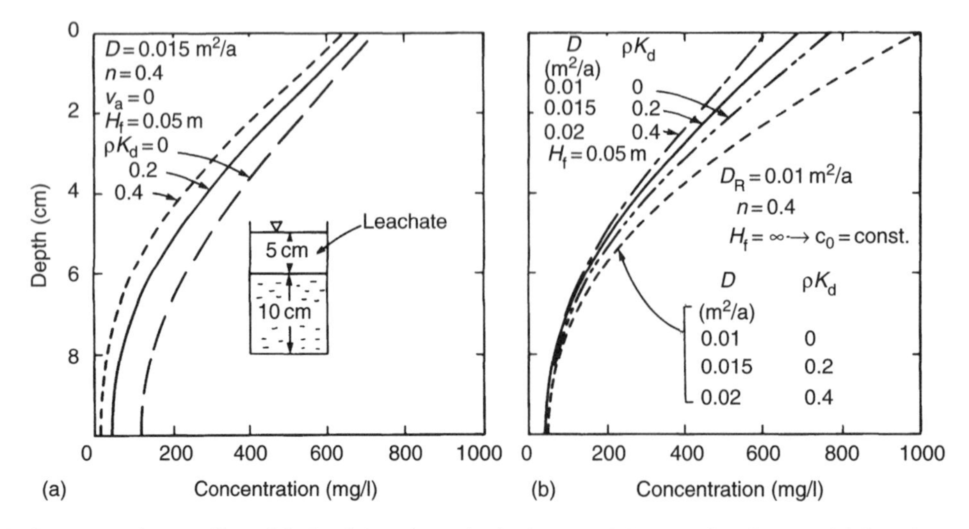

Figure 8.5 Concentration profiles with depth in a hypothetical test at 0.1 years showing sensitivity of concentration profiles (a) to small change in ρK_d and (b) to different combinations of D and ρK_d giving the same value of D_R.

($c_T(t) = c_0$) in these tests, the three sets of parameters would give identical concentration profiles at any time. However, as is evident from Figure 8.5b, in the test where the source concentration is allowed to drop with time ($H_f = 0.05$ m here), these three sets of parameters give rise to different concentration profiles illustrating the different effects of D and ρK_d. It is for this reason that both parameters can be evaluated. The effect is even more pronounced for smaller values of leachate height H_f.

8.3 Example tests for obtaining diffusion and distribution coefficients for inorganic contaminants

8.3.1 Advective–diffusive tests

A series of laboratory column tests were performed on samples of unweathered grey clay taken from beneath the Confederation Road landfill (near Sarnia, Ontario). The basic geotechnical properties and mineralogy of the soil are summarized in Table 8.1.

A schematic diagram of the apparatus used to obtain diffusion and distribution parameters in the presence of advection is shown in Figure 8.6. Details are given by Rowe *et al.* (1988). The objective of this test was to simulate field conditions as realistically as possible. This involved having an applied stress and downward advective flow. As will be demonstrated, this elaborate test turned out not to be necessary and similar results were obtained using a simple pure diffusion test.

A hanger weight system was set up and a pressure of 87 kPa was initially applied to the

Table 8.1 Soil properties for the Confederation Road landfill (adapted from Crooks and Quigley, 1984)

Property	*Below landfill waste*
Liquid limit (%)	~39
Plastic limit (%)	~12
Specific gravity	2.73
Moisture content	~23
Mineralogy (<74 μm) (%)	
Carbonates	~34
Quartz and feldspars	12–20
Illite	23–27
Chlorite	22–26
Smectite	~1
Cation exchange capacity (<2 μm) (meq/100 g)	10.5

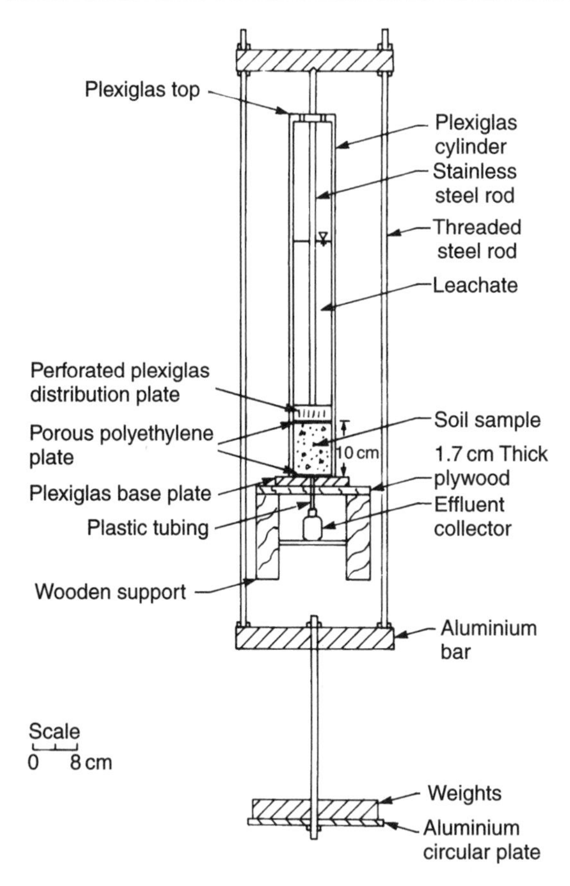

Figure 8.6 Schematic diagram of Plexiglas model (advective–diffusive test).

soil sample for 2 days. This pressure was considered to be large enough to provide good seating of the sample in the Plexiglass column while being well below the preconsolidation pressure of 172 kPa (Ogunbadejo, 1973). After the first 2 days, the soil sample was allowed to reconsolidate for two more days at an applied pressure of 30 kPa. To prevent drying of the clay surface, a small quantity of distilled water was maintained above and below the sample (zero hydraulic gradient) during the consolidation period. After consolidation, the distilled water was replaced by salt solution above the soil and drainage into a small polyethylene collection bottle was permitted at the bottom of the column. The source salt solution was mixed periodically to maintain a relatively uniform concentration throughout the reservoir depth. The models were maintained at a laboratory temperature of 22 ± 1°C. A similar procedure can be used with leachate as a source fluid.

The total fluid flow through the soil and into the collection bottle was monitored. To prevent a drop in height of solution in the reservoir due to seepage into the soil, a volume of distilled water equal to the volume of effluent collected was added after each monitoring period. Thus, the height of leachate in the reservoir remained relatively constant. The dilution resulting from the addition of distilled water is automatically considered by equation 8.1.

Six tests (referred to as models A–F) were conducted as described above. In each test, a specified salt solution (calcium chloride, sodium chloride, or potassium chloride) was placed into contact with the clay under a controlled total head for a predetermined period as indicated in Table 8.2. Models A–E involved a single cation source solution. Model F involved a source solution of both potassium and calcium chloride.

The effluent discharge volume was found to be linear with time over the entire test period for each of the six tests. Based on these discharge rates, the Darcy velocity was deduced as shown in Table 8.2. The calculated hydraulic conductivities of between 2×10^{-10} and 4×10^{-10} m/s are only marginally higher than a field value of 1.5×10^{-10} m/s obtained from falling head tests on piezometers installed in the clay below the landfill at Sarnia (Goodall and Quigley, 1977).

Figures 8.7 and 8.8 show the observed and typical best-fit theoretical matching curves for source concentration (c_T) of sodium chloride (NaCl) and potassium chloride (KCl) solutions, respectively.

The results presented in Figures 8.7 and 8.8 were obtained for the same volume of source leachate ($H_f = 0.3$ m) and it can be seen that the decrease in chloride concentration with time is fairly similar with the minor differences reflecting a small difference in Darcy velocity.

Table 8.2 Characteristics of various diffusion tests

		Average background concentration (mg/l)							*Source solution concentration of key species* (mg/l)						
Series	*Model*	Cl^-	Na^+	K^+	Ca^{2+}	Mg^{2+}	*Temperature* (°)	*Type*	Cl^-	Na^+	K^+	Ca^{2+}	Mg^{2+}	*Darcy velocity* (m/a)	*Duration of test* (days)
(i)	A	175	180	~20	345	255	22	Single salt	1,501	975	–	–	–	0.034	86
(i)	B	"	"	"	"	"	22	"	1,725	–	–	975	–	0.033	84
(i)	C	"	"	"	"	"	22	"	1,725	–	–	975	–	0.030	141
(i)	D	"	"	"	"	"	22	"	885	–	975	–	–	0.035	105
(i)	E	"	"	"	"	"	22	"	885	–	975	–	–	0.025	97
(i)	F	"	"	"	"	"	22	Two salt	2,609	–	975	975	–	0.035	108
(ii)	G	55	150	10	85	120	10	Single salt	1,480	955	–	–	–	0	15
(ii)	H	"	"	"	"	"	10	"	364	–	400	–	–	0	15
(ii)	I	"	"	"	"	"	10	"	450	–	–	250	–	0	15
(ii)	J	"	"	"	"	"	10	"	856	–	–	–	293	0	15
(iii)	K	"	"	"	"	"	10	Leachate	1,000	955	400	250	291	0	15

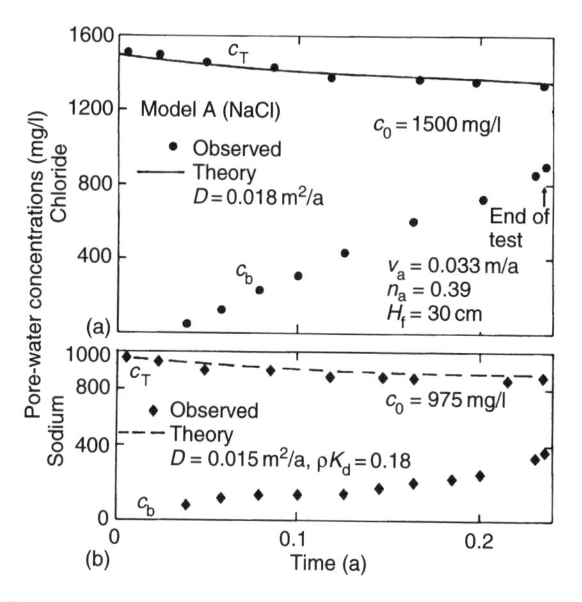

Figure 8.7 Source (c_T) and base (c_b) concentration changes over time in model A for (a) chloride and (b) sodium (modified from Rowe *et al.*, 1988).

However, comparing the rate of change in cation concentrations with time for three cations (Na^+, K^+), it is apparent that there is a significant difference in diffusion coefficient (D) and distribution coefficient (K_d) for the three cations. The values of the parameters D and ρK_d deduced by fitting the theoretical curve to the observed change in concentration are summarized in Table 8.3.

The variation in source concentration (c_T) with time provides an initial means of estimating the parameters D and ρK_d; however, the variation in the concentration throughout the sample at the termination of the test provides the primary data for estimating, or checking, these parameters. Figures 8.9–8.11 show the observed anion (Cl^-) and cation concentrations with depth for models A, D and E, respectively. Also shown are the theoretical curves obtained using the values of D and ρK_d deduced from the

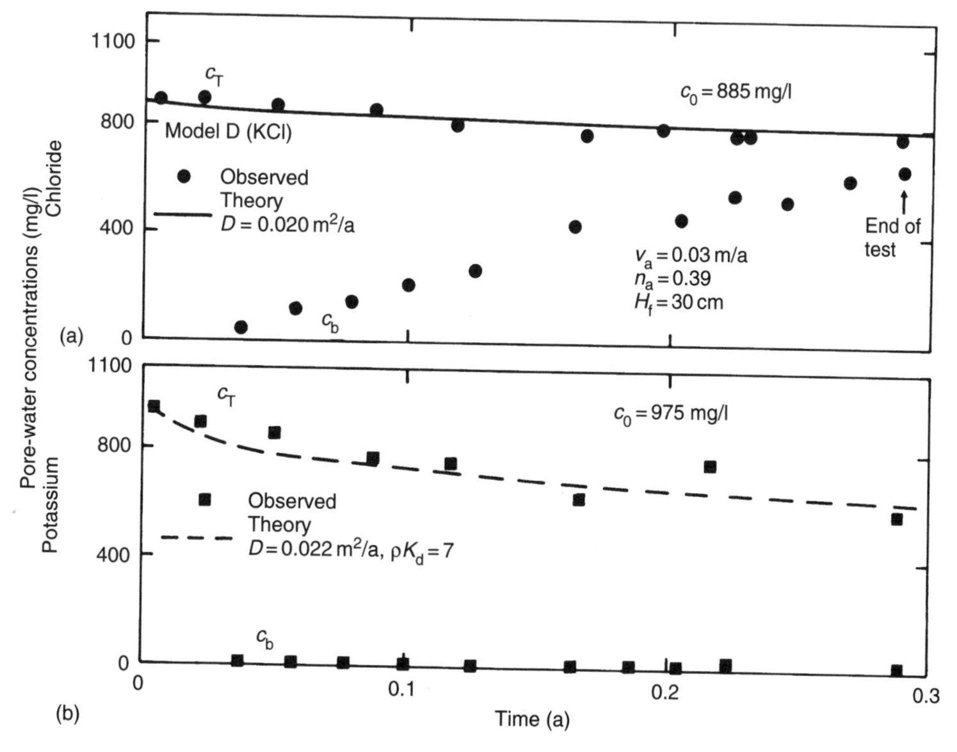

Figure 8.8 Source and base concentration changes over time in model D for (a) chloride and (b) potassium (modified from Rowe *et al.*, 1988).

Table 8.3 Comparisons of diffusion coefficient D (m^2/a) and sorption parameter ρK_d (–)

Series	*Model*	*Cl⁻*	*Na⁺*	*K⁺*	*Ca²⁺*	*Mg²⁺*
(i)	A	0.018, 0	0.015, 0.18	–	–	–
(i)	B	0.018, 0	–	–	0.012, 2	–
(i)	C	0.019, 0	–	–	0.012, 2	–
(i)	D	0.020, 0	–	0.022, 7	–	–
(i)	E	0.020, 0	–	0.020, 7	–	–
(i)	F	0.019, 0	–	0.022, 7	[a]	–
(ii)	G	0.018, 0	0.018, 0.75	–	–	–
(ii)	H	0.019, 0	–	0.024, 4.5	–	–
(ii)	I	0.020, 0	–	–	0.013, 4	–
(ii)	J	0.019, 0	–	–	–	0.012, 5
(iii)	K	0.024, 0	0.014, 0.25	0.019, 1.7	[a]	[a]

[a]No good fit could be obtained.

variation in source concentration with time and given in Table 8.3.

Allowing for some small experimental scatter of data points, inspection of Figures 8.7–8.11 indicates that in each case the theoretical curves provide a very good fit to both the decrease in the source fluids concentration with time and the

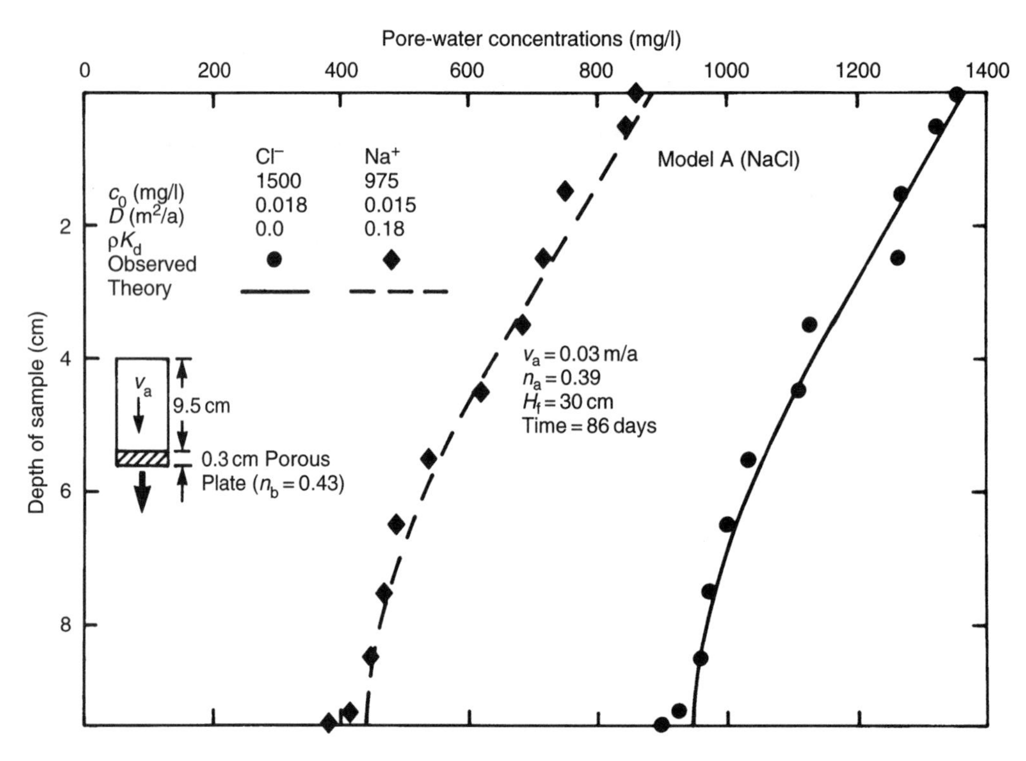

Figure 8.9 Chloride and sodium concentration versus depth in sample for model A (modified from Rowe *et al.*, 1988).

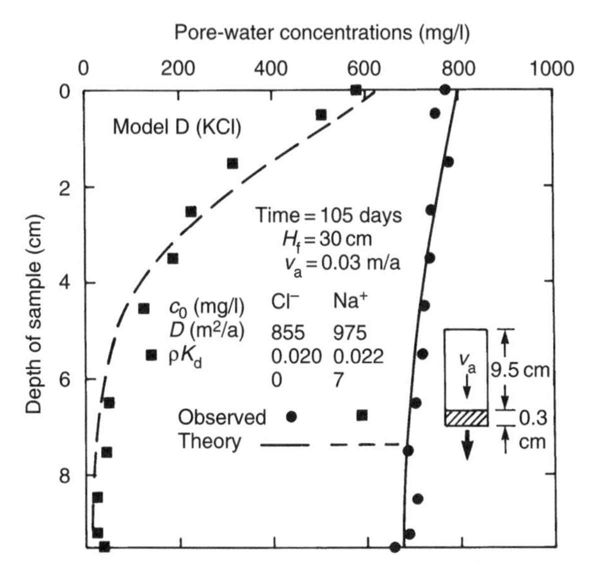

Figure 8.10 Chloride and potassium concentration versus depth in sample for model D (modified from Rowe *et al.*, 1988).

variation in concentration with depth in the soil at the end of each test. The consistency of results demonstrates the power of the analytical model (program POLLUTE) and provides some con-

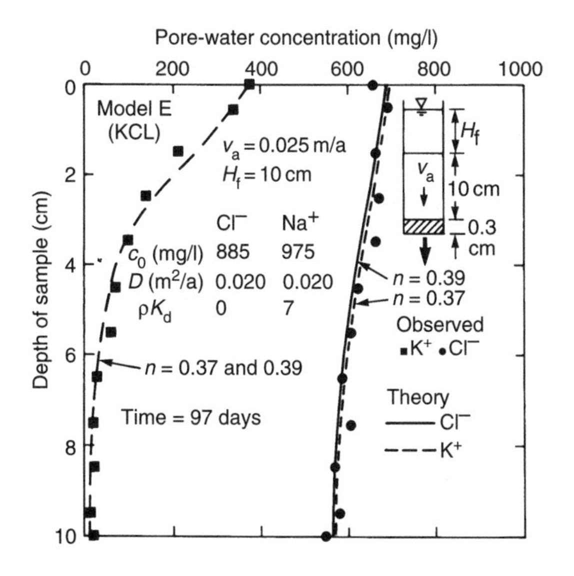

Figure 8.11 Chloride and potassium concentration versus depth in sample for model E (modified from Rowe *et al.*, 1988).

fidence in the parameters D and ρK_d for the clay and source fluids examined.

To provide an indication of parameter variation that might be expected for a given soil, a number of tests were duplicated. The diffusion coefficient, D, for chloride was deduced for each model and ranged between 0.018 and $0.02\,m^2/a$ with an average value of $0.019\,m^2/a$. This small variation in D does not appear to be related to small differences in Darcy velocity, nor does it appear to be particularly related to the nature of the associated cation (see Table 8.3). Rather, the variability from 0.018 to $0.02\,m^2/a$ is seen as an indication of the level of repeatability that may be achieved for this type of test.

The application of an effective stress to the soil sample adopted in these tests is not an essential part of the proposed technique for determining the parameters D and K_d. Tests performed for the particular combination of clay and permeants considered herein gave similar results both with and without the application of the effective stress. However, for some combinations of clay and permeant, shrinkage of the clay may occur in the absence of a confining stress and this can give quite misleading results (e.g., see Quigley and Fernandez, 1989). For these clays, and for GCLs (see Chapter 12), tests should be performed at an effective stress similar to that anticipated in the field.

8.3.2 Pure diffusion tests

In many cases, it is not necessary to perform an advection–diffusion test. Under these circumstances, a simple diffusion test can be performed for boundary conditions shown in Figure 8.2. In this test, the soil sample is placed in a Plexiglass cylinder by trimming the sample to a size marginally greater than the specimen and then pressing the specimen into the cylinder, using a cutting shoe attached to the cylinder, to perform the final trim. This procedure is found to work well for many clays. However, it does not work well for clays with a significant stone content because the

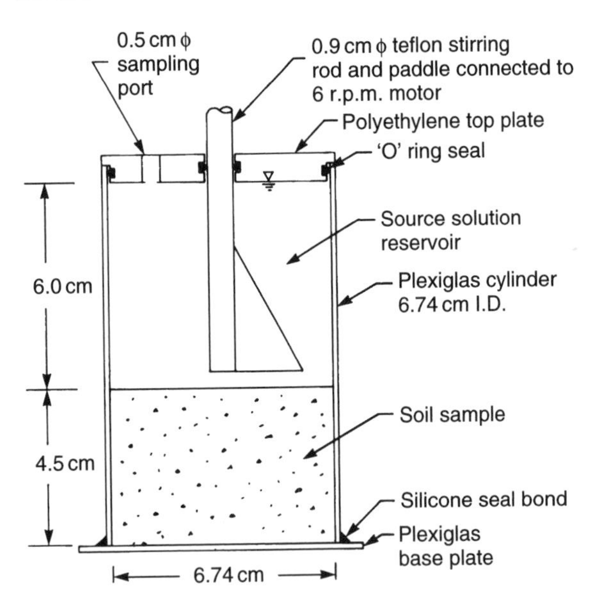

Figure 8.12 Schematic diagram of the diffusion model (a simple diffusion test).

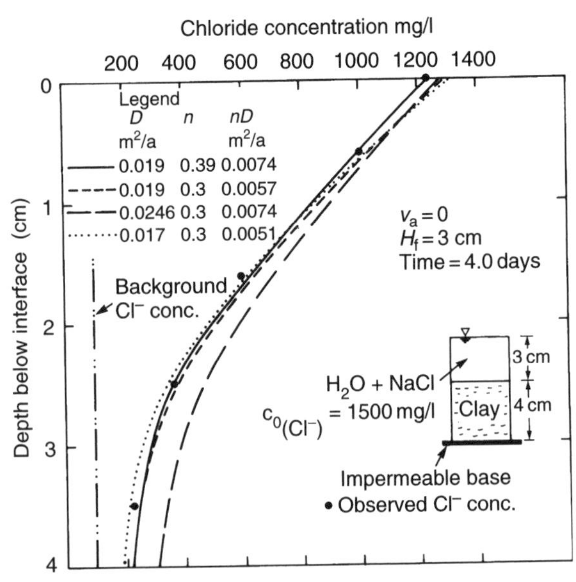

Figure 8.13 Effect of choice of effective porosity and diffusion coefficient on the comparison between observed and calculated concentration profile (modified from Rowe *et al.*, 1988).

stones tend to catch on the perspex and then tear the sample. In these cases, the modified procedure described in Section 8.3.3 may be adopted.

Once placed into the cylinder, a perspex base plate is placed below the sample (see Figure 8.12) and the source fluid (e.g., leachate) is placed above the specimen. The container is then sealed and left for diffusion to occur. The source leachate strength is monitored periodically. After some time t, the test is disassembled and the diffusion profile through the sample is established as shown in Figure 8.13. By adjusting parameters in a theoretical model, a fit to the data can be obtained, thereby yielding an estimate of the diffusion parameters.

8.3.3 Diffusion through soft clayey soils or soils with significant gravel size particles

For some very soft soils or some tills which have a significant gravel size, it is not practical to push the sample inside a Plexiglass cylinder because of the damage (e.g., compression or tearing) that may be done to the specimen. In these cases, an alternative procedure involving the use of a latex membrane around the specimen can be used (e.g., see Figure 8.14) to contain the specimen and the source fluid. Care is required to avoid leakage down the sides of the membrane. This procedure, while not as desirable as that described in Section 8.3.2, has been found to work well and, under similar circumstances, gives similar results to those obtained using the set-up shown in Figure 8.12.

In performing tests on samples that have significant gravel sizes, it is important to select samples which do not contain excessive gravel. Samples can be checked prior to testing using X-ray radiography.

8.3.4 Diffusion through sedimentary rock (shale, sandstone)

Techniques have been developed for performing diffusion tests through sedimentary rock as described by Barone *et al.* (1990, 1992a). Two

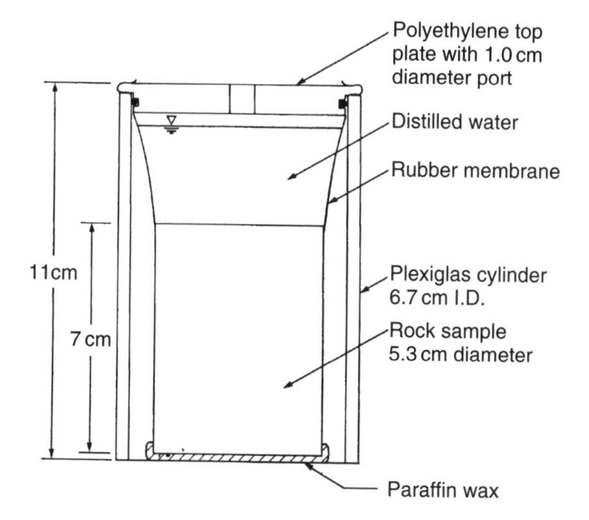

Figure 8.14 Schematic diagram of diffusion tests with latex membrane (for rock, stiff soils or soils with significant gravel size particles).

types of tests have been performed. If the sedimentary rock already has a relatively high concentration of the chemical species of interest (e.g., sodium, chloride, etc.), then a test can be performed by encasing the rock in a latex membrane and allowing the species of interest to diffuse out of the rock into a reservoir of fluid (e.g., distilled or reference water). Figure 8.15b shows a schematic illustration of one such test set-up and Figure 8.16 shows the diffusion profile obtained by Barone *et al.* (1990) after 45 days diffusion of chloride out of a sample of Queenston Shale. Based on the results of a number of these tests, the diffusion coefficient of chloride in this shale, which had a porosity of 10.6–11.3%, was between 1.4×10^{-10} m^2/s (0.0044 m^2/a) and 1.6×10^{-10} m^2/s (0.005 m^2/a) at 22 °C. Similar tests were also reported by Barone *et al.* (1992a) for Bison Mudstone, which has a porosity of 21.5–25.7%, and yielded a diffusion coefficient for chloride of between 1.5×10^{-10} m^2/s (0.0047 m^2/a) and 2.0×10^{-10} m^2/s (0.0063 m^2/a) at 10 °C.

If the sedimentary rock does not have a significant concentration of the chemical species of interest, then a diffusion test can be performed by placing a source "leachate" containing the species of interest in contact with the rock using an arrangement similar to that shown in Figure 8.15a. The flux into and out of the sample can be monitored and the concentration profile at the end of the test may be determined in a similar manner as that adopted for soil.

8.3.5 Diffusion in polymer gels

Polymer gels have the potential to be used as a barrier for containment of contaminants,

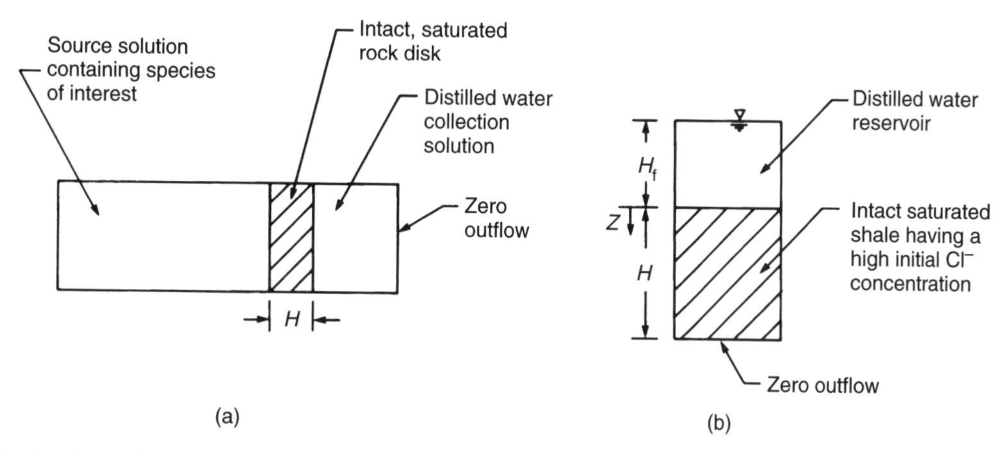

Figure 8.15 Schematic diagram of test set-ups for obtaining diffusion coefficients in porous rock: (a) diffusion from source through the rock to a receptor; (b) back-diffusion from rock with an initial high chloride concentration.

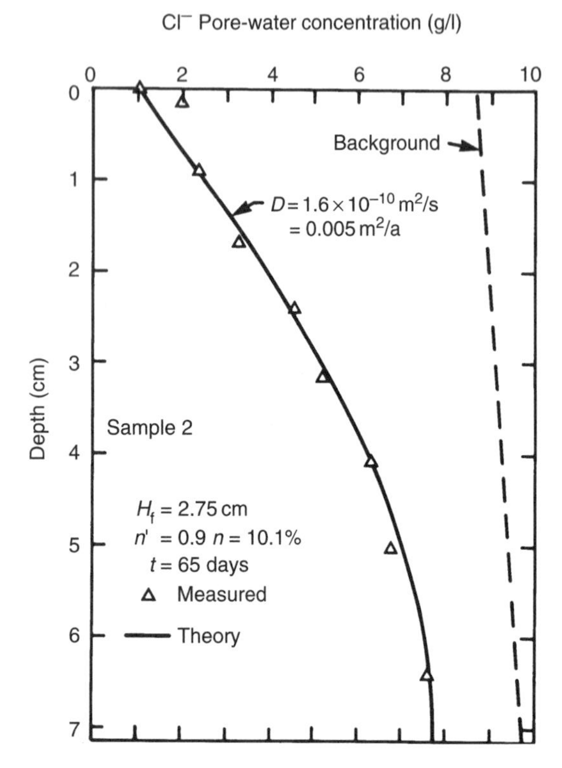

Figure 8.16 Diffusion profile for outward diffusion from a shale with an initial chloride concentration in the pore fluid (after Barone *et al.*, 1990; reproduced with permission, *Canadian Geotechnical Journal*).

however, they have received little attention in the literature. These macromolecules, consisting of a high number of units (monomers) bonded together, may be negatively or positively charged or neutral. When a chemical cross-linker is added, the individual chains are connected together at a number of points to form a 3D polymer chain network analogous to a porous medium with water with which the gel was prepared filling the pores between the polymer chains. The hydraulic conductivity of the gel is analogous to that of a porous medium and represents the resistance of the gel to water flow due to a hydraulic gradient and may be of the order of 10^{-12} m/s.

Using apparatus similar to that shown in Figure 8.2, tests can be performed to examine, diffusion of ions (e.g., Na^+ and H^+) through gels. For example, Darwish *et al.* (2004) examined the diffusion from an NaCl solution (170 mol/m^3) through a gel by using magnetic resonance imaging to measure the sodium and hydrogen ion concentration profiles in both the NaCl solution and the gel body throughout the test. These tests gave a gel diffusion coefficient similar to that of compacted clays but the partitioning coefficient for the gel was higher than that typically encountered for clays. These findings suggest that gels may have potential for use as a barrier, especially in situations where alternate systems are difficult to construct (e.g., in fractured media). More research is required to confirm the suitability of gels in this type of application.

8.3.6 Comments on boundary effects

When performing diffusion tests with a set-up such as that shown in Figures 8.2a, 8.12, 8.14 and 8.15, care needs to be taken not to run the test too long. For example, referring to the test reported in Figure 8.13, it can be seen that even though this test had only been run for 4 days, some chloride had diffused 4 cm to the bottom of the sample. When performing these tests using an impermeable base, it is desirable to terminate the test at about the time the most mobile species (typically chloride) reaches the bottom plate. If concentration is allowed to build up at the base (as seen in Figure 8.13) then accuracy may be lost; the greater the build-up, the greater the potential loss in accuracy. For example, in the worst case the test could be run long enough for steady-state conditions to develop. In this case, there would be a uniform concentration profile which is independent of the diffusion coefficient and hence it is not possible to obtain a unique diffusion coefficient. On the other hand, provided that the test is terminated while there is significant concentration gradient across the sample, the diffusion and distribution coefficients can be determined uniquely to within the experimental accuracy of the concentration determination. To avoid significant test over runs, an initial estimate of the time of termination can be made by estimating the diffusion coefficient and using a model (e.g., POLLUTE) to simulate the migration. This assumes that the diffusion coefficient can be reasonably estimated prior to the test, however, even if a good estimate cannot be made initially and a build-up in concentration does occur; the results of this test will usually give a fairly good estimate of the diffusion coefficient. This parameter can be used to estimate a better termination time for a second test, and the test can be repeated as a check. For the case shown in Figure 8.13, the diffusion coefficient was not affected by the build-up of concentration at the base.

It should be added that if one were concerned with the determination of parameters for a moderately to highly sorbed species, then it would be essential to have a length of sample such that there was not a significant build-up of either the associated anion (e.g., chloride) or desorbed cations (e.g., Ca^{2+}, Mg^{2+}) at the base of the sample. The problem can be avoided by either adopting a long sample or using a collection reservoir at the base of the sample (e.g., see Figure 8.2b).

The results presented in this section illustrate how the proposed technique can be used to estimate relevant parameters. The values of the parameters will depend, *inter alia*, on the mineralogy and structure of the clay, the pore water chemistry of the clay, the leachate composition and the temperature at which the test is conducted.

It should also be noted that a species that may be conservative in one environment (such as chloride discussed in this section) may not be conservative in a different environment (e.g., see Section 1.3.5(b)). Care is needed in assessing the parameters for conditions as close as possible to those expected in the field.

8.4 Dispersion at low velocities in clayey soils

The Darcy velocities of between 0.025 and 0.035 m/a used in the experiments described in Section 8.3.1 exceed that expected under operating conditions in most practical field applications involving clayey liners. The change in velocity from 0.025 to 0.035 did not give rise to a discernible difference in the coefficient D and this raises the question as to whether any dispersion is evident in these tests. To provide some indication regarding the effect of the advective velocity on the coefficient D, a pure diffusion test was conducted for chloride allowing for a concentration drop in the source leachate but zero flux at the base of the soil as described in Section 8.3.2. Tests at this scale can be performed very quickly and the concentration profile through the sample after 4 days is shown in Figure 8.13. Also shown is the predicted concentration profile using the average value of $D = 0.019\,m^2/a$ obtained from the advective–dispersive models A–E described in Section 8.3.2. The prediction is in excellent agreement with the observed profile and this suggests that the contribution of mechanical dispersion to the coefficient D at velocities of 0.035 m/a or less is negligible for this clay and hence it is appropriate here to refer to it as the diffusion coefficient.

In some circumstances, anion exclusion can give rise to significant dispersion and much faster movement of anions than water. Unlike traditional mechanical dispersion (see Section 1.3.4), which is related to variations in groundwater velocity about the mean, this type of dispersion is related to double-layer effects as discussed in the following section.

8.5 Effective porosity

It has been shown that for some intact saturated soils with predominantly active clay minerals (e.g., montmorillonite), the effective porosity can be significantly less than the total porosity (e.g., Thomas and Swoboda, 1970; Appelt *et al.*, 1975). This is a result of anion exclusion (as discussed in Section 6.4) from the highly structured double layer around negatively charged clay surfaces. For montmorillonitic soils compacted wet of optimum water content, the double layer may occupy a significant portion of the pore space and hence anion exclusion may result in an effective porosity for a negatively charged species, such as chloride, which is significantly less than the total porosity. Based on similar arguments, one would not expect there to be a significant difference between total and effective porosity for silty clays and silty clay tills of low activity since the double layer would only occupy a small portion of the pore space.

The authors' experience has been that the effective porosity, n, with respect to diffusion through saturated or near-saturated clayey barriers of low-activity clays is often reasonably estimated based on water content determined according to usual geotechnical practice (i.e., the effective porosity is essentially the same as the total porosity). This is also consistent with the findings of Kim *et al.* (1997) who found that the effective porosity in their test was essentially the same as the total porosity. However, as noted above, situations can be envisaged where the effective porosity could be less. The effective porosity can be estimated using the test procedure described in Section 8.3. For example, the tests described in Sections 8.3.2 and 8.3.3 were analysed assuming a value of porosity of 0.39, although some variability in water content was observed corresponding to a porosity ranging from 0.37 to 0.39. The calculated concentration profile using $D = 0.02\,m^2/a$ and both $n = 0.39$ and 0.37 are shown in Figure 8.11 for model E discussed in Section 8.3.2. For potassium, the difference in the curves is not plottable. For chloride, the difference is plottable but not significant. A similar conclusion is reached if the other model tests are reinterpreted using $n = 0.37$. Thus the uncertainty as to the precise value of porosity, n, due to the small variation in water content does not significantly influence the magnitude of the parameters D and ρK_d deduced using $n = 0.39$.

It could be argued that the effective porosity might be significantly less than the values calculated based on water content owing to the presence of immobile pore fluid or anion exclusion. To examine this possibility, an attempt was made to reinterpret the chloride tests using lower porosities (i.e., $n \leq 0.35$); however, it was not possible to obtain a good match to the experimental data for these parameters (see Figure 8.13). The discrepancies between the observed and calculated profiles increased with decreasing assumed porosity. Thus it would appear that the effective porosity of the soil tested is not significantly less than that deduced from the water content.

In some systems, anion exclusion may give rise to an average transport velocity of dissolved anions that is much larger than that of the accompanying water molecules due to the electrostatic repulsion by the negatively charged soil particles which forces the anions into the pore centres where the velocity is faster. Gvirtzman and Gorelick (1991) report a case where chloride and sulphate travelled about twice the velocity of tritium (effective porosities of 0.19, 0.11 and 0.09 for tritium, chloride and sulphate, respectively). The difference between chloride and sulphate was attributed to the greater charge of sulphate giving rise to greater repulsion from the charged surfaces of the soil particles. The migration of tritium was well described by a conventional advective–diffusive model using an effective porosity equal to the volumetric water content and a dispersion coefficient equal to the effective diffusion coefficient. However, the dispersion coefficient for anions was much greater than expected for molecular diffusion and at was about between 20 and 40 times greater than the effective diffusion coefficient.

8.6 Distribution coefficients and non-linearity

The approach described in the preceding sections is based on the assumption that the adsorption isopleth is linear and reversible. The species most affected by sorption in these tests is potassium. The tests involving potassium were performed at a concentration which could be expected to fully occupy the exchange sites on the clay. This represents the upper limit to which linearity of sorption could possibly occur and greatly exceeds concentrations found in many landfill leachates in Ontario.

A series of batch tests (see Chapters 1 and 6 for more details) were performed using this Sarnia clay and distilled water spiked with KCl (i.e., the same source solution as used in models D and E described in Section 8.3.2). The isopleth shown in Figure 8.17 is linear with $\rho K_d = 7.1$ up to a concentration of approximately 900 mg/l. This confirms that the assumption of linearity adopted in the interpretation of the model tests is reasonable. The value of ρK_d deduced from this batch test is in excellent agreement with the value back-figured from models D and E. The model test was also analysed using the finite element program SFIN (Rowe and Booker, 1983) which allows direct modelling of the isopleth obtained from the batch tests. The results from the non-linear analysis are not substantially different from those obtained using a linear isotherm for this case, since the initial concentration in the leachate is only marginally greater than the concentration at which non-linearity occurs.

The technique described in Section 8.3 provides a relatively simple means of estimating the diffusion and distribution coefficients of key contaminants as they diffuse through clayey soil. The interpretation of the test assumes that the sorption process can be reasonably approximated as being linear over the concentration range of interest. For problems where the concentration of contaminant is very high and where non-linearity becomes important, parameters back-figured using this procedure may

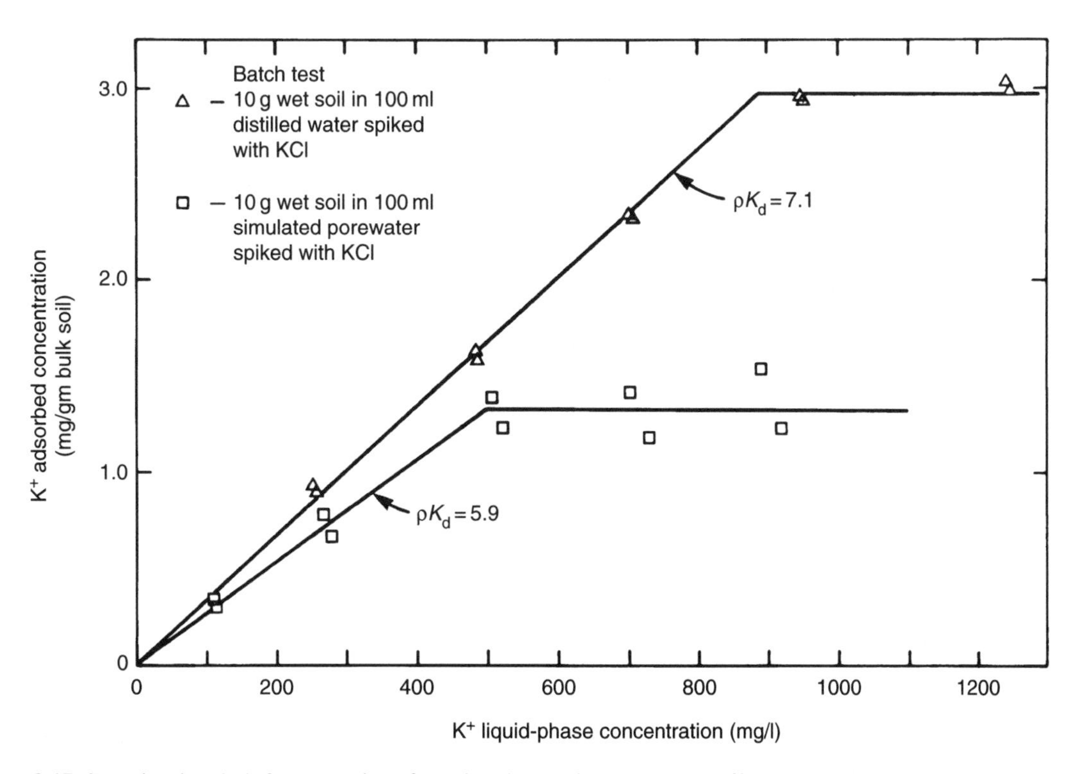

Figure 8.17 Sorption isopleth for potassium from batch tests indicating the effect of leachate chemistry on sorption of a given species (modified from Rowe *et al.*, 1988).

not be appropriate. The validity of the linearity assumption can be checked in one of two ways.

When dealing with inorganic contaminants, batch tests can be conducted to determine the range of linearity as discussed above and in Chapter 6. The diffusion test described in this chapter could then be performed within this range to provide values of both D and ρK_d for the soil of interest.

The alternative to performing batch tests in conjunction with a diffusion test is to perform at least two diffusion tests at different concentrations within the range of interest. If the linearity assumption is valid, then the values of D and ρK_d deduced from both tests will be the same. If the assumption is not reasonable, then markedly different values of ρK_d will be back-calculated for the different concentrations and, furthermore, it may be difficult to obtain a good fit to both the variation in leachate concentration with time and the variation in concentration with depth within the sample at the end of the test.

If sorption is found to be non-linear over practical concentration ranges, then in many cases either a Freundlich or a Langmuir isopleth may be established based on batch test data as discussed in Section 1.3.5. The diffusion coefficient can then be checked by performing a diffusion test and obtaining a theoretical fit to the experimental data as described in this chapter except that the sorption isopleth established from the batch test would be used as input for sorption parameters.

8.7 Effect of leachate composition, interaction and temperature

The migration of a particular contaminant species may be related to both the chemistry of the pore water in the soil or rock through which it will diffuse and the chemistry of the leachate. As discussed in Chapter 6, the diffusion coefficient can also be expected to be affected by temperature.

To illustrate the effect of chemistry, Barone *et al.* (1989) performed two series of tests (series (ii) and (iii)) using a similar Sarnia soil to that used for test series (i) which was discussed in Section 8.3.2. The essential differences between these three test series are summarized in Table 8.2 and the resulting diffusion coefficients and values of the sorption parameter ρK_d are summarized in Table 8.3.

Unless otherwise noted, a good fit could be obtained to the experimental data for all cases. For example, Figures 8.18 and 8.19 show the variation in concentration with time in the source reservoir, the variation in pore water concentration with depth in the sample and the adsorbed concentration with depth for both a leachate test (model K) and a single salt test (models G and H) for the Na^+ and K^+ cations, respectively. In these cases, a good theoretical fit could be obtained. Since Na^+ and K^+ were dominant, they were adsorbed onto the clay and other species (Mg^{2+}, Ca^{2+}) were desorbed.

The desorption of Ca^{2+} in the leachate test is evident in Figure 8.20 and as a result one cannot obtain a reasonable estimate of diffusion or sorption parameters for a simple single species model of the migration of these species. However, inspection of Figure. 8.20 does show that the single species model will give a reasonable fit to the data when these species are dominant as in the case in a single salt solution.

First inspection of Table 8.3 would suggest that the diffusion coefficient of chloride is relatively insensitive to temperature and leachate concentration; however, this is not the case. Figure 8.21 shows results from a different series of tests where the soil and chemistry were held constant and only the temperature was varied. As can be seen, the diffusion coefficient obtained for chloride at different temperatures does indeed vary as suggested by the theoretical relationship given in Chapter 6.

Thus, the fact that the diffusion coefficients for models A–F at 22 °C are similar to those for models G–J at 10 °C is coincidental and is a result of the counteracting effect of both different background chemistry and different temperatures. The effect of

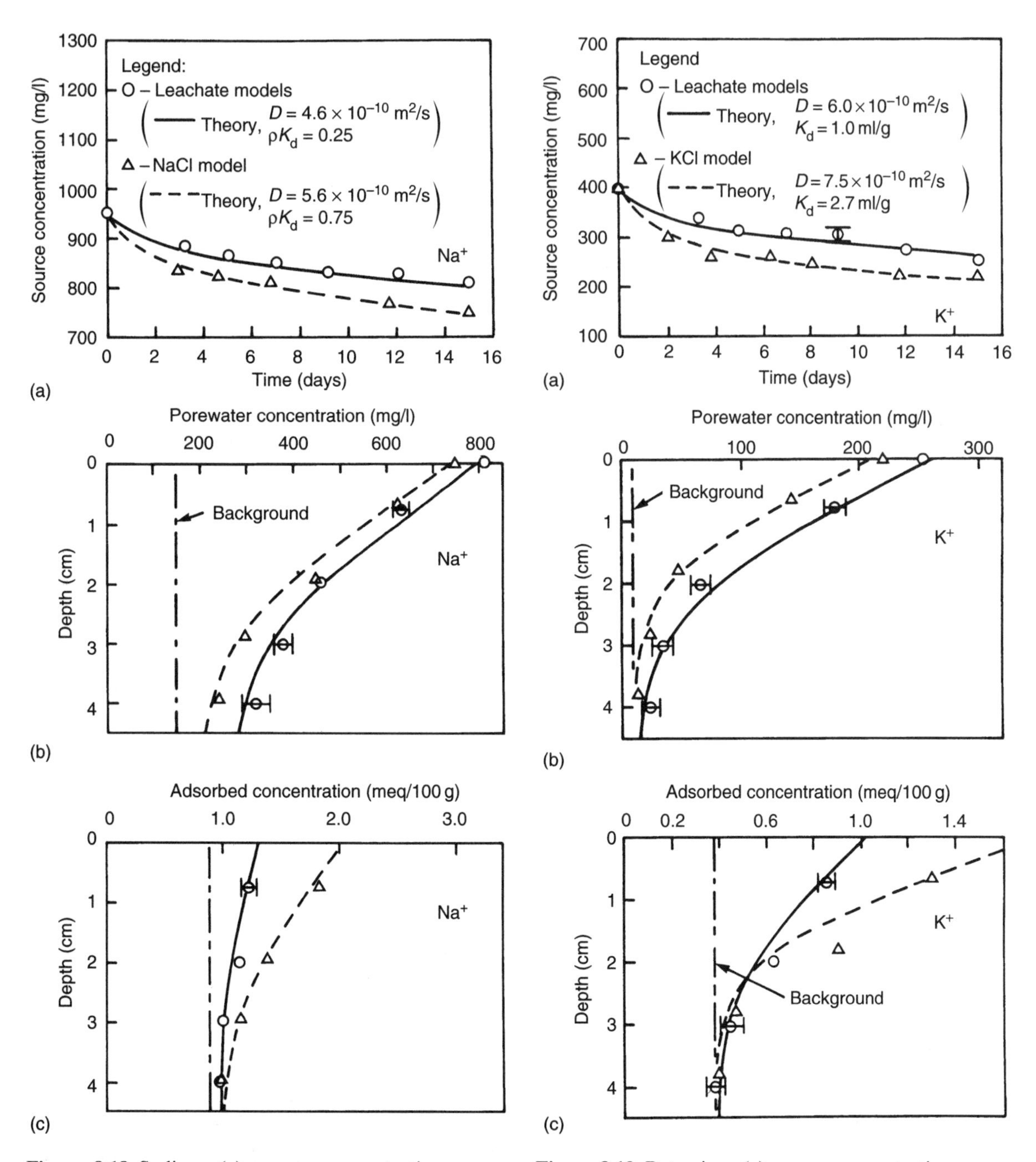

Figure 8.18 Sodium (a) source concentration versus time, (b) pore water concentration versus depth and (c) adsorbed concentration versus depth ($t = 15$ days) (modified from Barone *et al.*, 1989).

Figure 8.19 Potassium (a) source concentration versus time, (b) pore water concentration versus depth and (c) adsorbed concentration versus depth ($t = 15$ days) (modified from Barone *et al.*, 1989).

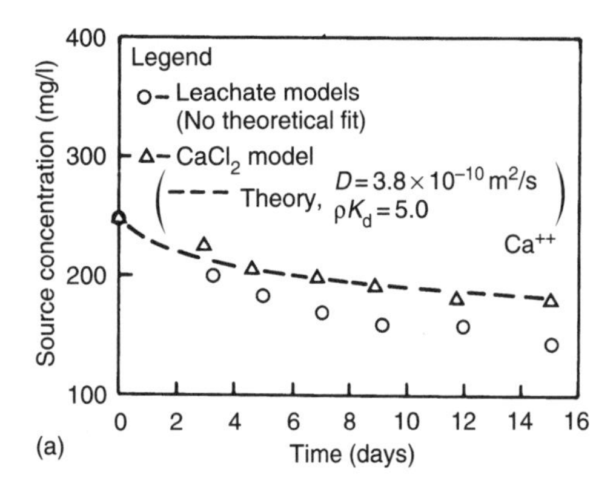

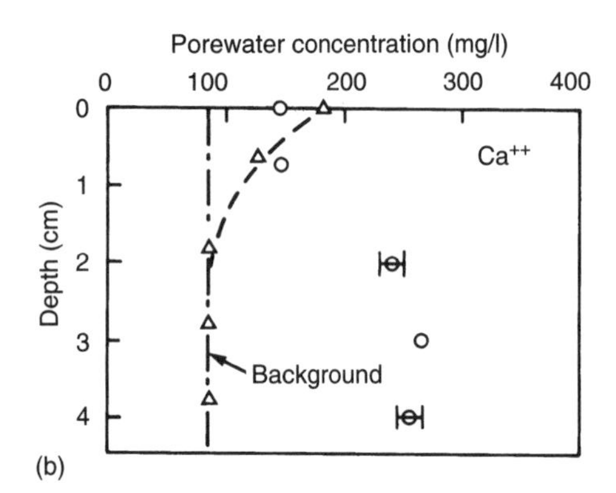

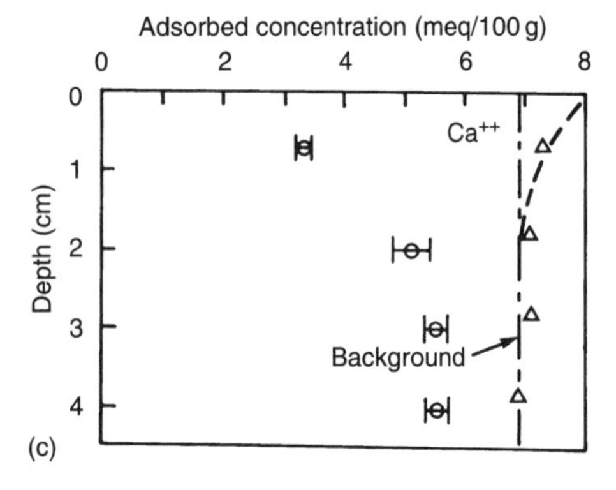

Figure 8.20 Calcium (a) source concentration versus time, (b) pore water concentration versus depth and (c) adsorbed concentration versus depth ($t = 15$ days) (modified from Barone *et al.*, 1989).

the difference in the clay is more evident when comparing the cation results in series (i) with the results in series (iii). For example, comparing the parameters for sodium in models A and G (both

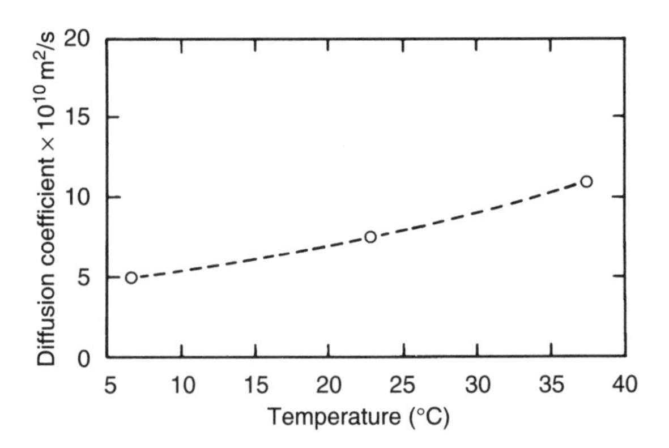

Figure 8.21 Effect of temperature on the effective diffusion coefficient, D_e, of chloride.

single salt tests), it can be seen that both the diffusion coefficient and sorption are higher for model G. The difference in temperature would be expected to result in a lower diffusion coefficient than that in model A. The fact that it is higher indicates that the physicochemistry of the soil has an important influence on the values of D and ρK_d.

Comparing the results of tests with the single salt solution in series (ii) with those obtained with the leachate in series (iii) further emphasizes the importance of chemistry in the apparent diffusion and distribution coefficients. The effect of the additional competition for sorption sites and the requirement of electroneutrality apparent in model K resulted in a higher chloride diffusion coefficient than that obtained in the single salt tests and reduced both the diffusion and distribution coefficients of the cations. Thus, even the diffusion coefficient for chloride, which is generally regarded as being conservative, does depend on the leachate and clay chemistry since, in order for there to be electroneutrality Cl^- must diffuse in association with cations, and the migration of cations is influenced by competition for sorption sites.

It should be emphasized that tests conducted on any particular soil with a given leachate are highly repeatable (e.g., compare models B and C, or D and E) or the two tests summarized by the "error bars" for tests with leachate shown in Figures 8.18–8.20 noting that where there are no error bars shown the results were identical to plotting accuracy. However, the results shown in Table 6.3 indicate that there are many complicating chemical–soil interactions influencing the rate of diffusion of a particular chemical species through a given clay. These factors are best captured empirically by performing tests using soil and leachate which most closely approximate that which will be used in the field application of interest.

For conditions where the diffusion and partitioning coefficients have been established at one temperature, an initial estimate of that parameter at another temperature can be obtained from the following relationship:

$$\frac{D_{T_2}}{D_{T_1}} = \frac{T_2 \eta_{T_1}}{T_1 \eta_{T_2}} \tag{8.2}$$

where D_{T_1} and D_{T_2} are the diffusion coefficients at the temperatures T_1 and T_2 K (273 + T in °C), and η_{T_1} and η_{T_2} are the viscosities at T_1 and T_2 (see Clark, 1996) and

$$\frac{K_{D_{T_1}}}{K_{D_{T_2}}} = \Lambda^{(T_1 - T_2)} \tag{8.3a}$$

$$\Lambda = \frac{E_a}{RT_1 T_2} \tag{8.3b}$$

where E_a is the activation energy (J/mol), T is an absolute temperature (K) and R is the universal gas constant ($R = 8.3143$ J/K/mol). The value of Λ is usually within the range 1.0–1.10 between 0 and 35 °C (see Schnoor, 1996). Rowe *et al.* (2003b) found that $\Lambda = 1.03$ provided a good description of sorption of hydrocarbons on a GCL over the temperature range 5–22 °C. The foregoing has potential application for sorption of organic compounds; however, more research is needed to establish the range of validity. In cases where the parameters are critical to performance, they should be established under conditions as close as possible to those anticipated in the field (including temperature).

8.8 Diffusion and sorption of organic contaminants

The techniques described in the previous sections can also be used to evaluate the diffusion and partitioning coefficients for organic chemicals; however, some modifications and considerable caution may be required. For example, with volatile organic compounds there may be difficulties in obtaining meaningful concentration profiles through the sample due to losses which occur when squeezing the pore water from the clay slices. In these cases, it is often necessary to rely on the source and collector concentrations for laboratory set-ups such as that shown in Figure 8.2b. The source and receptor concentrations can be used to infer both the diffusion coefficient D and the partitioning coefficient K_d, but care is also required to check that the results are meaningful. This has been discussed by Barone *et al.* (1992b), who examined the diffusion of acetone, 1,4-dioxane, aniline, chloroform and toluene through a natural clayey soil.

Barone *et al.* (1992b) showed that good results could be readily obtained for acetone, 1,4-dioxane and aniline (e.g., see Figure 8.22) by fitting a theoretical curve to the measured source and collector concentrations. However, they also showed that sorption onto the walls of the apparatus could give misleading results for chloroform and toluene unless the diffusion tests were complemented by a series of batch tests or unless the losses can be quantified. Barone *et al.* (1992b) performed control tests which could be used to model the time-dependent losses in the source reservoir. Using this data, together with the results of the diffusion tests, Smith *et al.* (1993) modified the Rowe and Booker (1985a) solution to allow for these losses and hence to directly infer the diffusion and sorption parameters from the results of the laboratory tests.

Other investigators have also performed diffusion or advection–diffusion tests for organic contaminants migrating through either clay liners or composite liners (geomembrane over clay).

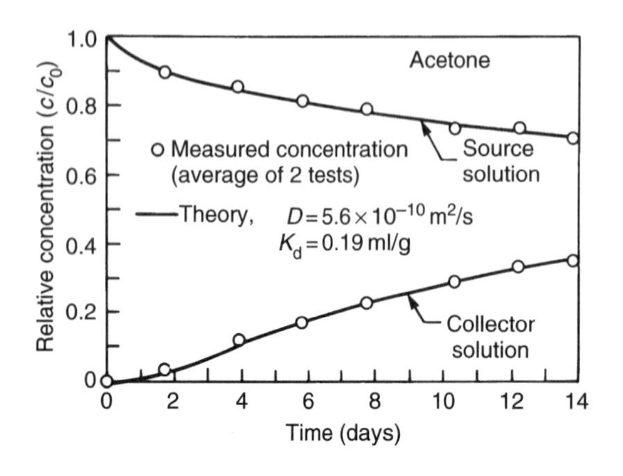

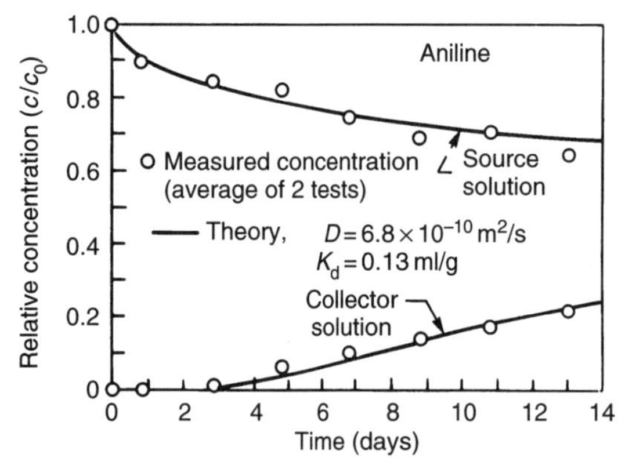

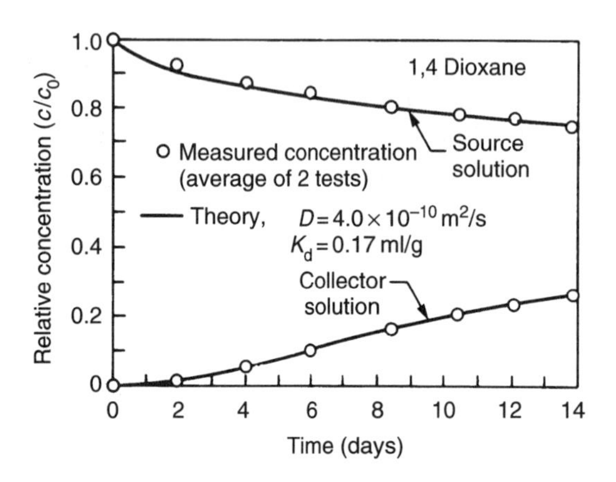

Figure 8.22 Source and collector solution concentration variation with time for three organic species (modified from Barone *et al.*, 1992a).

Examples include Millward (2000), Hrapovic (2001), Kim *et al.* (2001), Kalbe *et al.* (2002) and Krol and Rowe (2004), however, the results are not always consistent. For example, Kim *et al.* (2001) examined seven VOCs and obtained diffusion coefficients that varied from 2×10^{-9} to 1×10^{-10} m²/s corresponding to tortuosities ranging from 0.13 to 2.9! (some of these results are questionable). In contrast, Hrapovic (2001) examined diffusion of eight VOCs through compacted clay and reported a much narrower range of diffusion coefficients between 2.5×10^{-10} and 5×10^{-10} m²/s.

An alternative approach is to use the results of batch tests to estimate K_d. However, batch tests may also be subject to error and the results may depend on the solids to water ratio (Figure 8.23). Voice *et al.* (1983) attributed the sensitivity of K_d values to the solids/water ratio to microparticulate or macromolecular material being washed

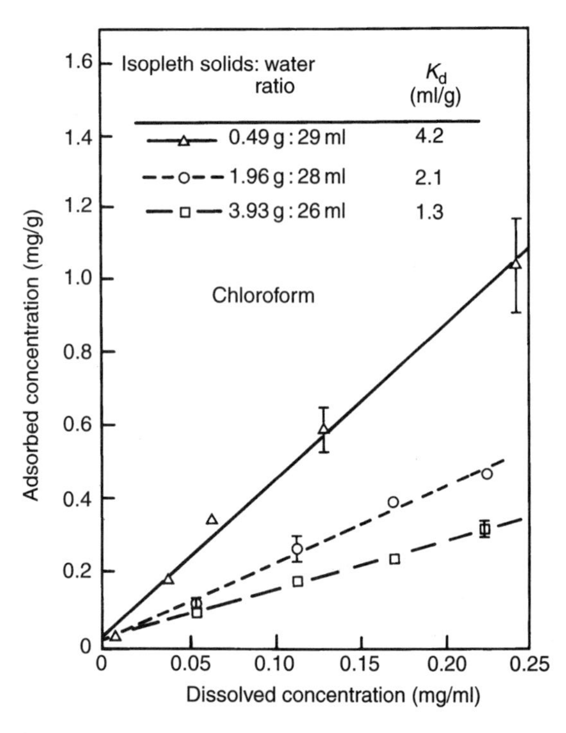

Figure 8.23 Batch test isopleths for chloroform at different solids/water ratios (modified from Barone *et al.*, 1992b).

Table 8.4 Summary of the diffusion and sorption coefficients at 22 °C as reported by Barone *et al.* (1992b)

Species	D (m^2/s)	D (m^2/a)	K_d (ml/g)	K_{oc} (ml/g)
Acetone	5.6×10^{-10}	0.018	0.19	33
1,4-Dioxane	4.0×10^{-10}	0.013	0.17	29
Aniline	6.8×10^{-10}	0.021	1.3	224
Chloroform	7.0×10^{-10}	0.022	4.2	724
Toluene	5.8×10^{-10}	0.018	11.3	1,950

off the soil particles during the batch test and then not being removed from the liquid phase during the separation procedure. These non-settling microparticles tended to increase the capacity of the liquid phase to accommodate solute and hence give rise to low apparent K_d values. Gschwend and Wu (1985) also studied this problem and based on their results it is evident that the lower the solids/water ratio, the less is the influence of the non-settling particles and hence the more representative the results. Combining the results of carefully performed batch tests at low solids/water ratios, with the results from diffusion tests, it is then possible to estimate the diffusion and adsorption characteristics of the organic contaminants for a given soil. Values inferred by Barone *et al.* (1992b) for a clayey soil with a porosity of 39%, and organic carbon content, f_{oc}, of 0.58% are given in Table 8.4.

The results reported by Barone *et al.* (1992b) were for single organic compounds dissolved in distilled water. Tests performed by Nkedi-Kissa *et al.* (1985) and by Quigley and Fernandez (1989, 1992) suggest that introduction of organic cosolvents may play a significant role on the interaction between an organic solute and clay minerals. This suggests that mixtures of organic liquids in water could behave differently than single species in water. Additional laboratory testing and field confirmation are needed for organic mixtures.

8.8.1 Diffusion from a DNAPL source

Pools of DNAPLs can act as a contaminant source as groundwater flows past these areas, leading to long-term contamination of the aquifer. Placing a soil–bentonite slurry wall around these pools and/or plumes is one means of limiting the migration of the contaminants. Krol and Rowe (2004) simulated a soil–bentonite slurry wall next to a pool of trichloroethylene (TCE) in a diffusion cell by having a system comprised of: a pure TCE as the source, a thin (3 mm) layer of clayey soil, a layer of glass beads initially saturated with water, and a 2% sodium bentonite–silty sand liner over a porous plate and water receptor. It was found that the concentration in the glass beads soon reached the solubility limit of TCE and that knowing the diffusion coefficient of the soil–bentonite wall (3.5×10^{-10} m^2/s from independent tests) a good prediction of migration through the system could be made by assuming a continuous source concentration equal to the solubility limit of the chemical of concern (TCE in this case).

The average sorption coefficient, $K_d = 0.2\,cm^3/g$, obtained by Krol and Rowe (2004) from batch tests with 2% bentonite–silty sand mixture was about a third of that obtained by Khandelwal *et al.* (1998) from the batch tests with 6% bentonite ($0.57\,cm^3/g$). However, Khandelwal *et al.* also found that the experimental K_d values obtained from diffusion tests ($K_d = 0.01\,cm^3/g$) were much lower than those from the batch tests ($0.57\,cm^3/g$). Gullick (1998) found that the partitioning coefficient of 4% bentonite backfill was negligible. Krol and Rowe (2004) also found that good predictions of the behaviour in the diffusion tests could be made with $K_d \sim 0\,cm^3/g$. Based on observations from these three investigators, it would appear that there is little sorption of TCE on typical soil–(sodium)bentonite mixtures.

8.9 Use of field profiles to estimate diffusion coefficients

Field profiles of diffusive migration of various chemical species have developed both by natural causes like the diffusion of Na^+ and Cl^- through clay from underlying salty bedrock or by diffusion from landfill leachate as will be discussed in more detail in Chapter 9. These profiles can often be used to back-figure diffusion coefficients. For example, the Confederation Road landfill near Sarnia, Ontario, has been the subject of many papers dealing with the diffusion of a variety of species into the underlying clays (Goodall and Quigley, 1977; Crooks and Quigley, 1984; Quigley *et al.*, 1984; Quigley and Rowe, 1986).

A summary plot of the chemical profiles for a variety of species is presented in Figure 8.24 for pore water squeezed from three close-spaced boreholes 15 years after placement of the domestic waste. The extrapolated curves are predicted from the lower 20% of the data points and extrapolated to the interface. The actual data points do not fit exactly because the source concentration was decreasing with time within the mass of waste. The D_R values used to back-calculate these curves are presented in Table 8.5. Based on the D_R value for chloride, it is possible to infer the tortuosity, τ, and hence estimate the effective diffusion coefficient D_e values for the other species as given in Table 8.5. Once the D_e value is obtained, the value of ρK_d can then be deduced from the value of D_R for these species as discussed in Chapter 6.

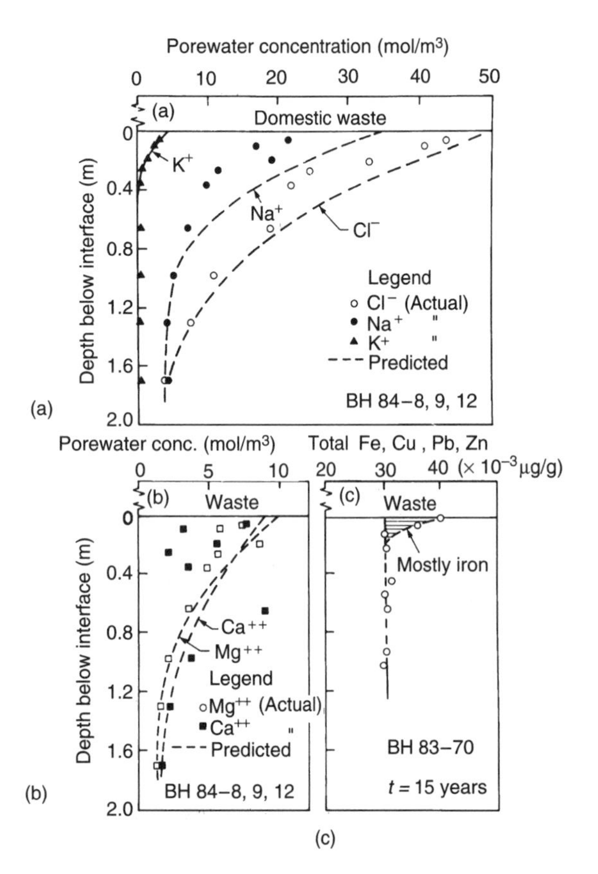

Figure 8.24 Concentration versus depth below waste/clay interface, Confederation Road landfill: (a) chloride, sodium and potassium; (b) calcium and magnesium; (c) heavy metals (modified from Quigley *et al.*, 1987b).

The sodium, potassium and chloride ion profiles (Figure 8.24a, actual and predicted) are in logical locations, the chloride furthest advanced, sodium slightly retarded and potassium greatly retarded. The relative positions of these three species are clearly reflected by the retarded diffusion coefficients, D_R.

Calcium and magnesium are much more difficult to deal with because, even though they are the dominant cations adsorbed on the clays, they are also intimately involved in heavy metal precipitation and dissolved CO_2 levels associated with bacterial respiration. The high D_R values used to calculate the location of the predicted curves may be more indicative of a hardness halo front than diffusion.

The distances to the position of diffusion fronts at $c/c_0 = 0.5$ are also presented in Table 8.5 yielding migration distances of 0.36, 0.30 and 0.11 m for Cl^-, Na^+ and K^+ respectively.

Finally, inferred values for K_d are also presented in Table 8.5. Conservative chloride by

Table 8.5 Tabulation of field diffusion data

Ion	Cl^-	Na^+	K^+	Ca^{2+}	Mg^{2+}
D_R (m²/a)	0.020	0.0072	0.0011	0.018	0.0115
D_e (m²/a)	0.020	0.012	0.016	0.009	0.008
K_d (ml/g)	0	0.16	3.23	−0.11	−0.07
Distance to $c/c_0 = 0.5$ (m)	0.36	0.3	0.11	–	–

definition has $K_d = 0.0$ mg/l, Na^+ yields a small K_d of 0.16 ml/g and K^+ (which tends to fix onto the clays) yields a larger value of 3.23 ml/g. The negative values for Ca^{2+} and Mg^{2+} imply desorption.

A comparison of the results back-calculated from these field profiles (Table 8.5) with values obtained from laboratory tests (Table 8.3) indicates reasonable agreement with the values back-figured from the field falling within the range of diffusion coefficients and sorption parameters obtained from the laboratory tests.

8.10 Summary of diffusion parameters for saturated soil and rock

The foregoing sections have discussed a number of key parameters to be considered in making predictions of contaminant migration through clayey barriers. With respect to diffusion, limited data would suggest that the effective porosity of clayey soils may be close to the values deduced based on water content. However, situations can be visualized where this may not be the case (e.g., due to anion exclusion). Diffusion tests such as those described in this chapter can be used to test the hypothesis that the effective porosity of active clays is less than the bulk porosity and, if so, to estimate the effective porosity.

Techniques for estimating the diffusion and distribution/partitioning coefficients of chemical species have been described and illustrated for a number of tests. The tests are relatively easy to perform and, provided that one can obtain accurate measures of concentration, they give repeatable results for a given soil and leachate chemistry. However, both the diffusion coefficient and distribution/partitioning coefficient for a given species will depend on the chemical composition of the source leachate and background chemistry and mineralogy of the soil. This can be true even for conservative ionic species such as chloride. Thus the diffusion and sorption coefficients should be obtained for the proposed soil using a leachate as near as practicable to that expected in the field situation. Thus the complex factors influencing the diffusion of chemical species are empirically captured in the laboratory parameters.

Fortunately, despite the potential variation in diffusion coefficients due to soil and leachate composition, the range of variation is relatively small compared to that of many other parameters (like hydraulic conductivity). In engineering terms, diffusion is in fact a very predictable process. As discussed in Chapter 9, field diffusion profiles and, indeed, natural diffusion processes established over thousands of years have been shown to be consistent with diffusion coefficients which can be determined from relatively simple short-term laboratory tests such as those discussed in this chapter.

With the foregoing caveats in mind, Table 8.6 summarizes diffusion and distribution coefficients either calculated from laboratory tests or back-figured from observed field profiles. This table may be a useful guide when planning a testing program but is not intended for use in the design of barriers. These values should be independently confirmed for any particular project.

8.11 Diffusion in unsaturated porous media

8.11.1 Solute diffusion through unsaturated porous media

The tests described in Section 8.3 have been successfully used for natural and compacted clayey soils, geosynthetic clay liners (GCLs) (see Section 12.6) and for sedimentary rock with greater than 90% saturation. For unsaturated soils or rock, care is required in the interpretation of results in order to separate the effects of matrix suction (which may induce flow into the sample) and diffusion. One means of minimizing the effect of matrix suction is to allow the soil to

Table 8.6(a) Summary of effective diffusion (D_e) and sorption (K_d) coefficients for various inorganic contaminant species n = porosity; PI = plasticity index; CEC = cation exchange capacity; c_0 = initial concentration; c_f = final concentration; θ_v = volumetric water content.

	Soil description												
Species	*n* (%)	*Silt content* (%)	*Clay content* (%)	*PI* (%)	*CEC* (meq/ 100 g)	*Soil pH*	D_e $\times 10^{10}$ (m²/s)	K_d (ml/g)	*Test method*	*Source solution*	*Temperature* (°C)	*Source solution pH*	*Reference*
Ammonium	32	42	23	11	–	–	5.7	1.2	Diffusion test on undisturbed soil	Leachate c_0 (NH_4) = 957 mg/l	10	–	Rowe (personal files)
Arsenic									One point batch test soil : solution = 1 : 20				
	–	25	30	–	17	6.1	–	5	c_f = 81 mg/l	Distilled deionized water spiked with KH_2AsO_4	22	–	Griffin *et al.* (1986)
	–	0	100	–	15	8.1	–	3	c_f = 87 mg/l				
	–	40	15	–	5	7.5	–	3	c_f = 88 mg/l				
Bromide	51–60		100	23	5	3.7	4.8–6.1[a]	0	–	Simulated leachate spiked with KBr c_0 (Br) = 645–1012 mg/l	23	3.7–6.7	Shackelford and Daniel (1991)
	45–47		82	42	25	6.9	18.2[a]	0					
Cadmium	–	–	86	26	8	–	–	8	Batch test soil:solution = 1: 20 by wt.; linear range 0–5 mg/l	MSW leachate spiked with $CdNO_3$	–	7.6	Yong and Sheremata (1991)
	51–60		100	23	5	3.7	3.2–4.2[a]	2	Batch test, soil:solution = 1: 4 by wt.; linear range 0–50 mg/l	Simulated leachate spiked with CdI_2	23	3.7–6.7	Shackelford and Daniel (1991)
	45–47		82	42	25	6.9	3.0–4.0[a]	35	Linear range 0–10 mg/l				
									One point batch test soil: solution=1:20	Distilled deionized water spiked with $CdCO_3$	22		Griffin *et al.* (1986)
	–	25	30	–	17	6.1	–	70	c_f = 22 mg/l				
	–	0	100	–	15	8.1	–	70	c_f = 22 mg/l				
	–	40	15	–	5	7.5	–	29	c_f = 41 mg/l			–	

Calcium	10	43	45	27	10	8.1	3.8	1.2	Diffusion test on undisturbed soil (with advection, $v_a = 0.033$ m/a)	$CaCl_2$ in deionized, distilled water c_0 (Cl) = 975 mg/l	22	7	Rowe *et al.* (1988)
Chloride	51–60 45–47	100 82		2.3 42	5 25	3.7 6.9	4.4–6.0[a] 1.5–1.8[a]	0 0	– –	Simulated leachate spiked with $ZnCl_2$ c_0 (Cl) = 231–448 mg/l	23	3.7–6.7	Shackelford and Daniel (1991)
	39	43	45	27	10	8.1	7.5 5.6 5.9 6.0 6.2	0	Diffusion test on undisturbed soil	MSW leachate c_0(Cl)=1000 mg/l Distilled, deionized water and NaCl KCl $MgCl_2$ $CaCl_2$	10	7.0	Barone *et al.* (1989)
	39	43	45	27	10	8.1	5.7 5.9 6.3	0	Diffusion test on undisturbed soil (with advection, $v_a = 0.025$–0.0035 m/a)	NaCl solution $CaCl_2$ solution KCl solution	22	7.0	Rowe *et al.* (1988)
									Diffusion test on undisturbed soil	MSW leachate	10	–	Rowe (personal files)
	28	38	28	–	–	–	3.0	0		c_0 (Cl)=1463 mg/l			
	23	44	19	–	–	–	4.0	0		c_0 (Cl) = 1463 mg/l			
	18	32	4	–	–	–	7.8	0		c_0 (Cl) = 580 mg/l			
	57	–	85	32	–	7.0	3.3	0		c_0 (Cl) = 1350 mg/l			
	47	–	81	21	–	–	4.0	0		c_0 (Cl) = 1350 mg/l			
	33	–	34	8	–	–	5.3	0		c_0 (Cl)= 14,500 mg/l			
	44	–	–	17	–	–	4.0	0		c_0 (Cl) = 1050 mg/l			
	37	–	–	–	–	–	5.0	0		c_0 (Cl) = 1050 mg/l			
	34	55	45	8	–	–	8.0	0		c_0 (Cl) = 1250 mg/l			
									Diffusion test on recompacted clay	MSW leachate	10	–	Rowe (personal files)
	40	–	–	10	–	–	5.0	0		c_0(Cl) = 1050 mg/l			
	38	–	39	11	–	–	5.0	0		c_0(Cl) = 14,500 mg/l			
	–	63	37	–	12	8.1	7.0	0	Diffusion test on recompacted clay	c_0(Cl) = 1500 mg/l	22 ± 2	7.2	Lake (2000)

Table 8.6(a) *Continued*

Species	*n* (%)	*Soil description* Silt content (%)	Clay content (%)	*PI* (%)	*CEC* (meq/100 g)	*Soil pH*	$D_e \times 10^{10}$ (m^2/s)	K_d (ml/g)	*Test method*	*Source solution*	*Temperature* (°C)	*Source solution pH*	*Reference*
Chloride	27	48	22	6	–	–	6.0	0	Diffusion test on undisturbed till	MSW leachate $c_0(Cl) = 967$ mg/l	10	–	Rowe (personal files)
	20	39	15	5	–	–	5.0	0					
	35	40	24	9	–	–	5.0	0	Diffusion test on recompacted till	MSW leachate $c_0(Cl) = 967$ mg/l	10	–	Rowe (personal files)
	30	48	22	6	–	–	7.0	0					
	30	39	15	5	–	–	6.0	0					
	29	50	23	6	16	–	5.7	0		Distilled water spiked with NaCl $c_0(Cl) = 1800$ mg/l		7.0 –	Rowe and Badv (1996a)
	32	42	23	11	–	–	4.7	0	Diffusion test on undisturbed clay	MSW leachate $c_0(Cl) = 2180$ mg/l	10	–	Rowe (personal files)
	43	46	54	18	–	–	4.5	0		$c_0(Cl) = 1250$ mg/l			
	41	70	30	6	–	–	7.5	0		$c_0(Cl) = 1250$ mg/l			
	47	53	47	19	–	–	5.0	0		$c_0(Cl) = 1000$ mg/l			
	36	–	40	–	–	–	2.0	0		$c_0(Cl) = 970$ mg/l	7		Barone (1990)
	37	3.5	0	–	1.4	–	9.8	0	Diffusion test on fine sand $\theta_v = 36–37\%$	Distilled water spiked with NaCl $c_0(Cl) = 1250$ mg/l	23 ± 2	–	Rowe and Badv (1996a)
	38	0	0	–	0.6	–	10.4	0	Diffusion test on coarse sand $\theta_v = 38\%$	Distilled water spiked with NaCl $c_0(Cl) = 1825$ mg/l	23 ± 2		Rowe and Badv (1996b)
							0.8–10.4	0	$\theta_v = 2.8–38\%$	$c_0(Cl) = 950$ mg/l			
	38	0	0	–	0.8	–	10.4	0	Diffusion test on fine gravel $\theta_v = 38\%$	Distilled water spiked with NaCl $c_0(Cl) = 1710$ mg/l	23 ± 2	–	Rowe and Badv (1996b)
							0.64–10.4	0	$\theta_v = 2.3–38\%$	$c_0(Cl) = 1040–1330$ mg/l			

Chloride									Diffusion test on				
	3.4	N/A	N/A	N/A	–	–	0.6–0.8	0	Sandstone	NaCl solution	22	–	Rowe (personal files)
	9.2	N/A	N/A	N/A	–	–	0.8–0.9	0	Mudstone	NaCl solution			
	10.8	N/A	N/A	N/A	12.5	–	0.15	0	Shale	Back-diffusion	22		Barone *et al.* (1990)
	23.8	N/A	N/A	N/A	38	–	0.15–0.2	0	Mudstone	Back-diffusion	10		Barone *et al.* (1992a)
Copper	–	43	45	27	10	8.2	–	400	Batch test, 1:200 soil:solution by wt. (linear range 0–20 mg/l)	MSW leachate spiked with $Cu(NO_3)_2$	22	7.8	Yanful *et al.* (1988b)
Iodide	51–60	100		23	5	3.7	3.5–14.7[a]	0	–	Simulated leachate spiked with CdI_2, $c_0(I) = 1089 - 1567$ mg/l	23	3.7–6.7	Shackelford and Daniel (1991)
	45–47	82		42	25	6.9	5.3[a]	0	–				
Lead	–	43	45	27	10	8.2	–	1900	Batch test, 1:200 soil:solution by wt. (linear range 0–5 mg/l)	MSW leachate spiked with $Pb(NO_3)_2$	22	7.8	Yanful *et al.* (1988b)
Potassium	51–60	100		23	5	3.7	11.7–17.7[a]	1.7	Batch test, 1:4 soil:solution by wt.	–	23	3.7–6.7	Shackelford and Daniel (1991)
	45–47	82		42	25	6.9	19.6[a]	1.1					
	39	43	45	27	10	8.1	6.0	1.0	Diffusion test on undisturbed soil	MSW leachate, $c_0(K^+) = 400$ mg/l	10	7	Barone *et al.* (1989)
							7.5	2.7		KCl solution in deionized distilled water			
	–	63	37	–	12	8.1	7.0	–		MSW leachate, $c_0(K^+) = 210$ mg/l	22 ± 2	7.2	Lake (2000)
	36	–	40	–	–	–	5.0	1.0	Diffusion test on undisturbed soil	MSW leachate, $c_0(K^+) = 280$ mg/l	7		Barone (1990)
Sodium	32	42	40	11	–	–	2.0	0.03	Diffusion test on undisturbed soil	$c_0(Na^+) = 565$ mg/l	10		Rowe (personal files)

Table 8.6(a) *Continued*

Species	*n* (%)	Soil description: *Silt content* (%)	*Clay content* (%)	*PI* (%)	*CEC* (meq/100 g)	*Soil pH*	$D_e \times 10^{10}$ (m^2/s)	K_d (ml/g)	*Test method*	*Source solution*	*Temperature* (°C)	*Source solution pH*	*Reference*
Sodium	39	43	45	27	10	8.1	4.6	0.15	Diffusion test on relatively undisturbed soil	MSW leachate, $c_0(Na^+) = 955$ mg/l	10	7.0	Barone *et al.* (1989)
	36	–	40	–	14.8	8.7	3.7	0.3	Diffusion test on relatively undisturbed soil	ISW leachate, $c_0(Na^+) = 5100$ mg/l	7	9.0	Barone (1990)
	39	43	45	27	10	8.1	4.8	0.18	Diffusion test on undisturbed soil (with advection, $v_a = 0.034$ m/a)	Distilled water spiked with NaCl, $c_0(Na^+) = 975$ mg/l	22	–	Rowe *et al.* (1988)
	–	63	37	–	12	8.1	–	0.2	Batch test, 1:10 soil:solution	MSW leachate with KCl	22 ± 2	7.2	Lake (2000)
							4.0	0.2	Diffusion test on recompacted clay	MSW leachate, $c_0(Na^+) = 1503$ mg/l			
	29	50	23	6	16	–	5.0	0.15	Diffusion test on recompacted till	Distilled water spiked with NaCl, $c_0(Na^+) = 1170$mg/l	23 ± 2	–	Rowe and Badv (1996a)
									Diffusion test on coarse sand	Distilled water spiked with NaCl	23 ± 2		Rowe and Badv (1996b)
	38	0	0	–	0.6	–	10.4	0	$\theta_v = 38\%$	$c_0(Na^+) = 1185$mg/l			
							0.87–10.4	0	$\theta_v = 3.2–38\%$	$c_0(Na^+) = 620$ mg/l			
									Diffusion test on fine gravel	Distilled water spiked with NaCl	23 ± 2	–	Rowe and Badv (1996b)
	38	0	0	–	0.8	–	10.4	0	$\theta_v = 38\%$	$c_0(Na^+) = 1710$ mg/l			
							0.64–10.4	0	$\theta_v = 2.3–38\%$	$c_0(Na^+) = 670$ mg/l			
Sulphate									Diffusion test on undisturbed soil	MSW leachate,			
	36	–	40	–	–	–	2.0	0		$c_0(SO_4^-) = 3300$ mg/l	7	–	Barone (1990)
	32	42	23	11	–	–	2.0	0		$c_0(SO_4^-) = 855$ mg/l	10	–	Rowe (personal files)

Zinc	51–60		100		23	5	3.7	3.5–4.5[a]	2	Batch test, 1:4 soil:solution by wt (linear range 0–50 mg/l)	Simulated leachate spiked with $ZnCl_2$, $c_0(Zn) = 301–374$ mg/l	23	3.7–6.7	Shackelford and Daniel (1991)
	45–47		82		42	25	6.9	1.5–2.8[a]	35	Batch test, 1:4 soil:solution by wt (linear range 0–10 mg/l)			7.8	
	0.4	73		27	18	12	7.0	4.1	13–60	Batch test, 1:10 soil:solution by wt and diffusion test on compacted clay	Artificial leachate from waste having high zinc, leached with MSW leachate acidified with acetic acid, $c_0(Zn) = 210$ mg/l	22	5.0	Quigley (personal files)
	–	25			–	7.3	9.4	0.54–6.5	0.43–0.76	Diffusion test Column test	Aqueous solution of $ZnCl_2$ $c_0 = 445–525$ mg/l	18–23	3	Shackelford *et al.* (1997)

[a]There were problems with mass balance in these tests, and the reported diffusion coefficient should be used with caution. Results are reported for completeness.

Table 8.6(b) Summary of effective diffusion (D_e) and sorption (K_d) coefficients for various organic species n = porosity; PI = plasticity index; f_{oc} = soil organic carbon content; $K_{oc} = K_d/f_{oc}$; c_0 = initial concentration; K_f and ε, Freundlich sorption parameters – concentration in ppm (Figure 1.19).

Species	Soil description: *n* (%)	Silt content (%)	Clay content (%)	PI (%)	Soil pH	f_{oc} (%)	$D_e \times 10^{10}$ (m^2/s)	K_d (ml/g)	K_{oc} (g/ml)	K_f, ε (ml/g), (–)	Test method	Source solution	Temperature (°c)	Reference
Acetic acid	26		45	15	7.6	0.14–0.45	–	0.06	–	–	Batch test, 1:3.6 soil:solution	Dilute aqueous solution, c_0 = 333–8118 mg/l	24 ± 2	Hrapovic (2001)
							3.5	–	–	–	Diffusion test, 1:3.6 soil:solution	Dilute aqueous solution		
							3.5	–	–	–	Diffusion test, 1:3.6 soil:solution	Synthetic leachate with acetic acid (7 mg/l)		
Acetone	39	43	45	27	8.1	0.58	5.6	0.19	33	–	Diffusion test on relatively undisturbed soil	Distilled, deionized, organic free water spiked with acetate, c_0 = 300 mg/l	22	Barone *et al.* (1992b)
Aniline	39	43	45	27	8.1	0.58	6.8	1.3	224	–	Diffusion test on relatively undisturbed soil	Distilled, deionized, organic free water spiked with aniline, c_0 = 300 mg/l	22	Barone *et al.* (1992b)
Benzene	–	54	42	–	–	0.60	3.6	13.8	2300	–	Batch test, 1:1.8 soil:solution, max. concn = 3.5 mg/l	Organic free water spiked with benzene	20	Myrand *et al.* (1987)
	–	100	0	–	–	2.78	–	2.3	83	–	Batch test, 1:50 soil:solution, max. concn = 900 mg/l	Organic free water spiked with benzene	25	Karickhoff *et al.* (1979)
	26		48	15	7.6	0.14–0.45		1.33	–	7.4, 1.538	Batch test, 1:3.6 soil:solution	Dilute aqueous solution, c_0 = 91–5597 mg/l	24 ± 2	Hrapovic (2001)
	26		48	15	7.6	0.14–0.45	2.5–5.0	0.3–0.7	–	–	Diffusion test, 1:3.6 soil:solution	Dilute aqueous solution, c_0 = 3–4.7 mg/l	24 ± 2	Hrapovic (2001)

Benzene	–	–	100	–	–	0.31	–	40.9, 37.1, 50	–	–	Batch test: 1:10, 1:5, 2:5 soil:solution ratios, max. concn = 110 mg/l	Distilled water spiked with benzene	20	Headley *et al.* (2001)
	–	–	100	370	–	0.01	–	37.6, 14.5, 3.7, 0.1	–	–	Batch test: 1:100, 1:50, 3:100, 1:20 soil:solution ratios			
	–	26	70	49	7.5	0.6	–	0.34–8.0	–	–	Batch test, 1:100 to 4:11 soil:solution	Deionized water spiked with 1–7 μg/l benzene	20	Donahue *et al.* (1999)
							3.2	0–1.0	–	–	Single reservoir diffusion test	Aqueous benzene solution, $c_0 = 215–265$ mg/l		
							3.2	1.0	–	–	Double reservoir diffusion test	Aqueous benzene solution, $c_0 = 241–261$ mg/l		
	–	34	66	51	–	0.85	–	3.2–7.2	–	–	Batch test, 1:20 soil: solution, max. concn = 300 mg/l	Distilled water spiked with benzene	–	Zhang *et al.* (1998)
	–	>43	7	–	7.8	0.02	1.8	0.23	1148	0.058, 0.64	Batch test, 1:2 soil: solution	Aqueous coal-tar solution, $c_0 = 44$ g/l	–	Broholm *et al.* (1999)
Butyric acid	26		48	15	7.6	0.14–0.45	–	0.08	–	–	Batch test, 1:3.6 soil: solution	Dilute aqueous solution, $c_0 = 26–992$ mg/l	24 ± 2	Hrapovic (2001)
							1.5	–	–	–	Diffusion test	Dilute aqueous solution		
							1.5	–	–	–	Diffusion test	Synthetic leachate with butyric acid (1 mg/l)		
Chloro-benzene	–	–	–	–	–	0.15	–	0.39	260	–	Batch test, 1:5 soil: solution, max. concn = 0.02 mg/l	$CaCO_3/CO_2$ water spiked with chlorobenzene		Schwarzenbach and Westall (1981)

Table 8.6(b) *Continued*

Species	Soil description						D_e $\times 10^{10}$ (m^2/s)	K_d (ml/g)	K_{oc} (g/ml)	K_f, ε (ml/g), (–)	*Test method*	*Source solution*	*Temperature* (°c)	*Reference*
	n (%)	*Silt content* (%)	*Clay content* (%)	*PI* (%)	*Soil pH*	f_{oc} (%)								
Chloroform	39	43	45	27	8.1	0.58	7.0	4.2	724–1034	–	Combination of diffusion test on relatively undisturbed soil and batch tests, 1:60 soil:solution	Distilled, deionized, organic free water spiked with chloroform, $c_0 = 300$ mg/l	22	Barone *et al.* (1992b)
1,4-Dichloro-benzene (DCB)	–	–	–	–	–	0.15	–	1.10	733	–	Batch test, 1:5 soil: solution, max. concn = 0.02 mg/l	$CaCO_3/CO_2$ water spiked with 1,4-DCB	22	Schwarzenbach and Westall (1981)
	–	8	2	–	–	2.55	–	–	–	40.5, 0.618	Batch test, 1:5 soil: solution, max. concn = 0.2 mg/l	Organic free water spiked with 1,4-DCB	21	Uchrin and Katz (1986)
	–	24	6	–	–	1.28	–	–	–	32, 0.178				
1,2-Dichloro-ethene (DCA)	26		48	15	7.6	0.14–0.45	–	0.45–0.77	–	16.6, 1.667	Batch test, 1:3.6 soil:solution	Dilute aqueous solution, $c_0 = 124–3323$ mg/l	24 ± 2	Hrapovic (2001)
							2.5–5.0	0.2–0.8	–	–	Diffusion test	Dilute aqueous solution, $c_0 = 5$ mg/l		
Dichloro-methane (DCM)	35	56	31	9.2	–	0.29	8.5	1.2	410	–	Diffusion test on recompacted soil	Distilled, deionized, organic free water spiked with DCM, $c_0 = 180$ mg/l	22	Rowe and Barone (1991)
	32	55	29	6.3	–	0.45	8.0	1.5	330	–				
	31	57	22	4.5	–	0.36	8.5	1.4	390	–				
	26		48	15	7.6	0.14–0.45	–	0.62–1.76	–	8.0, 1.471	Batch test, 1:3.6 soil:solution	Dilute aqueous solution, $c_0 = 74–3333$ mg/l	24 ± 2	Hrapovic (2001)
							2.5–5.0	0.06–0.7	–	–	Diffusion test	Dilute aqueous solution, $c_0 = 5$ mg/l		

2,4-Dichlorophenol (DCP)	–	8	2	–	–	2.55	–	–	–	26.4, 0.601	Batch test, 1:5 soil: solution, max. concn = 1 mg/l	Organic free water spiked with 2,4-DCP	21	Vchrin and Katz (1986)
	–	24	6	–	–	1.28	–	–	–	8.3, 0.247				
1,4-Dioxane	39	43	45	27	8.1	0.58	4.0	0.17	29	–	Diffusion test on relatively undisturbed soil	Distilled, deionized, organic free water spiked with 1,4-dioxane, $c_0 = 300$ mg/l	22	Barone *et al.* (1992b)
Ethylbenzene	26		48	15	7.6	0.14–0.45	–	1.32–2.36	–	17.2, 1.429	Batch test, 1:3.6 soil: solution	Dilute aqueous solution, $c_0 = 65-6301$ mg/l	24 ± 2	Hrapovic (2001)
							1.5–2.5	1.2–2.5	–	–	Diffusion test	Dilute aqueous solution, $c_0 = 1-2.5$ mg/l		
	–	>43	7	–	7.8	0.02	–	0.46	2300	0.13, 0.71	Batch test, 1:2 soil: solution	Aqueous coal-tar solution, $c_0 = 44$ g/l	–	Broholm *et al.* (1999)
2-Fluorotoluene	–	–	100	–	–	0.31	–	157, 114, 114	–	–	Batch test: 1:10, 1:5, 2:5 soil:solution ratios, max. concn = 47 mg/l	Distilled water spiked with 2-fluorotoluene	20	Headlay *et al.* (2001)
	–	–	100	370	–	0.01	–	51.2, 29.6, 33.6	–	–	Batch test: 1:50, 1:25, 1:20 soil:solution ratios			
Pentachlorophenol (PCP)	–	–	–	–	4.8	2.67	–	–	–	35, 0.79	Batch test, 1:10 soil: solution, max. concn = 0.004 mg/l	Organic free water spiked with PCP	20	Banerji *et al.* (1986)
Phenol	–	43	45	27	8.1	0.58	–	–	–	2.5, 0.628	Batch test, 1:1 soil: solution, max. concn = 50 mg/l	Distilled, deionized organic free water spiked with phenol	22	Mucklow (1990)

Table 8.6(b) *Continued*

	Soil description													
Species	*n* (%)	*Silt content* (%)	*Clay content* (%)	*PI* (%)	*Soil pH*	f_{oc} (%)	D_e $\times 10^{10}$ (m^2/s)	K_d (ml/g)	K_{oc} (g/ml)	K_f, ε (ml/g), (–)	*Test method*	*Source solution*	*Temperature* (°c)	*Reference*
Phenol	–	>43	7	–	7.8	0.02	–	0.08	400	0.0065, 0.27	Batch test, 1:2 soil:solution	Aqueous coal-tar solution, $c_0 = 44$ g/l	–	Broholm *et al.* (1999)
Propionic acid	26		48	15	7.6	0.14–0.45	–	0.11	–	–	Batch test, 1:3.6 soil: solution	Dilute aqueous solution, $c_0 = 208–5627$ mg/l	24 ± 2	Hrapovic (2001)
							2.5	–	–	–	Diffusion test	Dilute aqueous solution		
							1.5–2.0	–	–	–	Column diffusion test, 1:3.6 soil:solution	Synthetic leachate with propionic acid (5 mg/l)		
Tetrachloro-ethylene (PCE)	–	–	–	–	–	0.15	–	0.56	373	–	Batch test, 1:5 soil: solution, max. concn = 0.1 mg/l	$CaCO_3/CO_2$ water spiked with tetrachloroethylene	22	Schwarzenbach and Westall (1981)
	–	8	0.4	–	3.9	0.54	–	2.20–2.56	410–478	3.36, 0.92	Batch test, 1:2 to 1:4 soil:solution, max. concn = 100 mg/l	Aqueous solution	–	Kimani Njoroge *et al.* (1998)
	–	15	13	–	4.9	0.10–0.19	–	0.14–0.42	143–220	0.10–0.32, 1.01–1.07	Batch test, 1:3 to 1:4, soil:solution, max. concn = 100 mg/l			
	–	17	41	–	4.9	0.127	–	0.15	118	0.12, 1.09	Batch test, 1:3 soil: solution, max. concn = 100 mg/l			

Toluene	26		48	15	7.6	0.14–0.45	–	0.99–1.42	–	12.3, 1.449	Batch test, 1:3.6 soil: solution	Dilute aqueous solution, $c_0 = 95–4377$ mg/l	24 ± 2	Hrapovic (2001)
							2.0–3.0	0.9–1.5	–	–	Diffusion test	Dilute aqueous solution, $c_0 = 1.5–3.2$ mg/l		
	–	54	42	–	–	1.95	–	125	6410	–	Batch test, 1:5 soil: solution, max. concn = 0.02 mg/l	$CaCO_3/CO_2$ water spiked with toluene	22	Schwar-zenbach and Westall (1981)
	39	43	45	27	8.1	0.588	5.8	11.3	1950–4480	–	Combination of diffusion test on relatively undisturbed soil and batch tests, 1:90 soil:solution	Distilled, deionized, organic free water spiked with toluene, $c_0 = 200$ mg/l	22	Barone *et al.* (1992b)
	–	54	42	–	–	0.60	3.0	53.3	8883	–	Batch test, 1:1.8 soil: solution, max. concn = 3 mg/l	Organic free water spiked with toluene	20	Myrand *et al.* (1987)
	37		1.9	–	–	0.45	–	1.18 0.96 0.78	2.622 2.133 1.773	– – –	Advection dispersion column test	$c_0 = 5.2$ mg/l $c_0 = 5.2–15.5$ mg/l $c_0 = 10.3$ mg/l	22	Sleep and McClure (2001)
	–	–	100	–	–	0.31	–	154, 139, 129	–	–	Batch test: 1:10, 3:10, 2:5 soil: solution ratios, max. concn = 30 mg/l	Distilled water spiked with toluene	20	Headley *et al.* (2001)
	–	–	100	370	–	0.01	–	60.3, 34.9, 16.5	–	–	Batch test: 1:100, 1:25, 1:20 soil: solution ratios			
	–	>43	7	–	7.8	0.02	–	0.32	1585	0.072, 0.64	Batch test, 1:2 soil:solution	Aqueous coal-tar solution, $c_0 = 44$ g/l	–	Broholm *et al.* (1999)

Table 8.6(b) *Continued*

Species	Soil description						$D_e \times 10^{10}$ (m^2/s)	K_d (ml/g)	K_{oc} (g/ml)	K_f, ε (ml/g), (–)	Test method	Source solution	Temp-erature (°c)	Reference
	n (%)	Silt content (%)	Clay content (%)	PI (%)	Soil pH	f_{oc} (%)								
1,2,4-Trichloro-benzene (TCB)	–	8	0.4	–	3.9	0.54	–	12.9–15.3	2410–2850	18.2, 0.95	Batch test, 1:2 to 1:5.9 soil: solution, max. concn = 20 mg/l	Aqueous solution	–	Kimani Njoroge *et al.* (1998)
	–	15	13	–	4.9	0.10–0.19	–	1.33–4.26	1360–2230	1.34–5.33, 0.93–1.01	Batch test, 1:1.8–2.6 to 1:4, soil: solution, max. concn = 20 mg/l			
	–	17	41	–	4.9	0.127	–	1.59	1250	2.02, 0.9	Batch test, 1:3 soil: solution, max. concn = 20 mg/l			
1,2,4-Trichloro-benzene (TCB)	–	23	52	–	4.9	0.06–0.10					Batch test, 1:3 soil: solution, max. concn = 20 mg/l	Aqueous solution	–	Kimani Njoroge *et al.* (1998)
Trichloro-ethylene (TCE)	26		48	15	7.6	0.14–0.45	–	0.99–1.42	–	12.3, 1.449	Batch test, 1:3.6 soil: solution	Dilute aqueous solution, c_0 = 86-4605 mg/l	24 ± 2	Hrapovic (2001)
							2.0–3.0	1.26–1.4	–	–	Diffusion test	Dilute aqueous solution, c_0 = 1.5-3.5 mg/l		
	–	54	42	–	–	1.95	–	125	6410	–	Batch test, 1:1.8 soil: solution, max. concn = 0.1 mg/l	Organic free water spiked with TCE	20	McKay and Trudell (1989)
	–	54	42	–	–	0.68	–	58.6	8617	–				

Trichloro-ethylene(TCE)	–	54	42	–	–	0.60	3.5	15.5	2583	–	Batch test, 1:1.8 soil: solution, max. concn = 2 mg/l	Organic free water spiked with TCE	20	Myrand *et al.* (1987)
	–		100	–	–	1.0	–	8.3	830	–	Batch test, 1:50 soil: solution, max. concn = 10 mg/l	Organic free water spiked with TCE	27	Acar and Haider (1990)
m and *p*-Xylene	26		48	15	7.6	0.14–0.45	–	2.01–2.75	–	15.5, 1.333	Batch test, 1:3.6 soil: solution	Dilute aqueous solution, $c_0 = 63-3361$ mg/l	24 ± 2	Hrapovic (2001)
							1.5–2.5	1.9–3.0	–	–	Diffusion test	Dilute aqueous solution, $c_0 = 1.5-2.5$ mg/l		
	–	>43	7	–	7.8	0.02	–	0.4	2000	0.091, 0.66	Batch test, 1:2 soil: solution	Aqueous coal-tar solution, $c_0 = 44$ g/l	–	Broholm *et al.* (1999)
o-Xylene	26		48	15	7.6	0.14–0.45	–	1.67–2.24	–	12.0, 1.33	Batch test, 1:3.6 soil:solution	Dilute aqueous solution, $c_0 = 65-3447$ mg/l	24 ± 2	Hrapovic (2001)
							1.5–2.5	1.5–2.4	–	–	Diffusion test	Dilute aqueous solution, $c_0 = 1.0-2.7$ mg/l		
o-Xylene	–	>43	7	–	7.8	0.02	–	0.38	1900	0.093, 0.67	Batch test, 1:2 soil:solution	Aqueous coal-tar solution, $c_0 = 44$ g/l	–	Broholm *et al.* (1999)

equilibrate in contact with a small quantity of squeezed pore fluid from the same soil prior to performing the diffusion tests. In the case of rock, a simulated pore fluid may be used.

Liners are often constructed over an unsaturated zone that may include unsaturated silts, sands or gravels or an engineered secondary leachate collection system (e.g., see Figures 1.4 and 1.5). The effective solute diffusion in unsaturated soils may be substantially smaller than for saturated soil (Klute and Letey, 1958; Porter *et al.*, 1960). Rowe and Badv (1996a,b) found that the diffusion coefficient, D_e, for unsaturated silt, sand and fine gravel was given by

$$D_{e\theta} = \frac{(\theta - \theta_{min})}{(n - \theta_{min})} D_e \tag{8.4a}$$

where $D_{e\theta}$ is the effective diffusion coefficient in the unsaturated soil at a volumetric water content θ, n and D_e are the volumetric water content (total porosity) and effective diffusion coefficient in the same soil under saturated conditions and θ_{min} is the water content at which there is no interconnected water through which solute diffusion can occur. For their tests, Rowe and Badv (1996a,b) found that $\theta_{min} \approx 0$ and equation 8.4a reduced to

$$D_{e\theta} = \frac{\theta}{n} D_e \tag{8.4b}$$

(where $D_e \approx 0.028–0.033\,m^2/a$ in their test on silt, sand and gravel with $n = 0.34–0.38$).

Rowe and Badv (1996a,b; Badv and Rowe, 1996) demonstrated that advective–diffusive transport through a CCL underlain by unsaturated silt, sand or gravel could be well predicted using finite layer theory (Chapter 7) and program POLLUTE (Rowe and Booker, 1994, 2004). Mechanical dispersion was very small for clay, silts, sands and fine gravels at volumetric water contents greater than 0.04 and a Darcy flux less than $6 \times 10^-10\,m/s$ (i.e., 0.02 m/a). However, advective transport and mechanical dispersion were dominant over diffusion for an unsaturated coarse gravel drainage layer (20–45-mm particle size) with a volumetric water content of 0.027–0.035 at Darcy fluxes of between 6×10^{-10} and $8 \times 10^{-8}\,m/s$ (0.02–2.5 m/a). Under these conditions, the coefficient of hydrodynamic dispersion was given by

$$D = D_{e\theta} + \alpha V \quad (m^2/s) \tag{8.5a}$$

where $D_{e\theta}$ (m^2/s) is given by equations 8.4a and 8.4b, α is the dispersivity given by

$$\alpha = 35{,}200v + 0.013 \quad (m) \tag{8.5b}$$

and the average linearized groundwater velocity, v, is given by

$$v = \frac{v_a}{\theta} \quad (m/s) \tag{8.5c}$$

where v_a (m/s) is the Darcy flux through the liner and θ the volumetric water content of the coarse gravel. Equation 8.5b must be used with the units defined above.

The unsaturated diffusion coefficient as defined above will be important with respect to the transport of non-volatile contaminants. This is especially true when there is an air pocket (see Rowe *et al.*, 2000c). Unsaturated soil may also control transport of volatile contaminants when the degree of saturation is sufficiently high that there is no significant interconnected gas-filled void. However, for low degrees of saturation, the migration of volatile organic compounds will be dominated by diffusion in a gaseous phase as discussed in the following section.

8.11.2 Gaseous diffusion

In general terms, the mass flux of a gas (e.g., volatile organic compounds, methane, carbon dioxide, oxygen) through the liquid and gas phases of an unsaturated soil is given by (McCarthy and Johnson, 1995):

$$f = -\left[\theta \tau D_0 \frac{dc}{dx} + \theta_a \tau_a D_a \frac{dc_a}{dx}\right] \quad (8.6)$$

where θ is the volumetric water content, τ the water phase tortuosity, D_0 the diffusion coefficient in water (and where $\tau D_0 = D_{e\theta}$ as defined in the previous section), c the concentration in the dissolved phase, θ_a the effective air porosity, τ_a the gas-phase tortuosity, D_a the diffusion coefficient of the gas in air, c_a the concentration in the gaseous phase, and the relationship between the gaseous phase concentration, c_a, and the dissolved phase concentration, c, is given by Henry's law:

$$c_a = K'_H \, c \quad (8.7)$$

where K'_H is the dimensionless Henry's coefficient. Millington (1959) developed the following approximate relationship for the gas tortuosity:

$$\tau_a = \frac{\theta_a^{(7/3)}}{n^2} \quad (8.8)$$

where n is the total porosity of the soil ($n = \theta + \theta_a$). Karimi *et al.* (1987) indicated Millington's tortuosity adequately predicted benzene flux through soils at various water contents. The diffusion coefficient in air, D_a, and water, D_0, typically differ by about four orders of magnitude (see Table 8.7). As was the case for diffusion in the dissolved phase, the diffusion coefficient in air increases with increasing temperature but to a much lesser extent.

McCarthy and Johnson (1995) combined equations 8.6 and 8.7 and established an expression for the bulk diffusion coefficient, $D_{p\theta}$, viz.:

$$f = -D_{p\theta}\left[\frac{dc}{dx}\right] \quad (8.9a)$$

where

Table 8.7 Some published diffusion coefficients in air and water at 25 °C (adapted from Rowe, 2001)

	In air $D_a \times 10^5$ (m^2/s)	*In water* $D_0 \times 10^9$ (m^2/s)
Acetic acid	1.33	0.88[a]
Ammonia	2.8	1.76[a]
Aniline	0.72	0.98
Benzene	0.88	1.15
Butyric acid	0.76	0.92
Carbon dioxide	1.64	1.77[a]
p-Cresol	0.77	0.87
1,2-Dichloroethane	0.90	1.07
1,4-Dichlorobenzene	0.68	0.81
Dichloromethane	1.04	1.26
Ethylbenzene	0.75	0.91
Hydrogen	4.1	5.85
Methyl ethyl ketone	0.85	0.99
Propionic acid	0.95	1.01
Oxygen	2.06	2.35
Toluene	0.85	0.96
m-Xylene	1.23	0.79

[a]At 20 °C.

$$D_{p\theta} = \frac{\theta^{(10/3)} D_0 + K'_H \, \theta_a^{(10/3)} D_a}{K'_H \, n^2} \quad (8.9b)$$

Millward (2000) performed a series of diffusion tests to examine the migration of dichloromethane (DCM) through two different composite liner systems (a geomembrane over (a) compacted clay and (b) a GCL) overlaying unsaturated fine gravel and found that the diffusion of DCM through the unsaturated gravel could be reasonably predicted using equations 8.9a and 8.9b. It was also reported that the movement through the gravel was relatively fast and that the diffusive flux was largely controlled by the composite liners.

When unsaturated soils have continuous gas-filled pores, movement of gases will be predominantly through these pores and equation 8.6 can be approximated by

$$f = -\theta_a D_\theta \frac{dc_a}{dx} \quad (8.10)$$

where $D_\theta = \tau_a D_a$ is the diffusion coefficient of the gas of interest through a porous medium and the other terms are as previously defined.

The gaseous diffusion coefficient through the unsaturated soil, D_θ, is often related to the diffusion of gas through air, D_a. van Bavel (1952) found that for dry soil and sand mixtures with porosities between 0.1 and 0.6, $\tau_a = D_\theta/D_a = 0.58$. Studies examining the variation in the gas tortuosity $\tau_a = D_\theta/D_a$ with the degree of saturation for a range of materials (sands, silts, sand–bentonite, loams, till and clay) have shown that at low degrees of saturation $\tau_a = D_\theta/D_a$ approaches van Bavel's ratio of 0.58. As the degree of saturation increases, the value of $\tau_a = D_\theta/D_a$ decreases significantly and the scatter increases. The scatter at high degrees of saturation is attributed to variability in the effective porosity due to different grain size distributions and hence different amounts of connected gaseous pore space at the same degree of saturation. This arises because diffusion will be much faster for a sample with many continuous gaseous pores than for a sample with the same number of gaseous pores but where the paths between pockets of gaseous are blocked by water-filled pores.

MacKay (1997) reported that an empirical "best-fit" relationship between $\tau_a = D_\theta/D_a$ and the degree of saturation, S_r, was given by

$$\tau_a = \frac{D_\theta}{D_a} = \exp[-1.03\ \exp(0.017S_r)^{1.64}] \quad (8.11a)$$

with an upper bound given by

$$\tau_a = \frac{D_\theta}{D_a} = \exp[-1.03\ \exp(0.014S_r)^{1.64}] \quad (8.11b)$$

and a lower bound given by

$$\tau_a = \frac{D_\theta}{D_a} = \exp[-1.03\ \exp(0.019S_r)^{1.64}] \quad (8.11c)$$

where S_r is expressed in per cent. Equations 8.11a–8.11c can be used to obtain an initial estimate of the tortuosity and hence effective gas diffusion coefficient in unsaturated soils with continuous gas-filled voids. At high degrees of saturation, equations 8.9a and 8.9b may be used to estimate diffusion in both phases.

8.12 Diffusion and partition coefficients for geomembranes

Knowledge of diffusion and partition coefficients for a geomembrane is required to assess the potential contaminant migration through the liner system. As discussed in Chapter 6, the diffusion coefficient depends on the properties of the permeant and the polymer, and temperature. This section describes the laboratory methods used to estimate diffusion coefficient D_g and partition coefficient S_{gf}.

Based on the flux of permeant passing through an intact geomembrane, given by equation 6.18, the mass flux across a geomembrane of thickness t_{GM} is given by

$$f = S_{gf}\, D_g \frac{\Delta c_f}{t_{GM}} \quad (8.12)$$

or

$$f = P_g \frac{\Delta c_f}{t_{GM}} \quad (8.13)$$

where Δc_f is the difference in concentration in the fluid on either side of the geomembrane. It is possible to experimentally infer the value of P_g by performing a steady-state experiment (see Figure 6.16) in which the concentration is controlled at c_{f1}, c_{f2} in the fluid on either side of the geomembrane and the flux, f, is measured. Thus

$$P_g = \frac{f t_{GM}}{\Delta c_f} \tag{8.14}$$

It follows from this that while a steady-state test will allow the measurement of P_g, it does not allow the evaluation of either S_{gf} or D_g without additional tests. However, transient contaminant transport is controlled by the diffusion coefficient, D_g, and not by the permeability, P_g. This becomes evident when one substitutes equation 6.16 into equation 6.19 to give

$$S_{gf} \frac{\partial c_f}{\partial t} = S_{gf} D_g \frac{\partial^2 c_f}{\partial z^2} \tag{8.15}$$

and, on dividing by S_{gf}, this reduces to

$$\frac{\partial c_f}{\partial t} = D_g \frac{\partial^2 c_f}{\partial z^2} \tag{8.16}$$

which depends on D_g but not S_{gf} or P_g.

Thus it is important to distinguish between the partition (or Henry) coefficient S_{gf} and the diffusion coefficient D_g. The partition coefficient, S_{gf}, provides a measure of the discontinuity in concentration between the fluid and solid phases (as shown in Figure 6.16), whereas the diffusion coefficient quantifies the migration of permeant molecules through an intact polymer.

8.12.1 Laboratory tests to obtain diffusion and partition coefficients for geomembranes

Table 8.8 summarizes a number of available techniques for obtaining D_g and S_{gf}. Two classes of laboratory methods are available, namely (a) immersion tests and (b) permeation tests.

(a) Immersion tests

Immersion or sorption methods involve a geomembrane sample of known mass immersed in a container filled with permeant of known concentration. Normally both faces of the geomembrane are in contact with the permeant, allowing equilibrium to be reached relatively quickly. The increase in mass of the geomembrane can be recorded during the test. The value of S_{gf} can be calculated from consideration of mass balance. However, this weight gain technique is not well suited for testing volatile organic compounds given the mass loss during weighing. Sangam and Rowe (2001) developed a method that only requires measurement of the final mass of the geomembrane, thus preventing loss of contaminant due to volatilization, but records the decrease in solution concentration with time through a sampling port as shown in Figure 8.25a. Concentrations are monitored until equilibrium is reached. A mass balance at equilibrium can then be expressed as

$$M_{s0} = M_{sF} + M_{gF} + M_R \tag{8.17}$$

where M_{s0} is the initial mass of contaminant in the system, M_{sF} the final mass in the solution, M_{gF} the mass of contaminant sorbed by the geomembrane and M_R the mass of contaminant lost from sampling. Expressing equation 8.17 in terms of concentrations and volumes yields

$$c_{f0} V_{f0} = c_{fF} V_{fF} + \frac{M_g}{\rho_g} c_{gF} + \sum V_i c_i \tag{8.18}$$

where c_{f0} is the initial solution concentration [ML^{-3}], c_{fF} the final equilibrium solution concentration [ML^{-3}], V_{f0} and V_{fF} are the initial and final solution volumes [L^3], M_g the mass of the geomembrane [M], ρ_g the density of the geomembrane [ML^{-3}], c_{gF} the final equilibrium concentration in the geomembrane [ML^{-3}] and $\sum V_i c_i$ the mass removed by sampling events [M] (V_i and c_i being the volume and concentration removed at each sampling event).

Table 8.8 Some techniques for measuring the partitioning and diffusion coefficients for geomembranes (modified from Sangam and Rowe, 2001)

Technique	*Method*	*Comments*
Weight gain	Monitor increase in mass of geomembrane immersed in fluid of interest from initial value m_0 until mass of geomembrane becomes constant at m_∞ (plot $(m_t - m_0)/(m_\infty - m_0)$ versus $\sqrt{t}$) $S_{gf} = \frac{\rho_{GM}}{c_{fF}}\left(\frac{m_\infty}{m_0} - 1\right)$ $D_g = \frac{0.049 t_{GM}^2}{t_{1/2}}$	Faster than alternative tests but each chemical must be examined separately. Suitable for pure solvents. Weight gain must be corrected for sorbed water. Prone to error due to mass loss when weighing (especially for VOCs). c_{fF} = final equilibrium fluid concentration $t_{1/2}$ time to get; $(m_t - m_0)/(m_\infty - m_0) = 0.5$
Time lag	Monitor mass movement through geomembrane with time for test where c_{f1} = const, $c_{f2} \simeq 0$, plot cumulative mass, F, through geomembrane against time and extrapolate steady-state value to $F = 0$ to obtain the time lag τ. Then $D_g = \frac{t_{GM}^2}{6\tau}$	
Pouch method	Geomembrane pouch filled with test permeant and immersed in liquid of known composition. Weight change of pouch is monitored as diffusion occurs	Must avoid leaks in pouch; interpretation assumes diffusion through geomembrane is slow relative to that in fluid
Sorption test	Monitor decrease in solution concentration with time	Suitable for testing VOCs
Diffusion test	Diffusion from solution on one side of geomembrane to solution on other side. Change in source and receptor solution monitored with time. S_{gf} calculated from equation 8.19 at equilibrium, D_g inferred from variation in source and receptor concentration with time	May be used in conjunction with weight gain method to allow evaluation of parameters prior to equilibrium in the diffusion test

The partition coefficient can be obtained by substitution of equation 8.18 into equation 6.16 giving

$$S_{gf} = \frac{[c_{fo} V_{f0} - c_{fF} V_{fF} - \sum V_i c_i]\rho_g}{M_g c_{fF}} \qquad (8.19)$$

Results obtained from immersion tests conducted by Sangam and Rowe (2001) for dilute organic solutions and 2.0-mm thick HDPE geomembranes are shown in Figure 8.26. Selected properties of the organic contaminants are provided in Table 8.9. These tests were conducted at temperature of $22 \pm 2\,^\circ$C. Even though glass cells were used to minimize the chemical interactions with the cell walls (Rowe *et al.*, 1995b), decreases in source concentration attributed to sorption on to the glass was measured for some organic species – most noticeably for TCE and aromatic hydrocarbons. It is therefore important to consider sorption on to the cell when calculating transport parameters, which may be achieved in part by monitoring and modelling the concentration in control cells not containing geomembrane samples (Sangam and Rowe, 2001).

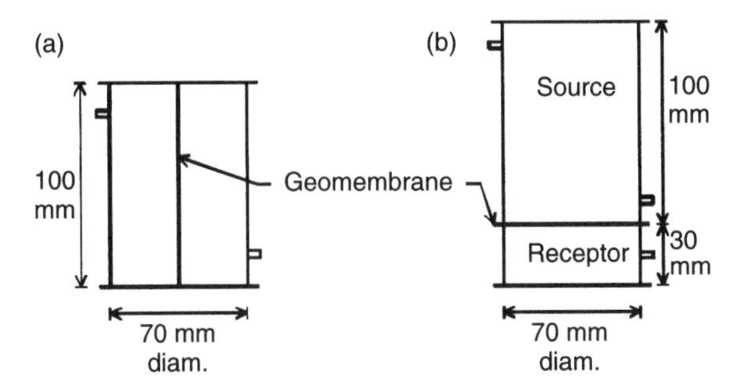

Figure 8.25 (a) Sorption and (b) diffusion cells (modified from Sangam and Rowe, 2001).

For the three chlorinated hydrocarbons tested (Figure 8.26a), TCE showed the fastest decrease in concentration, reducing to 30% of the original concentration after 20 days. More polar dichloromethane and 1,2-dichloroethane have less affinity for the geomembrane than TCE and consequently concentration reduced to 95 and 75% of the initial concentration, respectively. Concentrations of the aromatic hydrocarbons (Figure 8.26b) ethylbenzene, *m*- and *p*-xylene and *o*-xylene, showed greater decreases than the chlorinated hydrocarbons as noted in Chapter 6. Of the aromatic hydrocarbons, benzene ($\log K_{ow} = 2.13$) showed less concentration decreases than *m*-, *n*- or *o*-xylene (with $\log K_{ow}$ equal to 3.20, 3.13 and 3.18, respectively). Upper and lower bound partitioning coefficients estimated using equation 8.19 are provided in Figure 8.26.

(b) Permeation tests

Permeation or diffusion techniques are better suited for measuring contaminant transport through an intact geomembrane for landfill applications than the other methods listed in Table 8.8. In these diffusion tests, the geomembrane separates source and receptor solutions, each of known concentration, such that one side of the geomembrane is in contact with the permeant, thereby more closely simulating the transport conditions expected in a landfill. A schematic illustration of the cell used by Sangam and Rowe (2001) is shown in Figure 8.25b, and is similar to those used by August and Taztky (1984), Haxo and Lahey (1988) and Rowe *et al.* (1995b, 1996). During testing, the concentrations in both the source and receptor solutions are monitored. For these closed systems, the mass of contaminant in the source c_{ss} and the receptor c_{rs} solutions at any time t can be written as a slightly modified version of equation 8.1, as

$$c_{ss}(t) = c_{s0} - \frac{1}{H_{ss}} \int_0^t f_{ss}(\tau)\, d\tau \tag{8.20}$$

$$c_{rs}(t) = c_{r0} + \frac{1}{H_{rs}} \int_0^t f_{rs}(\tau)\, d\tau \tag{8.21}$$

Table 8.9 Selected properties of organic contaminants tested by Sangam and Rowe (2001)

Organic contaminant	*Molar weight* (g/mole)	*Density* (g/cm^3)	*Molar volume* (cm^3)	*Aqueous solubility at 20 °C* (mg/l)	*log* K_{ow}	*Boiling temperature* (°C)	*Dipole moment (debye)*
Dichloromethane	84.93	1.3266	64.02	20,000	1.25	40.2	1.6
1,2-Dichloroethane	98.96	1.253	78.98	8,690	1.45	83.5	1.44
Trichloroethylene	131.39	1.4642	89.74	1,100	2.53	87.2	0.77
Benzene	78.11	0.8765	89.11	1,780	2.13	80.1	0
Toluene	92.14	0.8669	106.28	515	2.79	110.6	0.3
Ethylbenzene	106.17	0.867	122.46	152	3.13	136.2	0.36
m-Xylene	106.17	0.8642	122.85	162	3.2	138	0.3
o-Xylene	106.17	0.8802	120.62	152	3.13	144	0.63
p-Xylene	106.17	0.8669	122.47	156	3.18	138.3	0

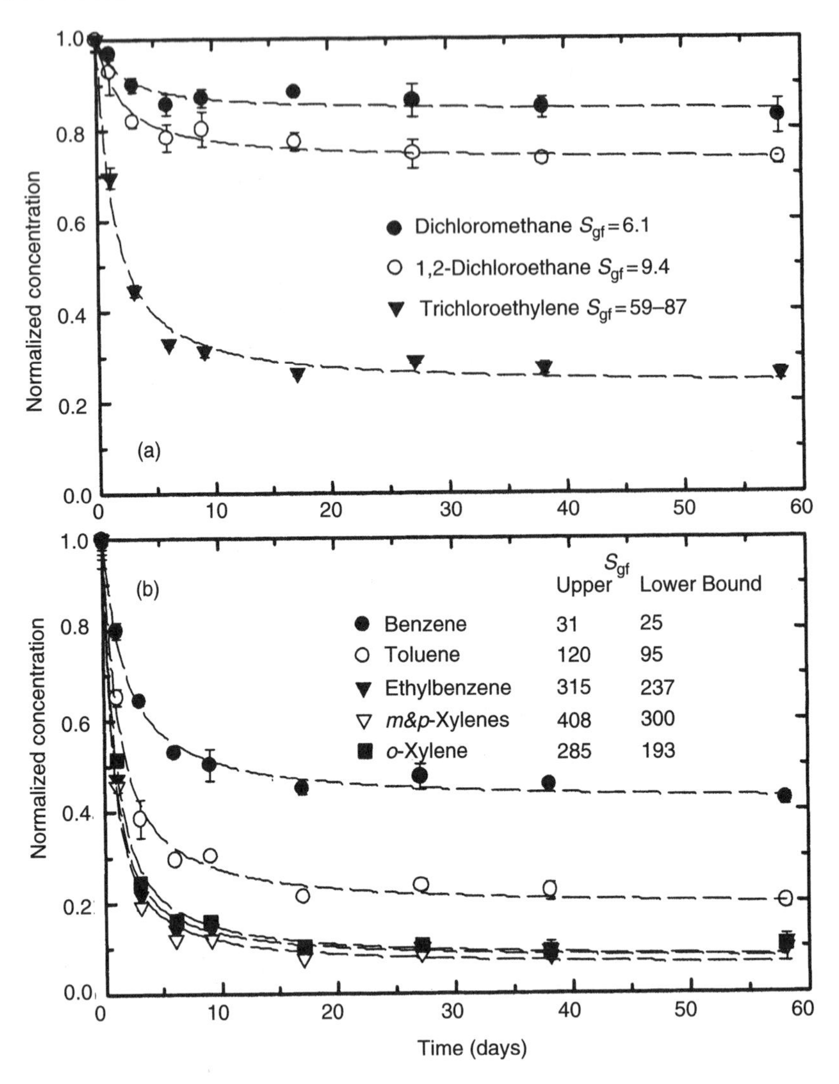

Figure 8.26 Results from sorption tests conducted with certain (a) chlorinated and (b) aromatic hydrocarbons concentration for a 2.0-mm thick HDPE geomembrane (modified from Sangam and Rowe, 2001).

where c_{s0} and c_{r0} are the initial concentrations in the source and receptor solutions (normally c_{r0} is set to be zero), H_{ss} and H_{rs} are the reference height of source and receptor solutions, f_{ss} is the mass flux of contaminant into the geomembrane and f_{rs} the mass flux of contaminant from the geomembrane into the receptor. The diffusion and partitioning coefficients can then be calculated fitting the variation of the source and effluent concentrations with time using computer software that models the boundary conditions, phase change and transport through the geomembrane (e.g., Rowe and Booker, 2004).

Results from diffusion tests also conducted by Sangam and Rowe (2001) are reported in Figure 8.27 for selected chlorinated hydrocarbons, and Figure 8.28 for certain aromatic hydrocarbons. For each test, the measured decrease in source

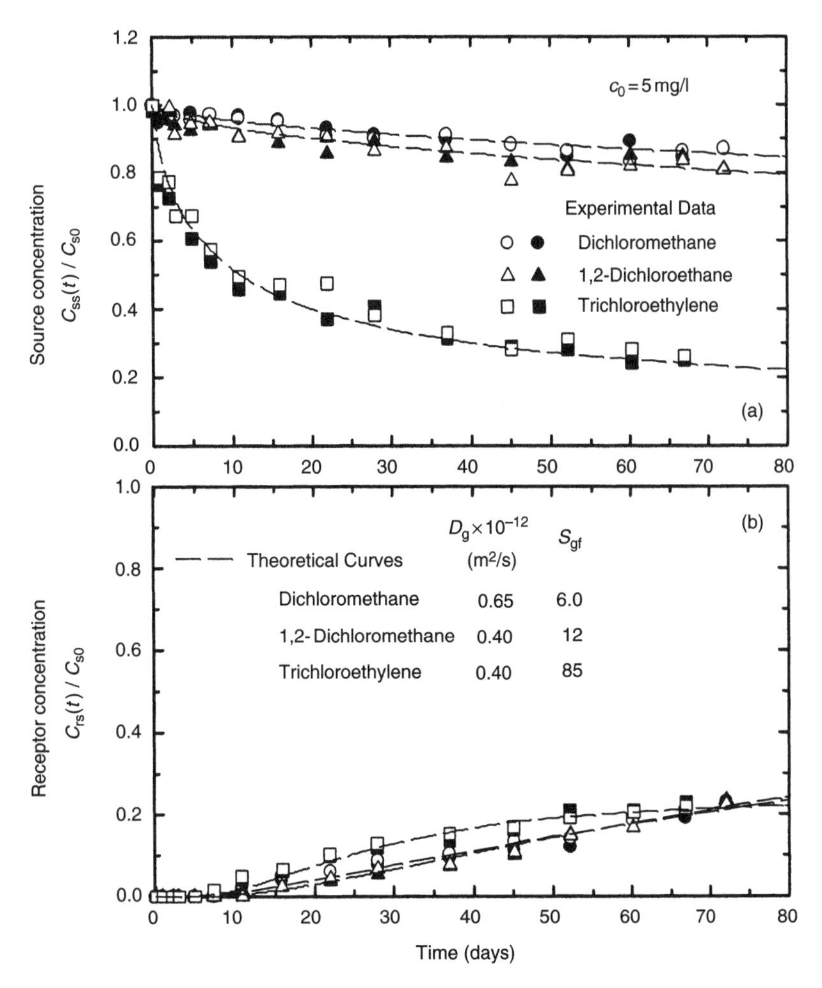

Figure 8.27 (a) Source and (b) receptor concentrations during diffusion tests for certain chlorinated hydrocarbons migrating through a 2.0-mm thick HDPE geomembrane (modified from Sangam and Rowe, 2001).

concentration and increase in receptor concentration are shown with each experimental data point representing the average of three measurements. Theoretical predictions using program POLLUTE (Rowe and Booker, 2004) are also plotted in Figures 8.27 and 8.28 and shows excellent agreement with the measured data. Inferred diffusion and partition coefficients for the organic solutions (Table 8.9) and particular conditions examined (2-mm thick HDPE geomembrane, temperature of $22 \pm 2\,^{\circ}$C) are given in the figures. The partitioning coefficient controlled the decrease in source concentration, and the diffusion coefficient governed the increase in receptor concentration.

A diffusion test examining the migration of sodium chloride from a source solution, through a geomembrane and into a receptor has been in progress for over 10 years. Figure 8.29 shows the concentrations monitored in the receptor over this period. After 10 years there is still minimal migration of chloride through the geomembrane, however, one can place bounds on the

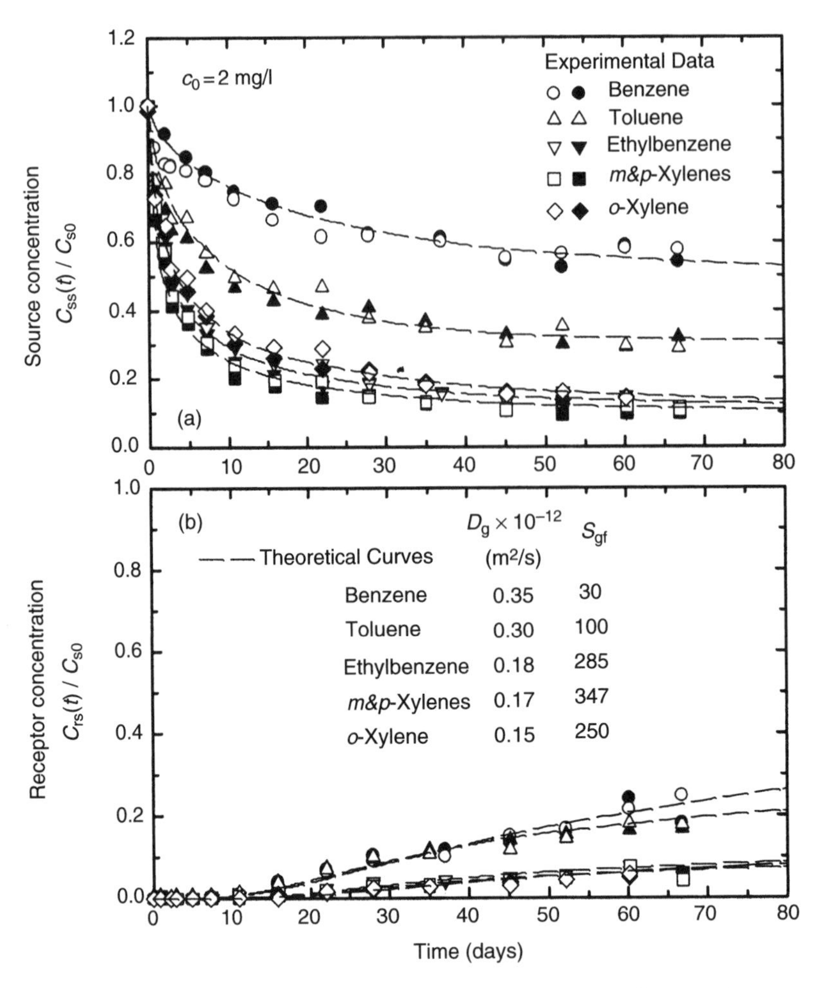

Figure 8.28 (a) Source and (b) receptor concentrations during diffusion tests for certain aromatic hydrocarbons migrating through a 2.0-mm thick HDPE geomembrane (modified from Sangam and Rowe, 2001).

diffusion coefficient based on this data. An upper bound estimate of the diffusion parameters for chloride is $S_{gf} = 8 \times 10^{-4}$ and $D_g = 4 \times 10^{-14}\,m^2/s$, based on 10.5 years of laboratory data at $23 \pm 2°C$ (Figure 8.29). This value is likely too high and a more appropriate estimate based on available data is $S_{gf} = 8 \times 10^{-4}$ and $D_g = 1 \times 10^{-14}\,m^2/s$. Even this value may subsequently decrease as more data become available. As discussed in Section 13.5.1, these data imply that a geomembrane is an excellent barrier to ionic diffusion.

8.12.2 Effect of fluorination of geomembranes

Surface fluorination involves the oxidation of HDPE by a mechanism wherein fluorine atoms exchange for hydrogen atoms on the surface of the geomembrane from a covalent C—F bond that creates a barrier layer to certain hydrocarbons. Surface-fluorinated HDPE has been routinely used in the polyolefin container industry to improve the shelf life storage of hydrocarbon fluids. The

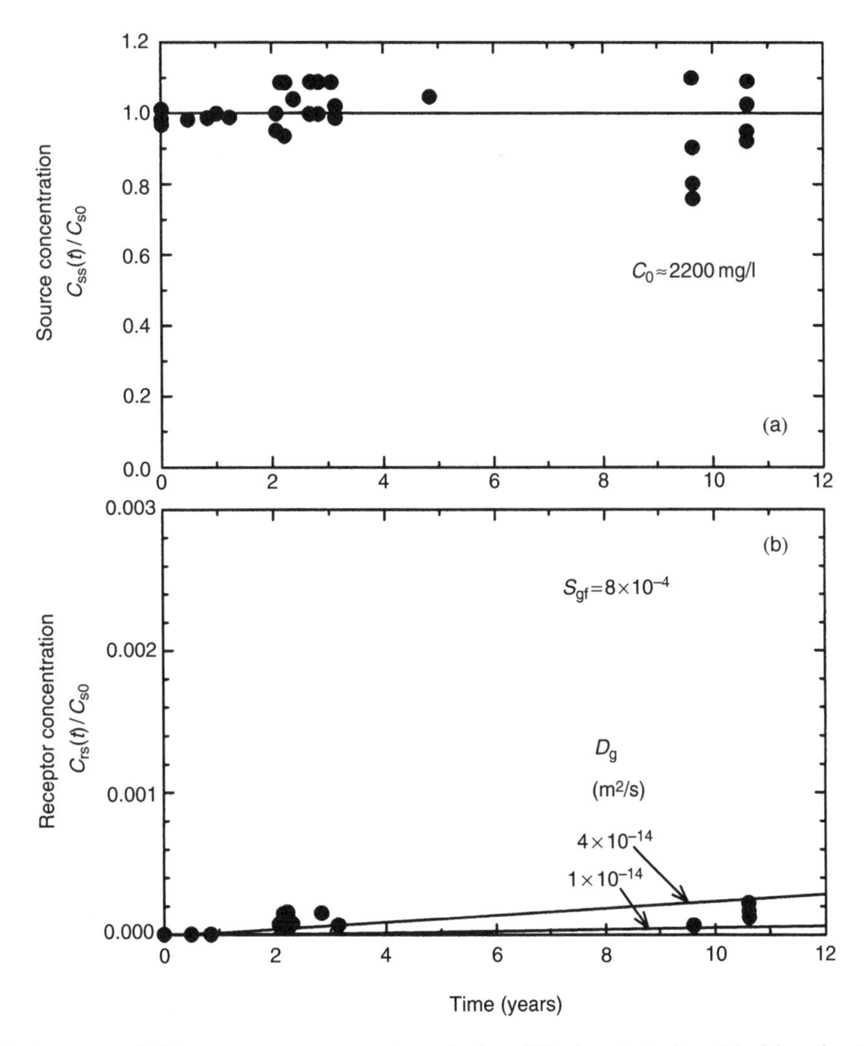

Figure 8.29 (a) Source and (b) receptor concentrations during diffusion tests for chloride migrating through a 2.0-mm thick HDPE geomembrane.

permeation resistance of HDPE to hydrocarbons may be increased by up to two orders of magnitude for thin HDPE membranes (Barsamyan and Sokolov, 1999) while the physical performance properties (e.g., tensile properties) remain practically unchanged since for surface fluorination, fluorine/hydrogen exchange occurs only in a thin layer along the exposed surfaces.

The fluorination of the surface increases in the permeation resistance to hydrocarbons due to (Anand *et al.*, 1994): (1) an increase in the surface energy, increased surface water-wettability and the consequent reduced solubility (or partitioning) of the non-polar organic liquids in the material; (2) the reduced free volume of the polymer through which the chemical can permeate and (3) cross-linking which reduces the segmental mobility of the polymer chains and hence reduces the diffusivity of larger penetrant molecules. The reduction in the permeation of hydrocarbons depends on the thickness of the fluorinated surface layer, which in turn depends on the nature of the treatment as illustrated by Carstens *et al.* (1999) who demonstrated 80-fold improvement associated with quadruple fluorination treatment as compared to twofold improvement provided by a single treatment.

Anand *et al.* (1994) describe two fundamental methods of surface fluorination, namely post-treatment (post-forming) and *in-situ* treatment. Post-treatment involves exposure of the polymer to fluorine after it has been formed. This approach gives operational flexibility and better economics for smaller productions. *In-situ* treatment makes use of existing polymer-processing equipment to apply fluorine to the polymer. In both cases, the effectiveness of the treatment depends on the fluorine partial pressure and dilution, reaction time, temperature, and the presence of co-reactants (Anand *et al.*, 1994).

Sangam and Rowe (2003) reported an evaluation of the effectiveness of fluorination in improving the diffusive barrier properties of a 1.5-mm HDPE with a crystallinity of 61% and a standard oxidative induction time (OIT) of 120 and 196 min, respectively, for the untreated and treated geomembranes. The treated geomembrane was fluorinated (post-treatment) on both surfaces. The fluorinated layer was found to have an average thickness of about 4 μm. The investigation involved diffusion and sorption tests on both the original untreated geomembrane and the fluorinated geomembrane. The diffusion, partitioning and permeation coefficients for chlorinated and aromatic hydrocarbons were compared for the two geomembranes.

Sorption tests indicated that the partition parameter (S_{gf}) was not significantly affected and remained unchanged except for dichloromethane for which S_{gf} appeared to have increased. The diffusion tests showed that, for the compounds examined, the diffusion coefficient decreased and, as a consequence, the permeation coefficients decreased by a factor varying from 1.7 to 2.9 when the geomembrane was fluorinated. The reduction factor was compound-dependent and increased exponentially with increasing log K_{ow}, indicating that the resistance to diffusion of highly hydrophobic contaminants is greatly improved by fluorination.

The results of the Sangam and Rowe (2003) study suggest that surface fluorination treatment may improve the effectiveness of HDPE geomembranes as a barrier to organic contaminants, especially aromatic hydrocarbons. They suggest that the performance of the treated geomembrane could be further improved by increasing the thickness of the fluorinated layer. However, it also follows that different fluorinated geomembranes may have quite different properties depending on the thickness of the fluorinated layer and that this should be specified and monitored. Additional research is required to assess how well the fluorinated layer stands up to scratching during installation. A field study examining the use of a fluorinated geomembrane to contain a jet fuel spill has been reported by Li *et al.* (2002). This geomembrane is presently being examined in terms of its performance.

8.12.3 Diffusion and partitioning coefficients for geomembranes

Values of the partition coefficient, S_{gf}, and diffusion coefficient, D_g, based on the literature, are summarized in Table 8.10. These values provide an initial estimate of parameters for a given contaminant; however, the values should be used with caution since they are temperature and maybe concentration-dependent. The values may also vary due to

Table 8.10 Partition and diffusion coefficients for various contaminants in polyethylene geomembrane

Chemical	*Thickness (mm)*$^{+}$, x	*Solution*	S_{gf} (–)	D_g ($\times 10^{12}$ m^2/s)	*Reference*
Acetic acid	2	4,000 mg/l	–	≤0.003 b	Rowe *et al.* (1996a)
	2.5	Pure	0.0086 ad	0.52–0.58 a	Müller *et al.* (1998)
	2.5	500–900 g/l	0.015 ac	0.11–0.29 a	Müller *et al.* (1998)
Acetic acid	1.0	Aqueous	0.021 ad	1.1 b	Müller *et al.* (1998)
Ethyl ester	2.0	Aqueous	0.023 ad		Müller *et al.* (1998)
Acetone	0.75	Pure	0.012 ad	0.51 a	Park and Nibras (1993)
	1.0	Aqueous Saturated	0.20 ac	<0.0001 a	Prasad *et al.* (1994)
	1.0	Pure	0.009 ad	–	Müller *et al.* (1998)
	2.0	Pure	0.009 ad	–	Müller *et al.* (1998)
	2.0	Pure	–	0.26–1.0 a	Müller *et al.* (1998)
	2.5	Pure	0.0099–0.0112 ad	0.87–0.91 a	Müller *et al.* (1998)
	2.5	10 vol. %	0.32 ac	0.66–0.84 a	Müller *et al.* (1998)
	2.5	50 vol. %	–	0.88 a	Müller *et al.* (1998)
	2.5	Aqueous Saturated	>0.013 ac	–	Haxo and Lahey (1988)
Benzene	0.15**	Pure	–	2.0 a	Saleem *et al.* (1989)
	1.0	Aqueous Saturated	57.2 ac	0.037 a	Prasad *et al.* (1994)
	1.5	Pure	0.08 ad	2.2–3.3 a	Ramsey (1993)
	0.75, 2.54	Pure	–	4.2 a	Britton *et al.* (1989)
	1.5, $x = 66\%$	2 mg/l	32 b 20–24 a	0.12 b	Rowe *et al.* (2003a)
	2.0, $x = 47\%$	2 mg/l	30 b 25–31 a	0.35 b	Sangam and Rowe (2001)
	2.5	Aqueous Saturated	54.3 ac	–	Haxo and Lahey (1988)
Carbon tetrachloride	0.15**	Pure	–	0.66 a	Saleem *et al.* (1989)
	1.0	Pure	0.180 ad	2.4 b	Müller *et al.* (1998)
	2.0	Pure	0.200 ad	–	Müller *et al.* (1998)
	2.0*	89.1 mg/l	–	57 b	Luber (1992)
	2.0	89.1 mg/l	–	48 b	Luber (1992)
Chlorobenzene	0.75	Pure	0.083 ad	2.4 a	Park and Nibras (1993)
	1.0	Pure	0.097 ad	3.6 b	Müller *et al.* (1998)
	1.5	Pure	0.110 ad	2.2–3.2 a	Ramsey (1993)
	2.0	Pure	0.108 ad	–	Müller *et al.* (1998)
Chloroform	0.15**	Pure	–	1.8 a	Saleem *et al.* (1989)
	1.0	Pure	0.134 ad	5.9 b	Müller *et al.* (1998)
	2.0	86.2 mg/l	–	17 b	Luber (1992)
	2.0*	86.2 mg/l	–	25 b	Luber (1992)
	2.0	Pure	0.153 ad	–	Müller *et al.* (1998)
Chloride	2.0	2,000–4,000 mg/l	–	0.002–0.005 b	Rowe *et al.* (1996a)
	2.0	2,000–4,000 mg/l	0.0008 b	0.1–0.3 b	Rowe (1998a)
Cyclohexane	0.15**	Pure	–	0.61 a	Saleem *et al.* (1989)
	1.0	Aqueous Saturated	2,378 ac	0.012 a	Prasad *et al.* (1994)
Dichloromethane	0.75	Pure	0.06 ad	4.9 a	Park and Nibras (1993)
	0.75	13.1–234.4 mmol/l	1.8–2.9 ac	0.58–2.28 a	Park and Nibras (1993)
	0.75–2.5	Pure	–	9.0 a	Britton *et al.* (1989)

Table 8.10 *Continued*

Chemical	*Thickness (mm)*$^{+}$, x	*Solution*	S_{gf} (–)	D_g ($\times 10^{12}$ m^2/s)	*Reference*
Dichloro methane	0.76–2.5	100 mg/l	1.8–5.6 bc	–	Sakti *et al.* (1992)
	1.5, $x = 66\%$	5 mg/l	3–5 ab	0.45–0.5 b	Rowe *et al.* (2003a)
	2.0	Pure	–	2–10 a	Durin *et al.* (1998)
	2.0	2–10 mg/l	–	1–3 b	Rowe *et al.* (1996a)
	2.0	2–10 mg/l	2.3 b	0.95–1.2 b	Rowe (1998a)
	2.0, $x = 47\%$	5 mg/l	6 ab	0.65 b	Sangam and Rowe (2001)
1,1-Dichloroethane	0.75–2.5	Pure	–	2.3 a	Britton *et al.* (1989)
	2.0	2–10 mg/l	–	1–2.5 b	Rowe *et al.* (1996a)
1,2-Dichloroethane	0.75–2.5	Pure	–	2.3 a	Britton *et al.* (1989)
	1.0	Aqueous Saturated	7.2 ac	6.8 a	Prasad *et al.* (1994)
	1.5, $x = 66\%$	5 mg/l	10 b 6–8 a	0.16–0.18 b	Rowe *et al.* (2003a)
	2.0	2–10 mg/l	–	3–6 b	Rowe *et al.* (1996a)
	2.0, $x = 47\%$	5 mg/l	9.4 a; 12 b	0.40 b	Sangam and Rowe (2001)
Ethyl acetate	0.75	Pure	0.024 ad	0.79 a	Park and Nibras (1993)
Ethyl benzene	0.75	Pure	0.1 ad	2.8 a	Park and Nibras (1993)
	2.0	2 mg/l	237–315 a 285 b	0.18 b	Sangam and Rowe (2001)
Formaldehyde	1.0	37 wt. %	0.004 ad	–	Müller *et al.* (1998)
	2.0	37 wt. %	0.003 ad	–	Müller *et al.* (1998)
Heptane	2.5	Pure	0.0646–0.0660 ad	1.52–1.74 a	Müller *et al.* (1998)
n-Hexane	0.75	Pure	0.006 ad	3.6 a	Park and Nibras (1993)
	2.5	Pure	0.0646–0.0663 ad	2.08–2.47 a	Müller *et al.* (1998)
Methyl ethyl ketone	2.0	2–10 mg/l	–	0.3–0.8 b	Rowe *et al.* (1996a)
	2.5	Pure	0.0179 ad	0.75 a	Müller *et al.* (1998)
	2.5	Pure	0.019 ad	0.86 a	Müller *et al.* (1998)
	2.5	Aqueous Saturated	>0.025 ac	–	Haxo and Lahey (1988)
n-Octane	0.75	Pure	0.08 ad	1.9 a	Park and Nibras (1993)
Propanoic acid	2.5	Pure	0.0212 ad	0.30 a	Müller *et al.* (1998)
	2.5	Pure	0.0209 ad	0.32 a	Müller *et al.* (1998)
Tetrachloroethane	2.0	61.0 mg/l	–	78 b	Luber (1992)
	2.0*	61.0 mg/l	–	87 b	Luber (1992)
Tetrachloroethylene	1.0	Pure	0.190 ad	3.8 b	Müller *et al.* (1998)
	2.0	Pure	0.217 ad	–	Müller *et al.* (1998)
Toluene	0.15**	Pure	–	1.8 a	Saleem *et al.* (1989)
	0.75	Pure	0.09 ad	4.4 a	Park and Nibras (1993)
	0.75	0.28–5.0 mmol/l	63.5–151 ac	0.35–0.56 a	Park and Nibras (1993)
	0.76–2.5	100 mg/l	115–125 bc	–	Sakti *et al.* (1992)
	0.76–2.5	Pure, 1.25 M	–	0.23 a	Britton *et al.* (1989)
	1.0	0.05 wt. %	–	0.23 b	August and Tatzky (1984)
	1.0	Aqueous Saturated	192 ac	0.51 a	Prasad *et al.* (1994)
	1.0	Pure	0.080 ad	6.1 b	Müller *et al.* (1998)

Toluene	1.0	Aqueous	160 ac	0.2 b	Müller *et al.* (1998)
	1.5	Pure	0.09 ad	2.6–4.0 a	Ramsey (1993)
	1.5, $x = 66\%$	2 mg/l	65–70 b 57–69 a	0.12–0.15 b	Rowe *et al.* (2003a)
	2.0	Pure	0.090 ad	–	Müller *et al.* (1998)
	2.0	Pure	–	0.18–4.0 a	Durin *et al.* (1998)
	2.0, $x = 47\%$	2 mg/l	100 b 95–120 a 100 b	0.30 b	Sangam and Rowe (2001)
	2.5	Aqueous Saturated	137 ac		Haxo and Lahey (1988)
1,1,1-Trichloroethane	2.5	Aqueous Saturated	78.2 ac	–	Haxo and Lahey (1988)
Trichloroethylene	0.75	Pure	0.11 ad	12 a	Park and Nibras (1993)
	0.75	0.42–8.4 mmol/l	–	0.44–0.76 a	Park and Nibras (1993)
	0.76–2.5	100 mg/l	94–98 bc	–	Sakti *et al.* (1992)
	1.0	Aqueous Saturated	134.5 ac	0.52 a	Prasad *et al.* (1994)
	1.0	Aqueous Saturated	134.5 ac	0.50 b	Prasad *et al.* (1994)
	1.0	0.1 wt. %	–	0.69 b	August and Tatzky (1984)
	1.0	Pure	0.168 ad	10.8 b	Müller *et al.* (1998)
	1.5, $x = 66\%$	5 mg/l	60–68 b 53–56 a	0.16–0.18 b	Rowe *et al.* (2003a)
	1.0	500 mg/l	–	0.6 b	Müller *et al.* (1998)
	2.0	Pure	0.190 ad	–	Müller *et al.* (1998)
	2.0	61.0 mg/l	–	73 b	Luber (1992)
	2.0, $x = 47\%$	5 mg/l	85 b 59–87 a	0.40 b	Sangam and Rowe (2001)
	2.0*	61.0 mg/l	–	85 b	Luber (1992)
	2.5	Pure	0.195–0.200 ad	7.70–8.40 a	Müller *et al.* (1998)
	2.5	500 mg/l	189 ac	0.20–0.30 a	Müller *et al.* (1998)
	2.5	Aqueous Saturated	131 ac	–	Haxo and Lahey (1988)
m-Xylene	0.15**	Pure	–	1.5 a	Saleem *et al.* (1989)
	0.75	Pure	0.093 ad	3.7 a	Park and Nibras (1993)
	0.75	0.2–1.6 mmol/l	192.7–310.2 ac	0.31–0.36 a	Park and Nibras (1993)
	0.76–2.54	100 mg/l	365–370 bc	–	Sakti *et al.* (1992)
	1.5, $x = 66\%$	2 mg/l	205–220 b 190–216 a	0.06–0.08 b	Rowe *et al.* (2003a)
	2.0, $x = 47\%$	2 mg/l	300–408 a 347 b	0.17 b	Sangam and Rowe (2001)
	2.5	Aqueous Saturated	366 ac	–	Haxo and Lahey (1988)
o-Xylene	0.15**	Pure	–	0.94 a	Saleem *et al.* (1989)
	1.5, $x = 66\%$	2 mg/l	180–200 b 170–180 a	0.04–0.06 b	Rowe *et al.* (2003a)
	2.0, $x = 47\%$	2 mg/l	193–285 a 240 b	0.15 b	Sangam and Rowe (2001)
	2.5	Aqueous	422 ac	–	Haxo and Lahey (1988)
p-Xylene	0.15**	Pure	–	1.6 a	Saleem *et al.* (1989)
	1.5, $x = 66\%$	2 mg/l	205–220 b 190–216 a	0.06–0.08 b	Rowe *et al.* (2003a)
	2.0, $x = 47\%$	2 mg/l	300–408 a 347 b	0.17 b	Sangam and Rowe (2001)
n-Xylene	2.5	Aqueous Saturated	387 ac	–	Haxo and Lahey (1988)

Table 8.10 *Continued*

Chemical	*Thickness (mm)*$^{+}$, x	*Solution*	S_{gf} (–)	D_g ($\times 10^{12}$ m^2/s)	*Reference*
Xylene	0.15**	Pure	–	1.8 a	Saleem *et al.* (1989)
	1.0	Pure	0.083 ad	4.7 b	Müller *et al.* (1998)
	1.0	Aqueous	556 ac	0.2 b	Müller *et al.* (1998)
	1.0	Aqueous	–	1.8 b	August and Tatzky (1984)
	1.0	Aqueous Saturated	498.5 a	1.0 ab	Prasad *et al.* (1994)
	2.0	Pure	0.095 ad	–	Müller *et al.* (1998)
Water	1.7	–	$\leq$0.0003 ad	$\geq$0.4 a	Eloy-Giorni *et al.* (1996)
	1.7	–	0.0008 bd	0.29 b	Eloy-Giorni *et al.* (1996)
	2.0	–	–	0.4–9.0 a	Durin *et al.* (1998)
	2.5	–	0.0008–0.001 ad	0.82–0.90 a	Müller *et al.* (1998)
	–	–		1.6 –	Lord *et al.* (1988)
	–	–	–	2.6 –	Hughes and Monteleone (1987)

Notes:
a Sorption/absorption/immersion.
b Permeation/diffusion.
c Solubility in GM/solubility in water.
d Mass of chemical/Mass of GM).
+ HDPE except where specified.
* MLDPE.
** LLDPE.
– Not reported/measured.
x Crystallinity.

the chemical composition of the contaminant source, the polymer crystallinity, additives, etc. Thus published values should only be used as an initial guide and do not replace experimentally determined values for the geomembrane of interest for projects where uncertainty regarding the diffusion coefficient or sorption could have a significant impact.

Chapter nine

Field studies of diffusion and hydraulic conductivity

9.1 Introduction

Examples of field diffusion are rather scarce in the literature, partly because barrier exhumations are very expensive once several metres of waste are in place, and partly because most research on pollutant migration has been devoted to plumes in granular deposits since these impact immediately on local groundwater resources. Fortunately, geological history has provided some long-term data regarding the process of diffusion.

The primary purpose of this chapter is to present a variety of concentration versus depth profiles documented for clay deposits. Both long-term (~10,000 years) and short-term (3–22 years) profiles will be discussed. A secondary objective is to examine a field case where a composite liner was used for a leachate lagoon which highlights the importance of appropriate design, construction and maintenance of composite liners.

9.2 Examples of long-term field diffusion

Three sites are discussed in this section; a 10,000-year-old salt profile in 30 m of saline Leda clay near Ottawa, Ontario; a 12,000-year-old set of profiles in 35 m of freshwater clay near Sarnia, Ontario; and a 12,000-year-old set of profiles in about 35 m of freshwater clay near Niagara Falls, Ontario, Canada.

9.2.1 Hawkesbury Leda clay

A geotechnical profile showing the preconsolidation state of the Hawkesbury Leda clay is presented in Figure 9.1 along with the present-day salinity profile. The site was described by Quigley *et al.* (1983) and the interrupted consolidation profile was attributed to removal of ~21 m of saturated sand about 70 years after deposition by a meander of the post-glacial Ottawa River.

The importance of the σ'_p curve is that it indicates double drainage from the centre of the clay at ~15 m. From this one would logically deduce that the salt content of the entire clay layer would have been constant at the time of sand erosion about 10,000 years BP. At the present time, the groundwater conditions are almost hydrostatic and it is difficult to see how they could ever have been much different over the past 10,000 years.

The salinity profile itself suggests that the original salt content was about 15 g/l and that salt has diffused towards surface and been removed by surface runoff.

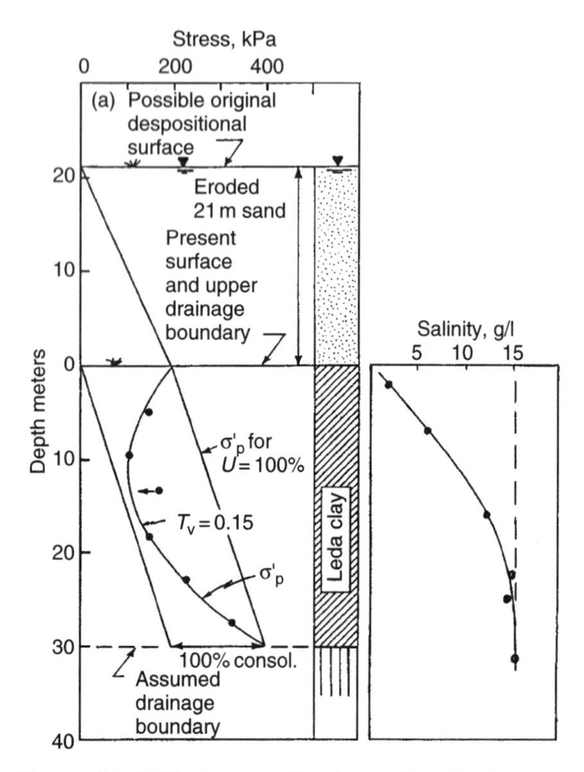

Figure 9.1 Diffusion profile for saline Leda clay, Hawkesbury, Ontario ($t \approx 10{,}000$ years) (modified from Quigley *et al.*, 1983).

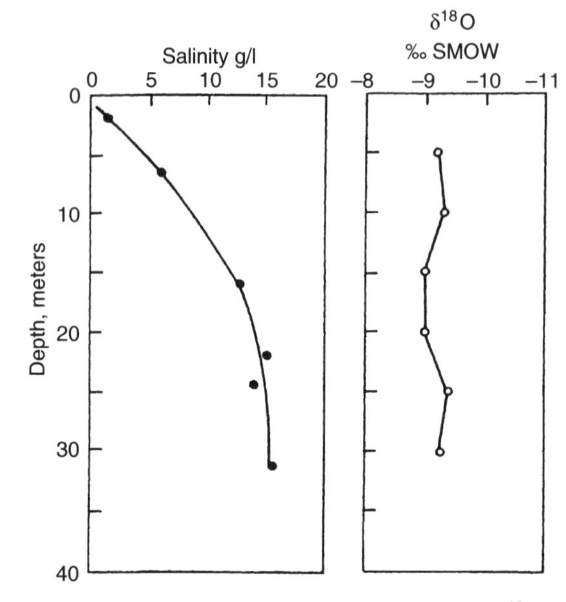

Figure 9.2 Profiles of pore water salinity and $\delta^{18}O$ in Leda clay at Hawkesbury.

In addition to the salt content profile, one other profile is presented in Figure 9.2 for $\delta^{18}O$ where

$$\delta^{18}O = \frac{(^{18}O/^{16}O)_{sample} - (^{18}O/^{16}O)_{SMOW}}{(^{18}O/^{16}O)_{SMOW}} \times 1{,}000 \qquad (9.1)$$

where SMOW is the standard mean ocean water.

$\delta^{18}O$ is a measure of the amount of oxygen isotope ^{16}O relative to the isotope ^{18}O. It is normally −20‰ or less for glacial water and close to zero for seawater. The $\delta^{18}O$ values in the clay are approximately constant at −9‰ and reflect several factors as follows:

1. The depositional water at the centre of the clay layer probably had a $\delta^{18}O$ of −9‰ due to mixing of seawater and glacial ice water.
2. At the time of erosion, $\delta^{18}O$ was constant throughout the layer due to double drainage.
3. Surface rainwater at the site has an average annual $\delta^{18}O$ of about −9‰ so that it is not possible to develop a diffusion profile similar to that for salinity.

$\delta^{18}O$ is quite important because it correlates with salinity in many ocean environments as shown in Figure 9.3. For the Champlain Sea clays, a $\delta^{18}O$ value of −9‰ would correlate with a salinity of 15 g/l, exactly that found at Hawkesbury.

Now that the reference salinity has been "verified" as 15 g/l, one has confidence to calculate a diffusion coefficient for bulk salinity assuming no advection and $t = 10{,}000$ years, the approximate time of isostatic emergence, Champlain Sea drainage and sand erosion. The result is a diffusion coefficient of 2×10^{-10} m^2/s

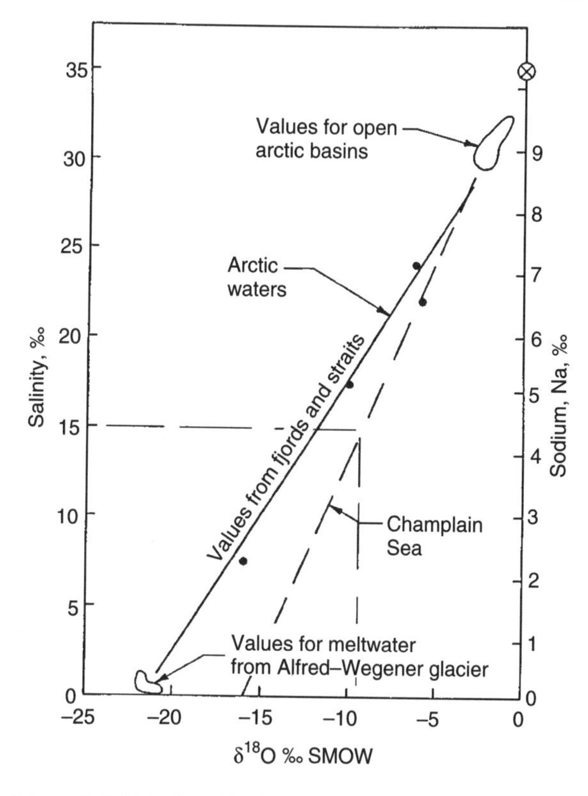

Figure 9.3 Relationship between salinity and oxygen-18 of present-day Arctic waters and predicted relationship for Champlain Seawater (modified from Hillaire-Marcel, 1979).

($0.0063\,m^2/a$) (Figure 9.4). This value is little lower than most of the other available data (e.g., see Table 8.6).

9.2.2 Sarnia water-laid clay till

Chloride and sodium diffusion profiles for the 35-m thick, freshwater, glaciolacustrine clays at Sarnia, Ontario, are presented in Figure 9.5. The important feature of these curves is an increasing concentration with depth; Cl^- increasing from 0 to about 400 mg/l at depth and sodium increasing to about 220 mg/l. Since these are freshwater clays, the source of this salt must have been by diffusion from the bedrock.

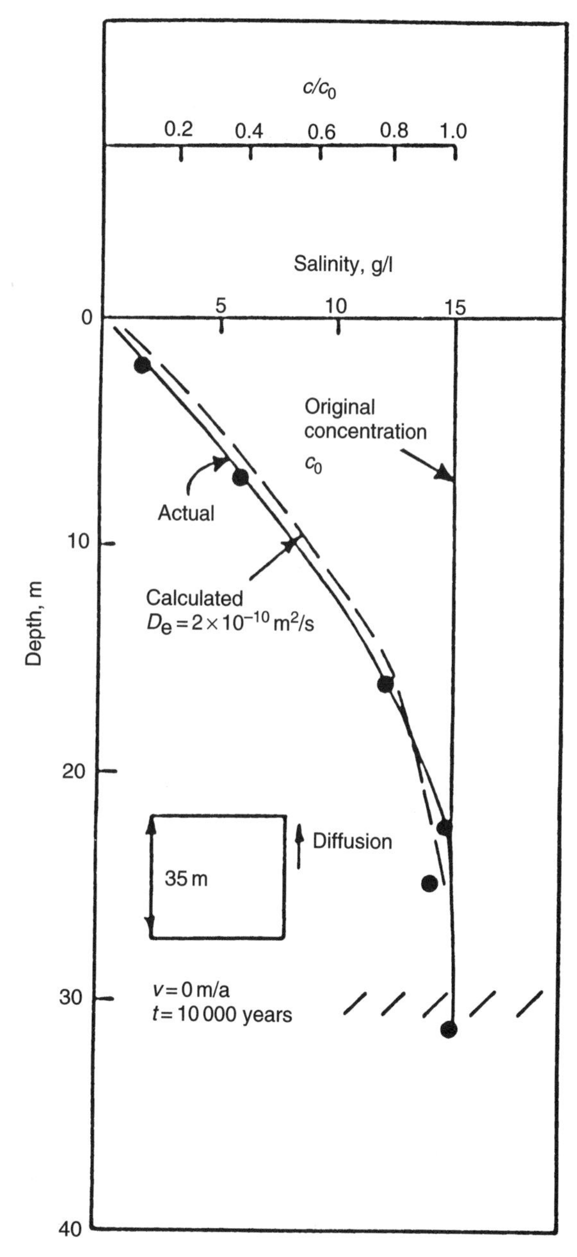

Figure 9.4 Actual and calculated diffusion profiles for bulk salinity, Hawkesbury Leda clay.

The chloride profiles in Figure 9.5 were analysed for upward diffusion of salts from the bedrock against a small, measured, downward regional flow of 0.0003 m/a (Desaulniers *et al.*,

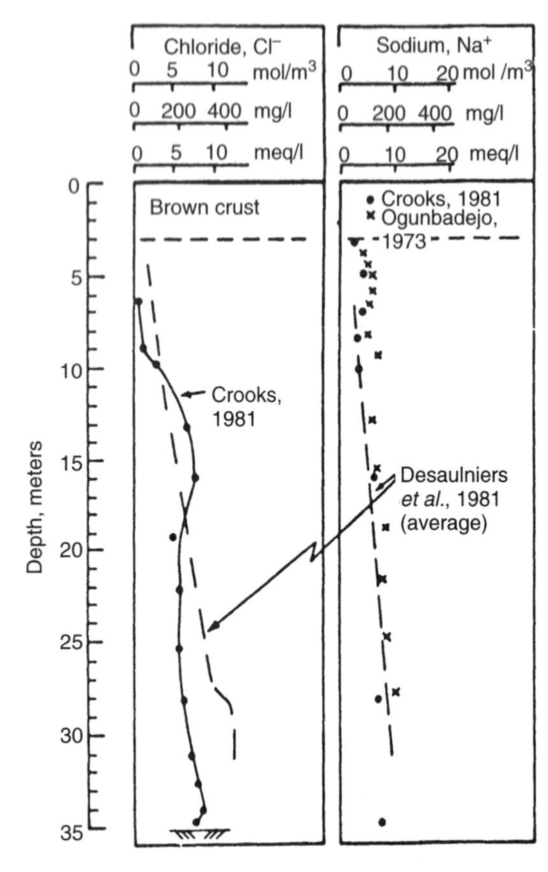

Figure 9.5 Na^+ and Cl^- profiles in thick freshwater glacial clays, Sarnia, Ontario (after Quigley and Crooks, 1983; reproduced with permission of *ASCE*).

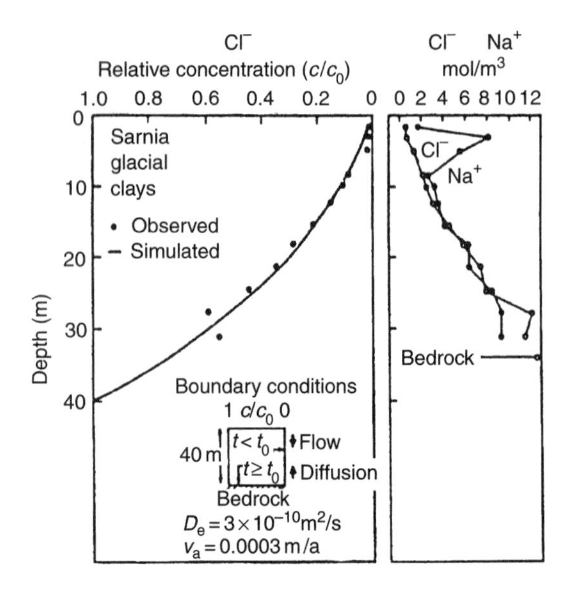

Figure 9.6 Chloride ion versus depth with concentrations calculated for upward diffusion from bedrock (modified from Desaulniers *et al.*, 1981).

1981). For a diffusion coefficient of 3×10^{-10} m^2/s ($0.01\,m^2/a$) and $t = 10{,}000$ years, a curve was drawn and the value calculated from that is a reasonable fit to the actual c/c_0 for the pore water chloride (Figure 9.6).

A $\delta^{18}O$ profile was also presented by Desaulniers *et al.* as shown in Figure 9.7 complete with calculated curves using a diffusion coefficient of $3 \times 10^{-10}\,m^2/s$ ($0.01\,m^2/a$). The $\delta^{18}O$ at surface (–10‰) represents the average annual rainfall value and the value of –18‰ at depth reflects glacial melt water. Originally the ice-derived pore water in the glaciolacustrine sediments probably had a $\delta^{18}O$ value of –20‰ or less, and 10,000 years of diffusion have

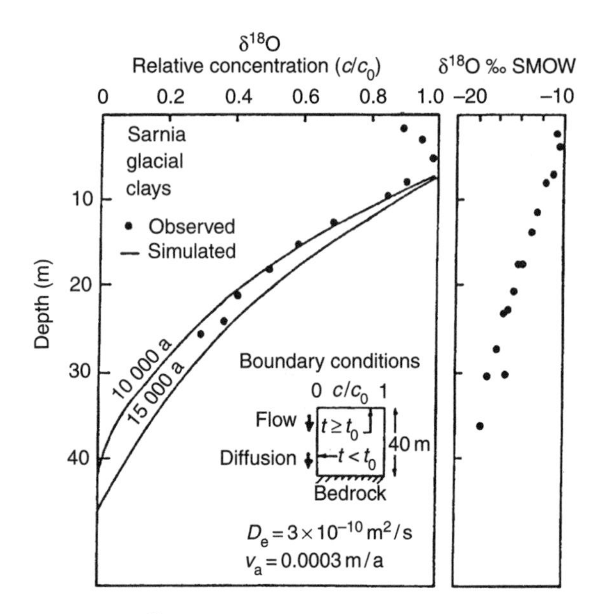

Figure 9.7 $\delta^{18}O$ versus depth and calculated dilution by downward diffusion of rainwater (modified from Desaulniers *et al.*, 1981).

resulted in a profile which ranges from surface value of ~–10 to –18‰ at the bottom.

Long-term diffusion profiles are difficult to find because of changing geologic and climatic conditions. Fortunately, assessment is not sensitive to a detailed knowledge of exactly when the geological changes occur. As shown in Figure 9.7, there is little difference for the calculated c/c_0 profiles for t values of 10,000 and 15,000 years.

An additional aspect of diffusion that can have important practical engineering consequences is that of long-term upward diffusion of dissolved methane from the bedrock. Dissolved gases in the groundwater, originating from the underlying sedimentary bedrock within which the oil and gas are located in the Sarnia area, exsolve in response to stress release and are observed venting in boreholes drilled into, or close to, the top of the confined basal till aquifer. Dittrich *et al.* (2002) have reported on the implication of these gases with respect to the construction and stability of approach slopes for two railway tunnels in the Sarnia area. Of greater direct relevance to the subject of this book are the implications for a landfill located near Sarnia, Ontario. The design and operations of this facility involve excavation of sub-cells, followed by placement of waste and a compacted clay cap with the objective of providing long-term diffusion-controlled contaminant migration through thick, low-permeability clays along the lines discussed in Section 1.2.1(a) and illustrated schematically in Figure 1.1. A geotechnical assessment and limit equilibrium stability analyses had been conducted to identify suitable excavation depths, slope lengths and duration of exposure. During excavation and waste filling operations, lateral slope movements, base heave and pore water pressures are monitored.

Following excavation to a depth of 25 m for a new sub-cell in September 1999, venting of gas and water were noted in three separate locations in the base of the excavation. Chemical analyses indicated that the vented water originated from the underlying confined aquifer, and the vented gas originated from the underlying bedrock. Pressure tests indicated gas pressures in excess of 70 kPa above atmospheric pressure at one venting location. The observed venting phenomena locally compromised the low-permeability till and the sub-cell was largely lost for waste disposal. The occurrence of the phenomena at this location, when it had not previously occurred for adjacent excavations to the same depth, was attributed to the combined effects of (a) a reduction in the mobilized shear strength of the soil (relative to that assumed in stability calculations) due to the exsolution of the dissolved gas in the pore water during unloading; and (b) combined with the presence of a local bedrock high beneath the venting area that reduced the thickness of the overlying till, resulting in a lower confining pressure for the gas and a shorter path length for gas and water movement.

This case highlights the implication of the diffusion of gas on the stability of excavation and highlights the importance of identifying the presence of dissolved gas and assessing its implications with respect to stability and barrier performance when considering excavations into gassy soil.

9.2.3 Freshwater clay

A third and final example of a natural diffusion profile for chloride is shown in Figure 9.8 (Rowe and Sawicki, 1992). At this site, 34–39 m of fine-grained glaciolacustrine deposit consisting primarily of silts and clays overlie bedrock of the Salina Formation which is composed of soft, erodible shales with evaporites (gypsum) and harder dolostone layers. Based on chemical analyses from four wells in the bedrock, the average bedrock chloride concentration is about 1,400 mg/l. The observed concentration extends upward from the bedrock with decreasing concentration towards the surface. Radiocarbon dates of 12,000–13,000 years BP have been reported (Lewis, 1969; Fullerton, 1980) for sediments deposited in glacial Lake Warren (the source of this clay unit). Based on the observed

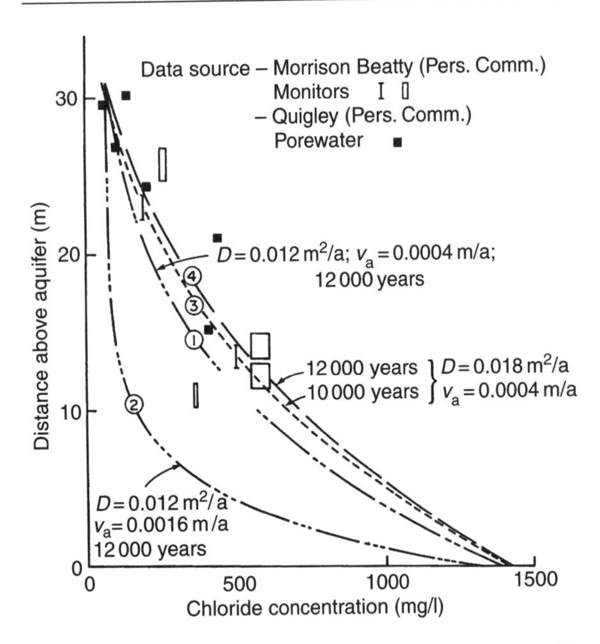

Figure 9.8 Modelling of an existing diffusion profile after 10,000–12,000 years diffusion (modified from Rowe and Sawicki, 1992).

gradients and measured hydraulic conductivity values, the downward Darcy velocity is estimated to be between 0.0004 and 0.0016 m/a. Laboratory diffusion tests give a diffusion coefficient of $3.8 \times 10^{-10}\,m^2/s$ ($0.012\,m^2/a$).

Figure 9.8 shows four calculated concentration profiles for three combinations of parameters and two times. A prediction based on the measured diffusion coefficient and a Darcy velocity of 0.0004 m/a (curve 1) gives a reasonable fit to the observed concentration profile for 12,000 years of diffusion. The potential importance of downward advection is shown by comparing curve 1 ($v_a = 0.0004\,m/a$) and curve 2 ($v_a = 0.0016\,m/a$) for a diffusion coefficient of $3.8 \times 10^{-10}\,m^2/s$ ($0.012\,m^2/a$). Curve 2 does not provide a good fit to the data and this implies that the downward advective velocity must have been very low for most of the past 12,000 years. To illustrate the sensitivity of results to diffusion coefficients and time, curves 3 and 4 show the predicted profile for a diffusion coefficient of $5.7 \times 10^{-10}\,m^2/s$ ($0.018\,m^2/a$) and times of 12,000 and 10,000 years. Based on these analyses, it appears that the diffusion is generally consistent with a diffusion coefficient of between $3.8\times 10^{-10}\,m^2/s$ ($0.012\,m^2/a$) and $5.7 \times 10^{-10}\,m^2/s$ ($0.018\,m^2/a$).

9.3 Examples of short-term field diffusion

Diffusion profiles presented in this section are restricted to extensive studies carried out at the Confederation Road landfill near Sarnia, Ontario, plus one profile for the Keele Valley Liner, Toronto. An additional profile is discussed for the case examined in Section 9.5.

9.3.1 Confederation Road Landfill, Sarnia

A summary of the Confederation Road landfill site at Sarnia was presented by Quigley and Rowe (1986). The nature of the Confederation Road site is shown by the geotechnical data in Figures 9.9 and 9.10. The site consists of ~7.5 m of waste (including cover) placed in a 5.5-m trench excavated through a brown desiccated crust for embankment borrow. The calculated effective stress profile incorporates the effect of a slight regional downward gradient which creates a downward average linearized groundwater velocity of ~0.0024 m/a (Goodall and Quigley, 1977). Figure 9.9 also shows the soil to be overconsolidated by ~90 kPa with a slight increase near the clay/waste interface where the moisture content decreases slightly from 23 to 21%. The nature of this crust is further illustrated in Figure 9.10. The desiccated crust is highly fissured in the upper 4 m of brown oxidized clay and much less fissured from 4 to 6.5 m in the lower grey portion of the crust. At the interface, which is probably within 1 m of the base of the desiccated crust, no fissures have ever been observed in the grey interfacial clay samples obtained in the 50 or more boreholes drilled at the site.

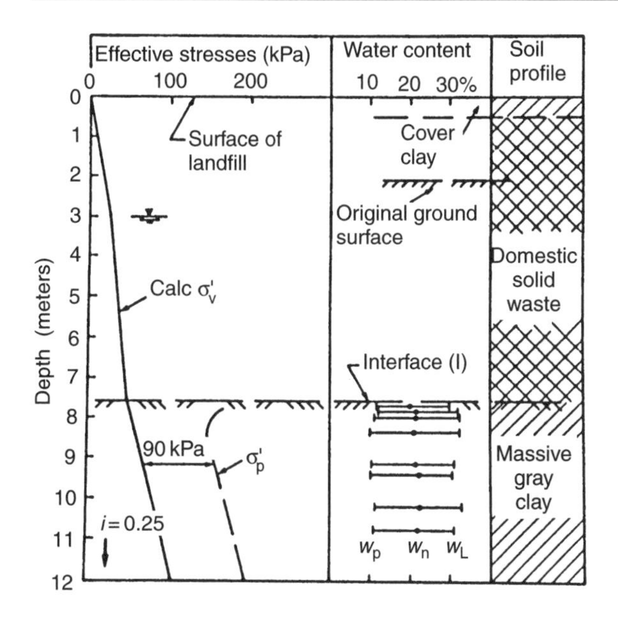

Figure 9.9 Soil conditions at Confederation Road landfill site, 1967 to present: σ'_v = vertical effective stress; σ'_p = preconsolidation pressure; w_P = plastic limit; w_n = natural water content; and w_L = liquid limit (after Quigley and Rowe, 1986; reproduced with permission of *ASTM*).

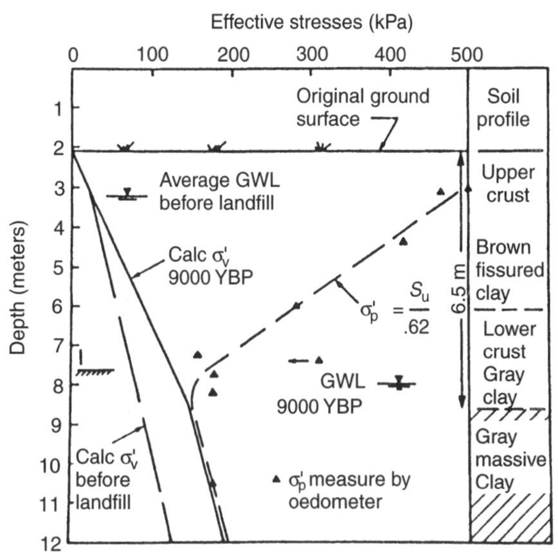

Figure 9.10 Soil conditions 9,000 years before present (YBP) and before cutting landfill trench in 1966 (I is waste/clay interface; s_u = undrained shear strength, GWL = groundwater level; see also the legend in Figure 9.9) (after Quigley and Rowe, 1986; reproduced with permission of *ASTM*).

(a) Sodium chloride

Chemical profiles for Na^+ and Cl^- are presented in Figures 9.11 and 9.12, respectively, for $t = 12$ years using units of mg/l and mol/m^3. Also plotted are the estimated seepage fronts for $t = 12$ and 100 years. Both figures indicate salt migration for a distance of about 1.5–2.0 m in 12 years which is far ahead of the estimated advection distance of 3 cm. This would seem to confirm the significant role that diffusion plays in the migration of chemicals. As noted in Chapter 8 (Section 8.9), diffusion and distribution coefficients can be inferred from these field profiles and yield parameters as given in Table 8.5 for Cl^-, Na^+ and K^+. For chloride, the diffusion coefficient was about 6.3×10^{-10} m^2/s (0.02 m/a) which is quite consistent with values obtained from laboratory tests on samples of the same clay (e.g., see Table 8.3).

The plot in mg/l suggests that Cl^- (stippled) has migrated significantly faster than Na^+; however, this impression is an artefact of the plotting related to the higher molecular weight of chloride (35) compared to sodium (23). If the data are plotted in chemical units (mol/m^3), as shown in Figure 9.12, the two species are seen to be migrating at approximately the same rate with Cl^- slightly in advance.

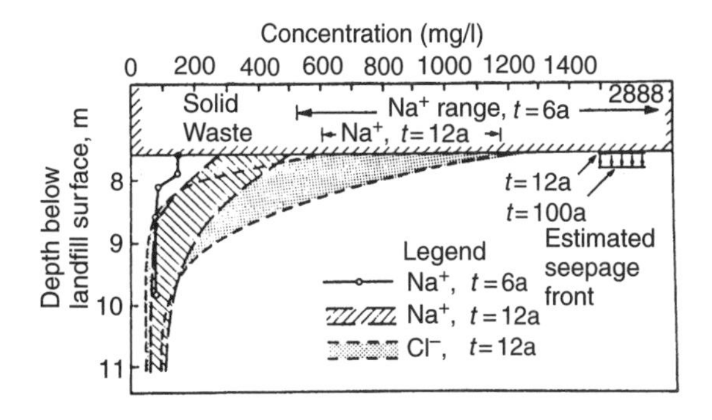

Figure 9.11 Pore fluid concentration of Na^+ and Cl^- (mg/l) in clay below waste after 6 and 12 years of diffusion (t = 6a and 12a) (after Quigley and Rowe, 1986; reproduced with permission of *ASTM*).

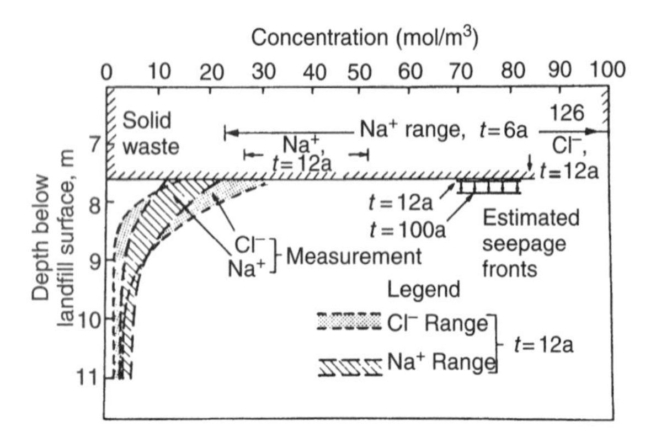

Figure 9.12 Pore fluid concentration of Na^+ and Cl^- (mol/m^3) in clay below waste at $t = 12$ years ($t = 12a$) (after Quigley and Rowe, 1986; reproduced with permission of *ASTM*).

(b) Heavy metals

Profiles of total metal concentration for lead, copper, zinc, iron and manganese are presented in Figure 9.13. The long vertical portions represent total background values and the short curved sections near the interface represent metals added to the soil by diffusion from the leachate. It seems quite clear that these metals have migrated a maximum of 20 cm and more likely only 10 cm in 16 years. The nature of these metal compounds is shown in Figure 9.14 (Yanful *et al.*, 1988b). The extractions suggest that all species in the background soil exist in soil carbonate and within the lattice structure of the minerals (primarily the clay minerals). An organic phase may also be present for iron, zinc and copper. In the elevated concentration zone near the interface, carbonate dominates the mineral species which is believed to have precipitated at the soil pH levels of ~8 and the low redox potentials of –150 mV at the interface. The interested reader is referred to Yanful *et al.* (1988a,b).

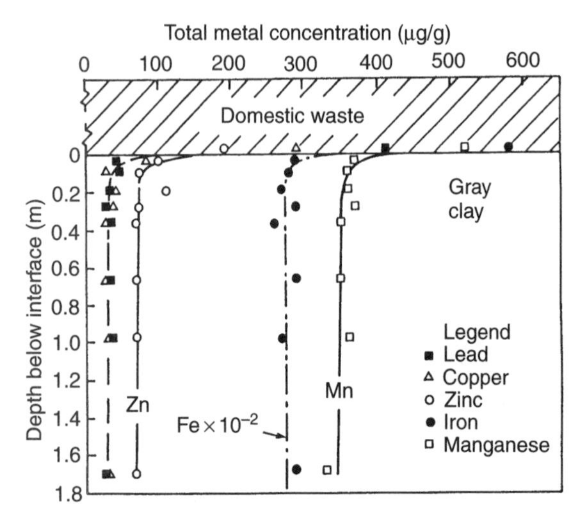

Figure 9.13 Profiles for total metals (BH 84–10, $t = 16a$) in clay below domestic waste (modified from Yanful and Quigley, 1986).

(c) Dissolved organic carbon (DOC)

Finally, Figure 9.15 presents a plot of DOC in pore fluid squeezed from the interfacial clays at $t = 16$ years. The curve indicates a migration distance of about 0.8 m compared to 1.5 m for chloride. This is encouraging since it suggests that natural biological activities are actively degrading any organics from the leachate in a zone very close to the waste.

Another DOC curve is presented in Figure 9.16 for Confederation Road at $t = 21$ years (Quigley *et al.*, 1990a). This curve also appears to be a diffusion profile with interface DOC values of ~100 mg/l decreasing to ~10 mg/l at 0.9–1.0 m depth below the interface. GC–MS analysis of the pore water yielded only trace amounts of xylene and bulk phenol values of 0.4–0.6 mg/l at the interface. No other EPA priority pollutants were identified, again suggesting that all have been biodegraded within the upper 1 m of clay in the 21 years since disposal.

The DOC curve itself remains a mystery in that its components are still not identified. It is possible that the curve does not represent migration of organic substances at all; rather it could represent biological activity that has broken down solid organics in the clay to produce lower molecular weight compounds that were removed

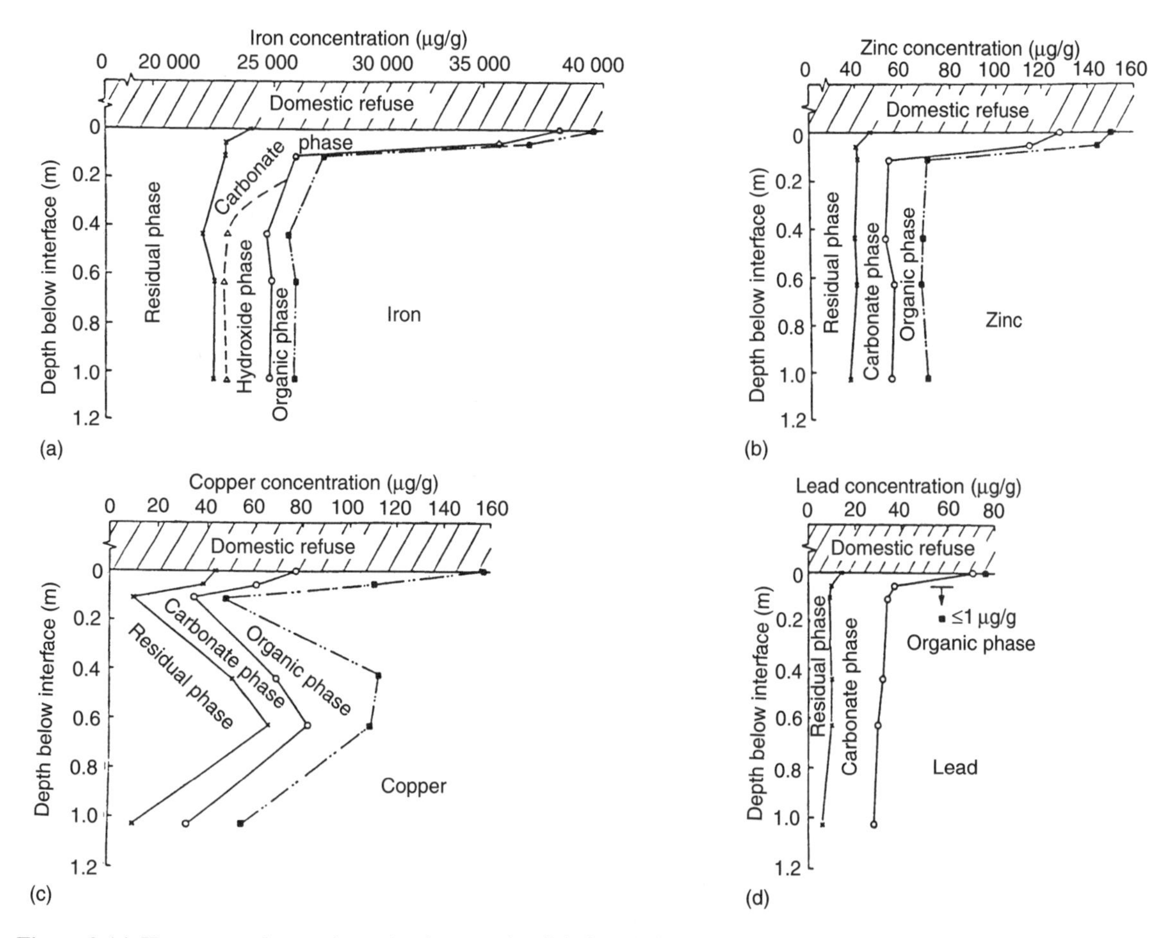

Figure 9.14 Heavy metal complexes in clayey subsoil inferred from sequential extraction, BH 83–7, $t = 15$ years: (a) iron; (b) zinc; (c) copper; (d) lead (modified from Yanful *et al.*, 1988b).

with the water by high-pressure squeezing. The background TOC available for such decay amounts to about 0.7% of the soil solids.

The phenol profile at the site is shown in Figure 9.17. Background values vary between 25 and 75 μg/l. Despite the amount of scatter generated by samples from three 1988 boreholes, there does appear to be a diffusion profile extending about 1 m below the interface. Since phenol levels are high in many MSW leachates, it is speculated that most of the phenol has biodegraded away in the 21 years since deposition of the waste. A half-life of about 6 years would be required to cause a decay from around 5,000 μg/l to present values. This half-life seems reasonable given available data on decay of phenol in soil. More research is required to better define the half-life of organics such as phenol in waste and in soil.

(d) Isotopes

In Figure 9.18, tritium analyses and fitted profiles are presented for squeeze water obtained in 1984. The central fitted curve, which also appears to be a good average curve, was calculated using a c_0 for tritium calculated for 1967 when the trench filled with rainwater. A radioactive decay was also applied to account for the

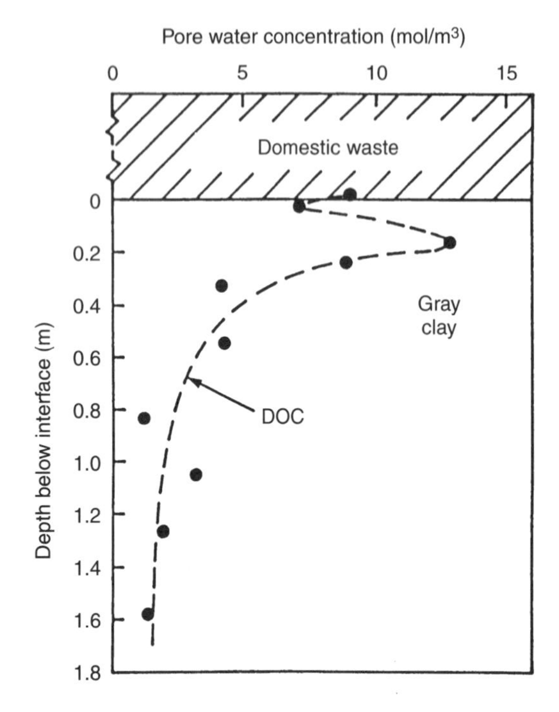

Figure 9.15 DOC in pore water squeezed from samples at $t = 16$a.

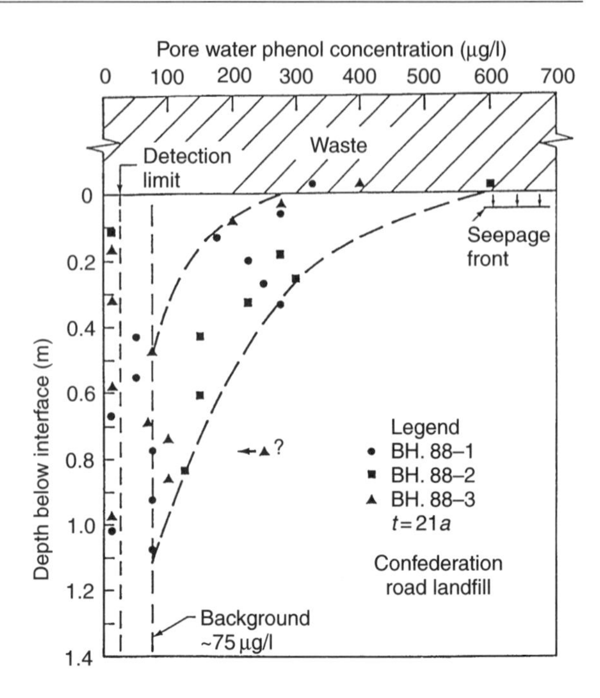

Figure 9.17 Bulk phenol versus depth below the waste/clay interface at $t = 21$a (modified from Quigley *et al.*, 1990a).

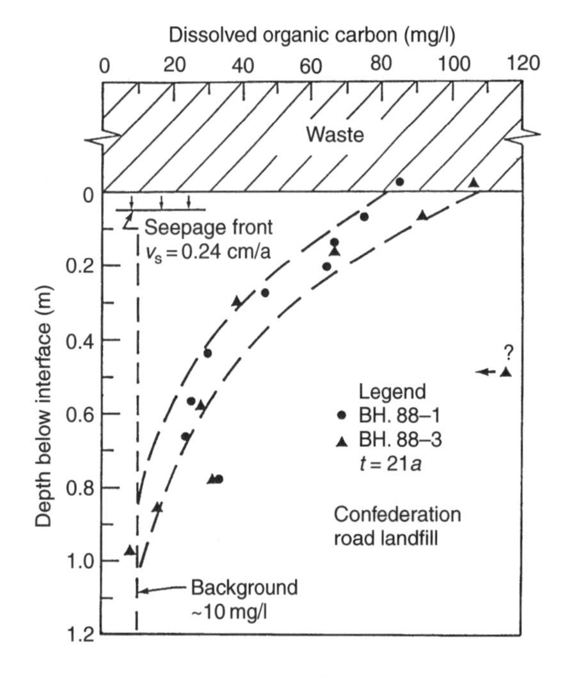

Figure 9.16 DOC versus depth below the waste/clay interface at $t = 21$a.

short half-life of tritium (12.4 years). The curve shows migration to ~ 2 m which is consistent with the distance of chloride migration.

Two curves are presented in Figure 9.19 for δ oxygen-18 and δ deuterium. Both curves describe diffusion of heavy water from the waste into the underlying clay which contains lighter water. The explanation for these profiles runs as follows:

(a) The background values in the soil are in equilibrium with the local rainwater at $\delta^{18}O \approx -11‰$ SMOW.
(b) Evapotranspiration from leaves results in escape of light water (i.e., light oxygen 16 and light hydrogen 1).
(c) On decay in the landfill, the heavy water left in the leaves is released from the waste and migrates by diffusion downwards creating a diffusion profile.

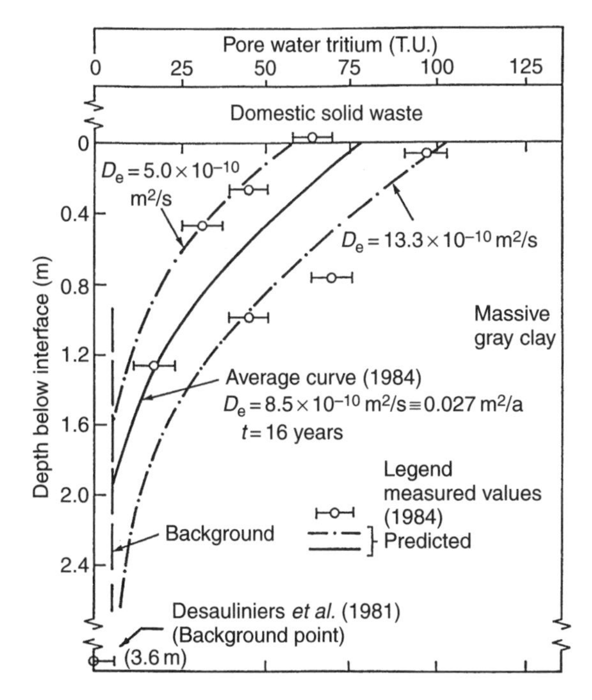

Figure 9.18 Pore water tritium profile in clay below the 1967 waste including range of calculated possible profiles using calculated values for c_0 in 1967.

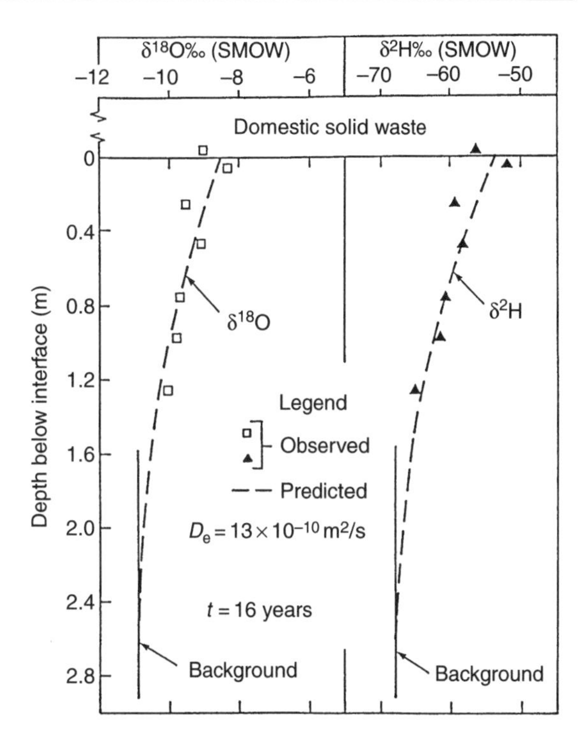

Figure 9.19 Profiles of measured δ oxygen-18 and δ deuterium in the clay below the 1967 waste (modified from Yanful and Quigley, 1990).

Another example of this phenomenon is described for a plume by Fritz *et al.* (1976). At the time of this last investigation, the Confederation Road site was 23 years old and in certain sections appears to have been flushed with surface water whose circulation has been aided by a railway ditch along the north side of the landfill. The Na^+ and Cl^- profiles shown in Figure 9.20 were obtained on samples taken in 1984 at $t = 16$ years. They indicate an abrupt back-diffusion into the source leachate from the clay soil. A simple analysis of the back-diffusion profile suggests that this back-diffusion event had been initiated about 6–9 months prior to the time of drilling at this location. This plot demonstrates again the speed of diffusion over very short periods of time when there is a high concentration gradient. On the other hand, the diffusion profiles at depth are now migrating so slowly at $t = 20$ years (~10 mm/a) that it is no longer practical to study the rate of advance.

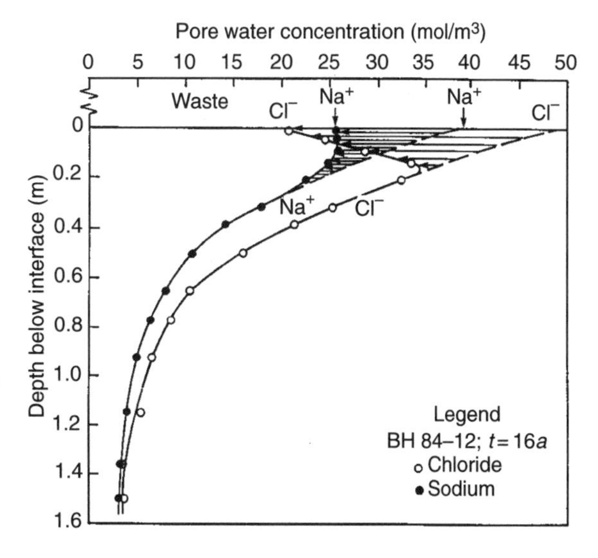

Figure 9.20 Chemical profiles for sodium and chloride showing diffusion from clay back into the waste due to site flushing.

9.3.2 Keele Valley landfill

A chloride ion diffusion profile and a series of cation diffusion profiles have been obtained at the Metropolitan Toronto and Keele Valley landfill site (King *et al.*, 1993).

The chloride data obtained from liner samples exposed by an exhumation are presented in Figure 9.21. The water samples for analysis were centrifuged from the sand layer and squeezed from the clay liner. The results for chloride show a smooth curve starting at the top of the sand layer and extending a total distance of 70–75 cm over a period of 4.25 years. A diffusion profile fitted to the data assuming negligible advection yielded a field value for $D_{Cl} \approx 6.5 \times 10^{-10}\,m^2/s$ ($0.02\,m^2/a$). This is nearly identical to values obtained in the field at the Confederation Road site (see Section 9.3.1) and on lab samples from Sarnia presented in Chapter 8 (Table 8.3).

The most important aspect of the chloride curve is that diffusion starts at the top of the partly clogged sand which appears to be acting as part of the barrier, adding an extra 30 cm of thickness with respect to diffusion. This is good from the perspective of diffusive transport but not so good from the perspective of leachate mounding. Fortunately, the hydraulic conductivity of the clay liner is so low (5×10^{-11} m/s) that advection is small even with the leachate mounding that has occurred to date.

Cation concentration profiles (Na^+, K^+, Mg^{2+} and Ca^{2+}) presented in Figure 9.22 also demonstrate a variety of important features. Na^+ has migrated about 65 cm and is thus only slightly retarded by adsorption onto the clays. The Na^+

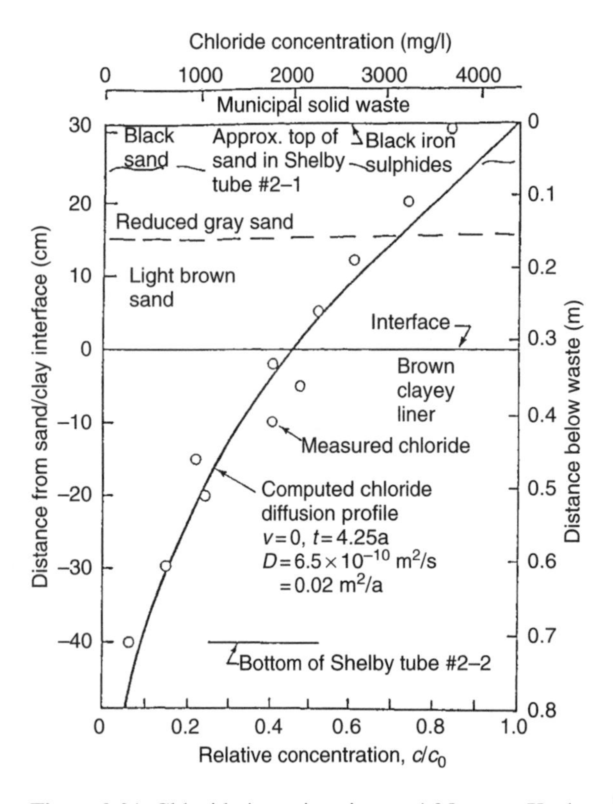

Figure 9.21 Chloride ion migration at 4.25 years, Keele Valley liner, Maple, Ontario (modified from Reades *et al.*, 1989).

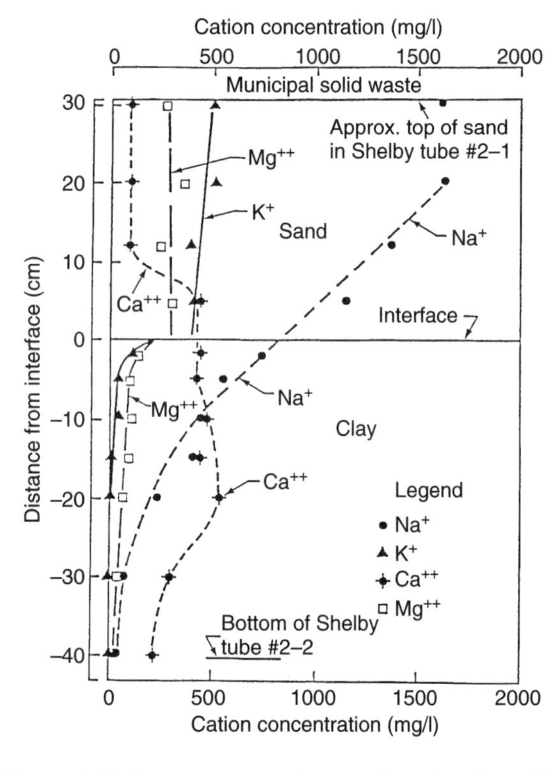

Figure 9.22 Pore water cation profiles for East Pit Shelby Tubes #2–1 and 2–2 (modified from Reades *et al.*, 1989).

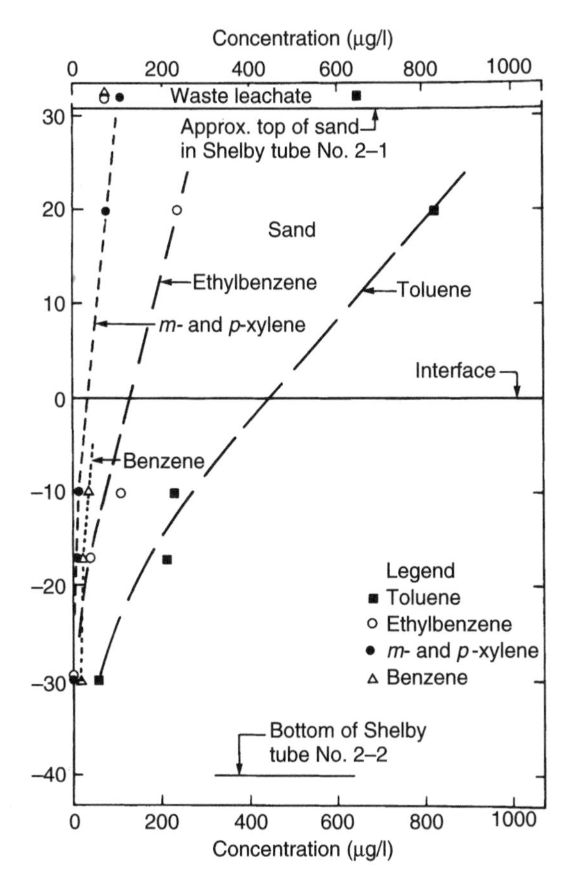

Figure 9.23 Pore water organic profiles at $t = 4.25$a in the Keele Valley landfill liner (modified from Barone *et al.*, 1993).

profile also starts at the top of the sand layer confirming the extra 30 cm of diffusion barrier thickness suggested by the chloride curve.

K^+ has migrated through the sand where it has not experienced significant retardation but only 5 cm into the clay demonstrating retardation by adsorption onto the liner clay. By implication, NH_4^+ has probably migrated about the same distance.

The Ca^{2+} profile forms a concentration hump throughout the depth examined. This hump is a hardness halo produced by desorption of Ca^{2+} that accompanies adsorption of Na^+, K^+ and possibly NH_4^+ and Mg^{2+}. If salinity probes are used to track the migration of salts, they should pick up the hardness halo and thus probably suggest migration further than it has actually progressed.

Finally, a series of organic profiles reported by Barone *et al.* (1993) are presented in Figure 9.23. These profiles indicate migration of low concentrations of volatile organic liquids to depths of ~60 cm in about 4.25 years.

The presence diffusion profiles have developed in the sand layer can be attributed to the effects of biological clogging of the upper portion on the sand layer (see Section 2.4.4) very shortly after waste placement began. The consequent reduction in hydraulic conductivity of the sand results in it behaving as part of the diffusion barrier system rather than as a granular drainage layer.

9.3.3 Other landfills

The writers have been involved with four field exhumations of clay liners varying in age from 2 to 10 years. Although these data cannot be presented in this book, it is noted that diffusion is clearly the dominant transport mechanism in each case. Furthermore, reasonable predictions of migration through these liners can be made using parameters obtained from the laboratory tests described in Chapter 8.

There is limited published data related to composite liner systems; however, as discussed more in Chapters 13 and 16, there is data indicating both leakage through holes in the geomembrane and diffusion through the barrier system. In particular, there is considerable potential for volatile organic compounds (VOCs) to diffuse through an intact geomembrane (see Section 6.8) and, experiencing relatively little retardation in clay (see Section 9.3.2 and Figure 9.23), to diffuse to the leak detection/secondary leachate collection systems. Although the pathways have not been unequivocally established, there is growing evidence of diffusive migration of organics through the primary liners into these collection systems. For example, Workman (1993) examined the chemistry of the fluid in the SLCS of a number of landfill cells and detected several VOCs including chloroethane, ethylbenzene and trichloroethylene at low concentrations. Workman hypothesized that these VOCs migrated as gases from the primary to secondary system at the side-slopes. While this may be a reasonable hypothesis, the alternative hypothesis of migration through the primary liner by a combination of diffusion and consolidation-induced advection warrants serious consideration. Other examples include (unpublished) detection of benzene, chloromethane, trichloroethylene and tetrahydrofuran (amongst others) in lysimeters at a number of US landfills.

9.4 Hydraulic conductivity of contaminated clay liners

9.4.1 The Confederation Road landfill

An interdisciplinary study published by Quigley *et al.* (1987a) appears to be one of the very few scientific field studies performed on the hydraulic conductivity of a contaminated clay beneath a domestic waste site. A brief review is presented here since it confirms that municipal waste leachate did *not* increase the hydraulic conductivity of inactive barrier clay confirming several laboratory studies which have employed MSW leachate as the permeant (Bowders *et al.*, 1986; Quigley *et al.*, 1988).

The site was described in Section 9.3.1. The hydraulic conductivity study consisted of oedometer k-testing of 3-inch (76 mm) Shelby tubes as Phase I and constant flow rate k-testing as Phase II.

A summary plot showing the oedometer k-test results along with water content and salt profiles for BH 83–2 representing $t = 15$ years is presented in Figure 9.24. The hydraulic conductivity profiles show approximately constant values except within about 20 cm of the clay/water interface where a slight decrease is observed. Of the two profiles presented in Figure 9.24, the

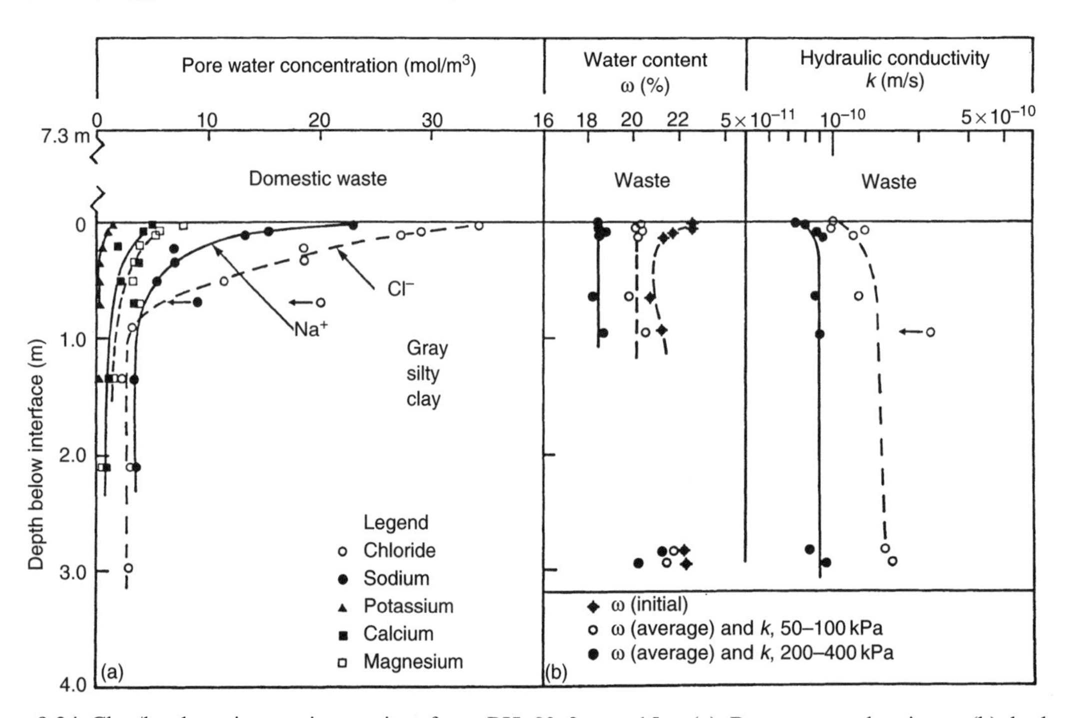

Figure 9.24 Clay/leachate interaction at interface, BH 83–2, $t = 15$a. (a) Pore water chemistry; (b) hydraulic conductivity and bulk water contents calculated from oedometer tests at pressures just below and above the preconsolidation pressure of 150 kPa.

hydraulic conductivity calculated for the stress range from 50 to 100 kPa (just below σ'_p) yielded values closest to the field k values of Goodall and Quigley (1977) and the directly measured k still to be discussed. Also shown in Figure 9.24 are profiles for Cl^- and the major cations. These profiles indicate contaminant migration to $\sim$1 m for this particular borehole (# 83–2). As a final comment, note that the average moisture of the clay for the two stress ranges used to calculate k is essentially constant. This is important since it suggests that the slight drop in hydraulic conductivity at the interface is chemically controlled.

If the pore water cation concentrations are used to calculate a value of sodium adsorption ratio (SAR), a profile like that shown in Figure 9.25 is obtained. This profile suggests up to 8% Na^+ on the interfacial clays decreasing to a background 3% Na^+ within 0.4 m of the interface. This should certainly cause at least a small reduction in hydraulic conductivity.

The results of constant flow rate k-testing on seven undisturbed tube samples from BH 84–10 (for $t = 16$ years) are presented in Figure 9.26. The k-test profile, obtained on samples with $\sigma'_v = 90$ kPa corresponding to the field stress, again implies a decrease in k in the zone within 20–30 cm of the interface. Unfortunately, the interface at this borehole location was 0.4 m higher than that at BH 83–2 used for the oedometer k-tests. This seems to include the lower part of the soil crust since the water content also decreased towards the interface suggesting complete ω_n control on k. Care was taken in doing these tests and pore water squeezed from adjacent soil samples was used as the permeant for the tests (Quigley *et al.*, 1987a).

The results of mercury intrusion porosimetry and bulk bacteriology run on the interfacial clays are presented in Figures 9.27 and 9.28. Both the modal and median pore diameters plotted in Figure 9.27 for freeze-dried specimens from BH 84–3 imply a decrease in pore size near the interface. Just how much of this must be apportioned to desiccation or heavy metal precipitation or freeze-drying damage is difficult to assess. Similarly, the bulk bacteriology which was done by culturing diluted wash extracts, not natural specimen counting, also suggests the possibility of bacterial clogging at the interface.

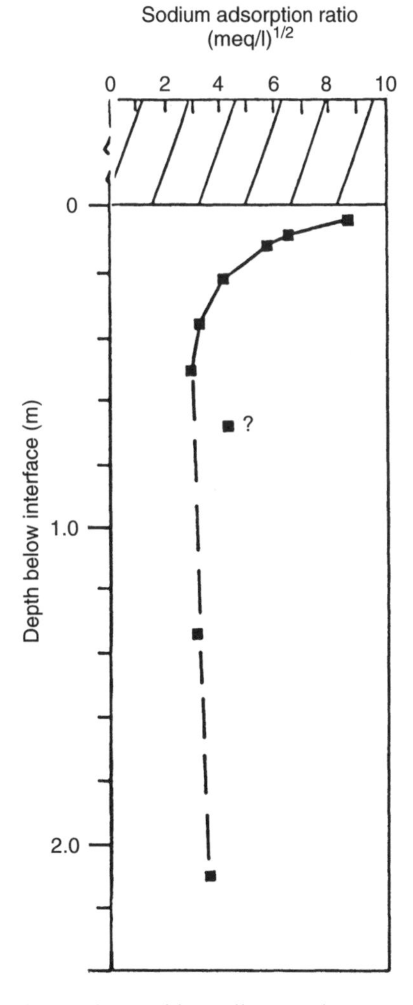

Figure 9.25 Exchangeable sodium cations on barrier clays, BH 83–2, $t = 15$ years; SAR $\approx$ % Na^+ (modified from Quigley *et al.*, 1984).

A summary plot of hydraulic conductivity versus bulk void ratio for all the oedometer k-tests and direct k-tests is presented as Figure 9.29. On the basis of this plot, one might conclude that municipal waste leachate has not

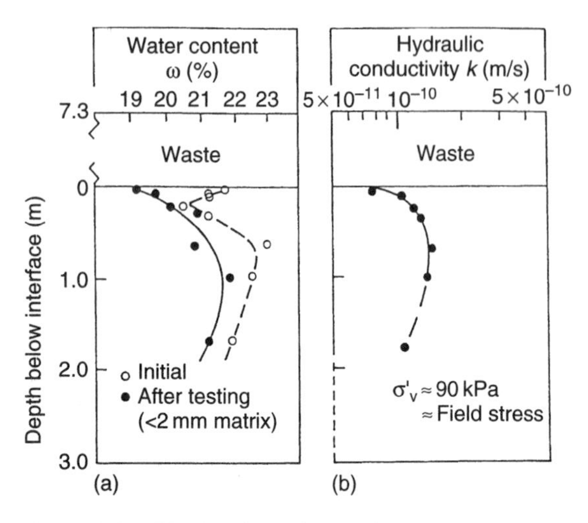

Figure 9.26 Clay/leachate interaction (BH 84–10): (a) initial (*in-situ*) and post-testing water contents of <2 mm soil matrix; (b) directly measured hydraulic conductivity using pore water as influent permeant.

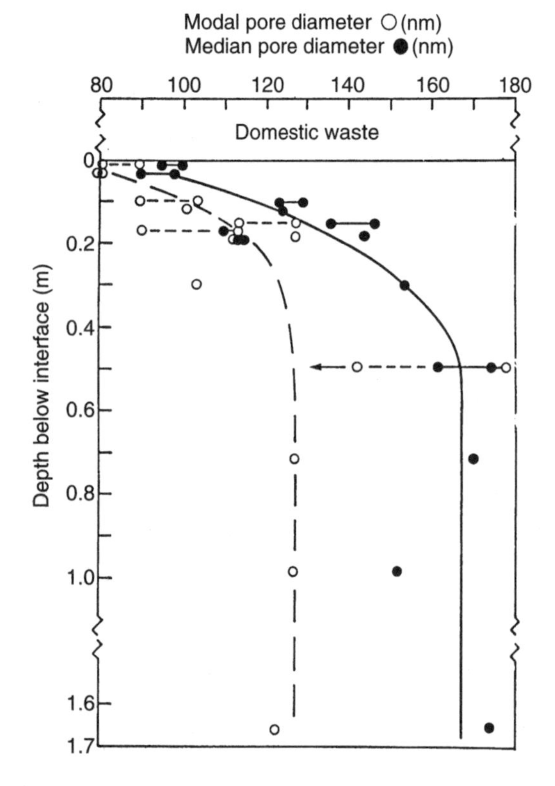

Figure 9.27 Median and modal pore size by mercury intrusion of freeze-dried samples (BH 84–3).

significantly altered the hydraulic conductivity of the inactive grey clays at Sarnia. It is possible that the range in values at constant void ratio could be assigned to chemical changes. As shown in the figure, this amounts to a decrease in k from about 1.4×10^{-10} to 1.0×10^{-10} m/s at a void ratio of 0.56.

9.4.2 Three Wisconsin liners

Gordon *et al.* (1989) have reported that clayey barriers at three Wisconsin landfill sites have significantly improved with time, based on data from field lysimeters. All three clays were residual soils derived from weathering of metamorphic rock, granite or dolomite and met the general criteria of: 50% < 200 mesh, liquid limit, $\omega_L \geq 30\%$; plasticity index, $I_p \geq 15$ and laboratory hydraulic conductivity of less than 1×10^{-9} m/s. The results of this study were previously summarized in Table 3.1.

9.4.3 Keele Valley landfill

Reades *et al.* (1989) and later King *et al.* (1993) have reported data which are reproduced in Figure 9.30 and Table 3.1 and which clearly show that once the waste was in place, the field (lysimeter) k-values decreased to $\sim 5 \times 10^{-11}$ m/s. The soil used at Keele Valley is a well-graded glacial till containing 20–28% <2 μm sizes.

9.4.4 Summary

On the basis of the above field studies and several other laboratory studies (Griffin *et al.*, 1976; Daniel and Liljestrand, 1984; Schubert *et al.*, 1984; Bowders *et al.*, 1986; Quigley *et al.*, 1987a), it appears that MSW leachate does not adversely affect the hydraulic conductivity of clayey barriers composed of inactive clay minerals illite and chlorite.

At least two important factors appear responsible for improving the performance of these clay liners with time. The first is replacement of

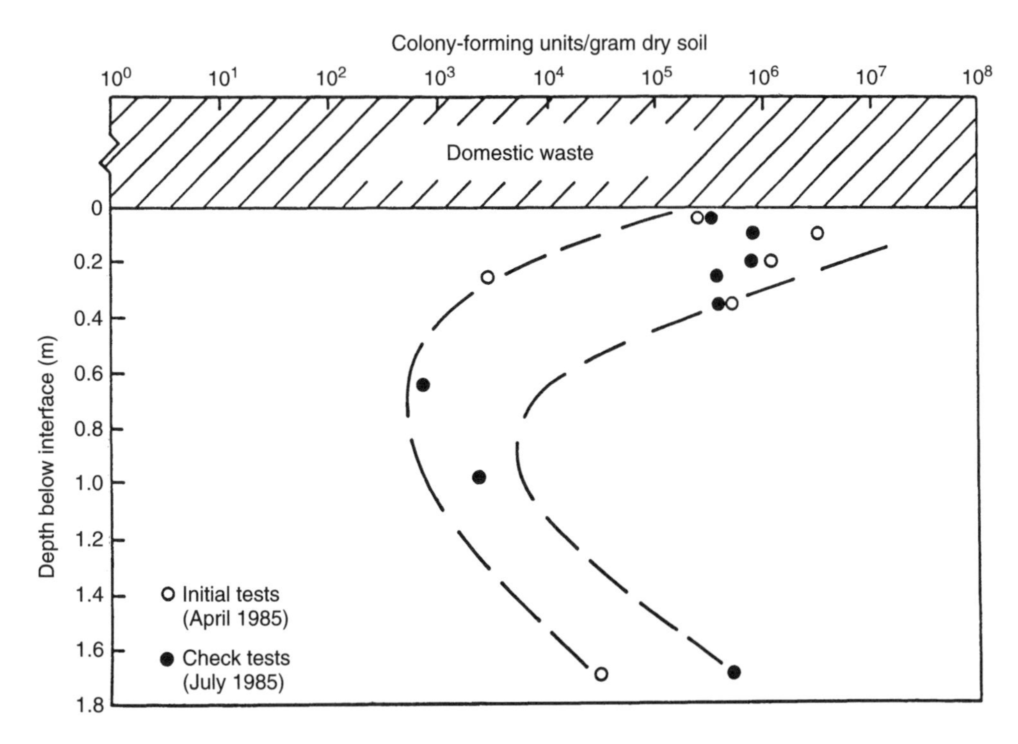

Figure 9.28 Total colony-forming bacterial units/gram of dry soil from hydraulic conductivity samples (BH 84–10) (after Quigley *et al.*, 1987a; reproduced with permission of *Canadian Geotechnical Journal*).

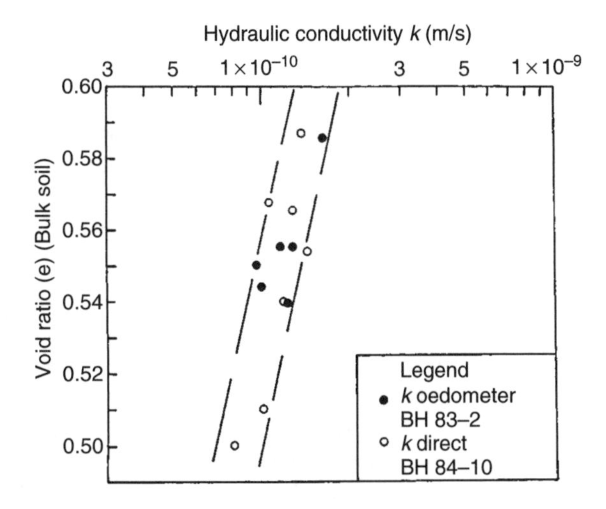

Figure 9.29 Hydraulic conductivity versus void ratio.

the Ca^{2+} and Mg^{2+} adsorbed on most natural clays by Na^+ in the MSW leachate. About 8% of the adsorption sites appear to pick up Na^+.

The second is consolidation caused by the increasing weight of waste applied to the wet-of-optimum liners. As noted earlier (Chapter 3), inert clayey soils placed about 2–3% wet-of-optimum have a preconsolidation pressure of about 120–150 kPa. Since they are heavily remolded, the clays appear to consolidate under the combined effects of stress increase and chemical migration by diffusion, resulting in high-quality liners over the long term.

It is suggested that compaction drier than 2% wet-of-Standard-Proctor optimum would produce a considerably stiffer clayey liner that would not be nearly as amenable to self-healing by consolidation.

Much work remains to assess the effects of various leachates on vermiculitic soils which are prone to *c*-axis contraction by K^+ and NH_4^+ fixation with resulting decreases in cation exchange capacity (CEC). Similar work has yet

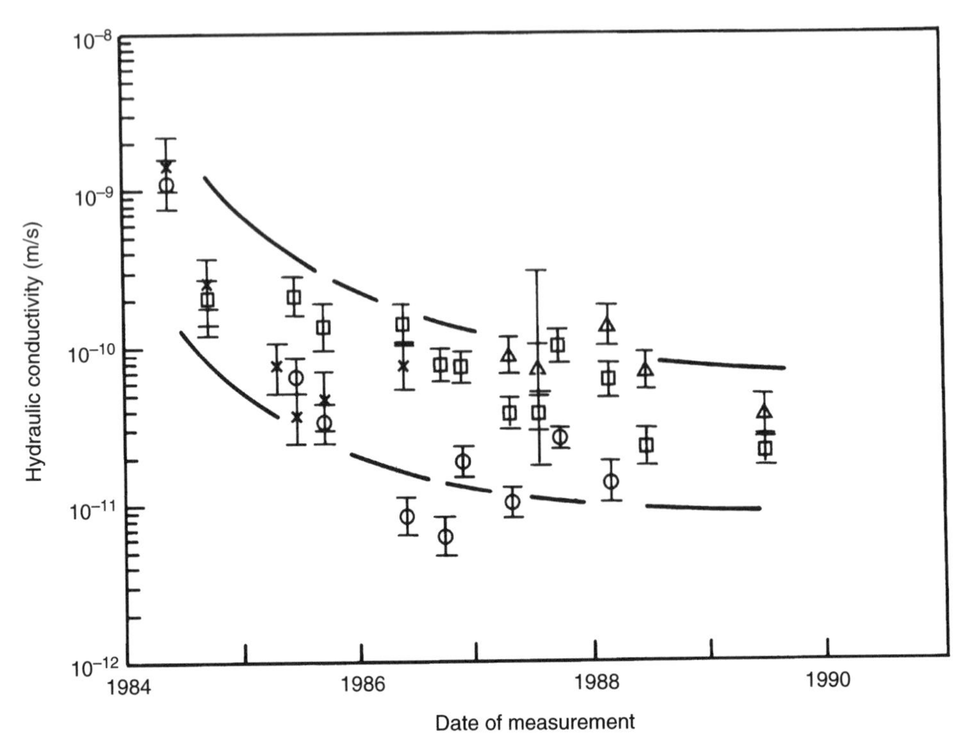

Figure 9.30 Estimated hydraulic conductivity based on shallow lysimeter effluent flow rates; Keele Valley landfill. Range bars represent possible error limits (modified from King *et al.*, 1993).

to be done on bentonites which are even more prone to *c*-axis contraction if highly peptized on placement. Fortunately, the bentonites are not likely to incur any decreases in CEC. Finally, much work remains to be done on bacterial activity within the mass of a barrier clay.

9.5 Performance of a composite lagoon liner during 14 years of operation

Due to the relatively short history of HDPE geomembrane usage in landfill applications, most of the reported case records are related to their use in liquid containment applications (Schmidt *et al.*, 1984; Hsuan *et al.*, 1991; Adams and Wagner, 2000). Field cases where HDPE geomembranes have been used in landfill bottom liner application have been reported by Rollin *et al.* (1994) and Eith and Koerner (1997).

This section summarizes a case record based on the exhumation of a 14-year-old composite liner from a leachate lagoon reported by Rowe *et al.* (2003a). The lagoon was placed into service in 1982 and was decommissioned after 14 years of service. The leachate contained typical inorganic constituents including transition and heavy metals.

The lagoon liner system consisted of a smooth 1.5-mm thick HDPE geomembrane overlying an approximately 3-m thick compacted clay liner. There was no protection layer above the geomembrane, which was directly exposed to the leachate and, above the leachate level, to the sunlight and atmosphere. At the time of decommissioning, there was a significant amount of liquid trapped between the geomembrane and the clay liner. According to the landfill operators, the lagoon had been drained several times during its lifespan to remove sludge from the lagoon and to patch geomembrane liner defects.

9.5.1 Defects in the geomembrane

Detailed mapping of the geomembrane revealed many defects (cracks, holes and patches) as summarized in Table 9.1. Cracks observed on the slopes where the HDPE geomembrane liner was not covered by the leachate and hence exposed to the sunlight and climate extremes were typically oriented down the slope and located near either seams or patches. The cracks observed at this site were similar to those reported by Peggs and Carlson (1989) and Hsuan (1999) who attributed the cracks to high thermal contraction stresses along the top of the slope adjacent to the trench where the geomembrane is completely restrained from contraction. Some of these cracks were up to 30-cm long suggesting high susceptibility of the geomembrane to cracking and hence that the geomembrane was relatively brittle.

Most of the holes were located on the slope portion above the leachate level (Table 9.1), with the majority being found on the east side where the service way was located. A total of 82 cracks, holes and patches having different forms, patterns and sizes were observed in the geomembrane, giving 528 defects per hectare over the 14-year period of operation. Of these, 70% (348 defects/ha) were above the leachate level and 30% (180 defects/ha) were below the leachate level.

It was evident that some of the defects were the result of maintenance activities. Protection of the geomembrane would have reduced damage during maintenance activities and provided protection from sunlight. Also the performance would likely have been enhanced if special care had been taken during sludge removal and if the travelled areas had been given additional protection.

9.5.2 Basic geomembrane properties after 14 years

Geomembrane samples were collected from a range of locations. All samples had very low values of oxidative induction time (OIT between 1.8 and 6 min compared to modern geomembranes which typically have a specified minimum

Table 9.1 Distribution of defects (cracks, holes and patches) on the geomembrane (modified from Rowe *et al.*, 2003a)

Defect	*Locations*[a]	*East*	*West*	*North*	*South*	*Bottom*	*Total*	*Defects*/ha
Cracks	Slope above leachate level	4	1	8	7	–	20	1,351
	Slope below leachate level	0	1	0	0	–	1	10
	Bottom	–	–	–	–	0	0	0
	Total cracks	4	2	8	7	0	21	
	Cracks/ha	136	69	277	241	0		135
Holes	Slope above leachate level	5	0	1	0	–	6	405
	Slope below leachate level	0	1	0	0	–	1	10
	Bottom	–	–	–	–	0	0	0
	Total holes	5	1	1	0	0	7	–
	Holes/ha	171	35	35	0	0	–	45
Patches	Slope above leachate level	17	4	5	5	–	31	2,095
	Slope below leachate level	5	4	5	3	–	17	168
	Bottom	–	–	–	–	6	6	154
	Total patches	22	8	10	8	6	54	–
	Patches/ha	751	277	346	275	154	–	348

[a]The "North" slope faces south, etc.

of 100 min). The crystallinity ranged from 65 to 67%, indicating a very high degree of crystallinity which was consistent with the high densities (0.965 and 0.967 g/cm^3) measured. One of the expected consequences of high crystallinity would be increased susceptibility of the geomembrane to stress cracking and indeed the stress cracking resistance (SCR; based on the single notched constant load test; ASTM D5397) was remarkably low with all specimens failing within 4 h compared to the 200 h specified for a new modern HDPE geomembrane (Hsuan and Koerner, 1997). Since the SCR data for the original geomembrane was not available, one cannot assess whether or not the low cracking resistance is due to ageing although this is likely a contributing factor given that the exposed geomembrane had a lower failure time (i.e., was more susceptible to stress cracking) than the covered samples. This finding is consistent with other properties that were measured (melt flow index, tensile properties and initial tear resistance) which also suggested that samples exposed to the sun had experienced more degradation than samples from other locations (e.g., bottom of the lagoon or anchor trench). These findings highlight the importance of proper geomembrane specifications (i.e., OIT, crystallinity and SCR). Typical values of Std-OIT of 100 min, CBC of 2–3%, crystallinity of 45–50% and SCR of 200 h as proposed by Hsuan and Koerner (1997) and Hsuan (2000b) should be a minimum for lagoon liners.

The importance of the compacted clay liner in minimizing advective and diffusive transport into the underlying groundwater system is emphasized from results presented herein. Without it, the lagoon would not have functioned properly.

9.5.3 Seams

Shear tests used to evaluate the quality and the integrity of the seams (ASTM D4437) indicated that, except for one sample, the shear strength ratios were well below today's typical quality control criteria (greater than 90%) with values of less than 70%. The peel tests also indicated very low seam–sheet strength ratios (<30%) compared to today's typical quality control criterion of 75% suggesting either very poor initial welding or a significant loss of peel strength during the 14 years of exposure.

9.5.4 Sorption and diffusion characteristics of the geomembrane

Sorption and diffusion tests (see Chapter 8 for a discussion) were conducted for seven organic chemicals representative of chlorinated hydrocarbons (dichloromethane, DCM; 1,2-dichloroethane, 1,2-DCA; and trichloroethylene, TCE) and aromatic hydrocarbons (BTEX: benzene, toluene, ethylbenzene and xylenes).

Partitioning coefficients, S_{gf} (Table 9.2), for the chlorinated aliphatic compounds ranged from a high of 60–68 for TCE, to about 10 for DCA and a low of 3–5 for DCM. For aromatic compounds, S_{gf} was the highest for *m*- and *p*-xylenes (205–220) and lowest for benzene (26–30). The S_{gf} values generally increased with increasing *n*-octanol/water coefficient ($\log K_{ow}$) (which indicates the chemical hydrophobicity, and hence the ability of the chemical to partition with organic material).

Although there was some (relatively small) variability in the results for a given compound, there was no significant or consistent difference between the results from the different sample locations. This may be attributed to the fact that crystallinity of the samples was similar at the three locations (with averages ranging from 65.5 to 67.5%).

Rowe *et al.* (2003a) compared the diffusion and sorption characteristics of the 14 years old geomembrane with those obtained by Sangam and Rowe (2001) for an unaged modern HDPE geomembrane (Table 9.2). The major difference between the two geomembranes is the crystallinity (65–68 and 47% for the old and new modern geomembranes, respectively). The permeation

Table 9.2 Comparison of permeation coefficients for the 14 years old geomembrane with the new modern geomembrane (modified from Rowe *et al.*, 2003a)

Chemicals	*New GM** ($\chi = 47\%$) S_{gf}	*New GM** ($\chi = 47\%$) $P_g (10^{-12}\ m^2/s)$	*Old GM* ($\chi = 65.5$–67.5%) S_{gf}	*Old GM* ($\chi = 65.5$–67.5%) $P_g (10^{-12}\ m^2/s)$	P_g (New GM)/P_g (Old GM) (–)
Chlorinated hydrocarbons					
Dichloromethane	6	3.9	3–5	1.4–2.3	1.7–2.8
1,2-Dichloroethane	12	4.8	10	1.6–1.8	2.7–3.0
Trichloroethylene	85	34.0	60–68	9.6–11.6	1.9–3.5
Benzene	30	10.5	26–32	3.1–4.2	2.5–3.4
Toluene	100	30.0	65–70	7.8–10.8	2.9–3.8
Ethylbenzene	285	51.3	150–160	9.8–12.0	3.6–4.6
m- and *p*-Xylene	347	59.0	205–220	12.3–17.6	3.4–4.8
o-Xylene	260	36.0	180–200	8.0–13.3	3.2–4.5

χ: crystallinity.
*From Sangam and Rowe (2001).

coefficient, P_g, was substantially higher for the new geomembrane than for the old geomembrane (by a factor of between 1.7 and 4.8). The difference in permeation coefficient between the two geomembranes is compound specific with the smallest difference being for DCM (which has the lowest molecular volume) and generally increases with increasing molar volume. Benzene and TCE have a similar molar volume and a similar difference in permeation coefficients between old and new geomembranes. The greatest difference was for ethylbenzene and xylenes (which had the largest molar volumes). The reduction in permeation coefficient may be attributed to the high crystallinity of the samples. As indicated by Naylor (1989) and Rogers (1985), the crystalline zones in semi-crystalline polymers act as relatively impermeable barriers to the migrating molecules by (1) reducing the sorptive and diffusive regions, and (2) restraining the mobility of the polymer molecules required for the accomplishment of the diffusive jump. As the segmental mobility of the chains is restrained, the diffusion process becomes more dependent on size and shape of the migrating molecule (Rogers, 1985; Naylor, 1989).

The results obtained by Rowe *et al.* (2003a) suggest that VOC diffusion coefficients will decrease with increasing crystallinity of the geomembrane. Thus if the crystallinity increases with geomembrane ageing, as reported by several investigators (e.g., Sangam, 2001), then it is conservative to use diffusion coefficients of the unaged material in design.

9.5.5 Compacted clay liner

Significant portions of the clay liner below the geomembrane and above the leachate level were desiccated. This phenomenon has been observed by others (Basnett and Brungard, 1992; Corser *et al.*, 1992). Below the leachate level, the clay appeared saturated and was covered by a thin layer of black sludge. Continuous core samples were taken from the clay liner at five locations. The water contents measured ranged from 38% at the top of the compacted clay liner to about 20–24% at a depth of about 0.8 m. The water contents of 20–24% are consistent with unpublished construction reports indicating that the clay liner was compacted at 2–4% wet-of-optimum with water contents ranging from 15.5 to 23%. The higher water contents obtained at the upper part of the clay liner were attributed to swelling of the clay in contact with leachate trapped between the geomembrane and clay liner combined with the negligible effective

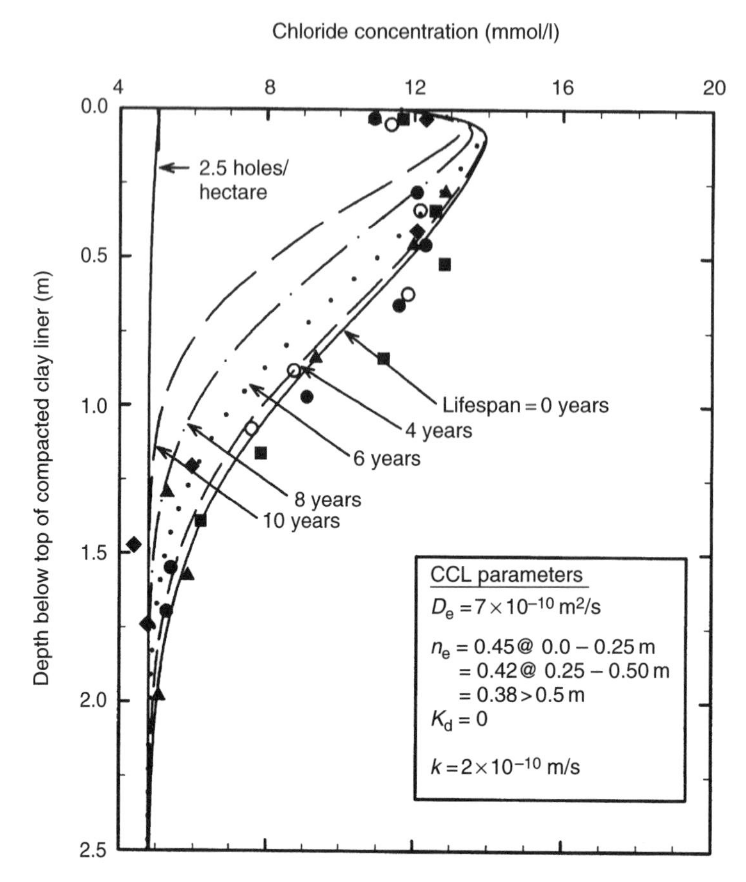

Figure 9.31 Chloride concentrations through a compacted clay liner used as part of a composite liner for a leachate lagoon together with prediction of pore fluid concentration for different assumed geomembrane service lives. Observed profiles based on samples from five boreholes (modified from Rowe *et al.*, 2003a).

stress at and near the clay surface. A contributing factor may have been biological activity taking place at and near the interface (associated with the black sludge noted earlier) similar to the observations by Hrapovic (2001).

Pore water concentration profiles (Figure 9.31) in the clay liner after 14 years of service showed that chloride had migrated approximately 1.7 m which is consistent with expectations based on previous field cases in which the leachate was in direct contact with a clay liner (Section 9.3.1). Although the clay liner appeared to have performed well (there was over 1.0 m of uncontaminated clay below the contaminated zone) in protecting the environment during the service period, the significant chloride profile raises questions regarding the effectiveness of this particular geomembrane liner in impeding the migration of chloride.

The average hydraulic conductivity of the clay liner (below the leachate level) was about 2.2×10^{-10} m/s. The chloride diffusion coefficient for the clay measured in the laboratory was about 7×10^{-10} m^2/s.

9.5.6 Effectiveness of the geomembrane liner

Intact HDPE geomembranes are an excellent barrier to advective migration and an excellent diffusive barrier against inorganic and polar contaminants (Chapter 13). Chloride is an inorganic ion that has a very low diffusion coefficient through HDPE geomembranes (Rowe *et al.*, 1995b, 1997a). If the HDPE geomembrane examined by Rowe *et al.* (2003a) had remained intact for a significant period of time, very little chloride would be expected in the compacted clay liner. Conversely, the fact that significant concentrations of chloride (above background levels) were found in the compacted clay suggests that the geomembrane had failed to perform its intended task.

The observation that a significant amount of leachate was trapped under the geomembrane and in direct contact with the compacted clay leads to the conclusion that, at some point during the lifetime of the lagoon, the geomembrane stopped functioning as an effective hydraulic barrier. A comparison of the observed chloride profiles with those calculated assuming that the geomembrane was ineffective at different times after construction (Figure 9.31) suggests that the geomembrane ceased functioning effectively somewhere between 0 and 4 years after construction.

9.5.7 Implications for composite liner design

The composite liner examined by Rowe *et al.* (2003a) was constructed in the early 1980s at a time when proper construction quality control was not ensured for small lagoons. However, the observations of poor seam quality and poor geomembrane performance highlight the need for good design (including the selection of good materials), the use of experienced installation personnel and good construction control/assurance procedures, and the need for appropriate maintenance procedures. The geomembrane did not perform its design function adequately for more than a few years. However, because the compacted clay liner was 3-m thick and well constructed, chloride had only diffused 1.7 m and failure of the geomembrane had not caused any adverse impact on the groundwater. This highlights the benefits of the clay liner as a backup to the geomembrane.

Chapter ten

Contaminant migration in intact porous media: analysis and design considerations

10.1 Introduction

The foregoing chapters have introduced the concepts of the various contaminant transport mechanisms, and have discussed the determination of parameters, the character and mounding of leachate, and methods of analysis.

The objective of this chapter is twofold. First, to discuss the mass of contaminant and its influence on the potential impact of a landfill on groundwater, and, second, to examine how some very simple finite layer models can be readily used to assess the impact of contaminant migration through clay barriers for a range of situations and combinations of parameters. Many of the examples relate to clayey barriers underlain by a thin natural aquifer (e.g., Figures 1.2–1.4) since this is a common occurrence in many areas; however, similar analyses can also be performed for compacted clay liners constructed over an engineered drainage layer (e.g., see Figures 1.5–1.8) or for designs involving cut off walls or permeable surrounds (e.g., see Figure 1.9).

10.2 Mass of contaminant, the reference height of leachate, H_r, and the equivalent height of leachate, H_f

For waste disposal sites such as municipal landfills, the mass of any potential contaminant within the landfill is finite. The process of collecting and treating leachate involves the removal of mass from the landfill and hence a decrease in the amount of contaminant which is available for transport through the barrier system and into the general groundwater system. Similarly, the migration of contaminant through the barrier also results in a decrease in the mass available within the landfill. For a situation where leachate is continually being generated by percolation of water through the landfill cover, the removal of mass by either leachate collection and/or contaminant migration will result in a decrease in leachate strength with time and there will be a decrease in concentration similar to that observed in the laboratory tests described in Chapter 8.

The impact of a waste disposal site upon groundwater quality is usually judged by monitoring the concentration of potential contaminants at a number of specific monitoring points. As will be demonstrated in this chapter, the variation in concentration with time at these points will be a function of the mass of contaminant in the system, the infiltration through the landfill cover, the proportion of leachate collected and the proportion of leachate passing into the hydrogeological system.

A reasonable estimate of peak concentration, c_0, of a given contaminant species can usually be obtained from past experience with similar landfills (e.g., see Section 2.3). The total mass of contaminant is more difficult to determine. Nevertheless, estimates can be made by considering the observed variation in concentration with time at landfills where leachate concentration has been monitored (e.g., see Section 2.3.4) or by considering the results of lysimeter tests on waste.

Until fairly recently, there has been a paucity of data concerning the available mass of contaminants within landfills; however, this situation is changing now that many landfills have leachate collection systems. Given that concentration is simply mass per unit volume, the mass of a given contaminant collected in a year is equal to the average concentration multiplied by the volume of leachate collected. By monitoring how this mass varies with time, it is then possible to estimate the total mass of that species of contaminant within the landfill. In the absence of this information, studies of the composition of waste (see Section 2.3.4) can be used to estimate the mass of given contaminant or groups of contaminants. For example, Table 2.12 summarized an estimate of refuse composition reported by Hughes *et al.* (1971). For contaminant species predominantly formed from breakdown or synthesis of other species (e.g., by biological action), an upper bound estimate of the mass of contaminant may be obtained from the estimates of the mass of chemicals which go to form the derived contaminant.

For the purposes of modelling the decrease in concentration in the leachate due to movement of contaminant into the collection system and through the barrier, it is convenient to represent the mass of a particular contaminant species in terms of parameters defined as the "reference height of leachate" H_r, or the "equivalent height of leachate", H_f, as described in the following sections.

10.2.1 Reference height of leachate, H_r

On the simplest level, suppose that the infiltration (percolation) into the landfill is q_0, the exfiltration through the liner is q_a and the leachate collected (per unit area) is q_c then, assuming the landfill is at field capacity, continuity of flow requires that

$$q_0 = q_c + q_a \tag{10.1}$$

If the initial mass of a contaminant species (e.g., chloride) can be estimated, m_{TC}, then the reference volume of leachate which would contain this mass at an initial concentrating c_0 is

$$V_{TC} = \frac{m_{TC}}{c_0} \tag{10.2}$$

In general, this volume will not correspond to the actual volume of leachate because it is based on the assumption that all this available mass can be quickly leached from the solid waste. It is convenient for both mathematical and physical reasons to express the volume V_{TC} in terms of a reference height of leachate, H_r, which is defined as the reference volume of leachate, V_{TC}, divided by the area, A_0, through which contaminant passes into the primary barrier, i.e.,

$$H_r = \frac{V_{TC}}{A_0} \tag{10.3a}$$

or

$$H_r = \frac{m_{TC}}{c_0 A_0} \tag{10.3b}$$

where H_r is the reference height of leachate [L], m_{TC} the total mass of a contaminant species of interest [M], c_0 the peak concentration of that species in the landfill [ML^{-3}] and A_0 the Area through which contaminant can migrate into the underlying layer [L^2].

An equation can then be written for conservation of mass as indicated in Figure 10.1 and, as shown, this can be reduced to

$$c_T(t) = c_0 + c_r t - \frac{1}{H_r}\int_0^t f_T(c,\tau)\,d\tau - \frac{q_c}{H_r}\int_0^t c_T(\tau)\,d\tau \tag{10.4a}$$

which, on substitution for the flux $f_T(c,\tau)$ (equation 1.8), becomes

$$c_T(t) = c_0 + c_r t - \frac{1}{H_r}\int_0^t \left(nvc_T(\tau) - nD\frac{\partial c}{\partial z}\right) d\tau - \frac{q_c}{H_r}\int c_T(\tau)\,d\tau \tag{10.4b}$$

Note that as H_r approaches infinity, $c_T = c_0 + c_r t$ and for no increase in mass ($c_r = 0$), c_T becomes constant at a value of c_0.

Equation 10.4b may be used for contaminants which do not experience any first-order decay (e.g., conservative contaminants such as chloride). However, some contaminant species may also experience first-order biological or radioactive decay and for these contaminants, equation 10.4b is replaced by

$$\begin{bmatrix}\text{Mass of contaminant}\\ \text{in the landfill at time } t\\ m_t = A_0 H_r c_T(t)\end{bmatrix} = \begin{bmatrix}\text{Initial mass}\\ \text{of contaminant}\\ m_{TC} = A_0 H_r c_0\end{bmatrix} + \begin{bmatrix}\text{Increase in}\\ \text{mass deposited}\\ m_{IC} = A_0 H_r c_r t\end{bmatrix} - \begin{bmatrix}\text{Mass which has}\\ \text{passed into the}\\ \text{soil up to time } t\\ A_0\int_0^t f_T(c,\tau)d\tau\end{bmatrix} - \begin{bmatrix}\text{Mass collected}\\ \text{by the leachate}\\ \text{collection system}\\ \text{up to time } t\\ A_0\int_0^t q_c c_T(\tau)d\tau\end{bmatrix}$$

$$\therefore\ m_t = m_{TC} + m_{IC} - A_0\int_0^t f_T(c,\tau)d\tau - q_C\int_0^t c_T(\tau)d\tau$$

and

$$A_0 H_r c_T(t) = A_0 H_r c_0 + A_0 H_r c_r t - A_0\int_0^t f_T(c,\tau)d\tau - A_0 q_c\int_0^t c_T(\tau)d\tau$$

Where A_0 = Area of landfill through which contaminant can pass into the soil
H_r = 'Reference height of leachate'
$A_0 H_r c_r$ = Increase in mass per unit time due to deposition
$f_T(c,\tau)$ = Surface flux (mass per unit area per unit time) passing into soil
$A_0 q_c$ = Volume of leachate collected per unit time

Thus dividing by the equivalent volume of leachate ($A_0 H_r$) gives

$$c_T(t) = c_0 + c_r t - \frac{1}{H_r}\int_0^t f_T(c,\tau)d\tau - \frac{q_c}{H_r}\int_0^t c_T(\tau)d\tau$$

Figure 10.1 Conservation of mass in a landfill with contaminant inputs and leachate collection.

$$c_T(t) = c_0 + c_r t - \frac{1}{H_r}\int_0^t \left(nvc_T(\tau) - nD\frac{\partial c}{\partial z}\right) d\tau$$
$$- \left(\frac{q_c}{H_r} + \lambda_T\right)\int_0^t c_T(\tau)\, d\tau \qquad (10.4c)$$

where λ_T is the first-order decay constant as defined in Sections 1.3.6 and 1.5.

It is a relatively simple matter to formulate a mathematical model (e.g., program POLLUTE; Rowe and Booker, 2003) to incorporate equations 10.4a–10.4c (above) for the surface boundary condition, where c_0, c_r, H_r and q_c are specified boundary parameters and D, n, and v are as defined in Chapter 1 (see also Appendix B) and are specified for the layers of the soil deposit.

10.2.2 Equivalent height of leachate, H_f

The equation for source concentration given by equations 10.4a–10.4c is the most rigorous for the case of a landfill with a leachate collection system. An alternative (approximate) formulation in place of equation 10.4b is given as

$$c_T(t) = c_0 - \frac{1}{H_f}\int_0^t \left(nvc(\tau) - nD\frac{\partial c}{\partial z}\right) d\tau \qquad (10.5)$$

can also be used, where the "equivalent height of leachate" H_f is given by

$$H_f = H_r\frac{q_a}{q_0} \qquad (10.6a)$$

or

$$H_f = \frac{m_{TC}}{c_0 A_0} \times \frac{q_a}{q_0} \qquad (10.6b)$$

Here, H_f can be derived as shown in Figure 10.2 and represents the mass of contaminant available for transport into the hydrogeologic system and excludes the mass that is collected by leachate collectors.

As noted above, q_0 represents the volume of leachate generated within the landfill (per unit area) and can usually be taken to be equal to the infiltration into the landfill due to percolation through the landfill cover. The quantity q_a may be defined as the average mass flux into the barrier normalized (divided) by the average concentration within the landfill and referred to here as the normalized average flux. For situations where advection is the dominant transport mechanism, q_a is approximately equal to the Darcy velocity, v_a (i.e., $v_a = nv = ki$ where k is the hydraulic conductivity of the barrier and i the outward hydraulic gradient in the barrier). However, for situations where diffusion is a significant transport mechanism, the determination of the normalized average flux q_a is a little more

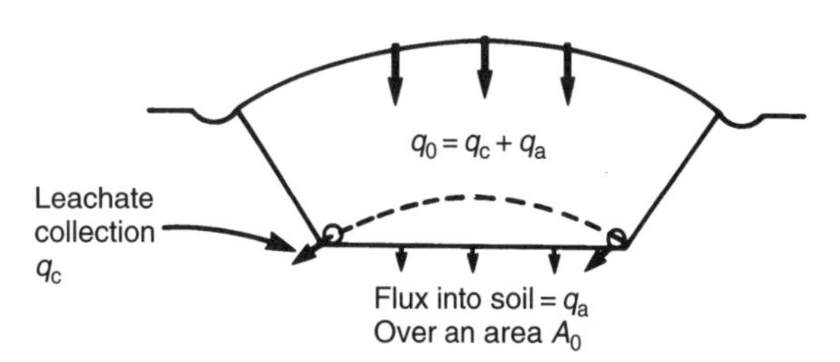

Proportion of contaminant which can pass into the soil	$= q_a/q_0$
Maximum 'initial' concentration of contaminant species	$= c_0$
Total mass of contaminant species in the waste	$= m_{TC}$
Mass of contaminant likely to be transported into the soil:	$m_0 = m_{TC} \cdot q_a/q_0$
Equivalent volume of leachate = mass/initial/concentration:	$V_0 = m_0/c_0$
Equivalent height of leachate = (volume)/(area ⊥ flow):	$H_f = V_0/A_0$
i.e.	$H_f = \frac{m_{TC}}{c_0 A_0} \cdot \frac{q_a}{q_0}$

Figure 10.2 Calculation of the equivalent height of leachate.

complicated but can be readily estimated as outlined below.

The mass flux $f(\tau)$ of contaminant into the barrier beneath the landfill at time τ is given by

$$f(\tau) = nvc(\tau) - nD\left(\frac{\partial c}{\partial z}\right) \tag{10.7}$$

where c_T and $\partial c/\partial z$, respectively, represent the concentration and concentration gradient at the boundary between the landfill and the barrier. Up to some time t, the total mass of contaminant passing into the soil is obtained by integrating equation 10.7. Thus, the average mass flux, f_a, is obtained by dividing the total mass by the period t to give

$$f_a = \int_0^t \frac{f(\tau)}{t} \mathrm{d}\tau \tag{10.8}$$

The normalized average flux into the barrier over this time period is then obtained by dividing the average mass flux by the average concentration c_a, viz.

$$c_a = \int_0^t \frac{c(\tau)}{t} \mathrm{d}\tau \tag{10.9}$$

giving

$$\begin{aligned} q_a &= \frac{f_a}{c_a} = \frac{\int_0^t f(\tau)\,\mathrm{d}\tau}{\int_0^t c(\tau)\,\mathrm{d}\tau} \\ &= \frac{nv \int_0^t c(\tau)\,\mathrm{d}\tau - nD \int_0^t \left(\frac{\partial c}{\partial z}\right) \mathrm{d}\tau}{\int_0^t c(\tau)\,\mathrm{d}\tau} \end{aligned} \tag{10.10a}$$

and as previously discussed, for the case where D/v approaches zero (i.e., when advection governs) equation 10.10a reduces to

$$q_a = nv = v_a \tag{10.10b}$$

where v_a is the Darcy velocity through the barrier.

Using mathematical modelling (e.g., Program POLLUTE), the contaminant flux, $f(\tau)$, into the clay barrier can be calculated for any combination of advection and diffusion. The average mass flux, f_a, into the liner can be automatically determined and the normalized average flux, q_a, can then be calculated from equation 10.10a.

Although in design cases where POLLUTE is used one would model the leachate collection explicitly (by specifying H_r and q_c, and the program would use equation 10.4 for the surface boundary condition), it is of interest to examine the impact in terms of the mass available for transport into the hydrogeological system (using H_f) especially for generic studies such as those reported in this and the following chapter. For these situations, one can readily obtain an estimate of H_f from a simple hand calculation as described below.

10.2.3 A simple estimate of q_a

Consider a barrier of thickness H as shown in the inset to Figure 10.3. For the purposes of estimating H_f, it is conservative to determine the normalized average flux assuming (i) the concentration in the landfill remains constant and (ii) the concentration in the aquifer is zero. The results obtained for this case are shown in Figure 10.3 for a barrier of thickness greater than or equal to 1 m. (For thicknesses less than 1 m, this plot may underestimate q_a.) Here, the normalized average flux (q_a) is plotted against the Darcy velocity through the liner for a range of values of the product nD where n is the porosity of the barrier and D the effective diffusion coefficient of the contaminant being considered.

The results presented in Figure 10.3 may be conservatively used for situations where the concentration in the leachate decreases with time and/or the concentration beneath the barrier is greater than zero.

Advection controls contaminant transport when the normalized average flux, q_a, is approximately equal to the Darcy velocity, v_a. For

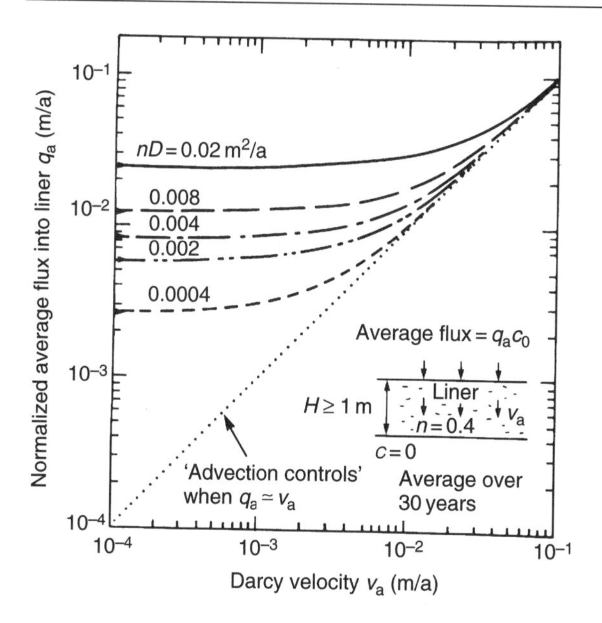

Figure 10.3 Relationship between normalized average flux into a liner and the Darcy velocity for a range of diffusion coefficients (modified from Rowe, 1988).

typical situations involving clayey barriers, this will be the case for Darcy velocities greater than 0.03 m/a. For velocities less than this, diffusion may noticeably affect the normalized average flux. Even if there was no flow ($v_a = 0$), contaminant would still diffuse into the barrier and hence there is a minimum value of q_a for a given diffusion coefficient as shown in Figure 10.3.

The results given in Figure 10.3 are based on an average flux over a 30-year time period. This is considered to be a reasonable averaging period for many practical situations. The normalized average flux will in fact decrease with increasing time period t. For example, putting aside the case where there is inward flow, the minimum value of the flux will correspond to steady-state diffusion ($v = 0$) and can be calculated from equation 10.10a, viz.

$$q_a(\text{minimum}) = \frac{nD}{H} \qquad (10.10c)$$

Clearly, if there is flow into the landfill, which opposes the outward diffusive flux, the normalized average flux would be even smaller than this (and can be calculated from equation 10.10a).

10.2.4 Example calculation of q_a/q_0

To illustrate the use of Figure 10.3, consider a clayey barrier with a porosity of 0.4 and a contaminant with an effective diffusion coefficient of $0.01\,m^2/a$ (i.e., $nD = 0.004\,m^2/a$).

(1) If the Darcy velocity into the barrier were 0.03 m/a, then from Figure 10.3 the normalized average flux q_a would also be approximately 0.032 m/a. If the infiltration (percolation) through the landfill cover, q_0, was 0.3 m/a, then the proportion of mass entering the system q_a/q_0 would be 0.1 (i.e., $m_0 = 0.1 m_{TC}$ or 10% of the total mass is available for transport into the barrier).

(2) If the Darcy velocity was only 0.003 m/a, then from Figure 10.3, $q_a \approx 0.01$ m/a (for $nD = 0.004\,m^2/a$). Thus the tenfold decrease in Darcy velocity compared to case (1) has only reduced the normalized flux by a factor of 3 and in this case diffusion has a significant influence on contaminant transport into the barrier. Again assuming $q_0 = 0.3$ m/a, this case would correspond to $m_0 = 0.03 m_{TC}$ or 3% of the total mass being available for transport into the barrier; the remaining 97% is being collected by the leachate collection system.

10.2.5 Example calculation of H_r and H_f

Suppose that an examination of the composition of typical municipal waste indicates that chloride represents less than 0.2% of the total mass of the waste in the landfill. For a proposed landfill of area A_0 of 50 ha with a total mass of waste of 2 Mt, this corresponds to a total mass of chloride in the waste $m_{TC} = 0.002 \times 2 \times 10^6\,t = 4{,}000\,t$. Supposing that the peak concentration of chloride $c_0 = 2{,}000$ mg/l = ($2{,}000\,g/m^3$) then from equation 10.3b,

$$H_r = \frac{m_{TC}}{c_0 A_0} = \frac{4,000 \times 10^6}{2,000 \times 50 \times 10^4} = 4\text{m}$$

and hence from equation 10.6a,

$$H_f = H_r \times \frac{q_a}{q_0}$$

The value of q_a may be evaluated as described above. The infiltration q_0 through the landfill cover must be estimated. It should be noted that when estimating contaminant impact it is important to be realistic in the estimation of the value of q_0. For example, it is *not* conservative to use a design value of $q_0 = 0.3$ m/a if the realistic infiltration is, say, 0.15 m/a.

Inspection of equation 10.3b or 10.6b shows that for landfills resting on a barrier, the reference height of leachate and equivalent height of leachate are proportional to the total mass of contaminant, m_{TC}, divided by the plan area, A_0. This represents the mass per unit area and is directly related to the height of the waste mound. Thus, for a given height of waste, H_r and H_f are the same for a landfill with an area of 50 ha and a total mass of 2 Mt as they are for a landfill with an area of 20 ha and a total mass of 0.8 Mt.

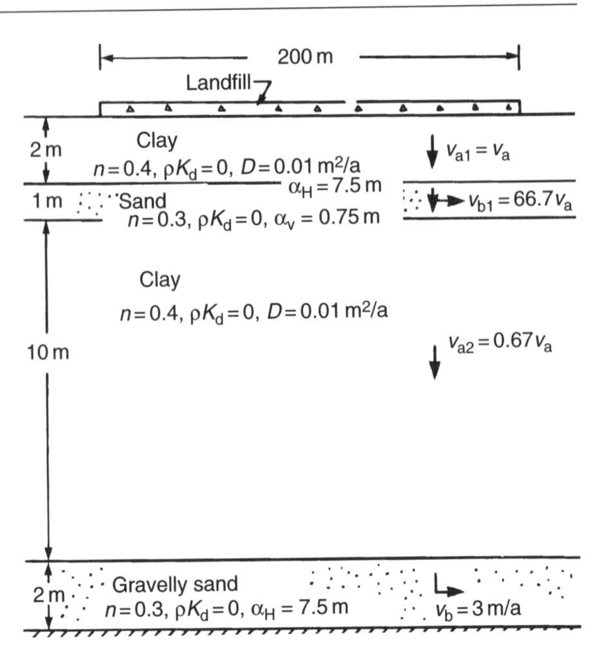

Figure 10.4 Soil profile considered in Example A.

10.2.6 Effect of considering the finite mass of contaminant

To illustrate the significance of parameters such as the equivalent height of leachate H_f and the downward Darcy velocity through the barrier, consideration will be given here to the potential impact of a hypothetical landfill on groundwater quality at the site boundary, taken to be 100 m downgradient from the landfill. For this case, denoted as "Example A", the assumed hydrostratigraphy is shown in Figure 10.4.

The migration of contaminant was modelled using the 2D finite layer solution to the 2D advection–dispersion equation for a multilayered system described in Chapter 7. The input to the model consists of the horizontal and vertical components of the Darcy velocity, the distribution coefficient and the coefficient of hydrodynamic dispersion for each layer, together with the density, porosity and thickness of each layer. In addition, it is necessary to specify the initial concentration of contaminant in the landfill and the equivalent height of leachate (which represents the mass of contaminant available for transport into the soil). It is noted that this model directly considers the variation in concentration with time within the leachate as contaminant is removed from the landfill. The model also considers the mechanism of attenuation due to diffusion of contaminant from the granular units into the clayey aquitard (above and below) as it moves from beneath the landfill towards the boundary of the site. This diffusion from the aquifer will reduce concentrations at the site boundary.

It was shown above that the mass of a given contaminant available for transport into the barrier can be expressed in terms of the equivalent height of leachate H_f. For the case shown in the inset to Figure 10.5, the concentration of a conservative contaminant in the leachate is plotted against time for three different assumed values of H_f and for a downward Darcy velocity, v_a, of 0.03 m/a. If the mass of contaminant is infinite ($H_f \to \infty$), then the concentration of contaminant within the landfill remains constant for all time. Conversely, the assumption of a constant source concentration is equivalent to assuming an infinite mass of contaminants. For the most realistic assumption of a finite mass of contaminant, the calculated concentration in the leachate decreases with time as contaminant is removed from the landfill. The rate of decline is related to the mass of contaminant and hence H_f. For example, for $H_f = 0.8$ m, the calculated concentration in the landfill has reduced to one-third the original value after 30 years whereas for $H_f = 0.4$ m it takes about 15 years for the same reduction to occur. It is also apparent from this that the mass of contaminant can be back-figured if the variation in concentration has been monitored in the leachate over a sufficiently long period.

The effect of the equivalent height of leachate upon the calculated impact of the landfill or water quality at a point "x" in the upper aquifer 100 m downgradient of the landfill is shown in Figure 10.6. If one assumes that the concentration of contaminant in the leachate remains constant (i.e., $H_f = \infty$), then the calculated concentration at point "x" increases until it reaches a steady-state value equal to 46% of the source concentration. When one considers a finite mass of contaminant (i.e., finite H_f), the concentration at point "x" increases to a peak value and then subsequently decreases. The magnitude of the peak and the time at which this occurs depend on the equivalent height of leachate H_f. The smaller the mass of contaminant available for transport, the smaller is the impact on a downgradient monitoring point. For the case of a downward Darcy velocity of $v_a = 0.03$ m/a and $H_f = 0.4$ m, the peak concentration at point "x" is approximately 11% of the original source concentration in the leachate compared with 46% of the source value obtained for $H_f = \infty$. Thus in this case consideration of a

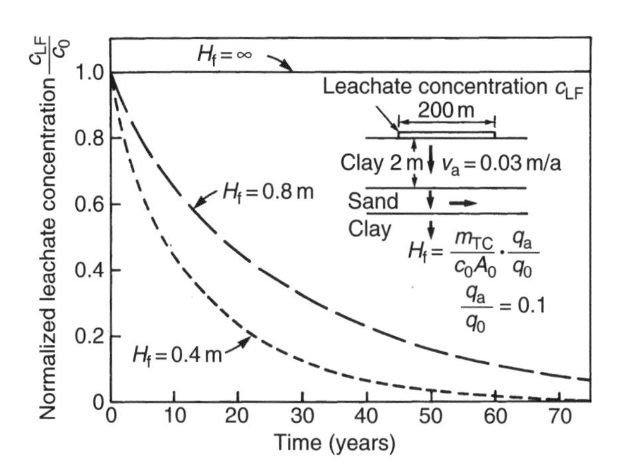

Figure 10.5 Effect of the equivalent height of leachate (H_f) on the variation in leachate concentration with time for Example A (after Rowe, 1988; reproduced with permission of the *Canadian Geotechnical Journal*).

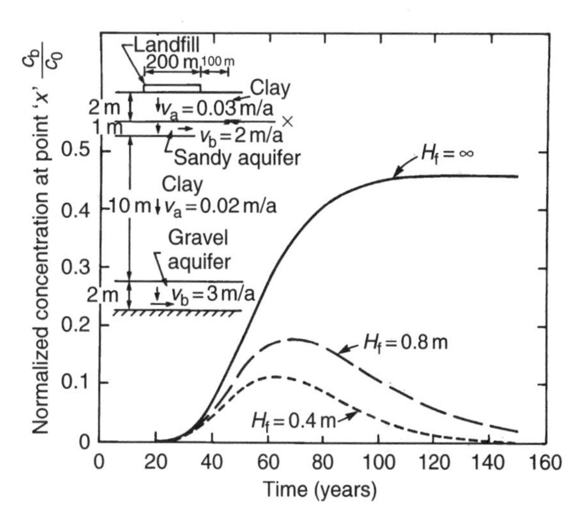

Figure 10.6 Effect of the equivalent height of leachate on the variation in concentration with time at a point within an aquifer at the site boundary for Example A (after Rowe, 1988; reproduced with permission of the *Canadian Geotechnical Journal*).

realistic mass of contaminant reduces the calculated potential impact on groundwater quality by a factor of 4, illustrating that the assumption of a constant source concentration may be unrealistically conservative.

Inspection of Figure 10.3 indicates that for the Darcy velocity of 0.03 m/a, contaminant transport through the barrier is being dominated by advection since $q_a \simeq v_a$. This is a situation which might occur if there were a failure of the leachate collection system. In many landfill designs, the Darcy velocity into the barrier will be substantially smaller than 0.03 m/a, particularly while the leachate collection system is functioning. To examine the effect of this Darcy velocity, analyses were also performed for $v_a = 0.003$ m/a. At this velocity, diffusion dominates contaminant transport and as previously discussed with respect to Figure 10.3, the tenfold reduction in velocity from $v_a = 0.03$ to 0.003 m/a only reduces the normalized average flux from about 0.03 to 0.01 m/a. Assuming an infiltration of 0.3 m/a, this corresponds to a threefold reduction in the proportion of mass available for transport into the barrier (i.e., from $q_a/q_0 = 0.1$–0.033).

The calculated variation in concentration with time at the downgradient monitoring point "x" is shown in Figure 10.7 for an infinite mass of contaminant ($H_f = \infty$; $q_a/q_0 = \infty$) and a finite mass of contaminant $H_f = 4q_a/q_0$ (m) (i.e., $H_r = 4$ m) for assumed downward Darcy velocities of 0.03 and 0.003 m/a. If one assumes that the concentration in the source remains constant for all time (i.e., $H_f = \infty$), then a tenfold decrease in Darcy velocity v_a only reduces the peak concentration by about 35% from $0.46c_0$ to $0.3c_0$. However, when one considers the finite mass of contaminant (specifically $H_f = 4q_a/q_0$), then this tenfold decrease in Darcy velocity gives rise to a more than tenfold decrease in peak concentration from $0.11c_0$ to $0.01c_0$. The corresponding increase in the time required to reach this peak was from a little over 60 years to about 700 years. For the Darcy velocity of 0.003 m/a considered here, the assumption of a constant source concentration results in an overestimate of the peak concentration by a factor of 30 if the mass of contaminant corresponds to $H_f = 4q_a/q_0$ m (e.g., 4,000 t over a site of area 50 ha at an initial source concentration of 2,000 mg/l).

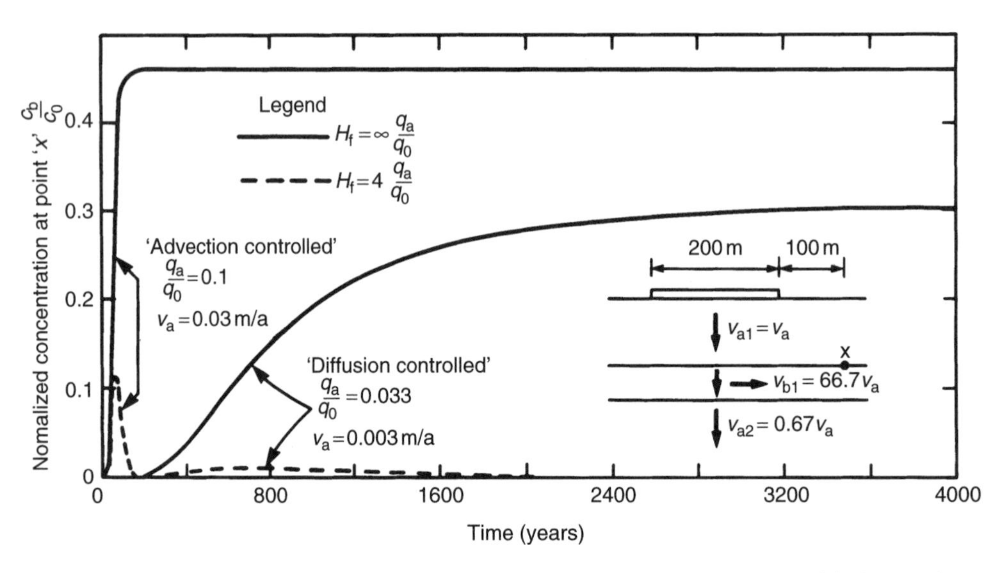

Figure 10.7 Effect of equivalent height of leachate (H_f) on the variation in concentration with time and contaminant impact at a point in the aquifer beneath the site boundary for Example A (after Rowe, 1988; reproduced with permission of the *Canadian Geotechnical Journal*).

It may be concluded that the mass of contaminants is an important parameter to be considered when calculating attenuation of contaminants as they move into the groundwater system.

10.3 Development of a contaminant plume

Proposed landfills are often sited above aquifers. This is particularly so when the geology consists of beds of clayey or silty soil separated by relatively thin granular units which are used for water supply (e.g., many glacial deposits). In these situations, it is necessary to evaluate the potential impact of the landfill on water quality in the aquifer(s) at the site boundary.

Because of factors such as the uncertainty associated with defining the groundwater system from limited data as well as the limitations regarding the adequacy of the data itself, it is not reasonable to expect that one could make an accurate prediction of the exact time at which contaminant would first migrate offsite no matter how sophisticated the theoretical model used. However, theoretical models can be particularly useful for examining the implications of different possible scenarios and different key parameters. The results from such a study can then be used in the formulation of an engineering opinion as to the potential for contamination of groundwater at the site boundaries due to a proposed landfill.

In the following sections, the effect of varying different key parameters will be examined. These results were obtained for the case of a clayey barrier underlain by a natural aquifer as shown in Figures 1.2–1.4; however, it should be emphasized again that similar calculations could be performed if there is a drainage system beneath a landfill liner (as shown in Figures 1.5–1.8).

Figures 10.8–10.10 examine various aspects of contaminant migration from a landfill with a length of 200 m parallel to the direction of flow in the underlying aquifer (see inset to Figure 10.8). The Darcy velocity, v_b, in the aquifer is assumed to be 1 m/a. The landfill is separated from the 1-m thick aquifer (which has porosity, $n_b = 0.3$) by 2 m of clayey soil (porosity, $n = 0.4$, diffusion coefficient, $D = 0.01\ \mathrm{m^2/a}$; downward

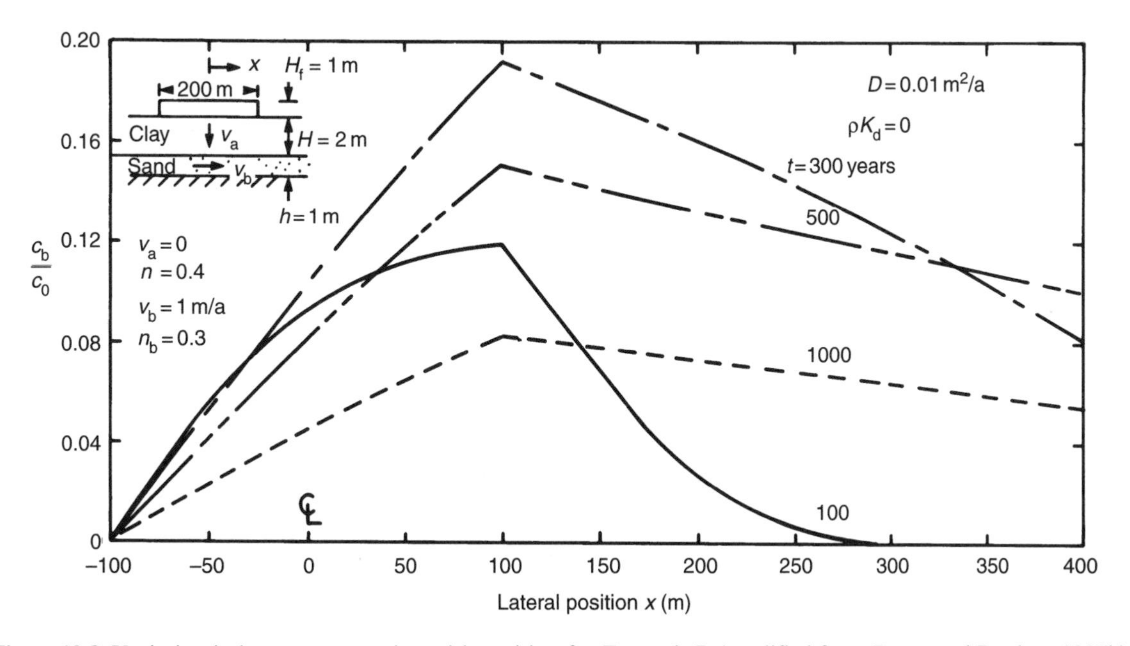

Figure 10.8 Variation in base concentration with position for Example B (modified from Rowe and Booker, 1985b).

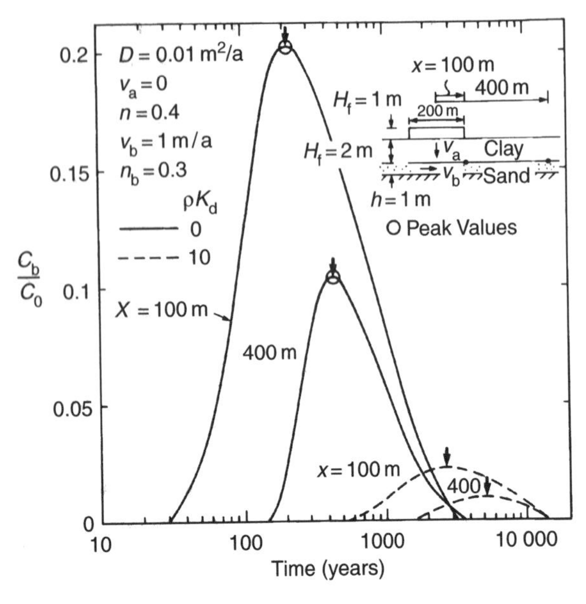

Figure 10.9 Effect of attenuation with distance from $x = 100$–400 m on concentration history in an aquifer for Example B (modified from Rowe and Booker, 1985b).

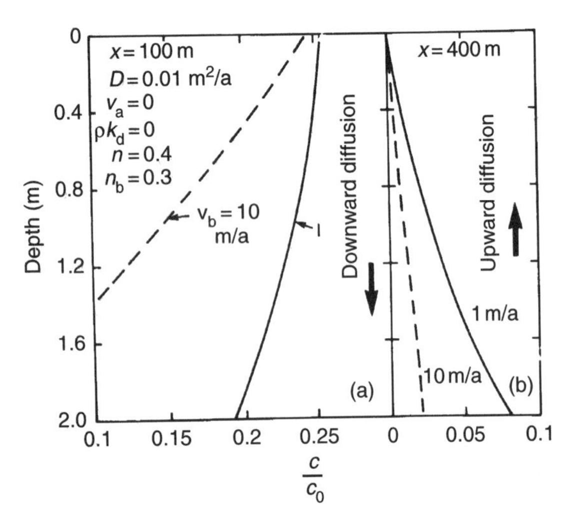

Figure 10.10 Concentration profile in the clay barrier at two positions ($t = 300$ years) (a) $x = 100$ m and (b) $x = 400$ m for Example B (modified from Rowe and Booker, 1985b).

Darcy flux, $v_a \simeq 0$ (i.e., zero gradient or very low hydraulic conductivity composite liner and a small gradient); no sorption for contaminant species of interest, $\rho K_d = 0$).

Figure 10.8 shows the variation in the concentration in the sand layer, c_b, with lateral position, at four times ($t = 100$, 300, 500 and 1,000 years). Except at smaller times, the concentration increases approximately linearly with lateral position beneath the landfill and in all cases attains a maximum value at the "downstream", in the sense of the direction of flow in the base strata, edge of the landfill.

Outside the landfill, the concentration decreases with increasing distance from the landfill. At smaller times, this decrease is primarily due to a time lag; however, at larger times it is primarily due to diffusion of contaminant back into the adjacent clay. Thus there is a natural attenuation mechanism in the system that will ensure that the maximum concentration reached at any point outside the boundaries of the landfill will never reach the maximum value calculated at the edge of the landfill.

The concentration of contaminant within the sandy aquifer beneath the clay can be reduced by increasing the thickness H of the clay. If this is not possible, the concentration can also be controlled by providing a buffer zone between the landfill and the areas where the concentration of contaminant in the groundwater must not exceed a specified level.

To illustrate the effect of the attenuation in a buffering zone, Figure 10.9 shows the variation in contaminant concentration in the sand with time at the edge of the landfill ($x = 100$ m) and at a point 300 m "downstream" from the landfill ($x = 400$ m) for two values of ρK_d. Considering first the curves for the case where there is no sorption ($\rho K_d = 0$), it is seen that the contaminant concentration at both positions increases with time until a peak value is reached and then decreases for subsequent time. There is a time lag between the times at which the peak values are reached at $x = 100$ and 400 m due to the time required for the contaminant peak to move the 300 m between these two points. This time lag is increased above that which would be expected for purely advective transport in the sand because of diffusion from the aquifer into the clay. Of greater interest is the fact that the magnitude of the peak concentration is substantially reduced even for this analysis where no horizontal dispersion in the sand layer is considered ($D_H = 0$).

Figures 10.8 and 10.9 have focussed on the concentrations within the aquifer, since this is of primary concern in the design of a landfill separated from an aquifer by a clay layer. However, the concentration profiles within the clay can also be determined as illustrated in Figure 10.10. Beneath the landfill (Figure 10.10a), the mass transport is predominantly downwards and the concentrations decrease with depth. For the case examined, the surface concentration within the landfill has reduced to about 25% of the original value after 300 years.

At locations remote from the landfill, the mass transport is also predominantly vertical but in this case it is upwards from the aquifer into the clay as shown in Figure 10.10b.

10.4 Effect of Darcy flux in a receptor aquifer

The results presented in Figure 10.9 were obtained for a specific value of the Darcy flux (velocity) within the aquifer ($v_b = 1$ m/a). This parameter is important; it is also difficult to determine in practice. What can be determined is a reasonable estimate of the range in which the velocity is expected to lie. Under these circumstances, finite layer techniques can be easily used to evaluate the effect of this uncertainty upon the expected impact. For example, Figure 10.11 shows the peak concentrations obtained at $x = 100$ and 400 m for analyses performed for a range of base velocities v_b.

Beneath the edge of the landfill ($x = 100$ m), the maximum concentration decreases monotonically with increasing base velocity due to the consequent increased dilution of the contaminant in high volumes of water. At points outside the landfill area, there is a critical velocity which gives rise to the greatest maximum concentration. As indicated by Rowe and Booker (1985b), this situation arises because of the interplay of two different attenuation mechanisms. The first of these, diffusion into the surrounding clayey soil, is dependent on the time required to reach the monitoring point. Generally, the lower the velocity v_b, the more time there is for contaminant to diffuse away and hence the lower the maximum concentration. The second mechanism, dilution, involves decreasing contaminant concentration due to higher volumes of water (i.e., higher v_b) with which contaminant migrating from the landfill can mix.

An important practical consequence of the foregoing is that it is not necessarily conservative to design only for the maximum and minimum expected velocities in the aquifer. In performing

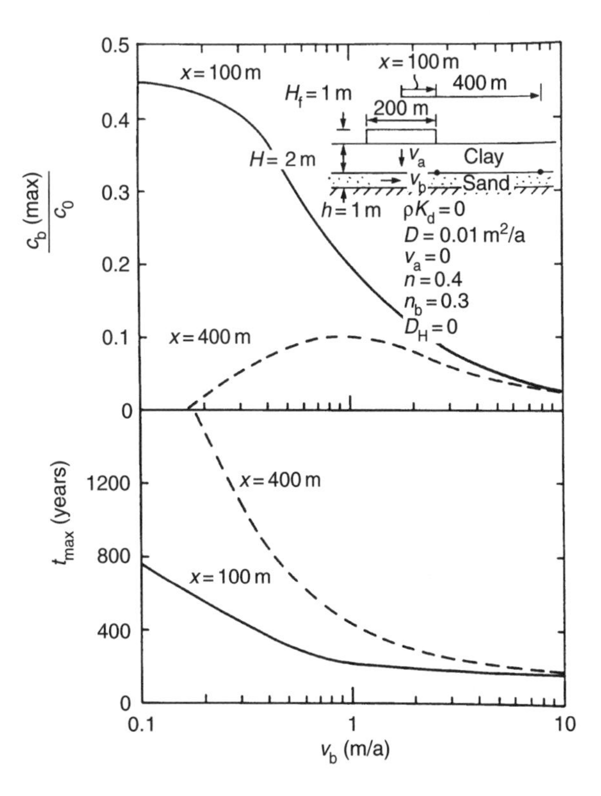

Figure 10.11 Variation in maximum base concentration with base velocity for Example B (after Rowe and Booker, 1985b; reproduced with permission of the *Canadian Geotechnical Journal*).

sensitivity studies, sufficient analyses should be performed either to determine the critical velocity or, alternatively, to show that the critical velocity does not lie within the practical range of velocities for the case being considered.

10.5 Effect of horizontal dispersivity in a receptor aquifer

In the analysis of the previous section, it was assumed that $D_H = 0$. Analyses were also performed for a range of values of the horizontal Darcy velocity, v_b, and coefficient of dispersion D_H. In general, it was found that for horizontal Darcy velocities, v_b, of 1 m/a or greater, dispersion tended to reduce the maximum base concentration, although the effect was relatively small even for high values of D_H as shown for $v_b = 1$ m/a in Figure 10.12.

For small advective velocities (i.e., less than 1 m/a), a high dispersion coefficient may have a significant effect on the peak concentration both beneath the landfill and at points outside the landfill. For example, Figure 10.12 shows the variation in the peak concentration in the aquifer with D_H at two points ($x = 100$ and 400 m) for a base velocity $v_b = 0.1$ m/a. For these low velocities, lateral dispersion in the aquifer reduces the peak base concentration beneath the edge of the landfill and the effect is quite significant for values of D_H between 0 and 50 m^2/a. At points remote from the landfill, lateral dispersion in the aquifer can lead to a modest increase in the predicted peak base concentration. This is because for low v_b, increasing D_H increases the rate of mass transport through the aquifer and reduces the amount of diffusion that can occur into the adjacent clay layers. The magnitude of D_H encountered in the field is highly variable, although for $v_b = 0.1$ m/a the expected range would be 0.01–20 m^2/a and most probably less than 3 m^2/a. These values could be an order of magnitude higher for $v_b = 1$ m/a.

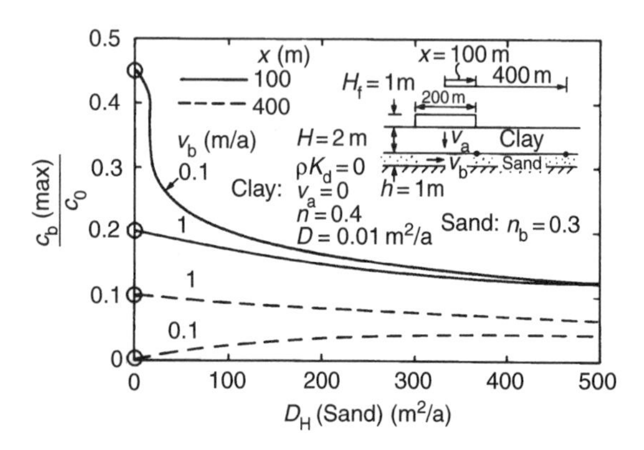

Figure 10.12 Variation in maximum base concentration with horizontal dispersion coefficient in the base layer for Example B (after Rowe and Booker, 1985b; reproduced with permission of the *Canadian Geotechnical Journal*).

As might be expected, the peak concentration at a point in the aquifer for $v_b = 0.1$ and 1 m/a tends towards a single value as D_H becomes very large and dominates over the effect of advection.

Because of non-homogeneities present within most aquifers, the coefficient of dispersion in the sand, D_H, is a difficult parameter to determine and there will always be considerable uncertainty regarding the precise value to use in a design. When considering concentration beneath the landfill, it is conservative to perform the analysis for $D_H = 0$. When considering points outside the landfill, analysis may be performed for $D_H = 0$ and the maximum reasonable value of D_H. For low horizontal velocities, the analysis for the high D_H may be critical; for moderate and high horizontal velocities, the analysis for $D_H = 0$ will be conservative when predicting peak impact.

10.6 Effect of aquifer thickness

The results to be discussed in this and in the following section were obtained using a 2D analysis for the case denoted as Example C as shown schematically in Figure 10.13, for parameters given in Table 10.1.

The thickness of the aquifer will have an effect on the concentration of contaminant at various

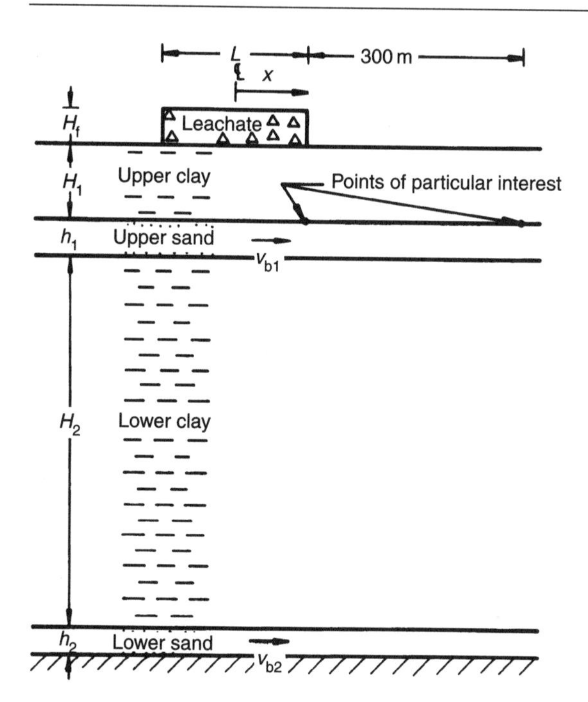

Figure 10.13 Multilayer problem analysed in Example C.

Table 10.1 Parameters relating to Example C (Figures 10.13–10.15)

Layer	*Symbol*	*Value*
Landfill	L (m)	200–1,000
	H_f (m)	1
Upper clay	D (m^2/a)	0.01
	n	0.4
	v_a (m/a)	0.0
	H_1 (m)	2.0
Upper sand	D_{H1} (m^2/a)	10.0
	D_{V1} (m^2/a)	0.2
	n	0.3
	v_{b1}(m/a)	1.0
	h_1 (m)	0.5–1.5
Lower clay	D (m^2/a)	0.01
	n	0.4
	v_a (m/a)	0.0
	H_2 (m)	10.0
Lower sand	D_{H2} (m^2/a)	10.0
	D_{V2} (m^2/a)	0.2
	n	0.3
	v_{b2} (m/a)	1.0
	h_2 (m)	0.3

locations along the aquifer. As shown in Figure 10.14, the concentration of contaminant at the edge of the landfill ($x = 100$ m) tends to decrease as the thickness of the aquifer is increased. This is primarily because of an increased dilution of contaminant that occurs for large values of h_1 (all other things being equal) due to the correspondingly higher flow in the aquifer. At points well away from the landfill (e.g., $x = 400$ m), the concentration of contaminant may increase with increasing thickness of aquifer (again, all other

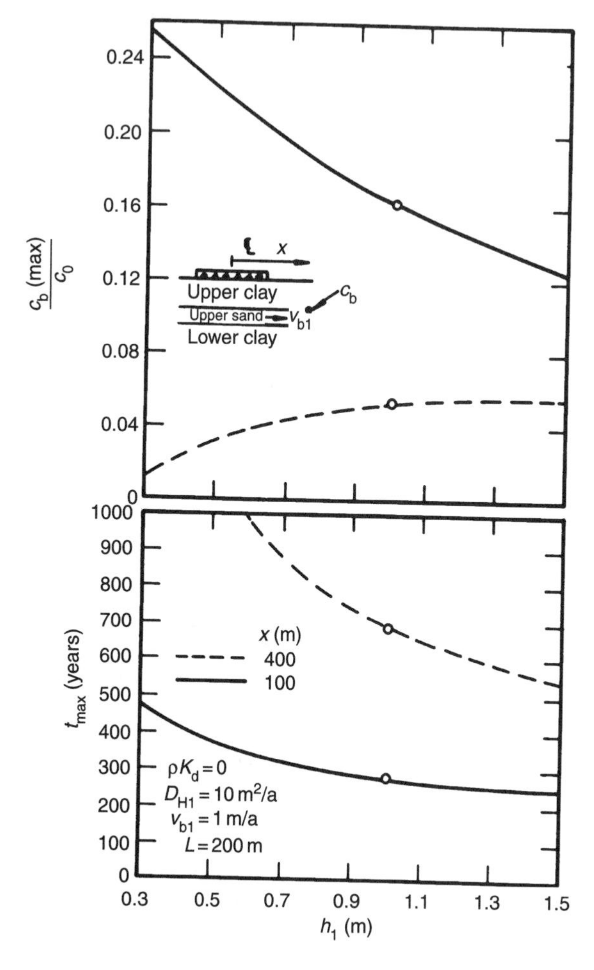

Figure 10.14 Variation in maximum concentration c_{bmax} in the upper sand layer with thickness h_1 of the sand layer for Example C (modified from Rowe and Booker, 1986).

things being equal) because the relative diffusion into the adjacent clayey soil is reduced. The result given in Figures 10.11 and 10.14 indicates that there is a fairly complex interaction between the effects of aquifer thickness, Darcy flux in the aquifer and the distance to the point of interest, on the maximum concentration expected to occur at that point.

10.7 Effect of landfill size on potential impact

All the foregoing results have been for a landfill 200-m long in the direction of groundwater flow (i.e., $L = 200$ m). For the other basic parameters given in Table 10.1, Figure 10.15 shows the variation in the maximum concentration at the edge and 300 m from the edge of a landfill of variable landfill length L. For the problem considered, increasing the length of the landfill increases the concentration at both points of interest although the maximum concentration tends to become asymptotic to a constant value for L approaching 1,000 m. The increase in concentration with L arises because of the increased mass loading of the aquifer which arises from a large total mass of contaminant within the landfill. The tendency of the asymptote to reach a constant value for very large L arises because significant diffusion can occur into the underlying clay between the time that the contaminant enters the aquifer near the upstream edge and the time that it approaches the downstream edge when the width of the landfill, L, is large. The value of L at which this occurs will depend on v_{b1}, h_1, H_f and the other parameters.

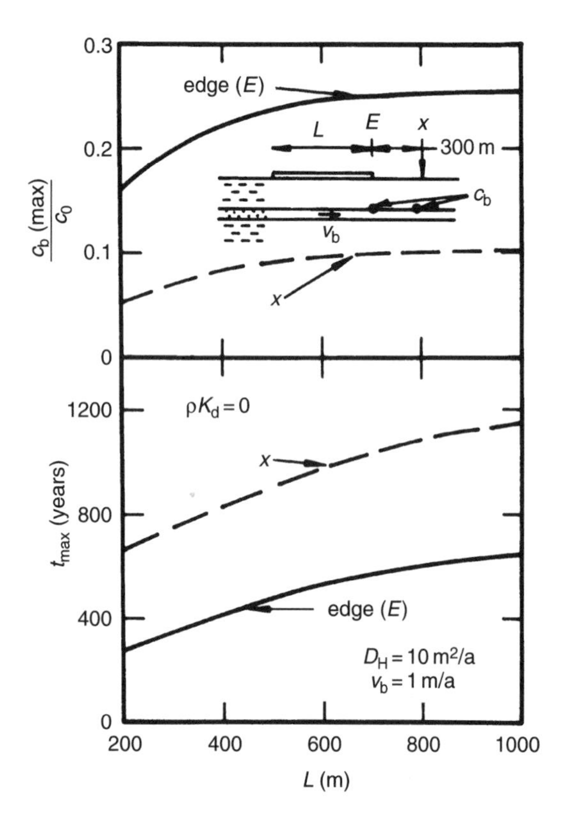

Figure 10.15 Variation in maximum concentration c_{bmax} in the upper sand layer with landfill width L for Example C (modified from Rowe and Booker, 1986).

10.8 $1\frac{1}{2}$D versus 2D analysis and modelling of the aquifer beneath a liner

The analyses reported in the previous section were all performed using a full 2D analysis. For 2D conditions there are, in fact, two ways in which the aquifer can be modelled:

1. As a boundary condition (see Section 1.6). This approach allows for spatial variations in concentration within the horizontal plane of the aquifer as well as advective–dispersive transport within the aquifer itself. Thus this advective–dispersive transport will depend on the horizontal velocity within the aquifer, v_b, and the coefficient of hydrodynamic dispersion in the aquifer, D_H. However, this approach does assume that the concentration in the aquifer is uniform in the vertical direction (i.e., $D_v = \infty$) and that the aquifer is underlain by an impenetrable boundary (i.e., zero mass flux across this boundary). This is the approach used to get the results shown in Example B (Figures 10.8–10.12).

2. As a physical layer having prescribed velocity components, v_b, and coefficients of hydrodynamic dispersion D_V, D_H, thereby allowing for spatial variations in concentration both vertically and horizontally within the aquifer. This approach of treating the aquifer as a physical layer in a manner similar to the clay, but with different parameters, permits us to examine two cases where the aquifer is underlain by:
   (a) an impenetrable boundary (as assumed in (1) above), or
   (b) an additional layer (or layers) of clay (and/or sand) as was the case in the analysis for Example A (Figures 10.4–10.7) and Example C (Figures 10.13–10.15).

To illustrate the effect of modelling the aquifer in different ways, a series of 2D analyses were performed for Example D using the parameters given in Table 10.2. Considering first case (a), where the aquifer is underlain by an impenetrable boundary, the concentration plume was calculated at $t = 300$ years using methods (1) and (2) above as shown in Figure 10.16. Method (1) implicitly assumes that the concentration c_{base} is vertically uniform within the aquifer. Method (2) makes no *a priori* assumption regarding the spatial variation of concentration and the calculated values at both the top (c_{b1}) and bottom (c_{b2}) of the aquifer are shown in Figure 10.16. For the parameters considered, there is relatively little vertical variation in concentration within the aquifer and the results obtained by treating the aquifer as a physical layer (Method (2)) closely bound the results from the computationally simpler approach where the aquifer is treated as a boundary condition (Method (1)).

Table 10.2 Parameters relating to Example D (Figures 10.16 and 10.17)

Layer	*Quantity*	*Symbol*	*Value*
Clay	Vertical Darcy velocity	v_a (m/a)	0.0
	Porosity	n	0.4
	Sorption potential	ρK_d	0.0
	Coefficient of hydrodynamic dispersion (horizontal and vertical)	D (m^2/a)	0.01
Sand	Horizontal Darcy velocity	v_b (m/a)	1.0
	Porosity	n_b	0.3
	Sorption potential	ρK_d	0.0
	Coefficient of hydrodynamic dispersion		
	Horizontal	D_H (m^2/a)	1.0
	Vertical (layered case)	D_V (m^2/a)	0.2
	Vertical (boundary condition)	D_V (m^2/a)	∞

Now consider case (b), where the aquifer is underlain by an additional 10-m thick layer of clay resting on an impenetrable base. It is found that (see Figure 10.16) the concentration plume is almost the same as that obtained for case (a) near the upstream edge of the landfill but for case (b) diffusion out of the aquifer into the underlying clay gives rise to smaller concentrations near the downstream edge of the landfill and at points outside the landfill ($x \geq 100$ m). The results given at the top and bottom of the aquifer indicate that there is a small concentration gradient in the vertical direction within the aquifer.

Analyses similar to those performed to obtain Figure 10.16 were repeated for different times to give the variation in concentration with time. Figure 10.17 summarizes the results for a point beneath the downstream edge of the landfill ($x = 100$ m) and a point well outside the landfill ($x = 400$ m). For the sake of clarity, only the concentrations calculated at the top of the aquifer, c_{b1}, are shown. A number of observations can be made regarding the results from this example.

First, diffusion of contaminant into the adjacent clay gives rise to natural attenuation of contaminant plume as it advances along the aquifer ($x > 100$ m). Thus the maximum concentration reached at $x = 400$ m is less than that obtained at the edge of the landfill ($x = 100$ m) by more than a factor of 2.

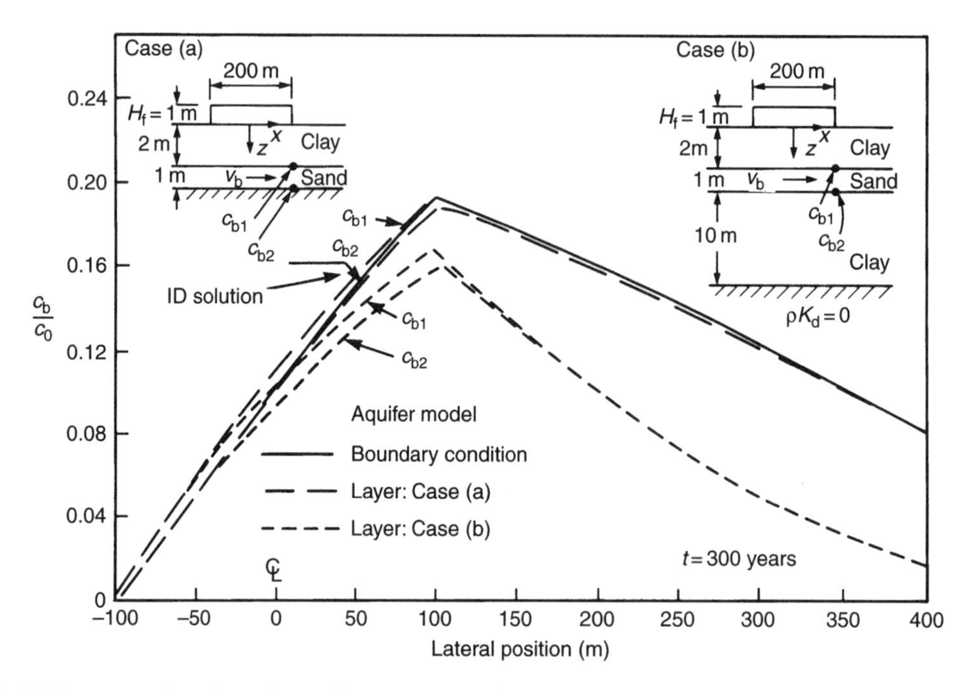

Figure 10.16 Concentration plume in aquifer at 300 years for Example D (modified from Rowe and Booker, 1987).

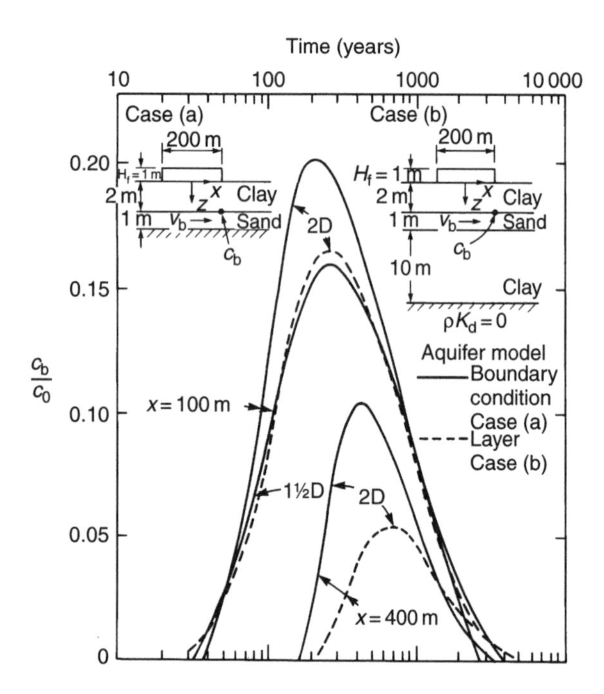

Figure 10.17 Concentration in aquifer beneath edge of landfill ($x = 100$ m): effect of 1D and 2D models for Example D (modified from Rowe and Booker, 1987).

Second, consideration of possible diffusion into a clay layer beneath the aquifer (case (b)) gives rise to additional attenuation of contaminant concentration. At the edge of the landfill ($x = 100$ m), diffusion into the lower clay layer reduces the maximum concentration (compared to case (a)) by approximately 20%. The effect increases with distance away from the landfill and at $x = 400$ m diffusion into the lower clay reduces the maximum concentration by almost a factor of 2. Thus the modelling of the aquifer as a boundary condition in a 2D analysis provides a conservative estimate of the contaminant concentrations within the aquifer.

The $1\frac{1}{2}$D analysis, denoted in the figure as the 1D solution, allows one to approximately analyse the 2D problems discussed above. To make the problem tractable, it is necessary to assume that the concentration within the aquifer directly beneath the landfill is spatially homogeneous at all times and that the only mechanism for transporting mass out from directly beneath the landfill is by advection in the aquifer. Since this assumption allows one to perform a very simple $1\frac{1}{2}$D analysis, it is of some interest to compare the results from the $1\frac{1}{2}$ D analysis with those of the more rigorous 2D analysis.

The results obtained for Example D are shown in Figures 10.16 and 10.17. Comparison of these results with those obtained at $x = 100$ m from the 2D analysis indicates that for this case the $1\frac{1}{2}$ D approach slightly overestimates the time required to attain the maximum concentration and underestimated the magnitude of the maximum concentration. However, the discrepancy, which was less than 30%, may be acceptable in preliminary calculations.

Figure 10.18 shows the variation in the maximum peak concentration determined from $1\frac{1}{2}$D and 2D analyses for a range of base velocities for case (a). It is seen that the $1\frac{1}{2}$D analysis consistently underestimates the maximum concentration but that in general the error is quite small. For this example, the maximum error of 30% occurs for base velocities between 0.3 and 0.5 m/a. Similarly, it was found that the $1\frac{1}{2}$D analysis also gives a reasonable estimate of the time required to attain the peak base concentration beneath the landfill.

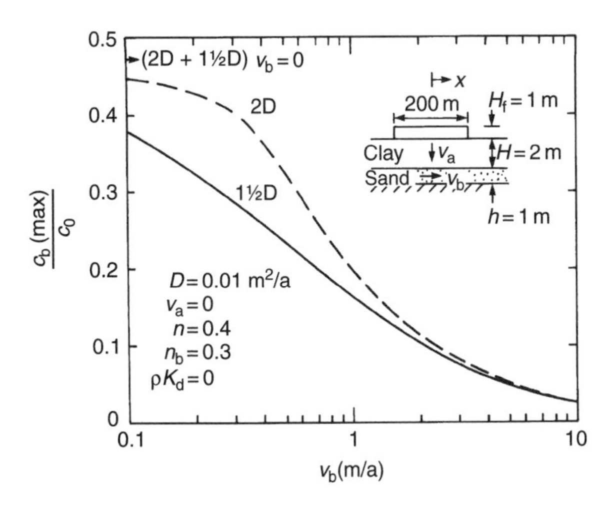

Figure 10.18 Comparison of peak concentration calculated from the $1\frac{1}{2}$D and 2D analyses for a range of base velocities (modified from Rowe and Booker, 1985b).

10.9 Effect of sorption

As indicated in the previous chapters, sorption can often be represented in terms of the dimensionless product ρK_d. The value of ρK_d will depend upon the soil properties, chemical reactions and their rates, and the range of concentration. Typical values lie between 0 and 100 although much higher values have been reported. As illustrated in Figure 10.9, even a modest amount of sorption can have a significant effect on the time at which the peak impact occurs. In addition, where there is a finite mass of contaminant, sorption (e.g., $\rho K_d = 10$ in Figure 10.9) can also significantly reduce the magnitude of peak impact. The effect of sorption on the magnitude of peak impact gets smaller as the mass of contaminant (i.e., H_f) increases (see Rowe and Booker, 1985a).

10.10 Effect of liner thickness

Having established that the $1\frac{1}{2}$D approach can give a reasonable indication of contaminant impact in the aquifer, the following results were obtained using a $1\frac{1}{2}$D analysis.

In design, the thickness and known attenuation potential (ρK_d) of the clay layer isolating the landfill from the underlying groundwater system may be used to control the maximum base concentration. Figure 10.19 shows the variation in maximum base concentration $c_{b(max)}$ with clay liner thickness H for $\rho K_d = 0$ and 10. Increasing the layer thickness substantially reduces the maximum concentration ever reached at the base and increases the time required to reach this maximum. For example, Figure 10.19 shows that with $\rho K_d = 0$, increasing the clay liner thickness from 0.5 to 4 m decreases $c_{b(max)}$ by up to an order of magnitude and increases the time t_{max} by up to an order of magnitude.

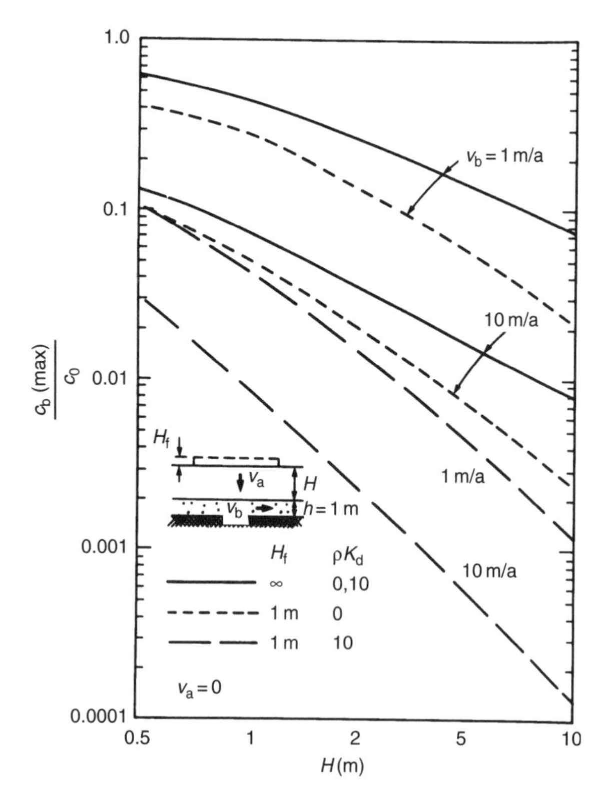

Figure 10.19 Effect of layer thickness, H, on maximum base concentration (modified from Rowe and Booker, 1985a).

The effect of layer thickness is increased by consideration of the available mass of contaminant (height of leachate) and the geochemical reaction. Thus, the thickness of liner required to ensure that a specified maximum base concentration is never exceeded may be significantly reduced by considering the interaction of all these factors.

10.11 Effect of Darcy flux in the barrier and design for negligible impact

The vertical Darcy velocity (flux), v_a, within the clayey barrier can be calculated knowing the hydraulic conductivity of the barrier and the difference between the head in the landfill and in the underlying aquifer. If the head in the landfill is greater than that in the aquifer, the velocity v_a is from the landfill towards the aquifer. If the head in the landfill is less than that in the aquifer, there will be upward flow into the landfill.

Although seepage may be minimized, it will rarely be precisely zero and it is of interest to examine the effect of the Darcy velocity (flux), v_a, through the liner upon the maximum base concentration. Analyses were performed for a range of Darcy velocities, and the general effects of base velocity, layer thickness, equivalent heights of leachate and geochemical reaction discussed in the previous sections for $v_a = 0$ were similar to those for $v_a > 0$. However, the Darcy velocity did increase the magnitude of the maximum base concentration and decrease the time required to reach this concentration as shown for a 2-m thick liner with $\rho K_d = 0$ in Figure 10.20. The results for other values of ρK_d are very similar except that the values of c_b and t_{max} differ in approximately the same ratio as they do for $v_a = 0$.

From Figure 10.20, it can be seen that for the range of parameters considered, the Darcy velocity could be neglected for $v_a < 0.0004$ m/a. For v_a up to 0.001 m/a, advection increases the base concentration by up to 33% although t_{max} is almost constant. Darcy velocities greater than 0.001 m/a may significantly affect the base concentration, with the values of $v_a = 0.01$ m/a being up to five times greater than those for $v_a = 0$. The time t_{max} is somewhat less affected by v_a, varying by less than a factor of 2 for the same range in Darcy velocities. This demonstrates the importance of minimizing the Darcy velocity in the design of the waste disposal site. For a clayey liner with a hydraulic conductivity of 10^{-9} m/s and a unit gradient of 1, the Darcy velocity would be about 0.03 m/a.

The foregoing discussion has considered outward flow from the landfill. As noted earlier, contaminant migration from a landfill can be minimized if the landfill is designed so that groundwater flow is into the landfill. However,

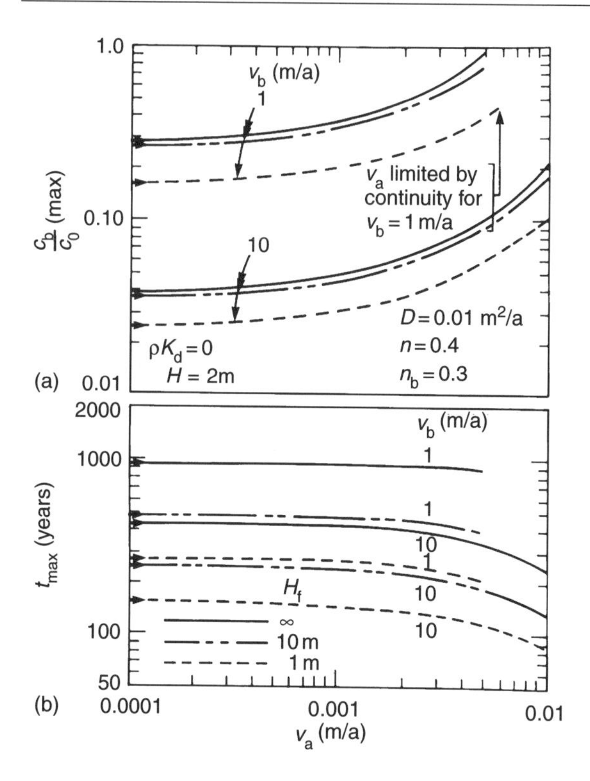

Figure 10.20 Effect of downward Darcy velocity on (a) maximum base concentration and (b) time to reach maximum concentration (modified from Rowe and Booker, 1985a).

the fact that there is inward flow does not necessarily mean that there will be no contaminant migration out of the barrier. The assessment of potential impact on groundwater quality involves two stages, viz.

(i) Calculate the Darcy velocity (flux) into the landfill.
(ii) Calculate the diffusive movement out of the barrier.

The construction of a landfill will usually change the hydraulic characteristics of a given site. For example, a functioning leachate collection system may give rise to an average head in the leachate which is below the original groundwater level, thereby inducing inward flow at the sides of the landfill and, if the head is lowered sufficiently, at the base of the landfill. On the other hand, if significant leachate mounding were to occur due to failure of the collection system, then the head in the landfill will increase and may exceed the head outside the barrier, thereby reversing the flow direction and resulting in outward flow. In both cases, the mounding of the leachate must be calculated as previously discussed and then the hydraulic gradient determined for the new flow system – this will require a seepage analysis.

Having calculated the Darcy velocities, contaminant transport modelling may then be used to assess the potential impact of contaminants moving through the barrier. If the flow is outward, then there can be no hydraulic trap and the problem is similar to that discussed in the previous section. If the flow is inward, then analysis may be used to estimate the rate of movement and maximum extent of the contaminant movement.

If it is assumed that the concentration of contaminant within the landfill remains constant at c_0 and that the base of the barrier is flushed by flowing water, such that the concentration is zero, then the steady-state solution is given by (Al-Niami and Rushton, 1977):

$$\frac{c}{c_0} = \frac{\exp[-v(H - z)/D] - 1}{\exp[-vH/D] - 1} \tag{10.11}$$

where z is the distance from the top of the liner (positive for movement into the liner), H the thickness of the liner, v the groundwater velocity which is positive for outward flow and negative for inward velocity (i.e., for the case of interest here), n the porosity and D the coefficient of hydrodynamic dispersion which, for the case of inward flow, may be taken to be equal to the effective diffusion coefficient. Thus the normalized flux into the aquifer ($z = H$) is given by

$$\frac{f}{c_0} = \frac{-nv}{\exp(-vH/D) - 1} \tag{10.12}$$

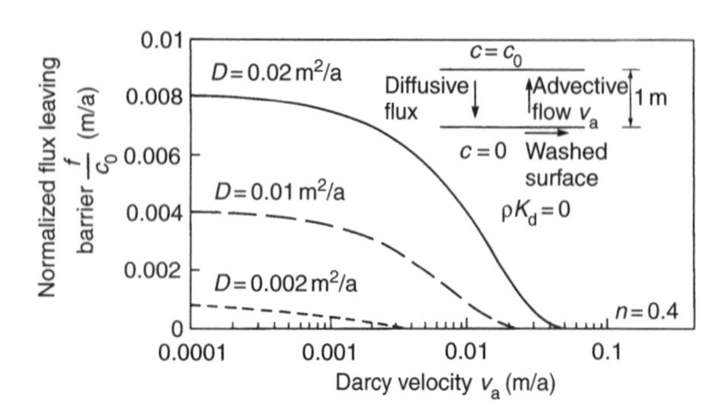

Figure 10.21 Diffusive flux of contaminant into an aquifer against an inward flow at a Darcy velocity v_a (steady-state conditions) (modified from Rowe, 1988).

Equation 10.11 may be useful for estimating the maximum extent of the contaminant plume. However, this equation *assumes* zero concentration at the base of the barrier and hence the calculation of a small concentration near the base does not necessarily imply that the impact on an underlying aquifer would be small. For example, Figure 10.21 shows the calculated steady-state contaminant flux passing into an aquifer beneath a 1-m thick clayey barrier for a range of inward Darcy velocities and assumed diffusion coefficients.

The flux f has been divided by the initial source concentration and the resulting normalized flux f/c_0 has units of velocity. The effective diffusion coefficient for many contaminants lies in the range $D = 0.01$–$0.02\,m^2/a$ (see Chapter 8). For these conditions, Figure 10.21 shows that the inward Darcy velocity would have to exceed 0.025 and 0.05 m/a for $D = 0.01$ and $0.02\,m^2/a$, respectively, before the outward flux was reduced to negligible levels for a 1-m thick liner.

As a rule of thumb, the impact on an underlying aquifer is likely to be negligible if the concentration calculated from equation 10.11 is negligible for depths $z \simeq 0.9H$. If the concentration is not negligible for $z \simeq 0.9H$, then it is necessary to assess the potential impact of the proposed landfill upon groundwater quality using a more realistic model that explicitly considers the aquifer and the finite mass of contaminant; this can be easily done using finite layer techniques (Chapter 7). If the calculated impact is unacceptable, then the barrier thickness must be increased and/or the inward Darcy velocity must be increased.

Chapter eleven

Migration in fractured media: analysis and design considerations

11.1 Introduction

Fractured porous media are frequently encountered adjacent to present or proposed waste disposal facilities. This fractured media may, for example, take the form of a fractured clay or till (e.g., with blocky fractures in the upper weathered zone and with predominantly vertical fractures in the unweathered clay or till), or a fractured rock (e.g., mudstone, siltstone, sandstone) which has significant horizontal and vertical fracturing. Typically, fractured media will involve a series of fractures separated by blocks of intact material which will be referred to as the matrix of the fractured soil or rock.

The primary transport mechanism for dissolved contaminants in fractured media is usually advective–dispersive transport along the fractures. In these systems, the average linearized groundwater velocity in the fractures, v_f, is related to the Darcy velocity v_a by the relationship

$$v_f = \frac{v_a}{n_b}$$

where n_b is the effective porosity through which flow is occurring. Thus, the fracture velocity v_f may be quite high because the fracture porosity of a fractured mass n_b is usually quite small with a typical range of between 0.1 and 0.001% based on hydraulic conductivity–fracture opening size–fracture frequency relationships (e.g., see Hoek and Bray, 1981). If conservative contaminants were to migrate along these fractures at the average linear groundwater velocity, v_f, then the rate of contaminant migration through the rock would be very fast. For example, the Burlington landfill is located in fractured Queenston shale in Southern Ontario, Canada. In the environs of the landfill, the Queenston shale has a typical fracture spacing between 0.05 and 0.35 m (Hewetson, 1985). At the time of monitoring, the Burlington landfill has been generating leachate for more than 15 years. If contaminant moved at the speed of the average linear groundwater velocity, and if, for example, the velocity were 50 m/a, then 15 years after leachate entered the rock, contaminants should be readily detected at distances of up to 750 m downgradient of the landfill. However, the available field data (e.g., the Burlington Landfill Plume Delineation Study, Gartner Lee Ltd, 1988) suggest that after 15 years this was not the case.

While it may be true that migration of conservative contaminants could occur at the rate close to that of the average linear groundwater velocity in porous media where the matrix of the media (e.g., the sand grains in a sandy aquifer) has a negligible effective porosity, it is not true that conservative contaminants (e.g., chloride) will migrate at the rate of the average linear groundwater velocity in fractured rock or soil systems where the matrix of the rock or soil between fractures has a significant effective porosity (see Grisak and Pickens, 1980). This is the case at the Burlington landfill where the Queenston shale has a matrix porosity of approximately 10%. In cases such as this, diffusion of contaminants from fractures into the adjacent matrix can control the migration of contaminants and represents an important attenuation mechanism. This phenomenon is known as matrix diffusion. The fact that diffusion can occur into or out of this type of rock was demonstrated in Section 8.3.4 for the Queenston shale, and the effective diffusion coefficient was found to be about 1.5×10^{-10} m^2/s for chloride.

The phenomenon of attenuation due to matrix diffusion is well recognized (Foster, 1975; Freeze and Cherry, 1979; Grisak and Pickens, 1980; Grisak *et al.*, 1980; Barker and Foster, 1981; Tang *et al.*, 1981; Sudicky and Frind, 1982) and Gillham and Cherry (1982) have reviewed a number of cases where matrix diffusion has been shown to decrease concentrations of species moving along fractures. For example, Foster (1975) showed that the diffusion of tritium from flowing groundwater in the fractures of porous chalk into the pore water of the porous rock matrix could account for a rapid decrease in tritium concentration in the fractures. Similarly, Day (1977) used this concept of matrix diffusion to account for a rapid decline in tritium concentration with depth in a fractured clay in the Winnipeg area. Finally, Grisak *et al.* (1980) have also demonstrated that a model that considers matrix diffusion can give relatively good agreement between theoretical simulations and laboratory experiments in which a tracer solution was passed through a large column of fractured, clayey glacial till. From these studies, Gillham and Cherry concluded that:

> molecular diffusion can exert a major influence on the rates and patterns of migration of contaminants in fractured argillaceous deposits. Diffusive loss of contaminants from paths of active flow in the fractures to the matrix can be a dominant mechanism of attenuation.

Matrix diffusion combined with sorption processes will result in even greater attenuation for reactive contaminants than for conservative contaminants. Some reactive contaminants may be removed from free solution by cation exchange either at the surface of the fracture or within the adjacent soil or rock. Organics may also be removed by preferential partitioning of contaminants on organic matter. In principle, partitioning could occur both at the face of the fracture and within the matrix; however relatively little is known about the effects of partitioning at the face, and it would be prudent and conservative to restrict consideration to partitioning that occurs within the soil or rock matrix.

Finite element techniques provide one means of modelling contaminant migration in fractured systems (e.g., Grisak and Pickens, 1980; Huyakorn *et al.*, 1983 and others). These approaches potentially make it possible to analyse quite complicated 2D and 3D fracture networks and may be useful when there is detailed data available concerning the distribution and characteristics of the fracture system. Frequently, however, this data is not available and it is necessary to assess potential impact based on knowledge of typical fracture spacings and orientations together with some knowledge of the hydraulic gradient and hydraulic conductivity in the system. Under these circumstances, analytic or semi-analytic solutions for contaminant transport in a fractured medium may be particularly useful for quickly assessing the potential effects of uncertainty regarding key parameters like fracture spacing, fracture opening size, etc. These techniques may also be useful for benchmarking more complex numerical procedures (e.g., finite element codes).

Various investigators have developed analytical or semi-analytical solutions for contaminant transport in idealized fracture media. For example, Neretnieks (1980) and Tang *et al.* (1981) developed a solution for 1D contaminant transport along a single fracture together with 1D diffusion of contaminant into the matrix of the rock adjacent to the fracture. Sudicky and Frind (1982) and Barker (1982) extended this approach to consider the case of multiple parallel fractures. The aforementioned researchers all considered constant concentration at the inlet of the fractures. Moreno and Rasmuson (1986) developed an analytical equation for a constant flux boundary condition at the inlet of a single fracture.

Rowe and Booker (1988a, 1989, 1990a,b, 1991a,b) considered the situation that is encountered with landfills wherein there is a finite mass of contaminant available for transport into the groundwater systems. They produced solutions for 1D, 2D and 3D fractured systems while allowing the concentration of contaminant in the source (landfill) to decrease with time as mass is transported into the fracture network. The theory behind this solution was presented in Chapter 7. The objective of this chapter is to examine the implications of modelling both matrix diffusion and the finite mass of contaminant within the landfill.

For the purposes of illustrating the importance of various parameters, the results presented in this chapter will generally be taken as those applicable to lateral migration in a fractured shale deposit in Section 11.3 and vertical migration in fractured clay or till in Section 11.4. It should, however, be noted that the same issues as discussed in Section 11.3 arise when considering lateral migration in fractured clay (e.g., out from the edge of a landfill) and similarly the issues raised with respect to predominantly vertical migration through fractured clay/till in Section 11.4 could equally well apply to vertical migration through fractured porous rock. A more detailed discussion of the issues examined in Section 11.3 can be found in Rowe and Booker (1989, 1990b). A detailed discussion of the issues raised in Section 11.4 is to be found in Rowe and Booker (1990a, 1991a,b).

11.2 Numerical considerations

The analysis used in this chapter was described in Chapter 7 and the evaluation of the parameter $\overline{\eta}$ which controls matrix diffusion from the fractures is given in Appendix C as:

For 1D conditions:

$$\overline{\eta}=n_{\mathrm{m}}R_{\mathrm{m}}\left(1-2\sum_{j=1}^{\infty}\frac{s}{[s+(D_{\mathrm{m}}/R_{\mathrm{m}})\alpha_j^2]}\times\frac{1}{(\alpha_jH_1)^2}\right)\tag{11.1a}$$

or

$$\overline{\eta}=n_{\mathrm{m}}R_{\mathrm{m}}\frac{\tanh\mu H_1}{\mu H_1}\tag{11.1b}$$

where $\mu^2=R_{\mathrm{m}}s/D_{\mathrm{m}}$. For 2D conditions:

$$\overline{\eta}=n_{\mathrm{m}}R_{\mathrm{m}}\left(1-4\sum_{j,k}^{\infty}\frac{s}{s+(D_{\mathrm{m}}/R_{\mathrm{m}})(\alpha_j^2+\beta_k^2)}\times\frac{1}{(\alpha_jH_1)^2}\times\frac{1}{(\beta_kH_2)^2}\right)\tag{11.2}$$

For 3D conditions:

$$\overline{\eta}=n_{\mathrm{m}}R_{\mathrm{m}}\left(1-8\sum_{j,k,l}^{\infty}\frac{s}{(s+(D/R_{\mathrm{m}})(\alpha_j^2+\beta_k^2+\gamma_l^2)}\times\frac{1}{(\alpha_jH_1)^2}\times\frac{1}{(\beta_kH_2)^2}\times\frac{1}{(\gamma_lH_3)^2}\right)\tag{11.3}$$

where

$$\alpha_j = (j - 1/2)\pi/H_1, j = 1, 2, \ldots, \infty$$
$$\beta_k = (k - 1/2)\pi/H_2, k = 1, 2, \ldots, \infty$$
$$\gamma = (l - 1/2)\pi/H_3, l = 1, 2, \ldots, \infty$$

and the definitions of H_1, H_2, H_3, h_1, h_2, h_3 are as shown in Figure 11.1, D_m, n_m and R_m are the diffusion coefficient, porosity and retardation coefficients for the matrix material between the fractures and s is the Laplace transform parameter (see Chapter 7).

These equations each contain infinite sums ($\sum_j^\infty$ etc.) in which the spacing between fractures plays an important role. Intuitively, one would expect the 3D and 2D solutions to reduce to the 1D solution as the spacing in the Y- and Z-directions (i.e., $2H_2$ and $2H_3$) is increased. Indeed, this is so, and for the case examined in Figure 11.2 the 2D and 3D solutions tend to approach the 1D solution (to less than 1.5% difference in concentration) as the spacings $2H_2$ and $2H_3$ are increased up to about 1 m. Further increasing the spacing from 1 to 10 m results in further slow convergence to the 1D solution.

Inspection of equations 11.2 and 11.3 suggests that although the 2D and 3D solutions will reduce to the 1D case (e.g., equation 11.1a), as noted above, the number of terms in the series required to obtain a given accuracy will also increase for increasing fracture spacing. Generally, good solutions can be obtained using only 8–10 terms in each series. Nevertheless, Figure 11.3 shows that the number of terms required to reach a high level of accuracy (i.e., better than 0.01%) can become quite large for widely spaced fractures. It can be seen that for spacings of $2H_2 \leq 1$ m (i.e., where 2D effects are the greatest), less than 30 terms are required to attain a high accuracy. For wider spacings (i.e., as the 2D solution reduces to the 1D solution, as shown in Figure 11.2), the number of terms in the series increases rapidly. This generally does not pose a practical problem since (a) the computation is still quite fast even on a microcomputer and (b) it is generally not necessary to evaluate the 2D or 3D solution for fracture spacings $2H_2$ or $2H_3$ exceeding ten times $2H_1$ since the corresponding 1D or 2D solution can be used instead. Obviously, one can program each of the equations 11.1, 11.2 and 11.3 and select the equation considered most appropriate for a given problem.

In general, relatively few terms are required in the infinite series and in the case of the 1D solution, the summation can be avoided altogether by defining $\bar{\eta}$ as given in equation 11.1b rather than as in 11.1a.

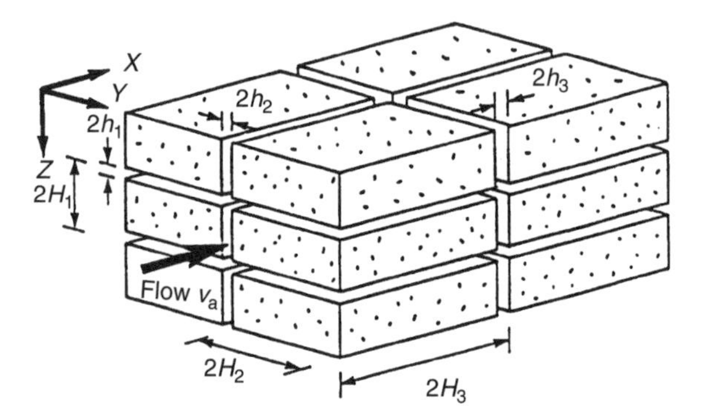

Figure 11.1 Definition of fracture geometry (after Rowe and Booker, 1989; reproduced with permission of the *International Journal for Numerical and Analytical Methods in Geomechanics*; © John Wiley & Sons Limited).

11.3 Lateral migration through fractured media

11.3.1 Evolution of a contaminant plume: the role of matrix diffusion

To illustrate the development of a contaminant plume under 1D and 3D conditions, consider a slightly reactive contaminant (e.g., 1,1-dichloroethane) migrating through fractured Queenston shale. The diffusion coefficient, matrix (intact shale) porosity and retardation coefficient are estimated to be $D_m = 0.0027\,\text{m}^2/\text{a}$; $n_m = 0.1$; and $R_m = (1 + \rho K_m/n_m) = 2.3$, respectively. The dispersivity is assumed to be $\alpha = 3$ m and the

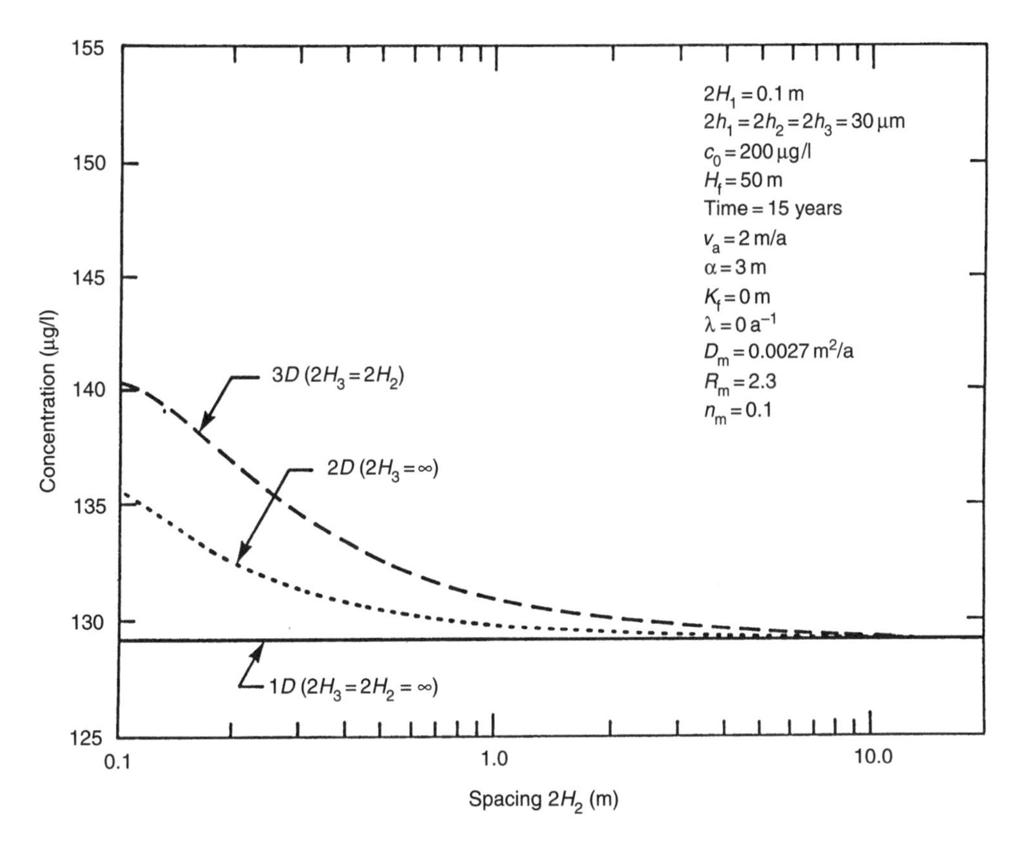

Figure 11.2 Variation in peak concentration at a point 100 m from the source as the spacing between secondary fractures ($2H_2$ and $2H_3$) is varied (after Rowe and Booker, 1990b; reproduced with permission of the *International Journal for Numerical and Analytical Methods in Geomechanics*; © John Wiley & Sons Limited).

equivalent height of leachate H_f is taken to be a rather high value of 50 m. The initial concentration of 1,1-dichloroethane is assumed to be 200 μg/l.

The reference case involves fractured shale with an assumed hydraulic conductivity of 2×10^{-6} m/s and a lateral gradient of 0.032 giving a Darcy velocity of 2 m/a. Based on the hydraulic conductivity of 2×10^{-6} m/s, the fracture opening size was estimated for a single set of parallel fractures (based on Hoek and Bray, 1981) to be approximately 30, 60 and 130 μm for fracture spacings of 0.01, 0.1 and 1 m, respectively. These correspond to fracture porosities $n_b = 0.003$, 0.0006 and 0.00013 and groundwater velocities of 667, 3,333 and 15,385 m/a for these three spacings.

Figures 11.4 and 11.5 show the evolution of a contaminant plume for 1D and 3D fracture systems, respectively. In both cases, the plume is plotted at four different times. Figure 11.6 shows the variation in concentration, with time, at a monitoring point 100 m from the source for the two cases.

First, inspecting Figure 11.4, it is evident that at small times (e.g., 3.9 years), the contaminant plume is restricted to a zone extending about 1,000 m downgradient of the landfill. Inspecting Figure 11.6, it can be seen that the concentration at a monitoring point 100 m downgradient increases with time until it reaches a peak value after 3.9 years. The concentration at this point then decreases for subsequent times as

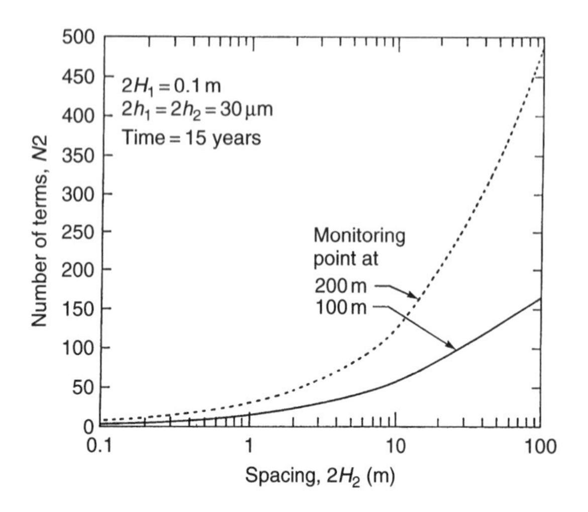

Figure 11.3 Number of terms in series required to obtain an accuracy of 0.01% (after Rowe and Booker, 1990b; reproduced with permission of the *International Journal for Numerical and Analytical Methods in Geomechanics*; © John Wiley & Sons Limited).

contaminant is flushed out of the landfill and also from the fractured rock near the landfill. This evolution of the plume is evident from Figure 11.4 where it can be seen that after 8.4 years the plume extends beyond 1,000 m and the concentration at locations between the source and the 100-m "monitoring" point has continuously dropped below the values that were present at 3.9 years. Between points at 100 and 160 m the concentration has also decreased below the values present at 3.9 years. Further than 160 m from the landfill there has been an increase in concentration between 3.9 and 8.4 years and the concentration at 250 m from the source reaches its peak after 8.4 years and decreases for subsequent times (e.g., see plumes for 16 and 24 years). This peak is less than the peak which occurred at 100 m from the source. Thus it can be seen that as the plume spreads out there is a decrease in the peak concentration which

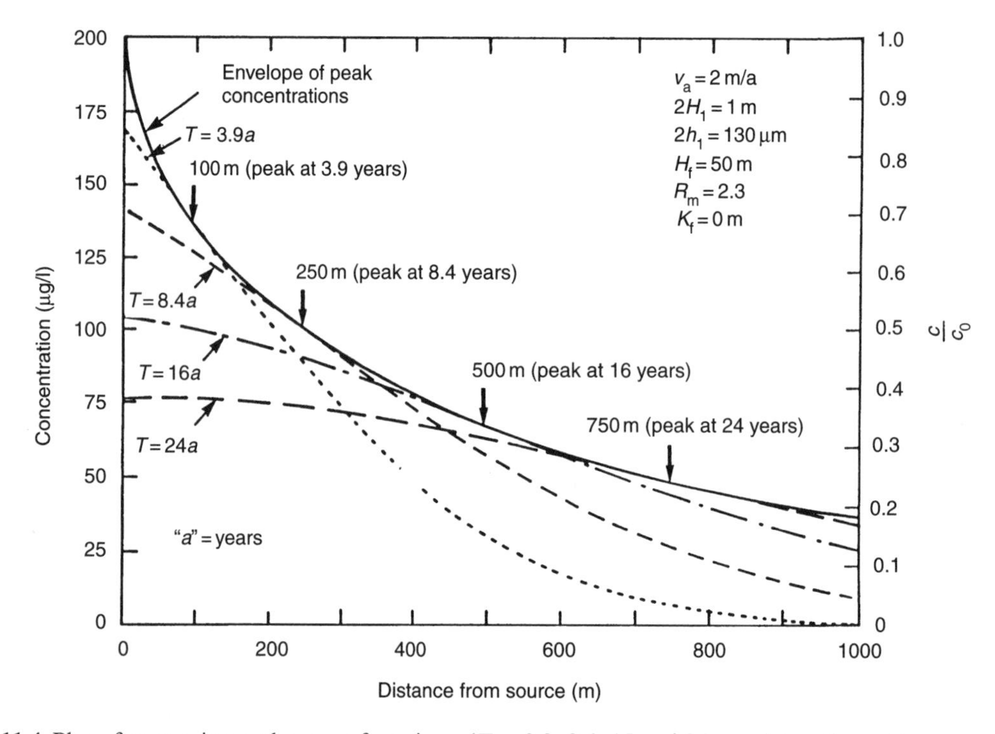

Figure 11.4 Plot of contaminant plumes at four times ($T = 3.9$, 8.4, 15 and 26 years) together with the envelope of peak concentrations. Fracture spacing 1 m, Darcy velocity 2 m/a, 1D fracture network (after Rowe and Booker, 1990b; reproduced with permission of the *International Journal for Numerical and Analytical Methods in Geomechanics*; © John Wiley & Sons Limited).

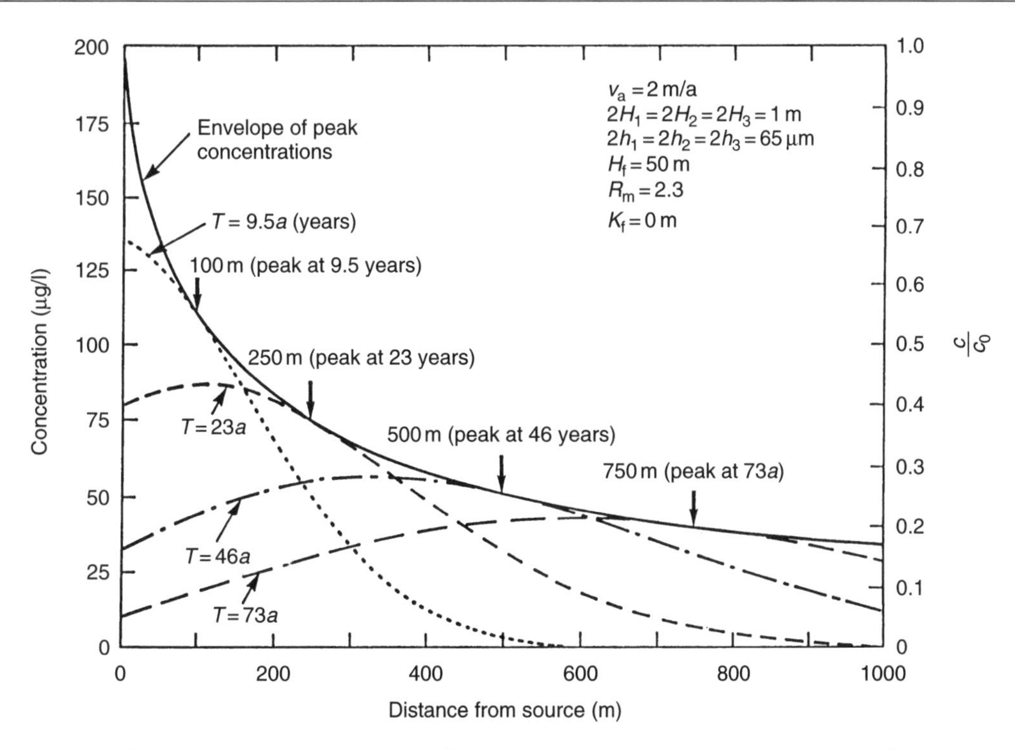

Figure 11.5 Plot of contaminant plumes at four times (9.5, 23, 46 and 73 years) together with envelope of peak concentrations, 3D analysis. Fracture spacing 1 m, Darcy velocity 2 m/a, 3D fracture network (after Rowe and Booker, 1990b; reproduced with permission of the *International Journal for Numerical and Analytical Methods in Geomechanics*; © John Wiley & Sons Limited).

is attained at points away from the landfill. The peak values at points 100, 250, 500 and 750 m are indicated in Figure 11.4 together with an envelope of peak concentration at all points up to 1,000 m downgradient of the landfill. There is significant attenuation of contaminant with distance from the source and for this combination of parameters, the peak impact at 1,000 m is less than 20% of the initial concentration in the leachate.

The results presented in Figures 11.4–11.6 also serve as a warning that one cannot infer the rate of movement of contaminant from isolated observations of the contaminant plume. To illustrate this, consider the rate of advance of a point where $c/c_0 = 0.5$, which is often taken as an indication of the speed of a contaminant plume. If observations were made after 3.9 years, the point where $c/c_0 = 0.5$ would be about 210 m from the source, implying a velocity of about 53 m/a. After 8.4 years, the location at which $c/c_0 = 0.5$ has moved to 250 m implying a velocity of about 30 m/a. Thus, the velocity of the plume (defined in terms of the movement of a point where $c/c_0 = 0.5$) is not a constant. Rather, it decreases with time. Furthermore, examination of Figure 11.6 shows that when the mass of contaminant is finite, there will be up to two times at which $c/c_0 = 0.5$ at a given point. For 1D condition it takes less than 1 year for c/c_0 to reach 0.5 at 100 m from the source. The concentration continues to rise at this point until it peaks after 3.9 years and then decreases to a value of $c/c_0 = 0.5$ again after 16 years. Thus one must be very cautious in interpreting the rate of contaminant migration from a few field observations.

The results presented in Figure 11.4 were obtained assuming a single fracture set at 1 m spacing (1D analysis). Figure 11.5 gives the

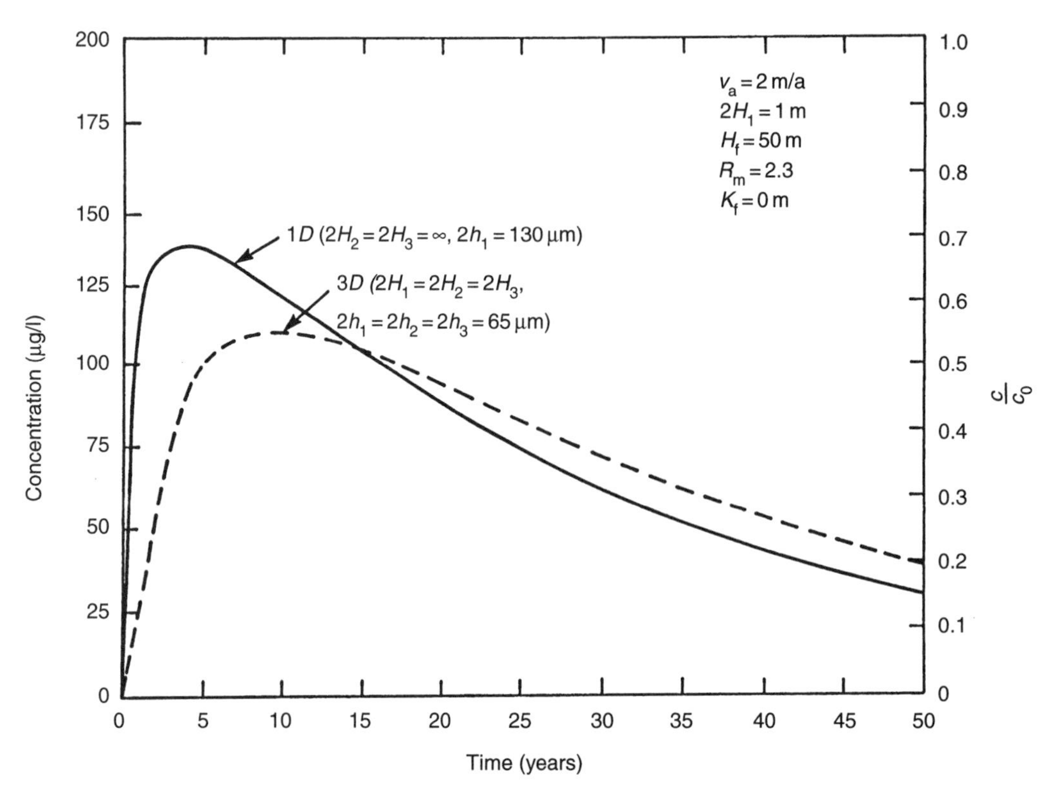

Figure 11.6 Variation in concentration with time at a point 100 m from the source for a 1D and 3D fracture network (after Rowe and Booker, 1990b; reproduced with permission of the *International Journal for Numerical and Analytical Methods in Geomechanics*; © John Wiley & Sons Limited).

corresponding results for a 3D fracture system with 1-m spacings in each direction. A comparison of Figures 11.4 and 11.5 shows that the additional matrix diffusion which can occur with the 3D fracture system has a significant effect on the contaminant plume and the rate of contaminant migration. Since more contaminant is "captured" by the matrix of the rock, with the 3D fracture system, the contaminant plume is smaller; for example, compare the plume at 8.4 years in Figure 11.4 with that after 9.5 years in Figure 11.5. Furthermore, the plume moves slower. This is evident from a comparison of the time required to reach the peak concentration at a given point (e.g., 100, 250, 500 or 750 m) for the 1D and 3D cases (see Figures 11.4 and 11.5) or from Figure 11.6 which shows the variation in concentration with time at the 100-m point.

An additional consequence of the greater movement into the matrix for 3D conditions is a much slower eventual decay of the concentration at points near the landfill which is evident from Figures 11.4–11.6. Referring to Figure 11.6, it is seen that for the first 15 years the 1D solution gives higher concentration than the 3D solution. After 15 years, the reverse is true because of the greater release of contaminant back into the fracture system (by diffusion from the matrix of the rock back into the fractures) which is possible in the 3D case. Similarly, a comparison of Figures 11.4 and 11.5 shows that the 3D plume after 23 years gives the higher concentration for the first 250 m from the source than does the 1D solution after 24 years.

A related issue is the comparison of the envelope of peak concentration obtained from the 1D and 3D analyses (see Figures 11.4 and 11.5). For distances less than 1,000 m, the 3D solution gives lower peak concentrations than the 1D solution. It is, however, evident that the difference between the two solutions is quite small at the 1,000-m point. The reason for this is that the effect of 1D, 2D and 3D matrix diffusion is scale-dependent. The differences are greatest when the spacing between fractures and the time of interest is such that much more diffusion can occur if 3D migration is possible than if only 1D migration is possible. For a spacing of 1 m, this difference in the effects of diffusion will be greatest at small to moderate times (i.e., times less than the time for diffusion to fully penetrate the matrix). Thus the difference between 1D and 3D cases will not be as significant for the times required for the peak contaminant to migrate 1,000 m as it is, say, for the times required to migrate 100 m. For similar reasons, it is found that there is no practical difference between the 1D and 3D results obtained from similar calculations performed with a fracture spacing, $2H_1$, of 0.1 m.

Referring to Figures 11.4 and 11.5, it can be seen that when there is a high Darcy velocity ($v_a = 2\,\text{m/a}$), the contaminant plume is quite extensive after a short period of time. Here, advective–dispersive transport is dominant but matrix diffusion still does play a significant role. Referring to the curves for similar times (24 years in Figure 11.4, 23 years in Figure 11.5), it can be seen that 3D diffusion considerably reduces the extent of the plume. In addition, the maximum concentration in the plume at 23 years (3D analysis) exceeds the maximum at 24 years (1D analysis) because a similar mass of contaminant is contained within a smaller volume of rock. For lower velocities (e.g., $v_a = 0.2\,\text{m/a}$), diffusion has a more dominant role and the time at which the peak impact reaches a given point is similar for the 1D and 3D cases (e.g., at 250 m it is 350 years 1D and 381 years 3D). However since the 3D plume is contained within a smaller area, the mass within the volume of rock where the peak impact is being monitored is greater and consequently the impact is greater.

11.3.2 Effect of matrix diffusion coefficient and porosity

The results presented in the previous section provide some insight concerning the effect of fracture arrangement for 1D and 3D conditions and a reactive contaminant. Similar observations can be made for conservative contaminants which do not interact with the matrix or fracture surface (i.e., $R_m = 1$, $K_f = 0$).

Figure 11.7 shows the calculated variation in concentration with position after 30 years migration of a conservative contaminant for a number of combinations of matrix diffusion coefficient, D_m and matrix porosity, n_m. For a given fracture spacing and Darcy velocity, the velocity at which the contaminant plume moves away from the source is primarily related to the matrix porosity. For this case it can be seen that for both diffusion coefficients considered, reducing the matrix porosity by a factor of two, from 0.1 to 0.05, increased the velocity of the peak of the contaminant plume by a factor of two from about 0.7 to about 1.4 m/a.

It is evident from Figure 11.7 that the peak concentration is smaller for a porosity of 0.1 than that for 0.05; however these plots do not fully illustrate the effect that porosity has on attenuation. To illustrate this, Figure 11.8 shows the variation in concentration with time at a point 30 m from the source for the four combinations of parameters considered in Figure 11.7. This figure reinforces the earlier observation that a higher porosity reduces the speed of the contaminant plume and reduces the magnitude of the peak impact at a given point. This is because the higher matrix porosity provides more fluid volume into which contaminant can diffuse from the fracture thereby reducing the mass of contaminant within the fracture system.

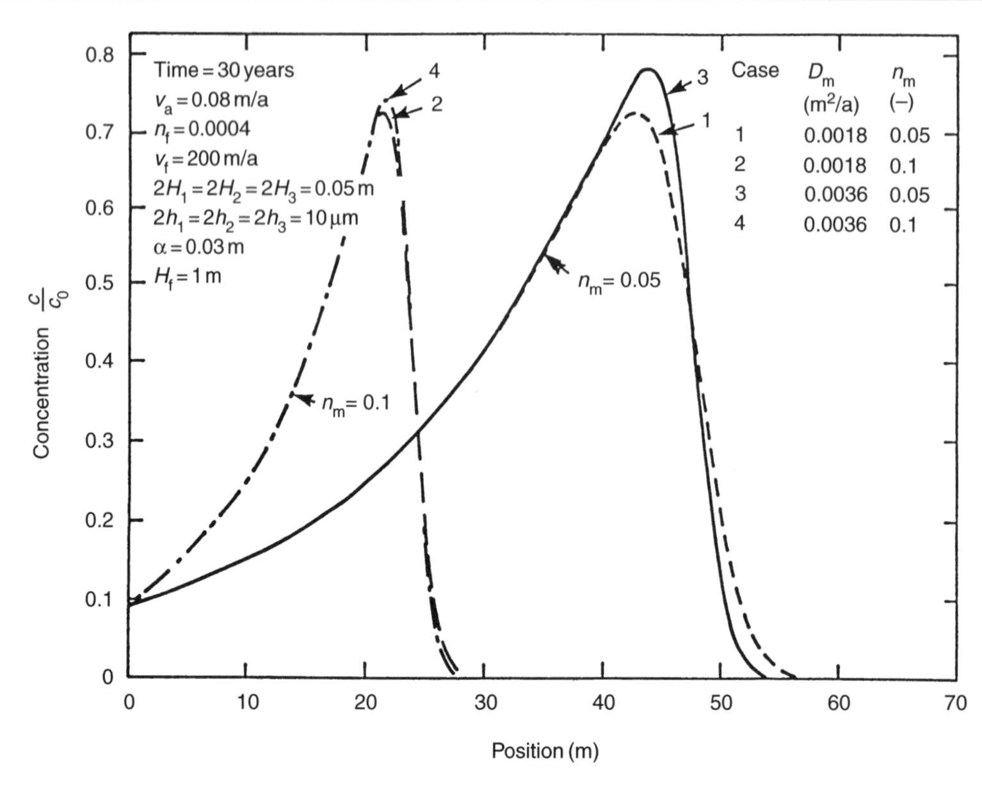

Figure 11.7 Effect of diffusion coefficient and matrix porosity on the contaminant plume: 3D fracture system (after Rowe and Booker, 1989; reproduced with permission of the *International Journal for Numerical and Analytical Methods in Geomechanics*; © John Wiley & Sons Limited).

The effect of diffusion coefficient on contaminant migration is related to the time required to "chemically saturate" the pore fluid between fractures with contaminant. In the following discussion, this time is defined as the time required to reach steady state for diffusion into the matrix assuming a constant concentration in the fracture at a given point. A higher diffusion coefficient allows more rapid movement of contaminant into and out of the rock matrix. Thus as the contaminant moves along the fracture, the extent of the plume will be limited by diffusion into the rock matrix until the concentration in the rock between fractures reaches a value equal to the concentration in the fracture. No more contaminant can then enter the rock at this point and, in fact, contaminant will diffuse out of the matrix once the concentration in the fracture begins to decrease. The time required to chemically saturate the matrix with contaminant at a given point is about 1 year for a single set of fractures at a spacing of 0.05 m and a diffusion coefficient of $0.0018\,m^2/a$ and about 0.5 years for a diffusion coefficient of $0.0036\,m^2/a$. Since these time scales are small compared with the 30-year time period being examined here, much of the rock is chemically saturated with contaminant in both cases. When chemical saturation can occur over a significant extent of the rock in the timeframe of interest, the maximum peak concentration and minimum extent of the plume will be obtained for the case where the contaminant can most quickly chemically saturate the rock between fractures. Thus, increasing the diffusion coefficient by a factor of 2 from 0.0018 to $0.0036\,m^2/a$ results in a slightly smaller plume

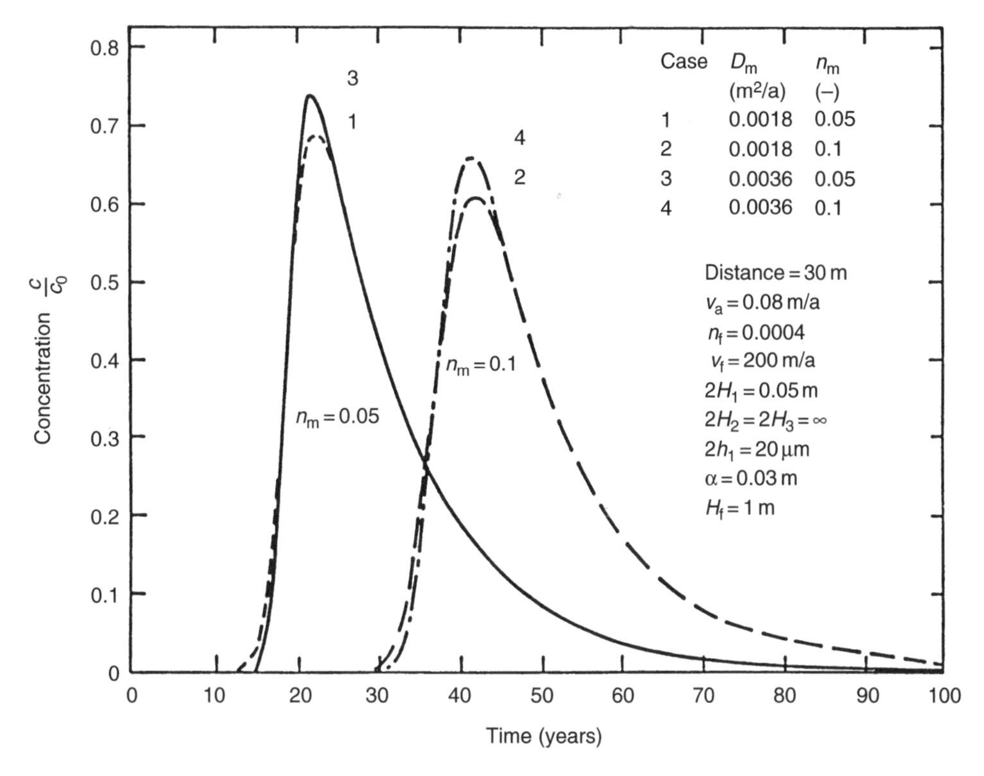

Figure 11.8 Effect of diffusion coefficient and matrix porosity on contaminant arrival times at a point 30 m from the source (after Rowe and Booker, 1989; reproduced with permission of the *International Journal for Numerical and Analytical Methods in Geomechanics*; © John Wiley & Sons Limited).

for the higher diffusion coefficient but, since essentially the same amount of mass is involved, the peak concentration is consequently higher as shown in Figure 11.7 or 11.8.

11.3.3 Effect of fracture spacing

As illustrated in Figures 11.7 and 11.8, the twofold reduction in diffusion coefficient from 0.0036 to 0.0018 m^2/a does, in this case, result in a modest reduction in the magnitude of the peak concentration but very little change in the velocity of the contaminant plume. However, this finding should not be extrapolated to other situations. For example, as the spacing between fractures increases, the time required to chemically saturate the rock with contaminant also increases. When the time required to saturate the rock becomes significant compared to the timeframe of interest, then a higher diffusion coefficient may be expected to exert a different influence on the shape of the contaminant plume since, by definition, there will not have been time for the contaminant to fully diffuse into the rock along most of the distance between the source and the point of interest. To illustrate this, Figure 11.9 shows the calculated contaminant plume for fracture spacings of 0.05 and 0.5 m for two values of matrix diffusion coefficient (i.e., $D_m = 0.0018\,m^2/a$ and $D_m = 0.0036\,m^2/a$).

It is evident from Figure 11.9 that the spacing between the primary set of fractures is an important parameter influencing contaminant movement. As previously noted, the time required for diffusion to chemically saturate the matrix with contaminant is about 1 year and 0.5 years for $D_m = 0.0018$ and

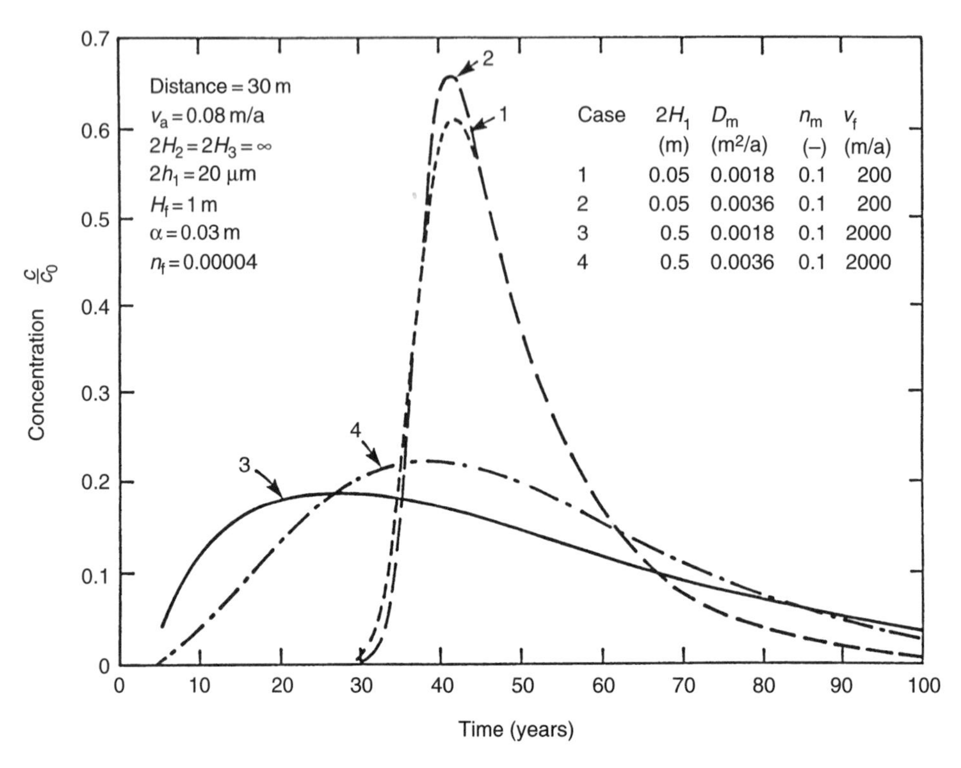

Figure 11.9 Effect of fracture spacing, $2H_1$, on arrival times at a point 30 m from the source (after Rowe and Booker, 1989; reproduced with permission of the *International Journal for Numerical and Analytical Methods in Geomechanics*; © John Wiley & Sons Limited).

$0.0036\,m^2/a$ respectively at a spacing of $2H_1 = 0.05$ m and about 100 years and 50 years respectively at a spacing of $2H = 0.5$ m.

In this case, the larger spacing between fractures ($2H_1 = 0.5$ m) also gives rise to a substantially smaller peak impact and earlier arrival of this impact than was calculated for the narrower fracture spacing ($2H_1 = 0.05$ m). This is primarily because there has not been sufficient time for chemical saturation of the rock matrix on the time scale being considered and so the contaminant is much more spread out along the fracture for the wider spacing. Since essentially the same mass is contained in both the plumes for 24 years – 0.05 and 0.5 m, the fact that it is spread over a larger volume of rock for the wider spacing means that the peak impact is correspondingly reduced.

The effect of fracture spacing on the contaminant plume is dependent on both the mass of contaminant and the ratio of the time of interest to the time required to chemically saturate the matrix between fractures. Because of the complex interaction between the effect of different parameters, care should be taken not to over generalize from the results of any one set of analyses. Fracture spacing is clearly an important parameter, and its effects and interaction with other parameters such as retardation coefficient will be discussed in more detail in Section 11.3.6.

11.3.4 Effect of dispersivity

Figure 11.10 shows the calculated variation in concentration with time at a point 30 m from the contaminant source for a 3D fracture set and

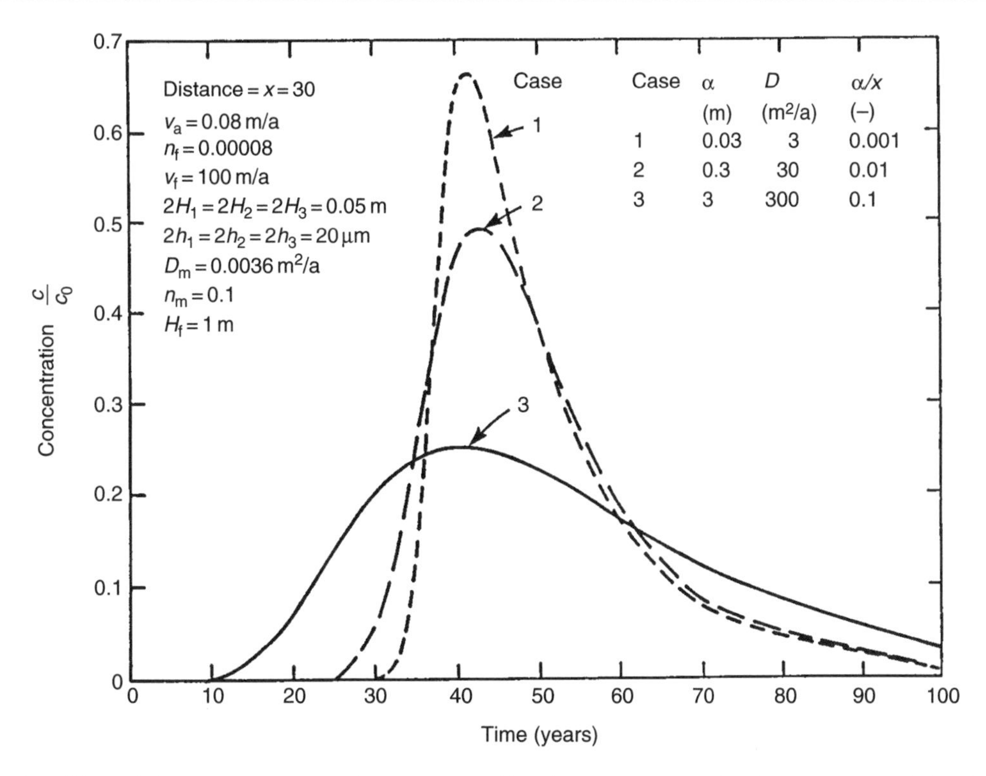

Figure 11.10 Effect of dispersivity, α, on arrival at a point 30 m from the source: 3D fracture system (after Rowe and Booker, 1989; reproduced with permission of the *International Journal for Numerical and Analytical Methods in Geomechanics*; © John Wiley & Sons Limited).

three different values of dispersivity α (i.e., 0.03, 0.3 and 3 m) which correspond to 0.1, 1 and 10% of the travel distance. All other things being equal, a small dispersivity, implying very little mechanical dispersion due to irregularities in the fracture system, gives the greatest impact. Higher dispersivities give rise to a more extensive contaminant plume, since the mass of contaminant is spread over a greater volume of rock, and a smaller peak impact at a given point.

Dispersivity is generally considered to increase with distance from the source, at least until the distance involved is large compared to the scale of the non-homogeneities of the flow system which give rise to the dispersion process (e.g., see Frind *et al.*, 1987). If one considered the dispersivity to be a linear function of distance (say 1% of the travel distance: $\alpha/x = 0.01$) then it is found that the peak impact at the monitoring point is reduced by more than a factor of five between the monitoring points at 30 and 300 m.

The fracture opening size is one of the most difficult parameters to determine and it is of interest to explore the implication of uncertainty regarding this parameter since the fracture porosity and, hence, groundwater velocity both depend on this parameter.

Figure 11.11 shows the calculated contaminant plume after 30 years migration for five cases involving a 3D fracture system. All cases assume the same Darcy velocity. In cases 1, 2 and 3, the fracture opening size is varied between 10 and 40 m giving a consequent fourfold variation in groundwater velocity from 200 to 50 m/a. For a given dispersivity of 0.03 m, it can be seen that this fourfold variation in opening size and

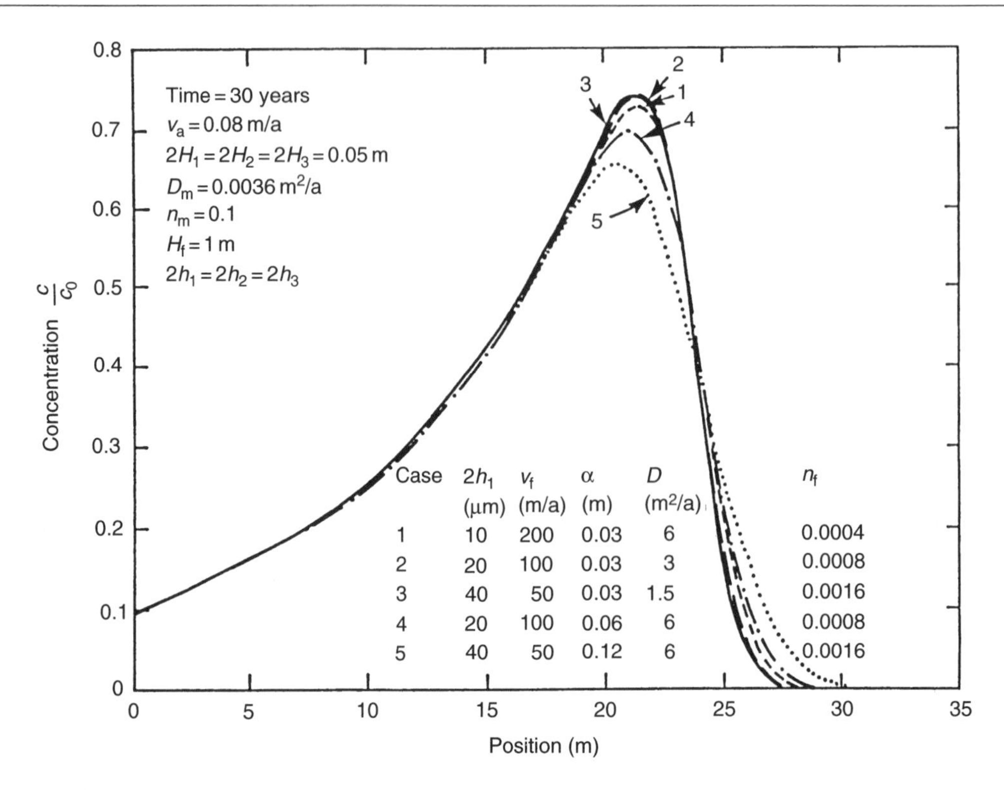

Figure 11.11 Effect of fracture opening size, $2h_1$, and dispersivity, α, on the contaminant plume (after Rowe and Booker, 1989; reproduced with permission of the *International Journal for Numerical and Analytical Methods in Geomechanics*; © John Wiley & Sons Limited).

groundwater velocity does slightly change the contaminant plume; however, this change is of no practical significance. In each case, the peak of the contaminant plume is moving at a velocity of 0.7 m/a irrespective of whether the groundwater velocity is 50 or 200 m/a.

The coefficient of hydrodynamic dispersion, D, is often assumed to be linearly proportional to the groundwater velocity v_f (i.e., $D = \alpha v_f$, where α is the dispersivity). In the comparison of cases 1, 2 and 3 above, it has been assumed that the dispersivity remained constant ($\alpha = 0.03$ m) and hence the coefficient of hydrodynamic dispersion did vary by a factor of 4 due to the variation in groundwater velocity. However, it may also be reasonable to argue that really it is the coefficient of hydrodynamic dispersion that remains constant. To examine the implications of this assumption, cases 1, 4 and 5 (Figure 11.11) represent results for fracture opening sizes of 10, 20 and 40 μm and a single coefficient of hydrodynamic dispersion of 6 m^2/a. In this case, it can be seen that increasing the assumed fracture opening size from 10 to 40 μm does result in a change in the contaminant plume primarily due to the fact that for the high opening size there is (relatively speaking) a fourfold increase in the ratio of the coefficient of hydrodynamic dispersion to groundwater velocity. Nevertheless, from a practical standpoint, it can be seen that the variation in average fracture opening size and hence groundwater velocity considered here does not significantly influence the speed of contaminant transport or the peak impact which varies by less than 10% for a fourfold variation in opening size. This implies that uncertainty as to the precise details concerning fracture opening size is not going to have a significant effect on prediction of contaminant transport for the range of cases considered. This conclusion is valid for typical ranges of fracture opening size and spacing in a fractured media. The assumptions used in developing these results may not be applicable to situations where the fracture porosity n_b is less than 10^{-5} or greater than 0.1. The results are only applicable to the situation where almost all of the advective contaminant transport is along the fractures rather than through the matrix.

11.3.5 Effect of Darcy velocity (flux)

It is evident from the preceding discussion that the groundwater velocity is of secondary significance when predicting the rate of contaminant migration through fractured porous media. This is fortunate since the groundwater velocity is very sensitive to changes in opening size and hence is difficult to estimate and may vary substantially within a given mass of rock. A more readily determined parameter is the Darcy velocity (Darcy flux) which represents the volume of water moving through the fractured rock mass per unit area per unit time. This quantity can be deduced from the measured bulk hydraulic conductivity and gradient within the fractured rock mass.

Figure 11.12 shows the effect of varying the Darcy velocity on the contaminant plume after 30 years of migration. Results are given for Darcy velocities of 0.04 and 0.08 m/a and it can be seen that doubling the Darcy velocity results in an approximate doubling of the velocity of the

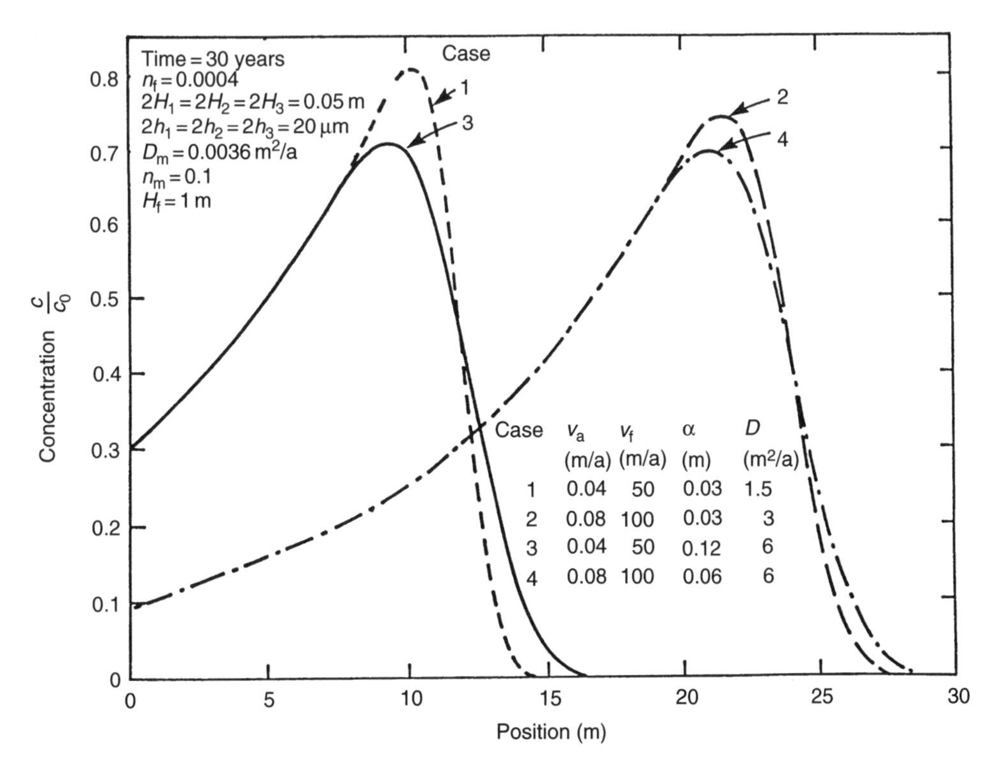

Figure 11.12 Effect of Darcy velocity on contaminant plume (constant fracture porosity) (after Rowe and Booker, 1989; reproduced with permission of the *International Journal for Numerical and Analytical Methods in Geomechanics*; © John Wiley & Sons Limited).

peak of the plume. Curves 1 and 2 show the effect of increasing the Darcy velocity while maintaining a constant dispersivity, α. Curves 3 and 4 show the effect of increasing the Darcy velocity while maintaining a constant coefficient of hydrodynamic dispersion. The assumption regarding what happens to dispersion when the Darcy velocity changes does have some impact on the contaminant plume; however, the primary variable is the Darcy velocity and dispersion is a secondary consideration.

The results presented in Figure 11.12 were obtained by varying the Darcy velocity for a given fracture distribution and fracture porosity. Thus the groundwater velocity varies in proportion to the Darcy velocity. To emphasize the point that it is the Darcy velocity (flux) and not the groundwater velocity which is the primary control on contaminant transport, the results in Figure 11.13 were obtained by varying the Darcy velocity and fracture porosity such that the groundwater velocity remained constant at 100 m/a. Comparing case 1 in Figures 11.12 and 11.13, it is seen that for a given dispersivity and Darcy velocity the difference in groundwater velocity between 50 and 100 m/a has no significant effect on the contaminant plume whereas it is seen from Figure 11.13 that for a given dispersivity and groundwater velocity, the difference in Darcy velocity between 0.04 and 0.08 m/a approximately doubles the movement of the peak of the contaminant plume. Thus, it is the Darcy velocity and not the groundwater velocity which controls the movement of the

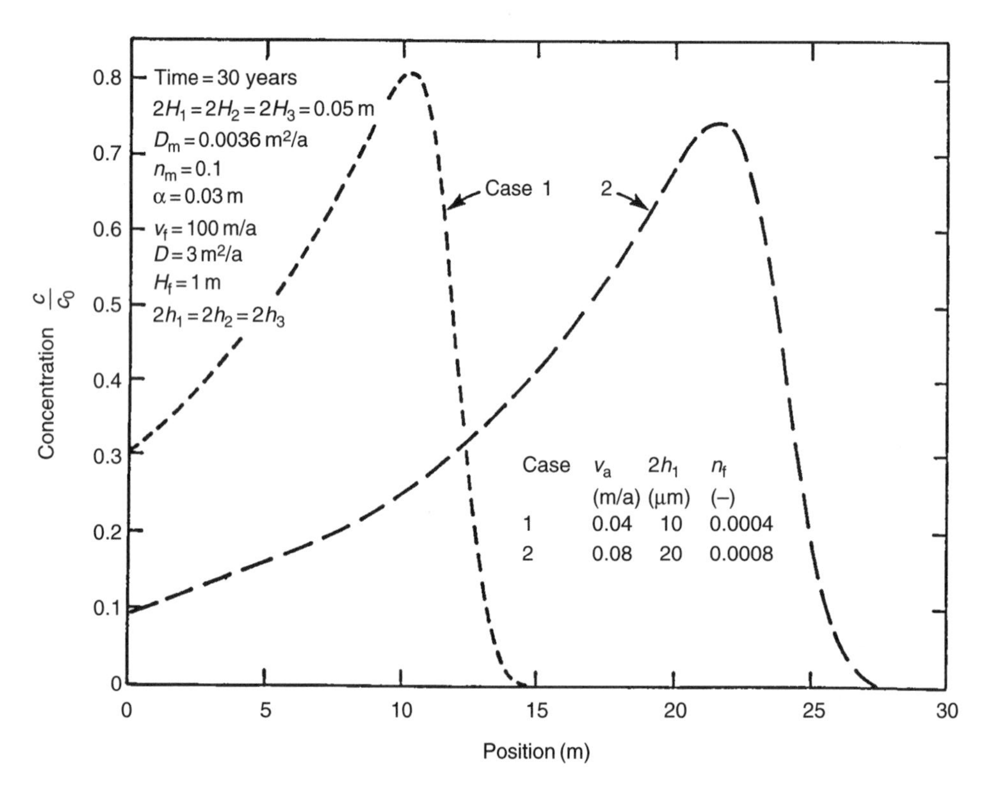

Figure 11.13 Effect of Darcy velocity on the calculated contaminant plume (constant groundwater velocity, v_f) (after Rowe and Booker, 1989; reproduced with permission of the *International Journal for Numerical and Analytical Methods in Geomechanics*; © John Wiley & Sons Limited).

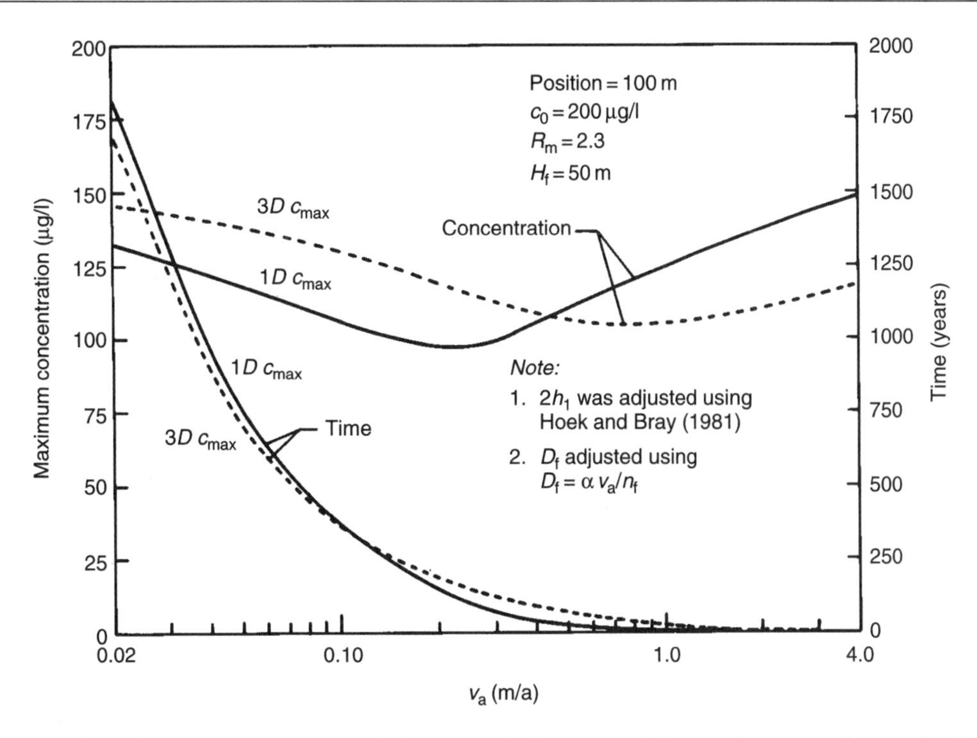

Figure 11.14 Effect of Darcy velocity on peak impact and time to peak impact for a reactive species ($R_m = 2.3$) (modified from Hammoud, 1989).

contaminant plume and the impact on groundwater quality in fractured porous media where matrix diffusion is a significant attenuation mechanism.

The effect of Darcy velocity on the peak impact at a point 100 m downgradient from a source is illustrated in Figure 11.14 for a mildly reactive contaminant ($R_m = 2.3$). Notice that the time at which the peak impact occurs decreases monotonically with increasing Darcy velocity; however, the magnitude of the peak impact decreases initially with increasing velocity and then increases for higher Darcy velocities. The velocity at which the minimum impact occurs is different for the 1D and 3D cases, but the trends are the same. Notice also that at low velocities the peak impact is greater for the 3D than for the 1D case but for higher velocities the reverse is true. Thus there is an optimal velocity which, for a given situation, gives minimal impact. It is not normally practical to design to take advantage of this optimal situation but it is important to recognize that an analysis for a single value of v_a could be quite misleading in that one may, fortuitously, be for an optimum or near optimal velocity. Where uncertainty exists regarding Darcy velocity as it invariably will, it is necessary to perform analyses for a number of different velocities which define a range of values in which the actual velocity is most likely to lie.

11.3.6 Effect of sorption in the rock matrix (R_m)

The results discussed in the previous sections have separately considered a contaminant which was either conservative (e.g., chloride with $R_m = 1$) or slightly retarded (1,1-dichloroethane with $R_m = 2.3$ for the shale being considered). Sorption can be an important attenuation mechanism and its effects are illustrated for $R_m = 1$ (a conservative species), $R_m = 2.3$ and $R_m = 10$ (a reactive species) in Figure 11.15. This figure shows the calculated variation in concentration with time at a monitoring point 100 m downgradient of the contaminant source. Since R_m represents sorption which occurs within the rock matrix, the effects of R_m will be greatest when there is significant diffusion into the rock.

In Figure 11.15, the peak concentration is decreased by a factor of 2 and the time required to reach the peak is increased by about 12 years due to a change in R_m from 1 to 10.

The effect of sorption is to remove contaminant from solution at times when the concentration is increasing. This reduces the peak impact at any given point away from the source. It is assumed here that sorption is reversible. A consequence of this assumption is that contaminant will be desorbed into solution when the contaminant concentration drops. Thus at times after the peak has passed, the concentration at a given point will be higher for $R_m = 10$ than for lower values of R_m because the contaminant which had been sorbed is now being released back into the groundwater. This assumption of full reversible sorption is conservative; it may not happen in practice since the sorption processes may not be reversible or there may be a much lower desorption partitioning coefficient than sorption partitioning coefficient.

Figure 11.16 shows the effect of retardation, R_m, upon the peak concentration expected at a point 100 m from the source. Results are given for $v_a = 0.2$ m/a and two different fracture spacings ($2H_1 = 1$ and 0.1 m). For each case, the peak impact at the 100-m monitoring point is

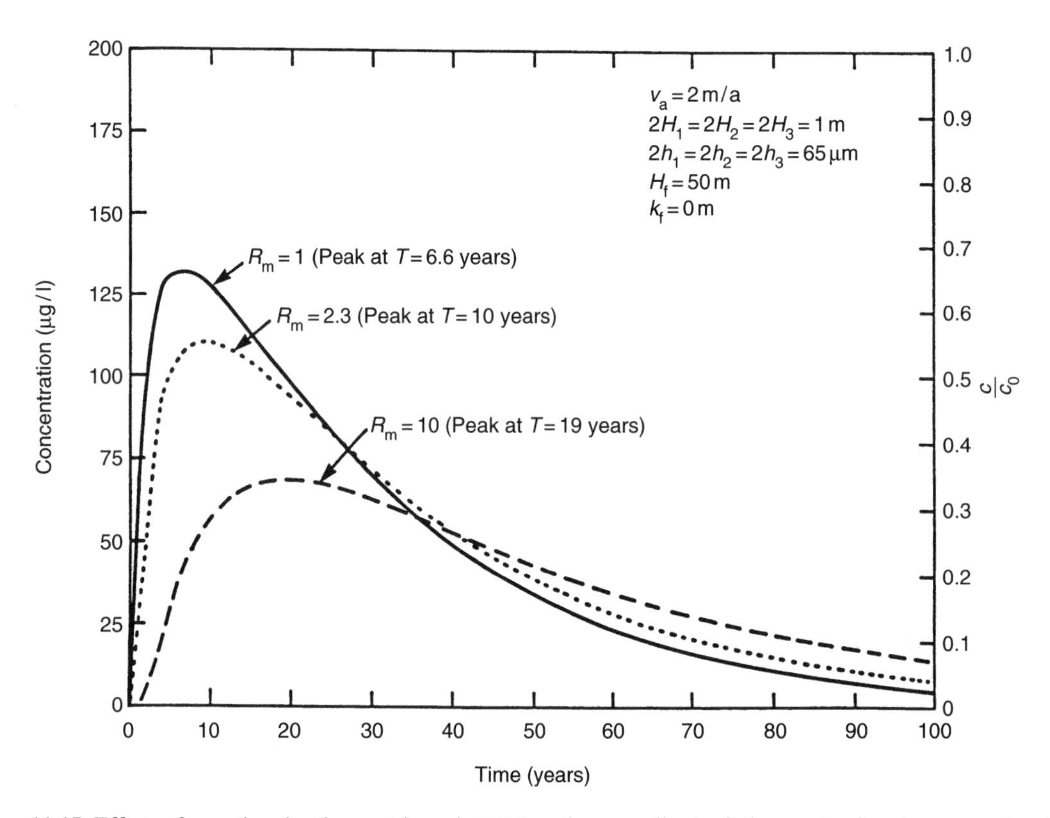

Figure 11.15 Effect of sorption in the matrix point 100 m downgradient of the contaminant source. Fractured spacing 1 m, 3D fracture network (after Rowe and Booker, 1990b; reproduced with permission of the *International Journal for Numerical and Analytical Methods in Geomechanics*; © John Wiley & Sons Limited).

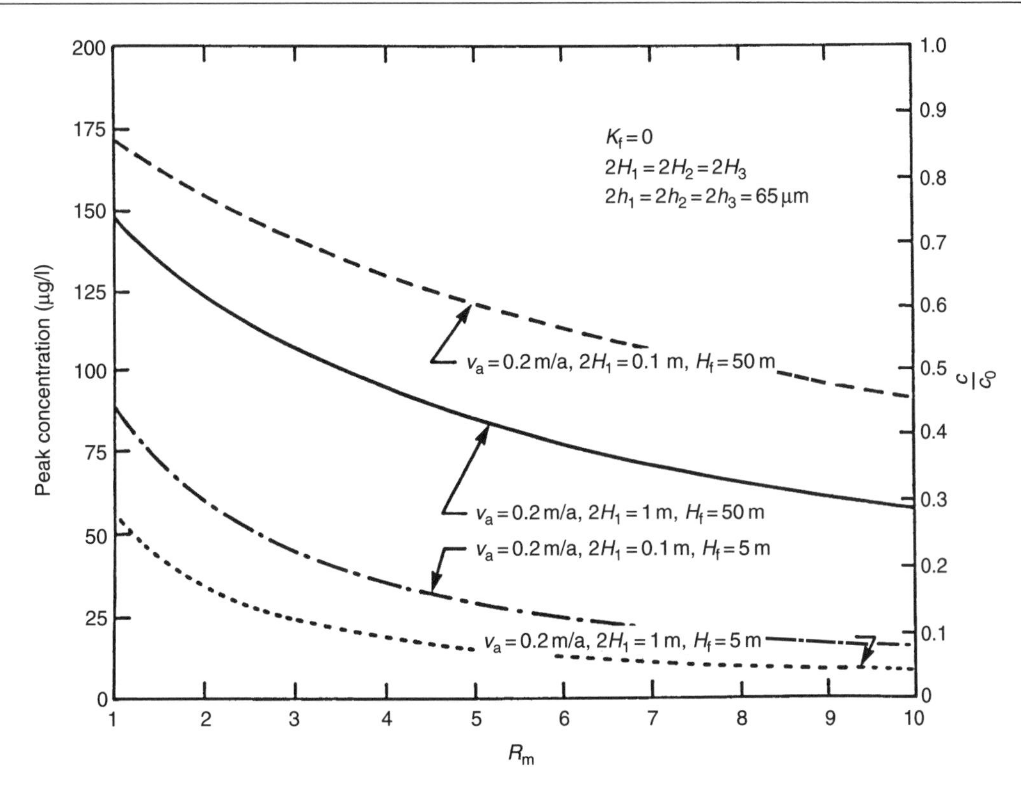

Figure 11.16 Effect of sorption in the matrix on calculated peak impact at a monitoring point 100 m downgradient of the source (after Rowe and Booker, 1990b; reproduced with permission of the *International Journal for Numerical and Analytical Methods in Geomechanics*; © John Wiley & Sons Limited).

greatest for a conservative species ($R_m = 1$) and least for a highly retarded species ($R_m = 10$). Curves are presented for two values of the equivalent height of leachate H_f (viz. 50 and 5 m) and it can be seen that sorption has a greater impact when the mass of contaminant originally in the landfill (defined in terms of H_f) is smaller.

11.3.7 Effect of the equivalent height of leachate H_f (mass of contaminant)

The results shown in Figure 11.16 indicated that the equivalent height of leachate is an important parameter and that attenuation will be considerably enhanced if the mass of contaminant available for transport into the fracture media can be minimized (e.g., by installing an effective leachate collection system). This is further demonstrated by Figure 11.17 which shows the effect of the equivalent height of leachate on the peak contaminant impact at the 100-m monitoring point. This shows that it is important to obtain a realistic, albeit conservative estimate of the equivalent height of leachate in the landfill if reasonable predictions of contaminant impact are to be made.

11.3.8 Sorption onto the fracture surface

Little is known about the sorption of contaminants onto fracture surfaces. Recognizing this, Figure 11.18 shows the effect sorption onto the fracture surface could have for an assumed range of values of distribution coefficient K_f. Since K_f is

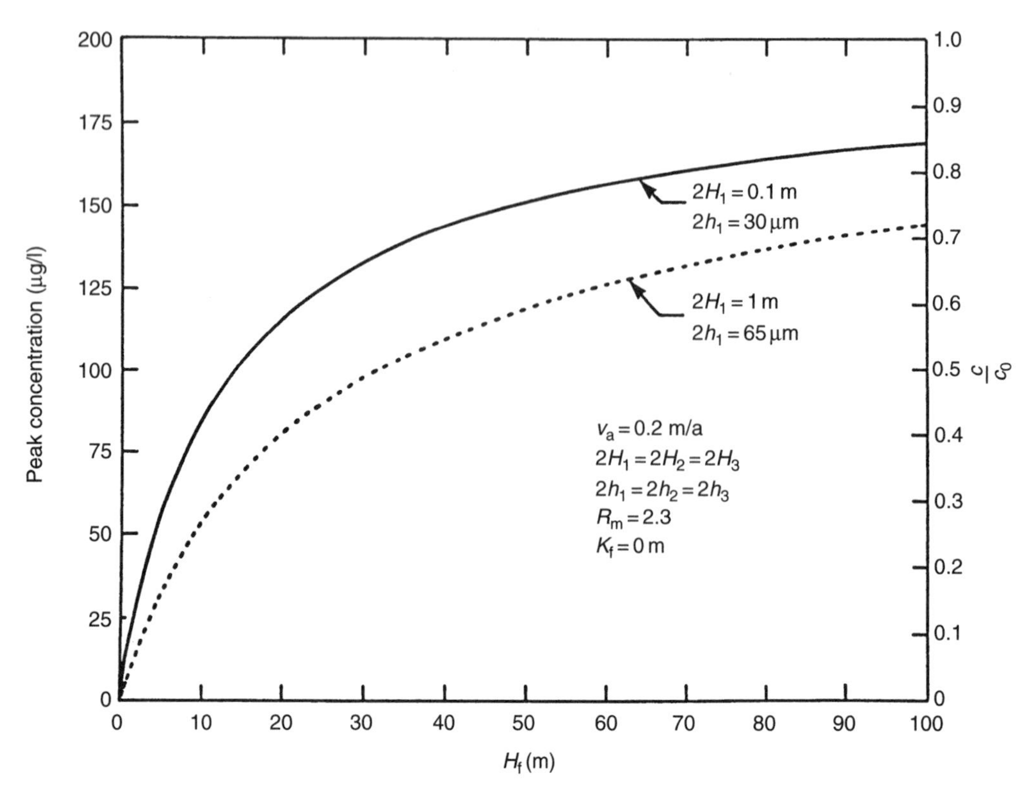

Figure 11.17 Effect of mass of contaminant (expressed in terms of the equivalent height of leachate, H_f) on contaminant impact at 100 m (after Rowe and Booker, 1990b; reproduced with permission of the *International Journal for Numerical and Analytical Methods in Geomechanics*; © John Wiley & Sons Limited).

defined (Freeze and Cherry, 1979) as the mass of solute adsorbed per unit area (M/L^2) of surface divided by the concentration of solute (M/L^3) it has units of length (L). It is noted that a range of values of K_f from 0 to 1 result in a decrease of up to a factor of four on the impact at a point 100 m downgradient for the landfill. More research is required to determine relevant and appropriate values of K_f which could be used in a given situation. It is, however, evident from Figure 11.18 that it is conservative to neglect sorption on fracture surfaces when predicting impact.

11.3.9 Case study

Two different landfills, known as the Bayview Park and Burlington landfills, were constructed directly on fractured shale in the city of Burlington, Ontario. Neither landfill was originally constructed with a leachate collection system although a toe drain has now been retrofitted to both landfills. A subsequent northern expansion of the Burlington landfill is located directly on clayey till and has a leachate underdrain system. This expansion is not expected to have lead to any significant leachate egress into the shale.

Although chloride is normally used as an indicator of the extent of contaminant migration from landfills, it does not provide a good leachate diagnostic in this case because the natural high salinity of the unpotable groundwater is such that the concentrations of chloride in the groundwater exceed values in the leachate. Diffusion tests on the shale (Rowe *et al.*, 1988;

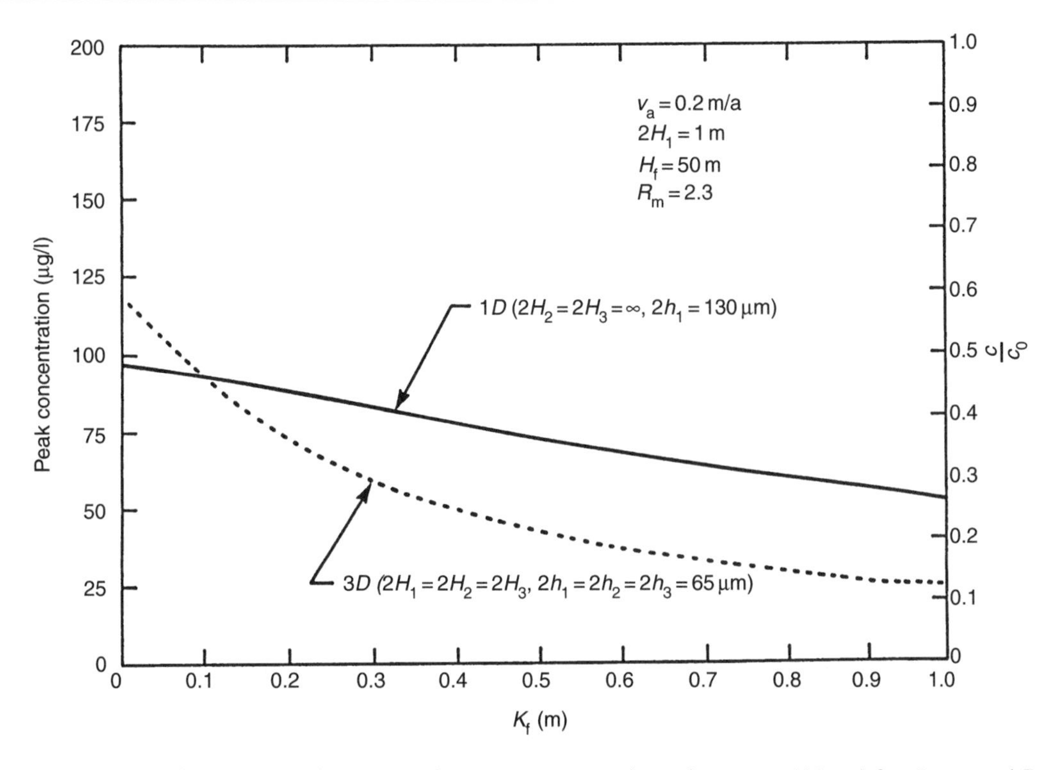

Figure 11.18 Effect of sorption on fracture surface, K_f, on contaminant impact at 100 m (after Rowe and Booker, 1990b; reproduced with permission of the *International Journal for Numerical and Analytical Methods in Geomechanics*; © John Wiley & Sons Limited).

see also Section 8.3.4) indicate that chloride can readily diffuse out of the shale which has a matrix porosity of between 0.1 and 0.11, with a diffusion coefficient of about 1.5×10^{-10} m^2/s (0.0047 m^2/a).

Since chloride cannot be used as a reliable indicator of the presence of contaminants at this site, investigations and monitoring have focused on organic contaminants. A number of stainless steel monitors ranging in distance from 25 to about 320 m downgradient of the Burlington landfill (Gartner Lee Ltd, 1988) have been monitored. Four organic chemicals (1,1-dichloroethane at 4 µg/l; vinyl chloride at 12 µg/l; trichloroethylene at 12 µg/l; and *cis*-1,2-dichloroethylene at 33 µg/l) have been detected at the monitor 25 m downgradient of the landfill (Gartner Lee Ltd, 1988). Rowe and Booker (1989) modelled the migration of the two most mobile species (vinyl chloride and 1,1-dichloroethane) and compared the calculated results with the field observations. They showed that the calculated contaminant plume which was based on a reasonable hydrogeologic evaluation of the site (a Darcy velocity of between 0.05 and 0.125 m/a and an average fracture spacing of between 0.05 and 0.3 m) was consistent with the observed very limited extent of the plume. These analyses indicate that contaminants would be expected to have migrated about 25 m in 15 years. On the other hand, calculations based on simple advective transport at groundwater velocities ranging from 50 to 1,071 m/a are not consistent with the field observations. This case demonstrates the significant effect which matrix diffusion can have on the attenuation of contaminants migrating through fractured porous media.

11.3.10 Summary

For the range of parameters considered in the examination of contaminant migration through fractured shale presented in this section, it may be concluded that:

1. The rate of contaminant transport did not significantly depend on the groundwater velocity. The rate of movement and the degree of attenuation were, however, greatly dependent on the fracture spacing, matrix porosity, dispersivity, Darcy velocity (i.e., Darcy flux) and the mass of contaminant available for transport.
2. For a given spacing between the primary fractures, consideration of the spacing between the secondary fractures was found to reduce the calculated velocity of the leading edge of the contaminant plume; however, it may either increase or decrease the calculated maximum impact of the plume depending on the specific case.
3. Any reasonable estimate of secondary fracture spacing gave results which would be sufficiently accurate for engineering purposes.
4. The larger the matrix porosity, the slower was the contaminant transport velocity and the greater the attenuation.
5. The magnitude of the matrix diffusion coefficient had a significant effect on the predicted plume when the time scale of interest was similar to or greater than the time required to reach steady state due to diffusion into the rock matrix at a point.
6. Fracture opening size was only important insofar as it influences the hydraulic conductivity of the rock mass and the Darcy velocity. Any combination of fracture opening sizes which gave rise to a given hydraulic conductivity also gave rise to a contaminant impact which, for all practical purposes, was the same. The effect of uncertainty regarding the magnitude of opening size upon the groundwater velocity did not have any significant effect on the prediction of contaminant migration.
7. A theoretical examination of the evolution of a contaminant plume also demonstrated that one may not be able to reliably infer the rate of advance of a contaminant plume based on isolated observations of the contaminant plume if the source of contaminant is finite.
8. The rate of contaminant movement in a fracture system was not linearly proportional to the Darcy velocity and may be almost independent of the groundwater velocity. The effect of changes in velocity may not be intuitively obvious and will greatly depend on other parameters such as fracture spacing, the diffusion coefficient, retardation factor, matrix porosity and mass of contaminant. The theory presented in Chapter 7 allows the designer to readily and quickly examine these interactions.
9. The effectiveness of sorption as an attenuation mechanism was dependent on the primary, secondary and tertiary fracture spacings; however in each case, higher retardation coefficients for sorption in the rock matrix were found to decrease the maximum contaminant impact at any specified monitoring point downgradient of the source.
10. The mass of contaminant available for transport into the fractured soil/rock was a major factor affecting the impact downgradient of the leaking landfill. By removing contaminant from the landfill, leachate collection systems reduce the mass available for contaminating the underlying soil/rock and can substantially reduce the downgradient impact of those contaminants which do escape the leachate collection system.

This section has considered lateral migration in a fractured porous media. In the following section, consideration is given to similar factors but this time relative to vertical migration from a contaminant source (e.g., a landfill) through a fractured clay and into an underlying aquifer.

11.4 Vertical migration through fractured media and into an underlying aquifer

Rapid migration of contaminant through what was originally thought to be relatively low hydraulic conductivity clayey soils, both in the USA (e.g., at Wilsonville, IL, USA) and Canada, has awakened concern regarding the potential fracturing of stiff–very stiff clays and clayey tills. The Wilsonville problem prompted research by Herzog *et al.* (Herzog and Morse, 1986; Herzog *et al.*, 1989) which has provided evidence to suggest that the unweathered till below the obviously weathered and fractured zone at Wilsonville was also fractured to extensive depths. Similarly, investigations in Southern Ontario (Ruland, 1988; D'Astous *et al.*, 1989; and the authors' own investigations) have indicated that clayey till which appears unweathered and unfractured in conventional borehole investigations may be fractured to depths of as much as 13 m below ground surface (e.g. to depths of up to 9 m below the weathered crust). In some cases, this may mean that there is little or no unfractured clayey till between the base of the landfill and an underlying aquifer (see Rowe *et al.*, 2000c). Careful examination of deep test pits has demonstrated that at some locations fracturing at 1–3 m spacings may exist both above and below an upper "permeable" zone (which could potentially act as a conduit for contaminant transport if a landfill were constructed above this unit). Similarly, hydrogeologic investigations at other sites have identified that even though there may be 4–5 m of till separating the base of a proposed landfill from an underlying permeable zone, only 1–2 m of the material may be unfractured. When fractures are identified, the question then arises as to what effect they will have on contaminant transport from a proposed landfill facility.

11.4.1 Steps to be followed in modelling fractured systems

In order to use the theory described in Chapter 7, it is necessary to follow the following steps:

1. Define the hydrostratigraphy of the site in terms of
   (a) the number of layers with different properties;
   (b) the level of fracturing of each layer (i.e., the fracture spacings $2H_1$, $2H_2$, $2H_3$ and fracture opening sizes $2h_1$, $2h_2$, $2h_3$) where the subscripts 1, 2, 3 correspond to the x-, y- and z-directions, respectively;
   (c) the vertical and horizontal hydraulic conductivity k_z, k_x of each layer;
   (d) the hydraulic boundary conditions.
2. Estimate the horizontal and vertical Darcy velocity (v_{ax}, v_{az}) for each layer. Depending on the complexity of the problem, this may be done by means of a hand calculation or using a computer flow model – see Chapter 5. The vertical velocities in any layer should be representative of the velocity beneath the landfill. The horizontal velocities should satisfy continuity of flow at the downgradient edge of the landfill.
3. Estimate the matrix porosity (n), the dry density (ρ) and the distribution/partitioning coefficient (K_d) for each layer.
4. Estimate the coefficient of hydrodynamic dispersion D_{zz}, D_{xx} in the vertical and horizontal directions. For clay layers, this is often equal to the diffusion coefficient. For fractured layers, this is the diffusion coefficient within the matrix between fractures. For aquifer layers, this will depend on the assumed dispersivity.
5. For fractured layers, estimate fracture distribution coefficient (K_f) and the coefficient of hydrodynamic dispersion along the fractures in the vertical and horizontal directions (D_{xf}, D_{zf}). This is usually equal to the dispersivity, α, times the average groundwater velocity in the fractures (e.g., $D_{xf} \simeq \alpha_x v_{xf}$; $D_{zf} \simeq \alpha_z v_{zf}$).
6. Estimate the peak source concentration, c_0, the equivalent height of leachate, H_f, or the reference height of leachate, H_r, and the volume of leachate collected by the collection system per unit area, q_c (e.g., see Section 10.2).

7. Define the boundary condition at the bottom of the deposit. Typically, a bottom aquifer is modelled as a boundary condition if it is thin and by a layer(s) if it is thick. If the lower aquifer is modelled by layers, the bottom boundary is usually impermeable.

Once the foregoing have been estimated, the contaminant transport model can be run to calculate concentration at any specific times and locations of interest. Typically, there will be uncertainty regarding values of many of the input parameters. The implications regarding this uncertainty can be evaluated by performing a sensitivity study.

11.4.2 Definition of example case

To illustrate the potential application of the theory presented in Chapter 7, consideration will be given to the situation where the base of a hypothetical landfill is to be located in the unweathered portion of a clayey aquitard which is fractured with an average fracture spacing of 1 m. The base of the landfill is assumed to be 4 m above a 1-m thick upper permeable unit (aquifer 1) and 7 m above a 2-m thick lower permeable unit (aquifer 2) as shown in Figure 11.19.

The proposed model can readily consider situations where there is horizontal flow in aquifers beneath a landfill. Under these circumstances, flow in the aquifer provides a potential for dilution of contaminants migrating from the landfill, and the horizontal flow at the downgradient edge of the landfill is equal to the flow at the upgradient edge plus the change in flow, as required by consideration of continuity of flow, which occurs beneath the landfill. From the perspective of contaminant impact, the "worst case" situation involves the landfill being located on a groundwater divide. This divide may have existed prior to the landfill being constructed. However, even if under natural conditions there is no groundwater divide, a landfill can change flow conditions if significant leachate mounding occurs within the landfill. The likelihood of this occurring should be evaluated for each landfill being considered.

For the purposes of this particular example, it is assumed that the landfill is 400 m wide and is to be located in a recharge zone directly over a groundwater divide. This represents worst case

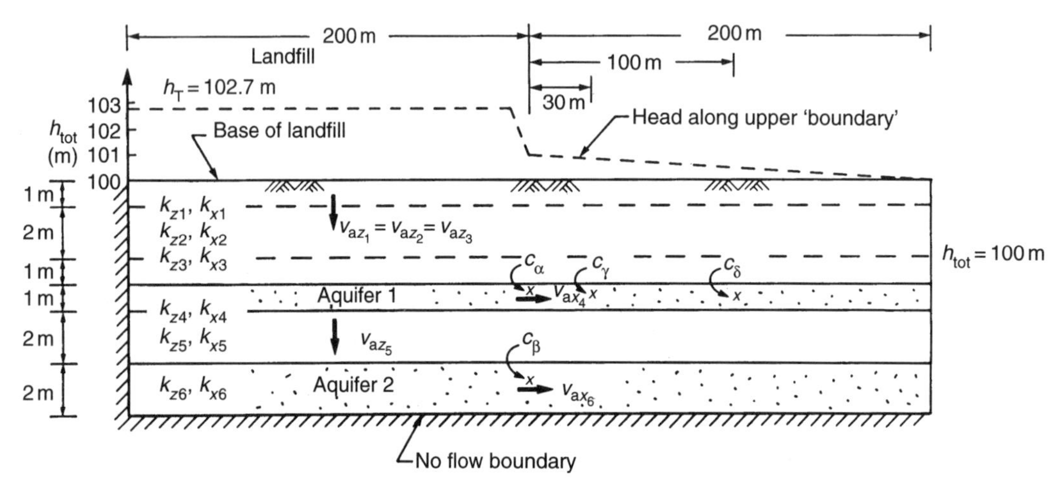

Figure 11.19 Schematic of general hydrostratigraphy considered (after Rowe and Booker, 1991a; reproduced with permission of the *Canadian Geotechnical Journal*).

Table 11.1 Assumed hydraulic conductivities k (m/s) (Figure 11.19)

Case	k_{z1}	k_{x1}	k_{z2}	k_{x2}	k_{z3}	k_{x3}	k_{z4}	k_{x4}	k_{z5}	k_{x5}	k_{z6}	k_{x6}
1	10^{-9}	10^{-8}	10^{-9}	10^{-8}	10^{-9}	10^{-8}	10^{-6}	10^{-6}	10^{-9}	10^{-8}	10^{-5}	10^{-5}
2	10^{-9}	10^{-8}	10^{-9}	10^{-8}	10^{-9}	10^{-8}	10^{-6}	10^{-6}	10^{-10}	10^{-9}	10^{-5}	10^{-5}
3	2×10^{-10}	2×10^{-10}	10^{-9}	10^{-8}	10^{-9}	10^{-8}	10^{-6}	10^{-6}	10^{-9}	10^{-8}	10^{-5}	10^{-5}
4	2×10^{-10}	2×10^{-10}	10^{-9}	10^{-8}	10^{-9}	10^{-8}	10^{-6}	10^{-6}	10^{-9}	10^{-8}	10^{-5}	10^{-5}
5	2×10^{-10}	2×10^{-10}	10^{-9}	10^{-8}	2×10^{-10}	2×10^{-10}	10^{-6}	10^{-6}	10^{-9}	10^{-8}	10^{-5}	10^{-5}
6	2×10^{-10}	2×10^{-10}	10^{-9}	10^{-8}	2×10^{-10}	2×10^{-10}	10^{-6}	10^{-6}	10^{-10}	10^{-9}	10^{-5}	10^{-5}
7	2×10^{-10}	2×10^{-10}	10^{-9}	10^{-8}	2×10^{-10}	2×10^{-10}	10^{-5}	10^{-5}	10^{-10}	10^{-9}	10^{-7}	10^{-7}

conditions. For the sake of definiteness, it is assumed that the average (long term) height of leachate mounding in the landfill is 2.7 m above a reference datum located 200 m downgradient of the proposed landfill and that there are boundary heads as shown in Figure 11.19. Other heights of mounding and boundary heads could have been considered equally as well. Greater levels of leachate mounding would increase the potential impact compared with the cases considered here, and lesser levels of leachate mounding would decrease the potential impact.

The symmetry associated with the assumed hydrogeologic conditions implies that the centre of the landfill will be a no-flow boundary and that only half (i.e., 200 m of the landfill) needs to be considered in flow and contaminant transport modelling. This also implies that any flow in the aquifer at the edge of the landfill arises from the landfill and that there is no dilution due to flow entering the aquifer upgradient of the landfill.

Figure 11.19 shows the general hydrostratigraphy considered. There are potentially six different hydrostratigraphic units with vertical and horizontal hydraulic conductivities of $k_{z\ell}$, $k_{x\ell}$. Table 11.1 summarizes the values of $k_{z\ell}$ and $k_{x\ell}$ assumed for the seven cases examined. Where the aquitard is assumed to be fractured, it has $k_z = 10^{-9}$ m/s, $k_x = 10^{-8}$ m/s. Where it is not fractured, it is assumed to have $k_z = 10^{-10}$ m/s, $k_x = 10^{-9}$ m/s. The clay and clayey liner are assumed to be isotropic with $k_z = k_x = 2\times 10^{-10}$ m/s.

In order to perform the contaminant transport analyses, it is necessary to first determine the Darcy velocities in each of the layers. Depending on the complexity of the problem, this can be done by hand calculation or by using computer flow models. In this case, a finite element computer flow model was used to obtain the flow field for each combination of hydraulic conductivity given in Table 11.1. Based on these flow analyses, representative Darcy velocities were selected for each case as summarized in Table 11.2. The velocities v_{az1} and v_{az5} represent average downward velocities beneath the landfill. The velocities v_{ax4}, v_{ax6} are the horizontal velocities in the upper and lower aquifer at the edge of the landfill. Figure 11.20 schematically illustrates the basic cases considered. The details will be discussed in subsequent paragraphs. Table 11.3 summarizes the contaminant transport parameters adopted for the modelling using program MIGRATE v8.

Table 11.2 Flow parameters (Darcy velocities, Figures 11.19 and 11.20)

Case	v_{az1} (m/a)	v_{az5} (m/a)	v_{ax4} (m/a)	v_{ax6} (m/a)
1	0.0085	0.0074	0.23	0.74
2	0.004	0.0024	0.32	0.24
3	0.0057	0.005	0.14	0.5
4	0.00325	0.002	0.24	0.2
5	0.0043	0.0038	0.10	0.38
6	0.00273	0.0017	0.19	0.17
7	0.0039	0.00007	0.79	0.007

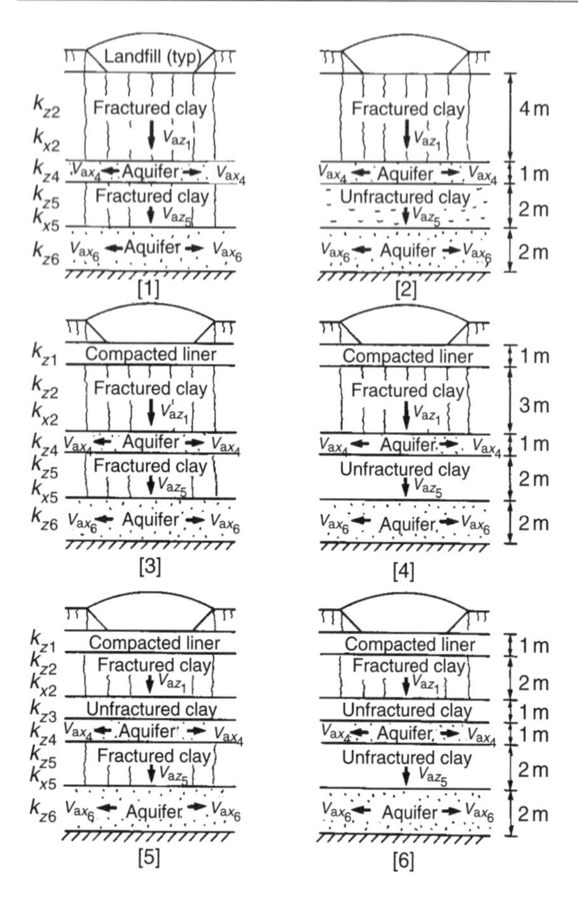

Figure 11.20 Schematic of the hydrostratigraphy for the cases considered (after Rowe and Booker, 1991a; reproduced with permission of the *Canadian Geotechnical Journal*).

The "reference height of leachate" H_r was calculated assuming an average thickness, H_w, and density, ρ_w, of waste of 8 m and 625 kg/m^3, respectively. Assuming that chloride is the contaminant species to be considered and that it represents 0.2% ($p = 0.002$) of the waste and that the peak concentration, c_0, is 1,000 mg/l (1 kg/m^3) then the "reference height of leachate" H_r is given by:

$$H_r = \frac{m_{TC}}{A_0 c_0} = \frac{pH_w\rho_w}{c_0} = \frac{0.002 \times 8 \times 625}{1} = 10\,\text{m}$$

The average volume of leachate collected (or escaping as surface seeps) per unit per unit time is assumed to be 0.15 m^3/a/m^2. The fractured layers were considered to have orthogonal fractures at an average spacing $2h_1 = 2h_2 = 8.5\,\mu$m.

The discussion in the following subsections relates to the particular example defined above. It should be noted that the interaction between different levels of fracturing and hydraulic conductivities in the different layers can be quite complex, as will become evident from the following subsections. Thus, while this example serves to illustrate the application of the proposed theory and some important factors to be considered, care should be taken not to overgeneralize the results. Each landfill must be carefully evaluated as a unique situation.

11.4.3 Monitoring period and delay in impact

As illustrated in Figure 11.20 (and Table 11.1), cases 1 and 2 both consider a 4-m fractured zone with $k_z = 10^{-9}$ m/s between the landfill and the upper aquifer. Case 1 also assumes that the aquitard between the upper and lower aquifers is fractured whereas case 2 assumes that it is not. Because of the higher transmissivity of the lower aquifer compared with the upper aquifer and hence its ability to remove fluid from the system, a reduction in lower aquitard permeability has a significant effect on the Darcy velocity, v_{az1}, leaving the landfill (see Table 11.2) which reduces from 0.0085 (case 1) to 0.004 m/a (case 2). The Darcy velocity in the upper aquifer, v_{ax4}, is increased from 0.23 (case 1) to 0.32 m/a (case 2) because more fluid must now be removed from this zone. The flow in the lower aquifer is only reduced by about a factor of three (i.e., from 0.74 (case 1) to 0.24 m/a (case 2)) even though the hydraulic conductivity of the lower aquitard was reduced by one order of magnitude. This reduction results from the lower flux which can be passed through the unfractured aquitard.

Figure 11.21 shows the decrease in leachate strength with time calculated for case 1. Also shown is a simple hand calculation of the

Table 11.3 Common parameters used in transport model

Quantity	*Value*
Reference height of leachate, H_r (m)	10
Volume of leachate collected/area, q_c (m/a)	0.15
Initial concentration, c_0 (mg/l)	1000
Porosity of till/clay matrix (−)	0.4
Diffusion coefficient in matrix (m^2/a)	0.02
Distribution coefficient (m^3/kg)	0
Fracture spacing, $2H_x = 2H_y$ (m)	1
Fracture opening size, $2h_x = 2h_y$ (μm)	8.5
Coefficient of hydrodynamic dispersion along fracture (m)	0
Porosity of aquifer (−)	0.3
Longitudinal dispersivity (m)	1
Transverse dispersivity (m)	0.1
Darcy velocity (flux) (m/a)	Table 11.2

leachate concentration based on the volume of leachate collected, q_c, viz.

$$\frac{c_{0L}}{c_0} = \exp\left(\frac{-q_c t}{H_r}\right) \tag{11.4}$$

where c_{0L} is the concentration of the leachate at the time of interest, t; c_0 the maximum leachate concentration and H_r the "reference height of leachate". Since the volume of leachate collected is large compared to the mass flux into the underlying soil, the simple hand calculation gives a good estimate of the concentration in the landfill compared with the more rigorous calculation performed using MIGRATE v8. Similar results to those for case 1 were obtained for each of the seven cases considered.

Reference to Figure 11.21 shows that for the assumed conditions, the average concentration of chloride in the leachate decreases to approximately half strength after about 45 years. After about 145 years, the leachate strength has reduced to 100 mg/l (assuming c_0 of 1,000 mg/l). Assuming that the background concentration chloride in the groundwater is less than 50 mg/l, then according to policies such as Ontario's "Reasonable Use Policy" (MoEE, 1993a), an increase in chloride concentration of up to 100 mg/l in the groundwater would be environmentally acceptable and hence one might be tempted to infer from this that there would be no need to monitor the landfill or groundwater after 145 years. That this is *not* the case is evident from Figure 11.22 which shows the calculated variation in concentration with time at two monitoring points. Assuming that for thin aquifers

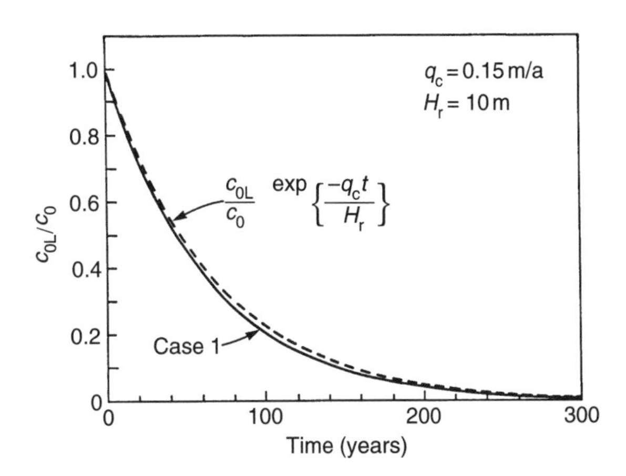

Figure 11.21 Variation in leachate with time (after Rowe and Booker, 1991a; reproduced with permission of the *Canadian Geotechnical Journal*).

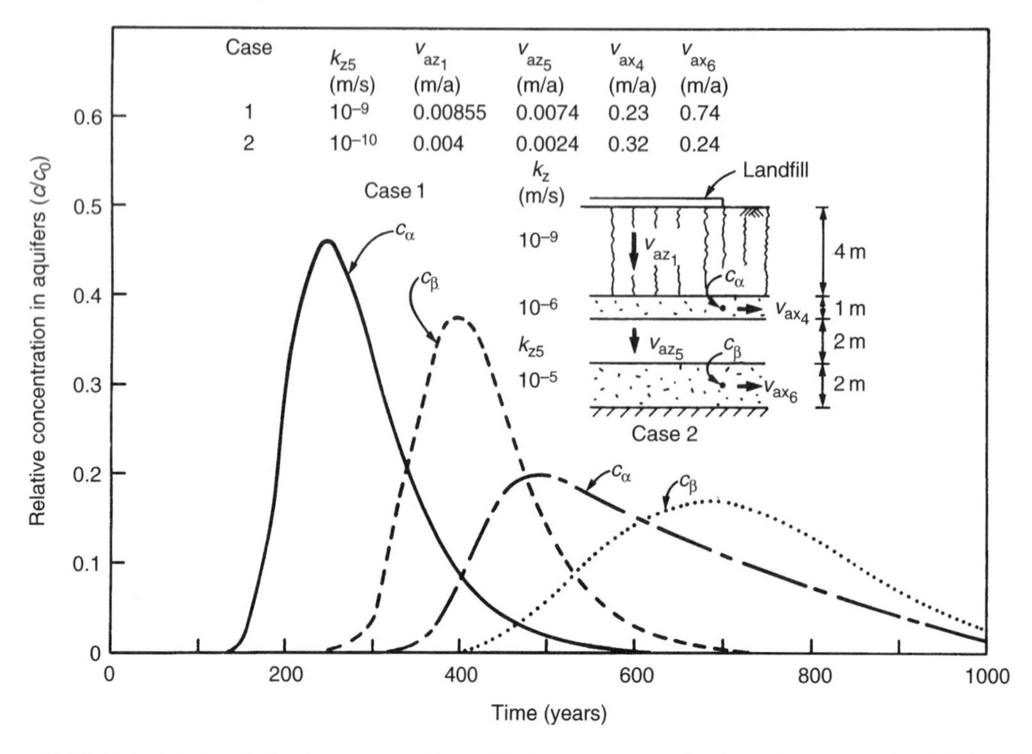

Figure 11.22 Calculated variation in concentration with time at two monitoring points, cases 1 and 2 (modified from Rowe and Booker, 1991a).

the monitoring wells would be screened across the entire thickness of the permeable unit, the concentrations c_α, c_β represent the average calculated concentration in the upper and lower aquifer respectively (i.e., the average of the calculated concentration at the top and bottom of each aquifer) at the edge of the landfill.

Inspection of Figure 11.22 shows that if the mounding of leachate to the design elevation 102.7 m occurred quickly then there would be no impact on either aquifer for at least 100 years. This would meet any regulatory requirements for no impact within 100 years but, eventually, there would be a significant impact. For case 1 (assuming both the upper and lower aquitards are fractured with $k_z = 10^{-9}$ m/s), the increase in chloride concentration in the upper aquifer is about 275 mg/l ($c/c_0 = 0.275$) after 200 years even though the concentration in the leachate is only 50 mg/l at this time ($c/c_0 = 0.05$). The chloride concentration in the aquifers continues to increase until a peak impact of about 460 mg/l ($c/c_0 = 0.46$) is reached after approximately 250 years at which time the concentration of chloride in the leachate is predicted to be less than 25 mg/l. This shows that even a fractured clayey aquitard can act as a buffer and can provide attenuation of leachate strength since the peak concentration in the upper aquifer is only 46% of the peak value originally in the leachate. However, it is equally clear that the delay in the impact should also be considered when assessing the suitability of a site and required monitoring period; once contaminant enters the aquitard it may take a long time before it impacts on an underlying aquifer, in this case well beyond the time when the leachate strength itself has reduced to a level where it does not impose an environmental risk.

The analyses for case 1 assume relatively minor fracturing of both aquitard layers (fractures with an opening size of 8.5 μm at 1 m spacing; $k_z = 10^{-9}$ m/s); however, even this minor fracturing can have a significant long-term effect as shown in Figure 11.22. As discussed above, the peak impact in the upper aquifer of about 460 mg/l occurs after approximately 250 years. At this time there is negligible impact on the lower aquifer, but after another 100 years the lower aquifer is also significantly impacted with a peak increase in chloride of 375 mg/l at about 400 years.

The only difference between cases 1 and 2 is the absence of fracturing of the lower aquitard for case 2. As discussed, this reduces the flow through both aquitards and leads to an increase to the time of impact and a reduction in the magnitude of impact compared with case 1. Even though the average Darcy velocity through the upper aquitard is only 0.004 m/a, the peak impact of about 200 mg/l at approximately 500 years is twice that permitted in the Province of Ontario based on the "Reasonable Use Policy" assuming a background concentration of 50 mg/l. It is also noted that there is significant impact on the lower aquifer even though the lower aquitard is not fractured and has a hydraulic conductivity of 10^{-10} m/s.

11.4.4 The effects of fracture spacing, opening size, matrix porosity, matrix diffusion coefficient, and sorption into the matrix and onto fracture surfaces

There will always be some uncertainty regarding fracture spacing and opening size for fractured media. The bulk hydraulic conductivity will depend on both the spacing and opening size and will tend to be dominated by those fractures with a larger opening. Based on Hoek and Bray (1981), a theoretical relationship between hydraulic conductivity, k (m/s), fracture spacing, $2H$ (m) and opening size, $2h$ (m), is given by:

$$k = \frac{8.1 \times 10^5 2h^3}{(2H)} \tag{11.5}$$

for one parallel set of fractures and

$$k = \frac{1.6 \times 10^4 2h^3}{(2H)} \tag{11.6}$$

for orthogonal fractures.

The fracture spacing can be reasonably estimated (or bounded) based on observations from test pits. The fracture opening size cannot usually be estimated from test pits since the process of excavation can open the fractures, and even small changes (e.g., from 5 to 15 μm) can have a significant effect on hydraulic conductivity as can be appreciated from equations 11.5 and 11.6. Thus the bulk hydraulic conductivity is usually obtained independently from field tests (e.g., see Section 3.2). Once the range of reasonable hydraulic conductivities is known for a given aquitard, one can then estimate fracture opening size from equation 11.5 or 11.6 based on the estimated fracture spacing.

Rowe and Booker (1990a, 1991a,b) have examined the effect of fracture spacing and matrix porosity. Based on these studies, it may be concluded that:

- Even if the bulk hydraulic conductivity is known, variations in fracture spacing will have a significant effect on the time at which contaminant impact occurs within an underlying aquifer. Often the greatest impact will occur for very closely spaced fractures; however, this is not always the case. For a given Darcy velocity, the effects of reasonable uncertainty regarding fracture spacing should be assessed by means of a sensitivity study which includes the assumption of no fractures as one limiting case.
- For a given fracture spacing, increasing the thickness of the clayey aquitard between the landfill and the aquifer increases the time required for a given impact to reach the aquifer and decreases the maximum impact. However, for practical ranges of fracture spacing, significant impact on the aquifer is

possible even if the fractured aquitard is up to 10 m thick.

- The effect of uncertainty regarding Darcy velocity (e.g., as a result of uncertainty concerning hydraulic conductivity) is a critical consideration in modelling contaminant migration through the fracture system (e.g., see also Section 11.3.5).
- The value of effective matrix porosity used in the analysis can affect the magnitude and time of occurrence of impact on the aquifer even if contaminant transport is through fractures.

As discussed in Section 11.3.5, if the hydraulic conductivity is known from field tests, then the assumed fracture opening sizes used in the contaminant transport analysis does not have a significant effect on calculated impact for typical opening sizes in fractured aquitards.

As discussed in Section 11.3.2, the matrix diffusion coefficient can also be an important parameter affecting contaminant transport through fractured systems. Often, the calculated impact will be greatest for the highest reasonable diffusion coefficient used; however, this is not always the case. When there is uncertainty concerning diffusion coefficient, the effect of that uncertainty can be assessed by a sensitivity study.

In summary, it has been found that uncertainty regarding fracture spacing, diffusion coefficient and effective porosity will influence the impact on the underlying aquifer. Because of symbiotic interaction, the effects of changing these parameters are not intuitively obvious and sensitivity analyses should be performed for each project to assess the impact of uncertainty concerning these parameters.

As also discussed in Section 11.3, sorption into the matrix and onto the surface of the fractures can significantly retard the movement of contaminants and reduce potential impact. The modelling of sorption into the matrix requires knowledge of sorption characteristics as discussed in Section 8.6. At present, there is very little data concerning sorption onto the surface of fractures.

11.4.5 Evaluating the effects of a clay liner

Given that the impact for cases 1 and 2 is likely to be unacceptable based on requirements for long-term protection of these aquifers, the question then arises as to what could be done to minimize impact. The magnitude of the impact could be reduced by reducing the height of leachate mounding which is the cause of downward advective transport. For example, if the leachate collection system were to maintain a leachate level at an elevation of less than 100 m then there would be inward flow into the landfill (a hydraulic trap), and the impact could be minimized even though the aquitard is fractured. For case 1, the results shown in Figures 11.21 and 11.22 can be used to determine an attenuation factor $c_\alpha/c_0 = 0.46$ which is the ratio of maximum impact in the aquifer, c_α, divided by the peak source concentration c_0. Assuming that to ensure that the impact on the aquifer is not to exceed 100 mg/l, it is evident that the leachate collection system must function until the source concentration has reduced to a value c_{0L} which will not cause unacceptable impact. In this case, the concentration must reduce to $c_{0L} = 100/(c_\alpha/c_0) = 100/0.46 = 217$ mg/l. Using equation 11.4, one can then infer that the leachate collection system would have to function for about 100 years for case 1 (i.e., until the chloride concentration in the leachate c_{0L} reduced to $c_{0L}/c_0 = 217/1{,}000 = 0.217$). For case 2, the corresponding period of operation would be about 45 years. This issue of longevity of leachate collection systems is discussed in more detail in Sections 2.4, 2.5, 16.4 and 16.5.

Another means of reducing impact would be to remove the upper 1 m of clayey aquitard below the proposed landfill base and rework it as a compacted clay liner. For the purposes of this example, it is assumed that the liner has a hydraulic conductivity of 2×10^{-10} m/s. Cases 3 and 4 consider a 1-m thick liner and either a fractured or unfractured lower aquitard respectively.

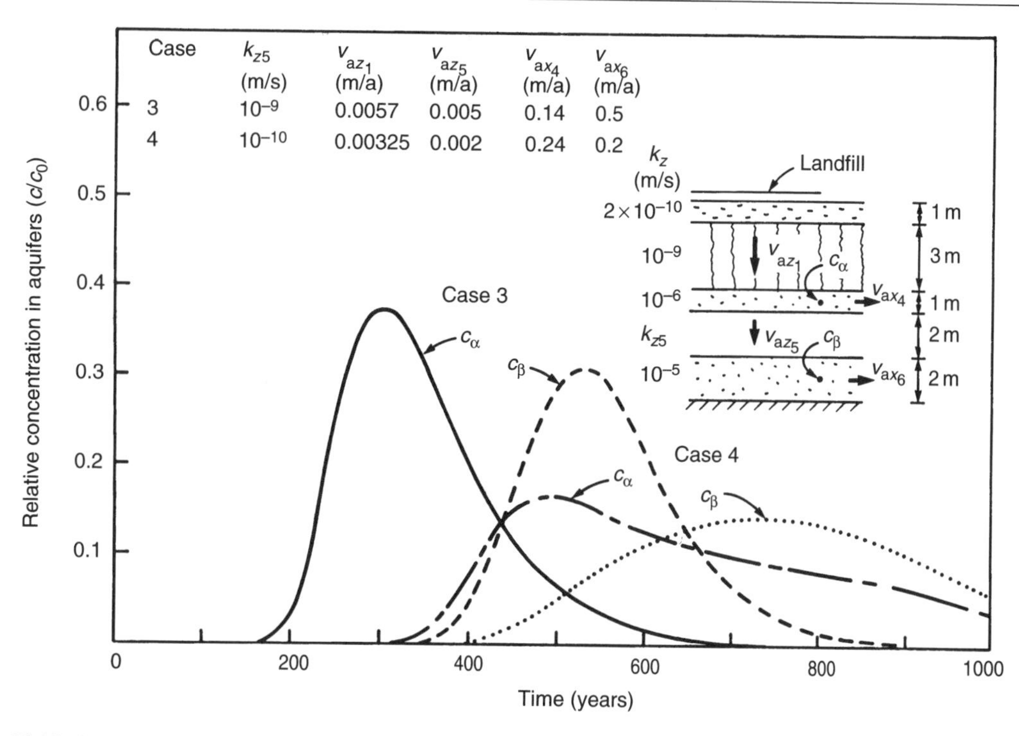

Figure 11.23 Calculated variation in concentration with time at two monitoring points, cases 3 and 4 (modified from Rowe and Booker, 1991a).

As might be expected, installation of the liner decreases the Darcy flow from the landfill and, consequently, the impact on the aquifer (see Figure 11.23). This landfill would readily meet regulatory requirements of no impact in 100 years. However, it is equally evident that this liner is not, of itself, enough since the peak impacts shown in Figure 11.23 are still quite significant even though the time required for impact to occur is large.

11.4.6 Evaluating multiple-component aquitards

Cases 5 and 6 (see Figures 11.20 and 11.24) consider the situation where the upper aquitard consists of 3 m of fractured clay or till ($k_{z2} = 1 \times 10^{-9}$ m/s) and 1 m of unfractured lacustrian clay ($k_{z3} = 2 \times 10^{-10}$ m/s). It is further assumed that the top 1 m of the fractured aquitard beneath the landfill is removed and recompacted as a liner with $k_{z1} = 2 \times 10^{-10}$ m/s. In case 5 the lower aquitard is assumed to be fractured ($k_{z5} = 1 \times 10^{-9}$ m/s); in case 6 it is not fractured ($k_{z5} = 1 \times 10^{-10}$ m/s). The presence of the unfractured clay layers in the upper aquitard serves to reduce flow and calculated impact as shown in Figure 11.24. For these cases, there is no impact within the first 100 years. The impact on the lower aquifer is probably acceptable based on "Reasonable Use Criteria" (MoEE, 1993a); however, the impact on the upper aquifer is still unacceptable based on these calculations.

At this point it is worth emphasizing the assumption made in modelling fractured flow, that contaminant is transported along the fractures rather than through the matrix of the fractured media. This may be a reasonable assumption when the fractures control the bulk hydraulic conductivity of the aquitard and where there is an easy path for contaminant to pass into the

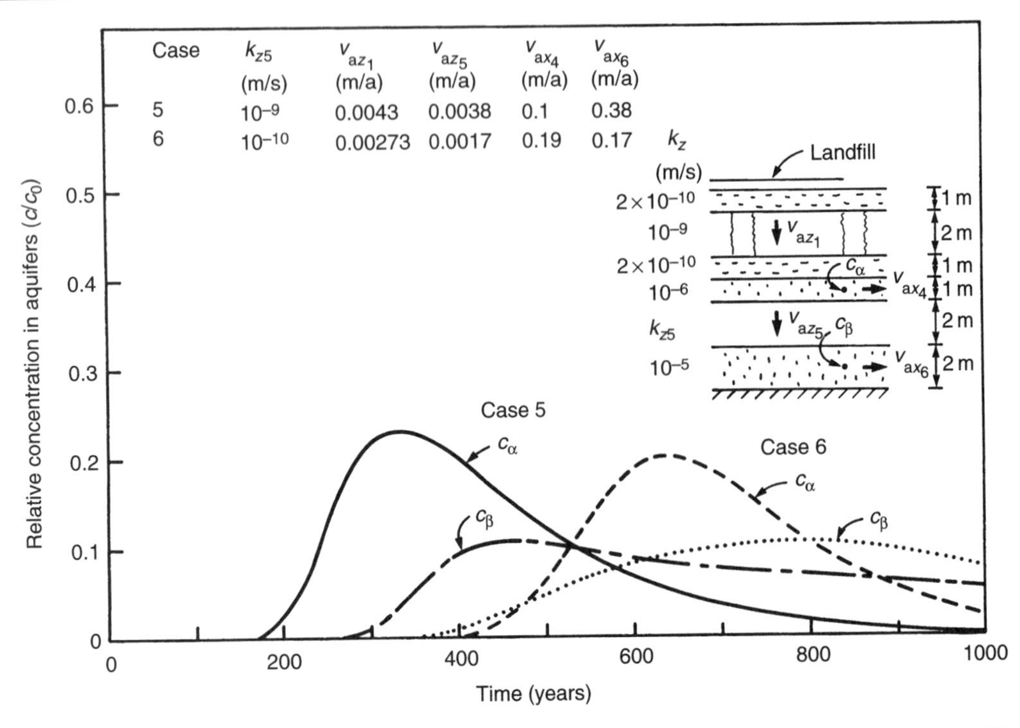

Figure 11.24 Calculated variation in concentration with time at two monitoring points, cases 5 and 6 (modified from Rowe and Booker, 1991a).

fractures (e.g., cases 1 and 2). However, this assumption may not be valid when the fractured portion of the aquitard is sandwiched between two unfractured layers of clay as for in the upper aquitard for cases 5 and 6. For these cases, the modelling of migration primarily through the fractures assumes that once leachate passes through the liner it spreads out and passes through the underlying fractures; this is equivalent to assuming that there is a thin sand layer between the liner and the fractured till which serves to distribute the "leachate" to the fractures.

In case 6, discussed above, it was assumed that the fractures were the primary conduits for contaminant movement and the hydraulic conductivity used to estimate the Darcy velocities used in the contaminant migration analyses are based on this assumption. Alternatively, if there is a good contact between the fractured till and the unfractured soil above and below it, then the contaminant will not have an easy path to the fractures, and much of it is likely to pass through the matrix of the clayey soil between the fractures rather than along the fractures. It is not a simple matter to determine what the true hydraulic conductivity would be for this case; clearly, it will lie between the limits of that obtained considering that the fractures are the primary conduit for fluid flow and the limit of the mass hydraulic conductivity of matrix material between the fractures. One could adjust the flows corresponding to these hydraulic conductivities associated with these two limiting cases and then perform contaminant migration analyses for each case. This would represent the extremes of potential arrival times and impact; however, the matrix flow condition would be unconservative since it totally neglects the presence of the fractures and hence could not justifiably be used in estimating potential impact in an environmental assessment.

Given the uncertainty regarding what the bulk hydraulic conductivity would in fact be for this case, a conservative approach would be to assume the worst-case hydraulic conductivity arising from the fractures being the primary conduit and then use the corresponding flows to estimate potential impact by performing contaminant transport analyses for the two cases where: (a) this flow is directed through the fractures as assumed in Figure 11.24; or (b) this flow is through the clay matrix. Option (b) will give earlier arrival times and impact than if the contaminant really did flow entirely through the matrix because the flow has been estimated based on fractured flow conditions. Figure 11.25 shows the effect of these two different assumptions on the calculated impact for case 6. The results obtained for the "fractured" assumption are the same as those shown in Figure 11.24. The results obtained for the "unfractured" assumption show a faster arrival time but a lower impact on the upper aquifer than those obtained for the "fractured" assumption. This counter-intuitive result arises primarily because matrix diffusion into the till as contaminant moves along the fractures provides a very effective retardation mechanism at the low Darcy velocity corresponding to case 6. If matrix diffusion were neglected and only advective–dispersive transport along the fractures was considered, then the "intuitive" result of very rapid arrival times would be obtained. In fact, it would only take about two days for contaminant to move through the fractured layer if matrix diffusion were neglected. However, matrix diffusion cannot be justifiably neglected, and it can provide very substantial attenuation as is evident from Figure 11.25. As noted earlier, a contributing factor to the early arrival times for the "no fracture" analysis is the conservative assumption that the flow through the material is equal to the flow through the fractures in the "fractured" case. Thus, based on these two limiting cases, it is possible to make a conservative

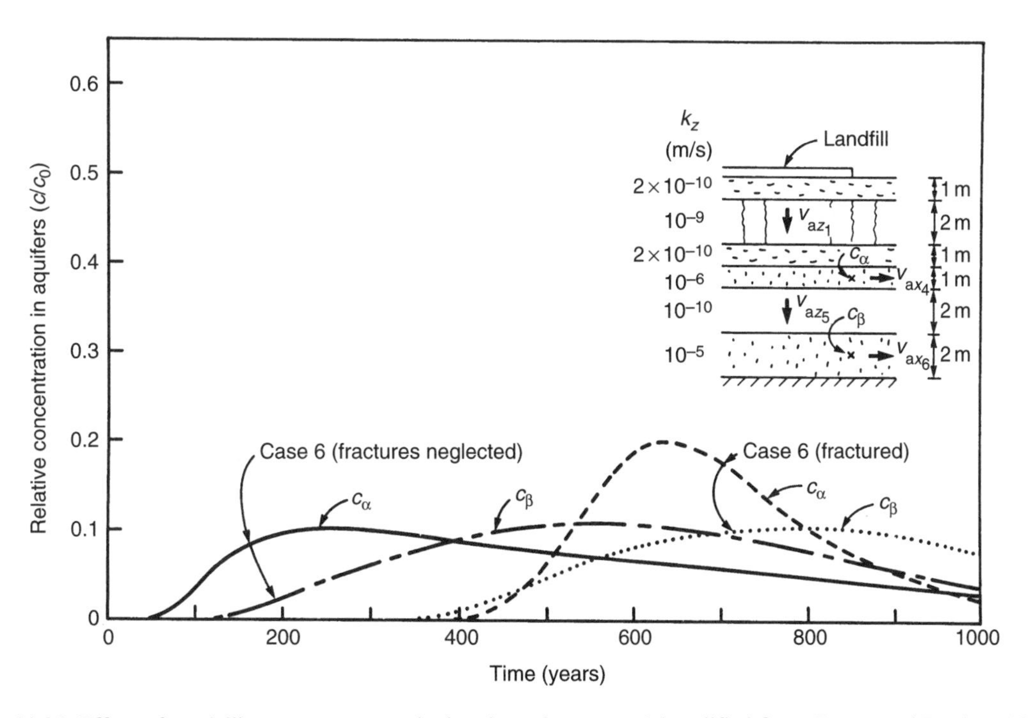

Figure 11.25 Effect of modelling strategy on calculated results – case 6 (modified from Rowe and Booker, 1991a).

estimate of both the first arrival time based on the "no fracture" case and of the peak impact based on the "fractured" case. Neither represents a "prediction" of impact since it is expected that reality lies somewhere between the results obtained from examining these limiting cases.

11.4.7 Evaluating hydraulic conductivity relationships between aquifers

The cases considered to this point have examined variations in the properties of the aquitards and have assumed constant properties of the aquifer. Case 7 shows that the hydraulic conductivity of the aquifers is also important. Here the aquitards have the same properties as considered for case 6 but the hydraulic conductivity of the upper aquifer is increased from 10^{-6} to 10^{-5} m/s and the hydraulic conductivity of the lower aquifer is decreased from 10^{-5} to 10^{-7} m/s. Thus in case 6 the lower aquifer is a significant drain for the system taking almost 50% of the flow from the landfill whereas in case 7 the upper aquifer is the major drain taking almost all the flow and this increase in the hydraulic conductivity of the upper aquifer increases the flow from the landfill compared to case 6. Results were obtained assuming migration through fractures for the fractured till and migration through the matrix in a similar fashion to that already discussed for case 6, and the results are shown in Figure 11.26.

Comparison of the results obtained for cases 6 and 7 shows that the higher Darcy velocity in the upper aquitard associated with case 7 gives rise to earlier impacts compared to case 6. For the case of migration through the intact soil, the peak impact in the upper aquifer is also greater; however, this is not the case when one considers migration through the fractures. This situation arises because of the buffering effect that results from matrix diffusion from the fractures into the intact material. Thus, as has previously been discussed in Section 11.3, there is a complex symbiotic relationship between the effects of

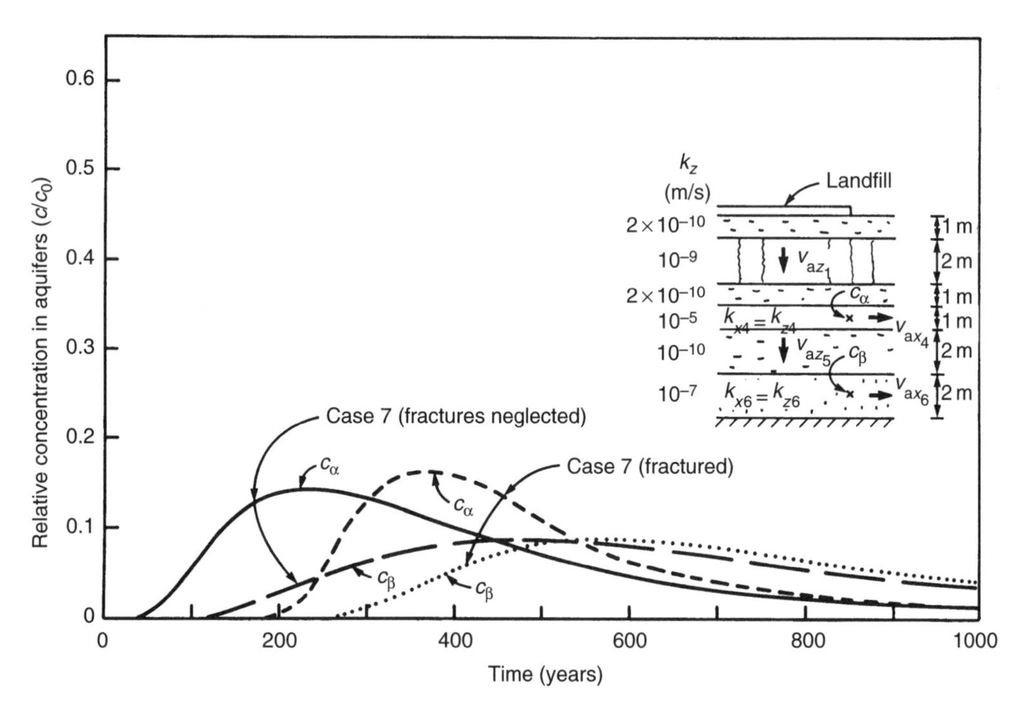

Figure 11.26 Effect of modelling strategy on calculated results – case 7 (modified from Rowe and Booker, 1991a).

Darcy velocity and matrix diffusion, and a higher Darcy velocity does not necessarily result in higher impact at a monitoring point when dealing with fractured media. When modelling contaminant migration in fractured media, it is important to examine a range of realistic possible situations in order to determine which combination of parameters provides the peak impact. This combination is not always obvious.

Assuming that regulatory requirements specify that one cannot increase chloride in the aquifer by more than about 100 mg/l, it is evident that for case 7 this requirement would be met in the lower aquifer but not in the upper aquifer at the edge of the landfill. One means of allowing additional reduction in impact is to acquire a natural attenuation zone. The impact of the landfill on the upper aquifer decreases with distance away from the landfill and in this particular case one could meet the 100-mg/l maximum increase in impact postulated here at a little over 100 m from the landfill.

When landfills are constructed on aquitards with a relatively low hydraulic conductivity (i.e., less than 10^{-9} m/s), the impact of that landfill upon underlying aquifers may not occur until a time long after closure of the landfill and indeed after the leachate itself has reached benign levels (e.g., see Figures 11.22–11.26). The clay acts as a buffer, slowing movement into the underlying soil but the impact, although long in coming, can be significant and can last for decades to centuries. Regulations which only consider a limited (e.g., 100 years) period in considering potential impact fail to recognize the long-term implications of designing landfills with clay liners. For example, in case 7, if the site boundaries were 30 m from the edge of the landfill, the increased impact would exceed 100 mg/l for a period of about 225 years.

11.5 Summary

This chapter has examined contaminant migration through fractured media. Particular consideration has been given to lateral migration in fractured porous rock and vertical migration through fractured clayey aquitards although the factors discussed are equally appropriate to lateral migration through fractured clayey aquitards (e.g., from the edge of a landfill) or vertical migration through a fractured porous rock. Attention has been focused on the migration of contaminants at concentrations typical of that encountered in domestic waste leachate. The migration of concentrated (dense non-aqueous phase liquids) (DNAPLs) has not been considered and it should be noted that these may move rapidly through fractures with little or no attenuation. Concentrated DNAPLs should not be stored or disposed of in (or adjacent to) fractured media without the provision of a multiple-component engineered barrier system (likely involving geomembranes and compacted clay – see also Section 4.3).

It has been shown that diffusion of contaminants from fractures into adjacent matrix of porous rock like shale and sandstone, or soil such as clay and till can result in significant retardation of contaminant movement and attenuation of peak impact compared to that which would be predicted neglecting matrix diffusion. It has also been shown that there is a symbiotic relationship between Darcy velocity, fracture spacing, matrix porosity and matrix diffusion coefficient which needs to be carefully examined on each particular project.

The development of any model invariably involves making idealizations of what, in reality, are complex situations. The model examined herein is no exception. To date, there is very limited field data to allow verification of models which consider contaminant migration through fractured systems such as those considered in this chapter. The reader should be aware of the assumptions that have been made in developing the model and should use engineering judgement when assessing the applicability of idealized models in practical projects.

It is worth emphasizing again that when landfills are constructed on relatively "tight" aquitards (i.e., hydraulic conductivity of 10^{-9} m/s or less), the clay acts as a buffer, slowing movement into the underlying soil but the impact, although long in coming, can be significant and can last for significant periods of time. Regulations which only consider a limited (e.g., 30 or 100 years) post-closure period for assessing potential impact fail to recognize the long-term implications of advective–diffusive transport. It also follows from this that one cannot terminate monitoring of a receptor aquifer simply because the leachate strength has reached benign levels. The monitoring programme should be linked to predictive modelling which, together, can provide an indication of when monitoring can be terminated.

Chapter twelve

Geosynthetic clay liners (GCLs)

12.1 Introduction

Geosynthetic clay liners (also sometimes known as geosynthetic clay barriers or GCBs) are thin (typically 5–10 mm) manufactured hydraulic barriers comprised of a layer of bentonite supported by geotextiles and/or a plastic sheet. GCLs are usually placed on a prepared foundation layer.

Bentonite, a weathered volcanic ash, is predominantly smectite minerals (montmorillonite typically constituting 75–90% of the mass of the bentonite) with exchange sites primarily occupied with sodium or calcium ions (see Section 3.4). Typically, natural sodium bentonite is used although "sodium-activated" calcium bentonite (in which the exchangeable calcium cations have been replaced with sodium ions to increase the swell capability and decrease the hydraulic conductivity of the bentonite in the GCL) also have been used. The other constituents of the bentonite may include quartz, carbonates (calcite, dolomite, siderite), feldspars, mica, organic matter, illite, kaolinite, chlorite, zeolites, volcanic glass, apatite, haematite, limonite, and heavy minerals such as pyrite, magnetite and zircon. The typical distribution of cations in sodium bentonite used for GCLs is (Egloffstein, 2001): 50–90% Na^+, 5–25% Ca^{2+}, 3–15% Mg^{2+}, 0.1–0.5% K^+. Calcium bentonite has also been used in GCLs but since the hydraulic conductivity of calcium bentonite is about one order of magnitude higher than that for sodium bentonite it is not commonly specified. Where calcium bentonite is used, the resulting GCL will typically have a much larger mass per unit area than a GCL manufactured from sodium bentonite.

There are many different types of GCL. Typically the bentonite is enclosed (sandwiched) between two geotextiles that are held together by adhesives, stitching or needle-punching. The needle-punching process involves punching many small needles through the entire GCL, causing some fibres from the top needle-punched nonwoven geotextile (called the "cover" geotextile) to extend through the bentonite and bottom geotextile (called the "carrier" geotextile), bonding the entire structure together (von Maubeuge and Heerten, 1994). The fibres that are punched through the bottom geotextile rely on natural entanglement and friction to keep the GCL together. In some cases, an additional bond is achieved by heating the carrier geotextile causing the fibres to fuse together and/or attach to the carrier (bottom) geotextile, creating a stronger bond between the two geotextiles and bentonite (these GCLs are sometimes referred to as "thermally treated", "thermal locked" or "heat burnished"). The cover geotextile is usually needle-punched nonwoven while the carrier geotextile may be either woven or nonwoven depending on the product. In some cases, the carrier is a combined woven and nonwoven geotextile (sometimes referred to as "scrim reinforced" or "nonwoven, woven reinforced"). As will be discussed in more detail in this chapter, the method of construction and the choice of type and weight of cover and carrier geotextile can have a significant effect on the engineering properties of GCLs and this should be carefully considered when selecting a GCL for a given application. Most commonly the bentonite in a GCL is air dried, although some GCLs have prehydrated bentonite. Another type of GCL involves bentonite bonded to a thin plastic sheet using adhesive.

GCLs are a manufactured product and hence a high level of quality control can be achieved. However, the fact that they come in thin sheets that are seamed by overlapping does mean that care is required during construction to avoid tearing the GCL sheets or opening the seams – especially when cover soil is being placed over the GCL.

The engineering design of barrier systems (be they liners, covers or lateral containment walls – e.g., see Li *et al.*, 2002) which includes GCLs typically requires evaluation of water flow ("leakage"), contaminant transport and stability. These, in turn, require consideration of hydraulic conductivity, clay-leachate compatibility, diffusion, sorption and shear strength. Careful consideration must be given to the stability of liner systems involving GCLs and other geosynthetics (see Chapter 15). Since the manufacturing process will typically have an effect on the engineering characteristics of the GCL (Lake and Rowe, 2000a), this must be considered when comparing the engineering behaviour of different GCLs.

12.2 Basic properties

The GCL bentonite properties of most common interest are swell index (ASTM D5890), moisture content (ASTM D4643) and fluid loss (ASTM D5891). The clay type is established by either X-ray diffraction (see Chapter 4) or the use of methylene blue (Heerten *et al.*, 1993). The basic GCL properties of interest include mass per unit area (ASTM D5993), grab (ASTM D4632) or wide width strength and elongation (ASTM D4595), peel strength (ASTM D6496) and internal shear strength (ASTM D6243). The flow rate (index flux) and hence the hydraulic conductivity ("permeability") of a GCL are commonly measured in a flexible wall permeameter (ASTM D5887), although a fixed wall permeameter may have advantages for assessing hydraulic conductivity after interaction with leachate (see Petrov *et al.*, 1997a; Kodikara *et al.*, 2002).

12.3 Swelling behaviour

Both the hydraulic conductivity, k, and the diffusion coefficient, D, of a GCL are highly dependent on the bulk GCL void ratio, e_B (as defined in Appendix D). This may depend on (a) the method of manufacture and (b) the confining stress at which hydration occurs. Petrov *et al.* (1997b) demonstrated that the final bulk GCL void ratio of a needle-punched GCL and an otherwise similar GCL without needle punching (i.e., fibre-free) varied substantially, as indicated in columns 2 and 3 of Table 12.1. Thus, needle punching has a significant effect, especially at lower confining stresses.

Thermal treatment of needle-punched fibres can also significantly influence the bulk GCL void ratio. For example, columns 3 and 4 of Table 12.1 compare the final bulk GCL void ratio for two otherwise similar needle-punched GCLs except that in one case there is no thermal treatment (column 3) and in the other there is thermal treatment (column 4). The effect of thermal treatment is greatest at low confining stress when the combination of needle punching and thermal treatment may be expected to reduce the amount of swelling (thereby reducing the bulk void ratio) and thus improve the performance of a GCL (other things being equal) in terms of both hydraulic conductivity and diffusion since both are related to the bulk void ratio for a given

Table 12.1 Final bulk GCL void ratios obtained in confined swell tests

	Final bulk void ratio (e_B)		
		Needle-punched	
Confining stress (kPa) (1)	*No needle-punching*[1] (2)	*No thermal treatment*[1] (3)	*Thermal treatment*[2] (4)
6	7.58	5.12	3.98
25	4.04	3.23	2.97
100	2.58	2.26	2.25
200	1.96	1.68	1.69
400	1.50	1.24	1.19

[1]Petrov *et al.* (1997b).
[2]Lake and Rowe (2000a).

GCL. Thermal treatment can also be expected to provide improved internal shear resistance.

The results presented in Table 12.1 are for tests where the stress is applied prior to hydration. The effect of thermal treatment is particularly evident in post-hydration confinement where the sample is allowed to hydrate at a low confining stress (e.g., 6 kPa) and then, after hydration, the stress is applied. This might correspond to the case where a GCL used as part of a composite liner is left to hydrate (e.g., from the underlying soil) with only the leachate collection system in place and with little or no waste being placed until after hydration had occurred. As shown by Lake and Rowe (2000a), in this case the effect of thermal locking is manifest throughout the stress history. This is because the thermal treated (locked) fibres are anchored to the carrier geotextile and hence are less likely to pull out of the carrier geotextile during hydration at low confining stress than when they are simply needle-punched.

From the preceding discussion, it follows that the relationship between the effective confining stress, σ'_v, and the bulk GCL void ratio depends on the method of GCL manufacture. The relationship between the bulk void ratio and confining stress needs to be developed for each type of GCL, and for a given GCL will vary with hydrating conditions (e.g., the chemical composition of the fluid used to hydrate the GCL and whether the stress was applied before or after hydration). Lake and Rowe (2000a) fitted regression curves to the data obtained for the three GCLs indicated in Table 12.2, yielding the following relationships for confined swell condition and hydration with distilled water:

Table 12.2 Typical properties of three thermally treated GCLs tested by Lake and Rowe (2000a,b)

Symbol	*Sodium bentonite layer*	*Polypropylene carrier geotextile*	*Polypropylene cover geotextile*	*Typical geotextile mass/area* M_{GEO} (g/m^2)	*Minimum GCL mass/area* M_{GCL} (g/m^2)	*Mean, std dev. of* M_{GCL} (g/m^2)
NWNWT	Granular	Nonwoven, woven reinforced	Nonwoven	620	5,270	5,898, 312
WNWT	Granular	Woven	Nonwoven	390	5,665	5,795, 113
WNWBT	Powder	Slit-film, woven	Nonwoven	500	5,481*	5,578, 85

*Includes 500 g/m^2 of powder bentonite in cover geotextile.

$$\log_{10}\sigma'_v \cong 3.25 - 0.57e_B \quad \text{for the NWNWT GCL} \tag{12.1}$$

$$\log_{10}\sigma'_v \cong 3.5 - 0.69e_B \quad \text{for the WNWT GCL} \tag{12.2}$$

$$\log_{10}\sigma'_v \cong 2.9 - 0.32e_B \quad \text{for the WNWBT GCL} \tag{12.3}$$

where σ'_v is in kPa. Even for these products, the variability associated with the manufacturing process may result in some deviation from these relationships. These relationships serve to illustrate general trends and should not be used for detailed design without independent verification.

12.4 Hydraulic conductivity

Test procedures (e.g., ASTM D5887) give the water flux (flow per unit area per unit time) through a GCL for a given head difference. The hydraulic conductivity can be deduced using the thickness of the hydrated test GCL specimen after the conclusion of the test. Since some test specimens will have a slightly curved surface while others (i.e., the stitch-bonded products) exhibit a pillow like pattern, an average thickness value is used. The effect of thickness can be eliminated by expressing the results in terms of permittivity, ψ (i.e., the flow rate per unit area divided by the head difference across the GCL giving rise to this flow rate).

As noted earlier, some GCLs are manufactured by bonding the clay to a plastic sheet. These plastic sheets may be manufactured from polymers similar to those used in geomembrane liners but are typically much thinner. In laboratory tests, these thin plastic sheets may make the GCL essentially impermeable to fluids. However, because they are thin (typically less than 0.5 mm thick), one cannot rely on the sheet remaining intact (and hence acting like a geomembrane) in the field. Thus, irrespective of whether or not a GCL has a plastic backing, a proper geomembrane (typically 1.5–2.5 mm thick, high-density polyethylene (HDPE)) is still required to form a composite liner, and only the bentonite component of GCLs with a plastic membrane backing can be relied upon to provide hydraulic resistance in the field. Thus, the bentonite component of these GCLs should be examined in the same way as other GCLs that do not have a plastic sheet/membrane backing.

Using average values of thickness, typical ranges of hydraulic conductivity of GCLs with respect to water are from 5×10^{-12} to 5×10^{-11} m/s, although values in the range 2×10^{-12} to 2×10^{-10} m/s have been reported (Bouazza, 2002). The hydraulic conductivity of a GCL is highly dependent on the hydrating conditions and the applied effective stress during permeation. These factors combined with the method of manufacture, water content prior to hydration and mass of bentonite all significantly influence the GCL thickness, and while there is some correlation between hydraulic conductivity, k, and thickness, H, there is a great deal of scatter. Petrov *et al.* (1997b) showed that by plotting the bulk void ratio, e_B, rather than thickness, H, against hydraulic conductivity, k, one obtains a much better correlation and much less scatter of the data for a given permeant (e.g., see Figure 12.1). This is consistent with the conventional geotechnical engineering concept of a strong correlation between the void ratio and hydraulic conductivity of a soil. For example, for the needle-punched NWNWT GCL and conditions examined by Petrov and Rowe (1997), the hydraulic conductivity with respect to distilled water is given by:

$$\log_{10} k \ (\text{m/s}) \cong -11.8 + 0.29e_B \tag{12.4}$$

The coefficients in equation 12.4 can be expected to vary from one GCL to another; however, in each case the hydraulic conductivity may be expected to decrease with decreasing bulk GCL

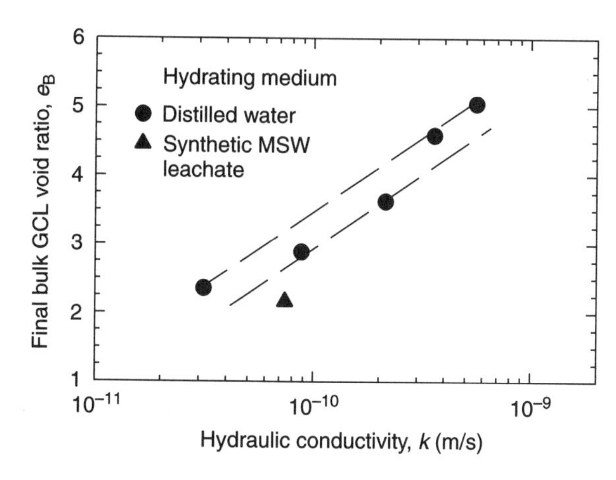

Figure 12.1 Final bulk GCL void ratio versus hydraulic conductivity for permeation of synthetic MSW leachate (modified from Petrov and Rowe, 1997).

void ratio (which is related to stress as implied by equations 12.1–12.3). For example, the hydraulic conductivity of an NWNWT GCL under a confining stress of 40 kPa may be estimated by using equation 12.1 to calculate e_B for $\sigma'_v = 40$ kPa (giving $e_B = 2.89$) and then substituting this into equation 12.4 to give $k = 1.1 \times 10^{-11}$ m/s.

12.5 Factors influencing hydraulic performance

12.5.1 Leachate compatibility

As discussed in Chapter 4, a permeant with a different chemical composition to the pore water may cause the double layers of the bentonite to contract, increasing the size of flow channels in the clay and hence increasing the hydraulic conductivity of the clay (i.e., GCL). A number of researchers (Shan and Daniel, 1991; Petrov and Rowe, 1997; Ruhl and Daniel, 1997; Petrov *et al.*, 1997a,b; Mazzieri *et al.*, 2000; Shackelford *et al.*, 2000; Jo *et al.*, 2001) have examined the issue of GCL hydraulic conductivity and GCL compatibility with various permeants, and many of the results are summarized by Rowe (1998a) and Shackelford *et al.* (2000).

Hydraulic conductivity tests typically employ a flexible wall, double ring or a fixed wall permeameter to assess the GCL hydraulic conductivity (see Petrov *et al.*, 1997a; Koerner, 1998; Orsini and Rowe, 2001). Koerner (1998) advocates the use of flexible wall devices in order to prevent misinterpretation of test results arising from potential sidewall leakage. The concern regarding sidewall leakage in fixed wall tests has been discussed by Petrov *et al.* (1997a) and Orsini and Rowe (2001). Petrov *et al.* (1997a) conducted tests using all three commonly used devices and found no significant discrepancies arising from the test method used provided that appropriate care is taken in traditional hydraulic conductivity and compatibility tests. However, Orsini and Rowe (2001) showed that even without any interaction with leachate, sidewall leakage from traditional fixed wall permeameters may give misleading results in non-traditional tests where the GCL is underlain by gravel or a geonet; in these cases a modified fixed wall permeameter was required to successfully prevent sidewall leakage.

As noted in Chapter 4, it is important that the tests for leachate compatibility be conducted until chemical equilibrium (i.e., the effluent and influent are essentially identical) is reached. As indicated by Rowe (1998a) and Shackelford *et al.* (2000) chemical equilibrium was not achieved for many of the test results reported in the literature. Significant pore volumes of flow (as high as 18 as reported by Petrov and Rowe, 1997) may be required to achieve chemical equilibrium – this tends to be difficult to achieve in reasonable time periods using a flexible wall permeameter.

From an examination of the compatibility of a GCL with salt solutions (NaCl) of different concentrations, Petrov and Rowe (1997) found that:

1. The hydraulic conductivity increases as the final bulk GCL void ratio increases.
2. As the NaCl concentration increases, the hydraulic conductivity increases for a particular void ratio.

3. The samples initially hydrated with water had lower hydraulic conductivities than those initially hydrated with the NaCl for a particular void ratio.
4. Relative to dilute 0.01 M NaCl concentrations, the hydraulic conductivity at a particular void ratio after leachate permeation increased by no more than an order of magnitude for the range of conditions examined.

The hydraulic conductivity, k, of a GCL after permeation with NaCl and ethanol is given in Table 12.3. There is some increase in k with increasing concentration of NaCl. For ethanol there is a decrease in hydraulic conductivity for a 25 and 50% ethanol/water mix due to decreased viscosity (see Chapter 4).

Table 12.4 summarizes changes in hydraulic conductivity of a GCL permeated with different real and simulated leachates. The hydraulic conductivity with respect to real leachate obtained by Ruhl and Daniel (1997; Table 12.4) was lower than that obtained with tap water. Grant *et al.* (1997) also reported low GCL hydraulic conductivity values for tests with a real municipal solid waste leachate (MSWL). It may be hypothesized that this decrease is due to the effects of biological processes that can change the characteristics of the leachate during the test, typically causing an increase in pH and a consequent "dumping" or precipitation of some inorganic constituents (see Section 2.3.3). This in turn can reduce the hydraulic conductivity due to (a) a reduction in pore space due to precipitation and bacterial clogging and (b) the formation of gas bubbles (typically methane or carbon dioxide) during anaerobic degradation of leachate. These gas bubbles can lodge in pore space and reduce the hydraulic conductivity. Both these mechanisms can occur in field applications; however, it is not known to what extent this will occur and hence it is more conservative to use a chemically similar synthetic leachate. Since at typical laboratory temperatures it usually takes about 100 days for significant biological activity to be induced in synthetic leachate (Rowe *et al.*, 2002b), there is also the potential for changes in leachate characteristics and gas formation even using synthetic leachate in very long-term tests.

Not all "synthetic" leachates are similar. The synthetic leachate used by Petrov and Rowe (1997) was modelled on MSW landfill leachate from the Keele Valley Landfill (KVL) (near Toronto, Canada) and hence has a very similar chemical composition to the real leachate. In contrast, the synthetic MSW leachate used by Ruhl and Daniel (1997) was not modelled on any specific leachate but was deliberately "designed" to have a high Na^+ and Ca^{2+} content knowing that this would have a significant impact on sodium bentonite. Where there is an interaction, the effect is to change the relationship between hydraulic conductivity and void ratio, e_B. For example, for the same NWNWT GCL used

Table 12.3 Hydraulic conductivity of a GCL after permeation with different concentrations of NaCl and ethanol/water (at 33–36 kPa)

NaCl concentration	*Hydraulic conductivity* (m/s)	*Proportion ethanol* % *(by mass)*	*Hydraulic conductivity* (m/s)
0	1.3×10^{-11}	0	1.6×10^{-11}
0.1N	2.0×10^{-11}	25	7.3×10^{-12}
SKVLL* (≅0.13N)	7.3×10^{-11}	50	6.0×10^{-12}
0.6N	9.3×10^{-11}	75	4.1×10^{-11}
2N	4.7×10^{-10}	100	2.0×10^{-9}

Source: Petrov and Rowe (1997); Petrov *et al.* (1997b). *Synthetic Keele Valley Landfill Leachate (no bacteria).

Table 12.4 Hydraulic conductivity of a particular GCL product (at 35 kPa) after permeation (adapted from Ruhl and Daniel, 1997)

Permeating fluid	*Hydraulic conductivity* (m/s)
Tap water	7×10^{-12}
Real leachate	$<1 \times 10^{-12}$
Synthetic leachate	1×10^{-11}
High strength synthetic leachate	2×10^{-8}

to obtain the results summarized in equation 12.4, the hydraulic conductivity was bounded by:

$$-11.4 + 0.42e_B < \log_{10} k(\text{m/s}) < -11.2 + 0.42e_B \quad (12.5)$$

when hydrated with water and subsequently permeated using a synthetic leachate based on the typical leachate chemistry at the KVL (see Petrov and Rowe, 1997).

In general, the changes in hydraulic conductivity for modest strength leachates (Table 12.4), salt solutions (<0.6 N; Table 12.3) or ethanol/water mix (<25% ethanol; Table 12.3) are all modest, and the hydraulic conductivity is still low ($<10^{-10}$ m/s). A permeant with a high salt or organic concentration can, however, result in very significant changes in hydraulic conductivity and the potential of interaction should be assessed on a case-by-case basis for a given leachate and GCL. This assessment should consider the hydrating stress, the expected applied stress and whether the GCL will be hydrated with water or the leachate of interest.

Notwithstanding the fact that there is typically some interaction between the bentonite and leachate, as noted above, GCLs may be used provided that design value for hydraulic conductivity is selected giving due consideration to this interaction and based on tests that have a stress history, hydrating conditions and permeating fluid which closely resemble those expected in the field. When performing these compatibility tests, it is important to monitor differences in permeant influent and effluent to ensure sufficient pore volumes of the permeant passed through the sample to reach chemical equilibrium. Results obtained prior to a sample reaching chemical equilibrium with the leachate (i.e., before the effluent concentration is essentially equal to the influent concentration) underestimate the effects of cation exchange and *c*-axis contraction on the hydraulic conductivity (see Chapter 4 and Petrov *et al.*, 1997a).

12.5.2 Interaction with adjacent soil-groundwater

In typical field applications, the sodium bentonite used in GCLs may change to a calcium bentonite due to the exchange of the sodium ions for calcium ions or possibly magnesium ions over a period of a few years (Egloffstein, 2001). This natural process can be expected to occur in cases where the pore water in the soil adjacent to the GCL contains calcium or magnesium. Under some circumstances, this can also occur due to calcite in the bentonite itself. Ion exchange due to calcite in the bentonite has been described by James *et al.* (1997) and confirmed by the excavation of a landfill cover reported by Egloffstein (2001) which showed that the calcite had dissolved completely (probably due to acidic landfill gas condensate). Activated calcium bentonite is likely to have more calcite than other bentonites and hence to be particularly prone to this mechanism. Egloffstein (2001) indicated that dissolving only 3% of the calcite in the bentonite would be sufficient to exchange the total amount of sodium in the bentonite for calcium. He also indicated that over time the sodium-activated calcium bentonite sometimes used in Europe tends to turn back into a calcium bentonite, because of exchange with calcium that remains in the bentonite. Dry–wet cycles encourage this ion exchange, since evaporation of the water causes the ion concentration in the remaining pore solution to rise.

It might be argued that leachate with high sodium but low calcium and magnesium content would represent a better fluid for contact than typical calcium-rich groundwaters. However, even in this case (a) the GCL is likely to be separated from the leachate by a geomembrane and/or (b) it is likely to take years before the leachate characteristics have these "desirable" characteristics. During this period, the GCL can be expected to experience significant uptake of available calcium and magnesium if the pore water in the soil adjacent to the GCL has significant concentrations of these ions. Egloffstein (2001) suggests that the maximum natural calcium concentrations in soil pore water are likely to be about 0.015 M (i.e., the solubility of gypsum). The permeation of GCLs with a permeant at around this level of calcium has been reported to increase hydraulic conductivity by a factor of 3–5 under water-saturated conditions (Lin and Benson, 2000).

The potential for change of a sodium bentonite to a calcium bentonite over time in many field situations begs the question as to whether one should not simply use calcium bentonite from the outset (Egloffstein, 1997; Gleason *et al.*, 1997). Egloffstein (2001) suggests that sodium bentonite is still preferable since much more calcium bentonite would be required for GCLs to achieve a similar permittivity as sodium bentonite. An example is quoted where a 4700-g/m^2 sodium bentonite (D4000) and 8,000-g/m^2 calcium bentonite (D8000) had permittivities with respect to water of $\psi = 0.47 \times 10^{-8}\,s^{-1}$ and $\psi = 2.3 \times 10^{-8}\,s^{-1}$, respectively. This permittivity data might be expected to apply during construction and for a short time afterwards (i.e., until significant ion exchange had occurred). During this time, the sodium bentonite is clearly far superior. After permeation with a 0.3-M $CaCl_2$ permeant (10,800 mg/Ca^{2+}/l), the permittivity of D4000 increased to $\psi = 2.2 \times 10^{-8}\,s^{-1}$ which is similar to, but still lower than, that of $\psi = 2.8 \times 10^{-8}\,s^{-1}$ obtained for the D8000 calcium GCL after permeation with the same solution. Thus after ion exchange, a medium-weight GCL (approximately 4–5 kg sodium bentonite/m^2) gave permittivity comparable to that of a heavy (8-kg bentonite/m^2) calcium GCL. It should be noted that even after ion exchange, a sodium bentonite has a different, and more favourable, structure than a calcium bentonite.

From the foregoing, it follows that in designing systems involving GCLs, the long-term hydraulic conductivity is unlikely to be the same as that typically provided by the manufacturer (these values are only likely to be valid during construction and shortly afterwards). Rather, the hydraulic conductivity to be used should be that for the expected field conditions after the appropriate ion exchange has occurred. This would need to be evaluated on a case-by-case basis but based on available data this cation exchange could give rise to an increase in hydraulic conductivity of between a factor of 3 and 10 (and possibly more in certain circumstances).

Since a calcium bentonite (no matter whether primarily a calcium bentonite or a former sodium bentonite after ion exchange) has relatively low self-healing capacity, desiccation of the bentonite layer should be minimized by providing adequate protection and stress to allow closing of those cracks that may form. This was not the case in the field tests at the Hamburg–Georgswerder landfill where the GCL was only covered with 300-mm soil layer and 150-mm drainage layer (for a detailed discussion, see Heerten, 2000). According to Egloffstein (2001), field tests in Germany suggest that a minimum overburden pressure of 15 kPa (and preferably 20 kPa) is required to achieve adequate performance (i.e., the GCL should be covered by 0.75–1 m of soil).

12.5.3 Partial hydration

The level of GCL hydration that can be achieved was examined by Daniel and Shan (1992) who indicated that sodium bentonite in contact with soil at the wilting point can be expected to have a water content rise to about 50%. They performed a number of tests where the bentonite from a GCL was in contact with sand at different water contents, and the bentonite–water content is given in Table 12.5 (after 40–45 days). Under these circumstances, the bentonite is very efficient in hydrating from the underlying moist soil. Eberle and von Maubeuge (1997) have reported that a GCL installed "dry" ($\omega \sim 9\%$) over a sand with a moisture content of 8–10% was hydrated to over 100% in less than 24 h and the water content increased to 140% after 60 days.

Daniel *et al.* (1993) examined the effect of partial hydration on the hydraulic conductivity, *k*, of a GCL. They found that when the partially saturated GCL was permeated with concentrated hydrocarbons (benzene, gasoline, methanol, trichloroethylene), the hydraulic conductivity was high for a low initial water content (i.e., tests with initial water contents of 17 and 50%). At a water content of 100%, *k* for benzene, gasoline and methanol was close to that of water. Thus, it was important that there be significant hydration prior to contact with the hydrocarbon but it was not essential for the GCL to be fully saturated to have a low hydraulic conductivity in these cases. The water uptake by the bentonite will be a function of applied stress and the method of manufacture, and the implications of this need to be considered for each particular application.

The effect of the prehydration water content on the hydraulic conductivity of GCLs permeated with divalent salt solutions has been examined by Vasko *et al.* (2001) and found (Figure 12.2) not to have any apparent effects on hydraulic conductivity for the intermediate (0.025 M) and weaker solutions (<0.01 M). For higher prehydration water contents, lower hydraulic conductivity was obtained with the stronger solutions (0.1–1 M). The hydraulic conductivity decreased from 1×10^{-6} to 3×10^{-9} m/s as the prehydration water content increased from 9 to 150% and then remained constant as the prehydration water content increased.

Vasko *et al.* (2001) showed that the benefits accrued by hydration with water followed by permeation with a non-wetting organic liquid (as obtained by Daniel *et al.*, 1993) are not obtained when the permeant liquid is a wetting aqueous solution. This difference was attributed to the different hydration mechanisms involved when the GCL is in contact with wetting and non-wetting liquids. Another possible explanation is that the tests conducted by Daniel *et al.* (1993) were terminated before equilibrium was established.

GCLs manufactured by gluing bentonite to a plastic sheet may pose a particular challenge with

Table 12.5 Uptake of water by bentonite (interpreted from Daniel and Shan, 1992)

Sand initial water content (%)	*Bentonite water content at 40–45 days (%)*
1	50
2	75
3	88
5	128
10	156
17	193

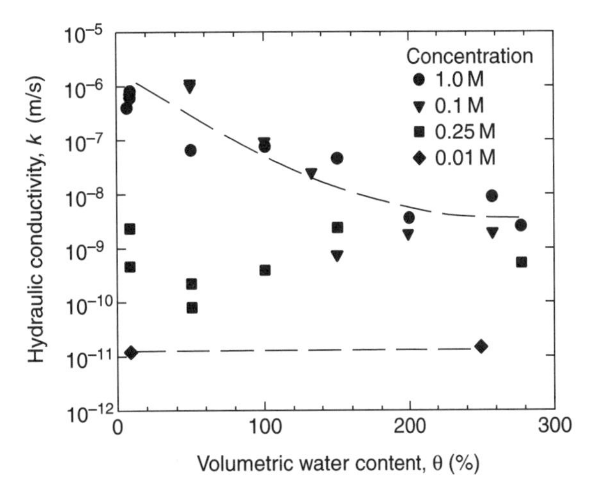

Figure 12.2 Hydraulic conductivity versus prehydration water content for unconfined GCL samples (modified from Vasko *et al.*, 2001).

respect to hydration when used as part of a composite liner with a normal geomembrane above the GCL since the only opportunity for hydration by water (as opposed to leachate at locations of holes in the upper geomembranes) is by holes forming in the lower plastic sheet (which will be highly variable) or at the location of overlaps of the GCL. The modelling of the uptake of water at overlaps has been examined by Giroud *et al.* (2002).

12.5.4 Damage, holes and/or thinning of the GCL

GCLs have the capacity to seal around imbedded objects (e.g., see Didier *et al.*, 2000b). They also have the capacity to seal holes under certain circumstances. The effect of holes in GCL has been examined by various authors (Shan and Daniel, 1991; Bouazza *et al.*, 1996; Mazzieri and Pasqualini, 2000; Sivakumar Babu *et al.*, 2001). Shan and Daniel (1991) cut three different size holes (12, 25 and 75 mm diameter) in three GCL samples. The hydraulic conductivity to water was measured following hydration at a confining stress of 14 kPa; the results are given in Table 12.6. Under the conditions examined, the bentonite swelled to completely fill the 12- and 25-mm diameter holes. Nevertheless, two out of the three 75-mm holes were reported to have not sealed and left an opening of 12 mm diameter giving a hydraulic conductivity greater than 2×10^{-6} m/s (the upper limit measurable with the equipment used) indicating that there is a limit to the capacity to self-heal. Didier *et al.* (2000b) showed that open holes up to 30 mm diameter totally healed after a relatively short time (15 days) although they also found that the stability of the self-healing area depended on the hydraulic head, and failure of the self-healed area occurred when the hydraulic head was higher than 1 m under a confining stress of 10 kPa. Sivakumar Babu *et al.* (2001) examined self-healing of GCLs with 6, 15, 30 and 55 mm diameter holes as well as tests on specimens dried at 20 and 60 °C under an applied stress of 20 kPa and 0.3 m differential head. These tests showed that needle-punched GCLs performed better than stitch-bonded GCLs but also that the GCLs tested had good self-healing properties. Thus, the self-healing capacity of sodium bentonite GCLs is high but this can be impeded if there is significant ion exchange (see Section 12.5.2; Lin and Benson, 2000).

Table 12.6 Effect of punctures in a GCL on hydraulic conductivity (data from Shan and Daniel, 1991)

Diameter of punctures	*Hydraulic conductivity to water* (m/s)
No punctures	2×10^{-11}
12 mm	3×10^{-11}
25 mm	5×10^{-11}
75 mm	$>2 \times 10^{-6}$

The susceptibility of a GCL to damage differs depending on the properties of the geosynthetics used (e.g., GCLs with a needle-punched nonwoven carrier geotextile are less susceptible to damage than those which have a woven carrier geotextile) and that the self-sealing ability also depends on the method of GCL construction. The potential for damage to the GCL can be minimized by appropriate choice of subgrade and material above the GCL. The effect that damage to the GCL can have in terms of increasing hydraulic conductivity has been illustrated by the field case reported by Peggs and Olsta (1998). Mazzieri and Pasqualini (2000) also indicated that the self-healing of GCLs can be compromised, and bentonite loss can occur if the damaged GCLs are placed on a coarse subgrade with large pores. Even in the absence of holes due to damage by the subgrade, the choice of subgrade and damage may be quite important in terms of the potential for internal erosion (see Section 12.5.6).

Damage to GCLs can take a form other than a simple hole or penetration. For example, Mazzieri and Pasqualini (1997) reported a case where an adhesive-bonded GCL punctured by plant roots gave rise to a significant increase in hydraulic conductivity. This was probably due to flow of water through the root itself (Daniel, 2000).

The distribution of bentonite mass per unit area may also affect the hydraulic conductivity of a GCL. Since the bentonite in a hydrated GCL has a very low shear strength, differential applied stresses arising from the pressure due, for example, to gravel particles (Fox *et al.*, 1998b, 2000) or a void due to a wrinkle in a geomembrane (Stark, 1998) may cause the bentonite to move laterally from areas of higher stress to those of lower stress. This can result in a local reduction in bentonite thickness which, in its turn, can cause a higher flow through the GCL (Koerner and Narejo, 1995; Fox *et al.*, 1996). Local bentonite displacement can be minimized by placing a suitable cover soil over a GCL before hydration. The presence of coarse-grained material, such as gravel, overlying a GCL can also be another cause of bentonite migration due to stress concentration. Daniel (2000) has discussed some steps that may be used to minimize bentonite thinning in GCLs.

12.5.5 Desiccation

Desiccation of a hydrated GCL can arise from exposure to either the sun or wind; however, in most instances the installed GCL is buried and hence not directly exposed. In these cases, desiccation can still occur if the GCL is subjected to prolonged heat or cold (freezing/frost action). Desiccation has the potential to increase the hydraulic conductivity of a GCL; however, preliminary work by Boardman and Daniel (1996) indicates that while the bentonite in GCLs did form open cracks upon drying, these cracks closed due to swelling upon re-wetting. Thus, after a few days, the initial high ($\sim 10^{-7}$ m/s) hydraulic conductivity had reduced to a value similar to the original value ($\sim 7 \times 10^{-11}$ m/s). While these results are encouraging, they were performed under a limited range of conditions and more data are required to verify these findings.

GCLs subjected to limited freeze-thaw conditions (Hewitt and Daniel, 1997; Kraus *et al.*, 1997) were found to perform well with no evidence of cracking, and no significant increase in hydraulic conductivity was observed related to the freeze-thaw of the GCL sheets. There is a need for more field data to verify these encouraging findings which are very different to those reported for a conventional compacted clay liner (CCL). For example, Othman and Benson (1993) observed a significant (one to two or more orders of magnitude) increase in the hydraulic conductivity of a CCL after being subjected to freeze-thaw cycles. Thus the, albeit limited, available evidence would suggest that GCLs may be preferable to CCLs where the liner cannot be protected from desiccation and freeze-thaw (e.g., in covers).

It should be noted that the positive results from laboratory freeze-thaw tests represent the best circumstances. As discussed in Section 12.5.2, ion exchange over a period of several years after installation in the field might result in much poorer self-healing characteristics (reduced swell potential), and adequate applied stress would be required to encourage self-healing. In this context, Reuter and Egloffstein (2000) and Blümel *et al.* (2002) reported good performance over a period of 4 and 3 years respectively for samples subject to ion exchange and severe drying when subject to 15–20 kPa confining stress.

12.5.6 Piping and internal erosion

Since GCLs are relatively thin, significant hydraulic gradients may occur if leachate is mounded above a landfill liner or if the GCL is used in pond, lagoon or reservoir applications. The existence of large hydraulic gradients in conjunction with fine-grained soils that may be inadequately filtered creates the potential for internal erosion (i.e., the movement of fine particles) and possible failure due to piping. Clay soils with a dispersed structure are particularly susceptible to internal erosion. The problem is most severe at low stress when the gradients capable of moving soil particles are coupled with a situation where the repulsive forces between the clay particles exceed attractive forces and cause individual particles to detach from each other, enter into suspension and be carried away by the flowing water. Depending on hydrating conditions, GCLs have the potential to exhibit this behaviour. Loss of bentonite from the GCL core by this mechanism may increase the hydraulic conductivity of the layer and may have detrimental consequences on the performance of underlying drainage systems due to increased clogging potential. Giroud and Soderman (2000) proposed a limit for acceptable bentonite migration into a geonet drainage layer of 10 g/m^2. A field example involving internal erosion has been reported by Stam (2000) for a case where abnormal leakage was observed in a GCL-lined lake. Excavation revealed "patchy" bentonite piping through the lightweight nonwoven bottom geotextile of the GCL into the coarse sand subgrade.

Rowe and Orsini (2003) present results of an investigation into the performance of a GCL over a 6-mm gravel subgrade, a geonet subgrade and a sand subgrade under high heads (hydraulic gradients). Using a specially modified constant-flow rigid-wall permeameter (Orsini and Rowe, 2001), they observed that once a threshold head (hydraulic gradient) was reached, there was a sudden increase in hydraulic conductivity and drop in head and hydraulic gradient. Following failure, when gradients were low in the constant flow rate apparatus, the bentonite was able to self-heal the channels created by piping at the high head. Subsequent to such healing, the gradient gradually increased until failure again occurred and the process was repeated.

Tests were conducted using five different types of GCLs (see Table 12.7). The results obtained for the GCLs tested over the 6-mm gravel subgrade showed that the BSNWD and SNWD GCLs, with a scrim reinforced nonwoven lower

Table 12.7 GCLs used in internal erosion tests by Rowe and Orsini (2003)

GCL	*Product descriptor*	*Polypropylene upper geotextile*	*Core sodium bentonite*	*Polypropylene lower geotextile*	*Total mass/unit area* (g/m^2)	*Bentonite moisture content* (%)
BWD[1]	BFG5000 (NWWBT)[3]	Bentonite filled (800 g/m^2) Nonwoven 300 g/m^2	Powder 4,200 g/m^2	Woven 200 g/m^2	5,500	<15
BSNWD[1]	B4000	Nonwoven 300 g/m^2	Powder 4,700 g/m^2	Nonwoven (250 g/m^2) Slit-film woven (100 g/m^2) Composite	5,350	<15
SNWD[1]	NW (NWNWT)[3]	Staple fibre nonwoven 200 g/m^2	Granular 4,340 g/m^2	Scrim reinforced nonwoven 305 g/m^2	4,845	<12
WD[1]	NS (WNWT)[3]	Staple fibre nonwoven 200 g/m^2	Granular 4,340 g/m^2	Slit-film woven 105 g/m^2	4,645	<12
NWD[2]		Woven 100 g/m^2	Granular 4,800 g/m^2	Nonwoven 220 g/m^2	5,100	22

[1] Bentofix, thermally locked and needle-punched.
[2] Bentomat, needle-punched.
[3] Notation used in Table 12.2 for a similar type of product (but samples from different rolls).

(carrier) geotextile on top of the subgrade, remained intact up to hydraulic gradients of approximately 5,500–6,500 (over 50 m water head). The NWD GCL remained intact up to gradients of 3,000–3,500 (25–30 m water head) while the WD and BWD GCLs sustained gradients of 2,500–3,500 (20–30 m water head). The difference in performance can be directly related to the type of geotextile in contact with the subgrade for each particular GCL.

In the tests on the GCLs over the geonet subgrade, the BSNWD GCL remained intact with no failure observed for gradients reaching as high as 6,300 (over 50 m water head). The SNWD GCL reached hydraulic gradients of 4,500–6,000 (40–50 m water head). The NWD GCL reached a final gradient of approximately 3,500 (~30 m water head) before a failure was observed. The WD and BWD GCLs remained intact for hydraulic gradients of up to 3,000 and 2,700 (20–25 m water head), respectively. While this is a relatively large head, it is likely that failures at lower heads would have occurred if more tests had been conducted and hence considerable care is needed in the selection of GCL in situations where it is underlain by a geonet and significant heads could develop over the GCL.

Three GCLs (SNWD, BWD, NWD) were tested over the sand subgrade. All GCLs performed very well, remaining intact at over 70 m water head. The results from these tests indicate that care is needed in the selection of a suitable GCL product and that problems could occur with an inappropriate product over a severe (e.g., gravel or geonet) subgrade.

The work of Rowe and Orsini (2003) shows the need for close scrutiny of designs where GCLs are to be used over coarse (e.g., gravel or geonet) subgrades. Water heads as low as 8 m caused failure and an increase in the hydraulic conductivity by one to two orders of magnitude. Since the gravel and geonet used in these tests meet the subgrade criteria of ASTM D6102, it follows that GCL installations that meet the requirements of this standard could experience internal piping and failure underwater heads that may be experienced in reservoirs, lagoons or even landfills (if leachate mounding occurs). The sand used for testing does not quite meet the specifications set by certain manufacturers for applications where the GCL is the sole barrier and subjected to large hydraulic heads. However, in these tests it did prove to be an adequate subgrade for the specific GCLs and applied heads examined.

Rowe and Orsini (2003) indicated that caution is needed in designing with GCLs under conditions involving high heads. However, their tests also showed that the choice of GCL in contact with the subgrade could have a significant effect on barrier performance and that some GCLs performed extremely well even under the severe conditions examined. Thus, the design for these conditions needs to be carefully examined on a case-by-case basis, matching the choice of product to the conditions (combined with laboratory testing appropriate to the conditions being examined).

12.5.7 Differential settlement

One of the problems encountered with the post-closure performance of landfills is internal cap distress due to subsidence. The heterogeneous waste composition and ageing process (waste biodegradation) can lead to substantial differential settlement of the cover system which in turn may lead to zones of tension cracking. For the case of earthen covers, it has been shown (Cheng *et al.*, 1994; Bouazza *et al.*, 1996) that the hydraulic conductivity of the cover increases when tension cracks start to appear in the soil, with the rate of increase becoming greater with the loosening and cracking of the soil cover. However, there is some evidence to suggest that GCLs can withstand reasonable distortion and distress while maintaining low hydraulic conductivity (Bouazza *et al.*, 1996; LaGatta *et al.*, 1997).

Bouazza *et al.* (1996) studied the effect of differential settlement on the hydraulic conductivity of a GCL. At a simulated 15% straining of the GCL, an increase in hydraulic conductivity of about one order of magnitude was observed (relative to the intact, unstrained, GCL). Nevertheless, the increased hydraulic conductivity after the imposition of a 15% strain still met the requirements of most regulations that limit advective flow through the landfill cover. LaGatta *et al.* (1997) also showed that GCLs can maintain a hydraulic conductivity of 1×10^{-9} m/s or less when subjected to tensile strains of 1–10%. The ability to withstand differential settlement may also depend on the method of construction of the GCL with those using nonwoven carrier geotextiles being able to sustain higher strains without significant loss of performance than those with woven carrier geotextiles. In situations where large strains can develop, particular care is needed in design and construction to ensure that the overlap at the seams is sufficient to prevent opening of the seams due to the lateral movements associated with the differential settlement.

12.6 Solute transport

GCLs represent an alternative to a conventional CCL, and considerations with respect to solute (e.g., contaminant) transport through GCLs are similar to those for clay liners as discussed elsewhere in this book (see especially the general concepts in Chapter 1 and modelling details in Chapter 7). The advective component of solute transport will depend on the hydraulic gradient and the hydraulic conductivity (see Sections 12.4 and 12.5). However, in landfill applications involving composite liners, the key factor controlling contaminant migration may be the rate of diffusion through the liner, and consideration of "equivalency" of a GCL with respect to a traditional CCL (see Chapter 16) may require consideration of both diffusion (Section 1.3.2) and sorption (Section 1.3.5) in addition to advective transport. Diffusion and sorption with respect to GCLs will be discussed in the following subsections.

12.6.1 Diffusion

Although the diffusion of contaminants through compacted bentonite has been the subject of some research (e.g., Choi and Oscarson, 1991; Madsen and Kahr, 1996), to date relatively little has been published regarding diffusion through GCLs. Rowe *et al.* (1997b, 2000d) describe three types of tests used to assess the diffusion of contaminants through GCLs. All of the tests involve subjecting a GCL specimen to a contaminant source on one side and monitoring chemical concentration within a receptor solution on the other side along the lines described in Section 8.2 (see also Figure 8.2b). However, the details of the test differ as noted below:

1. Specified volume diffusion (SVD) tests are used to obtain diffusion coefficients for a GCL where the height to which the GCL can swell is specified. This type of test allows control of the total porosity (and hence void ratio) during testing. The corresponding stress can be deduced but will vary from product to product for a given swell volume.
2. Constant stress diffusion (CSD) tests apply a known stress (rather than controlling the swell volume). The sample is allowed to swell to an equilibrium volume under the specified constant stress before diffusion begins. The height of the sample is monitored throughout hydration and diffusion testing.
3. The bentonite plug diffusion test uses a long sample of bentonite, rather than a single GCL, to examine whether GCL results for a thin plug of clay could be extended to a longer plug of clay.

Following collection of chemical concentration data, the diffusion coefficient is deduced as generally described in Section 8.2. With no advective transport through the sample, the mass flux, f, at any point in the GCL can be expressed in terms of the porous media diffusion coefficient, $D_p = n_e D_e$,

$$f = -n_e D_e \frac{dc}{dz} = -D_p \frac{dc}{dz} \tag{12.6}$$

In principle, the diffusion through a porous media is a function of both the effective porosity, n_e, and the effective diffusion coefficient, D_e. However, Lake and Rowe (2000b) presented results of numerous diffusion tests performed for the range of GCL materials described in Table 12.2 and found that except at very small times (of no practical consequence), the precise values of the effective porosity, n_e, and effective diffusion coefficient, D_e, are not critical to predicting transport through a single GCL provided the range of values of n_e and D_e correspond to the same value of $D_p = n_e D_e$. There was very little latitude in the selection of a value of D_p that would fit the experimental data (Rowe *et al.*, 2000d). This approach of using D_p is convenient since generally only the total porosity, n_t, is known for a GCL, and the effective porosity, n_e, may vary from one contaminant to another or with other factors such as the void ratio. For example, Figure 12.3 shows typical results for an SVD test involving the migration of chloride from the source to the receptor. By adjusting the diffusion coefficient for the bentonite porosity, $n_t = 0.73$, a diffusion coefficient of $D_t = 1.5 \times 10^{-10}\,m^2/s$ was inferred, giving a value of $D_p = 1.05 \times 10^{-10}\,m^2/s$. Therefore, the diffusion coefficients reported in Table 12.8, D_t, are the values deduced from the experimental data for $n = n_t$ (i.e., $D_p = n_t D_t$) and are not effective diffusion coefficients, D_e. The diffusion coefficients range between 3.5×10^{-11} and $3.2 \times 10^{-10}\,m^2/s$ for chloride and 6.0×10^{-11}–$5.0 \times 10^{-10}\,m^2/s$ for sodium. Various factors were found to influence the observed diffusion coefficient. The porous media diffusion coefficient, D_p for both sodium and chloride was found to be linearly related to the final bulk GCL void ratio, e_B. For example, for a simple 3–5 g/l NaCl solution, the relationship between D_p and e_B for chloride ($1 \le e_B \le 3.5$) was given by:

$$D_p \cong (1.02 e_B - 0.89) \times 10^{-10} \quad (m^2/s) \tag{12.7}$$

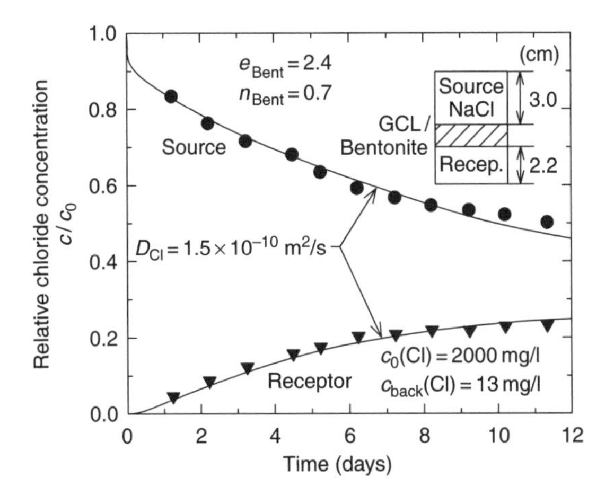

Figure 12.3 Variation in source and receptor concentration with time for a diffusion test on a needle-punched GCL (modified from Rowe *et al.*, 1997b).

This type of relationship arises because as the void ratio increases, the amount of free pore space through which contaminants can migrate increases, thus increasing the effective diffusion coefficient. For samples allowed to hydrate at a low confining stress (and thus high initial bulk void ratio) and subsequently consolidated to a lower void ratio, the void ratio at the time of diffusion (i.e., the consolidated void ratio) was found to govern the diffusion process. These results agree with the findings of Petrov and Rowe (1997) who also found that post-confinement resulted in a higher final bulk GCL void ratio than would be obtained if the same stress had been applied prior to hydration. Low stress hydration allows the clay fabric to arrange into a more flocculated structure, thus increasing the potential diffusive mass transport.

The diffusion coefficient also depends on the chemical species, the other chemical constituents in the leachate and the background chemistry of

Table 12.8 Results of diffusion tests (performed with 4.9 g/l NaCl solution unless otherwise noted) (data from Lake and Rowe, 2000b)

Test number	*Test type*	*Chloride diffusion coefficient,* D_t (m^2/s)	*Sodium diffusion coefficient,*[4] D_t (m^2/s)	*Final bulk GCL void ratio,* $e_B(-)$	*Bentonite total porosity,* $n_t(-)$
NWNWT1-SV	SVD (7.1 mm)[1]	1.5×10^{-10}	2.5×10^{-10}	2.1	0.73
NWNWT2-SV	SVD (11.1 mm)	3.0×10^{-10}	5.0×10^{-10}	3.2	0.80
NWNWT3-SV	SVD (5.6 mm)[1]	3.5×10^{-11}	6.0×10^{-11}	1.1	0.56
NWNWT4-SV	SVD (9.1 mm)	2.8×10^{-10}	4.4×10^{-10}	2.7	0.78
NWNWT5-SV	SVD (7.1 mm)	1.4×10^{-10}	2.2×10^{-10}	1.7	0.67
NWNWT6-SV	SVD (7.1 mm)[2]	2.2×10^{-10}	N/A	1.9	0.70
WNWT1-SV	SVD (7.1 mm)[1]	1.6×10^{-10}	2.7×10^{-10}	2.0	0.69
WNWT2-SV	SVD (9.1 mm)	2.7×10^{-10}	4.0×10^{-10}	2.9	0.77
WNWT3-SV	SVD (5.6 mm)	7.2×10^{-11}	1.1×10^{-10}	1.3	0.59
WNWBT1-SV	SVD (7.1 mm)[1]	1.3×10^{-10}	2.5×10^{-10}	2.1	0.74
WNWBT2-SV	SVD (11.1 mm)	2.9×10^{-10}	4.3×10^{-10}	3.6	0.83
WNWBT3-SV	SVD (9.1 mm)	2.8×10^{-10}	4.4×10^{-10}	2.8	0.79
NWNWTA-CS	CSD (22 kPa)	2.9×10^{-10}	4.8×10^{-10}	2.8	0.72
NWNWTB-CS	CSD (145 kPa)[3]	1.4×10^{-10}	2.2×10^{-10}	1.8	0.69
NWNWTC-CS	CSD (145 kPa)	2.3×10^{-10}	4.0×10^{-10}	2.6	0.77

Note: SVD = specified volume diffusion test, CSD = constant stress diffusion Test. All samples hydrated with deonized, distilled water.
[1] Source was 3.3 g/l NaCl.
[2] Source was synthetic leachate.
[3] Hydrated at 3 kPa and then consolidated to 145 kPa before diffusion test.
[4] Based on $K_d = 0$ ml/g.

the soil below the GCL. For example, Lake and Rowe (2000b) showed that the chloride porous media diffusion coefficient obtained in a constant volume test increased from 1×10^{-10} m^2/s for a 4.6-g/l NaCl solution to 3×10^{-10} m^2/s for a 114-g/l solution. For a given contaminant, the relationship between bulk GCL void ratio and diffusion coefficient was not significantly dependent on the test type (constant volume or constant stress) or hydrating conditions (i.e., pre- or post-confinement), or method of GCL manufacture (at least for the three GCLs examined, see Table 12.2). However, the relationship between applied stress and bulk GCL void ratio did depend on the type of manufacturing process and hydrating conditions (see Section 12.3). The greater the applied stress during hydration, the lower the bulk GCL void ratio, and hence the lower the diffusion coefficient (all other things being equal).

Lake (2000) developed a glass diffusion test apparatus and procedures for obtaining GCL diffusion coefficients for volatile organic compounds (VOCs). Test results (Lake and Rowe, 2004) showed that the rate of contaminant migration through the hydrated GCL was fastest for DCM and followed the sequence DCM > DCA > benzene > TCE and toluene. The difference was attributed to sorption of DCA, benzene, TCE and toluene to the geotextile component of the GCL more than to the bentonite present in the GCL. Diffusion coefficients (D_t) deduced from VOC diffusion testing ranged from approximately 2×10^{-10} to 3×10^{-10} m^2/s which were lower than that of chloride and sodium at similar bulk GCL void ratios ($e_B = 4.1$–4.6). These VOC diffusion coefficients were obtained at relatively high bulk void ratios compared to those expected for a GCL used as part of a base liner system in a

Table 12.9 Diffusion coefficients inferred by Lo (1992)

Liner material	*Effective diffusion coefficient, D_e (m^2/s)*		
	Chloride	*Lead*	*1,2 DCB*
CL	2.4×10^{-10}	5.9×10^{-10}	9.8×10^{-11}
Organo-clay	4.9×10^{-10}	9.0×10^{-10}	1.5×10^{-10}
HA-AlOH-clay	3.6×10^{-10}	7.6×10^{-10}	1.2×10^{-10}

landfill. Other estimates for BTEX range from about 2.5×10^{-10} to $4 \times 10^{-10}\,m^2/s$.

Lo (1992) used the program POLLUTE (Rowe and Booker, 1983–2004) to fit the concentration profile in the effluent data for tests performed using an ordinary sodium bentonite and two modified bentonites to obtain the diffusion coefficients summarized in Table 12.9. It was inferred that the diffusion coefficient for lead was greater than that for chloride which, in turn, was greater than for 1,2 dichlorobenzene. Diffusion coefficients for the modified clays were substantially greater than for conventional bentonite, with a difference of up to a factor of two for chloride.

12.6.2 Sorption

The third primary mechanism controlling contaminant transport is the retardation due to sorption. In conjunction with diffusion testing discussed in the previous section, Lake and Rowe (2000b) conducted experiments to evaluate the sorption characteristics for a number of ions commonly found in MSW leachate. Sorption was examined using a mass balance procedure on the influent and the effluent solutions used in diffusion testing. The results obtained from this procedure were then compared to results from modified batch tests using synthetic leachate in contact with the bentonite component of GCLs in centrifuge bottles at a range of soil-leachate ratios. The results, presented in Table 12.10, indicate that sodium (Na^+) and calcium (Ca^{2+}) ions desorbed during the diffusion testing, while potassium (K^+) and magnesium (Mg^{2+}) were adsorbed onto the clay to varying degrees. Linear distribution coefficients (K_d) were fitted to the experimental data and the results presented in Table 12.10 indicate generally good agreement between the two methods for N_a^+.

(a) Organic sorption on sodium bentonite and organoclays

Due to the hydration energy of exchangeable cations such as Na^+ and Ca^{+2}, the surface of the montmorillonite is quite hydrophilic (i.e., water loving) and this is not conducive for sorption of non-polar organic compounds to mineral. Usually the sodium bentonite used in GCLs has a very low fraction of organic carbon and hence partitioning of non-polar organic compounds would be expected to be small. Smith and Jaffe (1994) reported a K_d value of 0.2 ml/g for benzene for a

Table 12.10 Summary of sorption parameters for GCLs (adapted from Lake, 2000; Lake and Rowe, 2000b)

		NWNWTE-CS[2]			*NWNWTF-CS*[2]		
Cation	*CEC before testing*[1] (meq/100 g)	*CEC after testing* (meq/100 g)	K_d *(adsorb. data)* (ml/g)	K_d *(batch tests)* (ml/g)	*CEC after testing* (meq/100 g)	K_d *(adsorb. data)* (ml/g)	K_d *(batch tests)* (ml/g)
Na^+	55.9	23.3	–	–	23.3	–	–
K^+	1.9	3.2	3.0	3.1	3.0	2.3	3.1
Mg^{2+}	5.7	11.9	6.5	–	13.6	7.1	–
Ca^{2+}	27.3	22.1	–	–	20.6	–	–

[1] From Petrov and Rowe (1997).
[2] GCLs comprised of two nonwoven geotextiles, test conditions representing constant stress diffusion tests.

Table 12.11 Linear sorption coefficient K_d inferred by Lo (1992)

	K_d (ml/g)			
Liner material	*Lead (at pH=7)*	*1,2 DCB*	*1,2,4 TCB*	*1,2,4,5 TECB*
Sodium bentonite	6,000	1.4	2.2	10
Organo-clay	140	609.0	1,320.0	4,500
HA-AlOH-clay	417	20.0	38.0	254

bentonite while Lo *et al.* (1997) reported "insignificant" sorption of BTEX compounds with bentonite (see Table 12.11). Modified clays (also called organoclays or organophilic clays) may be used to provide greater sorption of VOCs than can be obtained from normal sodium bentonite (Boyd *et al.*, 1988a,b; Smith and Jaffe, 1994). One form of organoclay is created by exchanging the hydrated exchangeable cations of bentonite with various types of quaternary ammonium cations (Xu *et al.*, 1997). Within this group of modified bentonites, there may be differences in the sorption mechanism depending on the type of organoclay considered. Modified bentonites may have K_d values at least 2–3 orders of magnitude higher than unmodified bentonite usually used in GCLs.

"Organoclays" are not limited to clays modified with quaternary ammonium cations. Lo (1992) examined (see Table 12.11) modified bentonite in which humic acids are bound to the clay surface via a "cation bridge" which consists of "a formation of complexes with aluminium hydroxide species at clay surfaces". Another type is an "inorgano–organoclay" in which exchange sites are blocked by polycations, and a cationic surfactant acts as the source of surface organic carbon (Srinivasan and Fogler, 1990).

(b) Activated carbon sorption of VOCs

Activated carbon (granular and/or powdered) is a charred material (typically wood, coal or coconut shell) with a highly porous structure and a large internal surface area which has a high capacity for sorption of dissolved VOCs (Heilshorn, 1991). According to Voice (1988), the amount of removal of dissolved VOCs from solution by activated carbon will depend on the properties of the VOCs as well as the type of fluid in which they are dissolved.

When 2% of powdered activated carbon (PAC) was mixed by air-dried weight with powdered bentonite (PB), the sorption of the mixture was much higher than organoclays A, B, C and D for TCE, benzene and toluene (Table 12.12). DCM sorption to the 2% PAC/PB mixture was lower than that of organoclays A, B, C and D, while DCA sorption was similar between the 2% PAC/PB mixture and organoclays A to D.

(c) Sorption onto geotextile forming part of a GCL and sorptive GCLs

The tests reported above were for clay alone. When dealing with a GCL, there is also considerable potential for sorption of organics onto the geotextile component of the GCL. For example, Table 12.13 gives results obtained by Lake and Rowe (2004), and it can be seen that the sorption of TCE, benzene and toluene onto the geotextile component of the GCL is quite significant and can influence interpretation of diffusion testing if not considered correctly. The amount of sorption will be product-specific (depending on the type and mass of geotextile used); however, VOC sorption for conventional GCLs containing sodium bentonite will be predominately due to the geotextile (rather than the clay).

12.7 Gas migration

Although GCLs are usually installed to limit advection of fluids (e.g., water through a cover or lagoon or canal liner, or leachate through a landfill liner) they may also serve another important

Table 12.12 Summary of sorption results obtained for soil component alone by Lake (2000)

Soil	*Organic Content OC (%)*	*Soil: solution ratio*	K_d (ml/g) *and (log* K_{oc}*)*				
			DCM	*1,2 DCA*	*TCE*	*Benzene*	*Toluene*
Granular bentonite	0.3	1 g:11 ml	0.3	0.3	0.9	0.4	1.2
Powdered bentonite	0.4	1 g:11 ml	0.5	0.6	1.2	0.5	1.6
Organoclay A	26.0	1 g:80 ml	11 (1.63)	44 (2.23)	50 (2.28)	63 (2.38)	148 (2.76)
Organoclay A	26.0	1 g:25 ml	14 (1.73)	45 (2.24)	50 (2.28)	69 (2.42)	169 (2.81)
Organoclay B	27.0	1 g:25 ml	13 (1.68)	38 (2.15)	42 (2.19)	70 (2.41)	174 (2.81)
Organoclay C	28.0	1 g:25 ml	17 (1.78)	54 (2.29)	72 (2.41)	104 (2.57)	264 (2.97)
Organoclay D	38.0	1 g:25 ml	20 (1.72)	62 (2.21)	94 (2.39)	79 (2.32)	196 (2.71)
2% PAC/PB mixture	2.5	1 g:80 ml	7 (2.44)	41 (3.21)	491 (4.29)	369 (4.17)	3,354 (5.13)
2% PAC/PB mixture	2.5	1 g:25 ml	5 (2.30)	37 (3.17)	~889 (4.55)	774 (4.49)	~7,969 (5.50)

role in covers as a gas barrier. Gas migration, through landfill cover systems can be driven by the pressure differentials due to natural fluctuations in atmospheric pressure (barometric pumping). Elevated leachate/water levels and temperature gradients can also give rise to pressure differences that can cause gas migration. Research by Didier *et al.* (2000a), Bouazza and Vangpaisal (2000), Vangpaisal and Bouazza (2001) and Bouazza *et al.* (2002a) has shown that the gas permeability (or permittivity) of GCLs may vary depending on the manufacturing process, volumetric water content and the overburden pressure during the hydration phase (Figure 12.4).

Concentration gradients arising from differences in the concentration of a given gas on either side of the cover can give rise to gas diffusion. For example, the outward diffusion of methane and carbon dioxide through a landfill cover or inward diffusion of oxygen in a cover for certain mine wastes. Under these circumstances, a GCL with a high water content has the potential to provide a good barrier to gas diffusion. The diffusion of gases, such as oxygen, will be predominately through the air-filled pores of the soil with only a small amount occurring in the dissolved phase through the water-filled pores. Following from the discussion in Section 8.11.2 and considering diffusion in terms of dissolved gas concentrations for a soil with a volumetric air content θ_a (where $\theta_a = n(1-S_r)$, n is the total porosity and S_r the

Table 12.13 Distribution of sorption coefficients of geotextile (K_{dGEO}) for GCL studied by Lake and Rowe (2004)

DCM (ml/g)	*DCA* (ml/g)	*TCE* (ml/g)	*Benzene* (ml/g)	*Toluene* (ml/g)
0.0	5–6	53–77	18.7–22.4	66

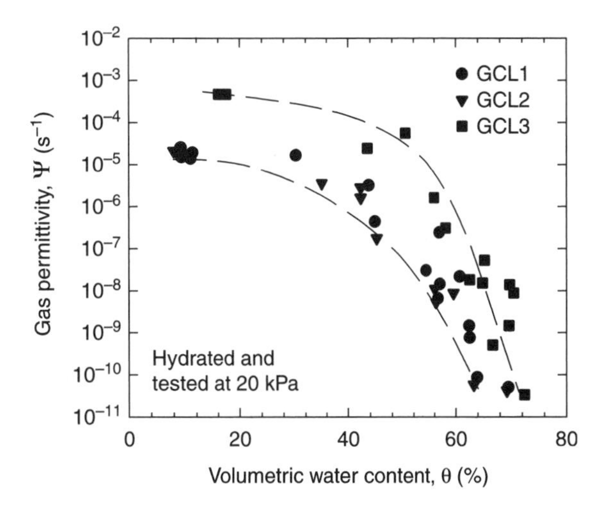

Figure 12.4 Variation of gas permittivity with volumetric water content for confined hydration. GCL1 : Bentofix X2000; GCL2 : Bentofix BFG; GCL3 : Bentomat (modified from Vangpaisal and Bouazza, 2001).

degree of saturation) and volumetric water content θ (where $\theta = nS_r$), the porous media diffusion coefficient, $D_{p\theta}$, can be expressed as a sum of both components as:

$$D_{p\theta} = \theta D_{e\theta} + K'_H \theta_a D_\theta \quad (12.8a)$$

or alternatively, if considering diffusion in terms of the gaseous concentration:

$$D_{\theta p} = \frac{\theta D_{e\theta}}{K'_H} + \theta_a D_\theta \quad (12.8b)$$

where D_θ, $D_{e\theta}$ are the diffusion coefficients for the air and water phases respectively in the GCL, and K'_H is the dimensionless Henry's law coefficient. D_θ and $D_{e\theta}$ are functions of the tortuously of the diffusion path and the diffusion coefficients in air, D_a, and water, D_0, respectively. It can be seen from Table 8.7 that the diffusion coefficient in air, D_a, and water, D_0, typically differ by about four orders of magnitude. The second term in equation 12.8 will dominate except for very high degrees of saturation. Based on data for wide range of unsaturated soils (MacKay, 1997), Rowe (2001) presented the best fit relationship and upper and lower bounds shown in Figure 12.5. The best fit relationship between the diffusion coefficient in the gaseous phase, D_θ (normalized with respect to the diffusion coefficient in air at the relevant temperature, D_a) and the degree of saturation S_r is given by:

$$\frac{D_\theta}{D_a} = \exp[-1.03 \exp(0.017 S_r)^{1.64}] \quad (12.9)$$

where S_r is expressed in per cent. Laboratory test results specifically for GCLs (as summarized by Bouazza *et al.*, 2002b) are also presented in Figure 12.5. The GCL results are significantly lower than those obtained for more conventional soils and most values fall close to the lower bound for other soils, suggesting that GCLs will have a lower gas diffusion coefficient than conventional soil covers. This relationship is given by:

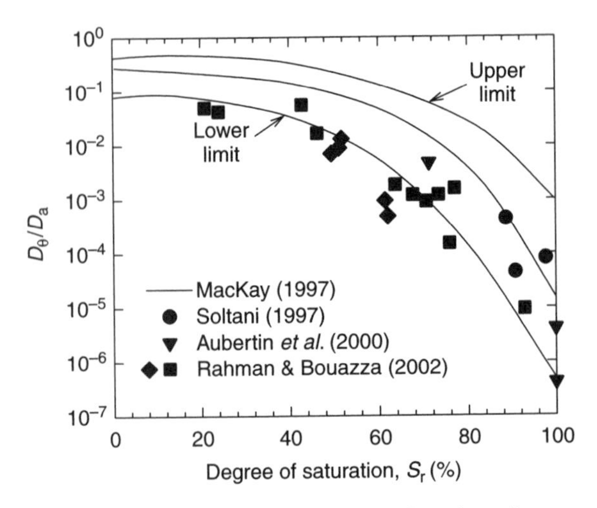

Figure 12.5 Diffusion coefficient as a function of saturation (modified from Bouazza *et al.*, 2002b).

$$\frac{D_\theta}{D_a} = \exp[-1.03 \exp(0.019 S_r)^{1.64}] \quad (12.10)$$

The general trend for GCLs is similar to that for the other soils with the ratio $D_\theta//D_a$ decreasing significantly as the degree of saturation increases. Soltani (1997) tested three different types of materials at different degree of saturation. It can be seen (Figure 12.5) that the results vary significantly at a similar degree of saturation, although they fall within the band drawn from the data sets reported by MacKay (1997). This is probably due to variability in effective air porosity at high degrees of saturation and product characteristics. Results by Aubertin *et al.* (2000) seem to agree with the results obtained by Rahman and Bouazza (2002) for the same type of GCL. Based on the available data, it appears that the relationship given by Rowe (2001) reproduced above as equation 12.9 can be used as a generally conservative estimate of the gaseous phase diffusion coefficient for GCLs, and the relationship given by equation 12.10 can be used to obtain an estimate of a reasonable value for use in calculating gaseous diffusive flux through unsaturated GCLs.

12.8 Installation

The installation procedure followed for construction of GCLs is an important factor in determining long-term field performance. If proper construction methods are not followed, damage of the GCL may result, thus impairing its ability to fulfil design requirements. Recommended installation procedures have been standardized in the form of ASTM D6102 – *Standard Guide for Installation of Geosynthetic Clay Liners*. Details of this specification include requirements that the subgrade surface be "firm and unyielding, with no abrupt changes, voids and cracks, ice or standing water... smooth and free of vegetation, sharp-edged rocks, stones, sticks, construction debris and other foreign matter that could contact the GCL" (ASTM D6102). Tests by Shan and Daniel (1991) and Didier *et al.* (2000b) have shown that GCLs are capable of self-healing holes or tears up to about 30 mm in size (see Section 12.5.4). Reuter and Egloffstein (2000) have also demonstrated good self-healing characteristics under severe drying condition when the GCL was subject to applied stress of 15–20 kPa. However, cases have been reported where improper subgrade preparation has lead to failure of GCLs (Peggs and Olsta, 1998; Stam, 2000). An uneven subgrade may lead to the lateral migration of bentonite, resulting in thinning of the liner and a reduction in hydraulic barrier performance. It is also essential to provide cover for GCLs as soon as possible after placement in order to minimize the potential for premature hydration and the uneven swelling. Care must be taken during cover placement to minimize damage from construction operations (Daniel and Koerner, 1995). Fox *et al.* (1998b) examined the installation damage of adhesive-bonded and needle-punched GCLs laid on a prepared subgrade, covered with varying thicknesses of angular sand and gravel. After hydration, bulldozers were driven over the test plots, and the samples were then exhumed and examined. It was reported that damage to the GCLs was minor for a cover soil thickness greater than 300 mm. Bentonite migration was insignificant except for the adhesive-bonded GCL covered with gravel. No failures were reported for installation conditions that met both the manufacturer's guidelines and ASTM D6102.

12.9 Internal and interface shear strength

The stability of landfill liner systems must be ensured before, during and after placement of the waste. Several failures involving geomembrane-lined landfills (Byrne *et al.*, 1992; Ouvry *et al.*, 1995) highlight the importance of this (see Rowe, 1998a; Chapter 15). The sodium montmorillonite in GCLs has very low shear strength (Mesri and Olsen, 1970) and hence the stability implications of the introduction of this potentially weak layer into a liner or cover system must be carefully considered. Not all GCLs have the same shear resistance, and the manufacturing process and techniques such as stitching, needle-punching and thermal treatment of needle-punched GCLs can greatly increase the internal shear strength of the GCL.

Factors that need to be assessed when considering the stability of liner (or cover) systems involving a GCL include the potential for: (1) internal failure of the GCL (in the bentonite or at the interface between the bentonite and geosynthetics in the GCL) and (2) failure at one of the interfaces in contact with the GCL (GCL in contact with a geomembrane, another geosynthetic or soil). Evaluation of these two potential failure mechanisms requires geotechnical laboratory test data and stability analyses.

Several researchers (e.g., Stark and Eid, 1996; Eid and Stark, 1997; Fox *et al.*, 1998a; Eid *et al.*, 1999) have studied the internal shear strength of reinforced and unreinforced GCLs. Table 12.14 presents some shear strength results. In general, unreinforced GCLs exhibit very low shear strength parameters upon hydration, making them largely unacceptable for use on slopes (Richardson, 1997a). The reinforced GCLs

Table 12.14 Peak internal shear strength parameters (cohesion c and angle of friction ϕ) for various GCLs with upper geotextiles all polpropylene (PP) slit-film woven (adapted from Fox *et al.*, 1998a)

GCL product	*Lower geotextile*	*Reinforcement*	c (kPa)	ϕ (deg)
GCL-1	PP slit-film woven	None	2.4	10.2
GCL-2	PP slit-film woven with a 0.1-mm PE geomembrane laminate	Stitch-bonded, 102 mm on centre	71.6	4.3
GCL-3	PP needle-punched nonwoven	Needle-punched throughout	98.2	32.6

(stitched, needle-punched) have significantly better internal shear strength than unreinforced GCLs; however, they have also yielded variable laboratory results depending on the type of GCL tested and the test method employed (Heerten *et al.*, 1995; Gilbert *et al.*, 1996; Stark and Eid, 1996; von Maubeuge and Eberle, 1998; Fox *et al.*, 1998a; Eichenauer and Reither, 2002). The internal shear strength of one particular type of GCL cannot be used for another GCL due to differences in the manufacturing processes. Some needle-punched GCLs have quite high peak strengths; however, if this value is to be used in design, care is required to either establish that large deformations cannot occur or will be dealt with by other mechanisms since there is some evidence that pull-out of the reinforcement after peak shear stress results in residual values approaching residual values of unreinforced GCLs ($\tau_{residual}/\tau_{peak} \sim 0.05$–$0.5$; Fox *et al.*, 1998a) at large displacements.

To prevent the residual or post-peak GCL internal strength from being mobilized, liner systems involving GCLs in combination with other geosynthetics (geomembrane, geotextiles) are often designed such that interface strengths of one of the various components of the liner system are lower than the GCL's peak internal shear strength but is higher than the residual strength of the GCL (Gilbert *et al.*, 1996). This is possible for low normal stresses, but as the normal stress increases, the failure plane may move internally to the GCL (Gilbert *et al.*, 1996).

Considerable research has been conducted relating to laboratory interface shear strength between GCLs and soils or other geosynthetics (Bressi *et al.*, 1995; Garcin *et al.*, 1995; Gilbert *et al.*, 1996; Feki *et al.*, 1997; Von Maubeuge and Eberle, 1998; Eid *et al.*, 1999; Triplett and Fox, 2001). It has been found in laboratory tests that bentonite extruding through the woven and nonwoven (for products where the mass per unit area of the nonwoven is less than 220 g/m^2) geotextile component of the GCL can form a bentonite smear between the GCL and geomembrane which significantly reduces the frictional resistance along this interface and results in low interface strengths (e.g., see Gilbert *et al.*, 1996). Techniques such as using a GCL with the nonwoven geotextile having a mass per unit area equal to or greater than 250 g/m^2 in contact with textured geomembranes can help to reduce this problem. Due to the variability of materials and conditions, published values of interface friction should not be used for design of a specific project without independent verification for the project in question.

Tanays *et al.* (1994) and Feki *et al.* (1997) have described the field behaviour of a stitch-bonded GCL installed on 2H (horizantal):1V (vertical) and 1H:1V slopes, respectively. Displacements in the GCL were found to be very low on the 2H:1V slope and remained unchanged during the 500 days of observation. The GCL on the 1H:1V slope experienced an average strain of 5.5% one day after its installation.

Daniel *et al.* (1998) have reported the findings from a field evaluation of the internal and interface shear strength behaviour of many different types of GCLs configured with other liner components such as geomembranes, geotextiles and native soils. All geosynthetic configurations on slopes inclined at 3H:1V were reported to be performing satisfactorily. Three slides have occurred on the steeper 2H:1V slope. In one case, an unreinforced bentonite between two geomembranes became hydrated causing a slide to occur. The other two slides occurred 20 and 50 days after construction and were reported to have been caused by bentonite extruding from the GCL through a woven geotextile in contact with a textured geomembrane, thereby reducing the strength of the interface. The tests section using a GCL with both nonwoven carrier and cover geotextiles remained stable. Data from this investigation should also eventually provide valuable long-term data on the potential creep of the reinforcement of the GCLs and how it affects stability with time.

The issue of creep of GCLs subjected to shear loads has been examined by Koerner *et al.* (2001). They examined the three reinforced GCLs used in the trials described by Daniel *et al.* (1998) by subjecting them to shear stresses up to 60% of the short-term shear strength for 1,000 h. Based on extrapolation of this data, they concluded that deformations at 30% of the short-term strength would give rise to deformations of 5–10 mm over a 100-year period. They caution that this is a "bold" extrapolation (over three log cycles of time) and that additional investigation is required to formalize a GCL specification with a polymer formulation that will ensure good long-term performance.

In summary, when designing GCL-lined slopes it is important to recognize the differences between different types of GCLs and consequently differences in interface and internal shear strengths. It is also important to consider the implications of low post-peak internal shear strength of the GCL and the need to prevent bentonite extrusion through the geotextiles in the GCL that decreases the frictional resistance of the interface between GCLs and other geosynthetics.

12.10 GCLs in composite liners

12.10.1 Reduced leakage

GCLs are often utilized in combination with geomembranes to form composite liners at the base of MSW landfills (e.g., Giroud *et al.*, 1997a; Richardson, 1997b). The primary method of transport of ionic contaminants through composite liners is with the flow of leachate through defects (holes, tears, holes over wrinkles, etc.) in the geomembrane. Much has been written about leakage through these defects when a geomembrane is used in combination with a CCL for the composite liner system (Brown *et al.*, 1987; Giroud and Bonaparte, 1989; Giroud *et al.*, 1992; Rowe, 1998a). Equations have been developed empirically and analytically in an attempt to predict flow through these defects. Wilson-Fahmy and Koerner (1995) looked specifically at GM/GCL contact conditions and found that GCLs have the potential to limit leakage rates through a GM/GCL composite system to values less than that would be expected for a GM/CCL system. The implications of this will be discussed in Chapter 13 and 16.

12.10.2 Service life

As part of a composite liner, a GCL may be expected to have a very long service life (hundreds/thousands of years) provided that:

(a) There is no significant loss of bentonite from the GCL during placement. This means that care will be required to avoid loss of bentonite into underlying drainage layers. In addition, a filter may be required to avoid bentonite loss for some GCLs (e.g., see Estornell and Daniel, 1992). Care is also required to adopt construction procedures that will maintain a uniform distribution of the bentonite in the GCL (the potential for bentonite movement may vary from one product to another, and construction specifications should recognize this fact). This involves consideration of both losses of powdered or granular bentonite and scraping off of bentonite (e.g., due to physically or thermally induced movement).

(b) There is no significant lateral movement (thinning) of bentonite during and following hydration that would cause an uneven distribution of the bentonite in the GCL during the contaminating life of the facility (see Sections 12.5.4 and 12.8). For example, wrinkles in a geomembrane may create a void or area of reduced stress into which bentonite in an underlying GCL could migrate (see Stark, 1998). Likewise, care is needed on side slopes (especially steep side slopes) to avoid bentonite migration downslope both in the "dry" and hydrated state (as noted earlier, the potential for bentonite movement may vary from one product to another).

(c) The geosynthetic component of the GCL is not critical to the long-term performance of the bentonite component of the GCL. von Maubeuge and Ehrenberg (2000) have suggested "conservative" estimate service lives of 25–50 years for polypropylene fibres and greater than 100 years for HDPE fibres. The issue of durability and service life of polymer fibres used in GCLs has also been discussed by Hsuan and Koerner (2002) and Thomas (2002). See Chapter 13 for a more detailed discussion of service life issues related to polymers.

(d) Seams are installed to ensure intimate contact, and the design and the construction procedures are such that the seams do not open up during or following placement (e.g., due to the shear loading of construction equipment or differential settlement).

(e) The choice of GCL and the design are such that there is no significant long-term loss of bentonite due to migration (internal erosion) through the GCL under the hydraulic gradients that may occur either during or after termination of the operation of the leachate collection system (see Section 12.5.6).

(f) The design for hydraulic conductivity is selected based on tests that examine the equilibrium (i.e., long-term) hydraulic conductivity of the GCL to the proposed leachate under conditions that simulate likely field conditions (i.e., similar hydrating conditions, pore water chemistry of the pore fluid adjacent to the GCL, applied stress, GCL and leachate characteristics; see Sections 12.5.1–12.5.3).

(g) The GCL is not allowed to desiccate (see Section 12.5.5).

Items (a), (b), (c) and (g) can be addressed by appropriate design, specifications and construction procedures. If the geosynthetic component of the GCL is critical to the long-term performance of the GCL then the service life of the GCL cannot exceed the service life of the geosynthetic component. The service life of the geosynthetic component will depend on the polymer and additives used. Some polymers have the potential to degrade due to biological action. For example, Giroud (1996b) reports that there are fungi and bacteria that catalyse hydrolysis of polyester. In situations where the service life of the geosynthetic component is critical, it may be argued that HDPE should be used with additives (especially antioxidants) similar to that used in a geomembrane (see Chapter 13). However, if a long service life of the GCL is required, it would be better, where practical, to adopt a design such that the geosynthetic component of the GCL is not critical to the long-term performance of GCL. For example, one could avoid situations where the geotextile component is necessary to prevent movement of bentonite into underlying open void space in drainage stone. Likewise, one could avoid situations where one was relying on needle-punching to provide long-term shear resistance or where the geosynthetic component is required to prevent lateral migration of bentonite into areas of reduced stress (e.g., beneath a wrinkle). The level of control of wrinkles (waves) in the geomembrane may need to be higher when a GCL is used as part of the composite liner than when a CCL is used due to much thinner clay and the potential for clay movement within the GCL. If GCLs are going to be used in bottom liners (base seals) and are critical to system performance, it would seem prudent to develop and apply a technique for non-destructive checking of seam integrity following placement of the overlying layers to address item (d). It is noted that the geotextile overlaps have been observed to move (open up) more than 0.5 m in some cases (Rowe *et al.*, 1993) and hence it is certainly possible that movement could occur for GCLs (e.g., due to the construction practice adopted). Items (e) and (f) can be addressed by appropriate hydraulic conductivity tests.

12.11 Equivalency GCLs and CCLs

Any assessment of the "equivalency" of GCLs and CCLs must consider the total design performance rather than individual properties. Such an assessment of the environmental protection afforded by the two systems should involve calculations that include consideration of the: (1) hydraulic conductivity of GCLs permeated with the leachate of interest, (2) diffusion and sorption, (3) interface contact with any overlying geomembrane and (4) thickness of the entire barrier system. Rowe (1998a), Shackelford *et al.* (2000), Manassero *et al.* (2000) and Rowe and Lake (2000) have discussed these issues in detail. A qualitative comparison of GCLs and CCLs is given in Table 12.15. For many criteria, the performance of a GCL is either equivalent to

Table 12.15 Potential equivalency between GCLs and CCLs (modified from Manassero *et al.*, 2000)

		Equivalency of GCL to CCL			
Category	*Criterion for evaluation*	*GCL probably superior*	*GCL probably equivalent*	*GCL probably inferior*	*Site- or product-dependent*
Construction issues	Ease of placement	X			
	Material availability				X
	Puncture resistance			X	
	Quality assurance	X			
	Construction rate	X			
Contaminant transport issues	Sorption				X*
	Gas permeability				X*
	Solute flux				X*
	Breakthrough time				X*
Hydraulic issues	Compatibility				X
	Consolidation water	X			
	Water flux		X		
Physical/mechanical issues	Bearing capacity				X
	Erosion				X
	Freeze-thaw	X			
	Settlement-total		X		
	Settlement-differential	X			
	Slope stability				X
	Wet–dry cycles	X			

*Depends on contaminant, liner and associated attenuation layer properties.

or better than that of a CCL. However, in terms of liner applications, careful consideration of both the liner itself and the underlying attenuation layer must be made when examining contaminant transport and equivalence in terms of potential contaminant impact over the contaminating lifespan of the facility. This is examined further in Chapter 16.

Chapter thirteen

Geomembrane liners

13.1 Introduction

Geomembranes are relatively thin (normally less than 2.5 mm thick) polymeric materials that are commonly used together with a compacted clay (Chapter 3) and/or a geosynthetic clay liner (GCL) (Chapter 12) to form a composite liner. For landfill applications, the primary function of a geomembrane is to impede the advective flow of contaminants through the liner and provide a diffusive barrier to inorganic contaminants. This chapter focusses on the issues important for the design of high-density polyethylene (HDPE) geomembranes. The physical response of geomembrane liners including the stress–strain behaviour of polyethylene, the phenomenon of stress cracking, geomembrane protection and wrinkles are examined. Factors influencing the long-term durability of HDPE geomembranes are then discussed and current best estimates of the service life for HDPE geomembranes at the base of a landfill are presented. The two primary mechanisms for contaminant migration (i.e., leakage through holes in the geomembrane and molecular diffusion through the intact HDPE geomembrane) are also considered.

13.1.1 Geomembrane selection

The many different commercially available geomembrane products can be useful in a variety of applications. For example, high density polyethylene (HDPE), very low-density polyethylene (VLDPE), linear low-density polyethylene (LLDPE), polyvinyl chloride (PVC), flexible polypropylene (fPP), chlorosulphonated polyethylene (CSPE) and bituminous geomembranes each have advantages and limitations, depending on the application. Koerner (1998) provides a description of various types of geomembranes available in the marketplace. Geomembranes can be smooth or textured to improve interface shear strength for use on side slopes or even coextruded with different polymers to obtain a combination of properties (e.g., coextrusion of HDPE/VLDPE/HDPE).

The selection of a geomembrane product depends on its application in the waste containment facility, whether as a primary or secondary base liner, or as a liner in the final cover. Factors to consider when selecting a geomembrane for these applications are listed in Table 13.1. In addition to adequate material strength, flexibility and puncture resistance (i.e., to minimize the development of small holes and tears), it is also important for all applications that the geomembranes have sufficient durability. Durability requirements depend on the exposure conditions. Resistance to stress cracking, a brittle failure mechanism manifest by semi-crystalline polymers as discussed in Section 13.2.2, is an important durability issue for all applications. Oxidation will lead to breakdown of the polymer and eventually failure of the geomembrane. Thus, oxidative resistance is very important for all liner applications, including secondary liners which have the potential to be exposed to both atmospheric oxygen in the secondary leachate collection system and leachate, and which require a long service life. Chemical

Table 13.1 Relative importance of some factors to consider when selecting a geomembrane on a scale of 1–5 (1, less important; 5, more important) for primary, secondary and cover liner applications (modified significantly from Peggs and Thiel, 1998)

	Relative importance for		
Factor	*Primary liner*	*Secondary liner*	*Cover liner*
Leachate resistance	5	4	1
UV resistance	3	2	4
Oxidative resistance	5	5	4
Flexibility for subgrade contact	4	4	5
Biaxial stress and strain	4	3	5
Puncture strength	1	1	1
Puncture strain	5	5	5
Interface friction	4	3	5
Stress crack resistance	5	5	5
Loss of additives	5	5	4
Seamability	5	5	3
Number of field seams	3	2	2
Detail seamability	4	3	3

resistance in contact with leachate (see Section 13.3) is very important for primary liners, and UV resistance may be very important for cover geomembranes.

HDPE is most commonly used for geomembranes in the primary and secondary barrier systems in landfills given its excellent resistance to a wide range of chemicals, despite the fact that it is both much stiffer and more susceptible to stress cracking than other geomembrane products that are available. For cover systems, a wider range of geomembranes, often with much higher elongation at failure than HDPE, can be used (e.g., PVC, VLDPE), since there is likely to be much less chemical exposure in the cover relative to the base liners. The current prevalent use of HDPE as base liners does not preclude the development of new polymer materials that have both excellent chemical resistance and outstanding physical properties; however, the introduction of any new material would require extensive laboratory testing and years of field verification prior to acceptance as an alternate liner material. Given their dominant use in practice, the remainder of this chapter focusses on HDPE geomembranes for use as primary and secondary liners.

It is very important to specify the desired engineering properties of the geomembrane. Generic specifications for HDPE geomembranes proposed by Hsuan and Koerner (1997) and Hsuan (2000a) are given in Table 13.2. This specification controls the quantity and distribution of carbon black introduced into the polymer formulation to adsorb solar energy to limit UV degradation. Control on the amount of antioxidants added to the formulation to stop deleterious oxidative chemical reactions can be achieved by specifying a minimum oxidative induction time (OIT) for both virgin and over aged samples. The specified stress crack resistance is such that it should not fail in 200 hours in a single point notched constant load test (SP-NCLT; ASTM D5397) at 30% of the yield stress. Specified minimum tensile properties depend on the thickness of the geomembrane and, in the case of the ultimate break strength and elongation, whether the geomembrane is smooth or textured.

Selection of the minimum geomembrane thickness may be governed by regulations. For example, Ontario Regulation 232 (MoE, 1998) specifies a 1.5-mm thick HDPE geomembrane for use in the primary liner and a 2-mm thick HDPE geomembrane in any secondary liner. Other factors being equal, a thicker geomembrane will have greater strength and higher puncture resistance, and a longer service life. However, in addition to increased product cost, the thicker geomembrane may be more difficult to place and seam.

13.2 Physical response of HDPE geomembranes

A geomembrane must be able to withstand the forces that develop during installation and over its subsequent service life and, in some applications, may be required to experience large strains

Table 13.2 Standard specifications for HDPE geomembranes as proposed by Hsuan and Koerner (1997), Hsuan (2000a) and GRI-GM13

Property	*ASTM method*	*Specification*
Density	D1505/D792	0.940 g/cm^3
Carbon black content	D1603	2–3%
Carbon black dispersion	D5596	A1
Oxidative induction time		
As manufactured		
Standard OIT	D3895	100 min
High pressure OIT	D5885	400 min
After oven ageing at 85 °C for after 90 days	D5721	
Standard OIT	D3895	55% retained
High pressure OIT	D5885	80% retained
After UV exposure		
High pressure OIT	D5885	50% retained after 1600 h exposure
Stress cracking resistance	D5397 App. A	200 h
Puncture resistance	D4833	Smooth: 320, 480, 640, 800 N[a] Textured: 267, 400, 534, 667 N[a]
Tear resistance	D1004	Smooth and textured: 125, 187, 249, 311 N[a]
Tensile properties	D638	
Strength at yield		Smooth and textured: 15, 22, 29, 37 kN/m[a]
Strength at break		Smooth: 27, 40, 53, 67 kN/m[a] Textured: 10, 16, 21, 26 kN/m[a]
Strain at yield		12%
Strain at break		Smooth: 700% Textured: 100%

[a]For geomembrane thickness equal to 1, 1.5, 2 and 2.5 mm, respectively.

without rupturing. Thus, proper assessment of the physical response of geomembranes is important to ensure the long-term environmental protection provided by the geomembrane. In this section, the physical response of HDPE materials to tensile stresses is examined.

Installation can give rise to tension in the geomembrane. For example, tensile stresses can be induced by: (a) installation of the geomembrane at a temperature higher than its expected service temperature, (b) uplift pressures caused by wind (Giroud and Bonaparte, 2001), or (c) placement of cover materials over the geomembrane on slopes. Tensions from the latter can be limited by placing cover materials from the toe of the slope towards the top of the slope.

Tension can also be induced from stresses and deformations arising from the weight of overlying materials. Predominantly this involves tension along side slopes and at anchoring locations at the top of slopes (Figure 13.1a). Qian *et al.* (2002) provide a good summary of the simple limit equilibrium calculations that are often used to estimate these tensile stresses. Alternatively, the solution developed by Kodikara (2000) that does not assume the slope is at the verge of failure can be used. Although anchoring of the geomembrane at the top of the slopes (either by trenches or run-out lengths) is necessary during construction, Bonaparte *et al.* (2002) recommend that the geomembrane be released from anchor trenches after construction to reduce the potential for damage to the geomembrane during earthquakes. It should also be noted that, in addition to geomembrane tensions, the stability of materials overlying the geomembrane must

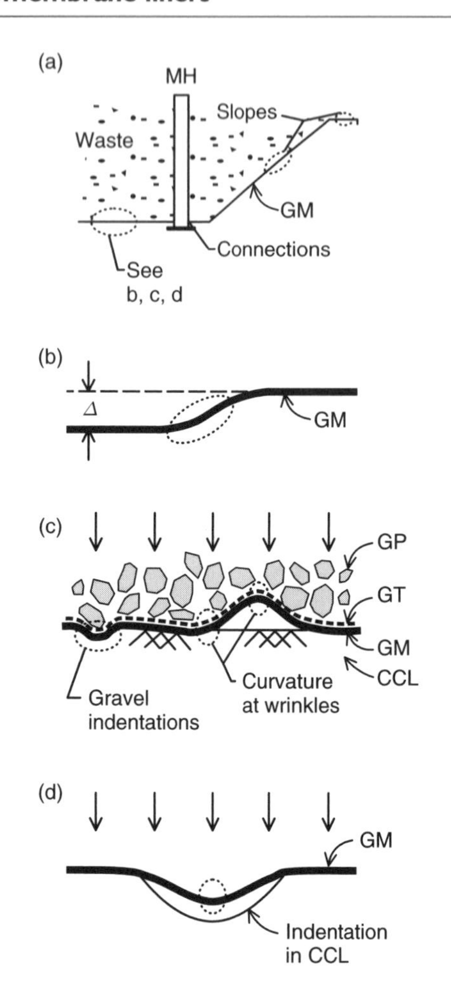

Figure 13.1 Illustration of possible sources of tension in geomembrane liners. Tension from: (a) side slopes and connections, (b) differential settlement, (c) gravel indentations and abrupt changes in curvature at wrinkles and (d) deformations into an indentation (e.g., tire track) (GM, geomembrane; GT, geotextile; GP, gravel; CCL, compacted clay liner; MH, manhole).

also be considered (e.g., sliding of materials on top of the geomembrane); this issue is discussed in Section 15.2.5.

Differential settlement of the material beneath the geomembrane (Figure 13.1b) may also introduce tension, and hence appropriate geotechnical design (see Section 15.2) is important to limit these tensions. Tension may also be induced by local deformations in geomembrane caused by indentations from overlying gravel and wrinkles in the geomembrane (Figure 13.1c). These issues are discussed further in Sections 13.2.3 and 13.2.4. Tension can occur at locations where the geomembrane spans an indentation in the underlying soil (e.g., a tire indentation, Figure 13.1d) and at connections between the geomembrane and structures (e.g., manholes) experiencing differential movements (Figure 13.1a). Giroud and Bonaparte (2001) provide equations for calculating the stresses and strains in the geomembrane for these cases.

13.2.1 Stress–strain behaviour

Polyethylene is a thermoplastic material with physical properties that depend on temperature; as the temperature increases, HDPE softens and has a lower strength. This thermoplastic behaviour also permits adjacent panels of HDPE geomembrane to be seamed by thermal fusion, where the geomembrane is heated and allowed to cool under pressure.

Thermoplastics also exhibit non-linear stress–strain behaviour at working strains that is influenced by time rate dependence at working temperatures. For example, the stress–strain response of HDPE depends on the rate at which the sample is tested, where in general a sample that is loaded faster will have a higher strength and stiffness than a sample of the same material loaded at a slower rate. Materials like HDPE also experience creep (i.e., the time-dependent increase in strain when the material is subject to a constant stress) and stress relaxation (i.e., the stresses in the material decrease under constant strain). The expected physical behaviour of HDPE in the field is difficult to quantify since loading rates are normally much lower than the slowest rates that can be reasonably simulated in the laboratory.

Uniaxial tension tests (ASTM D638) are useful for manufacturing quality control (Table 13.2) and as an index test in durability studies; however, their usefulness for obtaining strength and stiffness that can be used in design is limited because of the substantial "necking" of the sample that occurs prior to rupture. The effects of necking can be reduced by conducting wide strip tension tests (ASTM D4885). These tests conducted on a 200-mm wide specimen subject to uniaxial tension are normally used to assess strength and stiffness for geomembranes on side slopes.

Multiaxial tension tests (ASTM D5617) can also be conducted where a circular specimen of geomembrane is clamped around its perimeter and pressure is applied to one side while the other side is free to deform. The central section of the geomembrane specimen is subject to radially symmetric stresses. Results from multiaxial tension tests conducted on a 1-mm thick HDPE geomembrane by Merry and Bray (1997) are plotted in Figure 13.2. The large strains ($\geq 20\%$) experienced prior to rupture illustrate the ductile nature of the material response during short-term loading. These results also show the dependence of strength and stiffness for this particular geomembrane on strain rate. However, even the slowest strain rate

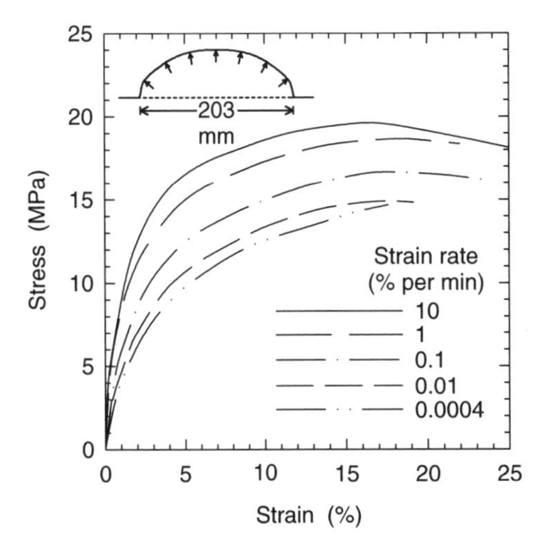

Figure 13.2 Calculated stress and strain in a 1.0-mm thick HDPE geomembrane from multiaxial tension tests reported by Merry and Bray (1997) conducted at various strain rates at a temperature of $21 \pm 1\,°C$.

reported in Figure 13.2 corresponds to failure in only 23 days and thus extrapolation is required to infer the material response over longer time frames.

13.2.2 Stress cracking

Although in many circumstances HDPE responds as a ductile material, HDPE is susceptible to a brittle failure mechanism referred to as stress cracking (or slow crack growth). Stress cracking occurs under sustained tensile stresses that may be much lower than the short-term strength. If the geomembrane cracks, it may no longer act as an effective contaminant barrier. Hence, to obtain long-term environmental protection, it is essential to minimize the potential for stress cracking.

Three conditions must exist for stress cracking to occur. The first requirement is a defect in the material which serves to initiate the crack. Defects may be induced by the seaming process, construction damage (e.g., scratches, punctures) and material flaws in the geomembrane. The second requirement is a microstructure that will allow the propagation of the crack. HDPE is a polymer that is particularly susceptible to this failure mechanism due to the semi-crystalline structure. Materials like VLDPE, which has a much lower crystallinity, are less susceptible to stress cracking as its structure retards crack propagation, although they are not as chemically resistant as HDPE. The third requirement is sustained tensile stresses. These stresses promote rapid crack propagation to the stage of failure. The magnitude and nature of the tensile stresses affect the extent of this failure.

Figure 13.3 illustrates the ductile to brittle transition for HDPE, plotting the tensile stress in the specimen (as a percentage of the short-term yield stress obtained from ASTM D638) versus the time to failure. At high stresses, the failure times are relatively short and failure occurs in a ductile manner. The failure times

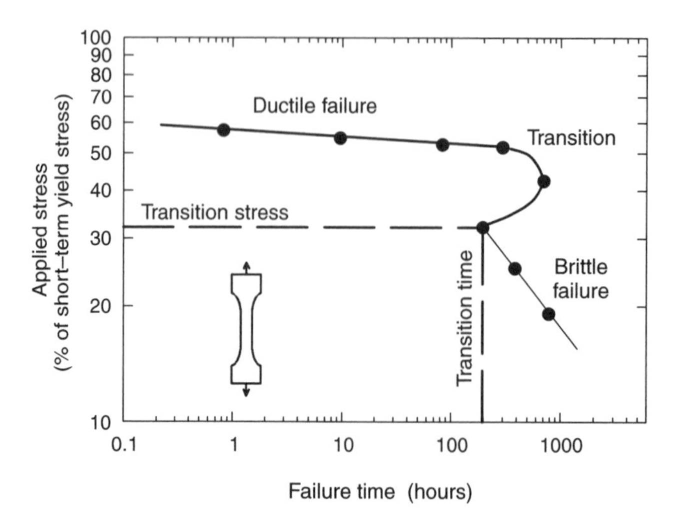

Figure 13.3 Illustration of experimental results from notched constant load tensile on specimens of HDPE geomembrane showing the transition from ductile to brittle failure (modified from Hsuan, 2000b).

are longer at lower stresses; however, these specimens fail in a brittle manner. The transition time between ductile and brittle failure depends on (Hsuan, 2000b) the crystallinity and molecular weight of the polymer, stress level, stress concentration factor, temperature and surrounding environment. Higher temperatures, sharper notches and/or chemical degradation will decrease the stress and time to the transition.

Stress crack resistance of a geomembrane can be evaluated by the notched constant tensile load (NCTL) test (ASTM D5397). In this test, a dogbone-shaped specimen of geomembrane is machined to contain a notch and is placed in a solution comprised of 10% Igepal and 90% water and at a temperature of 50 °C. The specimen is subject to a constant tensile stress and the time to failure is recorded. Hsuan *et al.* (1993) recommended that the time be greater than or equal to 100 h. As a simple manufacturer quality control test, Hsuan (2000b) recommended that geomembranes used as liners should not fail in 200 h in a single point notched constant load test (SP-NCLT; ASTM D5397, Appendix A) at 30% of the yield stress. These recommendations are based on chemically unaged specimens; the stress crack resistance may be expected to decrease with time as the geomembrane ages.

In summary, in order to reduce the potential of stress cracking it is important to: (a) have material with good stress crack resistance (as a minimum the geomembrane must meet the requirements listed in Table 13.2), (b) limit long-term tension in the geomembrane and (c) limit surface damage to geomembrane to the maximum practical extent (e.g., by providing appropriate geomembrane protection).

13.2.3 Geomembrane protection

Geomembranes can be damaged from contact with materials above or below either during construction or when subject to the weight of overlying materials. Use of coarse granular materials directly on top of the geomembrane (e.g., gravel in leachate collection systems, coarse crushed ore in heap leach pads) increases the likelihood of damage to the geomembrane from potentially large and irregularly spaced contact forces. Protection layers consisting of either sand, geotextiles, or other geosynthetic (i.e., rubber geomats)

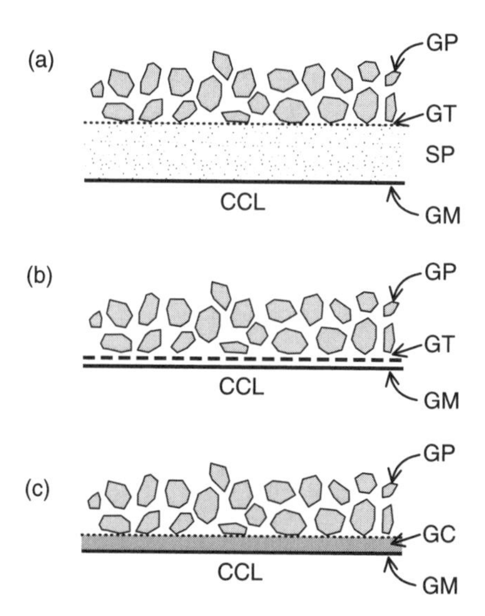

Figure 13.4 Examples of protection systems for geomembrane liners (GP, gravel; GT, geotextile; SP, sand; GM, geomembrane; GC, geocomposite; CCL, compacted clay liner).

and geocomposite materials are intended to limit the damage to the geomembrane (Figure 13.4). Protection from physical damage is required to: (a) minimize surface damage (e.g., scratches) to the geomembrane that, combined with a sufficiently large long-term tensile stress, may initiate stress cracking, (b) prevent puncture from coarse gravel particles that could result in substantial hydraulic transport through the geomembrane, and (c) limit the long-term stresses and strains in geomembrane (i.e., indentation may give rise to the long-term tensile stress required to cause stress cracking).

(a) Puncture resistance

For a given overburden pressure, the distribution and magnitude of contact forces reaching the geomembrane is related to: the size, particle gradation and particle shape of the overlying granular material; the thickness, stiffness, strength and durability of any protection layer; and, possibly, the particle size distribution and strength of the material beneath the geomembrane. The puncture resistance of a geomembrane depends on the stiffness, thickness, puncture and tensile characteristics of the geomembrane. Giroud *et al.* (1995) have developed a method of analysis for calculating geomembrane puncture resistance without a protection layer based on laboratory puncture resistance tests. Badu-Tweneboah *et al.* (1998) have developed an approach for evaluating the effectiveness of HDPE geomembrane liner protection. Reddy *et al.* (1996) evaluated the damage to a 1.5-mm HDPE geomembrane due to construction loads and concluded that "a geotextile as light as 270 g/m^2 ... completely protects the geomembrane from construction loading".

Wilson-Fahmy *et al.* (1996), Narejo *et al.* (1996) and Koerner *et al.* (1996b) examined the puncture protection of geomembranes provided by nonwoven, needle-punched geotextiles. Narejo *et al.* (1996) suggested that the mass per unit area μ (g/m^2) of a nonwoven needle-punched geotextile required to prevent puncture in a 1.5-mm thick HDPE geomembrane can be estimated using:

$$\mu = \frac{FS_{\mathrm{p}}\, p_{\mathrm{app}}\, H_{\mathrm{p}}^2}{450} MF_{\mathrm{S}}\, MF_{\mathrm{PD}}\, MF_{\mathrm{A}}\, FS_{\mathrm{CR}}\, FS_{\mathrm{CBD}} \tag{13.1}$$

where FS_{p} is the global factor of safety against puncture; p_{app} the vertical total stress acting on the geomembrane (in kPa); H_{p} the effective protrusion height (in mm, which may be taken as one-half of the maximum particle size of the gravel); MF_{S}, MF_{PD} and MF_{A} are modification factors for protrusion shape, packing density and soil "arching", respectively; and FS_{CR} and FS_{CBD} are partial factors of safety for creep and degradation, respectively. Modification and safety factors recommended by Narejo *et al.* (1996) and Koerner *et al.* (1996b) are summarized in Table 13.3. Factor MF_{A} was introduced to account for the stiffness of the material beneath the geomembrane. However, given the

Table 13.3 Modification and safety factors recommended by Narejo *et al.* (1996) and Koerner *et al.* (1996b) for use in equation 13.1 to obtain mass of geotextile (μ) to prevent puncture in a 1.5-mm thick HDPE geomembrane

Factor	*Value [–]*
Particle shape, MF_S	
Angular	1.0
Subrounded	0.5
Rounded	0.25
Packing density, MF_{PD}	
Isolated protrusions	1.0
Packed gravel	0.5
Soil "arching", MF_A	
None	1.0
Moderate	0.75
Maximum	0.5
Creep, FS_{CR}	
No geotextile	$\gg 1.5^a$
$\mu = 270\,g/m^2$	1.5^a, $>1.5^b$
$\mu = 550\,g/m^2$	1.2^a, 1.3^b, 1.5^c
$\mu = 1100\,g/m^2$	1.0^a, 1.1^b, 1.2^c, 1.3^d
Degradation, FS_{CBD}	1.5
Global factor of safety, FS_p^e	
Isolated protrusions	3^a, 4.5^b, 7^c, 10^d
Packed gravel ($H_p \leq 38$ mm)	3

[a, b, c, d]Protrusion height $H_p = 6$, 12, 25, 38 mm, respectively.
[e]Minimum recommended values.

uncertainty regarding the stresses actually acting around the protrusion in the tests of Narejo *et al.* (1996) arising from the effects of both the load being applied using a stiff plate and the friction between sand and the steel boundaries of their test cell, considerable caution is required if MF_A is chosen as less than 1.0. It should be noted that equation 13.1 is very sensitive to the effective protrusion height and this should be carefully considered in the context of the proposed grading curve for the drainage material to be used above the geomembrane.

To illustrate the use of equation 13.1, consider an example of a 1.5-mm HDPE geomembrane at the base of a landfill where the applied pressure acting above the geomembrane, p_{app}, is equal to 130 kPa. A leachate collection system consisting of angular coarse gravel is to be placed on top of the geomembrane. The maximum particle size of the overlying gravel is 76 mm, thus giving an effective protrusion height of $H_p = 38$ mm. From Table 13.3, with $MF_S = 1.0$ for angular particles, $MF_{PD} = 0.5$ with packed gravel (as opposed to isolated protrusions), $MF_A = 1$ since there will be no reduction from arching, $FS_{CR} = 1.3$, and $FS_{CBD} = 1.5$ would require a nonwoven needle-punched geotextile with mass per unit area greater than $850\,g/m^2$ to provide a global factor of safety of 3 in equation 13.1. However, Narejo *et al.* (1996) recommended that a geotextile no lighter than $1100\,g/m^2$ be used where the effective protrusion height is 38 mm. Thus in this case, a geotextile with mass per unit area of at least $1100\,g/m^2$ is required to prevent puncture, according to the design procedure of Narejo *et al.* (1996).

It must be recognized, however, that equation 13.1 was developed based on short-term puncture data and although there are factors of safety that may account for some long-term effects, it does not explicitly limit the deformations and thereby the long-term tensions in the geomembrane. This appears to be an important consideration in ensuring adequate long-term performance as a contaminant barrier, and more research is required to address this issue.

When using nonwoven needle-punched geotextiles as a geomembrane protection layer, the project specifications should require the geotextile to be free from needles that may be broken during manufacture, which could lead to a hole in the geomembrane.

(b) Limiting long-term strains

In addition to designing against puncture of the geomembrane, the long-term tensile stresses and strains in the geomembrane may need to be limited to low levels to reduce the potential for stress cracking. Brummermann *et al.* (1995), Saathoff and Sehrbrock (1995), Bishop (1996) and Seeger and Müller (1996) recommend minimizing the contribution of indentation due to coarse gravel under long-term loading condition to a very low (~0.25%) strain level.

Given both the unknown and highly variable contact forces from the gravel and the time-dependent constitutive response of most geomembrane materials, quantification of the long-term tensile stresses or strains in geomembranes is a complex task and has not yet been studied. Strains for a particular combination of gravel, protection layer, geomembrane and stress levels to be used for a given project are presently estimated from measured deformations of the geomembrane during relatively short-term laboratory tests. Deformations of the geomembrane are normally obtained by measuring the permanent indentations in a soft lead sheet placed beneath the geomembrane (BAM, 1995; Zanzinger, 1999; Gallagher *et al.*, 1999; Tognon *et al.*, 2000). However, considerable care is required in estimating geomembrane strains from the indentations made in the lead sheet. Significant errors can be introduced by the choice of method of calculating the strains from the measured deformations. The average longitudinal strain in the geomembrane is normally calculated from the measured deformations in the soft lead sheet using arch elongation theory (BAM, 1995). This approach assumes that the strain is uniform along the deformed profile and neglects bending strains. Tognon *et al.* (2000) found that strains estimated using arch elongation, in addition to being highly dependent on the selection of chord length and depth and hence subjective, may underestimate the strains by up to an order of magnitude. Consideration of both membrane and bending strains using the method of Tognon *et al.* (2000) is important for the calculation of strains in the geomembrane from measured deformations.

Results from a series of large-scale laboratory tests conducted to assess the effectiveness of different protection layers have been reported by Zanzinger (1999) and Tognon *et al.* (2000). The results of Zanzinger (1999) likely underestimate the strains in the geomembrane since arch elongation was used to calculate strains from the measured deflections. The tests of Tognon *et al.* (2000) were conducted in a 2-m wide, 2-m long and 1.6-m deep apparatus where vertical pressures were applied using an air bladder. A cross-section of the test apparatus is shown in Figure 15.8. The 1.5-mm thick HDPE geomembrane was underlain by 200 mm of compacted clay. A 600-mm thick layer of nominal 50 mm angular coarse gravel was placed over the protection layer and geomembrane. Gravel indentations into the geomembrane were quantified by measuring the permanent deformation of a soft lead sheet located beneath the geomembrane. Peak strains in the geomembrane were estimated from the measured indentations considering both membrane and bending strains. The tests were conducted with vertical pressures between 250 and 900 kPa, at room temperature (24 ± 1°C), and for relatively short times (200–720 min). The results are summarized in Table 13.4.

The smallest strains were obtained with the sand-filled cushion and grid-reinforced rubber and provided far superior protection than the nonwoven GT layers examined. For the two rubber protection layers, the one with larger tensile stiffness Geomat 2 (GT-rubber-GG) resulted in much smaller geomembrane strains relative to Geomat 1 (GT-rubber-GT). This indicates that the tensile stiffness of the protection layer may be important in reducing strain in the geomembrane. Thus, while mass per unit area (μ) may be an indirect indicator of tensile stiffness for a particular type of nonwoven geotextile, it should not be used as the sole measure of suitability of a geosynthetic protection layer.

The largest strains measured by Tognon *et al.* (2000) were obtained when using nonwoven geotextiles. The maximum strains of 13% calculated for 1,200 g/m^2 geotextile protection at a vertical pressure of 900 kPa were very close to the short-term yield strain. The geomembrane, however, did not puncture during these short-term tests in which the factor of safety against puncture was estimated to be about 1.3 using

Table 13.4 Measured maximum indentations and calculated peak strains for 1.5-mm thick HDPE geomembrane from nominal 50-mm coarse gravel backfill during short-term laboratory tests with different protection layers (data from Tognon *et al.*, 2000)

Protection layer	*Vertical pressure* (kPa)	*Test duration* (min)	*Number of indentations* (no./m^2)	*Maximum indentation* (mm)	*Peak strain* (%)
One layer of GT1[a]	250	200	350, 331[f]	5.05, 5.77[f]	8.0
Two layers of GT2[b]	900	720	338	7.63	13.0
Sand-filled geocushion[c]	650	520	69	3.8	0.8
Sand-filled geocushion[c]	900	720	78	2.86	0.9
Rubber geomat 1[d]	600	480	156	3.34	7.5
Rubber geomat 2[e]	600	480	38	1.70	1.2

[a] Nonwoven, needle-punched geotextile, mass per unit area $\mu = 435\,g/m^2$.
[b] Nonwoven, needle-punched geotextile, each layer $\mu = 600\,g/m^2$ (total $\mu = 1200\,g/m^2$).
[c] Depomat: polymer webbing filled with sand; upper non-woven, needle-punched geotextile, $\mu = 600\,g/m^2$; lower nonwoven, needle-punched geotextile, $\mu = 130\,g/m^2$.
[d] Rubber matrix $\mu = 5800\,g/m^2$; upper and lower nonwoven needle-punched geotextile with $\mu = 600\,g/m^2$.
[e] Rubber matrix $\mu = 5800\,g/m^2$; upper nonwoven needle-punched geotextile with $\mu = 600\,g/m^2$; lower polyester grid.
[f] Results from duplicate tests.

equation 13.1. This is consistent with what one would expect based on equation 13.1 (i.e., large strains for a factor of safety approaching 1) and highlights the need for an adequate factor of safety (Narejo *et al.* (1996) recommend a factor of safety of at least 3). While the use of equation 13.1 and Table 13.3 together with the factors of safety recommended by Narejo *et al.* (1996) appears to be adequate for controlling short-term puncture, it does not explicitly control the strains in the geomembrane and hence there is still some question regarding the level of protection required to ensure the long-term performance of the geomembrane.

There is some debate about the relatively arbitrary choice of 0.25% as a limiting strain due to indentations from overlying gravel (see Bishop, 1996). Furthermore, the allowable strain cannot be considered in isolation but, rather, must be considered in the context of the method of measurement used to assess whether this objective has been met for a given form of geomembrane protection. For example, Giroud (1996b) has quoted a case where the measured strain in a geomembrane that had been damaged, but not punctured, by stones was approximately 0.3% using a 5-mm scanning grid and an order of magnitude larger (3%) using a finer 0.5-mm scanning grid. Rather than relying on the results from scanning over a set grid, it is probably best to use the scan to find the deepest indentations, measure these indentations more carefully and then use combined membrane and bending action to estimate the strains in the geomembrane.

Thus, in addition to providing puncture resistance, it is essential to limit the strains due to gravel indentations to ensure the long-term performance of the geomembrane. The need to limit indentation strains to as low as 0.25% has not been firmly established; however, it is certainly desirable to keep the strain low, especially for geomembranes that require a long service life.

13.2.4 Wrinkles in geomembranes

Wrinkles, also sometimes called waves (see Figure 13.1c), may form in geomembranes during installation from material expansion by heating from the sun and by improper placement of overlying materials. The presence of wrinkles in the geomembrane raises two major concerns.

First, there is increased potential for contaminant migration through a hole in the geomembrane at or near the wrinkle (see Section 13.4.3). Second, there is increased potential for stress cracking due to the tensile strains that are induced by the wrinkle (Figure 13.1c), and these may be compounded by other tensile strains (e.g., from coarse gravel backfill).

This suggests the need for greater attention to reducing/avoiding wrinkles, possibly following the lines advocated in Germany (Averesch and Schicketanz, 1998). However, care is required to ensure that the elimination of wrinkles is not at the expense of inducing significant tensile strains in the geomembrane that could ultimately contribute to long-term stress cracking, particularly at the toe of slopes. One may expect that in this situation, stress relaxation will play an important role but the long-term stress in a geomembrane subject to tensile strain remains an open question and hence it would be prudent to minimize the tensile strains.

(a) Wrinkle formation, size and spacing

Giroud and Morel (1992) and Giroud (1995) showed that in addition to temperature, colour and the coefficient of thermal expansion, factors such as roughness and flexibility of the geomembrane can also influence the size of wrinkles. White geomembranes have been shown to lessen, but not eliminate, the problem (Koerner and Koerner, 1995a). Giroud and Peggs (1990) emphasized that good construction quality control is required to minimize geomembrane wrinkles. They reported that a temperature increase in the geomembrane of 50 °C could cause a wrinkle height of about 0.1 m for HDPE geomembranes and of about 0.01 m for PVC geomembranes.

Figure 13.5 is a sketch showing the distribution of wrinkles that developed on the base of a landfill cell during construction. The wrinkles were randomly distributed with no discernable patterns, thus making it challenging to quantify the length and spacing of wrinkles in the field. Wrinkles developed at locations parallel to the geomembrane seams as well at locations both perpendicular and inclined to the seams. Where wrinkles could be quantified from an aerial photograph, there were at least 81 wrinkles in the approximately 630 m^2 region marked on Figure 13.5. If extrapolated, this corresponds to over 1,200 wrinkles per hectare! The longest wrinkle was at least 17 m and may extend the

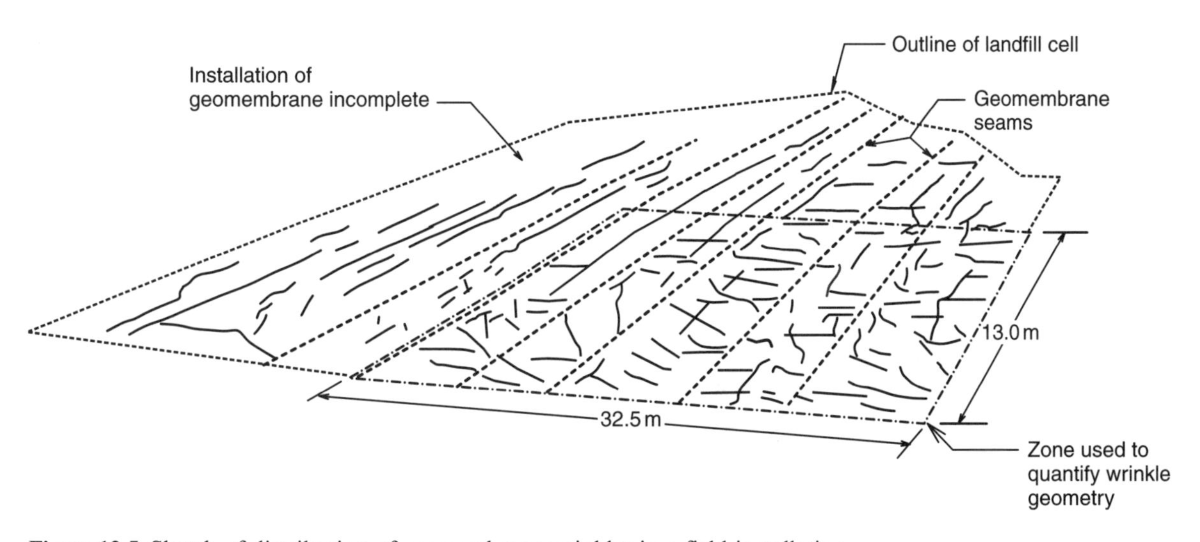

Figure 13.5 Sketch of distribution of geomembrane wrinkles in a field installation.

entire length of the cell (≈40 m). The shortest wrinkle was roughly 1 m long and the minimum spacing between wrinkles was about 0.5 m. The wrinkles inclined to the seams appear to be connected to the long wrinkles at each seam. Such a distribution of wrinkles would provide an extensive preferential pathway for liquid migration if there were any holes in the geomembrane (see case reported in Section 9.5).

Pelte *et al.* (1994) reported field observations of wrinkles in a 1.5-mm thick black HDPE geomembrane overlying clay in a 30-m by 30-m cell in a landfill in France. They observed that major wrinkles occurred parallel to the length of the geomembrane roll at the location of seams, and also perpendicular to the seam direction. They reported large wrinkles between 0.05–0.1 m high and 0.2–0.3 m wide, and had a spacing of 4–5 m and appeared to extend across a significant width of the cell. They also noticed small wrinkles (less than 0.05 m high and 0.2 m wide) appeared perpendicular to the seams.

Touze-Foltz *et al.* (2001a) used photogrammetric techniques to quantify field wrinkling of a 2-mm thick HDPE geomembrane over compacted clay. Wrinkles were also observed in two perpendicular directions. Touze-Foltz *et al.* (2001a) reported that wrinkle height varied between 0.05 and 0.13 m, wrinkle widths between 0.1 and 0.8 m, spacing between wrinkles ranged from 0.3 to 1.6 m, while the length of wrinkles was less than 4 m with most wrinkles 1–2 m long. The wrinkle height is similar, but wrinkle spacings are closer and wrinkle lengths are much shorter than those reported by Pelte *et al.* (1994). The limited size of the geomembrane (only 7.5 m by 7.5 m) examined by Touze-Foltz *et al.* (2001a) and the fact that it was anchored by sand berms around the perimeter most likely limited the length of wrinkle that could form. Consequently, these results are not representative for larger installations. The technique of Touze-Foltz *et al.* (2001a) represents a very useful way of quantifying wrinkles at larger sites since wrinkle width, length and spacing are important parameters needed to estimate leakage through geomembranes as discussed in Section 13.4.

(b) Physical response of wrinkles under overburden pressures

Some have thought that wrinkles will "go away" when subject to vertical pressures once the waste is placed; however, findings by Stone (1984), Soong and Koerner (1998) and Brachman and Gudina (2002) cast doubt on that assumption and suggest that wrinkles may exist under even high pressure. Soong and Koerner (1998) conducted a series of laboratory tests to quantify the effect of applied pressure, wrinkle height, geomembrane thickness and temperature on the physical response of HDPE geomembrane wrinkles with sand above and below the geomembrane. Their laboratory tests showed that although there was a reduction in height of the wrinkle, the wrinkles remained even with vertical stresses as high as 1.1 MPa. In each of their tests intimate contact between the geomembrane and the subgrade was not achieved even for the smallest initial wrinkle height studied (14 mm high) at a vertical pressure of 700 kPa. However, Soong and Koerner's results need to be carefully interpreted since it is possible that boundary effects from friction may have reduced the stress acting around the wrinkle in their tests.

Brachman and Gudina (2002) developed a 600-mm diameter test cell to study the physical response of composite liner systems (Figure 13.6). Multiple layers of lubricated plastic to reduce boundary friction to negligible (less than 5%) levels and stress-free axial boundary conditions provide a good simulation of the physical conditions expected for a geomembrane wrinkle in a landfill. Experiments conducted with sand backfill above and below the geomembrane showed that the wrinkles in all tests substantially reduced in size; however, out of the initial wrinkle height of about 65 mm, about 25-mm high wrinkles still remained even under a vertical pressure of 3 MPa. A substantial portion of the reduction in original wrinkle height took place at early stages of the tests (i.e., at the lower stress

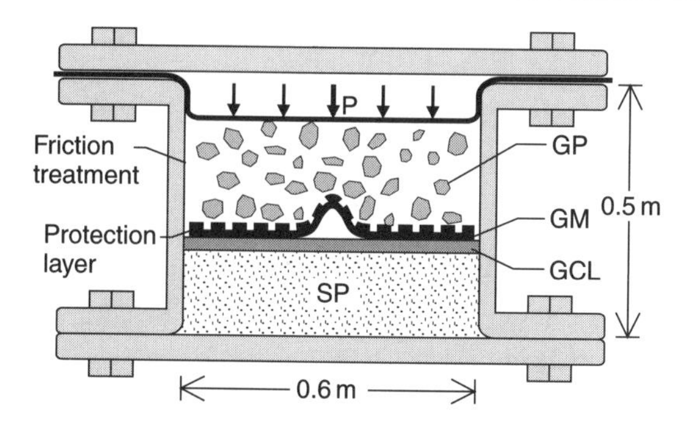

Figure 13.6 Experimental apparatus for measuring the physical response of composite liners (modified from Brachman and Gudina, 2002) (GP, gravel; GM, geomembrane; GCL, geosynthetic clay liner; SP, sand; p, applied pressure).

levels). Stiffening of sand under increased pressures as well as possible arching of earth pressures around the wrinkles may have resulted in only little additional wrinkle deflection as the applied pressure exceeded about 300–400 kPa.

Experiments have also been conducted for the more realistic case of 50 mm coarse gravel backfill on top of a 1.5-mm thick HDPE geomembrane with the apparatus developed by Brachman and Gudina (2002). The deformed shape of the wrinkle following application of 1,000 kPa (for 10 h) is plotted in Figure 13.7 for a test with sand above and below the geomembrane (SP) and a test with 50-mm gravel above and a GCL beneath the geomembrane (GP1 and GP2). The wrinkle had the same initial height of 70 mm and width of 200 mm for each test. With gravel backfill there is less vertical compression but more horizontal compression of the wrinkle relative to sand backfill. The wrinkle deformations are not symmetric with gravel backfill and vary along the wrinkle length from contact with the coarse gravel particles. At one section (GP1) pinching was measured at the top of the wrinkle, while at another section (GP2) the wrinkle flattened at the top. Clearly, much larger tensions can occur in the geomembrane with GP backfill relative to the experiments conducted with sand backfill.

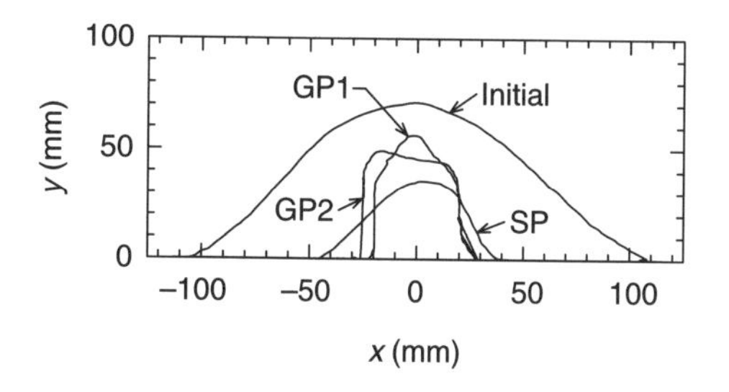

Figure 13.7 Measured wrinkle geometry in a 1.5-mm HDPE geomembrane before and after application of 1,000 kPa vertical pressure for 10 h. Results shown for sand above and below the geomembrane (SP) and 50-mm gravel above and a GCL beneath the geomembrane (GP1 and GP2).

Where there are wrinkles in a geomembrane over a GCL, there is potential for the bentonite to "extrude" laterally and establish close contact between the geomembrane and clay provided there is sufficient applied compressive stress (Giroud, 1997) and the wrinkle is small. However, this also creates potential thinning of the GCL (Stark, 1998) which could impact on the GCL performance if there was significant bentonite movement. Also, for large wrinkles there may be: (a) significant thinning of the GCL adjacent to the wrinkle and (b) failure to fill the gap between the GCL and geomembrane. Thus, minimization of wrinkles can be expected to improve the long-term performance of composite liners.

13.3 Durability of HDPE geomembranes

Even a well-designed and properly constructed geomembrane may be expected to experience some degradation or ageing over its lifetime. Eventually this degradation can lead to failure of the geomembrane. For a geomembrane liner used as part of a barrier at the base of a landfill, failure is said to have occurred when the geomembrane no longer acts as an effective hydraulic or diffusive barrier to contaminant migration. One possible failure mechanism is the deterioration or dissolution of the polymer when exposed to permeants. This extreme case is not expected for municipal solid waste landfills since the concentrations within the leachate are not sufficiently large to cause dissolution of HDPE geomembranes typically selected for this application because of its chemical stability with respect to landfill leachate. However, chemical compatibility does need to be considered when the geomembrane is expected to contain concentrated contaminants (e.g., hydrocarbons). Compatibility can be assessed by testing using the candidate geomembrane and the site-specific permeant. In another and more likely case, chemical ageing of the polymer with time may cause a decrease in stress crack resistance. Combined with tensile stresses this degradation may cause the geomembrane to crack and, eventually, to cease to be an effective contaminant barrier. The time required to cause failure of the geomembrane represents the service life of the geomembrane. The following sections describe the factors influencing the durability and service life of geomembrane liners and present current best estimates of the service life for HDPE geomembranes as barrier at the base of a landfill. The implications of a finite service life of a geomembrane on the contaminant migration from a landfill are examined in Section 16.6.

13.3.1 Factors influencing durability

The rate of degradation of a geomembrane liner in a landfill depends on the geomembrane properties including its thickness and the properties of the polymer such as the type, formulation, branching, tertiary hydrogen, crystallinity and additives (carbon black, antioxidants). It also depends on the exposure conditions including the temperature, surrounding medium (whether soil, fluid or gas), chemical concentration and applied mechanical stresses (particularly long-term tensile stresses). Rowe and Sangam (2002) have reviewed how these factors influence the durability of HDPE geomembranes.

The degradation of geomembranes consists of simultaneous physical and chemical ageing (Hsuan and Koerner, 1995) and will result in a reduction in the engineering properties of the geomembrane. Physical ageing involves a change in the crystallinity of the material with no breaking of covalent bonds. Chemical ageing involves degradation where there is breaking of covalent bonds.

Potential chemical degradation mechanisms for geomembranes include (Koerner *et al.*, 1990; Rowe and Sangam, 2002) swelling, extraction, biological degradation, photodegradation and oxidation. Swelling involves the increase in volume of the geomembrane by sorption of liquids (including leachate). Swelling is generally not a concern for HDPE geomembranes liners used to contain aqueous contaminants. Extraction involves the removal of some component of the plastic under chemical exposure. For HDPE geomembranes there can be extraction of protective antioxidants, increasing the susceptibility to oxidative degradation. The very high molecular weights of the resins used to produce typical geomembranes make the geomembrane a very poor substrate and hence biological degradation of geomembranes in a landfill is unlikely. Photodegradation may initiate oxidative degradation when exposed to visible or ultraviolet light, causing discolouration, surface cracks, brittleness and the deterioration of mechanical properties of the geomembrane. For geomembranes to be used at the base of the landfill, photodegradation is only a potential concern during construction. Carbon black (typically 2–3%) added to HDPE geomembranes to block and adsorb the light energy during periods of exposure to the sun, and the timely placement of a soil cover (e.g., protection system and leachate collection gravel), at least 0.15 m thick, reduce the likelihood of photodegradation during construction.

For HDPE geomembrane liners, oxidative degradation is the principal mechanism of chemical ageing. A schematic diagram illustrating oxidative degradation is shown in Figure 13.8.

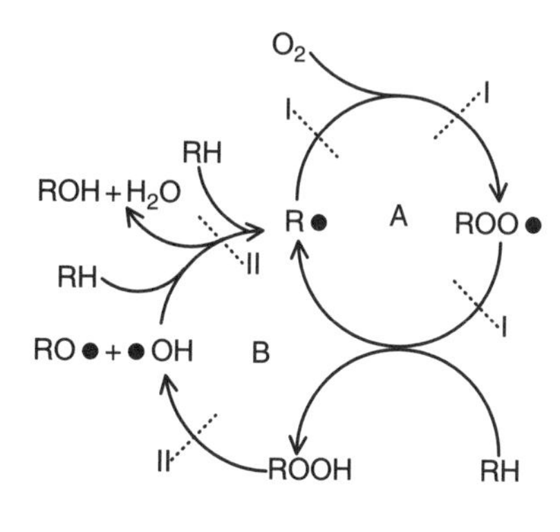

Figure 13.8 Polyethylene oxidation cycles A and B (modified from Grassie and Scott, 1985).

The initiation of oxidation in HDPE involves the formation of free radicals, *R*●, and requires an energy source such as heat or ultraviolet light. A free radical reacts with oxygen producing a hydroperoxy radical ROO● which can then react with another polymer chain RH to produce hydroperoxide ROOH and another free radical, denoted by cycle A in Figure 13.8 (Grassie and Scott, 1985). Oxidation propagates by cycle B as reactions involving the homolysis of hydroperoxide continue to degrade the polymer chain and form new free radicals. If oxidation is allowed to propagate, chain scission (i.e., breaking of bonds on the backbone of the polymer chain) leads to a decrease in molecular weight, making it more prone to stress cracking (Tisinger and Giroud, 1993). If any of the interactions between the two cycles are broken, oxidation can be retarded and even halted if all of the links are impeded. Antioxidants and stabilizers are added to the polymer to retard the oxidation of the polymer chain.

Two key issues with respect to the chemical ageing of HDPE geomembrane liners are: (a) the amount of antioxidant and the rate of consumption and/or removal (e.g., by diffusion or leaching) from the geomembrane and (b) the availability of oxygen. The availability of antioxidant in the geomembrane can be represented in terms of the oxidative induction time (OIT) which is a useful index for predicting the potential service life of a geomembrane, as discussed in the next section. Hsuan and Koerner (1997) and Hsuan (2000b) recommend specification of geomembranes that have a minimum OIT value as given in Table 13.2. The service life will also depend on the availability of oxygen. In this respect, the primary liner at the bottom of a landfill has relatively little exposure to oxygen since conditions at the base of a landfill are generally anaerobic except for very early in the operating period. There may be more persistent oxygen in the secondary leachate collection system (and hence above a secondary geomembrane) because of its separation from the biological processes ongoing in the landfill and the potential for air exchange due to atmospheric pumping resulting from changes in barometric pressure. Hsuan and Koerner (1997) note that the oxidation reaction of polyethylene can be increased in the presence of transition metals (e.g., Mn, Fe, Co, Ni, Cu, Zn and Pd). Since transition metals are usually present in leachate and oxygen is available, albeit in small quantities, in the air-filled voids of underlying unsaturated geotextiles (e.g., forming part of a GCL), soil and in wrinkles and other irregularities of the geomembrane-underlying soil interface, there is potential for oxidation to occur in the primary geomembrane even though mainly anaerobic conditions exist above the liner.

13.3.2 Existing field performance data

The relatively short history of geomembrane use in liquid or waste containment applications makes it rather difficult to predict a definitive service life for geomembranes. Schmidt *et al.* (1984) performed a series of index tests (i.e., to establish yield, elongation, break, and tear properties) on samples of polyethylene geomembranes from exposed and submerged geomembranes that had been in use for up to 16 years as canal/pond liners. The test results were compared with the original geomembrane material properties specified at the time of the geomembrane installation. They concluded that the geomembrane stiffened with time, but that this was less significant for buried (unexposed) material. The major causes of failure to the liners was from physical/mechanical damage, rather than weathering/ageing effects. Only minor changes in physical performance were observed when polyethylene geomembranes were buried (i.e., protected from UV radiation) over the observation period of 16 years.

Brady *et al.* (1994) examined the behaviour of HDPE in different environments over a period of 30 years. No substantial changes in density, water adsorption value and water extractable matter content were found between unaged and 30-year-old specimens. There was no significant change in impact resistance after 15.5 years; however, a reduction of about 50% was observed over a 30-year-period. The data from tensile tests showed that over the 30-year-period the tensile strength remained essentially constant but there was a reduction in the strain at the peak (yield) strength and the HDPE became stiffer and more brittle with time.

Hsuan *et al.* (1991) obtained samples of 1.5-mm thick HDPE geomembrane from a 7-year-old domestic solid waste leachate surface impoundment. The samples were obtained from four different locations in the lagoons, ranging from areas continuously and directly exposed to the atmosphere to those at the bottom of the impoundment, continuously and directly exposed to liquid. They concluded that: (1) there was no substantial macroscopic change in the geomembrane sheets or seams after 7 years exposure at the site; (2) an evaluation of the stress crack resistance of the materials indicated that constant outdoor exposure had not caused substantial changes in the internal structure of the material; and (3) changes in the geomembrane were observed only on the molecular level and they did not affect the engineering/hydraulic containment properties of the geomembranes.

Rollin *et al.* (1994) reported results obtained for samples of HDPE geomembrane that had been used to contain contaminated soil (including cyanide, phenolic compounds and heavy hydrocarbons) for 7 years. Ageing was found to increase the yield strength, decrease the tensile force required for rupture and lower the elongation at rupture. Ageing was more severe for samples from the bottom of the cells than for samples from the slopes or cover.

Eith and Koerner (1997) described a case in which an HDPE geomembrane was used as part of a double liner system for a landfill. During the 8 years of service, the primary geomembrane had been exposed to various concentrations of leachate constituents. The physical, mechanical and endurance test results indicated no apparent degradation of the HDPE geomembrane properties since they were still within the range of data generated for the original material at the time of installation.

Rowe *et al.* (1998, 2003a) examined an unprotected geomembrane after 14 years of exposure in a leachate lagoon liner (see Section 9.5 for more detail). For the geomembrane exposed to air and sun, very low standard OIT values were accompanied by low stress crack resistance (based on single point notched stress crack tests). The results of the melt flow index tests suggest that the degradation was induced by a chain scission reaction in the polymers. The geomembrane was severely cracked confirming that the material was highly susceptible to stress cracking. For geomembranes that were either covered by soil or leachate, the depletion of antioxidant was slower than for the UV exposed and partially UV exposed geomembranes. The amount of antioxidant present in these geomembranes seems to have been sufficient to protect the geomembrane from oxidation degradation over the 14-year exposure period. These results substantiate the importance of: (a) providing adequate geomembrane protection; (b) appropriate care during maintenance; and (c) the type and amount of antioxidant and stress crack resistance, to the longevity of the geomembrane. This case also suggests that OIT and stress crack resistance should be incorporated in the specification of an HDPE geomembrane and that they should be evaluated as part of the CQC/CQA procedures.

The observations from these field investigations suggest that there is a change in properties of geomembranes with time. However, given the short time frames involved (maximum 14 years for a geomembrane exposed to leachate) relative to the required lifetime of a geomembrane (likely much greater than 100 years) data from these limited field cases are insufficient for estimating the service life for geomembrane liners.

13.3.3 Predictions of service life

Several opinions have been provided regarding the service life of geomembrane liners for landfills. For example, a US Environmental Protection Agency (US EPA) ad hoc committee concluded (subject to numerous qualifications) that the service life of geomembranes was likely to be in the order of hundreds of years (Haxo and Haxo, 1988). Tisinger and Giroud (1993) echoed this opinion and stated "that in properly designed and constructed facilities, HDPE geomembranes should be able to protect groundwater from leachate for hundreds of years".

While the most reliable method of determining the service life of geomembranes would be from exposure under the actual field conditions, this is not presently feasible due to the length of time that would be required to obtain useful results. Consequently, several "accelerated ageing" tests using elevated temperatures have been developed which attempt to simulate long-term exposure of HDPE. The test results may then be extrapolated to a value representing the expected service conditions. Considerable care must be exercised such that the fundamental mechanisms being evaluated (e.g., oxidation, stress cracking) are not altered by the elevated exposure.

Accelerated ageing tests have been used for decades to quantify the service life of pressurized plastic pipes (e.g., natural gas pipelines). Hydrostatic burst tests are conducted on HDPE pipes (with or without notches) subject to an internal pressure (to induce tensile circumferential stresses in the pipe) and elevated temperatures. The time to failure is recorded for various internal pressures to detect the ductile to brittle transition in HDPE. Koch *et al.* (1988) reported that the application of the Arrhenius extrapolation technique to predict the service life of HDPE has been confirmed with over 30 years of testing on pressurized pipes. Although the stress levels are expected to be higher for pipes than geomembranes, Koch *et al.* (1988) indicated that the lifetime curves for pipes are applicable to geomembrane sheets with appropriate stress correction factors. Although the tests of Koch *et al.* (1988) were for samples submerged in water and not leachate, they state that the service life of HDPE geomembranes could be expected to be considerably greater than 100 years.

13.3.4 Predictions of service life using antioxidant depletion tests

The exposure conditions of a geomembrane may be expected to be different depending on whether it is part of a composite primary or secondary liner. As a component of the primary liner, the geomembrane is subject to direct contact with leachate and temperatures higher than normal groundwater temperatures. As part of a secondary liner, it has less contact with leachate than the primary liner and the leachate reaching the secondary geomembrane will likely have experienced some attenuation due to cation exchange, precipitation and biodegradation as it migrates through the primary clay liner. The secondary geomembrane is also likely to be closer to natural groundwater temperature. However, depending on the design and operation of the secondary leachate collection system, there is potential for significant exposure to atmospheric oxygen.

Since oxidation is the primary degradation mechanism that may limit the service life of HDPE geomembranes, and oxidation of the polymer is retarded by the presence of antioxidants, the time to complete chemical ageing of an HDPE geomembrane may be considered as the sum of (Hsuan and Koerner, 1998): (a) the time to deplete the antioxidants by consumption and/or extraction, (b) the induction time to the onset of polymer degradation and (c) the time for degradation of the polymer to decrease some property (or properties) to an arbitrary level (e.g., to 50% of the original value). If the times for these three stages as illustrated in Figure 13.9 can be measured or estimated, an estimate of the service life of a geomembrane can then be made for a given temperature and exposure conditions.

Sangam and Rowe (2002) reported the depletion times of antioxidants in samples of HDPE geomembranes in terms of the OIT. Tests were conduced at temperatures between 22 and 85 °C, and predictions were extrapolated to lower temperatures using Arrhenius modelling. A summary of extrapolated times for antioxidant depletion at a number of temperatures relevant to landfill liner design is presented in Table 13.5. The depletion times were longest for geomembranes in air and shortest for samples immersed in leachate, and all decreased with increasing temperatures. For example, the extrapolated antioxidant depletion time for a geomembrane immersed in leachate at 33 °C was estimated to be only 12 years. Since the geomembrane may be exposed to different conditions on either side, estimates of depletion times with leachate on one side and air, water and unsaturated soil on the other are also given in Table 13.5.

The results of Sangam and Rowe (2002) can be used to estimate the service lives of primary and secondary HDPE geomembranes under various exposure condition scenarios to which they may be subjected when used as bottom liners for MSW landfills. Referring back to Figure 13.9, the service life can be estimated using the antioxidant depletion times t_1 inferred from accelerated immersion tests (Sangam and Rowe, 2002), the induction time t_2 reported by Viebke

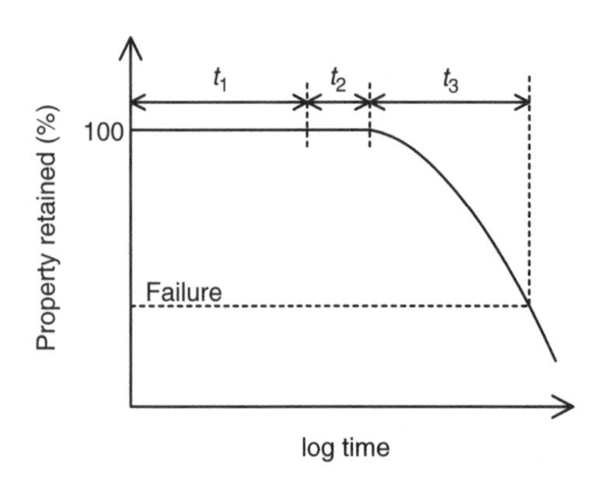

Figure 13.9 Illustration of the three conceptual stages of chemical ageing of HDPE geomembranes (modified from Hsuan and Koerner, 1998).

Table 13.5 Estimated antioxidant depletion time, t_1, (years) for a primary HDPE geomembrane based on immersion tests conducted in air, water and leachate by Sangam and Rowe (2002)

	Temperature (°C)				
Exposure medium	*13*	*15*	*20*	*25*	*33*
Air	390	330	230	160	90
Water	190	160	120	80	44
Leachate	40	36	26	22	12
Leachate and air[a]	210	180	130	90	50
Leachate and water[b]	110	100	70	50	28
"Unsaturated soil" [c]	290	250	180	120	70
Leachate and "unsaturated soil" [d]	160	140	100	70	40

[a]Average of exposure for air and leachate.
[b]Average of exposure for water and leachate.
[c]Average of exposure for water and air. Does not account for presence of soil particles.
[d]Average of exposure for leachate and unsaturated soil.

et al. (1994) for an unstabilized HDPE and an assumed degradation time t_3 of 25 years (Rowe, 1998a).

Current best estimates of service life for primary and secondary geomembrane liners are presented in Tables 13.6 and 13.7, respectively. The service life for a primary HDPE geomembrane liner is estimated to be greater than 200 years provided that the landfill is well maintained such that the temperature does not reach 15 °C for any significant period of time. The service life is estimated to be reduced to 70 years if the temperature at the base of the landfill is at 33 °C (note that this temperature is quite possible – see Section 2.6). For groundwater temperatures ranging between 7 and 10 °C, it is estimated that a secondary geomembrane should last about 400 years. All of these service life estimates assume that the geomembrane meets the specifications of Table 13.2 (and, in particular, has a suitable antioxidant package), is not subjected to significant tensile stress and is covered by an adequate protection layer, and that the degradation rates at elevated temperatures may be used to extrapolate antioxidant consumption at lower temperatures. Under these conditions, Tables 13.6 and 13.7 are considered to represent a lower bound estimate of the service life since: (a) the antioxidant depletion rates were obtained on samples immersed in either air, water or leachate and thus are likely to overestimate antioxidant depletion for the more realistic case with different conditions on both sides of the geomembrane and (b) the leachate strength was essentially constant during the immersion resulting in greater antioxidant depletion relative to that expected to occur in a landfill where the leachate concentrations may be expected to decrease with time. The authors are currently examining antioxidant depletion under more realistic exposure conditions.

Table 13.6 Estimated service life (years) for a primary HDPE geomembrane based on depletion time of antioxidants estimated by Sangam and Rowe (2002), induction time reported by Viebke *et al.* (1994) and an assumed degradation time of 25 years

	Temperature (°C)				
Exposure conditions	*13*	*15*	*20*	*25*	*33*
Leachate and air	270	230	170	130	80
Leachate and water	170	150	110	90	60
Leachate and unsaturated soil	220	190	140	110	70

Table 13.7 Estimated service life (years) for a secondary HDPE geomembrane based on depletion time of antioxidants estimated by Sangam and Rowe (2002), induction time reported by Viebke *et al.* (1994) and an assumed degradation time of 25 years

		Temperature (°C)		
Failure of primary geomembrane	*Exposure conditions*	*7*	*10*	*15*
220 years	Leachate and unsaturated soil	440	380	300
	Leachate and water	380	350	290
	Water and unsaturated soil	510	420	320
	Water and water	450	390	310
270 years	Leachate and unsaturated soil	470	410	330
	Leachate and water	420	380	330
	Water and unsaturated soil	520	440	330
	Water and water	480	410	330

13.4 Leakage through geomembranes

An intact geomembrane (i.e., one that is free of holes or tears) is essentially impermeable to water. However, fluid that flows through holes in the geomembrane is referred to as "leakage". Geomembranes are commonly used as part of a composite liner which also includes either a low permeability compacted clay liner or GCL to minimize leakage. The quantity of leakage through a geomembrane depends, inter alia, on the number and size of holes in the geomembrane, the hydraulic head acting above and below the composite liner, the hydraulic conductivity and thickness of any underlying soil and the nature of the interface between the geomembrane and underlying material. Techniques for calculating the leakage through composite liners were presented in Section 5.7. In this section, several practical cases will be examined to illustrate the potential magnitude of leakage through composite liners at the base of landfills.

At this point, it is instructive to consider the calculated leakage for the five cases shown in Figure 13.10 to illustrate a number of important points. Leakage rates per hole for the cases shown in Figure 13.10 are summarized in Table 13.8. Calculations were conducted for circular holes in the geomembrane with radius r_0 of 0.001 m and 0.01 m, a depth of fluid h_w of 0.3 m on top of the geomembrane, and a pressure head of zero at the base of the liner system ($h_a = 0$). The limiting case of a geomembrane with nothing below it (e.g., where the geomembrane is placed directly on a highly permeable drainage layer) is presented as a reference (Figure 13.10a). In this case, leakage is proportional to the square of the hole radius (see Section 5.7.1). The values given in Table 13.8 show that leakage through composite liners (Cases b, c, d and e) is much less than through a geomembrane resting directly on a highly permeable layer (Case a).

The leakage through a hole in a geomembrane that is in perfect contact with the underlying soil (i.e., interface transmissivity $\theta = 0$) can be calculated as described in Section 5.7.3. Thus, it can be shown that placing a 0.6-m thick layer of soil with $k = 1 \times 10^{-6}$ m/s (e.g., silty sand) beneath the geomembrane (Case b, Table 13.8 and Figure 13.10b) reduces the flow through the geomembrane by three to four orders of magnitude (i.e., relative to Case a). Replacing the silty sand with a 0.6-m thick CCL with $k = 1 \times 10^{-9}$ m/s reduces leakage by an additional three orders of magnitude (Case c, $\theta = 0$, Table 13.8 and Figure 13.10c). Leakage through the composite liner is even smaller if a GCL is used beneath the geomembrane (Case e, $\theta = 0$, Table 13.8 and Figure

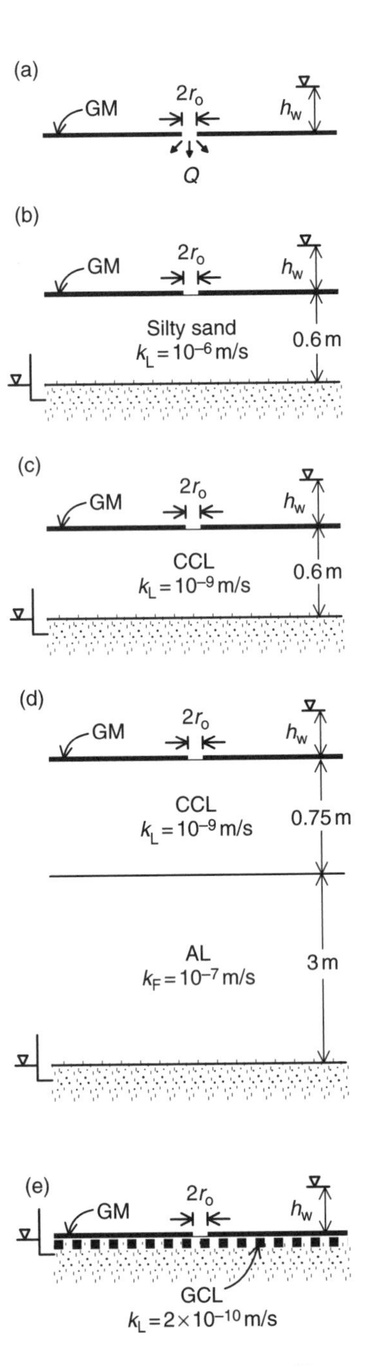

Figure 13.10 Cases considered to illustrate leakage through geomembranes.

Table 13.8 Calculated leakage through geomembrane (no holes in wrinkles)

Case in Figure 13.10	*Liner system*	k_L (m/s)	θ (m^2/s)	Q(m^3/s/hole)	
				$r_\theta = 0.001$ m	$r_\theta = 0.01$ m
a	GM alone	–	–	4.57×10^{-6}	4.57×10^{-4}
b	GM + 0.6-m soil	1×10^{-6}	0	3.60×10^{-9}	3.63×10^{-8}
c	GM + 0.6-m CCL	1×10^{-9}	0	3.60×10^{-12}	3.63×10^{-11}
			1×10^{-7}	2.54×10^{-8}	3.56×10^{-8}
d	GM + 0.75-m CCL + 3-m AL*	1×10^{-9}	0	7.79×10^{-11}	7.70×10^{-10}
			1×10^{-7}	2.81×10^{-8}	4.10×10^{-8}
e	GM + 0.01-m GCL	2×10^{-10}	0	2.59×10^{-13}	4.00×10^{-12}
			2×10^{-10}	8.46×10^{-11}	1.70×10^{-10}

*$k_F = 1 \times 10^{-7}$ m/s.

13.10e). The large reduction in leakage clearly demonstrates the benefit of using a geomembrane in conjunction with a low permeability layer (either a CCL or GCL) as a composite liner.

The nature of the interface between the geomembrane and any underlying material has a large influence on leakage. If flow occurs laterally along this interface (e.g., from gaps arising from unevenness in the geomembrane or the underlying soil, as discussed further in Section 13.4.2) then the head drop acts over a much larger area of the underlying soil (the wetted radius) and the leakage, calculated as described in Section 5.7.4, with interface transmissivity ($\theta > 0$) will be considerably larger than the case for perfect contact ($\theta = 0$). For example, the calculated leakage rate for the case shown in Figure 13.10c is 7000 times larger using an interface transmissivity θ of 1×10^{-7} m^2/s (corresponding to "poor" contact as defined by Giroud, 1997) than for perfect contact (i.e., no lateral flow) along the GM/CCL interface, but is still much smaller than the leakage through a geomembrane directly on a highly permeable layer. Lateral flow along the interface beneath the geomembrane also reduces the influence of hole size on leakage. For the case shown in Figure 13.10c and perfect contact, increasing the hole radius by a factor of ten results in increased leakage by a factor of ten, whereas with poor contact ($\theta = 1 \times 10^{-7}$ m^2/s) the same increase in hole radius increased leakage by a factor of 1.4.

The leakage through a GM/CCL composite liner is less than that for a CCL alone. For a given geometry this comparison depends on the hole radius, interface contact and the number of holes in the geomembrane (normally expressed as the number of holes per hectare). Flow through a 0.6-m thick clay liner with $k = 1 \times 10^{-9}$ m/s is 236 times greater than the leakage through the same clay liner with an overlying geomembrane (i.e., the composite liner of Figure 13.10c) with $r_0 = 0.001$m and poor contact ($\theta = 1 \times 10^{-7}$m^2/s), and assuming 2.5 circular holes/ha (a typical value reported by Giroud and Bonaparte (2001), and discussed further in Section 13.4.1). In fact for this combination of parameters, one would need a 590 holes/ha in the geomembrane to have the same flow as for the clay liner alone.

The calculated leakage rates in Table 13.8 demonstrate the benefit, in terms of minimizing leakage, of using a composite liner involving a geomembrane and a low permeability layer (either a CCL or GCL) compared to using either a geomembrane or a low permeability layer alone. As discussed in Section 16.6, provided that there are relatively few holes these composite liners may be so effective at controlling leakage that the contaminant impact through the composite liner is controlled by the service life of the geomembrane for inorganic and readily soluble organic contaminants, and the diffusive resistance and attenuation characteristics of the composite liner for low solubility organic contaminants such as volatile organic compounds (see Section 16.6). Although small, it is still important to include leakage rates in the contaminant migration assessment of any proposed barrier system. In the following sections, the effects of the number and size of holes in the geomembrane (Section 13.4.1) and the interface conditions between the geomembrane and the underlying material (Section 13.4.2) will be discussed. Calculated leakage rates for common composite liner systems are reported in Section 13.4.3. The concept of hydraulic equivalence is examined in Section 13.4.5, and finally comparisons are made between calculated and observed leakage rates through composite liners in Section 13.4.7.

13.4.1 Holes in geomembranes

Holes in a geomembrane may arise from: (a) flaws induced in the geomembrane during manufacturing (although with appropriate manufacturer quality control, geomembranes should leave the factory free of holes); (b) handling of the geomembrane rolls during delivery to the site and on-site placement and seaming; and (c) physical damage that occurs during the placement of material such as drainage gravel on top of the liner system (McQuade and Needham, 1999; Nosko and Touze-Foltz, 2000).

Visual inspection is required to detect holes prior to the placement of the overlying material. Once the drainage material is placed on top of the geomembrane, an electrical leak detection survey can be used to verify the condition of the geomembrane (e.g., Shultz *et al.*, 1984; Parra, 1988; McQuade and Needham, 1999). This involves inducing a voltage difference between an electrically conductive medium overlying the geomembrane (wet drainage gravel or protection geotextile) and the material beneath the geomembrane. Electrodes are then passed over the surface of the overlying material to measure electrical potential. Holes in the geomembrane are indicated by any anomalies in electrical potential caused by the flow of current along a conductive path through the hole.

The number of holes detected per hectare from published leak detection surveys (Laine and Darilek, 1993; Colucci and Lavagnolo, 1995; McQuade and Needham, 1999; Rollin *et al.*, 1999) is compiled in Figure 13.11. Many factors influence the number of holes including the type of geomembrane, type of the protection layer, characteristics of overlying and underlying materials, construction practices, the area that has been covered by geomembrane, whether the survey was conducted before or after placement of the cover material and the extent of a quality assurance programme at each site. Of the 205 results plotted in Figure 13.11: (a) no holes were detected for 30% of the cases and (b) less than 5 holes/ha were detected for half of these surveys. Of the 14 cases reported by McQuade and Needham (1999) that were known to have a thorough construction quality assurance (CQA) programme, no holes were detected in nine of these surveys and the maximum number of holes detected was just under 6 holes/ha. Also the studies reported in 1993 and 1995 typically detected more holes than the 1999 reports; the earlier studies represent a time with less stringent CQA than was commonly adopted toward the end of the decade. Thus, appropriate CQA practices assist in minimizing holes in geomembranes.

Holes detected during a leak detection survey are usually repaired; however, there may still be holes in the geomembrane undetected by the leak detection survey. Thus even following a leak detection survey and repairs to any detected

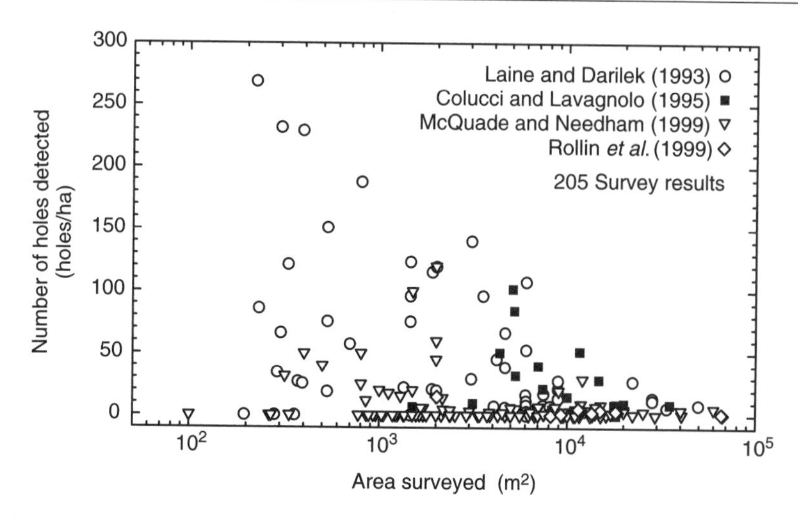

Figure 13.11 Number of holes detected in geomembranes from electrical leak detection surveys.

holes, it is prudent to assume a number of holes in geomembrane for design. As guidelines for the number and size of holes, Giroud and Bonaparte (2001) suggested that, for geomembranes installed with strict construction quality assurance, 2.5–5 holes/ha with a radius of 0.001 m can be used. Depending on the extent of CQA anticipated at a site, a range of hole sizes and number of holes should be considered during design.

Since the leak detection survey occurs shortly after construction of the liner system, it is uncertain how many holes may develop under combined overburden pressures, possibly elevated temperatures and chemical exposure years after construction. These holes may arise from: (a) indentations at gravel contacts following placement of the waste, (b) stress cracking at points of high tensile stress in wrinkles and (c) sub-standard seams subjected to tensile stresses. Thus, it is of interest to consider the number of visible gravel indentations in a geomembrane (e.g., from the large-scale physical testing conducted by Tognon *et al.* (2000) and as discussed in Section 13.2.3b) and ponder the implications on leakage if even a small proportion of the indentations, presumably those with the largest tensile strains, developed holes over time. For example, with a protection layer consisting of two layers of nonwoven geotextile yielding a total mass per unit area of 1,200 g/m^2 there were over 300 gravel indentations per square metre in the geomembrane (see Table 13.4). If only one out of every 10,000 indentations (i.e., 0.01%) eventually resulted in a defect, this would correspond to over 300 holes/ha. Thus, if one is to design with 2.5 holes/ha it is essential to have a protection layer that will limit the strains that develop in the geomembrane from overlying materials to a small value. This will typically require more than a layer of nonwoven needle-punched geotextile and may require a sand layer or special geocomposite protection layer in order to enhance the long-term performance of the geomembrane.

13.4.2 Contact between a geomembrane and underlying material

The quantity of leakage through a hole in a geomembrane is strongly influenced by the nature of the contact between the geomembrane and the underlying material. Lateral flow occurring along physical gaps between the geomembrane and underlying material increases the leakage through the composite liner relative to perfect contact conditions. One can envisage several sources of imperfect contact resulting from unevenness in the geomembrane or underlying material depending on construction practices and the extent of construction quality control.

One source relates to protrusions of gravel, small-scale irregularities or interclod voids that may exist in the underlying soil causing a gap between the geomembrane and the underlying soil (Figure 13.12a). A second source of imperfection arises from undulations or ruts which result in the unevenness of the surface of the underlying soil. When compacting clay liners to obtain low hydraulic conductivity, it is usually desirable to compact at a water content 2–4% above the standard Proctor optimum value. Since this is often close to the plastic limit of the soil and the soil is soft, it is difficult to obtain a smooth surface due to rutting from construction equipment (e.g., the overlap from a smooth drum roller, Figure 13.12b, or an imprint from a rubber tire, Figure 13.12c). Rutting can be reduced by compacting the upper layer of the clay liner at a lower water content thereby better allowing the preparation of a firm smooth surface. However, compaction at a lower water content will increase the hydraulic conductivity of this layer and increase the likelihood for interclod voids on the surface; both of which would increase leakage through holes in the geomembrane. Thus, a balance is needed between the desire to have the water content low enough to allow the preparation of a smooth surface but at the same time not so low as to increase the hydraulic conductivity of the layer above 10^{-9} m/s under field stress conditions (with landfill loading applied). In practice it will not be practical to have a liner surface that provides perfect contact and one should design with "poor" or "good" contact depending on what experience indicates can reasonably be achieved in a particular area.

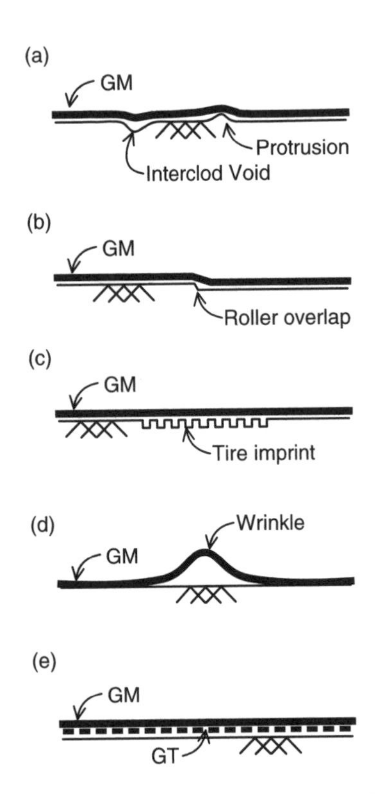

Figure 13.12 Illustration of possible sources of imperfect contact between a geomembrane and underlying clay soil.

A third source of imperfections affecting contact conditions is the presence of wrinkles (Section 13.2.4) in the geomembrane (Figure 13.12d). If a hole in the geomembrane occurs at or near the location of a wrinkle there would be essentially unimpeded lateral flow along gap between the geomembrane and underlying soil. As mentioned in Section 13.2.4, the size of the wrinkle may be reduced following application of overburden pressures; however, the wrinkle and thus the gap between the geomembrane and the soil will most likely remain. Attempts at reducing the number of wrinkles, other factors being equal, will reduce leakage through the geomembrane; however, it must be anticipated that there will be wrinkles and this should be accounted for in design.

A fourth source for imperfect contact can arise from a design where a needle-punched nonwoven geotextile is placed directly beneath the geomembrane as shown in Figure 13.12e.

The geotextile, which is intended to provide protection against puncture of the geomembrane from angular particles in the underlying soil, has a transmissivity much larger than that between a GM and CCL which will lead to larger lateral flow, and thus large leakage through the geomembrane, than would be realized if the geomembrane was in direct contact with the clay liner. This design detail needs to be carefully evaluated in terms of the additional protection afforded (plus any construction advantages) relative to the increased leakage rate through any hole in the geomembrane. It should only be used if the effect of this increased transmissivity has been considered in the design calculations and found acceptable. The alternative of using select soil material free of sharp stones as the final lift of the liner, or alternatively a GCL beneath the geomembrane is preferred to the detail shown in Figure 13.12e if puncture of the geomembrane from the underlying soil is a concern.

Quantifying the transmissivity between a geomembrane and underlying soil is challenging since even for apparently good contact conditions the flow pattern may be very localized with preferential channelling of flow along the interface. One possibility is to infer interface transmissivity from laboratory tests. For example, Fukuoka (1986) and Walton *et al.* (1997) conducted experiments on geomembrane specimens containing a known hole size with sand backfill placed above and below the geomembrane. Based on their tests, Walton *et al.* (1997) concluded that gaps between the geomembrane and underlying soil do not form at burial depths greater than 0.5 m. However, this conclusion is only valid for the particular conditions examined in their experiments and does not apply for geomembrane in contact with CCLs. Since the hydraulic conductivity of the sand in the tests of Walton *et al.* (1997) is much higher than that of compacted clay, lateral flow along the interface is not an issue for sand but it is an issue for clay beneath the geomembrane. The importance of lateral interface flow has been confirmed in the experiments conducted by Jayawickrama *et al.* (1988) with a 0.57-m diameter geomembrane specimen in contact with a compacted soil comprising a mixture of clay and sand. However, the usefulness of these results for inferring interface transmissivity is limited by the size of the apparatus and the boundary conditions.

Another approach to quantifying transmissivity is to conduct controlled experiments in the field reproducing actual construction practices. Field experiments measuring leakage through composite liners over an area of about 50 m^2 by Touze-Foltz (2001) showed that for known hole properties and applied head, there were highly variable leakage rates and substantial localized lateral flow along the interface between the compacted clay and geomembrane. Unfortunately, it is difficult to infer steady-state leakage rates from these experiments because of the short duration of testing. However, these field results demonstrated the sensitivity of leakage rates to the interface between the geomembrane and the clay, as low flows were attributed to very good contact and high flows attributed to extensive lateral flow along the GM/CCL interface that could be related to defects such as those shown in Figure 13.12(a–d).

Alternatively, one can examine the flow rates measured from operating landfills (e.g., Bonaparte *et al.*, 2002) as discussed in Section 13.4.4. However, even with this field data it is challenging to infer transmissivity since, in addition to differences between the field cases, the number and size of both holes and wrinkles and leachate levels at the base of the landfill are unknown.

Recognizing that quantifying leakage through a geomembrane is a complex problem involving highly variable and possibly unknown interface conditions, Giroud and Bonaparte (2001) provide simple equations to estimate leakage for what was defined as good and poor contact conditions between the geomembrane and an underlying CCL. Giroud (1997) described good contact "as conditions where the geomembrane has been installed, with as few wrinkles as possible, on a low-permeability soil layer that has been adequately compacted and has a smooth surface" and poor contact as conditions where the geomembrane "has been installed with a certain number of wrinkles, and/or placed on a low-permeability soil that has not been well compacted and does not appear smooth". Based on these equations, Rowe (1998a) inferred transmissivities of $1.6 \times 10^{-8}\,m^2/s$ and $1 \times 10^{-7}\,m^2/s$ for good and poor contact between a geomembrane and 0.6-m clay liner with hydraulic conductivity equal to 1×10^{-9} m/s, respectively. Until more data are published, these values represent the best available estimates for GM/CCL composite liners and are subsequently referred herein as Giroud's good and poor contact conditions.

For composite liners with a geomembrane overlying a GCL, there is greater potential for obtaining good contact since the GCL can be placed flat on a well-compacted, smooth and firm foundation. Even in the absence of construction-related gaps, the transmissivity of the interface between a GM and a GCL may be expected to be smaller than the interface with a CCL, particularity if the cover geotextile of the GCL is impregnated with bentonite. Swelling of the bentonite in the GCL upon hydration may greatly reduce small gaps that existed along the GM/GCL interface. Limited laboratory measurements by Harpur *et al.* (1993) indicate very low GM/GCL interface transmissivity as they reported values from 6×10^{-12} to 2×10^{-10} m^2/s from laboratory experiments for the interface between a GM and a GCL with woven and nonwoven cover geotextiles.

The available field data suggests that geomembrane interface conditions strongly influence the leakage rate and are highly variable. At some locations the contact may be expected to be good, while at other locations there will be small gaps, and possibly at others there may be large gaps. In calculations performed to assess leakage through composite liners, it is assumed that the interface can be quantified by transmissivity θ and that this value applies uniformly to the interface. Touze-Foltz *et al.* (2001b) have developed a solution that considers the effect of non-uniform transmissivity on leakage. While being a useful development, there is presently insufficient information to quantify geomembrane contact conditions to justify the use of this more complex solution for practical landfill calculations. The remainder of the leakage calculations presented in this chapter assume a uniform transmissivity for the interface between the geomembrane and underlying soil.

13.4.3 Leakage rates for common GM/CCL composite liners

(a) Leakage through circular holes with no wrinkles

Composite liners consisting of a geomembrane overlying a CCL are commonly specified as a minimum liner system in many regulatory jurisdictions. For example, composite liner designs developed to meet the requirements of the United States, Resource Conservation and Recovery Act (Subtitle D) typically comprise a 1.5-mm thick HDPE geomembrane and a minimum of a 0.6-m CCL with hydraulic conductivity of less than 1×10^{-9} m/s (or equivalent), as shown in Figure 13.10c. In Ontario, Canada, the prescriptive design permitted under Ontario Regulation 232/98 (MoE, 1998) involves either a single or double composite liner depending on the expected volume of waste (per unit area) and the background groundwater chemistry in the receptor aquifer. The single composite liner design, shown in Figure 13.10d, consists of a 1.5-mm thick HDPE geomembrane, 0.75-m thick clay liner with a hydraulic conductivity no greater than 1×10^{-9} m/s on a subgrade consisting of at least 3 m thick natural or engineered attenuation layer (AL) with a hydraulic conductivity less than 1×10^{-7} m/s. For illustrative purposes, it is useful to examine leakage rates for these two composite liner designs. In these calculations, the hydraulic conductivity of the clay liner is taken as 1×10^{-9} m/s, that of the attenuation layer for the Ontario liner is 1×10^{-7} m/s and for both cases the total head is assumed to be zero (i.e., $h_a = 0$) at the bottom of the liner system. The solution presented in Section 5.7.4 was used to calculate the leakage per hole as a function of hole radius r_o for both good ($\theta = 1.6\times10^{-8}$ m^2/s) and poor ($\theta = 1\times10^{-7}$ m^2/s) interface contact and for a typical design leachate head (h_w) equal to 0.3 m (see Section 2.4).

The results of these calculations (Figure 13.13) show that leakage increases with increasing hole size, with a greater increase in leakage occurring for large holes (r_o >0.1 m). In all cases, flow occurs within the area defined by the wetted radius, r_w (defined in Section 5.7.4 to be the radial extent of lateral flow along the interface). For a given liner system, the wetted radius is a function of interface contact (transmissivity θ), head drop across the liner and hole radius. The wetted radius calculated for a composite geomembrane liner with 0.6 m of compacted clay is plotted in Figure 13.14 for $h_w = 0.3$ m. For holes

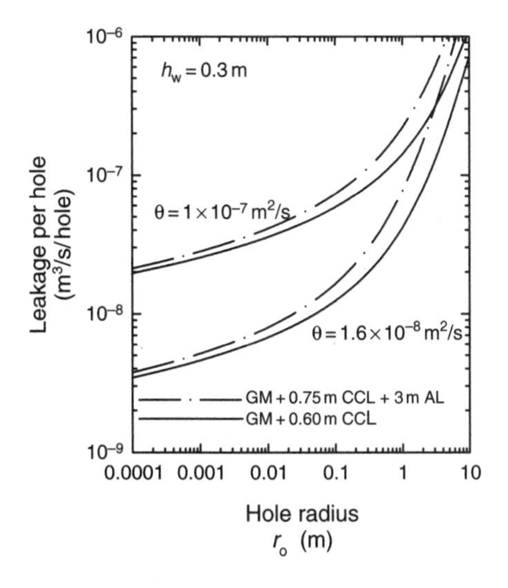

Figure 13.13 Calculated leakage rate per circular hole as a function of hole radius r_o and contact conditions θ for two GM/CCL composite liners (no wrinkles).

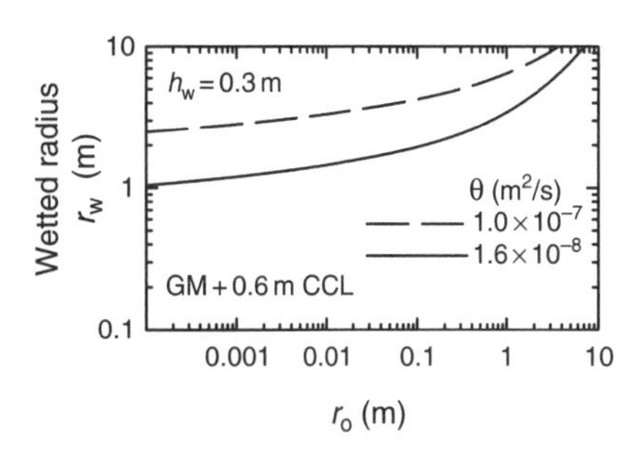

Figure 13.14 Calculated wetted radius as a function of hole radius r_o and contact conditions θ for a GM + 0.6-m CCL composite liner (no wrinkles).

with a radius less than 0.1 m, the wetted radius is much larger than the hole size and thus the leakage is less sensitive to hole size for small holes than for very large holes (r_o >1 m). Although hole size plays a role in the leakage and it is desirable to reduce the size of any holes in the geomembrane to minimize leakage, for the holes most likely to be encountered in the field (i.e., $r_o < 0.1$ m, in the absence of major tears), the interface contact has a much greater effect on leakage than does hole size. Figure 13.13 further shows that the leakage for both composite liners with poor contact is around 5–6 times greater than with good contact.

For situations where the leachate level (h_w) on the geomembrane is larger than 0.3 m, the calculated leakage per hole is plotted in Figure 13.15 for a range of hole sizes and good ($\theta = 1.6\times10^{-8}$ m^2/s) and poor ($\theta = 1\times10^{-7}$ m^2/s) contact conditions. Values of h_w greater than the typical design value of 0.3 m can arise from either the termination of leachate collection or a reduction in leachate collection efficiency from clogging (see Section 2.4.4). Predictably, the leakage increases with increasing head on the geomembrane liner. The calculated wetted radius (Figure 13.16) can be quite large.

The results plotted in Figures 13.13 and 13.15 also show that calculated leakage for the US liner shown in Figure 13.10c is consistently smaller than the leakage for the Ontario liner of Figure 13.10d. The leakage is greater for the

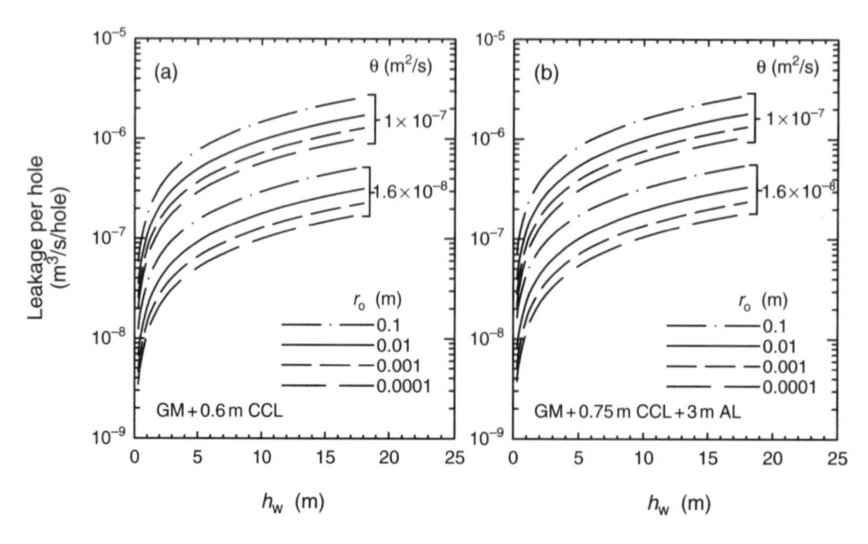

Figure 13.15 Calculated leakage rates per circular hole as a function of leachate head h_w, hole radius r_o and contact conditions θ for: (a) GM + 0.6-m CCL and (b) GM + 0.75-m CCL + 3-m AL composite liners (no wrinkles).

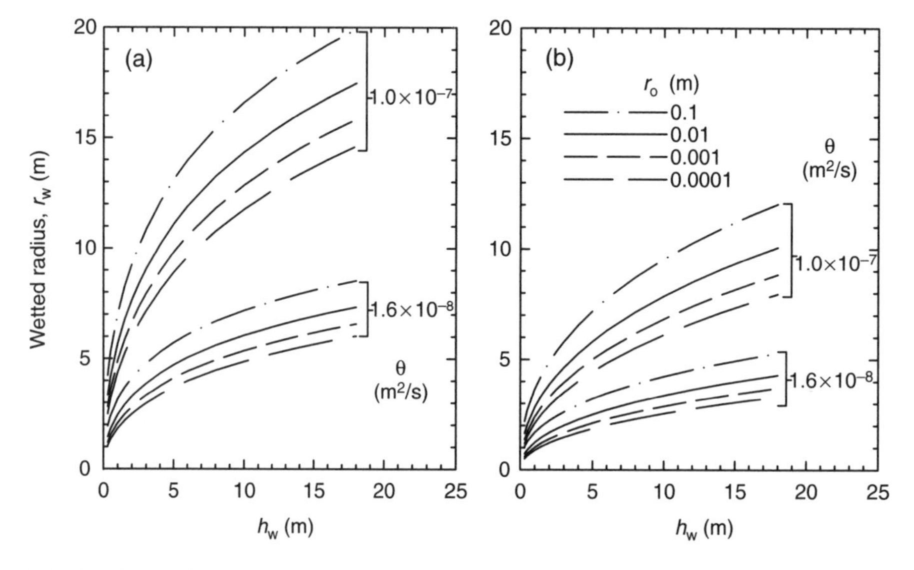

Figure 13.16 Calculated wetted radius r_w as a function of leachate head, hole radius r_o and contact conditions θ for: (a) GM + 0.6-m CCL and (b) GM + 0.75-m CCL + 3-m AL composite liners (no wrinkles).

thicker liner system due to the higher harmonic mean hydraulic conductivity of the soil between the geomembrane and the aquifer developed in the Ontario case. A similar increase would arise for the US system if, as would usually be the case, there was a subgrade (AL) between the compacted clay and the receptor aquifer. The maximum difference between these two designs occurs at low h_w and is less than 20% for r_o less than 0.01 m, and decreases as h_w increases or r_o

decreases. The AL serves as an important component of the Ontario composite liner as it acts as a diffusion barrier and provides a thicker medium for biodegradation of organic contaminants and sorption of reactive contaminants. The role of the AL in reducing the contaminant impact through composite liners is examined in Section 16.6.

(b) Leakage through holes in wrinkles

The results in the previous section correspond to the situation where a hole does not coincide with a wrinkle in the geomembrane. Given the local stress concentrations induced by a wrinkle, it is reasonable to hypothesize that holes would be likely to occur at the location of a wrinkle. The case where the hole coincides with a wrinkle in the geomembrane is now examined for the two composite liner designs examined in the previous section. The solution for leakage beneath a wrinkle given in Section 5.7.5 assumes that there is negligible flow at the ends of the wrinkle; however, it may be anticipated that there will in fact be flow at the ends. For a wrinkle of width $2b$ over a length L ($L > 2b$), this can be approximately accounted for by recognizing that generally the wrinkle does not stop immediately with a width $2b$ but, rather, has a transition back to the unwrinkled geomembrane that is approximately semi-circular. Taking this transition at the wrinkle ends to have radius b (see insert to Figure 13.17), the leakage through the wrinkle can then be approximated by the sum of: (a) the flow through a wrinkle of width $2b$ and length L (Section 5.7.5) and (b) the flow through a circular hole of radius b (Section 5.7.4). The calculated leakage rates using this approach are plotted in Figure 13.17 against wrinkle length for two different wrinkle widths, as well as for good ($\theta = 1.6 \times 10^{-8}$ m^2/s) and poor ($\theta = 1 \times 10^{-7}$ m^2/s) interface contacts. Values are given for typical US and Ontario minimum composite liners (Figures 13.17a and b, respectively) for a typical design leachate level, h_w, of 0.3 m above the geomembrane. Also plotted in Figure 13.17 are the calculaed flow rates if there were no wrinkles (shown by the symbols) for a range of holes sizes to illustrate the difference between leakage with and without wrinkles. For example, leakage through a hole with radius $r_o = 0.001$ m with good contact ($\theta = 1.6 \times 10^{-8}$ m^2/s) between the geomembrane and clay liner is increased by factor of 10 if that same hole is

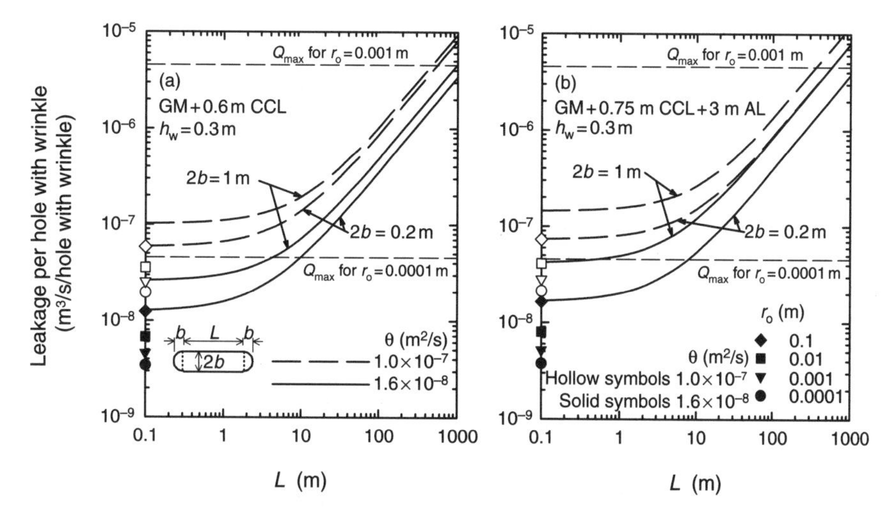

Figure 13.17 Calculated leakage rates per wrinkle with a hole as a function of wrinkle length L, wrinkle width $2b$ and contact conditions θ for: (a) GM + 0.6-m CCL and (b) GM + 0.75-m CCL + 3-m AL. Results independent of hole size until limit imposed by Bernoulli's equation as shown by Q_{max} for $r_o = 0.0001$ m and $r_o = 0.001$ m.

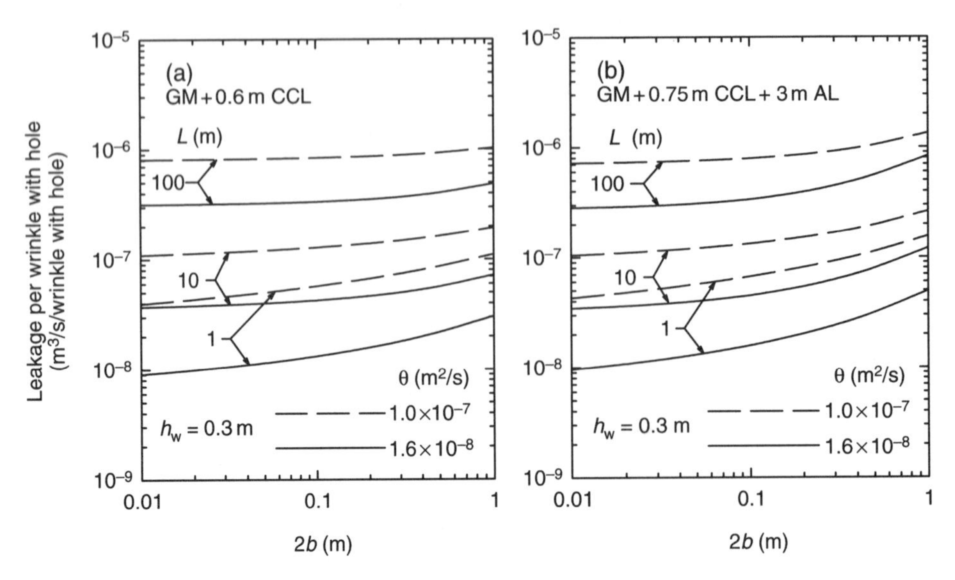

Figure 13.18 Calculated leakage rates per wrinkle with a hole as a function of wrinkle width $2b$, wrinkle length L and contact conditions θ for: (a) GM + 0.6-m CCL and (b) GM + 0.75-m CCL + 3-m AL. Results independent of hole size until limit imposed by Bernoulli's equation as shown by Q_{max} for $r_o = 0.0001$ m and $r_o = 0.001$ m in Figure 13.17.

coincident with a wrinkle in the geomembrane with width $2b = 0.2$ m and $L = 10$ m.

A number of observations can be made from Figure 13.17. First, the solution is not very sensitive to wrinkle length, L, for wrinkles of length less than about 2 m since end effects (controlled by b) dominate for low L. Second, end effects, and hence the effect of the wrinkle half width, b, decreases with increasing length L.

The upper limit to the leakage through a geomembrane, is controlled by the capacity of the hole to drain fluid (Section 5.7.6). This upper limit can be calculated using equation 5.53 and is plotted for wrinkles with single holes having radii of 0.0001 and 0.001 m in Figure 13.17 assuming an infiltration of $q_0 = 10^{-9}$ m/s $(= 0.03$ m/a $=$ 0.8 lphd) and hydraulic conductivity of the material overlying the geomembrane of $k_{om} = 10^{-2}$ m/s. For a single very small hole ($r_o = 0.0001$ m) in a wrinkle with $2b = 0.2$ m, flow is governed by hole size for $L > 10$ m with good contact ($\theta = 1.6 \times 10^{-8}$ m^2/s) and for all values of L with poor contact. For a single hole in a wrinkle with r_o greater than 0.001 m, the capacity of the hole may only be approached for extremely long (and unlikely) wrinkles ($L > 500$ m).

The sensitivity of the calculated leakage to variations in the wrinkle width is plotted in Figure 13.18. The leakage decreases as the wrinkle width is reduced; however, decreasing the width from 1 to 0.01 m reduces leakage by about 50% for the case in Figure 13.18a with good contact ($\theta = 1.6 \times 10^{-8}$ m^2/s) and a length of 10 m. For all situations where the length, L, is much greater than the width, $2b$, the width of the wrinkle does not have a great influence on leakage (as was the case for the influence of r_o on leakage without wrinkles) since lateral flow along the interface between the geomembrane and the underlying soil increases the area of clay liner that leachate can reach, and hence flow down through, the clay liner.

13.4.4 Leakage rates for a GM/GCL composite liner

Leakage calculated through a composite liner consisting of a geomembrane overlying a GCL is presented in Figures 13.19–13.22 for the geometry shown in Figure 13.10e. The hydraulic conductivity of the GCL was taken to be 2×10^{-10} m/s to account for possible interaction with leachate (e.g., see Section 12.5.1). The piezometric head at the bottom of the GCL was taken as zero ($h_a = 0$), which may correspond to the case where the GM/GCL liner is underlain by a secondary leachate collection system. Leakage is examined for the range of interface transmissivities reported by Harpur *et al.* (1993).

(a) Leakage through circular holes with no wrinkles

Figures 13.19 and 13.20 plot the calculated leakage for circular holes (but no wrinkles) in the geomembrane obtained using the solution in Section 5.7.4. Results are also shown for perfect interface conditions in Figure 13.19 that were obtained with the solution presented in Section 5.7.3. The leakage through the GM/GCL liner is very small, even though the GCL is thin (only 0.01 m thick) and the hydraulic gradients are large. The low hydraulic conductivity of the GCL and very low interface transmissivity result in small leakage through the liner. As for the GM/CCL liners, leakage through the GM/GCL liner increases with larger holes, higher interface transmissivity and greater leachate depth.

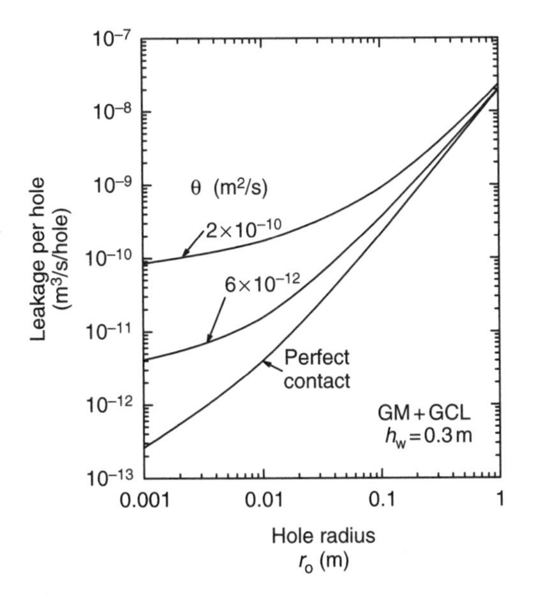

Figure 13.19 Calculated leakage rate per circular hole as a function of hole radius r_o and contact conditions θ for a GM/GCL composite liner (no wrinkles).

(b) Leakage through holes in wrinkles

Figures 13.21 and 13.22 plot the results for the GM/GCL liner with leakage at a wrinkle. Wrinkles result in a greater percentage increase in leakage for the GM/GCL liner than for GM/CCL liners. For the same hole and wrinkle geometry considered for the GM/CCL example, leakage through a circular hole with radius $r_o = 0.001$ m with good contact ($\theta = 1.6 \times 10^{-8}$ m^2/s) between the geomembrane and GCL is increased by factor of 3,600 if that same hole is coincident with a wrinkle in the geomembrane with width $2b = 0.2$ m and $L = 10$ m. The influence of both the radius of the hole in the case with no wrinkles and width of the wrinkle has a much greater effect on theoretical calculations of leakage for the GM/GCL liners than for the GM/CCL liners since the low GM/GCL interface transmissivity results in a smaller wetted radius. Wrinkles have a greater influence on leakage through the GM/GCL because the wrinkle results in a greater increase in wetted area for flow to occur. However, comparing Figure 13.17 and 13.21, it can be seen that while wrinkles are more important for a GCL than CCL, the leakage through the liners with GCLs is much lower.

13.4.5 Hydraulic equivalence of GM/CCL and GM/GCL composite liners

Most regulations will allow alternate liner designs to GM/CCL provided that one can demonstrate "equivalency" of the alternate design. Questions often arise as to whether a

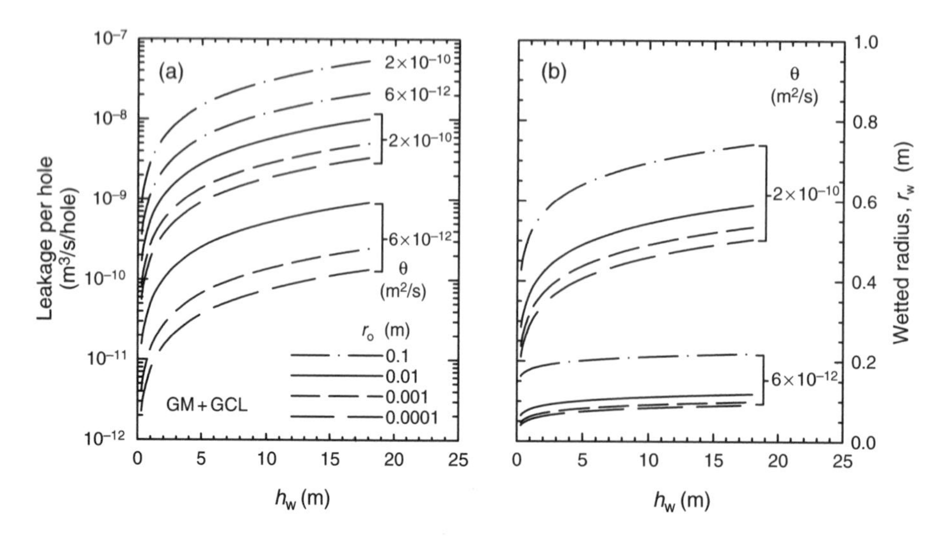

Figure 13.20 Calculated (a) leakage rate per circular hole and (b) wetted radius r_w as a function of leachate head h_w, hole radius r_o and contact conditions θ for a GM/GCL composite liner (no wrinkles).

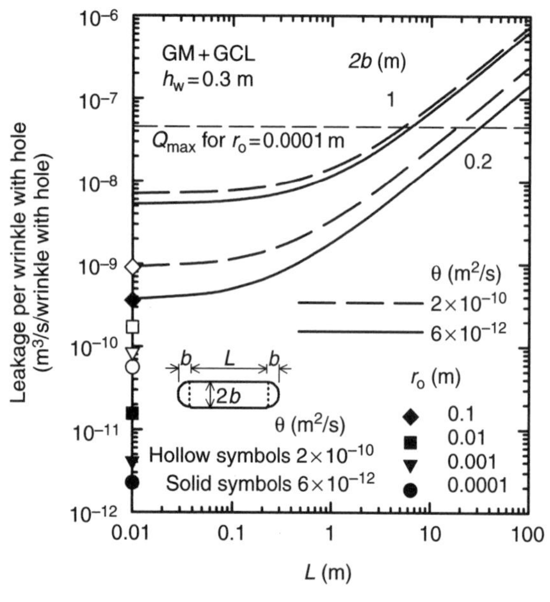

Figure 13.21 Calculated leakage rates per wrinkle with a hole as a function of wrinkle length L, wrinkle width $2b$ and contact conditions θ for a GM/GCL composite liner. Results independent of hole size until limit imposed by Bernoulli's equation as shown by Q_{max} for $r_o = 0.0001$ m.

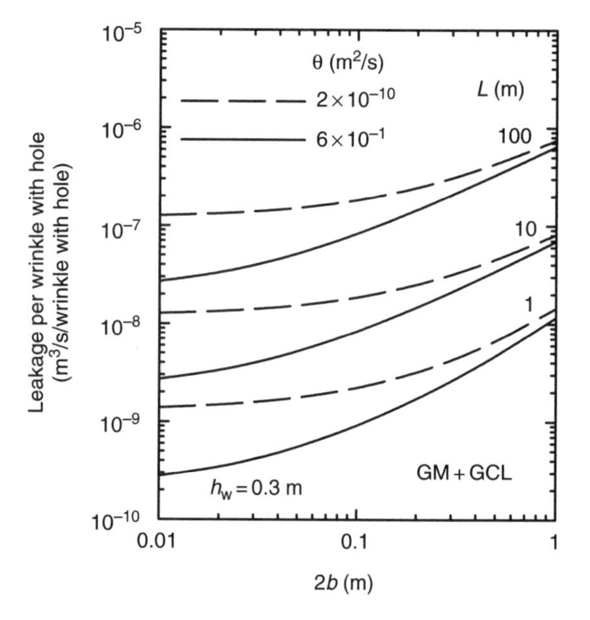

Figure 13.22 Calculated leakage rates per wrinkle with a hole as a function of wrinkle width $2b$, wrinkle length L and contact conditions θ for a GM/GCL composite liner. Results independent of hole size until limit imposed by Bernoulli's equation as shown by Q_{max} for $r_o = 0.0001$ m in Figure 13.21.

GM/GCL composite liner is equivalent to the standard GM/CCL composite liner. The hydraulic performance of the composite liner is one criterion that may permit comparison of different liner systems.

For example, the leakage rates for the GM + 0.6 m CCL composite liner can be compared with those for the GM + 0.01 m GCL composite liner to assess the hydraulic equivalence of the two composite liners. Without wrinkles, for good contact conditions for both interfaces, and taking $r_o = 0.001$ m and $h_w = 0.3$ m, the leakage through the GM/CCL is 1100 times greater than for the GM/GCL. For the same parameters but with the hole occurring in a 0.2-m wide by 10-m long wrinkle, the flow in the GM/CCL is still larger, but only three times larger than for the GM/GCL. The leakage with the GM/GCL is consistently lower than with the GM/CCL liner because of the much lower transmissivity between the GM and GCL compared to the GM and CCL. The difference between the leakage for the two composite liners decreases as the wrinkles become either wider or longer.

Thus, the hydraulic performance of the GM/GCL composite liner is certainly equivalent to, if not much better for many cases, than the GM/CCL composite liner for the conditions examined. However, this does not necessarily mean that the resistance to contaminant migration through the GM/GCL will be equivalent to that for the GM/CCL, as the diffusive migration through the composite liner and possible attenuation of contaminants needs to be considered to make this assessment. The equivalence of composite liners in terms of contaminant impact is examined further in Section 16.6.3, where it is shown that provided an AL is used along with the GM and GCL (with the thickness of the AL such that the total barrier thickness of the GM/GCL/AL system is approximately the same as the GM/CCL), the GM/GCL/AL liner may provide equivalent environmental protection to a GM/CCL.

13.4.6 Observed leakage rates through composite liners

Monitoring flow rates from a secondary leachate collection system (SLCS) may provide insight regarding the effectiveness of composite primary liners. However, the interpretation of the data from SLCS requires careful consideration of sources of fluid other than leakage from the landfill which may include (Gross *et al.*, 1990): (a) water that infiltrated into the SLCS layer during construction, (b) water arising from compression and consolidation of the clay liner and (c) groundwater seepage from outside of the landfill.

Leakage rates measured from the SLCS of landfills with primary and secondary liners have been reported by Bonaparte *et al.* (2002). Based on this data, average, maximum and minimum flows measured in the SLCS beneath GM, GM/CCL and GM/GCLs are plotted in Figure 13.23. These results are separated into those where sand or a geonet was used for the SLCS (indicated by the solid and hollow symbols, respectively, in Figure 13.23). Bonaparte *et al.* (2002) grouped the results for three life stages of the landfill during which there may be very different flows to the SLCS. The first stage is defined as the initial period of landfill operation where the thickness of the waste and any cover does not significantly impede the flow of rainfall to the PLCS. The second stage corresponds to the active period of landfill operation where the thickness of the waste and the presence of daily and intermediate cover reduce the infiltration to the PLCS. The third stage represents the period after closure of the landfill with a low permeability final cover placed over the entire cell.

Figure 13.23 shows that for the majority of cases involving a geomembrane over compacted clay, the fluid collected in the SLCS is considerably greater than that obtained for a geomembrane over a GCL. Much of the fluid collected in the SLCS for composite liners involving a CCL was attributed to consolidation water from the overlying CCL

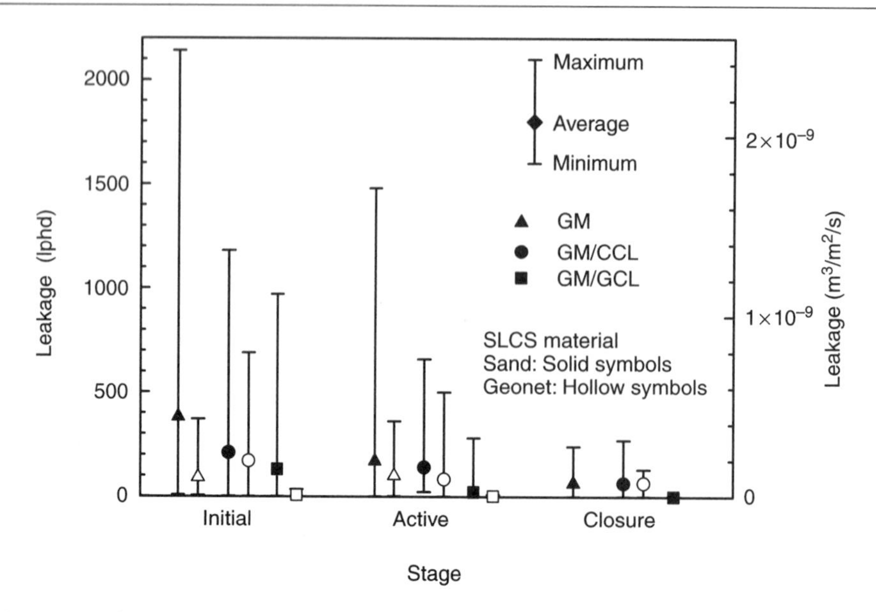

Figure 13.23 Measured flow rates in the secondary leachate collection system from landfills with primary and secondary liners (data from Bonaparte *et al.*, 2002).

(Bonaparte *et al.*, 2002). This hypothesis is supported by the observation that the chemistry of the leachate collected in the SLCS is not similar to the leachate collected in the primary leachate collection system (PLCS). However based on the available data, it is not possible to identify how much of the measured flow in the SLCS can be attributed to leakage through the composite liner.

Bonaparte *et al.* (2002) report data for 22 landfill cells containing GCLs as part of the composite primary liner from seven landfills that all had CQA programmes. Table 13.9 provides a statistical summary of the average and peak flows measured in PLCS and SLCS. Data in the active period are dominated by one landfill as 16 cells were obtained from that one landfill whereas the remaining six cells come from four other landfills. As might be expected, flow rates from the SLCS decrease as flow rates from the PLCS decrease. This is likely, at least in part, due to reduced head acting on the primary composite liner. The average SLCS flow rates are small during both the active and post-closure period (2 and 1 lphd, respectively). For 14 of the 22 data points in the active period, no flow was measured in the SLCS. Peak flow rates are 4–6 times larger than the average values but are still quite small. Although the SLCS flow rates reported by Bonaparte *et al.* (2002) are very low, there is still potential for diffusive transport of organic compounds through a composite liner involving a geomembrane over a GCL, as discussed in Section 16.6.

The results reported by Bonaparte *et al.* (2002) were for a maximum time period of about 10 years for the GM/CCL liners and about 7 years for the GM/GCL cases. Thus, this leakage data probably does not involve any mounding of leachate on the geomembrane since clogging of the leachate collection system may still be developing and not yet have occurred for these landfills. One would expect leakage rates to increase with increasing leachate head, as described in Section 13.4.3. Nonetheless, the data reported by Bonaparte *et al.* (2002) is an essential starting point to assess the hydraulic performance of composite liners. There is a need for continued monitoring of these sites and additional carefully documented case histories with a detailed record

Table 13.9 Statistical summary of average and peak flows measured in the primary and secondary leachate collection systems for GM/GCL composite liners from seven landfills (based on data reported by Bonaparte *et al.*, 2002)

Stage of landfill operation		*Average flow* (lphd)		*Peak flow* (lphd)	
		PLCS	*SLCS*	*PLCS*	*SLCS*
Initial	Mean	6000	35	15000	130
	Std dev.	4200	68	11000	250
	Data points	25	27	25	27
Active	Mean	1100	2	3500	11
	Std dev.	1400	3	4200	17
	Data points	22	22	22	22
Post-closure	Mean	140	1	370	4
	Std dev.	75	1	290	5
	Data points	5	5	5	5

of construction and operation records to permit the assessment of leakage through composite liners.

13.4.7 Comparison of calculated leakage with field measurements

The calculated leakage rates in the preceding sections were reported for either leakage per hole or leakage per wrinkle with a hole. In order to permit easier comparison with field leakage results, Figures 13.24–13.27 present calculated leakage rates through two primary composite liners (GM + 0.6-m CCL and GM + GCL) overlying an SLCS. Results are presented for either the number of holes per hectare (Figures 13.24 and 13.26) or the number of wrinkles with holes per hectare (Figures 13.25 and 13.27). In these comparisons, it is assumed that the leachate head on the geomembrane is 0.3 m.

The simplest comparison is for systems involving a geomembrane over a GCL since there should not be any significant consolidation water in the flows recorded in the SLCS. From Table 13.9, under active conditions, the observed mean flow is 2 lphd ($2.3 \times 10^{-12}\, m^3/m^2/s$) while under post-closure conditions it is 1 lphd ($1.2 \times 10^{-12}\, m^3/m^2/s$). These flows substantially exceed the calculated values in Figure 13.26 for most sizes of holes ($r_o \leq 0.01$ m) even for a very large number of holes per hectare. One would need 25–65 holes per hectare ($r_o = 0.1$ m) for $\theta = 2 \times 10^{-10}$ and $6 \times 10^{-12}\, m^2/s$, respectively to explain the average active period leakage. Since it is highly

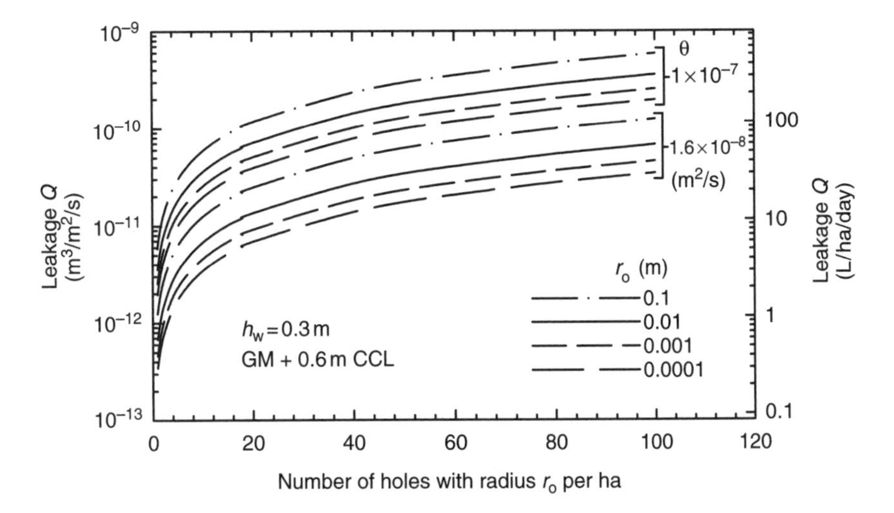

Figure 13.24 Calculated leakage rate for a GM/CCL composite liner as a function of hole size r_o, number of holes and contact conditions θ (no wrinkles).

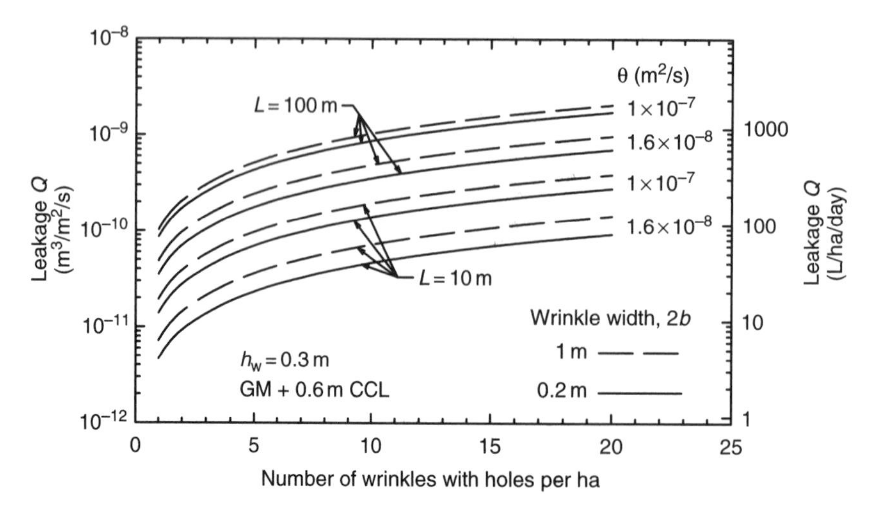

Figure 13.25 Calculated leakage rate for a GM/CCL composite liner as a function of wrinkle width $2b$, wrinkle length L, number of wrinkles with holes and contact conditions θ.

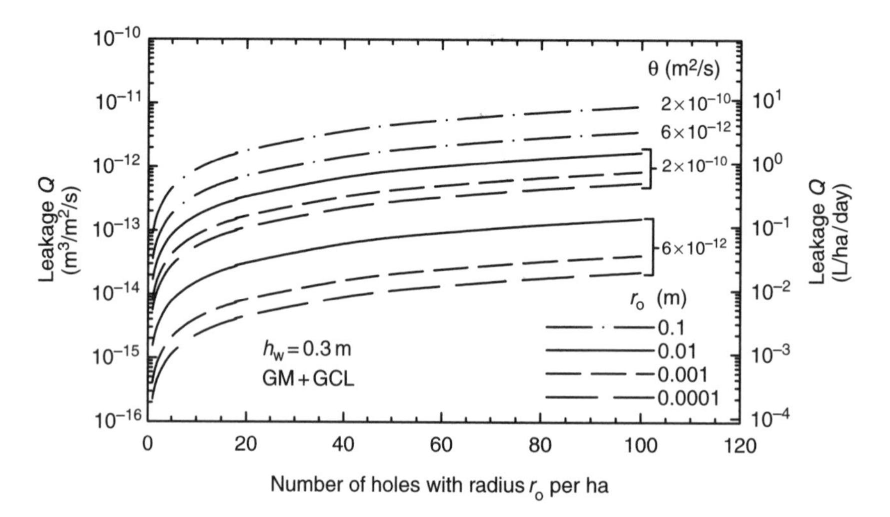

Figure 13.26 Calculated leakage rate for a GM/GCL composite liner as a function of hole size r_o, number of holes and contact conditions θ (no wrinkles).

unlikely that there are this many large holes, an alternate and more likely explanation is that there are wrinkles with holes in the geomembrane. Figure 13.27 shows that only 1–2 wrinkles with holes per hectare (with $2b = 0.2$ m and $L = 10$ m) are sufficient to explain a leakage rate of 2 lphd. The mean peak flow of 11 lphd and the maximum peak flow of 54 lphd reported by Bonaparte *et al.* (2002) possibly may have arisen from even longer or more frequent wrinkles.

For geomembranes over CCLs, the comparison is more difficult due to consolidation water contributing to the flow; however, the flows are large enough to lead one to believe that more than leakage through "large" holes in good contact with the clay is required to explain the flows.

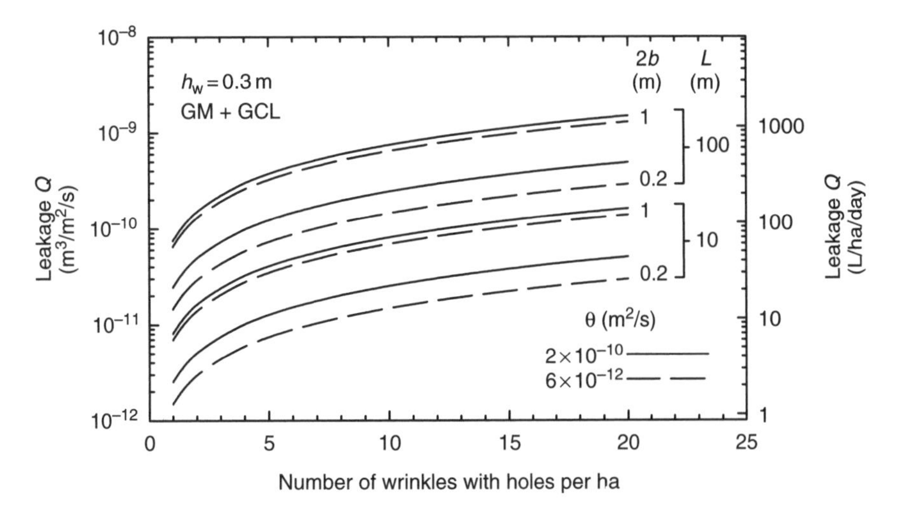

Figure 13.27 Calculated leakage rate for a GM/GCL composite liner as a function of wrinkle width $2b$, wrinkle length L, number of wrinkles with holes and contact conditions θ.

While some of the flow is invariably (unquantified) consolidation water from the clay liner, one can obtain an upper bound on the number of holes by neglecting this component of flow. Thus, if one was to explain a leakage rate of 100 lphd solely based on holes without wrinkles, an excess of 200 holes with $r_o < 0.01$ m would be required with good interface contact (Figure 13.24). With poor contact between geomembrane and clay between 20 and 60 holes (for $0.1 > r_o > 0.0001$ m) yield leakage of 100 lphd. Figure 13.25 shows that as few as 2.5 100-m-long wrinkles, or seventeen 10-m-long wrinkles with holes per hectare could account for a leakage rate of 100 lphd (for a wrinkle width $2b = 1$ m, and otherwise good contact between the geomembrane and clay $\theta = 1.6 \times 10^{-8}$ m^2/s).

In summary, it appears that wrinkles associated with holes in the geomembrane are a possible explanation to the observed leakage through composite liners (Table 13.9) involving a geomembrane over a GCL under both active and post-closure conditions. Likewise, holes in a wrinkle can explain the low leakages observed through a geomembrane over a CCL but probably not the high values of flow rate observed in SLCSs (Figure 13.23). This implies that other mechanisms such as consolidation of the CCL or groundwater seepage are required to explain the observed flow rates. Consolidation of the CCL may also give rise to earlier (and diluted) contaminant arrival times in the SLCS than would otherwise be expected.

The results presented in Figures 13.24–13.27 imply that it is desirable to minimize wrinkles if leakage is to be minimized. It may be hypothesized that the long-term performance of geomembranes with wrinkles may be even worse than implied by these calculations due to the potential for the development, with time, of stress cracking at the wrinkles and hence increased leakage through the geomembrane at locations that originally had no holes.

13.5 Diffusion of contaminants through HDPE geomembranes

Even in the absence of leakage, contaminants can migrate through geomembranes by the process of molecular diffusion. This process is thought to involve permeant adsorption on the inner surface, diffusion through the polymeric structure and desorption on the outer surface of the geomembrane (as described in Section 6.8) and can be characterized by partition coefficient S_{gf} and diffusion coefficient D_g. Methods that may be used to obtain these parameters were presented in Section 8.12. In this section, the importance of diffusion as a contaminant transport mechanism through composite geomembrane liners is examined.

Partitioning and diffusion coefficients for HDPE geomembranes and various contaminants were summarized in Table 8.9. Although there is considerable variability between diffusion coefficients reported in the literature (see Table 8.9), this variation is not as critical as it may initially appear. For a given contaminant, the geomembrane is either a good diffusion barrier (e.g., to water and hydrated ions such as Na^+, Cl^-, Zn^{2+}, Ni^{2+}, Mn^{2+}, Cu^{2+}, $Cd^{2+}$and Pb^{2+}) or it is not (e.g., to small chain organic compounds such as dichloromethane, toluene, etc.) with many orders of magnitude difference in the permeability of the geomembrane ($P_g = S_{gf} D_g$) for these two groups.

The available data on diffusion through geomembranes are sufficient to provide a basis for the rational analysis and design of barrier systems that include a geomembrane. The case of a composite liner consisting of a 1.5-mm thick HDPE geomembrane and a 0.6-m thick CCL overlying a 3-m thick aquifer is now considered to illustrate the diffusive migration of chloride, dichloromethane and toluene. It is assumed that there are no holes or defects in (and thus no leakage through) the geomembrane (i.e., $v_a = 0$), and that there is no substantial degradation or improvement in the diffusive resistance of the geomembrane over the time frame investigated.

13.5.1 Diffusion of ions through a GM/CCL composite liner

Figure 13.28 shows the predicted migration of chloride through the composite liner system for the two diffusion parameters deduced from Figure 8.29 (see Section 8.12.1). The assumed leachate chloride concentration is 2500 mg/l; however, it can be seen (Figure 13.28) that the geomembrane is an excellent diffusion barrier to chloride as demonstrated by the very small concentrations (less than 1 mg/l after 500 years, assuming the geomembrane lasts that long) predicted in the aquifer. These concentrations are well below the Ontario, Canada, aesthetic drinking water objective of 250 mg/l and thus the uncertainty regarding permeation parameters for chloride is of little practical consequence.

August *et al.* (1992) found that there was negligible diffusion of heavy metal salts (Zn^{2+}, Ni^{2+}, Mn^{2+}, Cu^{2+}, Cd^{2+}, Pb^{2+}) from a concentrated (0.5 M) acid solution (pH = 1 – 2) through HDPE over a 4-year test period. Thus, the evidence suggests very little diffusion of ions through HDPE geomembranes. For ionic contaminants, impact on an underlying aquifer will be controlled by the leakage through the liner and the service life of the geomembrane, as shown in Section 16.6.

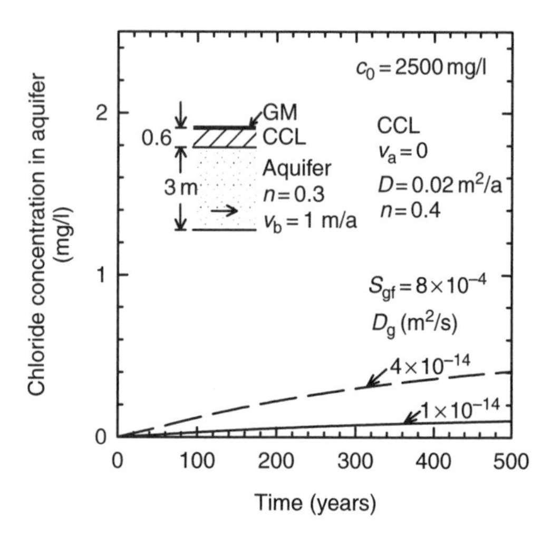

Figure 13.28 Calculated variation in chloride concentration in an aquifer due to diffusion through a 1.5-mm HDPE geomembrane and 0.6-m CCL.

13.5.2 Diffusion of organic compounds through a GM/CCL composite liner

Predicted concentrations of dichloromethane (DCM) in the aquifer using the parameters $S_{gf} = 6$ and $D_g = 0.65 \times 10^{-12}\,m^2/s$ deduced by Sangam and Rowe (2001) – see Figure 8.27 and Section 8.12.1 – are plotted in Figure 13.29. These calculations assume transport parameters in the clay as shown in Figure 13.29. The half-life of organic contaminants like DCM influences the concentrations calculated in the aquifer. A discussion on the selection of half-lives and the influence on calculated impact is presented in Sections 2.3.6 and 16.6, respectively. However, for the purposes of the current modelling, half-lives of 2 years in leachate and 50 years in soil were adopted for DCM. For this combination of parameters, the concentration of DCM in the aquifer would be expected to reach a maximum value of 50 μg/l after about 30 years. This calculated impact exceeds the US EPA's maximum concentration level (MCL) of 5 μg/l for DCM; the impact on the aquifer in this case can be reduced by either using a thicker clay liner or by adding an AL between the liner and the aquifer as will be illustrated in Section 16.6.

Figure 13.30 plots the predicted concentrations of toluene in the aquifer using the parameters $S_{gf} = 100$ and $D_g = 3.0 \times 10^{-13}\,m^2/s$ deduced by Sangam and Rowe (2001) for toluene in aqueous solution (see Figure 8.28). However, the parameters for sorption (K_d) and the half-life in the leachate and the soil are even more important than S_{gf} and D_g. Results are given in Figure 13.30 assuming no retardation ($K_d = 0$) and moderate sorption ($K_d = 3.9$) for a half-life in leachate of 10 years and in soil of 50 years. The Ontario aesthetic drinking water objective for toluene of 24 μg/l would be met only with moderate retardation and considering the toluene half-life as examined in Figure 13.30. The impact would be reduced if there was an AL between the CCL and aquifer.

For the organic contaminants examined here, the most critical liner parameters are the partitioning coefficient, K_d, and the half-life in the clay. As discussed by Rowe *et al.* (1995b) and Rowe and Weaver (1997), care is required in the selection of K_d values based on published correlations. The importance of diffusion as a

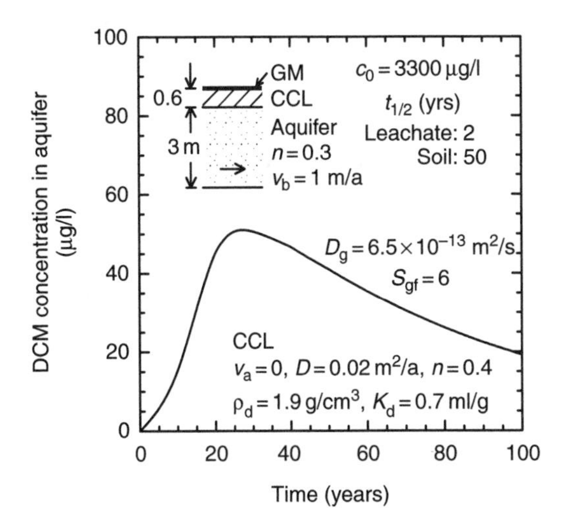

Figure 13.29 Calculated variation in DCM concentration in an aquifer due to diffusion through a 1.5-mm HDPE geomembrane and 0.6-m CCL.

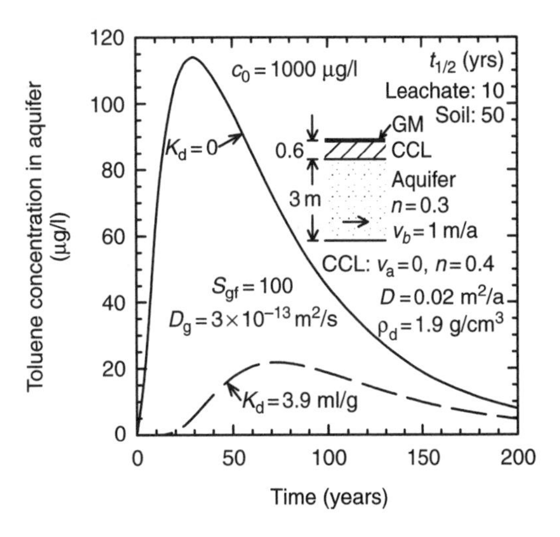

Figure 13.30 Calculated variation in toluene concentration in an aquifer due to diffusion through a 1.5-mm HDPE geomembrane and 0.6-m CCL.

transport mechanism for organic contaminant through a geomembrane can be seen from the results in Figures 13.29 and 13.30. Contaminant transport through composite liners considering leakage, diffusion, attenuation of contaminants, finite service life and operation of landfill is examined in Section 16.6.

13.6 Summary

Provided that adequate engineering design is carried out and appropriate CQC/CQA procedures are followed during manufacturing and installation, an HDPE geomembrane may be expected to perform well for a considerable period of time. In many landfills, the geomembrane may be required to act as a hydraulic and diffusive barrier for long periods of time (i.e., hundreds of years). The service life and long-term performance of the geomembrane is influenced primarily by the synergistic effects of physical stresses and chemical exposure over an extended period of time, possibly at elevated temperatures. A number of issues that need to be addressed in the design and installation of geomembrane liners to ensure sufficient long-term performance are summarized below.

1. The service life of an HDPE geomembrane depends on the type and formulation of the polymer (e.g., crystallinity, stress crack resistance and antioxidants), and the exposure conditions including temperature, surrounding medium, chemical concentration and long-term tensile stresses. Oxidative degradation is the principal mechanism of chemical ageing for HDPE geomembranes in landfills. The rate of oxidation is controlled by the amount of antioxidant in the geomembrane and the rate of antioxidant consumption and/or removal from the geomembrane. Geomembranes should have a specified minimum OIT as indicated in Table 13.2.
2. Tests examining antioxidant depletion rates from accelerated ageing tests at elevated temperatures can be used to obtain lower bound estimates of geomembrane service life. Based on presently available data, a primary HDPE geomembrane liner has a service life greater than 200 years provided that the temperature is not higher than 15 °C, whereas the service life is estimated to be reduced to 70 years if the temperature at the base of the landfill is at 33 °C.
3. The stress–strain response of HDPE geomembranes is influenced by loading rate and temperature, and the geomembrane may experience either creep or relaxation over time. When subject to tensile stresses, failure may be either ductile or brittle with the transition between brittle and ductile behaviour being affected by properties of the polymer, stress level, temperature and chemical exposure. To reduce the potential for brittle failure, the geomembrane must have sufficient stress crack resistance (Table 13.2) and the long-term tensile strains and surface damage should be minimized.
4. Sustained long-term tensions may occur due to differential movements, local deformations caused by indentations from overlying gravel and wrinkles in the geomembrane. It is necessary to provide adequate protection to the geomembrane to prevent puncture and limit long-term tensile strains. Limiting the number and size of wrinkles is also beneficial.
5. A geomembrane that is free of holes or tears is essentially impermeable to water. Leakage will occur through holes ranging from pinholes to large tears. Proper CQC/CQA can be expected to minimize the number of holes; however, some holes are to be expected and should be considered in design calculations. Leakage through a geomembrane overlying a low permeability layer (either CCL or GCL) is much lower than that for a geomembrane or a low permeability layer alone. The quantity of leakage depends on the number and size of holes in the geomembrane, the hydraulic head acting above and below the composite liner,

the hydraulic conductivity and thickness of any underlying soil, and the nature of the interface between the geomembrane and underlying material. Holes located at or near wrinkles will greatly influence the amount of leakage.

6. Some contaminants can migrate through geomembranes in the absence of leakage by the process of molecular diffusion. HDPE geomembranes act as a good diffusion barrier to water and hydrated ions such as chloride. For these contaminants, diffusion through the geomembrane is not significant and the potential contaminant impact is governed by leakage through holes in the geomembrane and service life of the geomembrane. However, for certain low solubility organic contaminants, diffusive migration may be quite significant for a well-constructed liner (i.e., one with only a few holes), and should be considered if one is to ensure a safe design.

Covers

14.1 Introduction

Covers are intended to serve several functions in the waste containment facility and may be required to: control infiltration of moisture into the waste; manage migration of gas to or from the waste; reduce nuisance from odours, litter, birds and rodents; prevent erosion; and facilitate post-closure activities. There are different types of covers encountered in landfills. For example, daily cover, that may consist of soil, wood chips, compost, geosynthetic materials or spray-on materials, is required at the end of each working day largely to reduce nuisances. Intermediate covers may also be required when a portion of the waste will be left for a period of time. However, in this chapter attention is focussed on the final cover (or top cap) for municipal solid waste landfills, although limiting the inward migration oxygen to sulphidic mine waste is also discussed in Section 14.5. The issue of whether to limit or promote infiltration of moisture into waste is first examined in Section 14.2. Common types of hydraulic cover systems are then examined and design issues for these covers are discussed (Sections 14.3 and 14.4). Finally the use of alternate cover systems is considered (Section 14.5).

14.2 Controlling infiltration

There are various philosophies to controlling the generation of leachate. One is to construct a low-permeability cover as soon as possible so as to minimize the generation of leachate. This approach has the benefits of minimizing the amount of leachate that must be collected and treated, and minimizing the mounding of leachate within the landfill. However, it also has the disadvantage of extending the contaminating lifespan (see Section 16.4.3).

Because of the heterogeneous nature of waste, some leachate will be generated almost immediately even for landfills designed with low infiltration covers. With low infiltration, it may take from decades to centuries before the field capacity of the waste is reached and full leachate generation occurs. This means that the full capacity of the leachate collection system may not be required for many decades after construction. However, during this period of time, degradation and biological clogging of the leachate collection system can be occurring unless the waste has been pre-treated to convert it to an essentially inert form. Furthermore, due to the variable nature of leachate generation, it may be difficult to assess whether there has been a failure of the leachate collection system by monitoring the volumes of leachate extracted.

An alternative philosophy is to allow as much infiltration as would practically occur. This would bring the landfill to field capacity quickly and allow the removal of a large proportion of contaminants (by the leachate collection system) during the period when the leachate collection system is most effective and is being carefully monitored (e.g., during landfill construction and, say, 30 years after closure). The disadvantages of this approach are twofold. First, larger volumes of leachate must be treated and this has economic consequences for the proponent. Second, once the leachate collection system fails, a high infiltration will result in significant leachate mounding.

The low infiltration philosophy is readily suited to meet environmental regulations that only require a limited period during which the landfill must not cause an unacceptable impact (be it 30 or 100 years). However, engineers have a moral responsibility to also consider the longer-term consequences. In some areas (e.g., Ontario, Canada), there is also a regulatory responsibility to consider environmental protection in perpetuity.

When long-term protection is considered, it may be desirable to find a balance between the low/high infiltration philosophies. Although it is not always practical (e.g., because of after use requirements), one option is to allow high infiltration during construction of the entire landfill (not just the cell) and for a period after landfilling ceases. Once the landfill has been largely stabilized or the leachate collection system starts to degrade noticeably, then a low infiltration cover would be constructed. This approach rapidly brings the landfill to field capacity and removes a substantial portion of the potential contaminants early in the life of the landfill when the engineered components are at their most reliable, thereby reducing the contaminating lifespan. This approach also reduces the infiltration and potential mounding of leachate after the low infiltration cover has been constructed and hence minimizes potential problems once the performance of key components of the engineered system (like the primary leachate collection system) have degraded.

The selection of whether to minimize infiltration or permit high infiltration not only influences the design of the final cover, but also should be considered in selecting the barrier system, in developing the monitoring and contingency plans, and in assessing the service life and contaminating lifespan of the facility.

14.3 Common final cover systems

Most final cover designs incorporate some sort of hydraulic barrier to control the infiltration into the waste. The final cover may be as simple as minimum of a 0.6-m thick cover material, underlying 0.15 m of topsoil and a vegetative cover, as shown in Figure 14.1a. This design corresponds to the minimum requirements specified for landfills in Ontario, Canada. The purpose of the topsoil (silt, silty-loam, loam, sandy-loam) is to sustain plant growth and thereby limit erosion of the cover soils by water and wind. Given a humid environment, the slope and type of cover soil can then be selected to

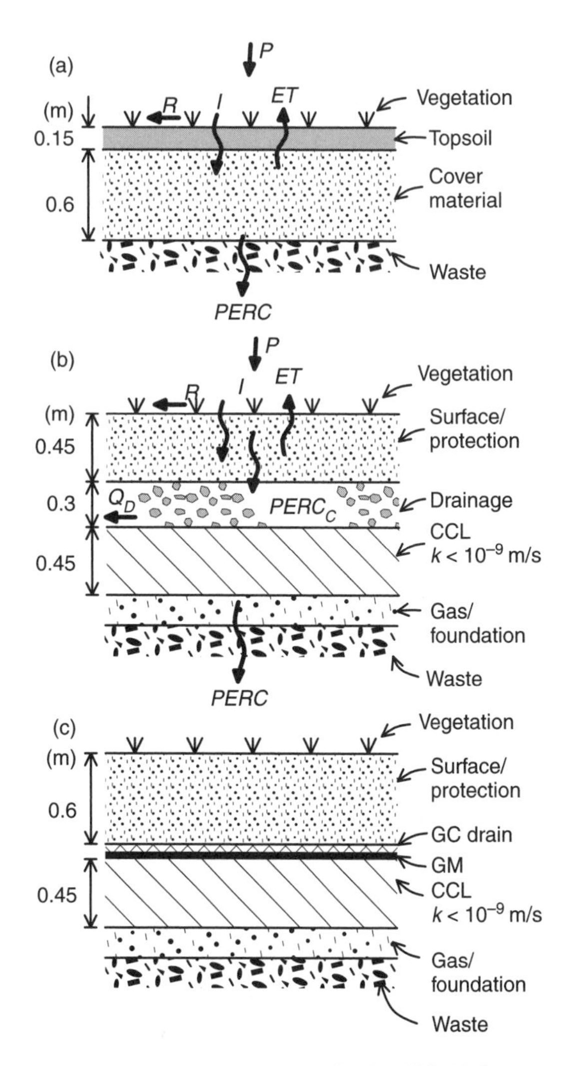

Figure 14.1 Final cover systems for landfills (CCL, compacted clay liner; GC, geocomposite; GM, geomembrane).

obtain an infiltration greater than or equal to $0.15\,m^3/m^2/a$ for the particular conditions at the site (i.e., precipitation and evapotransipitation – see Section 14.4), as required by Ontario Regulation 232/98 (MoE, 1998).

The final cover system required by RCRA Subtitle D, 40 CFR § 258.60(a), must be designed and constructed to: (a) minimize infiltration and erosion, (b) have a permeability less than or equal to that of any bottom liner system or natural subsoil present, or no greater than 1×10^{-7} m/s, (b) consist of a minimum of 0.45 soil cover and have an erosion layer that contains a minimum of 0.15 m of soil to sustain plant growth.

In cases where the rainfall infiltration into the waste is to be minimized, more complex cover systems consisting of multiple layers can be used. For example, Figure 14.1b illustrates a hydraulic barrier consisting of a CCL beneath a drainage layer and a protection/surface layer. The protection layer may store a portion of infiltrated water that may subsequently be removed from the cover through evapotranspiration and protect underlying layers from burrowing animals, roots, wetting–drying cycles and/or freezing (Daniel and Koerner, 1993b). The drainage layer reduces the hydraulic head acting on the hydraulic barrier, may increase the water-storage capacity of the protection/surface layer and is beneficial for cover stability on slopes (Bonaparte and Yanful, 2001). Proper drainage of fluid collected in this layer is important to reduce pore pressures that tend to reduce the stability of the cover slopes. Bonaparte *et al.* (2002) recommend the inclusion of a drainage layer above the hydraulic barrier when slopes of the cover exceed 5H:1V in order to prevent seepage pressures that may lead to cover instability. Either graded soil filters (e.g., see Cedergren, 1989) or geotextiles (Section 15.3) may be required to separate dissimilar materials in the cover. Figure 14.1b also shows a gas transmission layer beneath the compacted clay that is required where the waste generates gas (for municipal solid waste this normally arises from the degradation of organic matter in the waste). If constructed with granular soil, this layer may also act as a foundation for the cover system. In either case, particular care is needed to consider the potential degree of saturation of the gas transmission layer since the permeability with respect to gas flow can decrease significantly as the volumetric water content of the gas transmission layer increases (this has a negative effect when the objective of the layer is to encourage gas flow for collection and a positive effect when the objective is to minimize gas flow as discussed in Section 14.5.1).

The material presented in Chapter 3 applies equally to a clay liner in a cover or a base liner. However, in addition to the potential for desiccation cracking of the clay liners (Section 3.3.6), consideration must also be given to the possibility of tensile cracking induced in a clay liner from non-uniform settlement of the underlying waste (Section 15.2.1b). Allowable tensile limits for compacted clays will depend on the type of soil, water content and type of compaction. A summary of published data by LaGatta *et al.* (1997) suggests that the tensile strain at failure may vary from 0.1 to 4%, depending on the type of clay, plasticity index and water content. Proper geotechnical evaluation of the tensile limits of a clay liner is required on a case-by-case basis. If tensile strains in the clay may be anticipated to approach failure, consideration should be given to geosynthetic reinforcement beneath the liner (Figure 1.31) to limit tensile strains, or the use of an alternate composite liner to avoid an increase in hydraulic conductivity and thereby infiltration into the landfill.

Many variations on the design in Figure 14.1b are possible with the use of geosynthetics. For example, Figure 14.1c shows a cover system with a geocomposite drainage layer overlying a composite geomembrane/CCL. As discussed in Section 13.4, the leakage through a GM/CCL composite liner can be substantially less than a CCL alone. Consideration must be given to the flexibility, puncture strain and interface strength when selecting of a geomembrane product for use in a cover (refer to Table 13.1) because of the differential movements and tensile forces that the geomembrane will be exposed to in a cover above waste. Careful assessment of cover stability is required given the potential low interface strength between certain geomembranes and adjacent materials (Section 15.2.5). Work by Giroud and Beech (1989), Giroud *et al.* (1995a), Koerner and Soong (1998) and Bonaparte and Yanful (2001) can be consulted for more details regarding the stability of cover systems on slopes. Consideration of uplift pressures on the geomembrane from landfill gas is also required. Geosynthetics (geotextiles, geonets or geocomposites) can be used to transmit these gases to a higher vent or collection location. The use of geotextiles for gas transmission has been discussed by Koerner *et al.* (1984).

GCLs (Chapter 12) can be used instead of a thicker CCL in the cover, although generally the GCL should have about 1 m of soil above it to minimize the risk of desiccation. A schematic of a design that uses a GM/GCL composite liner in the cover system is shown in Figure 1.31. A GCL may be able to tolerate larger tensile strains than a CCL (LaGatta *et al.*, 1997); however, the tensile strain in GCLs still needs to be limited. The cover systems of Figures 14.1b,c and 1.31 can be quite effective in constructing a relatively tight cap, which will place a control on the infiltration into the landfill facility. A composite cap will also reduce the escape of gas (making for more effective gas collection if combined with a suitable gas collection system); however, as discussed in Section 14.2, it is important to recognize the interrelationship between the cover design (which controls infiltration), the effect of moisture availability on the degradation of organics (and hence gas generation), the contaminating lifespan of the landfill, and the service life of the components of the leachate collection, leak detection and leachate containment systems.

14.4 Calculating percolation through the cover

The percolation through the cover (and hence the infiltration into the waste) may be estimated by considering conservation of water mass between inflow, outflow and any storage. For the case where there is no drainage layer in the cover (Figure 14.1a), the percolation through the cover is:

$$PERC = P - R - ET - \Delta S \qquad (14.1)$$

where $PERC$ is the percolation through the cover; P the precipitation; R the surface water run-off; ET the net evapotranspiration (water loss to evaporation and plant use); ΔS the change in water storage at the cover surface, in plant foliage or in the soil materials of the cover system; and all terms have the units of L/T. When there is a drainage layer in the cover system (Figure 14.1b) that removes fluid at a rate of Q_D, the percolation through the cover becomes:

$$PERC = P - R - ET - \Delta S - Q_D \qquad (14.2)$$

Precipitation data for the site can normally be obtained from meteorological records. Any input of water from irrigation that may be required to sustain vegetation must also be included in the water balance. Surface run-off depends on the slope of the cover and the type of material on the cover surface. The net evapotranspiration is a function of the temperature and should be adjusted for the time of year and the latitude. The change in water storage depends on the infiltration, and the water content and the storage capacity of the cover soils.

For details on calculating the percolation into the waste with simple calculations, see Koerner and Daniel (1997). Bonaparte and Yanful (2001) have presented a useful review of commonly used computer programs that can also be used to estimate infiltration into the waste.

14.5 Alternate cover systems

Two alternate types of covers that may be used in arid and semi-arid climates involve capillary and evapotranspirative barriers (Benson and Khire, 1995; Bonaparte and Yanful, 2001).

14.5.1 Capillary barriers

A capillary barrier also acts as hydraulic barrier, but rather than using a fine-grained soil to provide low hydraulic conductivity, an unsaturated coarser-grained soil placed beneath a finer grained soil is relied upon to provide low hydraulic conductivity. This system is based on the observation that the hydraulic conductivity of an unsaturated soil may be substantially lower than that for a saturated soil because of matric suctions (e.g., Fredlund and Rahardjo, 1993). Thus, although a saturated coarse-grained soil (e.g., sand) will have a higher hydraulic conductivity than a saturated fine-grained soil (e.g., clay), at low degrees of saturation (water contents $<\sim 2\%$) the sand may have a hydraulic conductivity an order of magnitude lower than the clay, provided that the fine-grained material is able to sustain high matric suctions. A simple example of a capillary barrier may consist of a clayey soil overlying sandy soil. The clay layer serves as a protective cover, provides storage of infiltrated water and permits matric suctions to be developed in the underlying sand. However, if the water content of the lower coarse-grained soil were allowed to increase (e.g., due to high infiltration from a large precipitation event or cracking of the overlying clay layer), the hydraulic conductivity of the sand would greatly increase and the capillary barrier would be destroyed. As such, capillary barriers may best be suited for arid climates. The cover must have a slope that will promote proper drainage without erosion.

Capillary barriers also may be used to control acid production in sulphide mine waste. The objective in these designs is to construct a cover system with an upper finer-grained (relative to other layers in the system) soil layer that maintains a high degree of saturation and thus minimizes advective diffusive migration of oxygen through the cover (see Section 8.11.2). This is underlain by an unsaturated coarser layer that maintains a high matric suction at the soil interface and hence, does not allow any significant movement of moisture down from the overlying fine-grained layer thereby preventing the finer-grained layer from draining. A coarse-grained surface layer is frequently used over the finer-grained soil to increase percolation into, and reduce evapotranspiration from, the finer-grained soil layer. The reader is referred to Rasmuson and Eriksson (1986), Nicholson *et al.* (1989) and Yanful and Aubé (1993) for a more detailed discussion of capillary barriers used in mine applications, and to Yanful (1993), Yanful *et al.* (1993a,b), Woyshner and Yanful (1995) and O'Kane *et al.* (1995) for details regarding the documented performance of capillary barriers in field tests.

14.5.2 Evapotranspirative barriers

An evapotranspirative barrier attempts to control infiltration to the waste by relying on moisture storage and evapotranspiration of the cover soil. A thick (1–2 m) soil layer of fine-grained soil (e.g., silty sands, silt, and clayey silts) is placed over the waste. The thickness of the cover is then selected to result in no change in water content of the cover. There is little guidance available on the design of evaporative covers, although the thickness, unsaturated hydraulic conductivity and storage characteristics of the cover soil, as well as the precipitation and evapotranspiration at the site, are important design parameters. Zornberg *et al.* (2003) described the analysis and design of an evapotranspirative cover for a hazardous waste landfill in southern California.

14.5.3 Other options

Alternate materials to conventional soil and geosynthetic materials have also been used in limited cover applications. For example, McBean *et al.* (1995) describe the application of an asphalt barrier cover consisting of 5 mm of thick fluid asphalt membrane and 0.15 m hot mix asphalt concrete. Shallow and deep-rooted plants can also be used as a phytocover system to create a zone where plants extract moisture thereby minimizing the amount of water percolating into the landfill (e.g., Stack *et al.*, 1999). Recycled or waste materials have also been used including paper sludge, fly ash, incinerated sludge, dredged materials, street sweepings and incinerated tire slag.

Chapter fifteen

Geotechnical and related design issues

15.1 Introduction

In addition to the many issues related to contaminant transport from waste containment facilities detailed in the preceding chapters, consideration must also be given to the general stability of the facility and the performance of individual components within the facility. In this chapter, possible failure mechanisms and geotechnical design considerations related to the strength and stiffness of the soil and waste materials are examined. Design and construction issues for geotextiles and buried structures (leachate collection pipes and manholes) in waste containment facilities are also presented.

15.2 Geotechnical considerations

15.2.1 Properties of municipal solid waste (MSW)

The properties of MSW are highly variable. In part, they depend on the type of waste, method of placement, operation of landfill and age of waste. Even with these factors relatively constant, there can be substantial variations in properties given the heterogeneous nature of waste.

(a) Unit weight

The unit weight (or density) of MSW is used to estimate the volume of the landfill required for a given mass of waste and for estimating settlement and stability. This value may be obtained from excavated test pits in the waste. A detailed description of this approach, together with an assessment of the advantages and disadvantages of this and other methods for estimating unit weight, is given by Gachet *et al.* (1998). Other studies have used test fills (e.g., Marques *et al.*, 1998) and large-scale laboratory methods (e.g., Powrie and Beaven, 1999) to estimate unit weight.

The unit weight of MSW normally ranges between 6 and 12 kN/m^3 for an average degree of compaction. MSW that contains a significant proportion of metal, glass or rock can be expected to have a higher unit weight than MSW that is predominantly comprised of organic matter or textiles (Manassero *et al.*, 1996). Likewise, frequent placement of daily cover soil will influence the bulk unit weight of the waste mass (Knochenmus *et al.*, 1998). The unit weight is also sensitive to the moisture content, with reported values between 10 and 100%, depending on factors such as the organic content and age of the waste (Knochenmus *et al.*, 1998). Activities such as recirculation of leachate may be expected to increase moisture content and hence waste unit weight.

Given the likely range in unit weight (and hence density), the selection of a conservative value of unit weight to use in calculations depends on the situation being considered. For example, in obtaining the volume of landfill required to dispose of a given mass of waste it would be appropriate to select a lower bound value of density for the compaction and cover expected at the site. Whereas in assessing the stability of slopes in the waste pile (i.e., where the mass of the waste causes instability), it would be best to select an upper bound of the expected unit weight. Consideration should be given to the increase in unit weight with time (e.g., from waste compression and increased moisture). Sensitivity analysis should be performed over a range of likely unit weights to assess the sensitivity of the design to reasonable uncertainty regarding this (and other) parameter (e.g., see Stark *et al.*, 2000).

(b) Stiffness

Knowledge of the waste stiffness is required when estimating the settlement of the waste. These settlements need to be considered for the design of the final cover (see Chapter 14), manholes (see Section 15.5) and post-closure use at the site. This settlement occurs from self-weight compression of the waste, decomposition of the waste with time (e.g., biodegradation of organics, corrosion of inorganics) and any possible loading acting on the waste. Empirical procedures are, at present, best used to predict the long-term settlement of MSW. For example, Park *et al.* (2002) summarized four such predictive models and found that they were able to successfully calculate settlements when compared with available field measurements provided that the decomposition of the waste with time was considered.

Many investigators have examined the compression of MSW using parameters derived from laboratory or field tests. Manassero *et al.* (1996) advocated the use of the stiffness modulus to characterize MSW behaviour and reported values ranging from 0.5 to 3.0 MPa, depending on the vertical stress level. Powrie *et al.* (1999) reported compressibility values ranging from 7.45 to 1.07 MPa^{-1} under applied stresses of 34–463 kPa. There was a slight reduction for aged waste which gave compressibilities of between 7.38 and 0.66 MPa^{-1} over the same stress range.

(c) Shear strength

The shear strength of the waste is required to assess the stability of waste slopes. Similar to soil materials, the shear strength of waste is typically modelled using a Mohr-Coulomb failure criteria (Knochenmus *et al.*, 1998) which arises from a combination of interparticle friction, ϕ', and the reinforcing effect of sheet-like plastic and paper material, c' (Knochenmus *et al.*, 1998; Powrie *et al.*, 1999). The failure envelope is therefore a function of waste composition, particle size (Brandl, 1995) as well as density. The mobilized cohesion is reported to decrease with increasing moisture content (Gabr and Valero, 1995). The shear strength also can be expected to change due to time-dependent changes (Brandl, 1995; Knochenmus *et al.*, 1998) that may both lead to an increase in strength (e.g., densification of the waste) and a decrease in strength (e.g., biodegradation and breakdown of waste constituents) making it difficult to estimate the long-term strength.

Given the number of factors influencing inherent variability of the waste strength and the limited available data, it is difficult to generalize regarding the strength of MSW. In addition to the fundamental factors previously mentioned, the reported strength will also depend on factors related to its determination including the stress conditions and the size of the waste specimen tested. Eid *et al.* (2000) provided a summary of data from large-scale direct shear tests and back analysis of failed waste slopes and showed that a linear failure envelope with $\phi' = 35°$ and $c' = 25$ kPa (for normal stresses between 10 and 350 kPa) provided reasonable agreement with the available data. Alternatively using the same data, estimates of shear strength parameters can be obtained over smaller intervals of normal stress that would give lower ϕ' at higher normal stresses. For landfills where the waste slopes are no steeper than 4H (horizontal) to 1V (vertical), it may be appropriate to select parameters from published data for similar waste. For higher risk projects (e.g., where the waste slopes may exceed 4H:1V), published values may be used for an initial estimate, but large-scale direct shear tests on the expected waste may be required. Consideration of instability from sliding along weak planes (especially along interfaces involving geosynthetics – see Section 15.2.4) must be given when assessing the stability of the waste pile.

(d) Hydraulic conductivity

An assessment of waste hydraulic conductivity is needed to predict the rate and pattern of moisture movement within MSW, especially when practising leachate recirculation, and for assessing leachate mounding (Section 2.4.2). Hydraulic conductivity, like the other waste properties discussed in this section, is highly dependent on the composition of waste, degree of compaction (Manassero *et al.*, 1996), overburden pressure (Powrie and Beaven, 1999) and the age of the waste (Powrie and Beaven, 1999). Generally the higher the compaction and overburden stress and the greater the age of the waste, the lower the hydraulic conductivity.

Data summarized by Manassero *et al.* (1996) showed a wide range of hydraulic conductivity with values between 2×10^{-4} and 1×10^{-8} m/s being reported. Most of the published values fell between 10^{-4} and 10^{-6} m/s; however, much of this data were from tests performed at relatively low effective stresses in the uppermost portion of the waste. Large-scale test results from the Pittsea compression cell (Powrie and Beaven, 1999) gave hydraulic conductivity values that decreased from 1.5×10^{-4} m/s at an applied stress of 40 kPa to 3.7×10^{-8} m/s at 600 kPa. They reported the following best-fit line estimate of hydraulic conductivity, k_w (in m/s) in terms of the vertical effective stress, σ' (in kPa):

$$k_w(\text{m/s}) = 17[\sigma'(\text{kPa})]^{-3.26} \qquad (15.1)$$

Rowe and Nadarajah (1996c) reported the following correlation, based on field data at different depths, relating hydraulic conductivity, k_w (in m/s) to depth, z (in m):

$$k_w(\text{m/s}) = 1.8 \times 10^{-4} \exp[-0.269\, z(\text{m})] \qquad (15.2)$$

While both equations 15.1 and 15.2 represent best-fit curves to data, it must be recognized that there is considerable scatter of data around these curves as a consequence of the intrinsic variability of waste. Thus these, like any other empirical relationships, should be used with considerable caution.

The ratio between horizontal, k_{wh}, and vertical, k_{wv} hydraulic conductivity has been investigated by Hudson *et al.* (1999) and Landva *et al.* (1998). Hudson *et al.* (1999) reported that the anisotropy ratio, k_{wh}/k_{wv}, increased from about 2 at an applied vertical stress of 40 kPa to about 5 at an applied stress of 600 kPa. Landva *et al.* found the ratio to be relatively constant at $k_{wh} \approx 8k_{wv}$, regardless of the level of applied vertical stress in the range of 150–500 kPa. This anisotropy may be particularly important in designing leachate recirculation systems and has the potential to give rise to leachate seeps with the injection of leachate.

15.2.2 Settlement

Settlement of soils occurs because of increases in effective stresses; for waste containment facilities, the increase in effective stress normally arises from the weight of the waste. The magnitude of ground settlements can be estimated using standard procedures described in conventional Soil Mechanics texts (e.g., Lambe and Whitman, 1979). Settlements do not normally pose a major problem, provided that they are uniform across the site. Concern arises when the settlements are larger in one region relative to another. This can occur due to variations in the applied pressures (e.g., from variable thickness in waste), thickness of compressible layers and stiffness of underlying soil materials. Such differential settlements of the soil beneath a landfill and/or differential settlement of CCLs may reduce the effectiveness of leachate collection systems. Leachate will pond in areas where settlement occurs, increasing the hydraulic head and consequent flow through the underlying soil. Thus, the magnitude of differential ground settlements should be estimated when selecting the slope of the leachate collection system (Section 2.4), and whenever possible this slope should be selected to minimize leachate ponding after ground settlements have occurred.

Additionally, tensile stresses induced in geomembrane liners from differential settlement of underlying materials may initiate holes or ruptures. Giroud and Bonaparte (2001) present equations that can be used to calculate the strains in geomembranes arising from differential settlements. Differential settlements may also cause axial tensile stresses in leachate collection pipes, especially near connections with manholes. Existing solutions derived to obtain the stresses in laterally loaded piles (e.g., Poulos and Davis, 1980) may be used to estimate the axial stresses in the pipes in these cases. It is conceivable that under some extreme circumstances, differential ground settlements may lead to tension cracks in compacted or natural clay barriers.

Disposal of waste on top of an existing landfill (often referred to as vertical expansion) is becoming increasingly common to satisfy demand for landfill space and given the public aversion to the approval of new landfills. Vertical expansions may be expected to increase the occurrence of problems related to differential settlements. The use of geogrid reinforcement over areas of potential differential settlements may reduce such impacts. Since the settlements of the waste are expected to be highly variable (leading to large differential movements), the impact of these settlements on engineered components in the landfill needs to be carefully considered, and redundancy should be included in design if the consequences of excessive differential settlements could lead to unacceptable performance.

15.2.3 Bearing capacity

Consideration should be given to the overall bearing failure of the soil beneath the landfill. This may be especially important for deep landfills underlain by very soft soils where the strength of the soil may be insufficient to resist the pressures from the landfill above. Bearing failure may lead to large vertical and lateral displacements that can cause damage to the barrier system. The bearing resistance depends on the geometry of the landfill, the shear strength of the underlying soil and the nature of the underlying ground materials with depth (e.g., the presence of stiff or weak stratum) to a depth beneath the landfill approximately equal to the width of the landfill. Assessment of the potential of bearing failure can be made using conventional bearing capacity theory (that can be found in standard geotechnical engineering texts) treating the landfill as a flexible foundation resting on the underlying ground materials.

15.2.4 General stability

A landfill can represent a major loading on the underlying soil. Just as with the design of any structure on soil, care should be taken to ensure the overall stability of the landfill under both static and seismic conditions where appropriate. This involves an evaluation of the geotechnical conditions (e.g., type and strength of soil and waste materials, pore pressures in the soil and waste) and assessment of anticipated loadings followed by an appropriate stability analysis (e.g., see Leroueil *et al.*, 2001) to identify the most critical conditions in the development of the landfill. A factor of safety is used to keep the magnitude of the disturbing forces less than that of restoring forces along a potential rupture surface. With respect to global stability of a landfill, the factor of safety attempts to quantify uncertainty in material properties (both loads and resistances) and assumptions of the stability analysis, and the implications of failure with one single value. Thus, the required factor of safety for stability is not a unique value and depends on the situation being examined.

Stability analyses should consider potential instability: (1) during excavation of the landfill (e.g., side slopes, Figure 15.1a, or trenches); (2) during development of the landfill (e.g., waste slopes, Figure 15.1b) and (3) of the completed landfill (e.g., waste slopes, Figure 15.1c and around the landfill Figure 15.1d,e). Particular care is required not to underestimate the pressures that can be applied by partially saturated waste, and a conservative estimate of the unit weight of waste should be considered when evaluating landfill stability (see Section 15.2.1). This is particularly critical for landfills being constructed in soft soils, sensitive soil, near escarpments or on slopes. Proper assessment of pore pressures is essential, and the potential for increased pore pressures in the waste (Figure 15.1c) due to clogging of the leachate collection system or leachate recirculation should not be overlooked. Consideration must be given to non-circular failure surfaces either because of planes of weakness in underlying ground materials, in the waste, or along interfaces between geosynthetic components (Section 15.2.5). When existing landfills are being expanded particular care is required to integrate the new and old portions of the landfill and, in particular, to avoid stability problems when excavating at or near the toe of the existing landfill (Figure 15.1f).

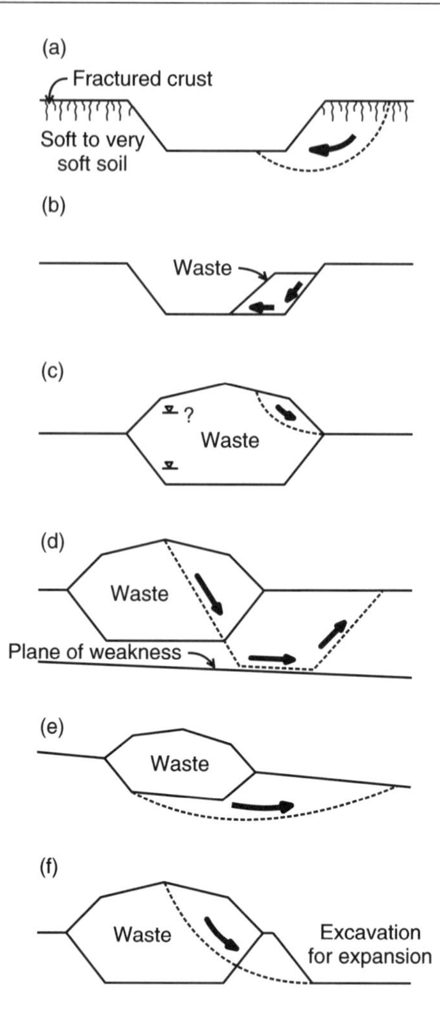

Figure 15.1 Illustration of a number of potential mechanisms to consider when examining landfill stability: (a) after excavation of cell, (b) during placement of waste, (c) waste slopes, (d) failure along planes of weakness, (e) general slope failure initiated by presence of landfill and (f) failure initiated by excavation for expansion.

Landfill failures have occurred due to failure to adequately assess overall stability. Selected case histories involving instability of landfills are presented in Table 15.1. In one such example reported by Reynolds (1991), failure occurred during the expansion of an existing 22-m high MSW landfill. Although there were a number of factors contributing to the failure, excavation of 1–3 m of surficial stiff clay and a trench at the toe of the slope of the old landfill were the principal causes of the instability. In this case, the density of the waste was also underestimated (by more than a factor of two) as greater compactive effort and more daily and intermediate cover (to control odour, birds and windblown waste) were used during placement of the waste than was assumed in the initial design calculations. Additionally, a stockpile of material that was placed near the crest of the landfill and heavy rains that occurred prior to the slide also

Table 15.1 Summary of selected case histories involving instability of landfills (based on reports in papers cited)

Case	*References*	*Barrier system*	*Description of instability*	*Likely cause*	*Lessons learnt*
Maine Slide	Reynolds (1991)	Existing landfill: unlined Expansion: GM + 0.6-m CCL	Large mass movements of waste (up to 50 m) and flow of remoulded soft clay (up to 120 m) from general shear failure during expansion of existing landfill	Several factors: excavation of stiff clay crust at toe of slope, density of waste was higher than expected, pile of excavated soil placed near landfill crest, and heavy rainfall	Must carefully consider how construction activities during expansion may influence stability
Kettleman Hills Slide	Mitchell *et al.* (1990a,b), Seed *et al.* (1990), Byrne *et al.* (1992), Stark and Poeppel (1994)	Triple composite liner (from top down): 1.5-mm GM + 0.45-m CCL + 0.3-m SLCS+ 1.5-mm GM + 1-m CCL +GT + drainage rock+ 2-mm GM + subgrade	Mass movement of waste with lateral displacement up to 11 m and vertical slump up to 4.3 m from slip along multiple geosynthetic interfaces	Low interface strength between secondary GM and CCL from undrained clay response	Need to carefully select interface strength considering field conditions, strain level, water content
French Slide	Ouvry *et al.* (1995)	Base: 2-mm smooth GM + GT + 3-m CCL Slopes: 2-mm smooth GM + 3-m CCL	Mass movement of waste up to 6.7 m at the top of the waste during waste placement	Slip between GM/CCL interface on slopes and GM/GT along base	Consider possible decrease in interface strength from heavy rainfall
Cincinnati Slide	Stark *et al.* (2000)	1.5-mm GM + 1.5-m CCL	Translation towards excavation near toe of existing landfill	Strength of ground beneath landfill exceeded	Proper geotechnical assessment of global stability essential during expansion

Table 15.1 *Continued*

Case	*References*	*Barrier system*	*Description of instability*	*Likely cause*	*Lessons learnt*
Dona Juana Landill	Hendron *et al.* (1999)	1-mm PVC GM + CCL or native soil	Large down slope movement of waste. Failure surface not along liner	Large pore pressures in waste from leachate recirculation system	Recirculating leachate can lead to large pore pressures that must be included in stability assessment
Bulbul Drive Landfill	Brink *et al.* (1999)	CCL liner below phase 1A, composite 1.5-mm fPP GM + CCL below phase 1B in valley landfill with a longitudinal slope of approximately 10% and side slopes of approximately 36%	Rapid translational slide when waste height reached 45 m in the phase 1B. About 160,000 m^3 of waste flowed into the valley below	Liquid waste deposited into trenches excavated into uppersurface of landfill near interface between phases 1A and 1B. A failure surface developed along the interface between the two phases of waste and then along the interface with the composite liner at the base	Interface between phases of waste disposal may be weaker than the waste mass. Injection of fluid raised pore pressures and reduced shear strength
Beirolas Slide Lisbon	Santayana and Pinto (1998)	1.5-mm GM + GCL+ CCL (on base) liner on the landfill base. Foundation soil consisted of 4–5 m of silty clay fill overlying a 20–35-m thick estuarine and alluvial soft clay deposit	Slide extended about 270 m along most of the area where 110,000 m^3 of contaminated soil had been placed. Nearly vertical failure scarp associated with 4 m of vertical and several metres of horizontal movement	Shear strength of the soft clay subsoil overestimated. Also actual failure mechanism not considered in the design calculations	Perform a proper site investigation. Examine all potential failure mechanisms. Avoid optimism in shear strength when faced with data that is inconsistent with that optimism

GM, geomen brone; GT, geotextile; CCL, compacted clay liner; SLCS, secondary leachate collection system; PVC, polyvinyl chloride; FPP, flexible polypropylene decreased stability. This case clearly highlights the need to consider many issues when assessing stability.

Careful consideration must also be given to the development of excess porewater pressures within a landfill, especially those practising leachate recirculation, since this has been identified as a principal factor in the failure at the Dona Juana Landfill (Hendron *et al.*, 1999). Finally, it is essential not to overlook traditional geotechnical issues related to the stability of the subsoil as illustrated by the Beirolas slide in Lisbon. Here the post-failure investigation (Santayana and Pinto, 1998) concluded that the failure occurred because the shear strength of the soft clay had been overestimated. Although it had been considered to be normally consolidated under a surcharge of 4 m from the existing fill, it was reported to be underconsolidated with very little dissipation of the excess pore pressures caused by placement of the fill in the 1970s and 1980s. Furthermore, the failure extended out into the river in an area where no fill had been placed (and hence no strength gain could have occurred due to consolidation) but, reportedly, the failure mechanism had not been considered in the design calculations.

15.2.5 Stability of engineered systems on side slopes

Figure 1.31 illustrates an engineered barrier system involving geomembranes, geonet drains, geotextiles and compacted clay all extending up a side slope. Tensile forces will be mobilized in the geosynthetic components lining the side slopes of the waste containment facility due to the waste overburden loads, waste settlement (down drag forces), and from the self-weight of the geosynthetic components themselves. The current method for the evaluation of these tensile forces and general stability is static equilibrium method (Richardson and Koerner, 1988). This evaluation requires knowledge of the interface strength characteristics between the various components of the lining system. Except possibly for only very low-risk projects, the interface properties should be measured for each specific project on a case-by-case basis (as opposed to relying on published test values) due to disparities of interface strengths between products from different manufactures, or even for otherwise identical materials from the same manufacturer (Bonaparte *et al.*, 2002). Bonaparte and Yanful (2001) summarize laboratory methods to obtain interface strength parameters. Most commonly direct shear friction testing (e.g., Bove, 1990) is used. It is essential that these tests be conducted with conditions representative of the actual field conditions (e.g., materials, stress conditions, water contents, stress and strain levels, and strain rates). From these experiments both peak and residual strengths can be obtained. The peak strength corresponds to the maximum strength obtained from the test while the residual strength is often much lower and occurs at large strains. The selection of the appropriate value for use in a stability calculation is the subject of much debate and depends on the strains (or displacements) expected in the field and the factors of safety to be applied. In the extremes, if only small displacements can be assured it is reasonable to use the peak strength, whereas the residual strength should be used if large displacements are likely to occur. However, in many practical situations it is not so straightforward to select the appropriate strength, and each case needs to be evaluated carefully by an experienced geotechnical engineer as illustrated by the cases summarized in Table 15.1 and highlighted in the following paragraphs.

A case involving failure to sliding along the interfaces within a composite, multilayered geotextile, geomembrane and clay liner system has been reported and studied by Mitchell *et al.* (1990a,b), Seed *et al.* (1990), Byrne *et al.* (1992) and Stark and Poeppel (1994). Down slope mass movements of about 11 m horizontally and up to 4.3 m vertically were observed when the waste was reaching its maximum height of 27 m above the base. Failure was attributed to sliding along multiple interfaces within the landfill liner system, primarily resulting from low interface shear strength between the geomembrane and clay in a secondary composite liner system. The strength of this interface was essentially undrained even one year following construction since the low hydraulic conductivity of the clay and the presence of the geomembrane limited the dissipation of excess pore pressures in the clay. Additionally, it was believed that the peak interface strength might have only been mobilized along a portion of the failure surface. Thus, an evaluation of stability based solely on peak strengths is inappropriate for the large mass movements that occurred in this particular case.

Another case that demonstrates the need to carefully consider stability of the barrier system has been described by Ouvry *et al.* (1995). At the base of this landfill, the barrier system consisted of a 2-mm thick smooth HDPE geomembrane placed on top of a 190 g/m^2 spun-bonded nonwoven geotextile overlying compacted clay, and on the side slopes the geomembrane was placed directly on top of clay. Down slope mass movements of waste (as large as 5–6.7 m) were caused by slip along the geomembrane/clay interface on the side slopes and along the geotextile/clay interface at the base. At the time of the slide, the waste had an average thickness of 12–15 m (maximum 20 m). The geomembrane was pulled out of the anchor trench over a length of 60 m. The moisture content of the clay from beneath the geomembrane was found to be 5–9% higher than it had been after compaction because of heavy rainfall during placing of the geomembrane on the clay in this case. Since the water content influences the shear strength developed at the interface between geomembrane and clay, the potential for a decrease in design interface strength due to events occurring during construction should be considered. Instability due to fluid pressures at the interface can also arise from landfilling operations that involve injection of fluid into the landfill. This has been illustrated by the failure at the Bulbul drive landfill (Brink *et al.*, 1999) where injection of fluid caused a reduction in the shear strength at the interface between two phases of waste placement and along the interface with the lining system.

When interface friction alone is not sufficient to prevent sliding, either the inclination of the slope must be reduced or tensile elements must be introduced to carry loads that would otherwise result in shear along a potential failure surface. The geosynthetics in tension must be anchored at the top of the slope and descriptions of anchoring methods and determination of the anchorage capacity are available (e.g., see Richardson and Koerner, 1988).

Figure 1.31 shows the inclusion of geogrids to steepen the side slopes, which will increase the available landfill airspace. Jewell (1991) has published design charts for geogrid reinforced slopes. Note that an increase in slope angle will be accompanied by an increase in the tensile forces mobilized in the geosynthetic lining the slope and an increased risk of sliding unless appropriate measures are taken.

The geosynthetics in tension must have sufficient strength to withstand the tensile forces to which they are subjected, and high safety factors should be used since strength losses due to installation damage, long-term creep and degradation mechanisms must be expected. Possible degradation mechanisms include UV exposure during construction on unprotected slopes, chemical reactions with leachate, swelling due to chemical adsorption, extraction, oxidation and biological attack (Koerner *et al.*, 1990).

15.2.6 Blowout or basal heave

In addition to considering general stability due to bearing capacity or slope failure, consideration should also be given to the potential for blowout (or basal heave) of the bottom of the excavation. This occurs when the uplift from water pressure in an underlying aquifer is similar to the self-weight of the overlying materials and is particularly critical for landfills being designed with hydraulic containment (see Section 1.2.1) where the water pressure in an underlying aquifer is to be used to minimize contaminant transport from the landfill. This same water pressure (if not controlled) can cause blowout of the base of the landfill. The factor of safety against blowout FS_{bh} is defined as:

$$FS_{bh} = \frac{\sum_{i=1}^{n} \gamma_i t_i}{h_p \gamma_w} \quad (15.3)$$

where γ_i is the unit weight and t_i the thickness of n layers of material overlying the aquifer, h_p the pressure head in the aquifer (see Section 5.2.1) and γ_w the unit weight of water. Blowout considerations may necessitate pumping of underlying aquifers to reduce water pressure, $h_p\gamma_w$ (and hence to ensure an adequate factor of safety against blowout) during excavation of a cell, placement of the barrier systems and the waste.

For example, consider the geometry shown in Figure 15.2. Assuming a unit weight of the clay till of $\gamma_t = 21\,\text{kN/m}^3$, of the engineered barrier system of $\gamma_e = 20\,\text{kN/m}^3$ and of the waste of $\gamma_{ws} = 6\,\text{kN/m}^3$, the maximum pressure head,

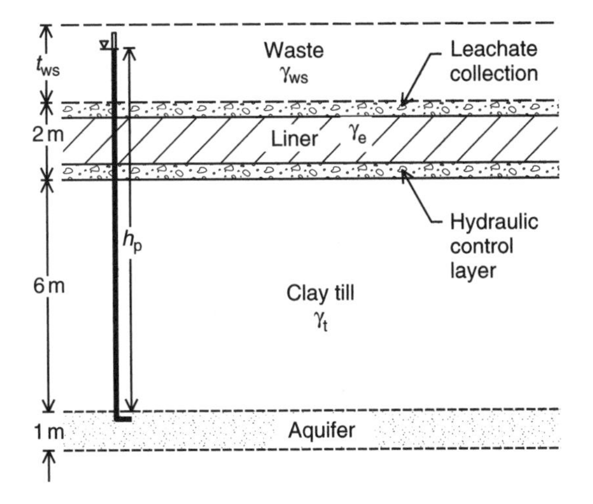

Figure 15.2 Schematic of natural soil, liner system and waste used for example blowout calculation.

h_p, in the aquifer for a factor of safety against blowout of $FS_{bh} = 1.4$ may be calculated using equation 15.3 for conditions during construction. The most critical condition with respect to blowout occurs after excavation of the till to the base of the landfill but before placing the engineered system, and the maximum allowable pressure head is equal to:

$$h_p = \frac{1}{\gamma_w}\left(\frac{6\gamma_t}{FS_{bh}}\right) = \frac{1}{9.8}\left(\frac{6 \times 21}{1.4}\right) = 9.2\,\text{m}$$

Placement of the 2-m thick engineered barrier system will increase the resistance to blowout; thus, the maximum allowable pressure head increases to:

$$\begin{aligned} h_p &= \frac{1}{\gamma_w}\left(\frac{6\gamma_t + 2\gamma_e}{FS_{bh}}\right) \\ &= \frac{1}{9.8}\left(\frac{(6 \times 21) + (2 \times 20)}{1.4}\right) = 12.1\,\text{m} \end{aligned}$$

Assuming that the natural pressure head in the aquifer h_p is 14 m, one could then estimate the thickness of waste t_{ws} that must be placed before pumping could be terminated and the aquifer is allowed to return to natural conditions from equation 15.3 as:

$$\begin{aligned} t_{ws} &= \frac{FS_{bh} h_p \gamma_w - 6\gamma_t - 2\gamma_e}{\gamma_{ws}} \\ &= \frac{(1.4 \times 14 \times 9.8) - (6 \times 21) - (2 \times 20)}{6} \\ &= 4.35\,\text{m} \end{aligned}$$

In this case, 4.35 m of waste would need to be placed in order to have a factor of safety against blowout of 1.4 with the natural hydraulic conditions in the underlying aquifer.

Basal stability requires special attention in areas where there is excavation into soil containing dissolved gas as illustrated by the cases reported by Rowe *et al.* (2002a). After excavation to a depth of about 25 m (in a 38–40-m thick clay deposit), venting of gas and water occurred at three separate locations. Additional details are given in Section 9.2.2.

15.2.7 Summary

Numerous landfills are safely constructed without stability problems. However, the issues discussed in this section illustrate the need for careful consideration to be given to the potential for instability during: (a) barrier construction, (b) placement of the waste, (c) the period of time after landfill closure and (d) expansion of existing landfills. The likelihood of failure occurring can be minimized by:

1. a proper geotechnical investigation of the subsoil properties;
2. carefully considering all potential failure mechanisms;
3. avoiding optimism regarding geotechnical properties;
4. taking account of the effect of a potential increase moisture content (e.g., due to leachate recirculation or injection of liquid waste) in increasing the unit weight of the waste and decreasing the shear strength of the waste and interfaces;
5. appropriate design and material selection (including appropriate laboratory tests and stability analyses);
6. good CQC/CQA to ensure that the barrier system is installed as designed;
7. taking account of the effect of excavation on stability (e.g., at the toe of existing waste);
8. development plans for expanded landfills that limit toe excavation and overfilling and define allowable conditions for construction of the expansion area and a means of monitoring adherence to the development plans;
9. operation plans that include consideration of stability as the waste is placed and means of monitoring adherence to the operation plans;
10. avoiding co-disposal of liquid waste or increasing the amount of liquid waste without fully assessing the potential impact on both stability and geoenvironmental protection;
11. contingency plans in the event of changed conditions occurring during construction (e.g., excessive rain, unexpected foundation conditions, etc.); and
12. disposal alternatives so that waste can be diverted if expansion schedules are not met.

15.3 Design of geotextiles

Geotextiles have found widespread use in modern waste containment facilities. They may be used as separators, filters, protection systems for geomembranes, reinforcement for soil and/or waste, or for drainage. Description of the different types and engineering properties of geotextiles can be found elsewhere (e.g., Koerner, 1998). The objective of this section is to discuss the design of geotextiles based on their intended function as separators and filters.

15.3.1 Geotextiles as separators

Geotextiles are used to separate dissimilar materials in the leachate collection system. For example, as shown in Figure 2.3, they may be used to separate waste from the leachate collection gravel, different gravels in the leachate collection system, and/or leachate gravel from an underlying clay liner. The design requirements of the separator geotextile between the waste and leachate collection system were discussed in Section 2.4.6. In all of these cases, the separator geotextile must have adequate strength to minimize damage during construction.

A separator geotextile is also required for barrier designs involving a secondary leachate collection system (SLCS) (or hydraulic control layer) beneath a primary CCL as shown in Figures 16.8, 16.10 and 16.11. This geotextile must also have adequate strength to survive construction, but in addition must have sufficient strength to span the voids in the underlying gravel when subject to overburden pressures from the overlying waste. At present, candidate geotextiles can only be evaluated based on laboratory testing like that conducted by Rowe and Badv (1996c).

15.3.2 Geotextiles as filters

When geotextiles are used as a filter between dissimilar materials they must have adequate soil retention capability, be sufficiently permeable and have sufficient resistance to clogging (Section 2.4.4c).

Adequate retention characteristics may be obtained by selecting the apparent opening size of the geotextile, O_{95}, from the guidelines given by Giroud and Bonaparte (2001) that are presented in Table 15.2. Selection of the opening size depends on the properties of the soil to be retained including soil density (characterized by density index I_D), particle size, d_{85} (i.e., the grain size for which 85% of the material is finer than by dry mass) and linear coefficient of uniformity, C^*_u, where:

$$C^*_u = \left(\frac{d'_{100}}{d'_0}\right)^{0.5} \tag{15.4}$$

and d'_{100} and d'_0 correspond to the grain sizes that intersect 100 and 0% passing obtained by extending the central portion of the grain size distribution curve (see Giroud and Bonaparte, 2001).

Consideration of permeability requires that excessive pore pressures do not develop in the soil and that the flow is not greatly reduced and may be satisfied provided that (Giroud and Bonaparte, 2001):

$$k_{FGT} > 10ki \tag{15.5}$$

and

$$k_{FGT} > k \tag{15.6}$$

where k_{FGT} is the hydraulic conductivity of the geotextile, k hydraulic conductivity of the soil to be retained and i the hydraulic gradient in the soil in near the filter. Giroud and Bonaparte (2001) provide some guidance on the selection of hydraulic gradient.

15.3.3 Construction damage

Defects arising from construction damage, poor seaming technique and rupture from excessive subgrade settlement may cause a geotextile to become discontinuous. The survivability of geotextiles during installation has been the subject of some study (e.g., Bonaparte *et al.*, 1988). Damage arising from construction may include puncture by coarse granular materials above and below the geotextile and tearing due to the action of heavy machinery. The geotextile

Table 15.2 Apparent opening size of geotextile O_{95} for adequate particle retention for geotextile filters (modified from Giroud and Bonaparte, 2001)

Density index of soil to be retained, I_D	*Linear coefficient of uniformity of the soil,* C^*_u	
	$1 < C^*_u < 3$	$C^*_u > 3$
Loose $I_D < 35\%$	$O_{95} < (C^*_u)^{0.3} d_{85}$	$O_{95} < \frac{9}{(C^*_u)^{1.7}} d_{85}$
Medium $35\% < I_D < 65\%$	$O_{95} < 1.5(C^*_u)^{0.3} d_{85}$	$O_{95} < \frac{13.5}{(C^*_u)^{1.7}} d_{85}$
Dense $I_D > 65\%$	$O_{95} < 2(C^*_u)^{0.3} d_{85}$	$O_{95} < \frac{18}{(C^*_u)^{1.7}} d_{85}$

should have sufficient strength and puncture resistance to survive normal construction. The construction specification should clearly emphasize the importance of maintaining the integrity of the geotextile. Rupture from subgrade settlements can be minimized by the use of low modulus geotextiles or by appropriate subgrade preparation.

For geotextiles with select waste over the geotextile (such that there are no coarse or hard materials directly in contact with the geotextile), Koerner and Koerner (1995b) proposed minimum values of geotextile strength to limit damage during construction. These guidelines listed in Table 2.25 may need to be increased if backfill or construction conditions are more severe than those considered by Koerner and Koerner (1995b). For example if the geotextile is in contact with coarse angular gravel (e.g., see Figure 2.3), a woven geotextile may be particularly susceptible to tearing even if it meets the requirements given in Table 2.25. Thus, the selection of a geotextile for an application involving harsh conditions such as this should involve a field trial to confirm that the proposed geotextile is suitable for the proposed conditions and construction procedures.

In one such field trial reported by Rowe *et al.* (1993), samples of geotextiles with the properties indicated in Table 15.3 were removed from interfaces between: (a) approximately 200 mm of angular, 50 mm crushed gravel and a CCL, and (b) a CCL and a rounded 10–30-mm gravel layer. No significant damage of the high survivability geotextile B was observed either in terms of observed holes or a change in tensile strength as obtained from the wide width tension test. The medium survivability geotextile A was indented but generally did not contain holes when placed between the liner and rounded 10–30-mm gravel; when placed between the 50-mm crushed gravel and a compacted clay, and subjected to many passes of a loaded water truck, some holes were observed. The reduction in wide width tensile strength was not significantly different for the two exposure conditions. The average reduction in tensile strength of geotextile A, due to construction damage, was about 15%. It is notable that the strain to failure decreased significantly (and the modulus increased) for both geotextiles, similar to findings of Bonaparte *et al.* (1988).

Table 15.3 Properties of nonwoven needle-punched geotextiles used in field trial reported by Rowe *et al.* (1993)

Property	*Geotextile A*	*Geotextile B*
Mass per unit area, M_A (g/m^2)	265	504
Filtration opening size, *FOS* (μm)	133	77
Wide width tensile strength (kN/m)	14	27

In addition to adequate strength to survive construction, careful attention to the details involving the seams between adjacent geotextile sheets is required. In the same field trial reported by Rowe *et al.* (1993), the geotextile seams were specified to have an overlap of 0.3–0.45 m and these specifications were met prior to the placement of the overlying material. However, excavation following construction showed that some of the seams had opened producing a gap up to 0.45 m wide between geotextile sheets. Thus, the use of overlapped seams is not recommended for any situation where a gap between geotextile sheets would have a significant effect on the performance of the system. If seams are sewn as an alternative to overlaps then the strength of the seam must be adequate to withstand the lateral forces that develop during construction.

15.4 Leachate collection pipes

Pipes are used in both PLCS and SLCS to promote collection and transmission of leachate. Leachate flows in these pipes under gravity to low points (i.e., sumps) and is then removed from the landfill. For example, Figure 15.3 shows a plan view outlining the network of drainage pipes in a PLCS. Design and construction issues for the collection pipes are presented in this section. First, the factors involving selection

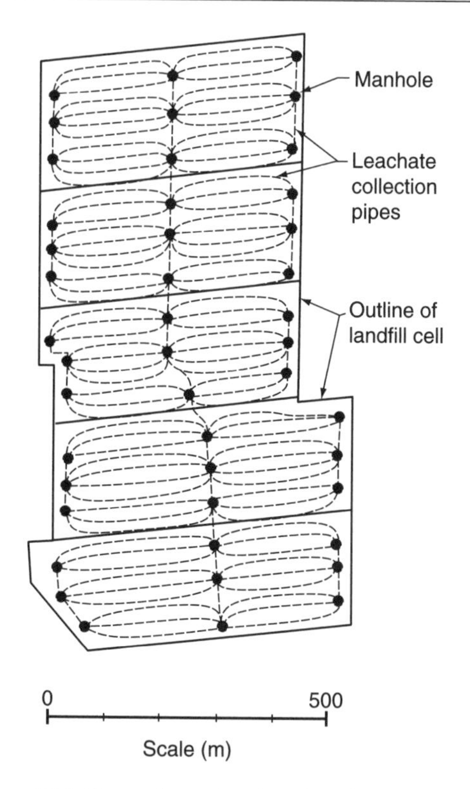

Figure 15.3 Plan view showing location of leachate collection pipes and manholes in a primary leachate collection system (modified from Rowe *et al.*, 2000c).

of the pipe are discussed. The structural performance of these pipes is then examined. Since the material (e.g., gravel) surrounding the pipe provides both loading and support to the pipe, the properties of this material and those of the structure, and the interactions between the two must be considered. The influence of coarse gravel backfill and circular perforations on the structural performance of pipes is also presented.

15.4.1 Pipe selection

HDPE pipes are normally used in landfills given their excellent chemical resistance (see Section 13.3). Either plane-wall pipes (with solid prismatic geometry) or profile-wall pipes may be used. However, given the desirability of using coarse gravel directly adjacent to the pipe, the local stiffness of profile-wall pipes may be insufficient to avoid excessive local bending stresses. Furthermore, deep burial can produce high wall stresses that lead to local bending and local buckling in profile-wall pipes (Dhar and Moore, 2001). Thick plane-wall pipes are therefore considered to be more suitable for landfill leachate collection systems.

The internal diameter, *ID*, of the pipe must be large enough to transmit the volume of leachate by gravity flow to a collection point (normally a sump in a manhole) where the leachate can be pumped out and removed from the collection system. The maximum flow rate in a circular pipe, Q_{max}, can be calculated using Manning's equation:

$$Q_{max} = 0.335 \frac{ID^{8/3} i^{1/2}}{n_M} \tag{15.7}$$

where i is the slope of the pipe [L/L] and n_M is Manning's coefficient [T/L$^{1/3}$] and depends on the roughness of the pipe. The average flow velocity in the pipe V (which should be large enough to minimize clogging (Section 2.4.4) and the deposition of suspended particles in the leachate) can be calculated using the equation derived by Giroud *et al.* (2000) knowing the flow rate Q in the pipe:

$$V = \left(1 - \frac{Qn}{2\ ID^{8/3}\ i^{1/2}}\right)\left(\frac{Q^4 i^{9/2}}{36\ ID^2 n_M^9}\right)^{1/13} \tag{15.8}$$

Since flows in the leachate collection system are often very small, the hydraulic capacity of these pipes is rarely a controlling factor. The selection of the internal diameter of the pipes is then based on maintenance requirements, such as being large enough to permit video camera inspection and hydraulic flushing. Internal diameters ranging from 150 to 300 mm are common for leachate collection pipes.

Adequate structural capacity of the pipe is obtained by selecting an appropriate wall thickness of the pipe for the specific overburden pressures and backfill conditions. For plane pipes, the pipe thickness is expressed using the dimension ratio, DR, where:

$$DR = \frac{OD}{t} \tag{15.9}$$

OD is the external diameter of the pipe and t is the minimum wall thickness of the pipe (thick pipes have low DR values). Common DR values for leachate collection pipes range between 17 and 6. Factors influencing the structural capacity and methods to select an appropriate pipe thickness are presented in the following sections.

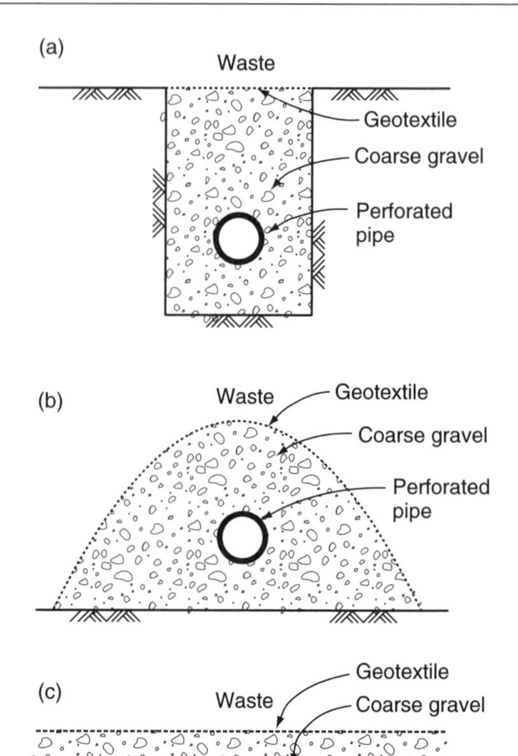

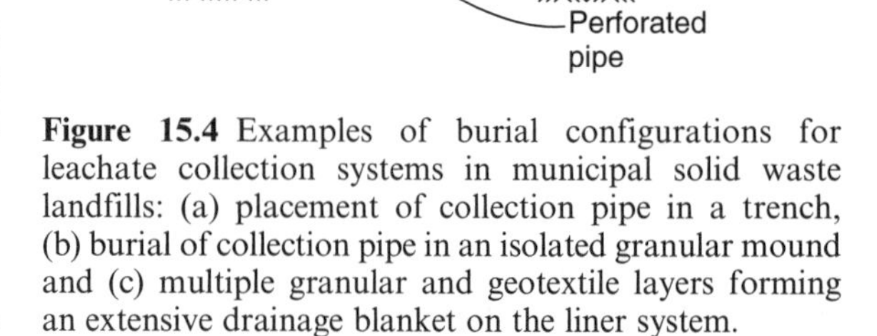

Figure 15.4 Examples of burial configurations for leachate collection systems in municipal solid waste landfills: (a) placement of collection pipe in a trench, (b) burial of collection pipe in an isolated granular mound and (c) multiple granular and geotextile layers forming an extensive drainage blanket on the liner system.

15.4.2 Backfill configuration

Leachate collection pipes may be buried within trenches, French drains (also called granular mounds) or laterally extensive blankets. Since for both trenches and mounds the flow of leachate into the pipe is over a restricted area, the use of laterally extensive blankets is preferable from the perspective of minimizing the implications of clogging of the leachate collection system, as discussed in Section 2.4. The backfill configuration also influences the vertical pressures acting on the pipe and lateral pressures confining the pipe, and thus the structural performance of the pipe.

(a) Trenches

Placement of pipes in trenches (Figure 15.4a) is common for gravity-flow and sanitary sewers. It is less common in leachate collection systems because of the susceptibility to clogging (Section 2.4.4). When used in these applications, only a proportion of vertical overburden stresses may reach the pipe if placed within a well-constructed trench. If vertical strains in the backfilled trench are greater than the supporting material, shear stresses mobilize along the trench walls, which lead to a reduction in vertical stresses acting on the pipe. This is illustrated in Figure 15.5, which shows that the vertical stress within the trench is less than applied pressure p if the stiffness of the trench is less than the surrounding material ($E_T = 0.4E_N$). The stresses in the trench are smaller as the trench becomes more slender (i.e., as H/B increases). Conventional trench-arching solutions based on the limiting shear stress mobilized along the trench wall can be used to estimate the stresses reaching the pipe for these cases (Moore, 2001).

In a trench at the base of a landfill however, vertical strains of the supporting material

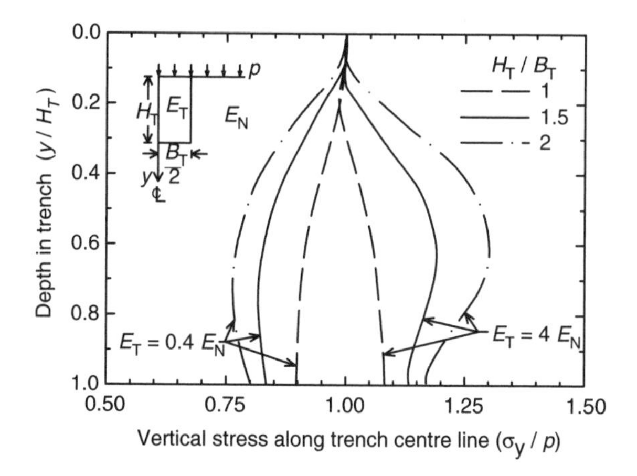

Figure 15.5 Calculated vertical stresses σ_y normalized by overburden pressure p in trenches with native soil (E_N) stiffer than trench (E_T) and trench material stiffer than native soil (modified from Krushelnitzky and Brachman, 2002).

(compacted clay liner) would probably be larger than the backfill (coarse gravel), due to consolidation of the clay liner and because the liner is not as stiff as coarse gravel. In this case, the trench will attract load and the pipe will experience greater stresses. For example, the results in Figure 15.5 also show that if the stiffness of the trench, E_T, is greater than the stiffness of the supporting material, E_N, the vertical stresses in the trench are larger than for the conventional trench. The maximum vertical stress in a trench with H/B equal to 2 is almost twice as large for a landfill trench (with $E_T = 4E_N$) than for a conventional trench (with $E_T = 0.4E_N$). Careful consideration of the stresses acting on the pipe is therefore required for the backfill configuration shown in Figure 15.4a. Numerical analysis is required to estimate the stresses reaching the pipe for these cases.

Lateral confinement provided to the gravel (and thus the pipe) in a trench may be better than in a mound (Figure 15.4b) or blanket (Figure 15.4c) because the coefficient of lateral earth pressure is larger for the clay than for waste or gravel. However, depending on construction practices, it is possible that lateral deflections of the clay could occur in the voids between coarse gravel particles along the trench wall. These deformations would decrease the lateral stresses in the clay and thereby decrease the lateral support for the pipe resulting in larger pipe deflections and stresses. This has been observed in a large-scale laboratory test and confirmed with numerical modelling (Krushelnitzky and Brachman, 2003). Thus, the design of a leachate collection pipe placed in a gravel trench surrounded by soft clay is not as straightforward as that for conventional municipal sewers and may require large-scale laboratory testing and numerical analysis, especially if the pipe is to be buried under large amounts (>25 m) of waste. Special attention should also be given to the potential for clogging and the service life of the gravel in the trench.

(b) Mounds

Placement of drainage pipes in granular mounds (Figure 15.4b) would have a slight advantage over trenches since the trench would no longer extend into the liner system. The mound would be subject to the entire weight of the overlying waste. However, the mound would be laterally supported by waste material, which is likely to provide low backfill stiffness. The stiffness of the backfill material is important since the pipe obtains essential support from the backfill. Soil stiffness is a function of confining stresses – the greater the confinement provided to the backfill soil, the stiffer the soil, resulting in smaller pipe deflections. If the gravel mound is surrounded by waste with potentially low and variable stiffness, large pipe deflections and greater variation in pipe behaviour would be expected. This is consistent with field measurements (Rogers, 1999) that showed increasing local pipe deformations as the size of granular mound decreased.

No current design method exists for pipes placed within granular mounds. Quantifying the lateral support provided by the waste is challenging since it depends on factors including the type of waste and how the waste was placed. If mounds are to be used there should be at least one pipe-diameter of gravel cover above the pipe, and three pipe-diameters of gravel on each side of the pipe.

(c) Blankets

The structural performance of the pipe will be better in a laterally extensive granular blanket (Figure 15.4c) compared with a mound installation. Here, the waste imposes only vertical loads on the drainage layer. With an extensive granular blanket, the gravel directly around the pipe is laterally confined by adjacent gravel. This can be expected to result in a stiffer ground response and therefore smaller pipe defections and stresses than would be obtained with a mound.

The vertical pressures from the weight of waste acting on the blanket probably vary because of the heterogeneous nature of MSW. This may cause localized zones of lower stiffness material due to the lower confinement, but these effects may be expected to be much less important in a granular blanket relative to a mound.

Laterally extensive granular blankets are therefore recommended for municipal soil waste landfills on the basis of: (1) reduced implications from biologically induced clogging (as described in Section 2.4) and (2) better backfill support provided to the pipe thereby improving the structural response of the pipe. Methods to estimate the pressures acting on the pipe and resulting pipe deflections and stresses for pipes placed in laterally extensive blankets are presented in Section 15.4.3.

15.4.3 Structural design of pipes in blankets

(a) Earth pressures acting on the pipe

For the deeply buried pipe shown in Figure 15.6a, vertical σ_v and horizontal σ_h earth pressures act at some distance from the pipe. The

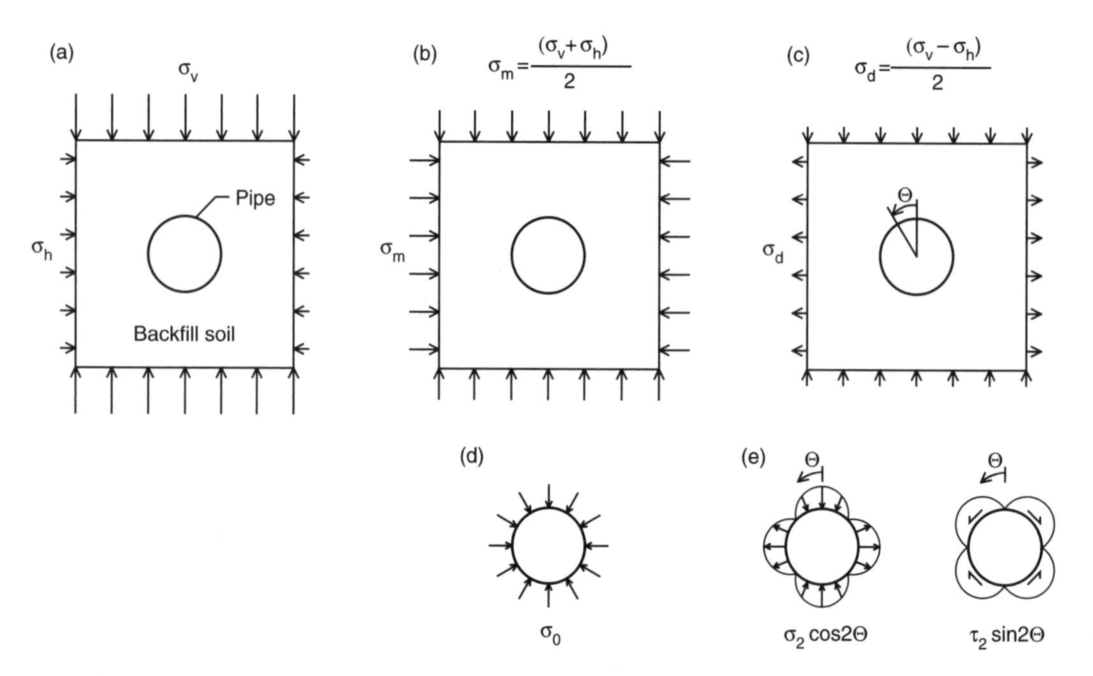

Figure 15.6 Illustration of earth pressures under deep burial: (a) idealized earth pressures acting distant to pipe, (b) mean component of earth pressures acting distance to pipe, (c) deviator component of earth pressures acting distant to pipe, (d) normal and shear stresses acting on pipe from mean component and (e) normal and shear stresses acting on pipe from deviator component.

vertical stress may be taken as the weight of the overburden material (including the waste, daily and intermediate cover):

$$\sigma_v = \gamma H \tag{15.10}$$

where γ is the unit weight of the overburden material and H the thickness of the overburden material above the pipe. Horizontal stresses develop from resistance to lateral movement and are given by:

$$\sigma_h = K\sigma_v \tag{15.11}$$

where K is the coefficient of lateral earth pressure.

To quantify the stresses that act on the pipe, it is mathematically convenient to first consider the distant earth pressures as the sum of two components. Using the notation of Moore (2001), these components consist of: (a) the mean stress σ_m (Figure 15.6b) which is the average of the vertical and horizontal earth pressures acting distant from the pipe:

$$\sigma_m = \frac{\sigma_v + \sigma_h}{2} \tag{15.12a}$$

and (b) the deviator stress σ_d (Figure 15.6c) which is one half the difference between the vertical and horizontal stresses acting distant from the pipe:

$$\sigma_d = \frac{\sigma_v - \sigma_h}{2} \tag{15.12b}$$

The normal and shear stresses acting on the exterior surface of the pipe can then be expressed as:

$$\sigma = \sigma_0 + \sigma_2 \cos 2\Theta \tag{15.13a}$$

$$\tau = \tau_2 \sin 2\Theta \tag{15.13b}$$

where σ_0 is the normal stress on the pipe from the mean boundary stresses (Figure 15.6d), σ_2 the normal stress on the pipe from the deviator boundary stress (Figure 15.6e), τ_2 the shear stress on the pipe from the deviator boundary stress (Figure 15.6e) and Θ the angle around the circumference of the pipe with respect to the crown (see Figure 15.6c).

The assumption of symmetric earth pressures acting on the pipe is a useful first approximation and provides suitable results for design of pipes with uniform backfill support. This approach may underestimate the stresses acting at the invert for pipes with poor backfill support at the haunches. Also, modification to pipe deflections and stresses is required when the pipe is surrounded by coarse gravel backfill as described in Section 15.4.4c.

The stresses acting on the pipe σ_0, σ_2 and τ_2 can be expressed in terms of the earth pressures acting distant to the pipe using:

$$\sigma_0 = A_m \sigma_m \tag{15.14a}$$

$$\sigma_2 = A_{d\sigma} \sigma_d \tag{15.14b}$$

$$\tau_2 = A_{d\tau} \sigma_d \tag{15.14c}$$

where A_m, $A_{d\sigma}$ and $A_{d\tau}$ are factors used to provide stresses acting on the pipe in terms of the mean and deviator field stresses obtained by Moore (2001) from the elastic arching solution of Hoeg (1968):

$$A_m = \frac{2(1 - \nu_s)}{1 + C(1 - 2\nu_s)} \tag{15.14d}$$

$$A_{d\sigma} = \frac{4(1 - \nu_s)(4 + 3C(1 - 2\nu_s) - 2F)}{\Delta}$$

for a bonded interface

(15.14e)

$$A_{d\sigma} = \frac{12(1 - \nu_s)}{2F + 5 - 6\nu_s} \quad \text{for a smooth interface}$$

(15.14f)

$$A_{d\tau} = \frac{16(1 - \nu_s)(F + 1)}{\Delta} \tag{15.14g}$$

for a bonded interface

$$A_{d\tau} = 0 \qquad \text{for a smooth interface} \tag{15.14h}$$

$$\Delta = C(1 - 2\nu_s)(5 - 6\nu_s + 2F) + 2F(3 - 2\nu_s) + 4(3 - 4\nu_s) \tag{15.14i}$$

$$C = \frac{E_s D_P}{2(1 + \nu_s)(1 - 2\nu_s)E_P A_P} \tag{15.14j}$$

$$F = \frac{E_s D_P^3}{48(1 + \nu_s)E_P I_P} \tag{15.14k}$$

in which E_s is the Young's modulus of the soil, ν_s the Poisson ratio of the soil, E_P the Young's modulus of the pipe, D_P the mid-surface diameter of the pipe, A_P the cross-sectional area per unit length of the pipe and I_P the second moment of area per unit length of the pipe. For a plain pipe with wall thickness t:

$$A_P = t \tag{15.15a}$$

$$I_P = \frac{t^3}{12} \tag{15.15b}$$

A secant value of soil modulus for the appropriate backfill density and stress conditions should be used. Typical parameters for 50 mm coarse gravel in a blanket configuration are given in Section 15.4.4c. The pipe manufacturer should be consulted for values of pipe modulus to use in design as the stiffness of HDPE decreases with increasing time, increasing temperatures and upon exposure to chemicals. For exposure to water at temperatures around 23 °C, the long-term modulus may vary from 140 to 210 MPa depending on the type of polyethylene (PPI, 1996). Data are not available for the effect of temperature on long-term modulus; however, short-term modulus decreases by nearly 2% for each 1 °C increase in temperature.

(b) Pipe deflections

Once the earth pressures acting on the pipe are known, the pipe deflections can be calculated. Pipe deflections are expressed as a change in the pipe diameter. Vertical ΔD_v and horizontal ΔD_h diameter changes are given by (Moore, 2001):

$$\Delta D_v = -2(w_0 + w_2) \tag{15.16a}$$

$$\Delta D_h = -2(w_0 - w_2) \tag{15.16b}$$

$$w_0 = \frac{\sigma_0 D_P^2}{4E_P A_P} \tag{15.16c}$$

$$w_2 = (2\sigma_2 + \tau_2)\frac{D_P^4}{288E_P I_P} \tag{15.16d}$$

For leachate collection pipes buried in laterally extensive blankets, the vertical diameter of the pipe decreases (i.e., ΔD_v is negative) while the horizontal diameter increases (ΔD_h is positive). Since the change in vertical diameter is larger than the change in horizontal diameter, ΔD_v normally controls pipe design. Diameter change can also be expressed as the per cent change in vertical pipe diameter divided by the initial pipe diameter D_P:

$$\%\Delta D = \frac{\Delta D_v}{D_P} \times 100 \tag{15.17}$$

The semi-empirical Modified Iowa equation (Howard, 1977) is another method that can be used to calculate pipe deflections; however, its use is not recommended for the design of leachate collection pipes since it: (a) neglects the soil-structure interaction between the backfill and the pipe, (b) predicts the wrong mode of deformation (Brachman, 1999) and (c) does not provide an opportunity to calculate the stresses in the pipe, whereas the method using equation 15.16 addresses all three of these items.

(c) Pipe stresses

The circumferential stresses σ_Θ at any location around the plane pipe (Θ) can be calculated using thin shell theory for each of the three loading cases:

$$\sigma_\Theta(\Theta) = (\sigma_\Theta)_{\sigma 0} + (\sigma_\Theta)_{\sigma 2} + (\sigma_\Theta)_{\tau 2} \quad (15.18a)$$

$$(\sigma_\Theta)_{\sigma 0} = \frac{\sigma_0 D_P}{2t} \quad (15.18b)$$

$$(\sigma_\Theta)_{\sigma 2} = \sigma_2\left(-\frac{D_P}{6t} - \frac{\alpha D_P^2}{2t^2}\right)\cos 2\Theta \quad (15.18c)$$

$$(\sigma_\Theta)_{\tau 2} = \tau_2\left(-\frac{D_P}{3t} - \frac{\alpha D_P^2}{4t^2}\right)\cos 2\Theta \quad (15.18d)$$

where $(\sigma_\Theta)_{\sigma 0}$, $(\sigma_\Theta)_{\sigma 2}$ and $(\sigma_\Theta)_{\tau 2}$ are the circumferential stresses in the pipe from the σ_0, σ_2 and τ_2 loading cases, respectively; and $\alpha = 1.0$ for locations at the pipe interior and -1.0 for the pipe exterior. Compressive stresses are taken as positive.

Compressive stresses are maximum at the interior springline ($\Theta = 90°$) and are given by:

$$(\sigma_\Theta)_{SPi} = \frac{\sigma_0 D_P}{2t} + \sigma_2\left(\frac{D_P}{6t} + \frac{D_P^2}{2t^2}\right) + \tau_2\left(\frac{D_P}{3t} + \frac{D_P^2}{4t^2}\right) \quad (15.19a)$$

while the stresses at the interior invert ($\Theta = 180°$) are given by:

$$(\sigma_\Theta)_{INi} = \frac{\sigma_0 D_P}{2t} - \sigma_2\left(\frac{D_P}{6t} + \frac{D_P^2}{2t^2}\right) - \tau_2\left(\frac{D_P}{3t} + \frac{D_P^2}{4t^2}\right) \quad (15.19b)$$

where this is the maximum tensile stress if the value of equation 15.19b is negative. The case of a bonded interface between the soil and the backfill provides the maximum stresses in the pipe. For leachate collection pipes in laterally extensive blankets, tensile stresses occur at the interior invert and crown locations given the low coefficient of lateral earth pressure (resulting in large σ_d) of coarse gravel backfills.

15.4.4 Large-scale testing of pipes in granular blankets

(a) Issues with coarse gravel backfill

The use of coarse gravel as backfill around the pipe is desirable from the perspective of minimizing clogging (see Section 2.4.4). However, concern has been raised as to whether coarse gravel has a detrimental effect on the structural performance of the pipe. Because the gravel particles are relatively large with respect to the size of the pipe, coarse gravel both loads and supports the pipe at discrete points around the outside surface. An imprint of contact points from 50-mm gravel on a 320-mm OD pipe shown in Figure 15.7 reveals that the gravel contacts have variable size, shape and spacing. This is quite different to the more continuous support provided by other conventional backfill materials (e.g., sand, well-graded gravel) that are much less suitable for use as the drainage

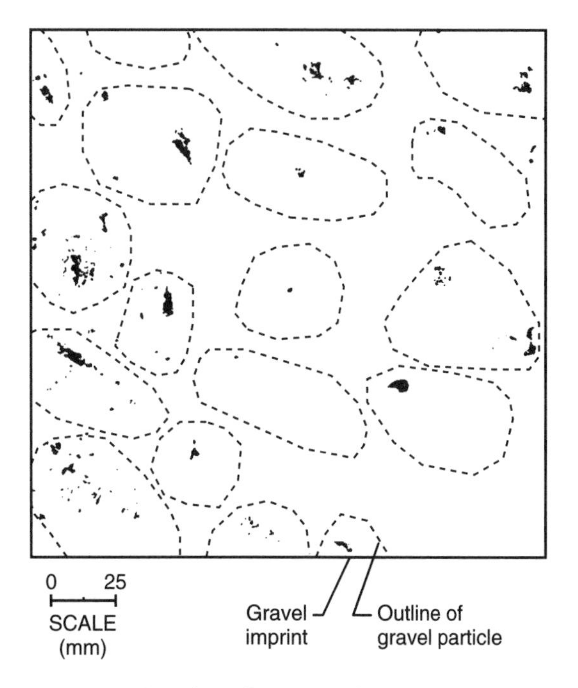

Figure 15.7 Imprint of contacts from nominal 50-mm gravel on the exterior surface of a 320-mm diameter pipe (modified from Brachman *et al.*, 2000b).

material in landfills because of concerns regarding clogging.

Discrete gravel contacts result in variations in pipe deflection and local bending stresses that must be recognized in the design of leachate collection pipes. However, quantifying the magnitude of local bending stresses in design calculations is very challenging given that the contact spacing is widely variable and contact forces are unknown. Calculated pipe deflections and stresses (from equations 15.16 and 15.18) can be modified for the effects of 50-mm coarse gravel using measured results from large-scale laboratory testing conducted by Brachman (1999) and Tognon (1999).

(b) Large-scale laboratory testing

Large-scale laboratory tests were conducted under biaxial earth pressures to assess the structural response of 220-mm *OD DR* 9 HDPE leachate collection pipes with 50-mm coarse gravel backfill. Figure 15.8 shows a cross-section through the test cell developed by Brachman *et al.* (2000a, 2001) to simulate large vertical

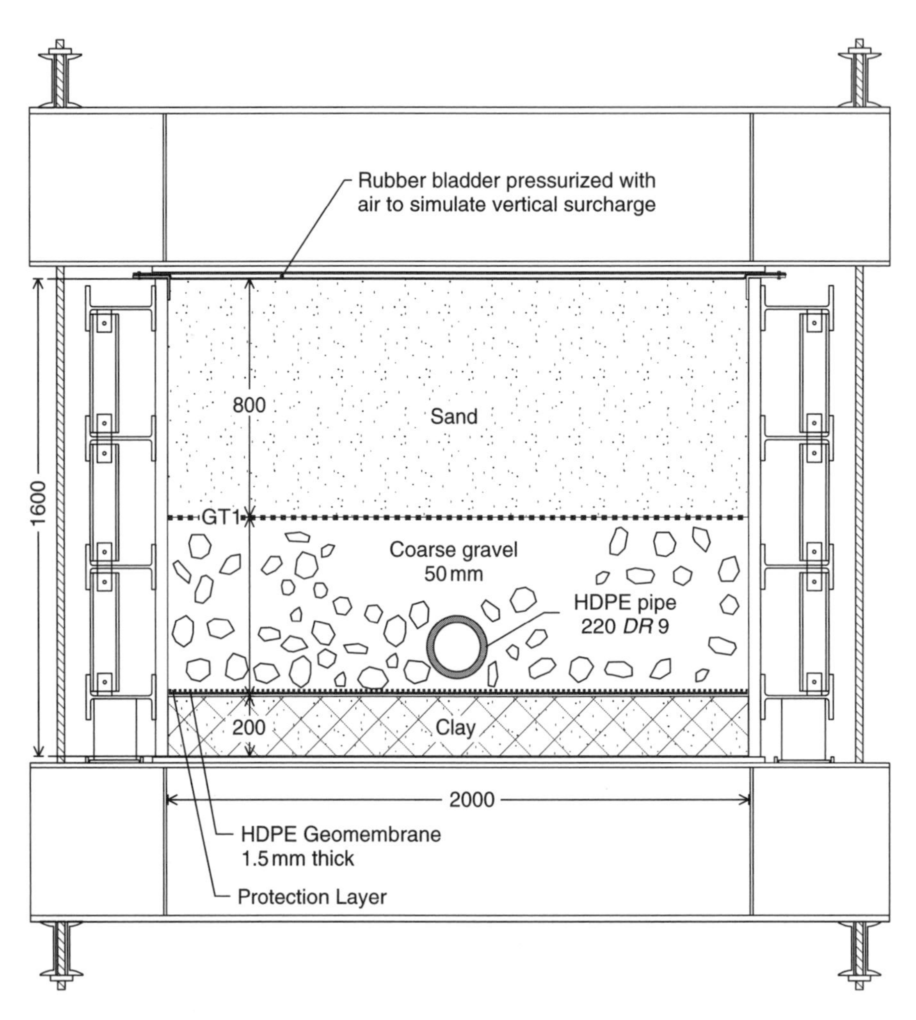

Figure 15.8 Longitudinal section through biaxial compression testing facility showing simulated landfill conditions (modified from Brachman *et al.*, 2001). Dimensions in mm.

(up to 1000 kPa) and horizontal stresses in the drainage layer containing the pipe. The pipe specimen was placed within a prism of soil with dimensions of 2.0 m wide × 2.0 m long × 1.6 m high. The soil was contained within a stiff steel structure. Vertical load was applied by a pressurized air bladder, which provided a uniform vertical pressure across the top surface. Horizontal stresses corresponding to essentially zero lateral strain conditions develop in the soil because deflection of the sidewalls is limited (Brachman *et al.*, 2001). Boundary effects from friction were minimized using polyethylene sheets lubricated with silicone grease along the sidewalls (Tognon *et al.*, 2000).

One set of specific conditions of a leachate collection pipe in a gravel blanket was examined. The pipe was placed within a 600-mm deep blanket layer of 50-mm coarse gravel (70% finer than 51 mm and 8% finer than 38 mm) to simulate the drainage layer. The pipe was placed with approximately 100 mm of gravel material between the pipe invert and the underlying 200-mm thick clay layer, included to simulate the effect of a compressible layer beneath the pipe. Poorly graded medium sand was used to fill the remainder of the test cell. Specific details of the materials tested have been reported by Brachman (1999).

The deflections of the pipe were measured as the vertical pressure in the bladder was increased. Vertical ΔD_v and horizontal ΔD_h diameter changes of the pipe measured at three locations along the pipe are plotted in Figure 15.9 for one test with a gravel blanket. For comparison, pipe deflections from a test conducted with the same pipe located in the middle of the test cell and backfilled with poorly graded medium-dense sand are also shown in Figure 15.9. Burial in the gravel blanket resulted in larger pipe deflections predominantly because the coefficient of lateral earth pressure is lower for the gravel than the sand. The pipe experiences much greater variations in deflections in the gravel blanket compared with sand backfill because of the discrete support provided by the coarse gravel. The increase in pipe deflection induced by the coarse gravel needs to be accounted for during pipe design as discussed in the next section.

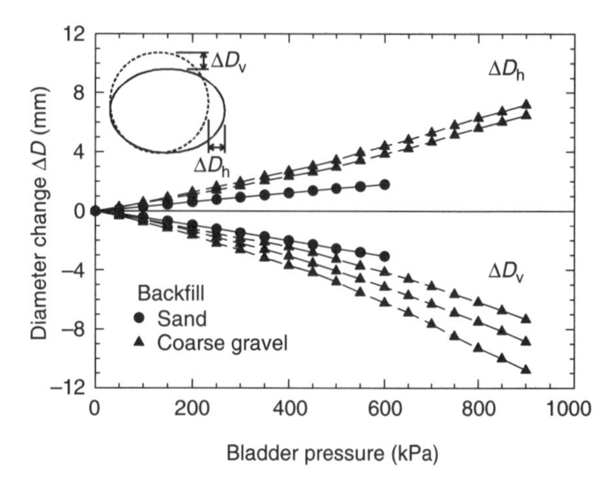

Figure 15.9 Measured vertical ΔD_v and horizontal ΔD_h diameter changes for HDPE pipe with outside diameter of 220 mm and *DR* 9 when tested with sand and gravel backfill (modified from Brachman *et al.*, 2001).

(c) Response of 50-mm coarse gravel

The stiffness of the particular 50-mm coarse gravel tested when subject to stress conditions expected in a blanket layer can be inferred by comparing the average measured deflections from the laboratory tests with values calculated using equation 15.16. Poisson's ratio $\nu_s = 0.3$ and lateral earth pressure coefficient $K = 0.15$ were used to characterize the coarse gravel. Up to applied vertical pressures of 500 kPa, the secant modulus of the gravel was estimated to be $E_s = 30$ MPa. The secant modulus of the coarse gravel backfill decreases (material softening) as the vertical pressures increase due to increasing shear strains in the gravel. For overburden pressures greater than 500 kPa, the secant modulus was estimated to be 25 MPa.

The diameter change and stresses of a landfill leachate collection pipe can be estimated by first calculating the average response (using the procedures of Section 15.4.3 and the gravel stiffness) and then multiplying those values by factors to account for variations induced by the coarse gravel. Empirical factors specifically for 50-mm gravel backfill and 220-mm *OD* pipes with a *DR* of 9 have been reported by Brachman (1999) based on a statistical analysis of the measured variations from the laboratory tests. For these conditions, the average diameter change should be increased by a factor of 1.3. Stresses in the pipe are more sensitive to local gravel effects. Maximum compressive stresses were 1.6 times the average response while maximum tensile stresses at the invert were 2.0 times the average response.

The gravel stiffness and magnification factors are applicable for the specific conditions tested by Brachman (1999), namely the particular 50-mm gravel in a 0.6-m deep blanket configuration and for 220-mm *OD DR*9 HDPE pipes. The stiffness may be different for other backfill materials, and at present the backfill stiffness can best be obtained from large-scale laboratory tests. The magnification factors may be even larger for thinner pipes ($DR > 9$), coarser gravel backfill or pipes with smaller outside diameters, and for these cases can only be obtained from a series of large-scale laboratory tests.

15.4.5 Allowable limits

Adequate structural performance is obtained by ensuring the pipe deflections and stresses are below allowable limits. Although pipe manufacturers should be consulted to obtain allowable limits for their specific pipe products, guidance on limits for pipe deflections and stresses is provided in the following sections.

(a) Limiting deflections

Deflection limits for HDPE pipes are listed in Table 15.4. The maximum allowable diameter change specified by ASTM F714-97 depends on the thickness (expressed as pipe *DR*) of the pipe. As thicker pipes (i.e., lower *DR*) would have higher bending stresses, these limits attempt to limit stresses (without explicitly calculating pipe stresses) by limiting pipe deflection. Alternatively, the maximum change in vertical diameter of 5% (PPI, 1996) can be used provided that the stresses in the pipe (calculated using equation 15.18) are below allowable limits.

Table 15.4 Recommended maximum allowable pipe deflections

Maximum allowable deflection % ΔD_P	*Source*
5.0 for $DR = 21$	ASTM F714 (1997)
4.0 for $DR = 17$	
3.3 for $DR = 11$	
5%	Plastic Pipes Institute (1996)

(b) Limiting stresses

Stresses in the pipe must be kept below yield stresses. Limits for both maximum compressive and tensile stresses must be checked. The compressive yield stress of polyethylene depends on the rate of loading and may be taken as 9 MPa. Applying a factor of safety of 2.0 against compressive yield would limit the maximum compressive stresses in the polyethylene pipes to less than 4.5 MPa.

For plane wall HDPE pipes, the maximum allowable tensile stresses can be obtained using the hydrostatic design basis (HDB) used to limit long-term tensile stresses in pressurized thermoplastic pipes. The HDB value is found by testing the pipe subject to uniform internal pressure for various times-to-failure. The hoop stress extrapolated to 10^5 h on a logarithmic plot of stress versus time-to-failure is defined as the HDB stress. The maximum allowable tensile stresses should be limited to one-half of the HDB. Values of the HDB should be obtained from the pipe manufacturer. ASTM F 714 specifies a minimum HDB value of 8.6 MPa for HDPE pipes, giving a maximum allowable tensile stress of 4.3 MPa. Use of the HDB to limit tensile stresses in landfill pipes appears to be satisfactory for medium size landfills (20–25 m waste), but may be overly conservative for pipes buried under large amounts of waste since the stress conditions and implications of failure are quite different for landfill pipes relative to pipes under internal pressure (Brachman, 2001). However, in the absence of data on the long-term response of leachate collection pipes, the HDB provides a mechanism for limiting tensile stresses in leachate collection pipes.

(c) Buckling

Buckling may occur if under deep burial if the compressive stresses are too large for the pipe thickness. The factor of safety against buckling FS_b should at least be equal to 2.0 and may be obtained from (Moore, 2001):

$$FS_b = \frac{N_b}{N_{sp}} \tag{15.20a}$$

$$N_b = 1.2 p_f (E_p I_p)^{1/3} \left(\frac{E_s}{1 - \nu_s^2} \right)^{2/3} R_h \tag{15.20b}$$

$$N_{sp} = \frac{\sigma_0 D_P}{2} + \left(\frac{\sigma_2}{3} + \frac{2\tau_2}{3} \right) \frac{D_P}{2} \tag{15.20c}$$

where N_b is the critical buckling thrust, N_{sp} the maximum thrust at the springline of the pipe (obtained assuming a bonded interface between the pipe and soil), p_f a performance factor and R_h a factor for burial depth and extent of backfill.

For granular backfill p_f may be taken as 0.55, while R_h is equal to 1.0 for gravel blankets in a landfill. Leachate collection pipes are usually thick (in order to minimize local bending effects from coarse gravel backfill) such that limiting deflections or tensile stresses and not buckling govern pipe design.

15.4.6 Perforations

(a) Design considerations

Perforations are holes in the wall of the pipe (Figure 15.10) that are essential for the purpose

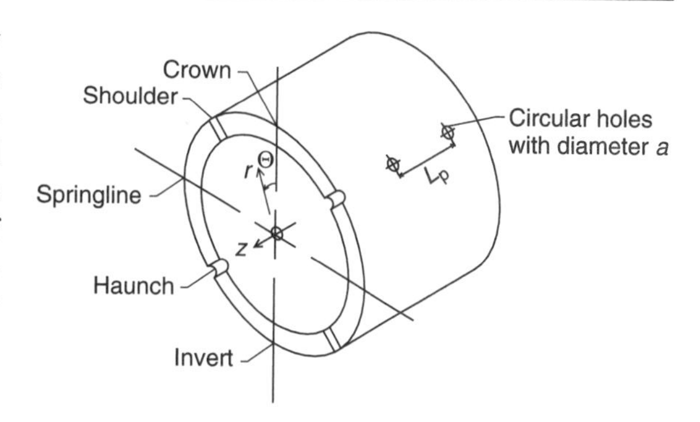

Figure 15.10 Perforated drainage pipe with four circular holes located at the quarter points (±45° from springlines).

of leachate collection. Perforations should be: (a) large enough to provide adequate flow, (b) small enough to minimize movement of fines from the backfill, (c) large enough to limit the extent of biologically induced clogging and also to maximize the effectiveness of cleaning (from pressurized hydraulic jets that pass through the pipe) and (d) small enough to limit increase of stress in the pipe.

The number and spacing of the perforations to provide adequate flow can be obtained using the technique developed by Panu and Filice (1992). Cedergren (1989) recommended that, to reduce the potential for loss of fines, the maximum diameter of circular perforations be less than ½ D_{85} of the backfill soil, where 85% of the backfill soil material is finer than D_{85} by mass.

Both circular holes and long narrow slots are used as perforations. Zanzinger and Gartung (1995) suggest the use of slots on the basis that they are hydraulically superior to round holes. However, since flow rates into the pipe may be very slow in landfills, the hydraulic performance of the perforation does not generally control design. Rather, concerns about the clogging of the perforation – which would be related to the size of the opening – are critical. It is postulated that a circular hole would clog less rapidly than a long and narrow slotted perforation of equivalent opening area. As discussed in Section 2.4, field observations by Fleming *et al.* (1999) revealed that 8-mm diameter circular holes might be too small for leachate collection pipes, since these holes were mostly clogged after only 1–4 years of exposure to landfill leachate. Thus, in landfill leachate collection systems, it is probably desirable to use fewer large diameter circular perforations instead of many small diameter circular holes or rectangular slots in order to minimize clogging of the openings.

(b) Stress concentrations around circular perforations

The principal effect that a perforation has on the structural performance of the pipe is a local perturbation in pipe stresses with its influence concentrated around the hole. The maximum stress acting around the perforation, σ_P, can be calculated by multiplying the stresses in the pipe where the perforation is to be located (obtained from equation 15.18) by stress concentration factors for each of the three loading cases σ_0, σ_2 and τ_2 viz.:

$$\sigma_P = \chi_{\sigma 0} \times (\sigma_\Theta)_{\sigma 0} + \chi_{\sigma 2} \times (\sigma_\Theta)_{\sigma 2} + \chi_{\tau 2} \times (\sigma_\Theta)_{\tau 2} \tag{15.21}$$

where $\chi_{\sigma 0}$, $\chi_{\sigma 2}$ and $\chi_{\tau 2}$ are stress concentration factors for pressures σ_0, σ_2 and τ_2, respectively.

Brachman and Krushelnitzky (2002) report stress concentration factors $\chi_{\sigma 0}$, $\chi_{\sigma 2}$ and $\chi_{\tau 2}$ obtained from 3D numerical analysis. These stress concentration factors depend on: (a) the circumferential location of the perforation, (b) the thickness and diameter of the pipe, (c) the diameter of the perforation and (d) the axial and circumferential spacing between perforations. Stress concentration factor, $\chi_{\sigma 0}$, for the pipe subject to the mean boundary stress σ_0 (Figure 15.6b) is plotted in Figure 15.11 against the ratio of pipe diameter D_P to perforation diameter a. Note that small values of D_P/a correspond to large perforations in the pipe. Values are shown

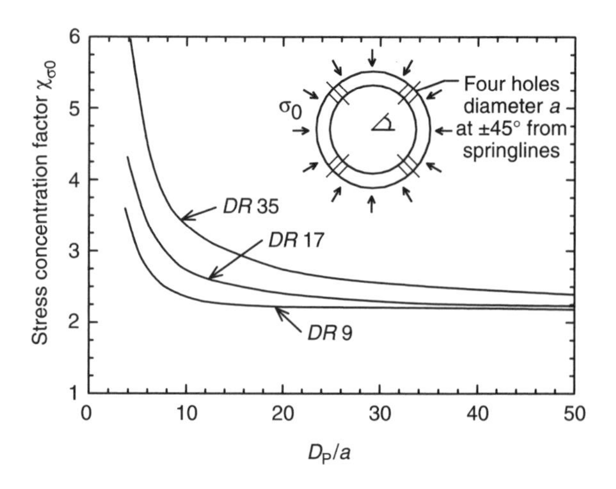

Figure 15.11 Stress concentration factor $\chi_{\sigma 0}$ around a circular perforation in a pipe subjected to uniform external pressure σ_0 (modified from Brachman and Krushelnitzky, 2002).

for pipes with DR varying from 9 (thick pipe) to 35 (thin pipe). The magnitude of $\chi_{\sigma 0}$ is independent of the position of the perforation around the circumference since the compression of the pipe subject to the mean boundary stress σ_0 is radially symmetric. The stress concentration factor, $\chi_{\sigma 0}$, increases with both decreases in pipe thickness and increases in perforation diameter.

Stress concentration factors $\chi_{\sigma 2}$ and $\chi_{\tau 2}$ for the deviator boundary stress depend on the location of the perforation around the pipe (see Brachman and Krushelnitzky, 2002). They have a maximum value of approximately 2.1 for perforations located at the crown, springline and invert. For perforations located at the quarter-points, factors $\chi_{\sigma 2}$ and $\chi_{\tau 2}$ are equal to zero and thus only stress concentration factor $\chi_{\sigma 0}$ need be considered.

The axial spacing between perforations also influences the magnitude of the stress concentration factors. Perforations should be located at a centre-to-centre spacing of at least four times the perforation diameter in the axial direction to avoid increases in the stress concentration factors.

Ideally, perforations should be located at the quarter-points of the pipe as shown in Figure 15.10 because the stresses are smallest near the shoulder and haunch locations of the pipe. The worst location to place perforations from the perspective of stresses in the pipe is at the crown, springline or invert, since the stresses are the highest at these locations. To illustrate the implications of the location of perforation on stresses in the pipe consider an example of a pipe with an outside diameter of 225 mm and *DR* of 9 buried in a laterally extensive granular blanket and subject to a vertical pressure of 600 kPa (equivalent to 50 m of waste for a unit weight of waste of 12 kN/m^3). Stresses calculated in the pipe are presented in Table 15.5 (and also taking $E_s = 25$ MPa, $\nu_s = 0.3$, $E_p = 150$ MPa, $\nu_p = 0.46$ and $K = 0.15$). Results are presented for the case with no perforations and for four circular holes ($a = 25$ mm, $D_P/a = 8$, $L_P/a = 150$ mm) when located at the quarter points and the crown, invert and springlines. Placement of the holes at the quarter points results in an increase in stress at those locations, but the maximum stresses remain at the crown, invert and springlines and are the same for the case with no perforations. Placing the holes at the crown, invert and springlines increases the maximum stresses in the pipe (relative to no perforations or with holes at the quarter points). In this case, the maximum compressive stress at the springlines exceeds the allowable compressive stress (see Section 15.4.5b) and note that increases in stress from coarse gravel have not yet been accounted for. The design would be acceptable provided that the perforations were at the quarter points.

15.4.7 Construction issues

(a) Pipe joints

Lengths of HDPE pipes can be joined in the field by thermal fusion butt joints. This process consists of first trimming the ends of the pipe to provide a clean and square surface for seaming. Heating both ends of the pipe to a certain temperature with a heating iron then softens the pipe material. The heating iron is then removed and the two pieces of pipe are held under axial pressure and allowed to cool for a certain time period. Pipe manufactures and ASTM D 2657 should be consulted for maximum heating temperatures and seaming pressures. The actual cooling time depends on the thickness of the pipe. For each seam, heating tool temperature, heating pressure, heating time, fusion pressure and fusion time should be recorded.

Seaming can introduce circumferential cracks that could propagate and may lead to stress cracking of the pipe in the presence of sustained axial tensile stresses (Parmar and Bowman, 1989). Qualified site supervision of the site-seaming process is therefore required. It is essential that the surfaces to be joined are clean and free of dirt. For seaming conducted in the field, it

Table 15.5 Calculated stresses at the crown and invert, quarter points and springlines of a 225-mm outside diameter *DR* 9 HDPE pipe with and without four circular perforations ($a = 25$ mm, $D_P/a = 8$, $L_P/a = 150$ mm, $E_s = 25$ MPa, $\nu_s = 0.3$, $E_p = 150$ MPa, $\nu_p = 0.46$, $\sigma_v = 600$ kPa and $K = 0.15$)

	Stresses at pipe interior, σ_Θ(MPa)		
Case	*Crown and invert*	*Quarter points*	*Springlines*
No perforations	−1.6	1.3	4.3
Four perforations at quarter points	−1.6	3.2	4.3
Four perforations at crown, invert and springlines	−3.3	1.3	9.7

may be necessary to provide a temporary work platform to prevent dirt being kicked up from the ground and possibly a temporary structure if dust is blowing in wind (e.g., from other construction activities on the site). Periodic verification of the surface temperature of the heating tool is required to ensure there are no cold spots on the fusion surface (e.g., due to a faulty heating iron). Each seam should be visually inspected for cracks, voids or other defects, and the seam should be redone if any defect is found.

Joints should be periodically evaluated to evaluate the quality of the seam. Samples of seamed pipe can be cut with 150 mm left on either side of the fusion joint from which samples for laboratory testing can be machined. Tensile elongation tests can be used to compare the short-term strength of the joint against that from control samples (i.e., unseamed polyethylene).

(b) Field inspection

Damage caused to the pipe during handling or installation is a source of flaws that in the presence of tensile stresses may lead to stress cracking. Damage can occur if the pipe is dragged on the ground surface during pipe seaming and installation. This can be avoided by sliding the pipe on top of closely spaced old rubber tyres to prevent the pipe from scraping on the ground.

Prior to placement of the pipe, it is important to prepare the bedding surface to be as even as possible with coarse gravel. The bedding surface should be free of very large gravel protrusions or voids beneath the pipe; otherwise the pipe may experience additional local bending effects.

The pipes should be inspected to verify that perforations are drilled in the specified circumferential location around the pipe and at the specified axial spacing. Prior to backfilling, the orientation of the perforated pipe should be checked to ensure that the perforations are located at the proper locations to avoid unanticipated larger stresses near the perforations. For example, maximum stresses in the pipe could be three times greater if the pipe was installed such that perforations intended to go at shoulder and haunch locations ended up at the pipe springline or invert (see Section 15.4.7). If there will not be close site supervision – such that perforations may end up at the springlines or invert – the increases in pipe stresses can be accounted for during pipe design (again see Section 15.4.7).

During backfilling, the density of the coarse gravel in the leachate collection system is not usually measured since the gravel is placed without compaction (given the marginal gain in stiffness by compacting poorly graded granular soils with angular particles). However, efforts should be made to ensure that gravel support at the haunches of the pipe is as continuous as possible with coarse gravel. This may require hand compaction of the gravel in the haunch regions with a steel rod, particularly for cases where pipe deflections are expected to be close to allowable deflection limits (e.g., in deep landfills). The size of construction equipment or the movement of equipment over the drainage system should be limited to avoid pipe damage or excessive pipe deformation during construction.

Pipe deflections that occur during installation must be accounted for in pipe design such that the total pipe deflection (i.e., those from installation and overburden pressures) is less than allowable limits. Installation deflections are normally negligible for thick pipes but may be important for pipes with *DR* greater than 11. Deflections can be monitored during installation, and construction practices should be adjusted if installation deflections are too large.

15.5 Manholes

Manholes are vertical shafts that extend through the waste to provide access to services buried below the surface (Figure 13.1). They are most commonly used at the connections between collection pipes (Figure 15.3) and provide access for maintenance (e.g., pipe cleaning and inspection), and at a sump (or low point) to remove leachate.

Manholes may also be connected to SLCS or leak detection system and may be used to both remove and introduce fluid (e.g., in the case of an engineered hydraulic trap).

The internal diameter of the manhole must be large enough to permit entry of personnel and maintenance equipment. The thickness of the manhole must be sufficient to ensure structural stability under loading from the waste. Radial pressures from the waste and fluids (if there is any leachate mounding), and settlement of waste around the manhole produce circumferential and axial compression of the manhole. Compressive stresses in the manhole must be kept below yield stresses; maximum circumferential and axial compression thrusts must be less than the critical buckling pressures.

The thickness of concrete manholes should be selected to account for deterioration of the concrete when exposed to leachate. Consequently, concrete manholes are often very thick and buckling rarely governs design. However, a thick concrete manhole may produce substantial vertical loading on the underlying soil and thus careful consideration to the design of the foundation for the manhole is required. Bearing failure or excessive settlement of the manhole may reduce the efficiency of the collection system and/or damage the barrier. The settlement of manholes occurs not only from the weight of the manhole itself but also the down drag forces that may develop due to the settlement of adjacent waste. The impact of waste settlement on the manhole can be estimated using techniques for the design of piles for down drag (Poulos and Davis, 1980).

Polymer manholes are also available for use in landfills. Although polymer manholes are more chemically resilient than concrete manholes, they are much more flexible (both circumferentially and axially) than concrete manholes. Careful consideration of their buckling capacity is required. For example, Gartung *et al.* (1989) and Biener and Sasse (1993) reported the failure of polymer manholes because of buckling, excessive lateral deformation and problems with connection details near the base. Gartung *et al.* (1993) reported that the field performance of two polymer manholes showed significant variations, particularly for radial stresses, when buried within relatively small (12 m) depths of waste.

A design approach for polymer manholes has been proposed by Petroff (1994). Alternatively, the circumferential thrust in the manhole can be calculated knowing the radial pressures acting on the manhole, which may be taken as the lateral stresses in the waste (although quantifying this in itself is challenging given the heterogeneous nature of the waste). The circumferential buckling resistance can be estimated using the method of Moore (2001) as given in equation 15.20b, provided that fluid pressures are small. However, leachate levels may lead to significant fluid pressures, depending on the efficiency of leachate collection. Laboratory experiments by Santamaria (2003) showed that fluid pressures significantly reduce the buckling capacity of a manhole below those expected when subjected to earth pressures alone. Given uncertainties in the radial pressures and axial forces acting on the manhole and the buckling capacity of polymer manholes under waste and fluid pressures, conservative factors of safety against circumferential and axial buckling should be used. Prior to use of polymer manholes in deep landfills, additional testing like that performed by Santamaria (2003) is required.

Another potential failure mechanism for landfill manholes is the connection details near the base where drainage pipes enter the manhole, and where geomembrane liners require a seal around the manhole. Differential settlements of the underlying soil materials (Section 15.2.1) may cause damage to the pipes and geomembrane. Giroud and Soderman (1995) have proposed design recommendations and simple calculations for the connections between the manhole and geomembrane liners.

Chapter sixteen

Integration of hydrogeology and engineering in barrier design and impact assessment

16.1 Introduction

In order to design safe waste disposal sites (e.g., landfills), it is essential that the interaction between the natural and engineered systems be considered. Thus, the objective of this final chapter is to discuss some of these interactions, paying particular attention to factors such as the evaluation of hydrogeology in the context of different engineering designs (Section 16.2), the different characteristics associated with a number of engineering designs (Section 16.3), the influence of concentration in the landfill on contaminant lifespan (Section 16.4), the long-term performance of leachate collection systems (Sections 16.5), the contaminant migration through composite liners when considering the finite service lives of the engineered components (Section 16.6) and the considerations involved in the expansion of landfills (Section 16.7).

16.2 Hydraulic conductivity of aquitards

It is well recognized that in the design of systems where there will be gradients and hence flow out of the waste disposal facility, the bulk hydraulic conductivity of the barrier and/or aquitard will be a key factor affecting potential impact of the landfill on an underlying aquifer. For situations involving outward gradients as shown in Figure 16.1, and all other things being equal, the greater the bulk hydraulic conductivity, the greater will be the outward advective flow, and hence the greater the potential impact on the aquifer. However, when dealing with inward gradient designs (i.e., hydraulic traps) as shown in Figure 16.2, since the inward advection from the aquifer to the landfill resists outward diffusion of contaminants, greater potential impact on the aquifer will occur for lower bulk hydraulic conductivity. Thus, although the hydrogeology may be exactly the same for Figures 16.1 and 16.2, with the only difference being the depth of the base contours adopted in the engineering design, the effect of uncertainty associated with the hydraulic conductivity of both the natural and engineered components of the barrier system, and hence what constitutes a "conservative" estimate of hydraulic conductivity, is quite different.

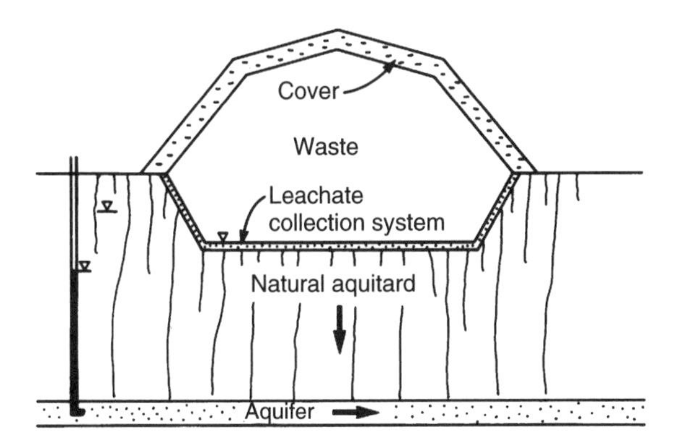

Figure 16.1 Schematic showing landfill located in a fractured aquitard with downward (i.e., outward) gradients.

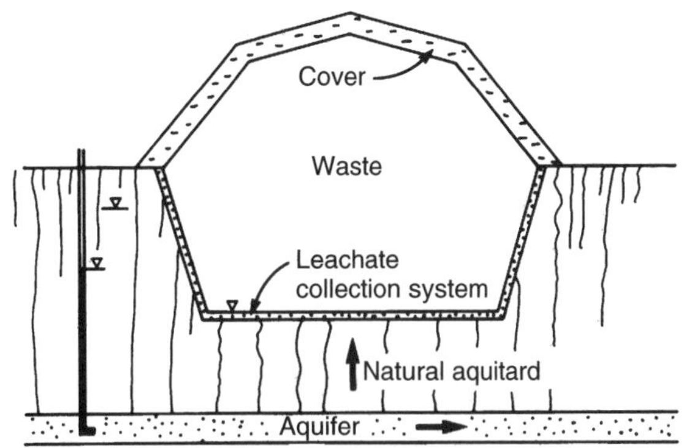

Figure 16.2 Schematic showing landfill located in a fractured aquitard with upward (i.e., inward) gradients (hydraulic trap design).

16.2.1 Outward gradient

To illustrate the potential significance of hydraulic conductivity, consideration will be given to contaminant migration from a 200-m wide ($L = 200$ m) landfill separated from a 1-m thick sand aquifer ($h = 1$ m, $n_b = 0.3$) by 4 m of clayey aquitard for three cases:

1. where there is 4 m of native fractured clay ($H_T = 4$ m, $H_L = 0$) between the landfill and the aquifer;
2. where the upper 1 m of fractured clay is removed and then recompacted to give a 1-m thick intact clayey liner (i.e., $H_L = 1$ m) underlain by 3 m ($H_T = 3$ m) of the original fractured clay; and
3. considering only the 1-m thick recompacted clayey liner ($H_L = 1$ m) in isolation and any retardation of contaminant as it moves through the fractured clay is neglected, although the hydraulic conductivity of the fractured clay is considered in determining the Darcy velocity which is, therefore, the same as for case 2.

The range of parameters considered for this example (example A) is summarized in Table 16.1. Here, it is assumed that there is a leachate collection system, that the difference in head between the leachate in the landfill and the water in the aquifer is 1 m and that the bulk hydraulic conductivity of the fractured clay k_T is 6×10^{-9} m/s. The Darcy velocity (Darcy flux), v_a, through the liner and fractured aquitard was then calculated by first determining the harmonic mean of the liner and fractured aquitard for a range of assumed hydraulic conductivities for the liner ($1 \times 10^{-10} \leq k_L \leq 1 \times 10^{-9}$ m/s) and then multiplying by the gradient which was held constant since it was controlled by the height of mounding in the landfill (assumed to be limited to 0.3 m) and the head distribution in the far more permeable aquifer (this technique is discussed in Section 5.2.2). As might be expected, the less permeable the liner, the smaller the Darcy flux through the system for a fixed head loss. The calculated concentration in the aquifer with time is plotted in Figure 16.3 for the three cases.

(a) Impact with fractured clay alone (no compacted clay liner)

For the assumed conditions in case 1 with a fractured aquitard alone, the fracture porosity is 0.00002 and corresponds to a groundwater velocity in the fractures of 2,500 m/a. If one were to consider simple "plug" (i.e., advective) flow in the fractures then full strength leachate would be expected to reach the aquifer in less than one day. However, if one considers migration along the fractures coupled with matrix diffusion into the adjacent clay, there is a substantially greater

Table 16.1 Summary of parameters considered for example A

Parameter	*Value*
Width of landfill, L (m)	200
Reference height of leachate, H_r (m)	5
Infiltration through cover, q_0 (m/a)	0.25
Initial concentration, c_0 (mg/l)	1,500
Downward Darcy velocity, v_a (m/a)	0.0001, $\underline{0.003}$, 0.0058, 0.0127, 0.0166, 0.021, 0.05
Thickness of fractured clay, H_T (m)	$\underline{3}$, 4
Porosity of clay matrix, n_m (−)	0.25, 0.3, 0.35, $\underline{0.4}$
Hydraulic conductivity of fractured clay, k_T (m/s)	6.3×10^{-9}
Thickness of clay liner, H_L (m)	0, $\underline{1}$
Hydraulic conductivity of liner, k_L (m/s)	$\underline{1 \times 10^{-10}}$, 2×10^{-10}, 5×10^{-10}, 7×10^{-10}, 1×10^{-9}
Thickness of underlying aquifer, h_b (m)	1
Porosity of aquifer, n_b (−)	0.3
Horizontal Darcy velocity in aquifer, v_b (m/a)	$v_b = v_a L + 2$
Diffusion coefficient in matrix, D_m (m^2/a)	0.02
Retardation coefficient for matrix, R_m (−)	1
Coefficient of hydrodynamic dispersion along fractures, D (m^2/a)	0.06
Fracture spacing, $2H_1 = 2H_2$ (m)	1
Fracture opening size, $2h_1 = 2h_2$ (μm)	10

Underlined quantities represent value used unless otherwise noted. a denotes annum (year).

time elapsed before contaminants reach the aquifer (see curve 1 in Figure 16.3). In this specific case, it takes between 16 and 17 years for contaminant to reach a concentration of 1% of the initial source leachate (i.e., $c/c_0 = 0.01$).

The important role of matrix diffusion (i.e., the migration of contaminants between the fractures and the pores of the adjacent intact material due to diffusion from high concentration to low concentration) in retarding contaminant migration through fractured media was discussed in detail in Chapter 11. However, it is evident that despite the attenuation due to matrix diffusion, there is a substantial increase in chloride concentration in the aquifer, to a peak value of about 540 mg/l ($c/c_0 \approx 0.36$) after about 45 years. This would generally be regarded as being unacceptable.

(b) Impact with a 1-m thick CCL ($k_L = 1 \times 10^{-9}$ m/s)

Case 2 (curve 2 in Figure 16.3) shows the calculated variation in concentration in the aquifer with time assuming that the top 1 m of fractured clay is excavated and recompacted as a liner having a hydraulic conductivity, k_L, of 1×10^{-9} m/s. This calculation involved modelling both the migration through the 1-m thick liner, and then along the fractures for the remaining 3 m of aquitard above the aquifer. This analysis considers both the reduced flow into the subsoil due to the presence of the liner and the attenuation due to diffusion into the matrix of the fractured aquitard. As a consequence, the time for contaminants to reach the aquifer at the 1% level is increased from about 17 years (case 1, no liner) to about 45 years (case 2). The peak impact is also reduced from 540 mg/l at 45 years (case 1) without a liner to 328 mg/l at 88 years (case 2) with this liner. Nevertheless, the impact is still excessive and would generally be unacceptable. This impact may be reduced by using a more effective liner (e.g., with a lower hydraulic conductivity as discussed momentarily).

As noted above, the results for case 2 were calculated considering attenuation due to matrix diffusion in the fractured aquitard below the

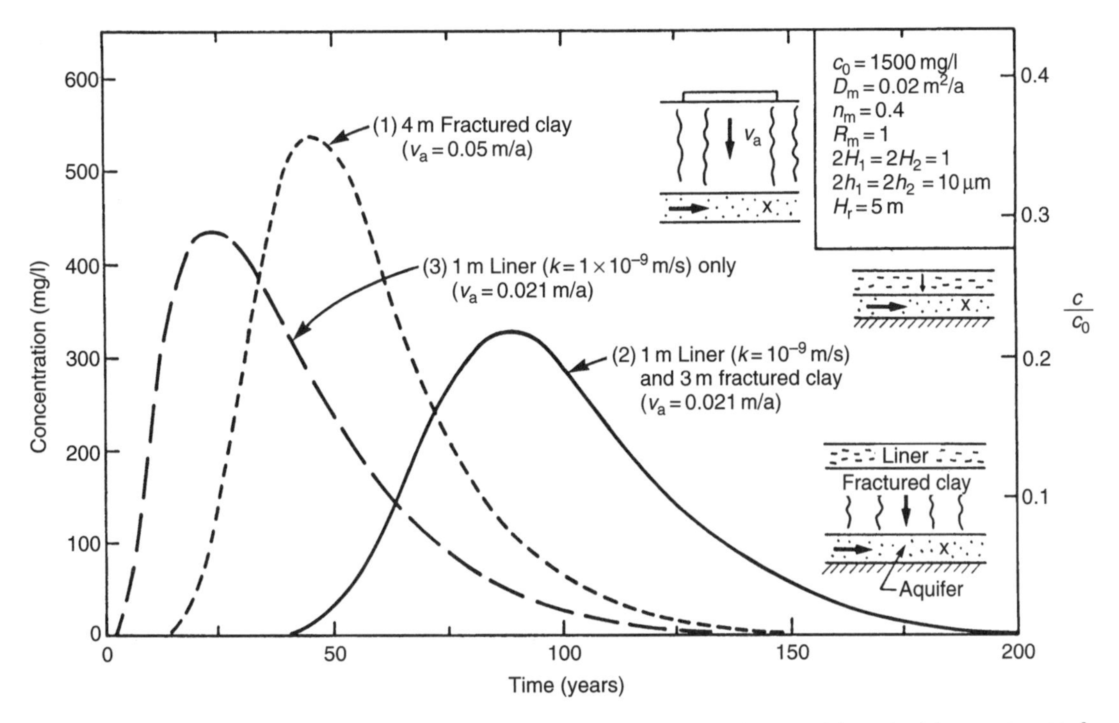

Figure 16.3 Impact on groundwater quality in the aquifer as a function of time with and without a 1×10^{-9}-m/s compacted clay liner, example A (after Rowe and Booker, 1991b; reproduced with permission of American Society of Civil Engineers).

liner. Case 3 (curve 3 in Figure 16.3) shows the results that are obtained if this attenuation is neglected. As can be seen, consideration of attenuation due to the liner alone (case 3) results in prediction of a contaminant impact which is much earlier than for case 2. (The issues of uncertainty regarding fracture spacing, opening size, etc. for fractured media are discussed in Chapter 11.)

(c) Impact with a 1-m thick (CCL) ($k_L = 1 \times 10^{-10}$ m/s)

The results presented in Figure 16.3 were obtained assuming a clay liner hydraulic conductivity of 10^{-9} m/s. As discussed in Chapter 9, there is substantial evidence to suggest that the bulk hydraulic conductivity of a well-designed and -constructed clay liner may be substantially less than 10^{-9} m/s and may be of the order of 10^{-10} m/s, or even lower. To illustrate the benefits of a lower hydraulic conductivity of a liner, Figure 16.4 shows the calculated variation in concentration with time for three cases. The results shown for case 1 (4 m of fractured aquitard) are precisely the same as those shown in Figure 16.3 (although the time scale has been changed) and the previous discussion applies to these results. Case 2 gives the results for a 1-m thick, $k_L = 1 \times 10^{-10}$ m/s liner overlying 3 m of fractured aquitard. It can be seen that the reduced flow and mass loading on the aquifer arising from the use of the 1×10^{-10}-m/s liner has a pronounced effect on the concentration with the aquifer. The time of arrival of contaminant at the 1% level in the aquifer is increased from about 17 years without the liner (case 1) to about 380 years (case 2). The maximum impact is reduced by an order of magnitude from 540 mg/l at 45 years with no liner (case 1), to 54 mg/l at 458 years with a 10^{-10}-m/s liner (case 2).

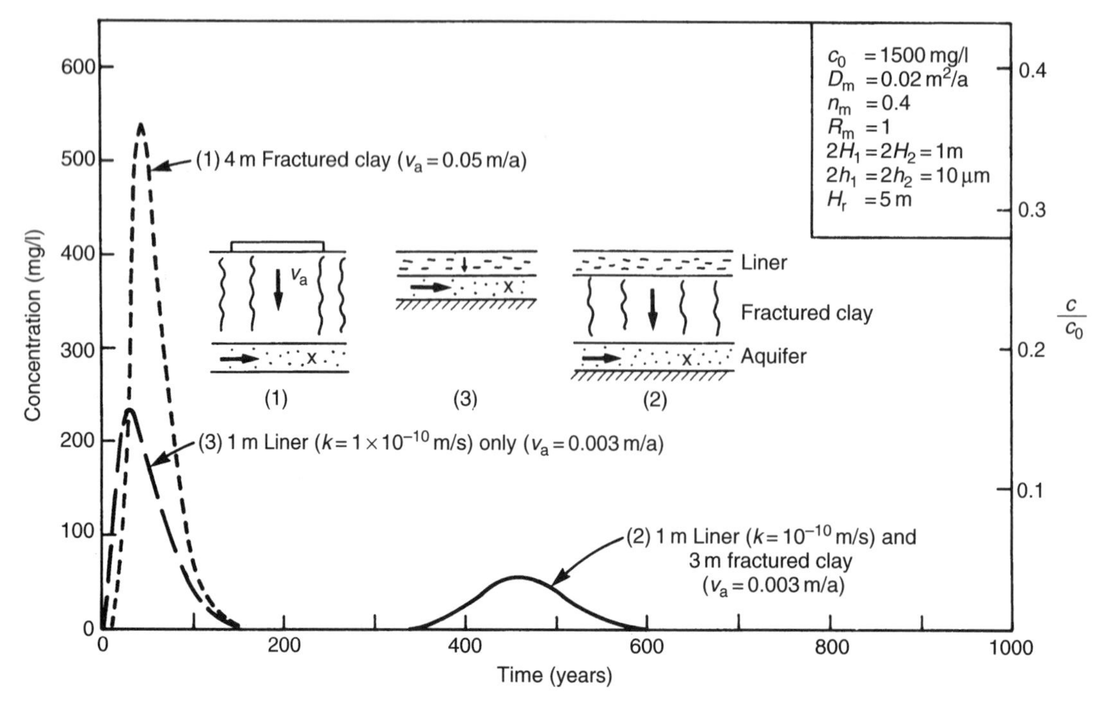

Figure 16.4 Impact on groundwater quality in an aquifer with and without a 1×10^{-10}-m/s compacted clay liner, example A (after Rowe and Booker, 1991b; reproduced with permission of American Society of Civil Engineers).

To illustrate the effect of considering the attenuation which occurs in the fractured aquitard, the results from case 2 may be compared with the results for case 3 which neglects any attenuation in the fractured aquitard. For the situation considered here, the analysis which neglects matrix diffusion gives arrival time much earlier and an impact much greater than that obtained when one considers matrix diffusion in the fractured aquitard.

(d) Effect of clay liner hydraulic conductivity

Comparison of the results given in Figures 16.3 and 16.4 demonstrates the importance of the hydraulic conductivity of the liner in controlling contaminant impact on the aquifer. Figure 16.5 summarizes the magnitude of peak impact for a range of values of liner hydraulic conductivity, k_L. As would be expected, a lower hydraulic conductivity results in smaller impact. The tenfold variation in k_L results in a more than sixfold decrease in impact on the aquifer and a fivefold increase, from 90 to 450 years, in the time to reach peak impact. A further reduction in impact could be achieved by increasing the thickness of the recompacted clayey liner.

16.2.2 Inward gradient

Section 16.2.1 considered the effect of varying hydraulic conductivity assuming that the potentiometric surface in the aquifer was 1 m below the level of leachate mounding in the aquifer and hence there were gradients outward from the landfill to the aquifer. To illustrate the different roles played by hydraulic conductivity with inward gradients, this section considers a similar thickness of aquitard and a CCL but assumes that the potentiometric surface in the aquifer is 1 m *above* the leachate level in the landfill (i.e.,

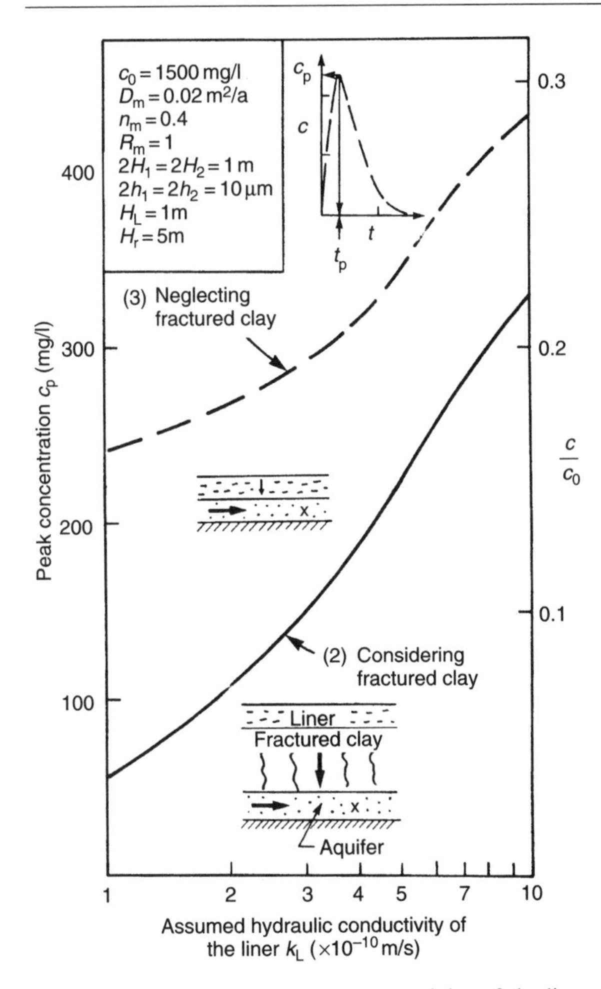

Figure 16.5 Effect of hydraulic conductivity of the liner on calculated impact in the underlying aquifer with outward gradients, example A (after Rowe and Booker, 1991b; reproduced with permission of American Society of Civil Engineers).

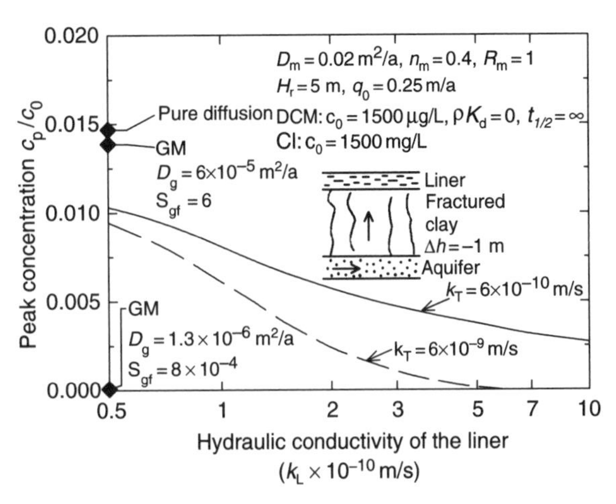

Figure 16.6 Effect of hydraulic conductivity of the liner on calculated impact in the underlying aquifer with inward gradients, example A.

the difference in head $\Delta h = -1$ m) thereby inducing an inward gradient. For the case of inward gradients, contaminant migration outwards from the landfill from diffusion is opposed by the inward flow of water to the landfill.

(a) Hydraulic trap with hydraulic conductivity of natural aquitard of 6×10^{-9} m/s

Figure 16.6 shows a plot of peak concentration in the aquifer as a function of liner hydraulic conductivity for a 1-m thick liner and 3 m of natural aquitard with a hydraulic conductivity of 6×10^{-9} m/s (the same hydraulic conductivity examined in the previous section). The peak increase in concentration in the aquifer is minimal for a liner hydraulic conductivity of 1×10^{-9} m/s (10×10^{-10} m/s) because the inward flow of 0.021 m/a overcomes most of the outward diffusion. As the hydraulic conductivity of the liner is reduced, so too is the inward flow and the calculated impact increases to a maximum value, for the range of k examined, for the hydraulic conductivity of 0.5×10^{-10} m/s. On the other hand, if there were no liner, and advection was controlled by flow through the fractures then the inward flow would not counteract diffusion through the matrix and, depending on the degree of fracturing, the impact would lie between that corresponding to the hydraulic conductivity of the intact clay and that shown for pure diffusion in Figure 16.6.

(b) Hydraulic trap with hydraulic conductivity of natural aquitard of 6×10^{-10} m/s

The results discussed above were obtained assuming that the fractured aquitard had a hydraulic conductivity of about 6×10^{-9} m/s. While this may have been a conservative interpretation of the bulk hydraulic conductivity of the aquitard for impact assessment based on outward flow, it is not conservative for assessing impact with inward flow. If the fractures were not hydraulically significant, and a reasonable lower bound bulk hydraulic conductivity of the clayey aquitard was 6×10^{-10} m/s, then greater contaminant impacts are predicted as shown by the curve for $k_T = 6 \times 10^{-10}$ m/s in Figure 16.6. This illustrates the need to consider the nature of the design being examined when assessing reasonable and conservative hydraulic conductivities to use in contaminant impact calculations.

16.2.3 Inward and outward gradients

In the previous two subsections, it was shown that the critical estimate of hydraulic conductivity corresponds to a high value for outward gradients and a low value for inward gradients. For designs with both inward and outward flow, such as that shown in Figure 16.7, it is not obvious as to what constitutes the critical hydraulic conductivity for the secondary liner and aquitard. With this design, the hydraulic control layer (HCL) is pressurized to induce an inward gradient from the HCL through the primary liner, thereby inducing an engineered hydraulic trap (see Section 1.2 for a more detailed discussion of these systems).

For the cases considered in this section, it is assumed that the potentiometric surface in the aquifer is below that in the HCL and hence there will be flow from the HCL to the aquifer. Thus contaminant, which can escape through the primary liner to the HCL by diffusion, will then be carried to the aquifer by advective–diffusive transport. If the hydraulic conductivity of the secondary liner and aquitard is low, then contaminant transport through the secondary liner-aquitard will be primarily by diffusion. This is slow, but there will be negligible dilution of contaminant in the HCL to reduce concentrations unless there is active pumping of this unit. If the hydraulic conductivity of the secondary liner and aquitard is high, then contaminants reaching the HCL by diffusion will be carried to the aquifer primarily by advection. This is fast but

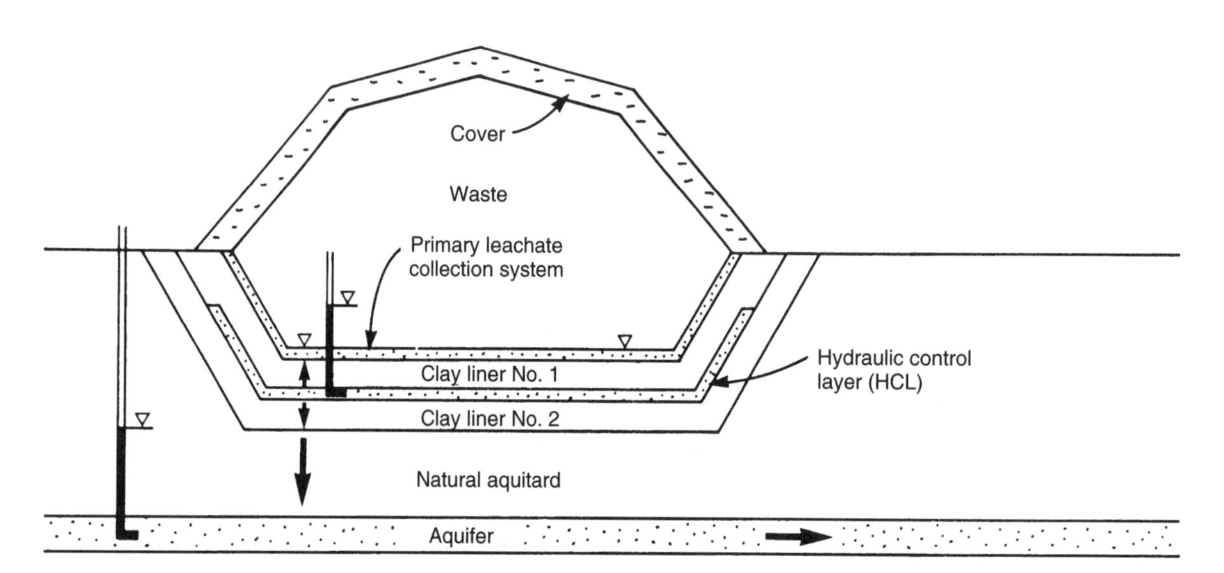

Figure 16.7 Schematic of a highly engineered system with a gravel layer beneath the primary liner that can be operated as an active hydraulic control layer.

there is also considerable potential for dilution of contaminant in the HCL because of the volume of water that must be injected to maintain the engineered hydraulic trap.

For this type of design, the hydraulic conductivity that would give rise to the maximum impact may not be either the highest or lowest reasonable hydraulic conductivity value. Rather, it may be some intermediate value that can only be established by performing a sensitivity study as illustrated below.

(a) Effect of secondary liner-aquitard hydraulic conductivity

Figure 16.7 shows a design involving a primary liner, a HCL and an aquitard that may include a man-made secondary liner. This case will be referred to as "example B". If it is assumed that the primary liner has a hydraulic conductivity $k_1 = 10^{-10}$ m/s, Figure 16.8 shows the calculated (using program POLLUTE; Rowe and Booker, 2004) maximum increase in concentration in the aquifer for a range of values of the harmonic mean hydraulic conductivity of the aquitard k_2, that incorporates the hydraulic conductivity of a secondary liner if present. As can be seen, the greatest impact occurs for a hydraulic conductivity that lies within the range considered and not at either end of the range.

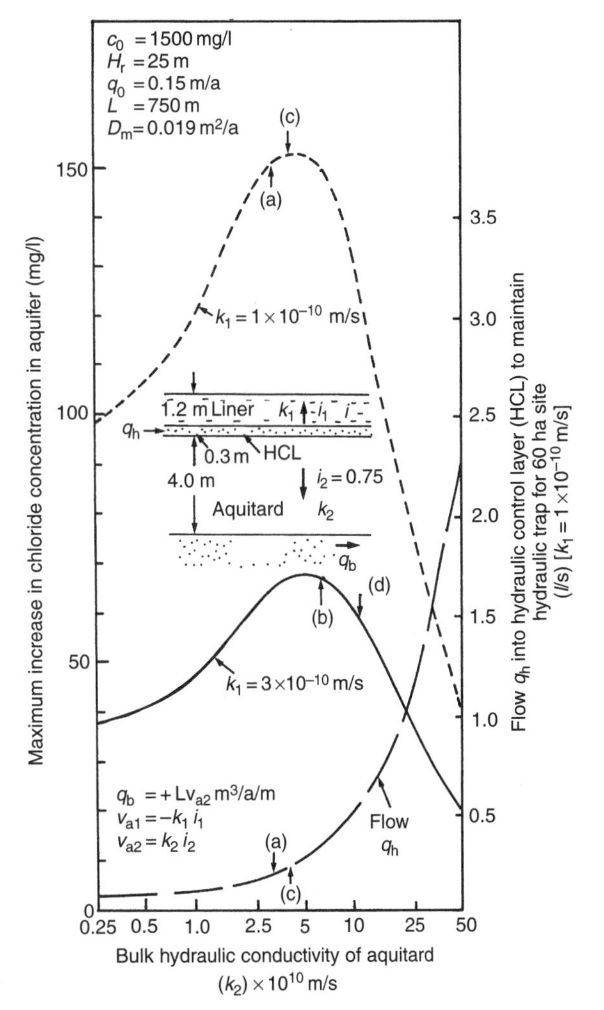

Figure 16.8 Effect of bulk hydraulic conductivity of aquitard on calculated impact in an underlying aquifer and flow required to maintain an engineered hydraulic trap, example B.

Assuming that the landfill has an area of 60 ha, Figure 16.8 also shows the volume of water (q_h) which must be injected to maintain the hydraulic trap as a function of the harmonic mean hydraulic conductivity of the aquitard. As might be expected, the flow increases with increasing hydraulic conductivity to a maximum of about 2.3 l/s for $k_2 = 50\times 10^{-10} = 5\times10^{-9}$ m/s. For the case giving rise to the greatest contaminant impact ($k_2 \approx 4\times 10^{-10}$ m/s), the flow rate is about 0.24 l/s.

Given that a CCL may consolidate with time giving rise to a reduction in hydraulic conductivity, consideration should be given to the likely range in hydraulic conductivity that might be expected. Suppose, for example, the likely range in liner hydraulic conductivity was 1×10^{-10}–3×10^{-10} m/s, inspection of Figure 16.8 shows that the lowest value of k_1 for the primary liner gives the greatest impact and thus for a likely range of between 1×10^{-10} and 3×10^{-10} m/s, one would use $k_1 = 1\times10^{-10}$ m/s in the contaminant impact calculations. For a 1-m thick secondary liner underlain by a 3-m aquitard having a hydraulic conductivity of 1×10^{-9} m/s, the harmonic mean hydraulic conductivity of the aquitard k_2 ranges between (a) 3×10^{-10} and (b) 6.3×10^{-10} m/s for liner hydraulic conductivities of 1×10^{-10} and 3×10^{-10} m/s, respectively. Assuming that the hydraulic conductivities of the primary and secondary liners are the same, the corresponding maximum impacts and flow required to maintain the hydraulic traps (see Figure 16.8) are 150 mg/l, 0.2 l/s for case (a) with $k_2 = 3\times10^{-10}$ m/s, and 67 mg/l, 0.46 l/s for case (b) with $k_2 = 6.3\times10^{-10}$ m/s. Thus, the upper and lower limits of the range of values of hydraulic conductivity of the liner give impacts that fall on either side of the maximum impact of about 153 mg/l that could be obtained.

If one examines the same range of liner hydraulic conductivities together with an assumed hydraulic conductivity of the natural aquitard of 10^{-8} m/s (rather than 10^{-9} m/s assumed above), one gets harmonic mean hydraulic conductivity k_2 of 3.9×10^{-10} m/s (for a liner of 1×10^{-10} m/s; case (c)) which gives a maximum impact of 153 mg/l and requires a flow $q_h = 0.24$ l/s to maintain the hydraulic trap at one end of the range of k for the liners and a value of k_2 of 11×10^{-10} m/s (for a liner of 3×10^{-10} m/s; case (d)) which gives a maximum impact of 60 mg/l and requires a flow $q_h = 0.68$ l/s at the other end of the range.

For the particular case examined here, a sensitivity study for reasonable uncertainty regarding the hydraulic conductivity of the liner (1×10^{-10}–3×10^{-10} m/s) and aquitard (10^{-9}–10^{-8} m/s) gives what would be judged unacceptable impacts in the Province of Ontario. On the other hand, a single calculation based on what might conventionally be regarded as conservative parameters (i.e., at the high end of the reasonable range) would correspond to $k_1 = 3\times10^{-10}$ m/s and $k_2 = 11\times10^{-10}$ m/s (for an aquitard with hydraulic conductivity of 10^{-8} m/s) giving a peak impact of only 60 mg/l (case (d) in Figure 16.8) which may well be judged acceptable for chloride. Thus, what might be regarded as "conservative" (case (d)) does in fact underestimate the potential peak impact by about 250%.

(b) Use of a composite secondary liner to minimize fluid flow required to maintain engineered hydraulic trap

The injection of water required to maintain the hydraulic trap could be reduced by including a geomembrane above the secondary liner. Assuming a well-designed and -constructed geomembrane, such that the average Darcy velocity from the HCL to the aquifer is reduced to 0.0001 m/a (i.e., less than 3 l/ha/day leakage through holes in the geomembrane), the injection rate required to maintain the hydraulic trap for a primary clay liner with $k_1 = 1\times10^{-10}$ m/s is reduced to about 0.06 l/s. The calculated impact on the aquifer would be negligible, allowing for a diffusion parameters of $D_g = 4\times10^{-14}$ m^2/s and $S_{gf} = 8\times10^{-4}$ for the geomembrane (see Section 13.5). This may be compared with the impacts calculated without a geomembrane (see Figure 16.8).

16.3 Some considerations in the design of highly engineered systems

In Section 16.2.3, consideration was given to the effect of the hydraulic conductivity of the secondary liner/natural aquitard system for the case where a granular layer beneath the primary liner (HCL) was used to create an engineered hydraulic trap in which water flowed from the HCL through the primary liner and into the landfill. This section examines a number of other ways in which an engineered permeable layer beneath the primary liner can be used and pays particular attention to the effect of the location of the potentiometric surface in an underlying aquifer to illustrate how the hydrogeology and engineered design interrelate.

For the purposes of quantitatively illustrating a number of points, consideration will be given to the design of a hypothetical landfill with an average waste thickness of 15 m which is constructed above a natural sand aquifer (example C), as illustrated in Figures 16.9 and 16.10. For simplicity of illustration, it is assumed that the design consists of (from the waste down) a PLCS, a 1.2-m thick CCL, a 0.3-m thick granular layer (for secondary leachate collection or hydraulic control) and a 1.5-m thick secondary (natural) clay liner which is underlain by a 1-m thick granular aquifer. For comparison purposes, consideration is also given to the case where the secondary liner/natural aquitard is 8-m thick below the landfill. The landfill is assumed to be 750 m long in the direction of groundwater flow and it is assumed that the primary piping and slope on the leachate collection system is out of the plane being considered (i.e., the cross-section being examined is the critical cross-section).

The basic parameters considered for this example are summarized in Table 16.2. Consideration is given to the migration of both chloride and dichloromethane. The reference height of leachate, H_r (see Section 16.4.2 for a discussion of the significance of H_r), represents a mass equal to 0.2% and 0.0002% of the total mass of waste for chloride and dichloromethane, respectively. For the purposes of this analysis, it is assumed that the diffusion coefficients for chloride and dichloromethane are the same; however, dichloromethane is retarded due to sorption ($\rho K_d = 2$). At this point, it is also assumed that the SLCS consists of coarse sand with porosity of $n = 0.3$ that remains fully saturated and provides no substantial resistance to diffusion ($D > 1\ m^2/a$). The influence of an unsaturated layer will be examined in Section 16.3.3.

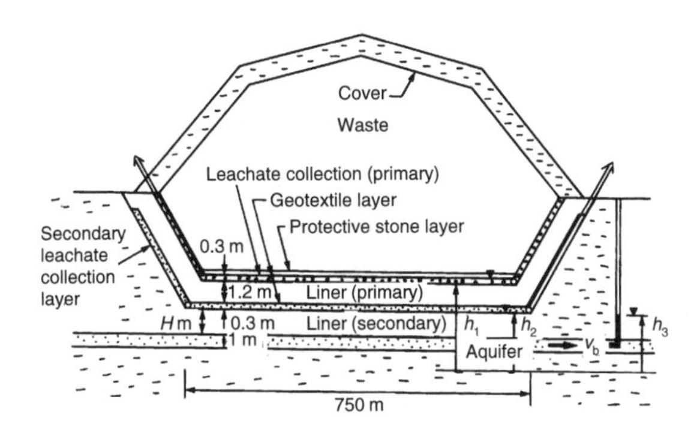

Figure 16.9 Schematic showing a primary liner underlain by a leak detection secondary leachate collection system. Advective flow is downward through the primary liner, example C.

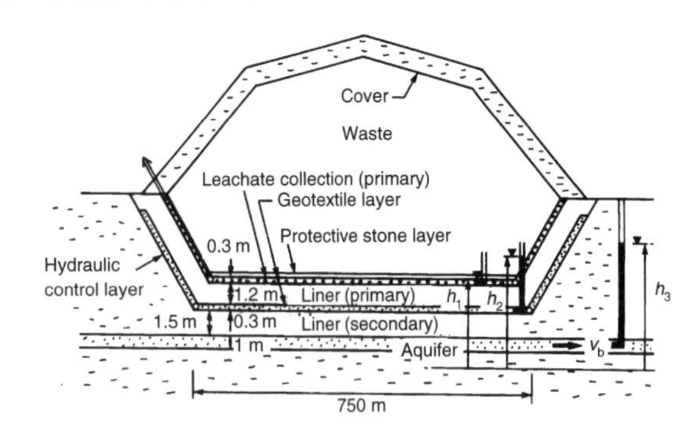

Figure 16.10 Schematic showing a primary liner underlain by a hydraulic control layer. The landfill is designed as a hydraulic trap with advective flow into the landfill, example C.

The leachate mound in the HCL is assumed to be 0.3 m above the top of the CCL (i.e., $h_1 = 10$ m, measuring head relative to an arbitrary datum and taking positive upwards). The heads in the secondary leachate collection/HCL (h_2) and in the aquifer (h_3) vary depending on the hydrogeologic conditions being considered (see Figures 16.9 and 16.10).

All the analyses reported herein were performed using a finite layer contaminant transport model described in Chapter 7, as implemented in computer program POLLUTE (Rowe and Booker, 2004).

16.3.1 Secondary leachate collection with a clay primary liner

For situations where the water table and potentiometric surface in the underlying aquifer are well below the base of the landfill (e.g., see Figure 16.9), the construction of a permeable drainage

Table 16.2 Summary of parameters considered for example C

Quantity	
Length of landfill, L (m)	750
Reference height of leachate, H_r (m)	12
Infiltration through cover, q_0 (m/a)	0.15
Initial chloride concentration, c_0 (mg/l)	1,500
Initial dichloromethane concentration, c_0 (μg/l)	1,500
Sorption parameter for dichloromethane, $\rho K_d(-)$	2
Primary clay liner	
Thickness, H_1 (m)	1.2
Porosity, n (−)	0.35
Hydraulic conductivity, k_L (m/s)	3×10^{-10}
Diffusion coefficient, D_m (m^2/a)	0.019
Engineered granular collection system	
Thickness, H_2 (m)	0.3
Porosity, n (m)	0.3
Hydraulic conductivity, k (m/s)	10^{-2}
Dispersion coefficient, D (m^2/a)	See text
Secondary liner/aquitard	
Thickness, H_3 (m)	Variable (1.5, 8)
Porosity, n (−)	0.25
Hydraulic conductivity, k (m/s)	Variable (10^{-9}, 10^{-10})
Diffusion coefficient, D (m^2/a)	0.015
Aquifer	
Thickness, h_b (m)	1
Porosity, n_b (−)	0.3
Hydraulic conductivity	10^{-5}
Horizontal Darcy velocity, v_b (m/a)	*

*Fixed by continuity of flow considerations. For a hydraulic trap $v_b = 1$ m/a is assumed; for outward flow $v_b = 1 + v_{a2}L$, v_{a2} given in Tables 16.3–16.4.

system, which is located beneath the CCL, serves two purposes. First, the drainage layer functions as an SLCS that can remove a portion of the leachate that escapes through the liner (and some escape is to be expected through any liner system where there are downward gradients). Second, this layer serves to reduce the hydraulic gradient through the underlying soil.

A key question in the design of these liner systems is "what will be the impact of the contaminant on groundwater in the underlying aquifer?". The answer to this question will depend on the properties of the soil, the drainage system and the liner, the geometry of the landfill, the design of PLCS and SLCS, and the properties of any underlying aquifer. Each case must be considered as a unique situation; however, it is important to recognize that some contaminant migration through the natural soil must be anticipated for designs such as that shown in Figure 16.9. Even in the limiting case where no leachate mounding occurred in the SLCS and all the leachate that escaped through the liner was collected, there would still be diffusion of contaminants into the natural soil from the secondary leachate collection layer. In most cases, there will also be downward advective contaminant transport since not all the leachate escaping through the liner is likely to be collected. Indeed, situations can readily occur where the majority of the leachate migrating through the primary liner also migrates down through the natural soil.

Ideally, the separation between the aquifer and the landfill would be as large as possible for a design involving an SLCS such as the one shown in Figure 16.9. However, in many practical situations the actual thickness may be quite thin. Under these circumstances, it cannot be assumed that even a perfectly operating liner, and PLCS and SLCS will necessarily prevent contamination of groundwater in an underlying aquifer. To illustrate this, consider case 1 where it is assumed that the base of the engineered landfill is chosen to correspond to the potentiometric surface of the underlying aquifer which, for this case, is assumed to correspond to the head of $h_3 = 8.2$ m at the downgradient edge of the landfill. This ensures an adequate factor of safety against "blowout" of the *in-situ* secondary clay liner (see Section 15.2.6) and, if there is no mounding of leachate in the secondary collection layer (i.e., $h_2 = h_3 = 8.2$ m), creates a situation where there is no inward or outward flow through the secondary liner. Thus for this case, there will be downward advective transport through the primary liner ($k = 3 \times 10^{-10}$ m/s, $i = 1.25$) corresponding to a Darcy velocity of 0.012 m/a. For this scenario, all leachate should be collected and contaminant transport through the secondary liner is by the process of molecular diffusion.

The results of analyses performed for case 1 are summarized in Table 16.3. Although this case represents perfect secondary leachate collection with no advective escape through the secondary liner, the process of molecular diffusion through the secondary liner results in significant impact in the aquifer for a thin secondary liner (case 1a). With a thick secondary liner/aquitard system (case 1b) the impact is much smaller, especially for dichloromethane which experiences attenuation due to sorption in the natural secondary liner aquitard. For this case, where contaminant migration is only by diffusion, dilution in the aquifer is an important attenuation mechanism. If the Darcy velocity, v_b, were ten times higher, the calculated impacts would be about ten times less. Thus for these systems, the hydrogeologic characterization of the head and flow in the aquifer can be quite important when assessing the potential impact due to a proposed design. As discussed in Chapter 10, the effects of reasonable uncertainty can be addressed by performing a sensitivity study.

The significance of the location of the potentiometric surface in the aquifer can be demonstrated by considering two different head conditions. In cases 2 and 3, it is assumed that there is an outward (downward) gradient of 0.33 from the secondary collection system to the underlying aquifer. Cases 4 and 5 examine the reverse situation where there is an inward (upward) gradient of 0.33 from the aquifer to the secondary collection system. This latter case represents the limiting inward gradient that can be achieved while ensuring a factor of safety against blowout of the base layer of 1.5 during construction of the secondary collection system for a unit weight of 20 kN/m^3 for the secondary barrier soil (see Section 15.2.6).

Table 16.3 Calculated impacts for clay primary liner and secondary leachate collection for example C (Figure 16.9)

	Secondary liner/aquitard			*Heads*[a, b]			*Darcy velocity*[c]			*Peak impact in aquifer*[d]			
										Chloride		*Dichloromethane*	
Case	k (m/s)	H (m)	i (–)	h_1 (m)	h_2 (m)	h_3 (m)	v_{a1} (m/a)	v_{a2} (m/a)	v_b (m/a)	*Conc.* (mg/l)	*Time* (a)	*Conc.* (μg/l)	*Time* (a)
1a	Any	1.5	0	10	8.2	8.2	0.012	0	1	240	160	75	560
1b	Any	8.0	0	10	8.2	8.2	0.012	0	1	20	800	3	5,320
2a	10^{-9}	1.5	0.33	10	8.2	7.7	0.012	0.0105	8.88	610	110	210	430
2b	10^{-9}	8.0	0.33	10	8.2	5.53	0.012	0.0105	8.88	470	270	90	1,740
3a	10^{-10}	1.5	0.33	10	8.2	7.7	0.012	0.001	1.79	280	160	85	540
3b	10^{-10}	8.0	0.33	10	8.2	5.53	0.012	0.001	1.79	50	730	5	4,880
4a	10^{-9}	1.5	−0.33	10	8.2	8.7	0.012	−0.01015	1	5	100	2	380
4b	10^{-9}	8	−0.33	10	8.2	10.86	0.012	−0.01015	1	<0.01	260	<0.01	1,710
5a	10^{-10}	1.5	−0.33	10	8.2	8.7	0.012	−0.001	1	175	160	55	540
5b	10^{-10}	8.0	−0.33	10	8.2	10.86	0.012	−0.001	1	5	730	<1	4,880

Notes: [a]Positive is downward; [b]Relative to fixed local datum, see Figure 16.9; [c]v_{a1} = vertical Darcy velocity (flux) through primary liner, v_{a2} = vertical Darcy velocity (flux) through secondary liner, v_b = horizontal Darcy velocity (flux) in aquifer at the downgradient end of the landfill; [d]All concentrations greater than 5 rounded to nearest 5 and all times rounded to nearest decade; see Table 16.2 for other parameters.

Relative to the pure diffusion case (case 1), an outward gradient through the secondary liner (aquitard) results in a substantial increase in potential impact and, in fact, if the hydraulic conductivity of the secondary liner is of the order of 10^{-9} m/s (case 2), then the SLCS is not very effective and collects only 13% of the contaminated water escaping through the primary liner, with the rest (87%) passing through the secondary liner. If the harmonic mean hydraulic conductivity of the secondary liner-aquitard is of the order of 10^{-10} m/s (case 3), then the secondary collection system is more effective and collects about 91% of the secondary leachate with about 9% still leaking through the secondary liner. One means of reducing impact would be to place a geomembrane above the CCL to create a composite primary liner, as examined in Section 16.6.

With inward gradients, the performance of the barrier system is substantially enhanced. For a hydraulic conductivity of the secondary liner of the order of 10^{-9} m/s (case 4), there is negligible impact on groundwater quality in the aquifer. However, it should be noted that this design extracts a substantial amount of water from the aquifer and the heads and flow considered here are those expected after consideration of landfill-hydrogeologic interaction as discussed in Section 16.3.4. For this case, the fluid to be collected from the secondary system is almost doubled. Thus for a 60-ha site, the volume of fluid collected from the secondary leachate system increases from about 0.23–0.43 l/s; however, this is still small compared with the volume of 2.63 l/s expected to be collected from the primary collection system. Case 5 involves a small (~10%) increase in the volume of leachate collected, but because of the low hydraulic conductivity of the secondary liner-aquitard there is still some diffusion of contaminant from the secondary collection system to the aquifer. In this case, there is a clear advantage to having a thicker aquitard, especially for attenuating organic chemicals such as dichloromethane as can be seen by comparing cases 5a and 5b.

16.3.2 Hydraulic control layer

As discussed in Section 1.2, an alternative to using the granular layer beneath the primary liner as an SLCS is to use it as a HCL. For example, suppose that the potentiometric surface in the aquifer shown in Figure 16.10 is such that $h_3 > h_2 > h_1$. In this case, there is both a natural hydraulic trap (i.e., water flows from the natural soil into the HCL) and an engineered hydraulic trap (i.e., water flows from the HCL into the landfill). Where practical, this design has the following advantages. First, since there is inward flow to the HCL and a relatively impermeable clay liner, it may be possible to design the system such that the engineered hydraulic trap is entirely passive. That is, all water required to maintain an inward gradient is provided by the natural hydrogeologic system and no injection of water to the HCL is required. Second, because of the two-level hydraulic trap, there will be substantially greater attenuation of any contaminants that do migrate through the primary liner. Third, since fluid can be injected and withdrawn from the HCL, it is possible to control the concentration of contaminant in the layer and hence minimize the impact at the site boundary in the event of a major failure of the leachate collection system (as discussed in Section 16.5).

There are three factors that must be considered in the design of this engineered hydraulic trap. First, the head in the HCL must be controlled such that blowout of either liner does not occur during or after construction (see Section 15.2.6). Second, the volumes of water collected by the hydraulic trap must be manageable and the hydrogeologic system must have the capacity to provide the water required to maintain the hydraulic trap; if this is not the case, then the head in the aquifer will drop and the effectiveness of the trap may deteriorate with time (see Section 16.3.4). Third, although there is a hydraulic trap, some outward diffusion of contaminants is to be expected in most cases.

Contaminant migration analyses are required to assess what impact may occur under these conditions. If the impact at the site boundary is not acceptable then it can be reduced by pumping water through the HCL (i.e., by injecting freshwater at one end and extracting contaminated water at the other end). The volume of fluid to be pumped can be assessed by appropriate modelling. Models are available (POLLUTE and MIGRATE) which readily allow the designer to estimate potential impact as a function of the flow in the HCL.

To illustrate the effect of a HCL, cases 6, 7 and 8 in Table 16.4 each examine the case where the total head, h_1, on top of the landfill liner is 10 m (i.e., 0.3 m of leachate mounding on the liner – see Figure 16.10) and the total head in the aquifer is 10.9 m. This induces an inward gradient across the liner system. In each case, the primary liner is assumed to have a hydraulic conductivity of 3×10^{-10} m/s. In cases 6 and 7, the secondary liner is assumed to have a hydraulic conductivity of 10^{-9} and 10^{-10} m/s, respectively and the HCL is assumed to be operating as a natural hydraulic trap (i.e., no human introduction or removal water from the HCL). Under these circumstances the head, h_2, in the HCL is controlled by the relative hydraulic conductivities and thicknesses of the primary and secondary liner and may be calculated as described in Section 5.2.

With the higher permeability secondary liner (case 6), the flows in the system (see Darcy velocities v_{a1}, v_{a2} in Table 16.4) are larger than for the lower permeability secondary liner (case 7) and hence the resistance to outward flow is also greater. The greater the inward flow, the greater the resistance to outward diffusion and, consequently, the impact for case 6 is less than that for case 7.

In case 6, both 1.5-m and 8-m thick secondary liner/aquitard are considered. For a given head in the aquifer, the thicker layer gives rise to lower gradients and hence (all other things being equal) lower inward Darcy velocities. Thus, case 6b represents a combination of less favourable inward gradients but more favourable thickness of the secondary liner-aquitard and the net effect of these two differences can be appreciated by comparing the results of cases 6a and b. It should be noted that cases 6a and b do not correspond to different designs for the same hydrogeology; rather, they represent two different sites where the head difference between the

Table 16.4 Calculated impacts for a clay primary liner and hydraulic control layer for example C (Figure 16.10)

	Secondary liner-aquitard			*Heads*[a, b]			*Darcy velocity*[c]			*Water added to HCL*[d]	*Peak impact in aquifer*[e]			
											Chloride		*Dichloromethane*	
Case	k (m/s)	H (m)	i (–)	h_1 (m)	h_2 (m)	h_3 (m)	v_{a1} (m/a)	v_{a2} (m/a)	v_b (m/a)	q_H (l/s)	*Conc.* (mg/l)	*Time* (a)	*Conc.* (μg/l)	*Time* (a)
6a	10^{-9}	1.5	−0.167	10	10.65	10.9	−0.005	−0.005	1	0	15	160	<5	580
6b	10^{-9}	8.0	−0.098	10	10.3	10.9	−0.00236	−0.00236	1	0	<1	660	<0.05	4,480
7	10^{-10}	1.5	−0.473	10	10.19	10.9	−0.0015	−0.0015	1	0	85	190	20	670
8	10^{-10}	1.5	0	10	10.9	10.9	−0.007	0	1	0.135	40	170	10	630
9	10^{-10}	1.5	0.667	10	11.9	10.9	−0.015	0.002	2.5	0.325	10	170	3	580

Notes: [a]Positive is downward; [b]Relative to fixed datum local datum, see Figure 16.9; [c]v_{a1} = vertical Darcy velocity (flux) through primary liner; [4]v_{a2} = vertical Darcy velocity (flux) through secondary liner; v_b = horizontal Darcy velocity (flux) in aquifer at the downgradient end of the landfill; [d]Water which must be added to maintain the hydraulic trap – assumes landfill area of 60 ha; [e]All concentrations greater than 5 rounded to nearest 5; all times rounded to nearest decade.

aquifer and the top of the primary liner just happen to be the same. If there were the same inward gradients for both clay thicknesses, then the impact would always be less for the case with the thicker secondary liner-aquitard.

Cases 8 and 9 examine the behaviour of an engineered hydraulic trap where water is introduced to increase the head in the HCL to 10.9 and 11.9 m, respectively. In case 8, there is the maximum gradient across the primary liner without creating an outward gradient across the secondary liner. This reduces the impact compared to the corresponding passive case (case 7) with a 10^{-10}-m/s secondary liner. Case 9 relies more heavily on the induced pressure in the HCL resisting outward movement of contaminant through the secondary liner and requiring a greater addition of fluid, q_H, to maintain the hydraulic trap.

In the design of HCLs, it is important that the transmissivity of the layer is high enough to ensure a relatively uniform head on the base of the primary and top of the secondary liners over the area of the landfill. This will often imply the need for a hydraulic conductivity greater than 0.1 m/s; generally the larger the landfill, the greater the transmissivity required. To minimize impact in the event of a failure of the PLCS, the HCL should be designed such that it can be used in an active mode, with injection and collection of fluid, without requiring a large head difference between the manholes where fluid is injected and removed.

16.3.3 Unsaturated granular layer

Both the volumetric water content and the coefficient of hydrodynamic dispersion need to be considered when modelling contaminant migration through the granular layer beneath the primary liner, whether it is providing secondary leachate collection or hydraulic control. The granular layer may be saturated even if it is operated as a collection layer (i.e., fluid is being removed but it is not pumped "dry"), or if acting as HCL. The volumetric water content is then equal to the porosity. Provided that advective flow is small (i.e., such that mechanical dispersion is negligible), the coefficient of hydrodynamic dispersion may be taken as the saturated diffusion coefficient (e.g., see Section 8.10 or Rowe and Badv, 1996b).

When operated as an SLCS such that the pressure head in this layer is equal to zero, the granular layer will be unsaturated. The volumetric water content is then used in contaminant transport modelling. The work of Badv and Rowe (1996; see Section 8.11.1) can be used to obtain estimates of the coefficient of hydrodynamic dispersion for unsaturated coarse gravel. They reported apparent coefficients of hydrodynamic dispersion of 0.009, 0.06 and 11 m^2/a for Darcy velocities of 0.017, 0.12 and 2.5 m/a, respectively. As the velocity increases, there is increased mechanical dispersion and hence an increase in the coefficient of hydrodynamic dispersion. Thus at very low velocities, for example in the case of a composite geomembrane and clay liner with small leakage (e.g., see Section 16.6), an unsaturated granular layer may act as an effective diffusion barrier to dissolved contaminants. In contrast, the unsaturated layer may readily allow migration of volatile organic contaminates (Section 8.11.2).

The influence of an unsaturated granular layer on the concentrations in the aquifer can be illustrated by re-examining case 2a from Table 16.3. If this case is modelled assuming that the SLCS is pumped dry and using a volumetric water content of $\theta = 0.03$ and coefficient of hydrodynamic dispersion of $D = 0.009$ m^2/a for coarse gravel in this layer yields a peak chloride impact of 640 mg/l at 105 years. This impact is about 5% larger than for case 2a primarily due to less dilution that occurs in the granular layer when it is unsaturated.

16.3.4 The effect on groundwater levels

A key consideration in hydrogeologic investigations is the understanding of the groundwater flow direction, natural variability in the water table and in the potentiometric surface in any underlying aquifers and flow in these aquifers. Landfills have been designed on the basis of this information; however, it is important to recognize that the construction of a landfill may change these conditions. This is particularly true when the landfill is of large areal extent.

If a landfill is located in a recharge area, the construction of an engineered landfill involving a leachate collection system and, frequently, one or more liners will often reduce the recharge to the underlying aquifer. Thus, the landfill casts a hydraulic shadow over the aquifer (and hence the term shadow effect), which may lead to a drop in water levels (i.e., a lowering of the potentiometric surface) within the aquifer. The extent to which this will occur will depend on the existing gradients and transmissivity of the aquifer, the hydraulic conductivity of the engineered barrier(s) and the elevation of the base contours relative to the existing potentiometric surface in the aquifer.

The greatest potential effect of the landfill on groundwater conditions occurs when a landfill of large areal extent is to be designed as a natural hydraulic trap above an aquifer of low transmissivity. In this case not only is recharge cut off over the area of the landfill but also water is being extracted from the aquifer to provide the inward flow necessary to have a natural hydraulic trap (e.g., see Figure 16.11). This can lead to a significant drop in water levels in the aquifer and, if not considered in the design, could lead to a failure of the hydraulic trap. The magnitude of the change in water levels may be estimated using numerical models (as discussed in Chapter 5).

An example of the potential significance of the shadow effect is found in the design of the Halton Landfill in Ontario, Canada (Rowe *et al.*, 2000c). The Certificate of Approval for this landfill required that it be designed as a natural hydraulic trap. Based on existing water level data, an initial set of landfill base contours could be readily established which provided the required hydraulic trap. However, flow modelling of this initial design which involved a landfill length of 750 m along the flow path in the aquifer revealed that the shadow effect could cause a lowering of the potentiometric surface in the aquifer in excess of 2 m. This would have caused a potential failure of the hydraulic trap over approximately half the landfill if the base of the landfill had been selected to provide a hydraulic trap (i.e., upward flow into the landfill) based only on the existing water levels (potentiometric surface) in the aquifer. This finding resulted in a lowering of the base contours by approximately 2 m such that the hydraulic trap would be maintained even after the consequent lowering of the potentiometric surface in the aquifer due to the construction of the landfill. It is noted that the modelling showed that the flow in the aquifer was reduced by about 50% due to the operation of the landfill as compared to the preconstruction flow.

The potential significance of the "shadow effect" can only be assessed by an integrated examination of both engineering design and existing hydrogeology. This will frequently involve flow modelling. In instances where the water levels in the aquifer are critical to the design, consideration should be given to the potential implications of long-term variability in water levels due to factors such as long-term changes in climate and potential developments in the recharge area for the aquifer. These effects are quite intangible and in these cases a contingency is required in the event of unexpected changes in the aquifer water levels. For example, in the case of the Halton Landfill, a HCL that is totally passive, referred to as the "subliner contingency layer", was installed (see Figure 16.11). Under operating conditions, no water is intended to be added or removed from this layer (i.e., the heads in the HCL will be maintained by flow from the underlying aquifer); however, in the event of some unexpected drop in the potentiometric surface in the aquifer, water may be added to the HCL to create an engineered

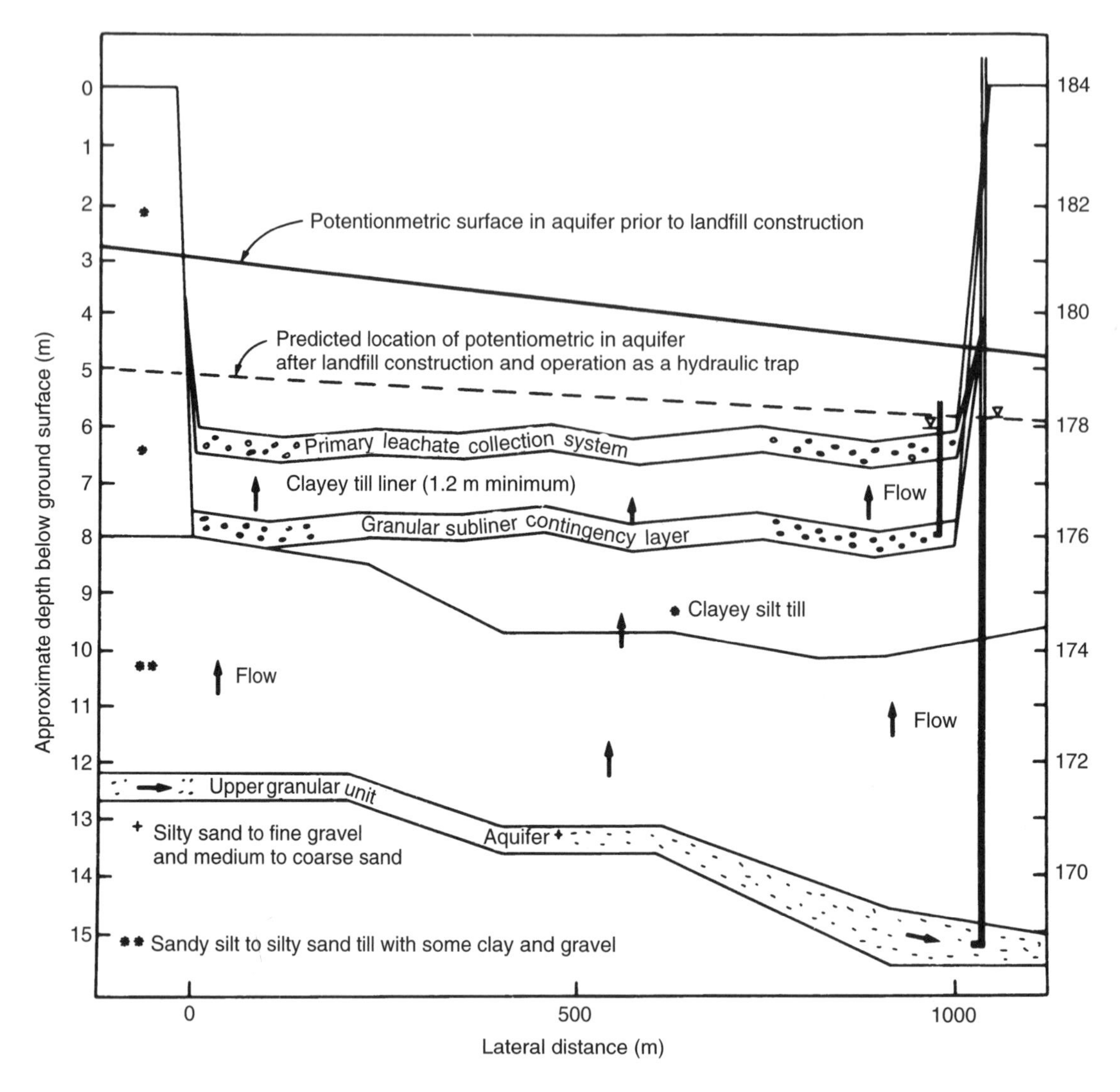

Figure 16.11 Hydraulic trap landfill (Halton) showing potential shadow effect due to landfill construction. Also shown is the subliner contingency layer.

hydraulic trap. Alternatively, if necessary, the layer may be pumped and operated as an SLCS. Thus, although it is not intended that this system be used, it is available for use in the event of an unexpected failure of the natural hydraulic trap due to either a change in natural conditions (e.g., lowering of the potentiometric surface due to climate change) or engineered conditions (e.g., a failure of the PLCS as discussed in Section 16.5).

16.4 Concentration of contaminant in the landfill

In addition to an estimate of the leachate head (discussed in Section 16.5), an estimate of the contaminant concentrations in the landfill is needed to solve the advection–diffusion equation as part of an impact assessment. These concentrations affect the calculated impact on an underlying aquifer, the contaminating lifespan of the landfill and hence the service lives of engineered components of the barrier system required to maintain negligible impact over the entire contaminating lifespan. In this section, the finite mass of contaminant available for transport and the contaminating lifespan are examined.

16.4.1 Modelling concentrations in the landfill

The actual concentration of contaminants in the landfill is highly variable both spatially within the landfill, and with time. These variations may arise from the heterogeneous nature of the waste, landfill practices, changes in chemical and biological activity, changes in precipitation, and removal of leachate. The measured concentration of leachate shows large variations that occur over short periods of time (e.g., see Figures 2.7–2.10). Rowe and Nadarajah (1996b), however, showed that such fluctuations in source concentration have negligible effect on concentrations at depths greater than 1 m (and hence impact on an underlying aquifer) below the source as diffusion dampens these effects. Thus to sufficient accuracy for engineering calculations, it is often assumed that a peak concentration (c_0) is reached early on in the life of the landfill and then this concentration will decrease with time given the finite mass of contaminant available for transport.

16.4.2 The finite mass of contaminant

For waste disposal sites such as municipal landfills, the mass of any potential contaminant within the landfill is finite. The process of collecting and treating leachate involves the removal of mass from the landfill and hence a decrease in the amount of contaminant that is available for transport into the general groundwater system. Similarly, the migration of contaminant through the underlying deposit also results in a decrease in the mass available within the landfill. For a situation where leachate is continually being generated (e.g., due to infiltration through the landfill cover), the removal of mass by either leachate collection and/or contaminant migration will result in a decrease in leachate strength with time.

The peak concentration, c_0, of a given contaminant species can usually be estimated from past experience with similar landfills. The total mass of contaminant is more difficult to determine. Nevertheless, estimates can be made by considering the observed variation in concentration with time at landfills where leachate concentration has been monitored, or by considering the composition of the waste.

Until fairly recently, there has been a paucity of data concerning the available mass of contaminants within landfills; however, this situation is changing now that many landfills have leachate collection systems. Given that concentration is simply mass per unit volume, the mass of a given contaminant collected in a given period is equal to the concentration multiplied by the volume of leachate collected. By monitoring how this mass varies with time, it is then possible to estimate the total mass of that species of contaminant within the landfill. In the absence of this information, studies of the composition of waste can be used to estimate the mass of given contaminant or groups of contaminants as discussed in Section 2.3. For contaminant species predominantly formed from breakdown or synthesis of other species (e.g., by biological action), an upper bound estimate of the mass of contaminant may be obtained from the estimates of the mass of chemicals that go to form the derived contaminant.

For the purposes of modelling the decrease in concentration in the leachate due to movement of contaminant into the collection system and through the barrier, it is convenient to represent the mass of a particular contaminant species in terms of the Reference Height of Leachate, H_r. Recall from Section 10.2 that H_r is obtained by dividing the reference volume of leachate (equal to the total leachable mass m_{TC} of a contaminant species, see Section 2.3.4, divided by the initial concentration c_0) by the area A_0 through which contaminant passes into the primary liner:

$$H_r = \frac{m_{TC}}{c_0 A_0} \tag{16.1}$$

H_r represents the mass of contaminant available for transport into the soil and/or collection by the leachate collection system and/or lost to surface waters through landfill seeps. The total mass of a contaminant species of interest m_{TC} can be obtained by multiplying the total mass of waste m_T by the mass of the species of interest as a proportion of total mass of waste p.

For the case where there is infiltration q_0 through the landfill cover and where all the leachate is collected (i.e., no migration into the underlying soils), it can be shown (Rowe, 1991a) that the process of dilution within the landfill gives rise to a decrease in the average annual leachate strength with time (recognizing that leachate concentrations may vary seasonally throughout a year) such that the concentration in the landfill at time t, $c_{0L}(t)$, is related to the peak concentration c_0 by the relationship:

$$c_{0L}(t) = c_0 \exp\left(\frac{-q_0 t}{H_r}\right) \tag{16.2a}$$

or, on rearranging terms, the time required for the leachate strength to reduce to some specified value, c_{0L}, is given by:

$$t = \frac{-H_r}{q_0} \ln\left(\frac{c_{0L}}{c_0}\right) \tag{16.2b}$$

For contaminant species that experience first order decay (e.g., due to biological and/or chemical processes), with a decay constant, Γ_{BT}, (T^{-1}), equations 16.2a and 16.2b can be rewritten as:

$$c_{0L}(t) = c_0 \exp\left[-\left(\frac{q_0}{H_r} + \Gamma_{BT}\right)t\right] \tag{16.3a}$$

$$t = -\left(\frac{q_0}{H_r} + \Gamma_{BT}\right)^{-1} \ln\left(\frac{c_{0L}}{c_0}\right) \tag{16.3b}$$

Decay of contaminants, as implied by equations 16.2a and 16.3a, has been observed for both conservative and non-conservative species (e.g., see Figure 2.6). For example, Figure 16.12 shows a summary of data for chloride obtained from leaching experiments together with a curve of best-fit developed by Reitzel (1990). Using equation 16.2a, it can be shown that Reitzel's (1990)

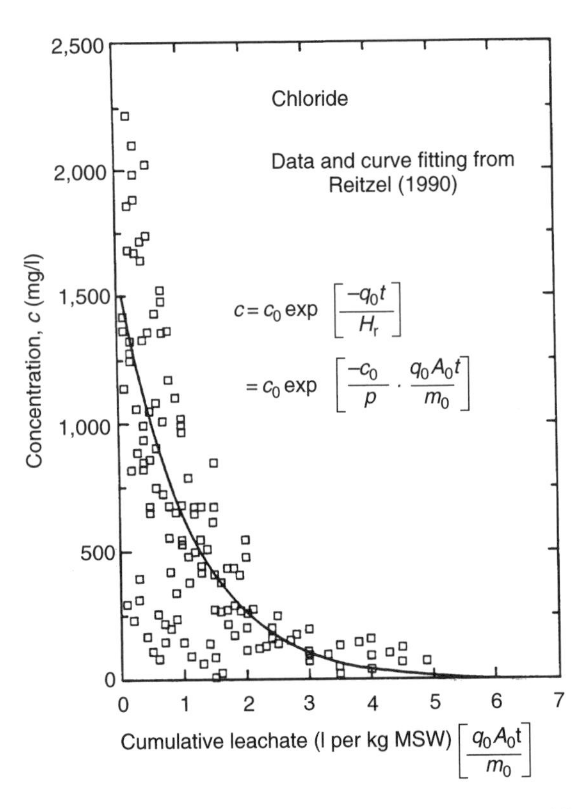

Figure 16.12 Decrease in chloride concentration with cumulative leaching.

curve of best-fit to the leaching data corresponds to $c_0 = 1{,}614$ mg/l and $p = 0.0018$. This implies that chloride represents, on average, 0.18% of the total mass of waste examined in the lysimeter tests summarized by Reitzel (1990). This proportion of mass is approximately twice that observed by Hughes *et al.* (1971). A similar back analysis of data reported by Ehrig and Scheelhaase (1993) gives $p \approx 0.0019$. These values may be rounded up to $p = 0.002$ and this value will be adopted for the following calculations.

Based on empirical work of Lu *et al.* (1981) and equation 16.3a, the first-order decay constant Γ_{BT}, and hence the first-order half-life $t_{1/2}$, may be deduced for a number of species as summarized in Table 16.5. It should be emphasized that Table 16.5 is based on limited data and that for design purposes, these half-lives should be increased if there may be significant consequences arising from the first-order decay not being as slow as implied in this table. Other information on decay rates is provided in Section 2.3.6.

To illustrate the implications of equation 16.2a, consider example D. Here, the average thickness of waste is taken to be $H_w = 10$ m, the average density of waste is $\rho_{dw} = 500$ kg/m^3 and chloride is assumed to represent 0.2% of the weight of the waste. The mass of chloride per unit area, (m_{TC}/A_0), is given by $H_w \rho_{dw} p$, which is equal to 10 kg/m^2. If the peak concentration of chloride is 1,000 mg/l (1 kg/m^3) then the reference height of leachate, H_r, is equal to 10 m.

Assuming an average infiltration through the landfill cover of 0.15 m/a, the decrease in chloride concentration with time, simply due to dilution in the landfill, can be calculated from equation 16.2a as indicated for case (i) in Figure 16.13. For this particular example, the chloride level would reduce to 250 mg/l after approximately 90 years and to 125 mg/l after about 140 years. If the infiltration was 0.3 m/a, all other factors being equal, the concentration would decay much faster (see case (ii), Figure 16.13) and would reduce from the peak value of 1,000 to 250 mg/l after about 45 years and to 125 mg/l after about 70 years.

If chloride represents 0.1% of the waste (Hughes *et al.*, 1971), rather than the 0.2% assumed above, then $m_{TC}/A_0 = 5$ kg/m^2 and thus $H_r = 5$ m. For an infiltration of 0.15 m/a, this gives case (iii) for which the results shown in

Table 16.5 Inferred first-order decay constant and half-life based on data by Lu *et al.* (1981)

Parameter	c_0 (mg/l)	Γ_{BT}(a^{-1})	*Inferred half-life (years)*
BOD_5	35,000	0.225	4.3
COD	89,000	0.192	5.5
TOC	14,000	0.260	3.6
NH_3—N	12,000	0.100	19.8
Cl	2,470	0.065	∞
SO_4	15,000	0.079	49.5
Cd	0.16	0.125	11.6
Cu	10	0.200	5.1
Cr	0.33	0.900	0.8

Note: Half-life of many of these parameters may be due to biologically induced (or accelerated) processes such as precipitation. (Adjusted for dilution effects)

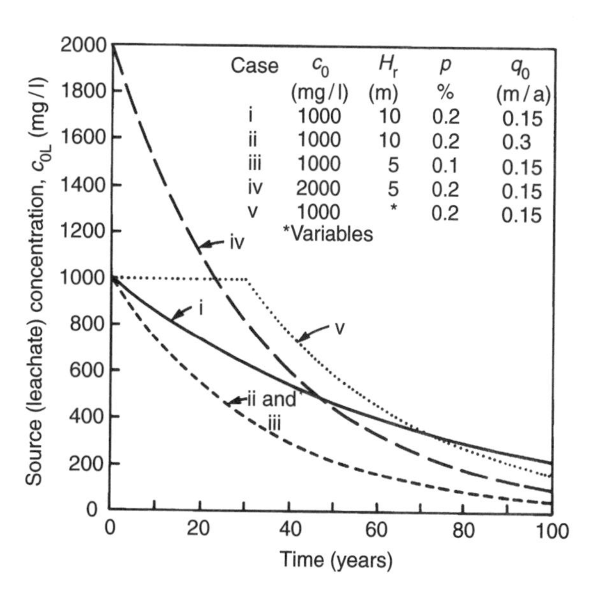

Figure 16.13 Decrease in chloride concentration with time, example D.

Figure 16.13 are precisely the same as those obtained for case (ii) (since the ratio H_r/q_0 is the same). The same result would also be obtained for 5 m of waste if chloride represented 0.2% of the waste.

If one were to assume the same total mass of chloride m_{TC} as in case (i) above, but if the peak concentration was equal to 2,000 mg/l then $H_r = 5$ m and the decrease in concentration with time for $q_0 = 0.15$ m/a is as shown by case (iv) in Figure 16.13. In this case, the concentration decreases from the peak value of 2,000–250 mg/l in about 70 years and to 125 mg/l in just over 90 years.

It may be argued that in some cases not all the contaminant in the source is immediately available for transport or collection. For example, some components of a contaminant may be released due to biological breakdown of the waste over a period of years. In these cases, the source concentration may increase to a peak value and then remain relatively constant (allowing for the usual seasonal variations) for a period of time because the removal of contaminant is being balanced by new contaminants becoming available (e.g., due to biodegradation). However, eventually all the contaminants that can be released are released, and at this point the concentration in the landfill must drop (due to consideration of conservation of mass). A similar phenomenon may be observed when the concentration in the waste is controlled by the solubility limit as examined by Rowe and San (1992).

If a reasonable estimate can be made of the time period over which the source concentration is likely to remain relatively constant, then the behaviour can be readily modelled. To illustrate this, case (v) in Figure 16.13 is presented where it is assumed that the mass of contaminant, peak concentration and infiltration are all the same as for case (i); however, it is also assumed that the source remains constant for a 30-year period (due to gradual release of chloride over this period balancing the removal of chloride by the collection system, etc.). After 30 years, the concentration will then decrease due to mass removal no longer being balanced by contaminant release. This rate of decrease will be controlled by a new value of H_r, which represents the mass of chloride still available in the landfill after 30 years. The mass removed, per unit area, over the first 30 years (for $t_c = 30$ years, $q_0 = 0.15$ m/a, $c_0 = 1{,}000$ mg/l) is given by:

$$m_c = t_c q_0 c_0 = 30 \times 0.13 \times 1\,\text{kg/m}^2 \\ = 4.5\,\text{kg/m}^2$$

thus the mass remaining at 30 years is:

$$m_{TC} - m_c = H_w \rho_{dw} p - m_c = 10 - 4.5 \\ = 5.5\,\text{kg/m}^2$$

and hence

$$H_r = \frac{m_{TC} - m_c}{q_0 c_0} = \frac{5.5}{1} = 5.5\,\text{m}$$

The corresponding rate of decrease in source concentration is shown by case (v) in Figure 16.13. This scenario maintains a higher source concentration for a period of time but, for the same mass as case (i), subsequently gives a more rapid decrease in concentration with time and, consequently, a shorter contaminating lifespan. For example, in case (v), the concentration in the source reduces to 250 mg/l in just over 80 years (compared with 90 years for case (i)) and to 125 mg/l in about 110 years (compared with 140 years for case (i)).

In the limiting case, if the concentration in the source remained constant until all the mass was removed, this would take 66.7 years for the concentration, mass and infiltration rate considered in case (i) (i.e., $c_0 = 1{,}000$ mg/l, $p = 0.002$, $H_w = 10$ m, $\rho_{dw} = 500\,\text{kg/m}^3$, $q_0 = 0.15$ m/a). Thus for this case, the minimum contaminating lifespan is about 67 years, assuming the mass is removed at a constant rate until it is all removed.

The effect of assumptions regarding the source concentrations on contaminant impact of an underlying aquifer is now considered for the barrier system shown in the inset to Figure 16.14. This case involves a 1m CCL, 2 m of fractured clay aquitard and 1 m of unfractured clay. Properties used in the contaminant transport modelling are summarized in Table 16.6 (cases 5i–5v). For all cases with varying source concentration, the time of the peak impact on the aquifer occurs around 400 years. The magnitude of the peak impact is greatest for the high initial concentration and low infiltration (case (5iv)) and lowest for the high infiltration or low proportion of chloride (cases (5ii and 5iii)) cases examined.

Case (5v) with the source concentration remaining constant for 30 years before it begins to decrease shows an earlier and somewhat higher impact on the aquifer than did case (5i). Thus in this case, the assumption of a constant source for 30 years followed by a decrease gives a shorter contaminating lifespan but, for the same flow conditions, a higher impact than case (5i) where the source concentration begins to decrease immediately. While this is true for this case, the conclusion should not be generalized. For example, if there was a SLCS, the landfill conditions modelled in case (5v) may result in smaller impacts than conditions (5i) because of the reduced contaminating lifespan. Thus, a sensitivity study is required on a site-/case-specific basis to assess which of a group of reasonable landfill assumptions gives the greatest impact.

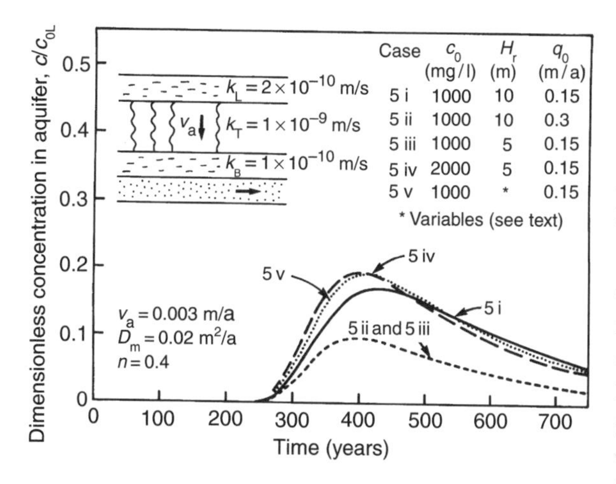

Figure 16.14 Effect of landfill source characterization on calculated impact in the underlying aquifer (case 5 – considering fractures), example D.

16.4.3 Contaminating lifespan

The contaminating lifespan of a landfill was defined in Section 1.9.5 to be the period of time during which the landfill will produce contaminants at levels that could have unacceptable impact if they were discharged into the surrounding environment. When dealing with groundwater contamination, it is necessary to consider the transport pathway (and consequent attenuation) when assessing the contaminating lifespan. This will vary from one landfill to another. The contaminating lifespan of a landfill will depend, inter alia, on the mass of contaminant per unit area (i.e., the thickness and density of waste), the infiltration through the cover, leachate characteristics and the pathway for contaminant release. The thicker the waste, the greater the mass of any given contaminant and, all other things being equal, the longer the contaminating lifespan. The greater the infiltration (and hence volume of leachate collected), the shorter will be the contaminating lifespan since there is greater opportunity for contaminant to be leached out and removed. The greater the potential for attenuation along the escape pathway, the shorter the contaminating lifespan.

When considering the contaminating lifespan of a landfill, it is necessary to define what is meant by unacceptable impact. In the Province of Ontario, the Ministry of the Environment and Energy has a policy (MoEE, 1993a) such that if a reasonable use for groundwater was as drinking water, then an unacceptable impact could be interpreted as an increase in contaminant which exceeds half the difference between the drinking water objective and background levels for

Table 16.6 Summary of parameters considered in example D

Quantity	*Value*					
Length of landfill, L (m)	200					
Porosity of clay matrix, n_m (−)	0.4					
Hydraulic conductivity of fractured clay, k_T (m/s)	1×10^{-9}					
Hydraulic conductivity of unfractured clay, k_B (m/s)	2×10^{-10}					
Hydraulic conductivity of liner, k_L (m/s)	2×10^{-10}					
Thickness of underlying aquifer, h (m)	1					
Porosity of aquifer, n_b (−)	0.3					
Horizontal Darcy velocity in aquifer, v_b (m/a)	$v_b = L \times v_a$					
Diffusion coefficient in matrix, D_m (m²/a)	0.02					
Retardation coefficient for matrix, R_m (−)	1					
Coefficient of hydrodynamic dispersion along fractures, D (m²/a)	0.06					
Fracture spacing[a], $2H_1 = 2H_2$ (m)	1					
Fracture opening size[a], $2h_1 = 2h_2$(μm)	10					
	Case					
	0	1	2	3	4	5
Downward Darcy velocity, v_a (m/a)	0	0.05	0.01	0.005	0.005	0.003
Thickness of fractured clay, H_T (m)	4	4	4	3	3	2
Thickness of unfractured clay, H_B (m)	0	0	0	0	1	1
Thickness of clay liner, H_L (m)	0	0	0	1	0	1
Total head drop between base of landfill and the aquifer, Δh (m)	0	6.35	1.27	1.27	1.27	1.27
	Case					
	i	*ii*	*iii*	*iv*	*v*	
Reference height of leachate, H_r (m)	10	10	5	5	variable	
Initial concentration, c_0 (mg/l)	1,000	1,000	1,000	2,000	1,000	

Note: Where only one value is given, it is the value used for all cases. [a]Assuming orthogonal fractures at equal spacings $2H_1$ and $2H_2$ and with equal fracture opening sizes $2h_1$ and $2h_2$.

aesthetic parameters (e.g., chloride) and a quarter of the difference between the drinking water objective and background for health-related parameters (e.g., dichloromethane). For chloride, this would mean a maximum increase at the site boundaries of 125 mg/l (or less if there are background levels of chloride in the groundwater).

Based on leachate strength decay, one can estimate the time required to reach a given concentration in the landfill using equation 16.2b. For example, the time required to reach a concentration of 125 mg/l for case (i) can be calculated from equation 16.2b:

$$t = \frac{-10}{0.15} \ln\left(\frac{125}{1000}\right) \approx 140 \text{ years}$$

If one adopts this definition of unacceptable impact, then for the examples considered in Figure 16.13 it would be necessary for the leachate collection system to operate for between a maximum of about 140 years for case (i) and a minimum of about 70 years (for cases (ii) and (iii)) before dilution of the leachate would reduce chloride to levels which are sufficiently low that they would not have an unacceptable impact if they were discharged to the environment after failure of the collection system.

This calculation (i.e., equation 16.2a) assumes that dilution of leachate is the only available attenuation mechanism. The question then arises as to how much attenuation may occur as contaminants pass through this barrier and into any underlying aquifer. Thus when considering contaminant impact on an underlying aquifer, the contaminating lifespan depends not only on the decrease in the leachate concentration but also on the potential attenuation in the soils between the landfill and the aquifer. This in turn will depend on the geometry of the landfill, the base elevation of the landfill, the head difference between the leachate and underlying aquifer, the properties of the aquitard and the properties of the underlying aquifer. Of these, the most important are the hydraulic conductivity of the aquitard (as discussed in Section 16.2) and the head difference between the landfill and the aquifer (to be discussed in Section 16.5).

The results (case (i) in Figure 16.14) show that the peak impact in the aquifer is 0.17 times the initial concentration in the landfill. These results can be used to define an attenuation factor a_t, with $a_t = 0.17$ for this case. Thus, if the barrier system can be counted on to reduce the concentration of contaminants from the landfill, the concentration in the landfill that would cause an unacceptable impact (c_{0L}) can be calculated using:

$$c_{0L} = \frac{c_a}{a_t} \quad (16.4)$$

where c_a is the allowable concentration level in the aquifer. Taking c_a equal to 125 mg/l for chloride, c_{0L} is equal to 658 mg/l for case (i), which when used in equation 16.2b yields a contaminating lifespan of 28 years.

The foregoing assessment of contaminating lifespan considered only chloride. Similar calculations may be performed for other critical contaminants (e.g., dichloromethane, benzene, etc.). The contaminant lifespan is taken as the length of time estimated for the most critical contaminant.

16.5 Failure of PLCS and SLCS

Sections 16.2 and 16.3 examined the performance of a number of combinations of engineered and hydrogeologic systems based on the assumption that the PLCS will maintain the leachate head on the primary liner to 0.3 m or less. As discussed in Section 2.4.4, clogging of the leachate collection system will decrease the effectiveness of leachate collection and will result in an increase in head acting on the liner. Clogging may result in increased impact on an underlying aquifer due to less mass removed from the landfill and greater advective transport though the barrier system. Additionally, as mentioned in Section 2.6, temperatures at the base of the landfill may be expected to increase as the leachate head increases, which may reduce the service life of geomembrane liners.

Based on available evidence, it would seem reasonable to project a service life for a well-designed multicomponent leachate system for municipal solid waste (MSW) (as discussed in Section 2.4) to be 100 years. The service life of systems with fine gravel or sand as the drainage material and/or inappropriately used geonets or geotextiles may be substantially less. It is noted that a collection blanket may be regarded to have failed once its hydraulic conductivity drops to about 10^{-6} m/s (or lower); at this hydraulic conductivity it may still be collecting leachate; however, mounding of leachate will occur increasing the head on the base of the landfill and hence increasing the potential for contaminant impact on underlying groundwater.

In new landfills, the objective normally is to design a barrier system such that the impact from the landfill does not exceed allowable limits even if the leachate collection system fails. In existing landfills with either no leachate collection or a poorly designed system (such that it clogs rapidly), the task may be more oriented towards controlling the height of mounding (e.g., using relief wells).

16.5.1 Estimating the required service life

An estimate of how long the leachate collection system must be operated (i.e., with active pumping) or last (i.e., without failure from clogging) can be obtained from the contaminating lifespan of the landfill. For example, an estimate of the contaminating lifespan can be made using equations 16.2b or 16.3b, assuming no attenuation in the underlying soil, by calculating the time required until the concentration in the leachate, c_{OL}, is less than the maximum increase in the critical contaminant (e.g., chloride) permitted in the aquifer or surface water. This value can be compared with expected service life of the collection system (which depends on the design and materials used in the leachate collection system). Thus if the contaminating lifespan of the landfill exceeds the expected service life of the primary collection system, consideration must be given to the implications of increased leachate mounding as the collection system fails. This is a simple but useful calculation prior to more elaborate modelling of the failure of the leachate collection system as discussed in next section.

16.5.2 Modelling the finite service life of leachate collection systems

Figure 16.15 illustrates an idealization of the leachate head acting on a primary liner with time. The actual leachate head early on in the operation of the landfill can be highly variable depending on the thickness of waste, the type and thickness of daily and intermediate cover, and local precipitation. Neglecting these temporal fluctuations, the head acting on the liner while the leachate collection system is operational (t < T1) is normally taken as the design head of 0.3 m (see Section 2.4.2). Provided that there are no significant changes in hydrogeological conditions during this period, advective flow

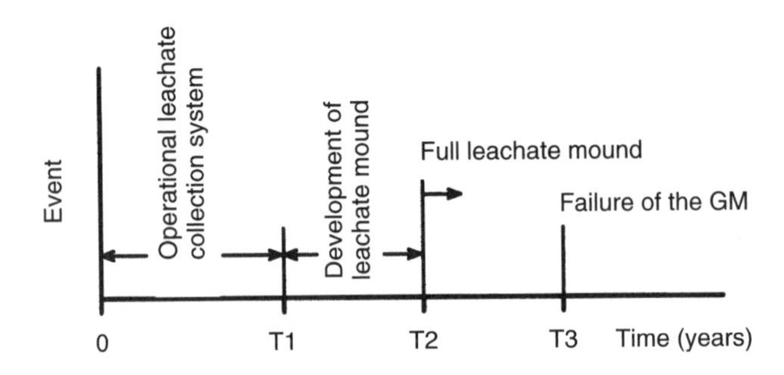

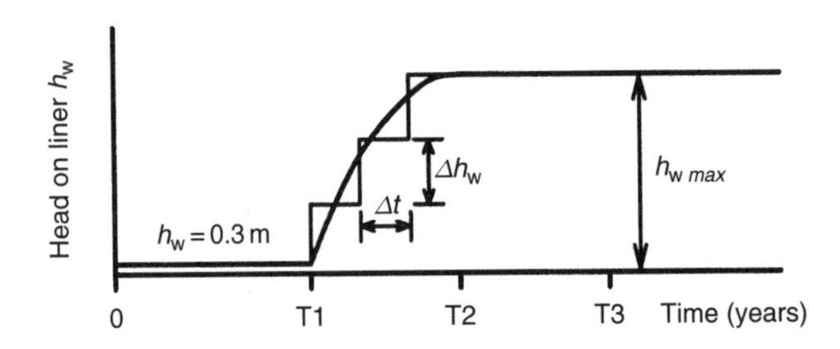

Figure 16.15 Illustration of leachate head acting at the base of the landfill and the development of a leachate mound with time.

through the barrier system may be calculated assuming steady-state conditions.

Following termination of leachate collection or once the effectiveness of the primary leachate collection system decreases, leachate will buildup in the waste until a steady-state balance is reached wherein the inflow through the landfill cover is balanced by the outflow through the liner system, leachate seeps from the sides of the landfill, and lateral migration above the intact liner system (if any). The magnitude of the leachate mound and the time needed to develop the mound depend on a number of factors including: the porosity of the waste, the water content of the waste at the time of landfill cover (cap) failure, the thickness and hydraulic conductivity of the waste, and the landfill geometry. These factors will vary from site to site. The height of the mound may also be limited by drains installed around the perimeter of the landfill (see Section 2.4.2) or relief wells installed in the waste (Rowe and Nadarajah, 1996c). Once the mound has fully developed (taken as time T2) steady-state flow conditions prevail, again, provided that hydrogeological conditions do not greatly change.

While the leachate mound is increasing, between times T1 and T2, transient flow conditions occur across the barrier, in part because of the change in head with time, but also because of the change in storage of fluid within the porous medium. For example, fluid can be released from storage due to compression under increasing effective stresses. Rowe and Nadarajah (1995) have shown that, to sufficient accuracy for practical calculations, the transient flow field during period ($T1 < t < T2$) when there is a long-term increase in leachate head can be modelled as a sequence of steady-state flows arising from an increase in leachate head on the liner without the need for a full consolidation analysis. Thus, either termination or failure of the leachate collection (and the subsequent increase in leachate head) can be modelled using the finite-layer approach presented in Chapter 7 (e.g., using program POLLUTE). The flowing sections examine the implications of failure of the leachate collection system modelled using this approach.

16.5.3 Natural barrier

Example A (Figure 16.3, Section 16.2.1) can be used to illustrate the influence of failure of the leachate collection system on the concentrations in the aquifer. The contaminating lifespan of the landfill examined in Figure 16.3 is estimated to be 50 years based on equation 16.2b (with $H_r = 5\,\text{m}$, $q_0 = 0.25\,\text{m/a}$, $c_0 = 1{,}500\,\text{mg/l}$ and c_{0L} taken to be the maximum allowable increase in concentration in the aquifer for chloride of 125 mg/l). If attenuation of contaminants in the barrier system is considered, c_{0L} is then taken to be the allowable chloride concentration divided by the attenuation factor for that barrier system. For example, for case 1 in Figure 16.3, the attenuation factor is 0.36, and the contaminating lifespan is calculated to be 30 years. Thus for case 1 failure of the leachate collection system after 30 years does not result in an increased impact relative to the assumption of indefinite operation of the leachate collection.

Greater attenuation provided by the barrier system will result in a shorter required service life of the leachate collection system, or reduced implications on the failure of the leachate collection system as illustrated in the following section.

16.5.4 Natural barrier and CCL

If one reanalyses example A, case 2 involving a 1-m thick CCL over a 3-m thick fractured aquitard (Section 16.2.1(b), Table 16.1, Figure 16.3) but now consider the effect of failure of the leachate collection system at 20 and 100 years on the concentration of chloride in the aquifer, one gets the results shown in Figure 16.16. In this situation with initial outward flow from the landfill, failure of the leachate collection system after 100 years has no effect on the peak concentration in the aquifer since the peak concentration occurs prior to failure at around 90

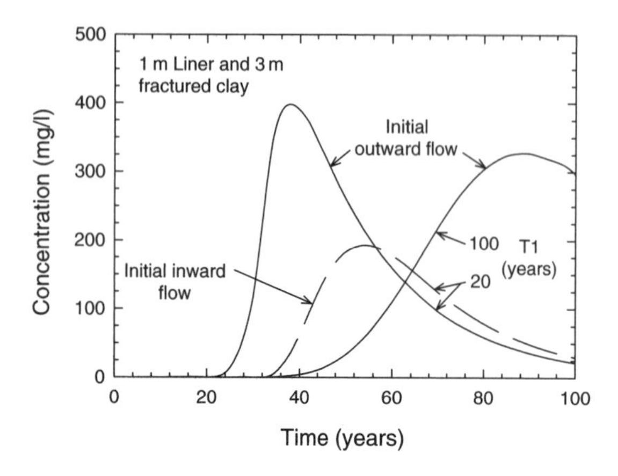

Figure 16.16 Calculated impact for chloride, example A, case 2 when the leachate collection system fails at time T1 with initial outward and inward flow.

years. However, for failure at 20 years the peak impact is 1.2 times larger than for an infinite service life, given the increased outward advection from the landfill and reduced mass of chloride removed from the landfill following failure of the leachate collection system.

For landfills designed with inward flow into the landfill, the increase in head in the landfill associated with the failure of the leachate collection system can result in a reversal of flow. Considering example A, case 2 again but with inward flow as discussed in Section 16.2.2, there was essentially no impact on the aquifer (Figure 16.6) with indefinite operation of the leachate collection system. If the collection system failed at 100 years, this would only cause a small increase in impact with a peak of less than 3 mg/l at 130 years. However, Figure 16.16 shows that if the collection system failed (or leachate collection was terminated) at 20 years, there would be a significant impact (peak concentration just less than 200 mg/l at 54 years) arising from the development of outward flow after the leachate mound begins to grow (since the total head in the landfill is greater than that in the aquifer). This emphasizes the importance of a properly designed leachate collection system when the landfill is designed as a hydraulic trap.

16.5.5 Primary and secondary liners with an SLCS

The inclusion of an SLCS (e.g., see Figure 16.9) provides an opportunity to collect some of the fluid that passes through the primary liner as discussed in Section 16.3.1. The SLCS may also reduce the implications of failure of the PLCS. Thus even if the head on the primary liner, h_w, causes an increased flow through the primary liner, the increase in h_w does not result in a direct increase in head on the secondary liner provided that the head in the SLCS is maintained at a low level by active pumping of the SLCS. Thus, inclusion of the SLCS hydraulically decouples the flow to the underlying aquifer from the head in the landfill. However, this does not necessarily mean that the impact on the aquifer will be unaffected by failure of the PLCS as illustrated by the example discussed in the next paragraph.

Considering example C, case 2a (Section 16.3.1, Table 16.3), but now assuming that the PLCS fails at time T1 and that over a period of 20 years (to time T2) a leachate mound of 12 m develops. A 6-m high mound was modelled between times T1 and T2. The SLCS is assumed to be properly designed such that it has a service life of 1,000 years (see Section 2.5). The secondary system is pumped such that the pressure head is equal to zero. A volumetric water content θ equal to 0.03 and apparent coefficients of hydrodynamic dispersion for the coarse gravel were selected depending on the flow in this layer (as discussed in Section 16.3.3) for the three time periods considered in the analysis. Figure 16.17 shows that this peak impact increases to 1,030 and 735 mg/l for service lives of 40 and 100 years of the PLCS.

16.5.6 Primary and secondary clay liners with an HCL

To illustrate the influence of failure of the PLCS on contaminant impact when the barrier includes a HCL (Figure 16.10), reconsider the

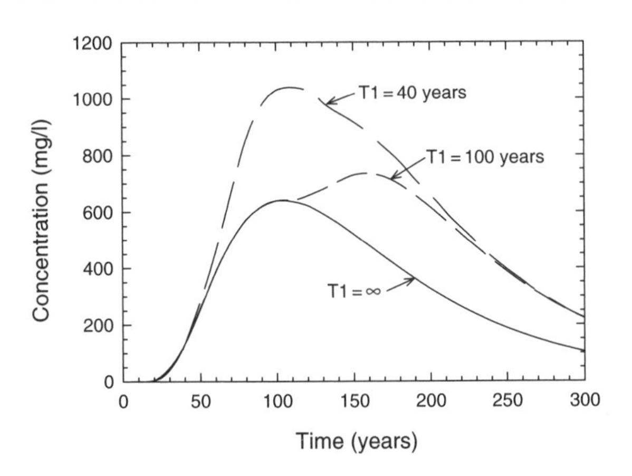

Figure 16.17 Calculation of contaminant impact for chloride, example C, case 2a with failure of the primary leachate collection system at time T1.

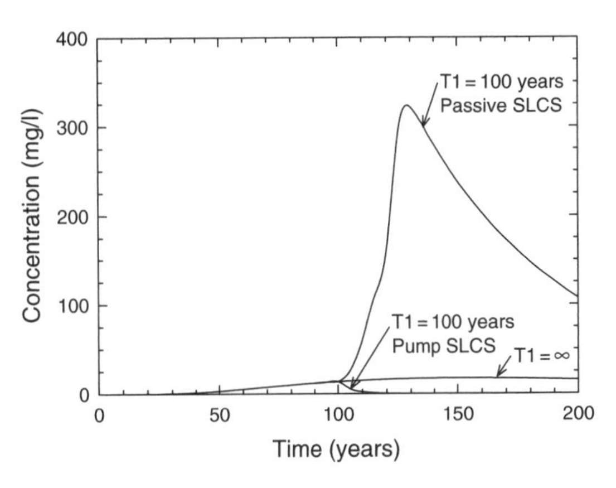

Figure 16.18 Calculation of contaminant impact for chloride, example C, case 6a with failure of the primary leachate collection system at time T1 with secondary leachate collection and passive control layer.

geometry from example C, case 6a from Table 16.4. During the initial operation of the landfill, where the PLCS functions as designed and the HCL is completely passive (no fluid added or removed from the HCL), a natural hydraulic trap exists with water flowing upward from the aquifer through the secondary liner, the HCL, primary liner and into the landfill. In this case, the water level in the HCL will be controlled by the hydraulics of the system and can be readily calculated as described in Chapter 5. The HCL was assumed to be coarse gravel with a porosity $n = 0.4$ and saturated diffusion coefficient $D = 0.03\,m^2/a$ (Rowe and Badv, 1996a).

With indefinite operation of the PLCS, the peak impact is less than 20 mg/l, as shown in Figure 16.18. When failure of the PLCS is modelled at time T1 = 100 years, the impact on the aquifer depends on the operation of the HCL. If it is operated as a completely passive system, the development of a 12-m leachate mound will cause a reversal of flow with downward flow through the primary and secondary liners. This situation would result in a peak impact of 320 mg/l (Figure 16.18). On the other hand if, following failure of the primary collection system, the HCL is actively pumped and operated as an SLCS (such that the leachate head acting on the secondary liner is very small) then this will induce a natural hydraulic trap from the aquifer to the secondary collection system. Thus, despite the strong downward gradient from the landfill through the primary liner and into the secondary system, there is only diffusion from the secondary system down to the aquifer. In this case, the secondary system will be unsaturated ($\theta = 0.03$) and the coefficient of hydrodynamic dispersion was obtained from Badv and Rowe (1996), as discussed in Section 16.3.3. When combined with the mass of contaminant removed from the HCL, the hydraulic containment over the secondary liner results in very small impact (even less than if $T1 = \infty$) as shown in Figure 16.18.

The SLCS (or HCL) also may be expected to have a finite service life or period of operation and this can also be modelled. In this particular case, failure of the SLCS could reverse the direction of flow across the secondary liner and hence contaminants would move down to the aquifer by advection as well as diffusion. However, inspection of the results in Figure 16.18 shows that the service life of the SLCS will not influence the impact on the aquifer provided that is greater than 200 years. Modelling the service life of the secondary collection system may be important for landfills with a very long contaminating lifespan or for barrier systems involving three liners and PLCS, SLCS and tertiary leachate collection system as examined by Rowe and Fraser (1993b).

16.5.7 Stochastic analysis of the failure of collection systems

The implications of failure of collection systems can be readily assessed using analyses like those described in the previous subsections. Such analyses are deterministic in nature since they model failure at definite times based on assumed service lives of the engineering systems. One challenge in this type of analysis is quantifying the service lives since there is inevitably some uncertainty regarding the service lives. One approach to address this uncertainty is to identify a likely range in service life for each system and then perform Monte Carlo simulations to establish the probability that the concentration in the aquifer will exceed some limiting allowable value.

Rowe and Fraser (1993a) have reported results from such calculations for a landfill similar to the one shown in Figure 16.10. The particular barrier system analysed consisted of (from the top down): a PLCS, 1-m primary clay liner, 0.3-m HCL, 1-m secondary clay liner, 7.7-m aquitard and a 1-m aquifer. The pressure head in the aquifer was equal to 14 m. While the PLCS was operational, the head on the top of the primary liner was assumed to be 0.3 m, the HCL was operated as a passive layer and there was hydraulic containment across the primary and secondary liners. Similar to the example considered in Section 16.5.6, the HCL was operated as an SLCS following failure of the primary system. The service lives of the PLCS and SLCS were modelled with triangular-shape distributions quantified by minimum, mode and maximum values of 25, 50 and 75 years for the PLCS, and 150, 300 and 700 years for the SCLS. Since the finite layer technique allows rapid modelling of any simulation, Monte Carlo analyses are quite feasible. Cumulative probability distributions for the peak concentrations in the aquifer generated based on 5,000 simulations are shown in Figure 16.19. For this particular case (see Rowe and Fraser (1993a) for the specific details), there is an 89% probability that the peak chloride concentration in the aquifer would be less than 125 mg/l and a 55% probability that the peak dichloromethane concentration would be below 12 μg/l. Similar analyses can be conducted to assess the implications of uncertainty in service lives on peak contaminant impact.

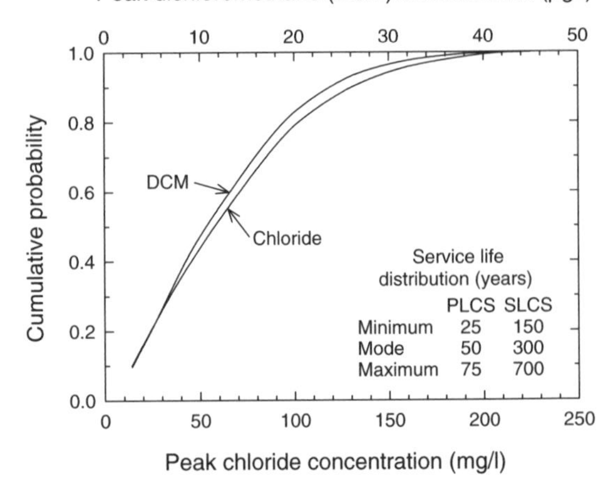

Figure 16.19 Cumulative probability curves for chloride and dichloromethane based on Monte-Carlo analysis considering the uncertainty in service lives of the primary and secondary collection systems (modified from Rowe and Fraser, 1993a).

16.5.8 Trigger concentration levels

The previous sections demonstrated that failure of the leachate collection system could lead to an increase in contaminant impact on an aquifer. Since there is inevitably a time difference (or lag) between when mounding occurs and when an unacceptable impact can occur in an aquifer, monitoring solely based on the aquifer may not be adequate, as once a problem is detected in the aquifer it may be too late to avoid unacceptable impact. If, however, conditions in the landfill are also monitored, a range of limiting concentration and head levels can be developed that based on these conditions in the landfill could lead to an unacceptable impact at some point in the future. If conditions in the landfill approach or exceed these limits, remedial action must then be initiated (or triggered) to avoid unacceptable impact. This is the premise behind developing trigger concentration levels for a given design.

Example D is now considered to illustrate the development of trigger concentration levels. The properties assumed in this example are summarized in Table 16.6. The barrier system considered here consists of a 4-m thick fractured clay layer. It is assumed that the landfill initially operates as a hydraulic trap. As mentioned in Section 16.2.2, under the most adverse conditions then, the advective flow into the landfill will be entirely through the fractures and so migration through the matrix would be by pure diffusion. This can be modelled as diffusive transport through this matrix, without the need to model fractures as given by case 0i in Figure 16.20. The peak impact on the aquifer would then be just under 100 mg/l after about 275 years. The contaminating lifespan can be estimated to be 140 years from equation 16.2b (assuming no attenuation in the aquitard and with $c_{0L}/c_0 = 0.125$, $H_r = 10\,\text{m}$ and $q_0 = 0.15\,\text{m/a}$). Since failure of the leachate collection system may be expected before this time, consideration must be given to the effect of leachate mounding on the impact on the aquifer.

Cases 1 and 2 are presented in Figures 16.20 and 16.21 to illustrate the impact on the aquifer if the head in the leachate was increased due to failure of the collection system. The change in head between the base of the landfill and the aquifer (Δh) is equal to 6.35 and 1.27 m for cases 1 and 2, respectively. In both of these cases, the development of the mound would cause a loss of the initial hydraulic containment and flow would be downward from the landfill to the aquifer. There are two possible bounding situations with regard to the effect of the fractures. On the one hand, the fractures may not be

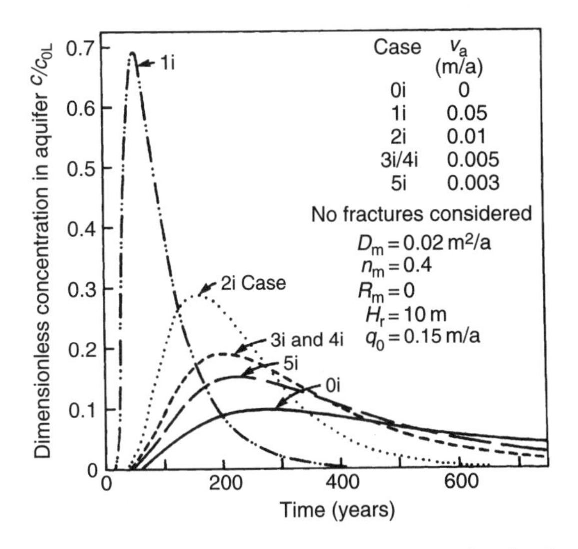

Figure 16.20 Variation in chloride concentration in the aquifer assuming migration through the matrix of the clay (neglecting fractures), example D (modified from Rowe, 1991a).

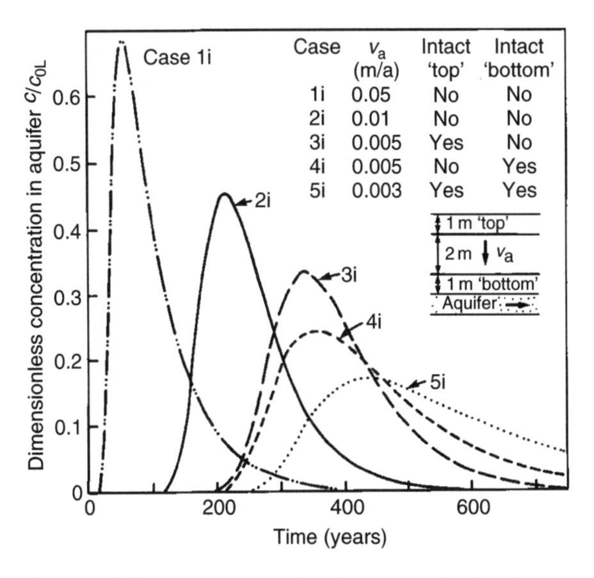

Figure 16.21 Variation in chloride concentration in the aquifer assuming migration along fractures, example D (modified from Rowe, 1991a).

significant conduits for contaminant transport and contaminant migration may simply occur through the matrix of the clay as given in Figure 16.20. On the other hand, the fractures may control migration and it can be assumed that all migration to the aquifer occurs through the fractures and none through the matrix (although attenuation may still occur due to matrix diffusion from the fractures into the adjacent clay – as discussed in Chapter 11). These results are given in Figure 16.21. Many situations will lie between these bounding cases (provided that the fracture spacing is not less than 1 m as assumed in this example), but by modelling these cases it is possible to obtain a reasonable engineering estimate of potential impact.

Very similar results are obtained for case 1 irrespective of whether it is assumed that migration is through the matrix or through the fractures. For either assumption, the attenuation factor is 0.69 (i.e., the peak concentration in the aquifer is 0.69 times the value in the leachate at the time of failure). The allowable concentration in the leachate at the time of failure c_{0L} for case 1i may be calculated by dividing the allowable concentration c_a (in this example it is taken to be 125 mg/l for chloride) by the attenuation factor giving $c_{0L} = 181$ mg/l. This value can be plotted against the difference in head at the base of the landfill and in the aquifer ($\Delta h = 6.35$ m in this case), shown as case 1 in Figure 16.22. This procedure can be repeated for different levels of leachate mounding to construct an envelope of trigger levels at which control measures would be required.

Equation 16.2a can be used to estimate how long the leachate collection system would have to work before failure and mounding to the level implied by case 1i could be allowed to occur. For the assumed landfill conditions ($c_0 = 1{,}000$ mg/l, $H_r = 10$ m, $q_0 = 0.15$ m/a), this time is equal to 114 years. If the failure occurs prior to this time then the eventual impact on the aquifer would be unacceptable based on the assumed conditions in this example.

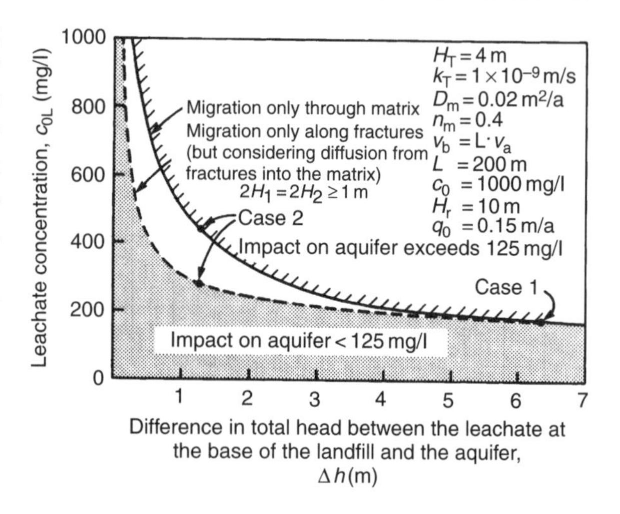

Figure 16.22 Trigger concentration levels for leachate pumping at various levels of mounding for the hypothetical case being considered, example D.

Case 2 in Figures 16.20 and 16.21 is also for the case of 4 m of fractured clay, but with a 1.27-m difference in head between the landfill and the aquifer (e.g., corresponding to a reduced level of leachate mounding). The assumption made concerning the mechanism of transport has a greater effect on the results than for case 1i. For migration purely through the matrix (Figure 16.20), the peak impact is $0.29c_{0L}$ (i.e., $a_T = 0.29$) about 160 years after failure of the leachate collection system. Whereas modelling migration through the fractures (Figure 16.21) gives a peak impact of $0.45c_{0L}$ (i.e., $a_T = 0.45$) about 215 years after failure. Thus, in order for the impact on the aquifer not to exceed 125 mg/l, the concentration in the leachate at failure would have to be less than 277 mg/l or less than 430 mg/l based on a_T equal to 0.45 and 0.29, respectively. These values are plotted for $\Delta h = 1.27$ m in Figure 16.22. Using equation 16.2b, it can be shown that to meet these requirements, the hydraulic trap must be maintained for between 56 and 85 years after which the mound would need to be controlled to give a head difference $\Delta h \leq 1.27$ m.

By monitoring the leachate levels and concentrations and comparing with the results shown in

Figure 16.21, it would be possible to determine whether supplementary leachate control (e.g., leachate wells) would be required for the case being considered here. If the combination of leachate mounding and concentration plots below the dotted line then the impact on the aquifer is expected to be less than 125 mg/l and to be acceptable for this case. If the combination of mounding and leachate concentration plots above the full line then future unacceptable impact may be anticipated unless some leachate control measures are taken. The zone between the dashed and full lines represents the range of variability associated with the extent to which contaminant migrates through the fractures and through the matrix of the fractured clay. It would be conservative to use the lower curve as the trigger for leachate control measures.

Consideration should also be given to the characteristics of the landfill when assessing potential impact (Section 16.4.2) when developing trigger plots.

The practicality of leachate sump wells controlling the leachate mound after failure of the collection system also needs to be carefully considered. The lower the hydraulic conductivity of the waste (see Section 15.2.1), the more wells that will be required and the less practical this option becomes. Generally, the larger the landfill (and, in particular, the thicker the waste), the lower will be the hydraulic conductivity of the waste and the less practical will be leachate wells.

Thus, although under operating conditions, this design may be successful, it relies heavily on the long-term maintenance of very low leachate levels. It may be argued that in terms of long-term potential impact and contaminating lifespan, this landfill design involving only a fractured clay barrier should not be accepted for the size of landfill considered here.

16.6 Composite liners

Composite liners as shown in Figure 16.23 consisting of either a geomembrane (Chapter 13) and a CCL (Chapter 3), or a geomembrane and a GCL (Chapter 12) can be very effective at controlling the migration of contaminants from landfills. Contaminant transport through the geomembrane is limited to leakage through small holes or defects in the geomembrane (Section 13.4) and diffusion through the intact geomembrane (Section 13.5) for many organic contaminants. The clay component of the composite liner systems reduces the leakage through holes in the geomembrane (Section 13.4) and resists diffusive migration of contaminants. The base upon which the liner is constructed also serves as an AL between the waste and a potential receptor aquifer, thereby acting as a partial diffusion barrier and providing a greater opportunity for a decrease in the concentration of many organic contaminants due to biodegradation (and possibly sorption).

16.6.1 Clay versus composite geomembrane and clay liners

For the purposes of illustrating the potential benefit of a composite geomembrane and CCL, it is assumed here that a well-designed and -constructed 1.5-mm thick HDPE geomembrane is placed over the barrier system examined in Figure 16.3, consisting of a 1-m thick CCL which overlies 3 m of fractured aquitard. With the geomembrane as part of the barrier system, the advective flow out of the landfill is controlled by leakage through holes/ha in the geomembrane. Assuming there are 2.5 "large" holes in the geomembrane with radius 5.64 mm (Giroud and Bonaparte, 2001) leakage is reduced to 4.4×10^{-5} m/a. The diffusion parameters through the 1.5-mm thick HDPE geomembrane for chloride are given in Table 16.7. The peak impact on the aquifer is reduced from 328 mg/l without the geomembrane (Figure 16.3) to much less than 1 mg/l with the geomembrane. In this case, contaminant transport is controlled by the relatively small leakage through the geomembrane with 2.5 holes/ha and the fact that the

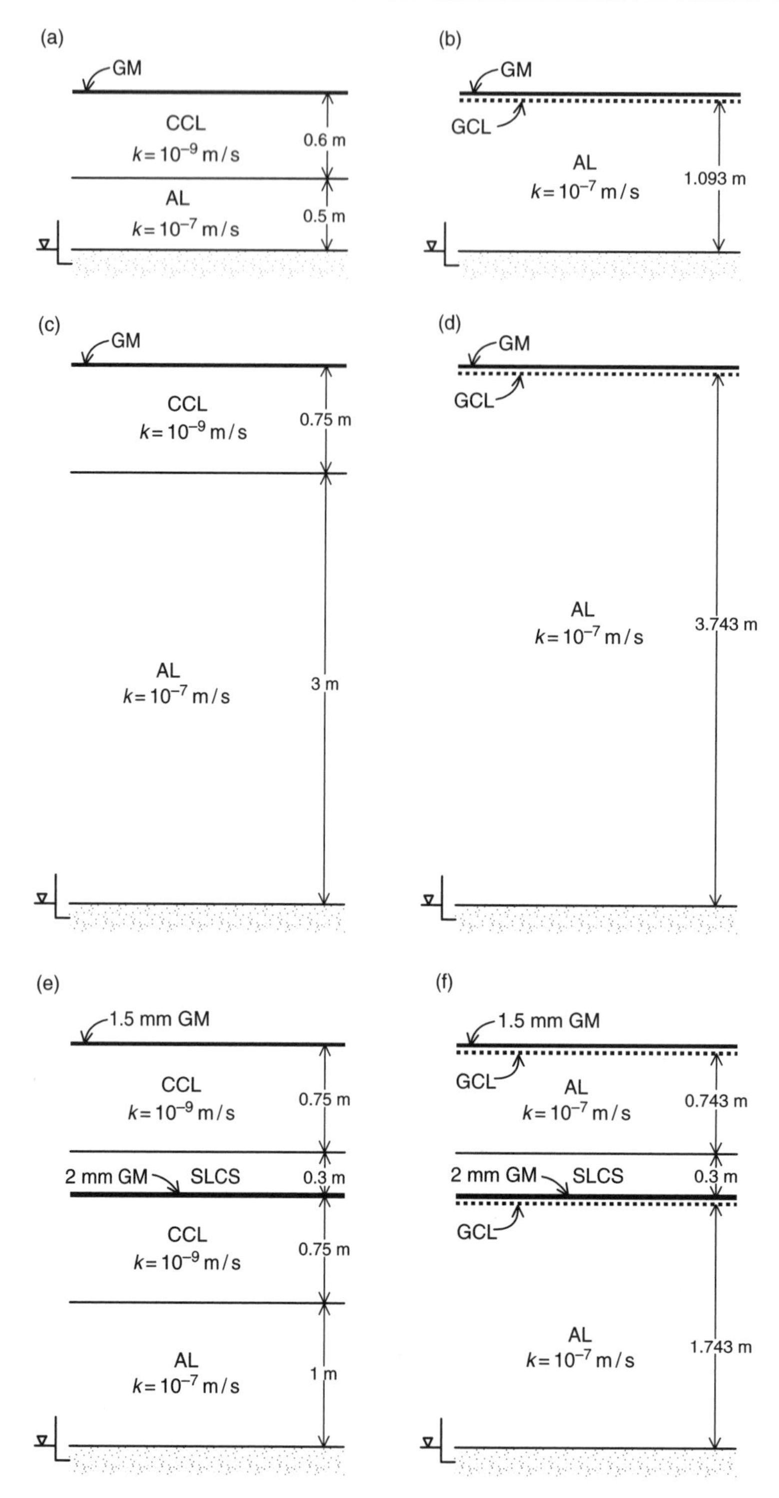

Figure 16.23 Schematic of the composite barrier systems examined in example E for cases a–f.

Table 16.7 Properties of the HDPE geomembrane used in the contaminant transport analyses

Property	*Value*	*Units*
Thickness		
Primary GM	1.5	mm
Secondary GM	2.0	mm
Number of holes	2.5	holes/ha
Hole radius	5.64	mm
Diffusion coefficient chloride, D_g	1.3×10^{-6}	m^2/a
Diffusion coefficient dichloromethane, D_g	2.0×10^{-5}	m^2/a
Partition coefficient chloride, S_{gf}	8×10^{-4}	–
Partition coefficient dichloromethane, S_{gf}	6	–
Service life of primary geomembrane		
20 °C	120	years
33 °C	70	years
Service life of secondary geomembrane		
10 °C	400	years
15 °C	300	years

geomembrane acts as an excellent diffusion barrier to chloride (see Section 13.5). The fractures play no significant role since the advective flow is so small as a result of the composite liner. Also, the precise leakage through the composite liner does not matter provided that it is less than 0.0001 m/a, since this flow is already so low that diffusion through the geomembrane controls. Thus, there is a substantial benefit to the inclusion of the geomembrane, for so long as the geomembrane continues to be a good barrier. Thus, the service life of the geomembrane is key to assessing the potential impact and the influence of the service life of the geomembrane on the concentrations in the aquifer is examined in Section 16.6.2.

If advective flow were inward towards the landfill from the aquifer (as examined in Figure 16.6), inclusion of a well-constructed geomembrane on top of the clay liner would effectively destroy what would otherwise be a hydraulic trap except at the location of holes in the geomembrane. The impact then would depend on the diffusion coefficient through the geomembrane. The maximum impact of 22 mg/l corresponds to the pure diffusion case in Figure 16.6 (i.e., assuming no diffusive resistance of the geomembrane). For chloride, the geomembrane provides excellent resistance to diffusion ($D_g = 4 \times 10^{-14}\,m^2/s = 1.3 \times 10^{-6}\,m^2/a$, $S_{gf} = 8 \times 10^{-4}$) and the impact with the geomembrane would be even less than if there were no geomembrane and a hydraulic trap. However for certain inorganic contaminants that can readily diffuse through HDPE geomembranes (e.g., dichloromethane with $D_g = 6.5 \times 10^{-13}\,m^2/s = 2 \times 10^{-5}\,m^2/a$, $S_{gf} = 6$), the installation of the geomembrane would degrade the performance of the system compared to the case where there was no geomembrane. Furthermore, the water pressures on the base of the geomembrane may lead to potential problems with the geomembrane unless they are controlled while the waste is being placed. Hence, careful consideration of the implications of using a geomembrane in situations with natural upward gradients is warranted.

16.6.2 Geomembrane and compacted clay composite liners

In this section, contaminant impact through composite liners consisting of a geomembrane on top of compacted clay is examined. The influence of the service lives of the leachate collection system and geomembrane liners is examined for a hypothetical landfill in this example, denoted as example E.

Properties of the geomembrane (assumed to be HDPE) used in the analysis are summarized in Table 16.7. It was assumed that the geomembrane had 2.5 undetected circular holes/ha with radius 5.64 mm (Giroud and Bonaparte, 2001) and that these holes do not coincide with wrinkles in the geomembrane. The diffusion parameters for the geomembrane were selected from the data presented in Section 8.12. The

Table 16.8 Properties of the compacted clay liner (CCL), geosynthetic clay liner (GCL) and attenuation layer (AL) used in the contaminant transport analysis, Example E

Property	*CCL*	*GCL*	*AL*	*Units*
Hydraulic conductivity to leachate	1×10^{-9}	2×10^{-10}	1×10^{-7}	m/s
Diffusion coefficient	0.02	0.005[a], 0.009[b]	0.022	m^2/a
Sorption	0	0	0	
Porosity	0.4	0.7	0.3	–
Transmissivity of geomembrane/clay interface	1.6×10^{-8}	2×10^{-10}	–	m^2/s
Half-life in soil				
Chloride	∞	∞	∞	year
Dichloromethane	50	50	50	year

[a]Chloride.
[b]Dichloromethane.

service life of the primary HDPE geomembrane liner was taken to be 120 and 70 years for temperatures at the base of the landfill of 20 and 33 °C (see Section 13.3.4). The properties of the CCL and AL used in the analyses are given in Table 16.8. The AL was assumed to be a sandy-silt material. Sorption of organics was not modelled in the materials beneath the geomembrane (i.e., $\rho K_d = 0$).

The properties of the landfill considered in this example are summarized in Table 16.9. The landfill was assumed to operate for a 20-year period. All time is expressed relative to the middle of this period. For the time period between 0 and T1, the landfill was assumed to have a low-permeability cover limiting infiltration to 3×10^{-4} m/a, and an operational leachate collection system limiting the hydraulic head on top of the geomembrane to 0.3 m. The operation of the leachate collection system was terminated after time T1 allowing the leachate head on the geomembrane to increase to a maximum of 12 m at time T2, taken 20 years following termination of leachate collection (i.e., T2 = T1 + 20 years). Between times T1 and T2, it was assumed that an average head of 6 m acts on the liner. After time T1, it was also assumed that the cover was no longer maintained and the infiltration was allowed to increase to 0.15 m/a. The primary geomembrane was assumed to fail at time T3, which is taken as time T1 plus the geomembrane service life estimates (from Section 13.3.4) of 120 and 70 years corresponding to 20 and 33 °C (i.e., the service life estimates are assumed to apply once the liner temperature starts to increase in response to leachate mounding, see Section 2.6).

Table 16.9 Properties of the landfill used in the contaminant transport analysis, Example E

Property	*Value*	*Units*
Landfill width	1,000	m
Mass of waste per unit area	250,000	t/ha
Proportion of waste		
Chloride	1,800	mg/kg
DCM	2.3	mg/kg
Initial concentration in waste		
Chloride	2,500	mg/l
DCM	3,300	μg/l
Half-life in landfill		
Chloride	∞	year
DCM	2	year

(a) Single composite liners

Results for the two single GM/CCL composite liners shown in Figure 16.23 are now considered. The liner of Figure 16.23a is similar to the minimum system specified by the US Environmental Protection Agency (Subtitle D of the Resource

Conservation and Recovery Act). The system shown in Figure 16.23c corresponds to the single composite liner option specified in Ontario, Canada (MoE, 1998). These are referred to as cases a and c, respectively.

Leakage rates for the different time periods corresponding to different heads acting on the GM are given in Table 16.10 and were obtained using the solution presented in Section 5.7.4. In all cases, the leakage is very low. These values also show that the thickness of the attenuation layer has no substantial effect on the leakage rate.

The calculated concentration of chloride in the aquifer is plotted in Figure 16.24 for the two single GM/CCL composite liner systems (cases a and c). Since the geomembrane provides an excellent diffusive barrier to chloride, there is essentially no arrival of chloride, in either case, prior to failure of the geomembrane at time T3. For times greater than T3, the concentration in the aquifer increases rapidly, reaches a peak value and then decreases because of the finite mass of chloride. With a finite service life of the geomembrane the ultimate impact is controlled by the thickness and hydraulic conductivity of the clay liner and natural system. The thicker single composite liner (case c) has a very similar peak impact with a slightly delayed arrival relative to the thinner system (case a).

It is also noted that a failure of the geomembrane after time T3 would not become evident from monitoring the aquifer until several years following T3. This raises questions as to the applicability of limited monitoring periods (e.g., 30 years post-closure – which in this case would only extend to time 40 years) when designing systems such as this. The effect of such a limited monitoring period may be even more pronounced for a larger landfill.

The chloride impact is larger for a shorter service life of the primary geomembrane as can be seen by comparing the results with T3 = 160 years to T3 = 110 years in Figure 16.24 (e.g., from a possible increase in temperature at the base of the landfill from 20 to 33 °C). Once the geomembrane fails essentially all of the chloride is available for transport throught the barrier. Since removal of mass from the landfill (via the leachate collection system) is the only way to reduce the concentrations of chloride in the landfill, if the geomembrane fails sooner there is less opportunity for the concentration of chloride to be reduced. Increasing the time that the low-permeability cover is maintained from T1 = 40 years to T1 = 110 years (both with T3 = 160 years to permit comparison of the effect of increased T1) results in larger concentrations of chloride in the aquifer. Although this may initially seem counter-intuitive, operation

Table 16.10 Calculated leakage rates through composite liners shown in Figure 16.23a–d for different leachate mounds (h_w) acting on the geomembrane. The geomembrane is assumed to have 2.5 undetected circular holes/ha of radius $r_o = 5.64$ mm, Example E

		Leakage rate (m/a)			
Figure 16.23	*Liner system*	*0–T1* $h_w = 0.3$ m	*T1 – T2* $h_w = 6$ m	*T2 – T3* $h_w = 12$ m	t > *T3* $h_w = 12$ m
a	GM + 0.6-m CCL + 0.5-m AL	5.0×10^{-5}	8.1×10^{-4}	1.6×10^{-3}	0.15*
b	GM + GCL + 1.093-m AL	1.7×10^{-6}	2.3×10^{-5}	4.3×10^{-5}	0.15*
c	GM + 0.75-m CCL + 3-m AL	5.5×10^{-5}	8.5×10^{-4}	1.7×10^{-3}	0.15*
d	GM + GCL + 3.743-m AL	2.0×10^{-6}	2.4×10^{-5}	4.4×10^{-5}	0.15*

*Flow limited by continuity.

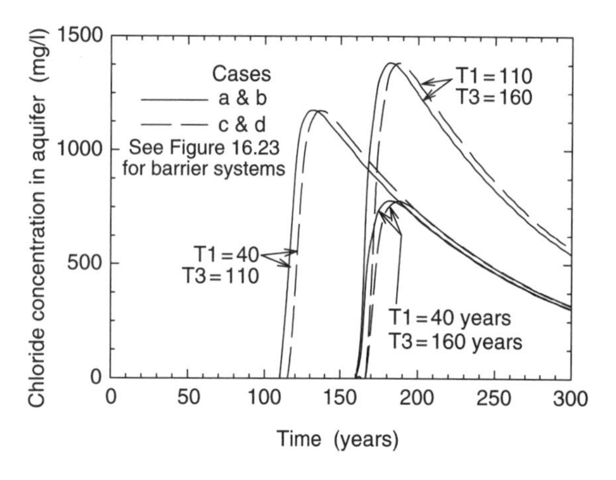

Figure 16.24 Calculated concentrations of chloride in the aquifer for single composite liners, example E.

of the low-permeability cover for a greater period of time results in less infiltration, less chloride removal by leachate collection and hence a larger chloride impact.

The calculated impacts of DCM on the aquifer underlying the single composite GM/CCL liner systems are shown in Figure 16.25 (cases a and c). For these cases, it was assumed that the leachate collection system was operational and low-permeability cover was maintained for

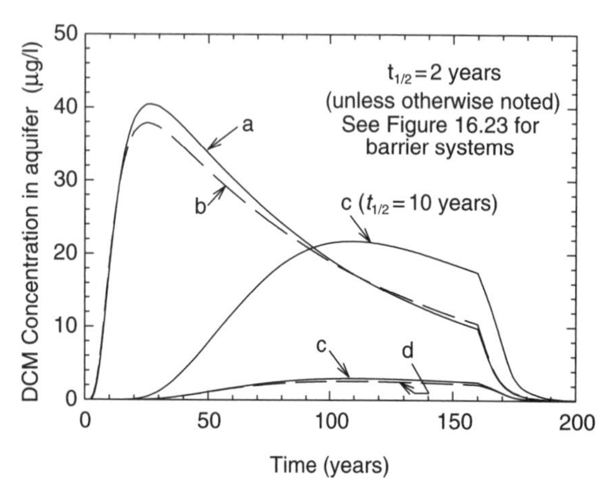

Figure 16.25 Calculated concentrations of dichloromethane in the aquifer for single composite liner systems with T1 = 40 years, T2 = 60 years and T3 = 160 years, example E.

T1 = 40 years, and the geomembrane was assumed to fail at T3 = 160 years. Provided the service life of the primary geomembrane was greater than 100 years, a shorter service life had no effect on the peak impact for DCM.

The half-life of DCM both in the landfill and soil beneath the geomembrane will influence the calculated contaminant impact. The presently available evidence suggests that the half-life of DCM in an MSW landfill is less than 10 years and may be as small as 2 years (Rowe, 1995; Rowe *et al.*, 1997b). The results in Figure 16.25 were obtained with a half-life of 2 years except for one case, c, where results are plotted for half-lives of 10 and 2 years. Over this range, the peak DCM impact decreases from 22 to less than 5 μg/l. There is a paucity of data on the biodegradation of DCM in the soil beneath a geomembrane liner. It is possible that the half-life of DCM in the soil beneath the geomembrane is longer than that in the landfill, since the geomembrane acts as a selective barrier by severely limiting the diffusion of volatile fatty acids (VFAs) to negligible levels while allowing diffusion of VOCs such as DCM. The very low levels of VFAs in the soil below the geomembrane will slow the biodegradation processes that degrade DCM and benzene (Hrapovic, 2001). The half-life of DCM in the soil beneath the geomembrane was conservatively taken as 50 years. Lower values would result in lower impacts.

The results in Figure 16.25 also show that the peak impact on the aquifer decreases as the total thickness of the soil barrier increases. Since an increased thickness of the soil beneath the geomembrane increases the time it takes for DCM to diffuse through the barrier system, this allows more time for the organic contaminant to degrade, thereby giving a lower impact on the aquifer for case c than for case a.

(b) Double composite liners

The impact with the single composite liners examined above exceeds a typical aesthetic drinking water objective for chloride of 250 mg/l.

The peak chloride impact can be reduced by using a double composite liner system. For example, consider the liner system shown in Figure 16.23e that consists, from the top down, of: 1.5-mm primary HDPE geomembrane, 0.75-m compacted clay primary liner, 0.3-m thick SLCS, a geomembrane protection layer (not shown), 2-mm secondary HDPE geomembrane, 0.75-m compacted clay secondary liner and 1-m AL. This liner system corresponds to the double composite liner option in Ontario, Canada (MoE, 1998).

Figure 16.26 shows the chloride impact on the aquifer for this double composite liner (case e) obtained with the parameters in Tables 16.7–16.9 and with T1 = 110 years and T3 = 160 years. Leakage rates through the primary and secondary composite liners are given in Table 16.11. The head acting above the secondary geomembrane was taken to be 0.03 m for service life of the SLCS (assumed to be 1,000 years – see Section 2.5). The unsaturated gravel of the secondary system was modelled using volumetric water content $\theta = 0.03$ and coefficient of hydrodynamic dispersion from Badv and Rowe (1996), as discussed in Section 16.3.3. The secondary geomembrane was assumed to fail at time T4. Results are given for T4 equal to 300 and 400 years, corresponding to temperatures of the secondary geomembrane liner of 15 and 10 °C, respectively (see Table 13.6).

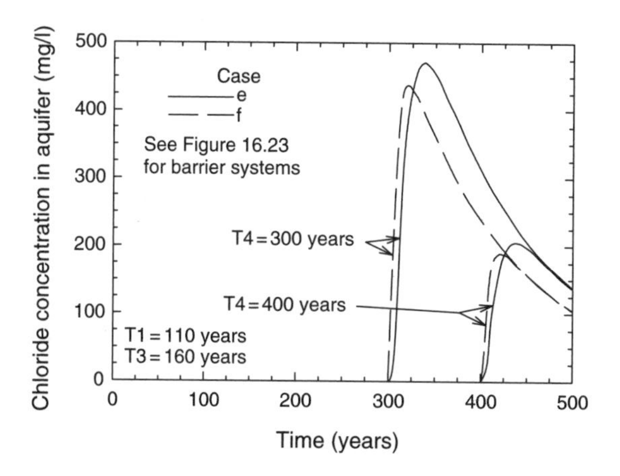

Figure 16.26 Calculated concentrations of chloride in the aquifer for double composite liner systems with T1 = 40 years, T2 = 60 years and T3 = 160 years, example E.

Even with failure of the primary geomembrane at 160 years there is negligible chloride that reaches the aquifer until failure of the secondary geomembrane at time T4. For this particular case, the secondary geomembrane would need a service life of 400 years to reduce the chloride impact to below 250 mg/l.

16.6.3 Geomembrane and GCL composite liners

Most regulations specify a minimum composite liner system consisting of a geomembrane and a CCL. Some regulations will permit different liner designs as an alternate to conventional GM/CCL systems provided that "equivalency" of the alternate can be demonstrated. Of practical interest is whether a GM/GCL composite liner is equivalent to a standard GM/CCL composite liner. Continuing with example E in this section, consideration is given to composite liners but now with a GCL as the clay component beneath the geomembrane.

The composite liners with a GCL shown in Figures 16.23b, d and f (referred to as cases b, d and f) are examined as alternate systems to the GM/CCL composite liners (Figures 16.23a, c and e). Since the GCL is thin (taken to be 7 mm thick in this example), the thickness of the AL beneath the GCL must be increased to provide similar diffusive resistance to the CCL it replaces. The thickness of the AL beneath the GCL was selected such that the total thickness between the geomembrane and the top of the aquifer for the GM/GCL alternative was the same as the corresponding GM/CCL system.

The properties of the GCL used in the analyses are given in Table 16.8. The hydraulic conductivity of the GCL was taken to be 2×10^{-10} m/s to account for exposure to leachate (see Section 12.5.1). The diffusion coefficient

Table 16.11 Calculated leakage rates through double composite liners shown in Figure 16.23e and f for different leachate mounds (h_w) acting on the geomembrane. The geomembrane is assumed to have 2.5 undetected circular holes/ha of radius $r_o = 5.64$ mm. The head acting on the secondary geomembrane is equal to 0.03 m

		Leakage rate (m/a)				
Figure 16.23	*Liner system*	*0–T1* $h_w = 0.3$ m	*T1–T2* $h_w = 6$ m	*T2–T3* $h_w = 12$ m	*T3–T4* $h_w = 12$ m	$t > T4$ $h_w = 12$ m
e	GM + 0.6-m CCL (primary)	4.8×10^{-5}	7.8×10^{-4}	1.5×10^{-3}	0.15*	0.15*
	GM + 0.6-m CCL + 1-m AL secondary	6.6×10^{-6}	6.6×10^{-6}	6.6×10^{-6}	6.6×10^{-6}	0.074
f	GM + GCL + 0.743-m AL primary	1.6×10^{-6}	2.2×10^{-5}	4.3×10^{-5}	0.15*	0.15*
	GM + GCL + 1.743-m AL secondary	3.6×10^{-7}	3.6×10^{-7}	3.6×10^{-7}	3.6×10^{-7}	0.15*

*Flow limited by continuity.

of the GCL was obtained from Lake and Rowe (2000b) for chloride and from Lake and Rowe (2004) for dichloromethane.

The hydraulic performance of the composite liner is one criterion that may permit comparison of different liner systems (e.g., Giroud *et al.*, 1997a; Richardson, 1997b). It was shown in Section 13.4.5 that the hydraulic performance of a GM/GCL composite liner is equivalent to, if not much better for many cases, than a GM/CCL composite liner (for the range of parameters examined) and this is certainly true for the cases examined here, as shown by the leakage rates given in Tables 16.10 and 16.11. However, this does not necessarily mean that any potential contaminant impact for GM/GCL composite liners will be less than that for GM/CCL composite liners, since assessment of equivalence solely on the basis of hydraulic performance neglects diffusion.

A more rational criterion to assess the equivalency of composite liners is the calculated contaminant impact on a receptor aquifer beneath a landfill. For example, results for the GM/GCL alternatives are also plotted in Figures 16.24–16.26 and can be compared with the GM/CCL systems. For all cases shown in Figure 16.24, there is negligible difference (no greater than 1%) between the peak impacts for the GM/CCL/AL and GM/GCL/AL systems. Thus for the parameters examined, the GM/GCL/AL composite liner system provides equivalent protection against chloride to the underlying aquifer as the GM/CCL composite liner system. Likewise, the double composite GM/GCL/AL (case f) liner provides equivalent environmental protection to (if not a little better than) the double composite GM/CCL liner (case e) as shown in Figure 16.26.

Comparison of results for DCM in Figure 16.25 shows that the impact for the GM/CCL composite liners is slightly greater than for the GM/GCL/AL alternatives. This largely occurs because the AL beneath the GCL acts as a better diffusive barrier than the CCL, since the product of the diffusion coefficient and porosity of the AL can be smaller than that for a clay liner. Under these circumstances, the use of the GM/GCL composite to control leakage and the AL to control diffusion of organic compounds proves to be quite effective and the GM/GCL/AL systems provide greater environmental protection than the GM/CCL/AL system. These GM/GCL/AL composite liner systems, therefore, are certainly equivalent to (and indeed a little better than) the GM/CCL/AL systems in terms of minimizing the impact of DCM on the aquifer.

Practical issues related to construction of the liner system may be considered in assessing the equivalence of composite liners. For example, it will generally be easier to construct the

GM/GCL/AL composite liner of Figure 16.23b than the GM/CCL liner of Figure 16.23a (due to the difference in the effort required to place a GCL compared to that required to construct a CCL). There is also greater potential to ensure field quality control for the composite liner with a GCL, provided that proper manufacturing quality control and assurance standards are achieved for the GCL. Better contact between the geomembrane and GCL (resulting in less leakage through holes in the geomembrane, see Section 13.4.2) may be achieved relative to a GM and a CCL since it is easier to obtain a good surface with a GCL. Depending on the availability of a local material suitable for a CCL, the GM/GCL system may be more economical. However, care is required to ensure that the integrity of the GCL is not compromised by internal erosion of bentonite under large hydraulic gradients (see Section 12.5.6) or by physical damage from overlying materials. Also, since not all GCL products are the same, equivalency must be assessed on a case-by-case basis.

16.7 Expansion of landfills

There is a growing trend to expand existing landfills rather than identifying and developing new landfill sites. These expansions may involve the construction of a separate but adjacent landfill cell and/or the placement of additional waste on top of an existing landfill. The influence of landfill expansion on the potential contaminant impact is examined in this section. The influence of construction, operation and post-closure of the expansion on the stability of both the new and existing landfill must also be considered, and is discussed in Section 15.2.

16.7.1 Horizontal expansion

If the expansion is perpendicular to the general direction of groundwater flow, the impact may be considered as that of two landfills that behave in isolation of each other, provided that construction of the expanded landfill does not significantly alter the hydrogeologic conditions. However, if the expansion occurs in the general direction of groundwater flow, interaction between the two landfills is to be expected and the combined impact may depend on the relative size of the cells (both geometry and mass loading), the operational period of each cell, the time between construction of the cells, in addition to the properties of the barrier and any underlying aquifer(s). It is possible that the combined impact could be dominated by the expansion (i.e., the original landfill has negligible contribution). This could occur, for example, if the second cell is larger (higher mass loading) than the first. It also is possible that the impact could be even larger than the sum of impacts of the individual landfills. Thus, it is desirable to check the interaction between landfills on the peak impact. Rowe and Booker (1998) have extended the 2D finite layer formulation presented in Section 7.5.2 to permit the modelling of the interactions between multiple landfill cells.

Consider the case illustrated in Figure 16.27 where landfill 2 is constructed 50 m down gradient from existing landfill 1, denoted as example F. The ground conditions and parameters used in the analysis are given in Figure 16.27. In this example, it is assumed that there is an operational leachate collection system with an indefinite service life that maintains the head on the liner such that there is no vertical flow from the

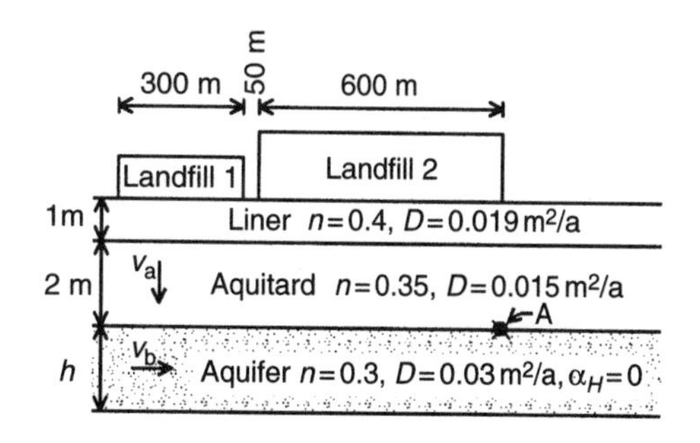

Figure 16.27 Geometry considered in example F involving lateral expansion of an existing landfill. Landfill 2 is constructed adjacent to existing landfill 1.

landfill ($v_a = 0$). Time $t = 0$ corresponds to the start of operation of landfill 1. Operation of landfill 2 commences at time T2. The source concentration of chloride in each landfill was increased by 500 mg/l every 5 years to its peak concentration (c_1 and c_2 for landfills 1 and 2, respectively) attained at the end of the operational period for each landfill (i.e., T1 and T3). Three cases are examined to show the effect of aquifer properties and landfill size (in terms of mass of waste) with the differences in the cases listed in Table 16.12.

The impact of chloride in the aquifer calculated at the down gradient edge of landfill 2 is plotted in Figure 16.28. For each case, results are shown for modelling landfill 1 and landfill 2 alone, and for modelling the interaction between landfills 1 and 2. For case 1, the peak impact when modelling both landfills is 168 mg/l. This is close to the impact estimated by adding the peak impacts of landfills 1 and 2 when modelled alone (which gives a combined impact about 170 mg/l at 190 years). For the same size of landfills but with a lower horizontal flux in the aquifer, the results from case 2 show that the combined effect of the two landfills is only marginally greater than that of landfill 2 alone (i.e., landfill 1 has only a small effect on the combined impact). The peak impact of landfill 2 when modelled alone is about 211 mg/l at 280 years while modelling the interaction between both landfills gives a peak impact of 220 mg/l at 300 years. With the aquifer properties held the same, the results from case 3 show that increasing the size of landfill 1 increases the combined impact (relative to case 2). As illustrated from these three examples, the peak impact of multiple landfills is not necessarily the sum of the individual peak impacts, or the maximum of the sum of concentrations from each landfill at corresponding times. The uncertainty introduced by simple superposition of separate landfills can only be established by conducting analysis that considers landfill interaction.

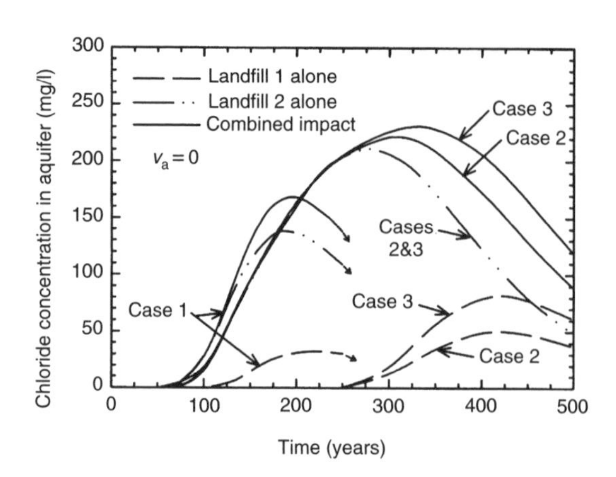

Figure 16.28 Contaminant impact on an aquifer (at point A in Figure 16.27) from lateral expansion (landfill 2) of an existing landfill (landfill 1), example F (modified from Rowe and Booker, 1998).

16.7.2 Vertical expansion

Since vertical expansion of an existing landfill results in a larger mass loading on the existing barrier system over the same footprint, careful

Table 16.12 Selected parameters for example F with landfill expansion

Parameter	*Case 1*	*Case 2*	*Case 3*
Aquifer thickness, h (m)	3	4	4
Horizontal Darcy velocity, v_b (m/a)	3	1	1
Maximum source concentration for landfill 1, c_1 (mg/l)	1,500	1,500	2,500
Time to maximum source concentration for landfill 1, *T1* (years)	15	15	25
Maximum source concentration for landfill 2, c_2 (mg/l)	2,500	2,500	2,500
Time to maximum source concentration for landfill 2, *T2* (years)	55	55	55

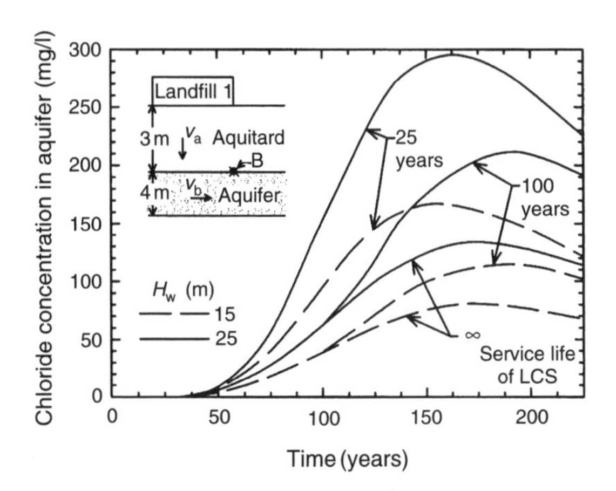

Figure 16.29 Contaminant impact on an aquifer (at point B) from vertical expansion of an existing landfill from waste height (H_w) of 15 to 25 m, example F (modified from Rowe and Booker, 1998).

consideration of the expected impact is required to avoid an unacceptable impact. It is possible that the existing landfill has an older and possibly less effective leachate collection system than a new design.

To illustrate the influence of vertical expansion of the landfill, consider only landfill 1 from Figure 16.27. Figure 16.29 shows the influence of increasing the thickness of the waste from 15 to 25 m (corresponding to the mass of waste increasing from 15,000 to 25,000 t/ha) with the same footprint ($L = 300$ m). Not surprisingly, the peak impact depends on the service life of the leachate collection system. For the cases with an infinite service life, it is assumed that the leachate collection system maintains the head on the liner such that $v_a = 0$ for all time. For the cases with a finite service life, the leachate collection system is assumed to fail (see Section 16.5) resulting in the increase in leachate acting on the liner (leading to outward advection in addition to outward diffusion) and less chloride removed from the system. The case with a service life of 25 years corresponds to a poor design of the leachate collection system that is more susceptible to clogging, whereas a service life of 100 years may be expected for a good design (see Section 2.4). For the larger landfill even an infinite service life of the leachate collection system exceeds that permitted in Ontario (maximum allowable is 125 mg/l for zero background concentration) but is well below the drinking water objective of 250 mg/l. With a service life of 25 years the peak impact would be almost 300 mg/l for the larger landfill, which well exceeds the drinking water objective. Engineering modifications would therefore be required to permit expansion, which could involve providing an increased attenuation zone, or horizontal expansion (perpendicular to the direction of groundwater flow, if possible) rather than vertical expansion.

16.8 Summary

This chapter has highlighted the fact that the safe design of waste disposal facilities requires consideration of the interaction between the natural system (hydrogeology) and the man-made system (engineering). In particular, it is important to:

1. assess the hydraulic conductivity of aquitards in the context of the proposed engineering design;
2. give consideration to the potential effect of the proposed facility on groundwater levels and long-term performance of the landfill system (i.e., the shadow effect);
3. give consideration to the service life of the engineered components of the barrier system (e.g., primary and secondary leachate collection, liners, HCLs, geomembranes, etc.) with respect to the contaminating lifespan of the landfill;
4. design a system such that there will be negligible impact on groundwater even after the service life of the engineered components of the barrier system have been reached;
5. monitor the performance of the system and match monitored performance against that expected. Trigger conditions may be

developed (e.g., level of leachate mounding and concentration of contaminants in the leachate; concentrations at volume of fluid collected from secondary collection systems) such that backup measures can be implemented once a failure of part of the system has been detected and before conditions have been established which will ultimately cause contamination of groundwater.

It is noted that in many systems there can be a considerable time delay (ranging from decades to centuries) between when a failure of an engineered component occurs and the subsequent contamination of an aquifer. If the failure is not detected and appropriate backup procedures initiated shortly after failure and long before contamination of the aquifer, the subsequent measures required to clean up the groundwater, when the aquifer is contaminated, may be extremely time-consuming and expensive.

In situations where the potentiometric surface in an aquifer is high relative to the base contours, a natural hydraulic trap can be designed to minimize impact on an underlying aquifer. In situations where the natural trap alone is not sufficient, the introduction of a HCL can permit long-term control of potential impact on an underlying aquifer.

One key point made in this chapter is that the design of highly engineered systems requires careful consideration of the likely range of hydraulic conductivities of barriers (including both natural aquitards and man-made liners) that can be expected in the field under the relevant stress conditions. Furthermore, it is important to recognize that when the primary design criterion is limiting long-term impact on groundwater, it is not necessarily conservative to consider either the highest or lowest reasonable hydraulic conductivity since the maximum impact may arise for some intermediate hydraulic conductivity; thus a sensitivity analysis is required in the design of these systems. The use of a composite secondary liner to minimize the demand for water to maintain the engineered hydraulic trap should be considered. In this case, the potential for impact on the underlying aquifer may be minimized by active removal of contaminated water from the HCL when significant levels of contamination eventually break through the primary clay liner.

While geomembranes act as excellent barriers to some contaminants (e.g., ions), there can be significant diffusion of some organic contaminants through even relatively thick (2 mm) HDPE geomembranes. Under these circumstances there can be significant impact on an aquifer due to diffusion through the geomembrane even if the volume of leachate physically leaking through defects in the geomembrane is negligible. Thus, the use of a geomembrane or composite geomembrane/clay liner to minimize physical leakage is not, in itself, enough to ensure negligible impact on an aquifer, and calculations should be performed to assess whether the liner combined with the AL is sufficient to provide adequate environmental protection. However, the geomembrane may be a very useful component of a barrier system and substantially reduces the migration of many contaminants of concern.

The performance of an SLCS as a barrier to contaminant transport can be enhanced by designing the system (where practical) to encourage inward flow from an underlying aquifer to the secondary layer. This may reduce the effectiveness of the secondary layer for the detection of leaks through the primary liner but it can considerably enhance the resistance to outward contaminant movement from the secondary collection system to an underlying aquifer. Thus in many cases, the disadvantages of this approach are outweighed by the advantages in terms of long-term environmental protection.

When there will be outward gradient from a secondary system to an underlying aquifer, there is an advantage to constructing a secondary composite liner below the secondary collection system. However, due to potential diffusive transport through the secondary composite liner, even a perfect SLCS which collects 100% of the volume of fluid escaping through the primary liner may not be sufficient to prevent impact on an underlying aquifer. Calculations should be performed to assess whether the barrier system, including the AL is sufficient to provide adequate environmental protection.

Equivalency of composite liners should be assessed in terms of practical construction issues, the hydraulic performance of the composite liner and the potential contaminant impact through the liner system. GM/GCL/AL composite liners can provide the same or even greater environmental protection to an underlying aquifer as GM/CCL composite liners provided that an AL with thickness equal to that of the CCL is included beneath the GCL.

Glossary

Acetic acid See volatile fatty acids.

Adsorption The process whereby molecular attraction holds solutes to the surface of solids such as rock or soil particles.

Advection A physical process whereby contaminants introduced into a groundwater flow system migrate in solution (as solutes) along with the movement of groundwater.

Aeration The process of exposing something to air or charging a liquid with gas.

Aerobic The biological state of living and growing in the presence of oxygen; requiring the presence of free oxygen.

After-use The use of a landfill site following completion and closure of the landfill. Typical examples of after-use include agricultural use, use for playing fields, and use for golf courses.

Agglomerate Together, form or grow into a ball or rounded mass, to cluster densely; particles of dust are brought together to form masses which are too heavy to be entrained in air.

Aliphatic A broad category of carbon compounds distinguished by a straight, or branched, open chain arrangement of the constituent carbon atoms. The carbon–carbon bonds may be either saturated or unsaturated. Alkanes, alkenes and alkynes are aliphatic hydrocarbons.

Alkanes Homologous group of linear saturated aliphatic hydrocarbons having the general formula C_nH_{2n+2}.

Alkenes Unsaturated hydrocarbons having the general formula C_nH_{2n} and characterized by being highly chemically reactive. Also referred to as olefins.

Alkynes Unsaturated hydrocarbons with a triple carbon–carbon bond having the general formula C_nH_{2n-2}.

Amorphous Arrangement of the atoms of a solid in space without regular packing or orientation of molecules.

Anaerobic The biological state of living and growing in the absence of oxygen; absence of the presence of free oxygen.

Anion A negatively charged ion (e.g., chloride, Cl^-).

Anisotropic The property of a material (e.g., hydraulic conductivity of soil) varies with the direction of measurement at a point in the medium (e.g., vertical versus horizontal). See also isotropic.

Anoxic Totally deprived of oxygen.

Antioxidant Additive in polymer formulation intended to halt oxidative reactions during manufacture and over the polymer's service life.

Apparent density The mass of solid waste placed in a landfill divided by the volume occupied by the waste in the landfill. This is not the actual density since the mass of cover soils is not included in the calculation. This may typically be used to determine the waste capacity of a landfill.

Aquifer A geologic formation that is capable of yielding usable quantities of groundwater to wells or springs. Movement is principally in a horizontal direction through porous underground strata.

Aquitard A relatively low-permeability stratum from which it is difficult to extract significant volumes of water.

Arithmetic mean Sum of a set of numbers, divided by the number of terms in the set.

Aromatic Organic compounds that resemble benzene in chemical behaviour. These compounds are unsaturated and are characterized by containing at least one 6-carbon benzene ring.

Artesian pressure Water pressure in a confined aquifer with a hydraulic head above the ground surface.

Ash Inorganic residue remaining after the ignition of combustible substances and can contain bottom and/or fly ash.

Assimilative capacity A measure of the ability of a receiving body (e.g., river, lake, etc.) to render innocuous, those substances deposited in it such that the quality of the receiving body does not degrade below a predetermined level.

Attenuation The process whereby the concentrations of chemical species in groundwater or leachate are reduced as they move throughout the subsurface.

Autotrophic Designating or typical of organisms that derive carbon for the manufacture of cell mass from inorganic carbon (carbon dioxide).

Background level Amounts of pollutants present in the environment prior to establishment, start-up and operation of a facility.

Bacteria Unicellular micro-organisms that exist either as free-living organisms or as parasites and have a broad range of biochemical and often pathogenic properties.

Base contour The contours of the bottom of the landfill.

Base flow The component of stream flow attributed to groundwater or spring contributions; the flow to which a stream will recede after a storm when surface runoff drops to zero.

Berm An artificial ridge of earth or other material used as a mitigative measure against visual and/or noise effects or, within a landfill, to contain leachate on an interim basis or provide stability to the toe of a landfill slope.

Bioassay Use of living organisms to assess the adverse effects of a potential contaminant or contaminants.

Bioavailability Availability of a compound for biodegradation.

Biochemical oxygen demand (BOD) The amount of oxygen used in the biochemical oxidation of organic matter in water over a specified time under specified conditions. BOD_5 is the BOD measured in a 5-day test. It is a standard test used in assessing wastewater strength.

Biocide Substance capable of destroying living organisms.

Biodegradable The ability of a substance to be broken down physically and/or chemically by micro-organisms.

Biodegradation Transformation (through metabolic or enzymatic action) of organic substances to smaller molecules via oxidation and reduction mechanisms induced by the metabolic activity of micro-organisms.

Biomass The amount of living matter in a given unit of the environment.

Blowout The upward movement of a low-permeability soil due to water pressures in an underlying aquifer which exceed the weight of the overlying aquitard. Blowout can cause fracturing of the aquitard and failure of the landfill.

Borehole A man-made hole in a geological formation which has been drilled, jetted, driven, or made by other similar techniques. It is used to determine soil and rock characteristics and also permits the installation of a water well or an observation well for groundwater-monitoring purposes.

Borrow area An area located on- or off-site from which material is extracted for use in earthworks construction (i.e., liner construction, berms, etc.) and/or operations (i.e., daily, final cover).

Breakthrough Contaminant arrival at an outflow or receptor.

BTEX The group of volatile aromatic hydrocarbons: benzene, toluene, ethylbenzene and xylenes.

Buffer area The area between an emission source(s) and nearby sensitive land uses, where land use controls are used to minimize any significant adverse effects.

Butyric acid See volatile fatty acids.

Capillary break layer Layer of high-permeability granular material used to stop upward capillary movement of soluble contaminants.

Capillary fringe Zone above the water table within which the porous medium is saturated with water which is at less than atmospheric pressure.

Capillary suction Process whereby water rises above the water table into the void spaces of a soil due to tension between the water and the soil particles.

Carbon black content (CBC) The proportion of carbon black in a plastic. Calculated according to (ASTM D 1603 – 76) as: carbon black(%) = $100^* W_r/W_z$, where W_r = weight of residue, g, and W_s = weight of sample, g.

Carbon dioxide (CO_2) One of the principal gases which comprises landfill gas.

Carcinogenic Capable of causing the cells of an organism to react in such a way as to produce cancer.

Cation exchange capacity (CEC) Reversible replacement of positively charged ions (cations) adsorbed on clays in an amount equivalent to the negative charge on the clay. Normally expressed in milliequivalents of cation/100 g of clay.

CCL Compacted clay liner.

Cell In respect to a landfill site, means a deposit of waste that has been sealed by cover material so that no waste deposited in the cell is exposed to the atmosphere.

Chemical oxygen demand (COD) The amount of organic substances in water. Includes non-biodegradable and recalcitrant (slowly degrading) compounds that are not included in BOD.

Clay Soil size particles smaller than 0.002 mm (2 μm).

Clod An artificially produced agglomeration of soil as in the process of tillage or coarse crushing prior to compaction.

Clogging A build-up of biofilm, chemical precipitates and small (e.g., silt and sand) particles that are deposited in pipes and the granular material (e.g., sand or gravel), and geotextiles that are used in drainage systems. This build-up progressively reduces the hydraulic conductivity of the system and hence its ability to drain fluids (e.g., leachate).

Closure The completion of a landfill facility including construction of the final cover, grassing, and preparation of the site for its after-use (including removal of all facilities used during the construction of the landfill and not required for the subsequent after-use).

Coarse-grained soil Soil materials with greater than 50% of the particles by mass larger than 0.075 mm (e.g., sands and gravels).

Coextruded Technique in polymer manufacture where two or more types of materials are joined under heat and pressure.

Cometabolism Simultaneous metabolism of two compounds in which the degradation of the second compound depends on the presence of the first compound.

Compaction Reduction in bulk of fill by rolling, tamping or other mechanical means.

Compaction curve The curve showing the relationship between dry unit weight (density) and the water content of a soil for a given compactive effort.

Compaction test A laboratory compaction procedure whereby a soil at a known water content is placed in a specific manner into a mold or given dimensions, subjected to a compactive effort of controlled magnitude, and the resulting unit weight determined. The procedure is repeated for various water contents sufficient to establish a relation between water content and unit weight (ASTM).

Complexation Reaction in which a metal ion and one or more anionic ligands chemically bond. Complexes may prevent the precipitation of metals.

Compliance point The point along the contaminant migration pathway where the target concentration is not to be exceeded.

Composite liner Barrier to contaminant migration composed of two or more liner materials (e.g., geomembrane and compacted clay composite liner; geomembrane and geosynthetic clay composite liner).

Compost A relatively stable mixture of decomposed organic waste materials, generally used to condition and fertilize soil.

Composting An aerobic process involving the biological stabilization of organic matter by micro-organisms. Generally comprises spreading or windrowing the organic waste which is sometimes mixed with a bulking agent to maximize air contact.

Concentration The relative fraction of one substance in another, normally expressed in mass per cent, volume per cent or as mass/volume.

Concentration gradient The change in chemical concentration per unit distance in a given direction.

Confined aquifer An aquifer which is overlain by an aquitard.

Confining layer A body of geologic materials (aquitard) in the subsurface which is of sufficiently low hydraulic conductivity to significantly limit the flow of groundwater into or out of an adjacent aquifer.

Conservative contaminant An unreactive contaminant that does not degrade and the movement of which is not retarded.

Constituent An essential component of a system or group such as an ingredient of a chemical mixture.

Containment The control of migration of gases, liquids and/or solids media from a site by use of containment measure(s).

Contaminant Any solid or liquid resulting directly or indirectly from human activities that may cause an adverse effect on the environment.

Contaminating lifespan The period of time during which the landfill will produce contaminants at levels that could have unacceptable impact if they were discharged into the surrounding environment.

Contingency plan An organized, planned and coordinated course of action to follow in case of any unexpected failure in the design of a waste-management facility. A contingency plan is considered to be a backup measure only, and proposed measures with a high level of probability of implementation are not contingency plans.

Continuous sampling The collection of cores of soil (overburden) during the drilling of a borehole. A continuous soil sample is collected from the surface to the bottom of the hole.

Correlation A statement of the kind and degree of a relationship between two or more variables.

Cover (daily and intermediate) Material that is placed on the waste during construction of the landfill to minimize impacts due to the blowing away of waste, birds, vermin and odour.

Cover (final) Soil (and sometimes geosynthetics) placed over the waste after completion (of a portion) of the landfill. This represents the final surface of the landfill and is intended to (a) control the infiltration of water into the landfill and (b) provide a “pleasing” appearance while containing the waste.

CQA Construction quality assurance.

CQC Construction quality control.

Creep The time-dependent increase in strain when the material is subject to a constant stress.

Crystalline A regular arrangement of the atoms of a solid in space. The crystalline regions (ordered regions) are submicroscopic volumes in which there is more or less regularity of arrangements of the component molecules (ASTM E1142 – 97).

Darcy velocity (flux) Volume of water flowing through a porous material (e.g., soil) per unit area, per unit time. Given by Darcy’s Law: $v_a = ki$, where k is hydraulic conductivity and $i = -d\phi/dz$ is the hydraulic gradient and the negative sign has been introduced in recognition that water flows from high head to low head and so $d\phi/dz$ will be negative in the positive z direction.

Decompose Separate into its elements.

Degradation potential Degree to which a substance is likely to be reduced to a simpler form by bacterial activity.

Denitrification Bacterial reduction of nitrate to gaseous nitrogen under anaerobic conditions.

Density The ratio of the mass of a substance to its volume.

Desiccation Cracking of fine-grained soils due to a decrease in water content.

Dichloromethane (DCM) A volatile chlorinated hydrocarbon used in paint removers and chemical processing. Also known as methylene chloride, CH_2Cl_2. Boiling point ~40 °C.

Dielectric constant (ε) A measure of the polarity of a medium, values include 1 for air, 5.9 for chlorobenzene, 80 for water.

Differential settlement Non-uniform displacement of ground or waste materials. May lead to ponding of surface water or leachate, damage to structures from excessive distortion, and/or cracking of fine-grained soils or rupture of geosynthetics from excessive tensions.

Diffusion Migration of molecules or ions in air or water as a result of their own random movements from a region of higher to a region of lower concentration. Diffusion can occur in the absence of any bulk air or water movement.

Dilution Increasing the proportion of solvent to solute in any solution and thereby decreasing the concentration of solute per unit volume.

Dispersion See hydrodynamic dispersion.

Disposal Means the discharge, deposit, injection, dumping, filling or placing of solid waste into or on any land or water.

Disposal facility A collection of equipment and associated land area which serves to receive waste and dispose of it. The facility may have available one, many, or all of a large number of disposal methods.

Disposal site Includes the fill area and the buffer area within the limit of solid waste.

Dissolved oxygen (DO) The quantity of oxygen dissolved in a certain volume of water.

Dissolved solids The anhydrous residues of the dissolved constituents in water.

Diurnal Occurring during a single day, and recurring daily.

DNAPL (dense non-aqueous phase liquid) A liquid immiscible with water and having a density greater than that of water.

Domestic waste Solid non-hazardous waste generated from households. Also referred to as residential waste or municipal solid waste (MSW). It does not include liquid waste or hazardous waste.

Domestic well A water well used for private household or farm supplies.

Dry density The mass of mineral matter divided by the total volume it is within.

Dry of optimum A soil compacted "dry of optimum" is compacted at a water content less than the optimum water content for that soil.

Dry unit weight The weight (force) of mineral matter per unit total volume.

Effective stresses Controls strength and stiffness of porous materials. For saturated soils, effective stresses are equal to the total stresses less the pore pressures.

Evaporation The physical transformation of a liquid to a gas at any temperature below its boiling point.

Evapotranspiration The combined loss of water from soil and plant surfaces by direct evaporation and by transpiration.

Exceedance An occurrence during which a regulatory limit is exceeded.

Fiber A linear element characterized by having a length at least 100 times its diameter or width, which can be spun into a yarn or otherwise made into a fabric.

Field capacity Quantity of water held by soil or compacted solid waste where application of additional water will cause it to drain to underlying material.

Filtration The process of retaining soil in place while allowing water to pass from soil.

Final closure Operational and engineering measures that are taken to ensure a former landfill operation will remain environmentally safe and acceptable.

Fine-grained soil Soil materials with greater than 50% of the particles by mass smaller than 0.075 mm (e.g., silts and clays).

Fissure A narrow opening, cleft or crevice.

Flux Rate of movement of mass through a unit cross-sectional area per unit time in response to a concentration gradient and/or advective force.

Footprint The area of the site within which solid waste will be placed.

fPP Flexible polypropylene.

GCB Geosynthetic clay barrier; alternative term for geosynthetic clay liner.

GCL See geosynthetic clay liner.

Geocomposite Manufactured, assembled material using at least one geosynthetic product among the components.

Geogrid Planar, polymeric structure consisting of a regular, open network of integrally connected tensile elements and whose openings are much larger than its constituents, used for reinforcement in geotechnical and civil engineering applications.

Geomembrane A relatively impermeable, polymeric sheet used as a liquid and vapour barrier in geotechnical and civil engineering applications.

Geomesh A geonet whose constituent elements are chemically or thermally bonded.

Geometric mean hydraulic conductivity (permeability) The average permeability of randomly distributed test results as defined by:

$$k_m = \sqrt[n]{k_1 x k_2 x k_3 x \ldots k_n}$$

where n is the number of test data and k_n the individual test result.

Geonet Planar, polymeric structure consisting of a regular, dense network of integrally connected overlapping ribs, used for liquid and vapour transmission in geotechnical and civil engineering applications.

Geospacer 3D, polymeric spacer layer, made of a cuspated sheet, monofilaments or any other structure, used in geotechnical and civil engineering applications.

Geosynthetic A polymeric material, synthetic or natural, used in geotechnical and civil engineering applications.

Geosynthetic clay barrier See geosynthetic clay liner.

Geosynthetic clay liner Factory-manufactured hydraulic barrier consisting of a layer of bentonite clay supported by geotextiles and/or geomembranes, and mechanically held together by needle-punching, stitching or chemical adhesives.

Geotextile Planar, polymeric (synthetic or natural) textile material, which may be woven, nonwoven or knitted, used in geotechnical and civil engineering applications.

Glaciolacustrine Fine-grained sediments that have settled from suspension in water bodies resulting from the melting of glaciers.

Grab test A tension test in which only a part of the width of the specimen is gripped in the clamps.

Groundwater Water occurring in a zone of saturation (complete or partial) in a soil or rock and which flows in response to gravitational forces.

Groundwater discharge area An area where water in the saturated zone discharges or flows out of the ground surfaces.

Groundwater flow path The directions in which groundwater flows within an aquifer.

Groundwater recharge area An area where precipitation infiltrates downward through the soil to the water table or saturated zone.

Half-life Time required for one-half of a chemical material to be degraded (e.g., biodegradation) or transformed (e.g., radioactive decay).

Halogenated Refers to compounds containing fluorine, chlorine, bromine, iodine or astatine.

Hardness A characteristic of water, imparted mainly by salts of calcium and magnesium that causes curdling of soap, deposition of scale in boilers and sometimes objectionable taste.

Harmonic mean hydraulic conductivity (permeability) Representative hydraulic conductivity, of a layered system for flow normal to a layered medium:

$$\bar{k} = \sum_{i=1}^{n} L_i / \sum_{i=1}^{n} (L_i/k_i)$$

where n is the number of layers, k_i the hydraulic conductivity of layer i and L_i the length of the flowpath through layer i.

HDPE High density polyethylene.

Head on liner The vertical distance between the top of the landfill liner and the top of the zone of saturation of leachate (also referred to as mound height).

Heat bonded In geosynthetics; thermally bonded by melting the fibers to form weld points.

Heavy metals Metallic elements such as mercury, chromium, lead, cadmium and arsenic with atomic weights greater than 23 (sodium).

Hectare A unit of area in the metric system equal to 10,000 square metres and equal to approximately 2.47 acres.

Heterogeneous In hydrogeology, the property (e.g., hydraulic conductivity) of the medium varies with the location within the medium.

High density polyethylene (HDPE) Linear polyethylene plastics having standard density of 0.941 g/cm^3 or greater.

Homogeneous In hydrogeology, the property (e.g., hydraulic conductivity) of the medium does not vary with the location within the medium.

Hot wedge A common method of heat seaming of thermoplastic geomembranes by a fusing process wherein heat is delivered by a hot wedge passing between the opposing surfaces to be bonded.

Hydraulic conductivity The ability of soil or rock to transmit water. The higher the hydraulic conductivity, the greater the ability to transmit water. Sometimes referred to as permeability.

Hydraulic control layer (HCL) A saturated, permeable (usually coarse gravel) engineered layer which may be pressurized (either naturally or by external introduction of water) to control the hydraulic gradients across a clayey barrier (liner). May be used to induce an inward hydraulic gradient across a clayey liner and hence create a hydraulic trap.

Hydraulic gradient The change in head per unit of distance in a given direction.

Hydraulic trap A term used to describe a landfill design where water flow is into the landfill and hence resists the outward movement of contaminants.

Hydrocarbon Any of the family of compounds containing hydrogen and carbon in various combinations.

Hydrodynamic dispersion Spreading of contaminant due to the combined effects of mechanical dispersion and diffusion.

Hydrogeology The study of the occurrence, movement and chemistry of groundwater in relation to the geologic environment.

Hydrology The science of dealing with the properties, distribution and movement of waters of the earth.

Hydrometer analysis A laboratory procedure for determining the distribution of particle sizes smaller than 75 mm in a soil sample by means of a sedimentation process which relates the rate of fall of soil particles in a water bath to the particle diameter.

Hydrophilic Having an affinity for water, or capable of dissolving in water; soluble or miscible in water.

Hydrophobic Tending not to combine with water or incapable of dissolving in water; insoluble or immiscible in water.

Impact The predicted effect or influence on public health and safety or the environment caused by the introduction of a proposed environmental undertaking. An impact may be positive or negative.

Impermeable Resistant to flow of, or penetration by water or other liquids.

Impervious Incapable of being passed through by moisture or chemicals.

***In-situ* density** The density of a soil sample in the field.

***In-situ* water content** The water content of a soil sample in the field.

Industrial waste Non-hazardous solid waste generated by business and industry. Collection (and sometimes disposal) is usually the responsibility of the industrial generator. It does not include liquid waste or hazardous waste.

Inert fill waste Uncontaminated earth or rock fill, also referred to as backfilling waste. These wastes typically contain no soluble or decomposable chemical substances.

Infiltration Water that penetrates the soil from the ground surface. Often used with reference to the water passing through a landfill cover (percolation).

Inorganic matter Chemical substances of mineral origin, not containing carbon to carbon bonding.

Instantaneous Happening with no delay; immediate.

Institutional waste Non-hazardous solid waste generated by schools, hospitals, nursing homes, etc. It does not include liquid waste or hazardous waste.

Interclod voids Pores in clayey soil that occur between soil clods that are held together by capillary forces.

Interface transmissivity The capacity of the space between a geomembrane and an underlying material to convey fluid. Depends on interface contact conditions and is required to estimate leakage through geomembranes.

Intrinsic permeability Measure of the relative ease with which a permeable medium can transmit a fluid. Intrinsic permeability is a property only of the medium and is independent of the nature of the fluid.

Isotropic The property of a material (e.g., hydraulic conductivity) is independent of the direction of measurement at a point in the medium. See also anisotropic.

Kneading compaction Compaction of clayey soil using a technique which remolds and works the clay and does not just attempt to compress it by pressure.

Landfill A land disposal site employing an engineered method of disposing wastes on land in a manner that minimizes environmental hazards by spreading wastes in thin layers, compacting the wastes to the smallest practical volume and applying cover materials at the end of each operating day.

Landfill gas (LFG) The mixture of gases generated by the decomposition of the organic wastes.

Landfilling Means the disposal of waste by deposit, under controlled conditions on land or on land covered by water, and includes compaction of the waste into a cell and covering the waste with cover materials at regular intervals.

LDLPE Low linear density polyethylene.

Leach To dissolve out by the action of a percolating liquid.

Leachable mass Mass of an element of compound that can be mobilized by water and removed from the landfill in leachate.

Leachate A liquid produced from a landfill that contains dissolved, suspended and/or microbial contaminants (see contaminant) from the solid waste.

Leachate collection pipes Network of perforated drains to collect and transmit leachate.

Leachate collection system (LCS) An engineered system designed to collect and remove leachate from the landfill.

Leachate mound The surface of gravity-controlled water (leachate) in a landfill. It generally corresponds to the top of the zone of saturation in the waste or leachate collection system. Generally represents the water pressure acting on the primary liner in a landfill (or the soil/rock below the waste when there is no liner).

Leak detection system Refers to geosynthetic or coarse grain drainage layer used to monitor volume of fluid and chemical concentrations passing through a liner system.

Leakage Refers to the movement of fluid through a hole in a geomembrane under a hydraulic gradient.

Lift Means the vertical thickness of a compacted volume of solid waste and the cover material immediately above it. Usually a lift represents the quantity of waste disposed in a landfilling cell in one day.

Limit of landfill/solid waste A line delineating the limit within which the municipal solid waste is contained.

Liner A relatively thin structure of compacted natural clayey soil or manufactured material (e.g., geomembranes, geosynthetic clay liners) which serves as barriers to control the amount of leachate that reaches or mixes with groundwater.

LLDPE Linear low-density polyethylene.

LNAPL (light non-aqueous phase liquid) A non-aqueous phase liquid that has a density less than that of water.

lphd Litres per hectare per day; used to quantify leakage through composite liners (1 lphd = 0.0000365 $m^3/m^2/a = 3.65 \times 10^{-5}$ m/a).

Lysimeter A sampling instrument used to monitor and measure the quantity or rate of water movement through soil, natural or artificial liners, or to collect percolated water for analyses.

Machine direction The direction in the plane of a geosynthetic product, parallel to the direction of manufacture.

Macropores Pore sizes larger than the dominant pore size of a soil. May include compaction-induced fractures and openings due to chemical shrinkage.

Manholes Vertical access structures to permit fluid collection and removal.

Matric suction Negative pore pressures that can be generated in unsaturated soils due to capillary effects at soil–water–air interfaces.

Mean A number or quantity contained within the range of a set of numbers or quantities and representative, by some method, of each of the set.

Mechanical dispersion Spreading of contaminant in groundwater due to small-scale variations in groundwater velocities.

Metabolic Refers to the exchange of matter and energy between an organism and its environment and the transformation of this matter and energy within the organism.

Methane (CH_4) An odourless, colourless, non-poisonous and explosive gas when mixed with air or oxygen in certain proportions. It is one of the two principal gases which comprise landfill gas.

Methanogenic conditions Environment under which methane is formed by obligate anaerobes that obtain energy by reducing carbon dioxide and oxidizing hydrogen: $CO_2 + 4H_2 -> CH_4 + 2H_2O$.

Micron A unit of measurement equal to one thousandth of a millimetre.

Mitigation Any action with the intent to lessen or moderate potential negative effects; refers to methods that may be used to prevent, avoid or reduce the severity of risks, impacts or service and cost concerns.

Modified Proctor compaction A laboratory compaction procedure used to determine the relationship between water content and dry unit weight of soils (compaction curve) compacted in a 101.6-mm or 152.4-mm (4–6 inches) diameter mold with a 44.5-N (10 lbf) rammer dropped from a height of 457 mm (18 inches) producing a compactive effort of 2,700 kNm/m^3 (56,000 ft.lbf/ft^3) (ASTM D1557).

Molecular diffusion See diffusion.

Monitoring programme Programmes designed to test on-site and off-site effects of landfills. Such programmes may be carried out over the operational life of a landfill and for several decades following closure.

Monitoring well A well used to obtain water samples for water quality analysis or to measure groundwater levels.

Moraine Commonly a ridge-like deposition of glacial material formed at the edge of a receding glacier.

Municipal solid waste (MSW) Consists of domestic waste and commercial and industrial wastes of similar composition in any combination or proportion but does not include liquid waste or hazardous waste.

Municipal waste landfill A waste disposal site operating as a landfill site and authorized to accept for landfill only domestic waste or only domestic waste in combination with either or both commercial waste or industrial waste but not including liquid waste or hazardous waste.

Natural attenuation Natural processes, including chemical, physical and biological processes, which lead to reduction in contaminant concentrations in the soil or groundwater. In the case of groundwater, natural attenuation may occur at the source and during migration of contaminants.

Necking Distortion of test specimen during physical loading.

Needle-punched Mechanically bonded by staple or filament fibers using barbed needles to form a compact fabric.

Non-aqueous phase liquid (NAPL) Liquids that are immiscible with water.

Nonwoven geotextile A textile structure produced by bonding or interlocking of fibers, or both, accomplished by mechanical, thermal or chemical means.

Observation well A well installed to enable the measurement of the groundwater level, the sampling of groundwater and testing of the characteristics of the surrounding overburden or bedrock.

Octanol/water partition coefficient (K_{ow}) A coefficient representing the ratio of the solubility of a compound in octanol (a non-polar solvent) to its solubility in water (a polar solvent). The higher the value of K_{ow} the more non-polar the compound.

Off-site Means any site that does not meet the definition of on-site.

On-site On the same property as the proposed or existing disposal facility.

Optimum water content The water content at the peak dry density for a given compaction energy.

Organic Containing carbon.

Overburden The soil and fragmented rock materials which lie above solid bedrock.

Oxidation A loss of electrons normally involving the combination of oxygen with another element to form one or more new substances.

Oxidative induction time (OIT) A relative measure of a material's resistance to oxidative decomposition as determined by the thermoanalytical measurement of the time interval to onset of exothermic oxidation of a material at a specified temperature in an oxygen atmosphere (ASTM D3895).

Partitioning A process by which a contaminant, released originally in one phase (e.g., dissolved in groundwater), becomes distributed between other phases (e.g., on organic matter in the soil).

Peds See soil peds.

Perched water table Groundwater lying above a low-permeability layer and separate from and above the water table.

Percolation A term applying to the downward movement of water through soil and especially through a landfill cover (see also infiltration).

Perforations Holes in drainage pipes to permit collection of leachate.

Permeability (1) The capacity of a porous medium to transmit a liquid or gas. See hydraulic conductivity.
(2) The capacity of a geomembrane to transmit a contaminant by diffusion; It is equal to the mass flux that will pass through a geomembrane under a unit concentration gradient between the two sides of the geomembrane.

Permeable Permitting the flow of water or other liquids. The property of a solid material which allows fluids to flow through it. Usually described as a rate of penetration at a defined pressure.

Permittivity The volumetric flow rate of water, normal to a geosynthetic, per unit cross-sectional area per unit head under laminar flow conditions.

pH The measure of the acidity or alkalinity measured on a logarithmic scale from 0 to 14. Neutral water, for example, has a pH value of 7, an acidic solution less than 7 and an alkaline solution greater than 7.

Phenols Organic compounds that contain a hydroxyl group (OH) bound directly to a carbon atom in a benzene ring.

Physiographic region Areas with recognizable local landform patterns.

Plume A plume refers to that portion of the groundwater beneath and in the vicinity of the landfill site where contaminant concentrations exceed certain specified limits. The limits may be defined on the basis of background water quality, drinking water quality standards or other appropriate standards.

Polyethylene A polyolefin formed by bulk polymerization (for low density) or solution polymerization (for high density) where the ethylene monomer is placed in a reactor under high pressure and temperature.

Polymer A macromolecular material formed by the chemical combination of monomers having either the same or different chemical composition.

Pore volume (PV) The volume of fluid required to occupy the pore space in a soil.

Porosity The ratio of the volume of pores of a material to the total volume.

Porous Permeable by fluids.

Potentiometric surface The contours of hydraulic head in a confined aquifer.

PPB/ppb Parts per billion (mass of substance (μg)/mass of solution (kg)).

PPM/ppm Parts per million (mass of substance (mg)/mass of solution (kg)).

Precipitation (1) A physical/chemical phenomenon in which dissolved chemical species in solution (e.g., metals) are transformed into a solid phase (precipitate) which can subsequently be separated from the solution by physical means; and (2) moisture which falls to the earth's surface as rain and snow.

Pressure head A measurement of pressure in a fluid system expressed as the height of an enclosed column of fluid which can be balanced by the pressure in the system.

Propanoic acid See volatile fatty acids.

Protection layer Geosynthetic or natural material placed on top of a geomembrane to prevent puncture and limit indentations in the geomembrane.

Pumping test A test performed on a well to determine characteristics of the aquifer and/or adjacent aquitard.

Purge well A pumping well installed for the purpose of extracting and controlling the movement of contaminated groundwater.

PVC Polyvinyl chloride.

Quality assurance (QA) A planned system of activities whose purpose is to provide a continuing evaluation of the quality control programme, initiating corrective action where necessary. It is applicable to both the manufactured product and its field installation.

Quality control (QC) Actions that provide a means of controlling and measuring the characteristics of (both) the manufactured and the field installed product.

Recharge The entry of infiltration into the saturated groundwater zone together with the associated flow away from the water table within the saturated zone. Generally, an area where water is added to the groundwater system by virtue of the infiltration of precipitation or surface water and subsequently moves downward to the water table is referred to a recharge area.

Recirculation of leachate The collection and subsequent re-injection of leachate into a landfill.

Reference height of leachate H_r, a measure of the mass of a contaminant of interest within a landfill. It is the ratio of the mass of a given contaminant per unit area divided by the peak leachate concentration for this contaminant and hence has dimensions of length. It generally does not correspond to an actual level of leachate in a landfill.

Refuse See solid waste.

Residential waste Non-hazardous solid waste generated by households.

Retardation A process that impedes the transport of contaminants by removing or immobilizing them from a free state.

Retention time The period of time that a contaminant remains in a system.

Risk The combination of the probability of occurrence of harm and the severity of that harm.

Runoff The portion of precipitation or irrigation water that drains over the land as surface flow.

Saturated zone The subsurface region below the water table where the soil pores are completely filled with water (saturated) and the moisture content equals the porosity.

Saturation The amount of moisture in the voids of a medium, equal to the volumetric moisture content divided by the porosity. The saturation ranges from 0 (dry) to 1 (or 100%) (completely saturated).

Scenario A possible course of action or events.

Semicrystalline Material comprised of amorphous and crystalline regions.

Sensitivity analysis An evaluation conducted to assess the impact of changes in the values of specific parameters.

Separation The use of geosynthetics as a partition between two adjacent materials (usually dissimilar) to prevent mixing of the two materials.

Site boundary criteria The groundwater quality criteria that are used to assess the hydrogeological performance of the site.

Soil peds Units of soil structure such as granules or crumbs. In this text considered to be smaller than clods.

Solid waste All non-hazardous solid materials which are discarded as waste and are not recyclable or reusable and includes daily and interim cover soil that is utilized as part of the landfilling operation (also referred to as refuse or waste).

Solid waste disposal area The area of a municipal solid waste disposal site set aside for landfilling (also referred to as limit of landfill/solid waste).

Solid waste management The systematic control of the storage, collection, transportation, processing and disposal of solid domestic/commercial and non-hazardous industrial waste.

Solvent A substance capable of, or used in, dissolving or dispersing one or more other substances.

Sorption Processes whereby a contaminant is removed from solution. These processes may include cation exchange or partitioning of organic compounds onto solid organic matter.

Standard Proctor compaction A laboratory compaction procedure used to determine the relationship between water content and dry unit weight of soils (compaction curve) compacted in a 101.6- or 152.4-mm (4–6 inches) diameter mold with a 24.4-N (5.5 lbf) rammer dropped from a height of 305 mm (12 inches) producing a compactive effort of 600 kNm/m^3 (12,400 $ft.lbf/ft^3$) (ASTM D698).

Staple fibers Fibers of short lengths; frequently used to make needle-punched nonwoven geotextiles.

Stress cracking An external or internal rupture in a plastic caused by a tensile stress less than its short-term mechanical strength (ASTM D883).

Stress relaxation The reduction in stress in a material under constant strain.

Subsidence Settling.

Surface drainage The overland movement of surface water.

Surface water Lakes, bays, ponds, springs, rivers, streams, creeks, estuaries, marshes, inlets, canals, and all other bodies of surface water, natural or artificial, public or private.

Surfactant A surface-active chemical agent.

Surficial At or near surface. Generally applied to soil and/or bedrock at or near the earth's surface.

Survivability The ability of a geosynthetic to be placed and to perform its intended function without undergoing degradation.

Suspension A mixture containing solid particles which are distributed throughout the fluid. The particles will ultimately settle out under the force of gravity.

Thermal fusion Process of joining or seaming polymer materials together under heat and pressure.

Thermoplastic Material that softens on heating and returns to original properties upon cooling; material that experiences permanent (plastic) strain beyond yield.

Till An unsorted mixture of clay, silt, sand, pebbles, cobbles and/or boulders deposited directly by glacial ice.

Tipping face Unloading area for vehicles that are delivering waste to the current active area of landfilling.

Topsoil The uppermost layer of organic rich soil which is capable of supporting good plant growth.

Total dissolved solids (TDS) The total amount of material dissolved in water (predominantly inorganic salts).

Total organic carbon (TOC) The amount or organic material suspended or dissolved in water.

Transition metals Metals which have a partially filled *d* shell. Examples include Cr, Mn, Fe, Co, Ni, Cu, Zn, Pd, Cd.

Transpiration A form of evaporation where growing plants give up water to the air.

Ultraviolet degradation The breakdown of polymeric structure when exposed to natural light.

Unit weight Weight per unit volume (with this definition, the use of the term weight means force).

Unsaturated zone Zone between the ground surface and the capillary fringe within which the moisture content is less than saturation and pressure is less than atmospheric. The zone excludes the capillary fringe.

Variability A quantity susceptible of fluctuating in value or magnitude under different conditions.

Vertical barrier In-ground vertical structure designed to contain contamination.

Vertical expansion The expansion of a landfill operation upward from the existing landscape (as opposed to horizontal expansion onto adjacent land not currently landfilled).

VFPE Very flexible polyethylene.

VLDPE Very low-density polyethylene.

VOC Volatile organic compound.

Volatile Any substance that evaporates at a low temperature.

Volatile fatty acids (VFA) Fatty acids contain only the elements of carbon, hydrogen and oxygen, and consist of an alkyl radical attached to a carboxyl group. The lower molecular weight, volatile, fatty acids are liquids that are soluble in water, volatile in steam and have a notable odour. Examples include ethanoic (acetic) acid (CH_3COOH), propanoic (propionic) acid (CH_3CH_2COOH) and butanoic (butyric) acid ($CH_3CH_2\,CH_2COOH$).

Volatility The property of a substance or substances to convert into vapour or gas without chemical damage.

Volatilization Transfer of a chemical substance from a liquid phase to a gaseous phase.

Volatilized Evaporated.

Volumetric water content The mass of moisture divided by the total mass in a given volume.

Waste Solid and non-hazardous garbage or refuse which no longer serves any useful purposes in its present form and is discarded by its owner.

Water level The measurement of the top of groundwater. The water level is reported as geodetic elevation to provide a common and comparative reference point.

Water table The surface of underground, gravity-controlled water; the surface of an unconfined aquifer at which pore water is at atmospheric pressure. It is generally located at the top of the zone of saturation in an unconfined aquifer.

Watershed The total land area above a given point on a stream or watershed that contributes runoff to that point.

Well nest A group of two or more observation wells found in one location and installed at different depths to monitor different geologic formations.

Wet of optimum A soil compacted "wet of optimum" is compacted at a water content higher than the optimum water content for that soil.

Wetlands Lands that are seasonally or permanently covered by shallow water as well as lands where the water table is close to or at the surface; in either case the presence of abundant water has caused the formation of hydric soils and has favoured the dominance of either hydrophytic or water-tolerant plants.

Wide-width strip tensile test A uniaxial tensile test for a geosynthetic in which the entire width of a 200-mm wide specimen is gripped in the clamps and the gauge length is 100 mm.

Woven geotextile A planar geotextile structure produced by interlacing two or more sets of elements such as yarns or fibers where the elements pass each other usually at right angles, and one set of elements are parallel to the geotextile axis.

Wrinkles Unevenness of a geomembrane when placed on a flat surface. Mostly occurs from solar exposure. Increases the leakage through holes in geomembranes and tensions in the geomembrane.

Xenobiotic Any substance foreign to living systems including drugs, pesticides and carcinogens. Detoxification of these substances occurs mainly in the liver.

Xenobiotic organic compounds (XOCs) Organic substances foreign to living systems including drugs, pesticides and carcinogens. Examples include halogenated hydrocarbons such as dichloromethane, trichloroethylene, etc.

Notation

a	diameter of circular perforation in a pipe [L]
A_0	area through which contaminant is being transported [L^2]
A_m, $A_{d\sigma}$, $A_{d\tau}$	arching factors [–]
A_p	cross-sectional area per unit length of the pipe [L]
A, B	constants used in solving equations
b	a parameter representing the rate of sorption [–]
$2b$	width of wrinkle in a geomembrane in Chapter 13 [L]
$c = c(t) = c(z,t)$ $= c(x,z,t)$ $= c(x,y,z,t)$	concentration in aqueous solution at depth z, time t [ML^{-3}]
$\bar{c} = \bar{c}(s)$	Laplace transform of concentration c
C	Fourier transform of concentration c
c_a	concentration in gaseous phase
c_r	rate of increase in concentration with time [$ML^{-3}T^{-1}$]
c_f	concentration of contaminant at a point (x,y,z) in a fracture at time t [ML^{-3}]
c_m	concentration of contaminant at a point (x,y,z) in the matrix of a fractured medium at time t [ML^{-3}]
c_I	initial (e.g., background) concentration in soil/rock [ML^{-3}]
c_0	initial source or maximum concentration in the source [ML^{-3}]
$c_{0L}(t)$	concentration in source at time t [ML^{-3}]
c_T	concentration at top of soil/rock mass [ML^{-3}]
$c_{b(max)}$	maximum value of impact at a specified point in an aquifer [ML^{-3}]
$c_b = c_b(x,t)$	concentration at bottom of soil/rock mass (e.g., in a base aquifer) [ML^{-3}]
C_u^*	linear coefficient of uniformity [–]
c_v	hydraulic diffusivity – coefficient of consolidation [L^2T^{-1}]
c'	effective cohesion intercept [$ML^{-1}T^{-2}$]
d	diameter of piezometer tube in Chapter 3
d_{85}	grain size for which 85% of the material is finer than by dry mass
d'_{100} and d'_0	grain sizes that intersect 100 and 0% passing, obtained by extending the central portion of the grain size distribution curve
D	coefficient of hydrodynamic dispersion ($D = D_e + D_{md}$)[L^2T^{-1}]
D_a	diffusion coefficient in air [L^2T^{-1}]
D_e	effective diffusion coefficient through soil/rock (coefficient of molecular diffusion through soil or rock) [L^2T^{-1}]

$D_{e\theta}$	effective solute diffusion coefficient in the unsaturated soil at a volumetric water content $\theta[L^2T^{-1}]$
D_θ	effective gas diffusion coefficient in the unsaturated soil at a volumetric water content $\theta[L^2T^{-1}]$
D_g	diffusion coefficient in a geomembrane $[L^2T^{-1}]$
D_m	diffusion coefficient of matrix material between fractures $[L^2T^{-1}]$
D_{md}	coefficient of mechanical dispersion ($D_{md} = \alpha v$)$[L^2T^{-1}]$
D_0	free solution diffusion coefficient $[L^2T^{-1}]$
D_p	porous media diffusion coefficient, $D_p = W_T D_0 = nD_e[L^2T^{-1}]$
D_P	mid-surface diameter of the pipe [L]
D_R	retarded coefficient of hydrodynamic dispersion ($D_R = D/R$)$[L^2T^{-1}]$
DR	dimension ratio for a pipe, (DR = outside diameter/wall thickness) [–]
$D_x = D_{xx}$, $D_y = D_{yy}$, $D_z = D_{zz}$	coefficient of hydrodynamic dispersion in the x, y and z Cartesian direction (principal components of the dispersivity tensor) $[L^2T^{-1}]$
D_{1x}, D_{1y}	coefficient of hydrodynamic dispersion in x- and y-directions, respectively for fracture set 1 $[L^2T^{-1}]$
D_{2x}, D_{2y}	coefficient of hydrodynamic dispersion in x- and y-directions, respectively for fracture set 2 $[L^2T^{-1}]$
D_{3x}, D_{3y}	coefficient of hydrodynamic dispersion in x- and y-directions, respectively for fracture set 3 $[L^2T^{-1}]$
D_H, D_v	horizontal and vertical coefficients of hydrodynamic dispersion in an aquifer ($D_H = D_x$; $D_v = D_z$)$[L^2T^{-1}]$
D_{H1}, D_{v1}, D_{H2}, D_{v2}	horizontal and vertical dispersivity in aquifers 1 and 2, respectively $[L^2T^{-1}]$
D_{ax}, D_{ay}, D_{az}	product of proportion of open area and coefficient of hydrodynamic dispersion in x-, y- and z-directions respectively (e.g., $D_{ax} = n_0 D_{1x}$) $[L^2T^{-1}]$
e	void ratio in Chapter 4 [–]
e_B	bulk void ratio of a GCL in Chapter 12 [–]
E_a	activation energy ($Jmol^{-1}$)
E_N	elastic modulus of native material outside trench $[ML^{-1}T^{-2}]$
E_p	elastic modulus of the pipe $[ML^{-1}T^{-2}]$
E_s	elastic modulus of the soil $[ML^{-1}T^{-2}]$
E_T	elastic modulus of material inside trench $[ML^{-1}T^{-2}]$
ET	net evapotranspiration $[LT^{-1}]$
erf(x)	error function
erfc(x)	complementary error function, erfc(x) = 1 – erf(x)
f	mass flux (mass transported per unit area per unit time) $[ML^{-2}T^{-1}]$
$\bar{f} = \bar{f}(s)$	Laplace transform of flux, f
F	Fourier transform of flux, f
$f_x = f_x(x,t) = f(x,y,z,t)$	mass flux transported in the x Cartesian direction at position (x,y,z) and time t $[ML^{-2}T^{-1}]$
$f_y = f_y(x,t) = f_y(x,y,z,t)$	mass flux transported in the y Cartesian direction at position (x,y,z) and time t $[ML^{-2}T^{-1}]$

$f_z = f_z(x,t)$ $= f_z(x,y,z,t)$	mass flux transported in the z Cartesian direction at position (x,y,z) and time t [$ML^{-2}T^{-1}$]
f_{1x}, f_{2x}	component of flux in first set of fissures [$ML^{-2}T^{-1}$]
f_{1y}, f_{2y}	component of flux in second set of fissures [$ML^{-2}T^{-1}$]
f_{av}	average mass flux transported over a time period t [ML^{-2}]
$f_b(\tau) = f_b(c,\tau)$	vertical flux into a base aquifer ($z = H$) at time τ [$ML^{-2}T^{-1}$]
f_{oc}	organic carbon content (e.g., of a soil) [–]
FOS	filtration opening size of a geotextile [L]
FS_b	factor of safety against pipe buckling [–]
FS_{bh}	factor of safety against blowout [–]
FS_p	global factor of safety against geomembrane puncture [–]
FS_{CR}, FS_{CBD}	partial factors of safety for creep and degradation of nonwoven protection geotextile [–]
g	rate at which contaminant is adsorbed onto fracture walls per unit volume of material (matrix and fissures)
h	thickness of base aquifer [L]
h_1, h_2	thickness of aquifers 1 and 2 [L]
h_1, h_2, h_3	half the fracture opening size (i.e., width of fracture opening) in fracture sets 1, 2 and 3, respectively [L]
h_a	pressure head of zero at the base of the liner system [L]
h_d	head loss across the geomembrane and the underlying layer [L]
h_{max}	maximum height of mounding between leachate drains [L]
h_{ave}	average height of mounding between leachate drains [L]
h_w	leachate height above the geomembrane [L]
h_p	pressure head [L]
H	thickness of soil deposit [L]
H_1, H_2	thickness of aquitards 1 and 2 [L]
H_1, H_2	piezometer head at times t_1, t_2 in Chapter 3 [L]
H_1, H_2, H_3	half the fracture spacing between fractures in fracture sets 1, 2 and 3, respectively [L]
H_f	equivalent height of leachate ($H_f = H_r \times q_a/q_0$) [L]
H_p	effective protrusion height [L]
H_r	reference height of leachate [L]
H_w	average height (thickness) of waste (excluding all cover) [L]
i	hydraulic gradient [–]
i, j, k	indices used in series summation or to denote one layer relative to another
I_D	density index of soil [–]
I_p	second moment of area per unit length of the pipe [L^3]
ID	internal diameter of pipe [L]
k	hydraulic conductivity (permeability) [LT^{-1}]
$\overline{k}$	harmonic mean hydraulic conductivity [LT^{-1}]
k_f	hydraulic conductivity of soil beneath a liner [LT^{-1}]
k_L	hydraulic conductivity of a liner [LT^{-1}]
k_h	hydraulic conductivity in the horizontal direction [LT^{-1}]
k_v	hydraulic conductivity in the vertical direction [LT^{-1}]

k_{om}	hydraulic conductivity of the material overlying the geomembrane [LT^{-1}]
k_w	hydraulic conductivity of waste [LT^{-1}]
k_{FGT}	hydraulic conductivity of the geotextile [LT^{-1}]
K, K_d	partitioning or distribution coefficient [L^3M^{-1}]
K	coefficient of lateral earth pressure in Chapter 15 [−]
K_f	fracture distribution coefficient (mass of solute adsorbed on fracture walls per unit area of surface per unit concentration of solution in the fracture) [L]
K'_H	dimensionless Henry's coefficient, $K'_H = c_a/c = K_H/(RT)$ where R is the universal gas constant and T is temperature [−]
K_m	partitioning or distribution coefficient in the matrix material of fractured media [L^3M^{-1}]
K_{oc}	partition coefficient of a compound between organic carbon and water [L^3M^{-1}]
K_{ow}	octanol-water partition coefficient [−]
L	length of landfill (in direction parallel to the horizontal flow in the underlying aquifer) [L]
L	length of wrinkle in a geomembrane in Chapter 13 [L]
L_P	centre-to-centre spacing of perforations in pipe in Chapter 15 [L]
L	length of piezometer intake in Chapter 3 [L]
ℓ	spacing between drains in a landfill
lphd	litres per hectare per day
m_{TC}	total initial mass of a contaminant species in waste [M]
m_{1c}	increase in mass deposited in landfill up to time t [M]
m_t	mass of contaminant in source at time t [M]
m_a	mass of contaminant transported from a contaminant source (e.g., landfill) by advection [M]
m_d	mass of contaminant transported from a contaminant source (e.g., landfill) by diffusion [M]
m_0	initial mass of a given contaminant in the source (e.g., landfill) that could be released for transport or collection. It includes contaminant in solid form that may eventually be dissolved but excludes contaminant in solid form which is never expected to be dissolved or contaminant that is or will be released in the gas phase [M]
m	mass of contaminant transported into the soil/rock [M]
m	$= (k_h/k_v)^{0.5}$ in Chapter 3 [−]
M_A	mass per unit area of a geosynthetic [ML^{-2}]
MF_S, MF_{PD}, MF_A	modification factors for protrusion shape, packing density and soil "arching" for geotextile protection layer [−]
n	porosity [−]
n_b	porosity of base aquifer [−]
n_m	matrix porosity (primary porosity) in a fracture system [−]
n_M	Manning's coefficient of pipe [$T/L^{1/3}$]
n_0	proportion of open area perpendicular to flow in a fracture system $n_0 = h_1/H_1 + h_2/H_2$[−]
n_f	fracture porosity (secondary porosity) [−]

N_b	critical buckling thrust of pipe $[MLT^{-2}L^{-1}]$
N_{sp}	maximum thrust of pipe $[MLT^{-2}L^{-1}]$
O_{95}	apparent opening size of a geotextile [L]
OD	outside diameter of pipe [L]
p	proportion of a contaminant species of the total dry mass of waste [–]
p_{app}	vertical total stress acting on the geomembrane [kPa]
p_f	buckling performance factor [–]
P	precipitation $[LT^{-1}]$
P_g	permeability (in polymer not geotechnical sense) or permeation coefficient of a geomembrane, $P_g = S_{gf}D_g[MT^{-2}]$
$PERC$	percolation through the cover $[LT^{-1}]$
q_a	normalized rate of contaminant flux into soil ($q_a = f/c$)$[LT^{-1}]$
q_c	volume of leachate collected per unit area of landfill $[LT^{-1}]$
q	rate at which contaminants are transported into the matrix per unit volume of material (matrix and fissures)
q_0	steady-state infiltration into a landfill (volume of leachate generated per unit area of landfill) $[LT^{-1}]$
Q	flow rate (leakage) through a hole in a geomembrane $[L^3 T^{-1}]$
Q_D	fluid removed by drainage layer in the cover system $[LT^{-1}]$
Q_{max}	maximum leakage rate through a hole in a geomembrane in Chapter 13 $[L^3 T^{-1}]$
Q_p	maximum flow rate in a circular pipe $[L^3 T^{-1}]$
r	distance from pumped well to piezometer in Chapter 3 [L]
r	rate at which contaminant is being "injected" into the fracture system per unit volume of matrix and fissures
r_0	radius of a circular hole in a geomembrane [L]
r_w	wetted radius defining lateral extent of flow along geomembrane/soil interface [L]
R	radius of screen or well in Chapter 3 [L]
R	universal gas constant ($R = 8.3143$ J K^{-1} mol^{-1}) in Chapter 8
R	surface water run-off in Chapter 14 $[LT^{-1}]$
$R = R_m$	retardation coefficient ($R = 1 + \rho K/n$) (R_m is for matrix material in a fracture system) [–]
R_h	buckling factor for burial depth and extent of backfill [–]
S	mass of solute removed from solution per unit mass of solid [–]
S	storativity in Chapter 3 [–]
S_{gf}	solubility or partitioning coefficient for a contaminant [–]
S_m	solid-phase concentration corresponding to all available sorption sites being occupied [–]
S'_s	specific storage ($S_s = m_v\gamma_w$)$[L^{-1}]$
s	Laplace transform variable
t	time [T]
t_p	thickness of plane wall pipe [L]
t_1	time to deplete antioxidants in polymer [T]
t_2	induction time to the onset of polymer degradation [T]

t_3	time for degradation of the polymer to decrease property to arbitrary level [T]
$t_{1/2}$	half-life of contaminant species [T]
t_D, t'_D	dimensionless time factor for aquifer and aquitard in Chapter 3 [−]
t_{max}	time at which maximum impact occurs at a specified point in an aquifer [T]
T	transmissivity [L^2T^{-1}]
T_0	basic time lag in Chapter 3 [T]
v	average linearized groundwater velocity (seepage velocity; groundwater velocity) [LT^{-1}]
v_a	Darcy velocity (also called discharge velocity and Darcy flux) ($v_a = nv$)[LT^{-1}]
v_R	retarded groundwater velocity ($v_R = v/R$)[LT^{-1}]
v_x, v_y, v_z	components of groundwater velocity in the x, y and z Cartesian directions [LT^{-1}]
v_{1x}, v_{1y}	groundwater velocity in x- and y-directions in fracture set 1 [LT^{-1}]
v_{2x}, v_{2y}	groundwater velocity in x- and y-directions in fracture set 2 [LT^{-1}]
v_{3x}, v_{3y}	groundwater velocity in x- and y-direction in fracture set 3 [LT^{-1}]
v_b	horizontal Darcy velocity (Darcy flux) in a base aquifer [LT^{-1}]
v_{b1}, v_{b2}	horizontal Darcy velocity in aquifers 1 and 2, respectively [LT^{-1}]
V	average flow velocity in the pipe [LT^{-1}]
W	width of landfill (e.g., perpendicular to the direction of flow in the base aquifer) [L]
W_T	complex tortuosity factor (Chapter 6) [−]
x, y, z	Cartesian directions
α	dispersivity [L]
α_H, α_v	dispersivity in horizontal and vertical directions [L]
α, β	variables used in derivation of solution to the contaminant transport equations
α_j, β_k, γ_l	variable used in derivation of solution to contaminant transport equation in the matrix of fracture media
Δ	surface area per unit volume of fracture media ($\Delta = 1/H_1 + 1/H_2 + 1/H_3$)[$L^{-1}$]
Δ	differential settlement in Chapter 13 [L]
ΔS	change in water storage [LT^{-1}]
ΔD_v, ΔD_h	vertical and horizontal diameter changes of the pipe [L]
ΔD	change in vertical pipe diameter [−]
η_T	viscosity at temperature T [$ML^{-1}T^{-1}$]
ϕ'	coefficient of internal friction [−]
ρ	dry density of soil/rock [ML^{-3}]
ρ_m	dry density of soil/rock forming the matrix material for a fractured medium [ML^{-3}]
ρ_{dw}	dry density of waste [ML^{-3}]
λ	first order decay constant which has components $\lambda = \Gamma_R + \Gamma_B + \Gamma_s$ due to radioactive and biological decay and fluid withdrawal, respectively [T^{-1}]
μ	mass per unit area of a geosynthetic [ML^{-2}]

Ω	dimensionless infiltration coefficient for waste, $\Omega = q_0/k$[−]
τ	dummy time variable – used in integrations with respect to time [T]
τ	tortuosity [−]
τ_a	gas-phase tortuosity [−]
θ	volumetric water content (= porosity of a saturated soil) [−]
θ	interface transmissivity between geomembrane and underlying material in Chapters 5 and 13 [L^2T^{-1}]
θ_a	effective gas/air porosity (= porosity of a dry soil) [−]
Θ	angle around the circumference of the pipe with respect to the crown [−]
Γ_R	$\ell n2$/(radioactive decay half-life) [T^{-1}]
Γ_B	$\ell n2$/(biological decay half-life) [T^{-1}]
Γ_s	sink term for removal of fluid. Equal to the volume of fluid removed per unit volume of soil per unit time [T^{-1}]
γ	unit weight of the overburden material [$ML^{-2}T^{-2}$]
γ_w	unit weight of water [$ML^{-2}T^{-2}$]
φ	total head [L]
σ	total stress [$ML^{-1}T^{-2}$]
σ'	effective stress [$ML^{-1}T^{-2}$]
σ'_p	preconsolidation pressure [$ML^{-1}T^{-2}$]
σ_v, σ_h	vertical and horizontal earth pressures distant to the pipe [$ML^{-1}T^{-2}$]
σ_m, σ_d	mean and deviator stresses distant to the pipe [$ML^{-1}T^{-2}$]
σ_0, σ_2	normal stress on the pipe from the mean and deviator boundary stresses [$ML^{-1}T^{-2}$]
$(\sigma_\Theta)_{\sigma 0}$, $(\sigma_\Theta)_{\sigma 2}$ and $(\sigma_\Theta)_{\tau 2}$	circumferential stresses in the pipe from the σ_0,σ_2 and τ_2 loading cases [$ML^{-1}T^{-2}$]
τ_2	shear stress on the pipe from the deviator boundary stress [$ML^{-1}T^{-2}$]
ν_p	Poisson ratio of the pipe [−]
ν_s	Poisson ratio of the soil [−]
ω	osmotic efficiency [−]
ω	water content in Chapter 3 [−]
$\chi_{\sigma 0}$, $\chi_{\sigma 2}$, $\chi_{\tau 2}$	stress concentration factors around circular perforations for pressures σ_0, σ_2 and τ_2, respectively [−]

Specific solution for matrix diffusion: one-dimensional, two-dimensional or three-dimensional conditions

The analysis presented in Section 7.6 is dependent on the quantity $\overline{\eta}$ introduced in equation 7.65. The quantity reflects the diffusion of contaminant from the fractures and into the matrix of the adjacent intact porous media. This quantity will depend on the nature of the fracturing and can be derived separately for 1D, 2D and 3D conditions as follows.

Consider a typical unit of the matrix. If advection in the matrix can be neglected, the matrix concentration, c_{m}, satisfies the equation:

$$n_{\mathrm{m}} D_{\mathrm{m}} \nabla^2 c_{\mathrm{m}} = (n_{\mathrm{m}} + \rho_{\mathrm{m}} K_{\mathrm{m}}) \frac{\partial c_{\mathrm{m}}}{\partial t} + \lambda_{\mathrm{m}} c_{\mathrm{m}} \quad \text{(C.1)}$$

where the subscript m indicates a property of the matrix and n, D, K and λ are the porosity, diffusion coefficient, distribution coefficient and first order decay constant (for radioactive or biological decay) in the matrix. The concentration at the surface of the matrix will be equal to that in the fracture system, c_{m}, so that $c_{\mathrm{m}} = c_{\mathrm{f}}$ on the surface Σ_{m} of the matrix.

It will be assumed that initially the matrix is uncontaminated so that:

$$c_{\mathrm{m}} = 0 \quad \text{when } t = 0 \quad \text{(C.2)}$$

The Laplace transform of equation C.3 is then:

$$n_{\mathrm{m}} D_{\mathrm{m}} \nabla^2 \overline{c}_{\mathrm{m}} = n_{\mathrm{m}} R_{\mathrm{m}} (s + \Lambda_{\mathrm{m}}) \overline{c}_{\mathrm{m}} \quad \text{(C.3)}$$

where

$$R_{\mathrm{m}} = 1 + (\rho K_{\mathrm{m}} / n_{\mathrm{m}}); \; \Lambda_{\mathrm{m}} = \lambda_{\mathrm{m}} / R_{\mathrm{m}}$$

Equation C.3 is to be solved subject to the boundary conditions that $c_{\mathrm{m}} = c_{\mathrm{f}}$ on the surface Σ_{m} of the matrix.

The quantity of primary interest is:

$$q = -\frac{\int f_{\mathrm{mn}} d\Sigma_{\mathrm{m}}}{\int dv_{\mathrm{m}}}$$

where f_{mn} denotes the component of flux normal to the surface and v_m the volume occupied by the matrix.

It thus follows from equation C.3 and Gauss's Divergence Theorem that:

$$\bar{q} = \frac{\int n_m D_m \nabla^2 \bar{c}_m dv_m}{\int dv_m} = n_m R_m (s + \Lambda_m) \frac{\int \bar{c}_m dv_m}{\int \bar{c}_f dv_m} \cdot \bar{c}_f \qquad (C.4)$$

Referring to equation 7.65, $\bar{\eta}$ is given by:

$$\bar{\eta} = n_m R_m \frac{\int \bar{c}_m \, dV_m}{\int \bar{c}_f dV_m} \qquad (C.5)$$

Case 1

First, suppose there is only a single set of fractures (set 1), then because of the assumed conditions in the landfill there is no variation of concentration with respect to y. Also, the variation of concentration is likely to be relatively slow along the fissures (x-direction) when compared with the variation between adjacent fissures (z-direction) and thus to sufficient accuracy, the concentration within the fissures satisfies the equation:

$$D_m = \frac{\partial^2 \bar{c}_m}{\partial z^2} = R_m (s + \Lambda_m) \bar{c}_m$$

where $\bar{c}_m = \bar{c}_f$; $z = \pm H_1$. It now follows that:

$$\bar{c}_m = \bar{c}_f \frac{\cosh \mu z}{\cosh \mu H_1}$$

where $\mu^2 = (R_m/D_m)(s + \Lambda_m)$.

Hence,

$$\bar{\eta} = n_m R_m \frac{\tanh \mu H_1}{\mu H_1} \qquad (C.6a)$$

If the solution for $\bar{c}_m$ is developed as a Fourier series, it is found that $\bar{\eta}$ can also be expressed in the form:

$$\bar{\eta} = n_m R_m \left[1 - 2 \sum_j \frac{(s + \lambda_m)}{[s + \lambda_m + \alpha_j^2 (D_m/R_m)]} \times \frac{1}{(\alpha_j H_1)^2} \right] \qquad (C.6b)$$

where $\alpha_j = (j - 1/2)\pi/H_1$; $j = 1,2,\ldots,\infty$.

Case 2

Suppose now that there are two sets of fractures (sets 1 and 2); and considering diffusive transport into the matrix in the z- and y-directions, the matrix concentration $\bar{c}_m$ can be written in the form:

$$\bar{c}_m = \bar{c}_f \left[1 + \sum_{j,k} X_{jk} \cos(\alpha_j z) \cos \beta_k y) \right]$$

with

$$\alpha_j = \left(j - \frac{1}{2} \right) \pi / H_1 \quad j = 1, 2, \ldots, \infty$$

$$\beta_k = \left(k - \frac{1}{2} \right) \pi / H_2 \quad k = 1, 2, \ldots, \infty$$

so that:

$$\bar{\eta} = n_m R_m \left[1 + \sum_{j,k} X_{jk} \frac{\sin(\alpha_j H_1)}{(\alpha_j H_1)} \frac{\sin(\beta_k H_2)}{(\beta_k H_2)} \right]$$

where

$$X_{jk} = -4 \frac{\sin(\alpha_j H_1)}{(\alpha_j H_1)} \frac{\sin(\beta_k H_2)}{(\beta_k H_2)} \times \frac{(s + \Lambda_m) R_m}{[R_m (s + \Lambda_m) + D_m (\alpha_j^2 + \beta_k^2)} \qquad (C.7a)$$

and so

$$\overline{\eta} = n_m R_m \left[1 - 4\sum_{j,k} \times \frac{(s + \Lambda_m)}{[s + \Lambda_m + (\alpha_j^2 + \beta_j^2)D_m/R_m]} \times \frac{1}{(\alpha_j H_1)^2} \frac{1}{(\beta_k H_2)^2} \right] \quad \text{(C.7b)}$$

Case 3

Finally, if there are three sets of fissures and it is observed that the transport distances of interest along the fracture system are large compared to the dimensions of the block then it is found that the solution has the form:

$$\overline{c}_m = \overline{c}_f \left(1 + \sum_{j,k,\ell} X_{jk\ell} \cos(\alpha_j z) \cos(\beta_k y) \cos(\gamma_\ell x) \right)$$

It follows in a similar manner to the previous case that:

$$\overline{\eta} = n_m R_m \left[1 - 8\sum_{j,k,\ell} \times \frac{(s + \Lambda_m)}{[s + \Lambda_m + (D_m/R_m)(\alpha_j^2 + \beta_j^2 + \gamma_\ell^2)]} \times \frac{1}{(\alpha_j H_1)^2} \cdot \frac{1}{(\beta_k H_2)^2} \cdot \frac{1}{(\gamma_\ell H_3)^2} \right]$$

where the definitions of α_j, β_k are as described previously and

$$\gamma_\ell = \left(\ell - \frac{1}{2} \right) \pi / H_3 \quad \text{(C.8)}$$

Bulk GCL void ratio

The final bulk GCL void ratio, e_B, is defined by (Petrov *et al.*, 1997a) as:

$$e_B = \frac{H_{GCL} - H_s}{H_s} \quad \text{(D.1)}$$

where H_{GCL} is the GCL height and H_s is the height of solids in the GCL and:

$$H_s = H_{s\,BENT} + H_{s\,GEO} \quad \text{(D.2)}$$

with

$$H_{sBENT} = \frac{M_{BENT}}{\rho_s(1 + w_0)} \quad \text{(D.3)}$$

$$H_{sGEO} = \frac{M_{GEO}}{\rho_{sg}} \quad \text{(D.4)}$$

where H_{sBENT} is the height of bentonite solids, H_{sGEO} the height of geotextile solids, $M_{GCL} = M_{BENT} + M_{GEO}$ is the mass per unit area of the GCL, M_{BENT} the mass of bentonite per unit area in the GCL, M_{GEO} the mass of geosynthetics per unit area in the GCL, ρ_s the density of bentonite solids (typical value of $2.61\,Mg/m^3$), ρ_{sg} is the density of geotextile solids (typical value of $0.91\,Mg/m^3$ for polypropylene), w_0 is the initial water content of the bentonite (typical value of 0.08).

The void ratio of the bentonite in the GCL is given by:

$$e_{BENT} = \frac{H_{BENT} - H_{sBENT}}{H_{sBENT}} \quad \text{(D.5)}$$

where H_{BENT} is the height of bentonite in the GCL and is given by:

$$H_{BENT} = H_{GCL} - H_{GEO} \quad \text{(D.6)}$$

where H_{GEO} is the height of geotextiles in the GCL.

For a given GCL, H_{GCL}, H_{GEO}, w_0, M_{GCL}, M_{GEO}, ρ_s and ρ_{sg} are either measured or known and hence e_{BENT} can be deduced from equations D.1–D.6 and the total porosity of the bentonite, n_t, can be calculated from:

$$n_t = \frac{e_{BENT}}{(1 + e_{BENT})} \quad \text{(D.7)}$$

In a consolidation or swell test, e_B is deduced from equations D.1 and D.2 based on known values of H_{GCL}, M_{BENT}, M_{GEO}, ρ_s, ρ_{sg} and w_0.

References

Abramowitz, M. and Stegun, I. (1964) *Handbook of Mathematical Functions*, National Bureau of Standards Applied Mathematics, US Government Printing Offices, Washington, DC, Ser. 55.

Acar, Y.B. and Haider, L. (1990) Transport of low concentration contaminants in saturated earthen barriers, *ASCE Journal of Geotechnical Engineering*, **116**(7), 1031–52.

Acar, Y.B. and Seals, R.K. (1984) Clay barrier technology for shallow land waste disposal facilities, *Hazardous Waste*, **1**(2), 167–81.

Acar, Y.B., Hamidon, A., Field, S.D. and Scott, L. (1985) The effects of organic fluids on hydraulic conductivity of compacted kaolinite, in *Hydraulic Barriers in Soil and Rock* (eds A.I. Johnson, R.K. Frobel, N.J. Cavalli and C.B. Pettersson), ASTM STP 874, American Society for Testing and Materials, Philadelphia, PA, pp. 171–87.

Adams, M.W. and Wagner, N. (2000) Evaluating physical property integrity of a geomembrane used at a wastewater treatment facility for 11 years, *Geotechnical Fabrics Report*, **18**(7), 36–9.

Allan, M.B. (1984) Why upwinding is reasonable, in *Proceedings 5th International Conference of Finite Elements in Water Resources*, pp. 13–23.

Al-Niami, A.N.S. and Rushton, K.R. (1977) Analysis of flow against dispersion in porous media, *Journal of Hydrology*, **33**, 87–97.

Al-Yousfi, A.B. and Pohland, F.G. (1998) Strategies for simulation, design and management of solid wastes disposal sites as landfill bioreactors, *Practice Periodical of Hazardous, Toxic and Radioactive Waste Management*, **2**, 13–21.

Anand, M., Hobbs, J.P. and Brass, I.J. (1994) Surface fluorination of polymers, in *Organofluorine Chemistry: Principles and Commercial Applications* (eds R.E. Banks, B.E. Smart and J.C. Tatlow), Plenum Press, New York, pp. 469–81.

Anderson, M.P. (1979) Using models to simulate the movement of contaminants through groundwater flow systems, *CRC Critical Reviews in Environmental Control*, **9**, 97–156.

Anderson, M.P. (1984) Movement of contaminants in groundwater: groundwater transport – advection and dispersion, *Groundwater Contamination*, NRC Studies in Geophysics. National Academy Press, Washington, DC, pp. 37–45.

Annex, R.P. (1996) Optimal waste decomposition – landfill as treatment process, *Journal of Environmental Engineering*, **122**, 964–74.

Appelt, H., Holtzclaw, K. and Pratt, P.F. (1975) Effect of anion exclusion on the movement of chloride through soils, *Soil Science Society of America Proceedings*, **39**, 264–7.

Armstrong, M.D. (1998) *Laboratory program to study clogging in a leachate collection system*, M.E.Sc. thesis, The University of Western Ontario, London, Ontario.

Armstrong, M.D. and Rowe, R.K. (1999) Effect of landfill operations on the quality of municipal solid waste leachate, in *Proceedings of 7th International Landfill Symposium*, S. Margherita di Pula, Cagliari, Sardinia, II, pp. 81–8.

Ashley, R.J. (1985) Permeability and plastics packaging, Chapter 7 in *Polymer Permeability* (ed. J. Comyn), Chapman & Hall, pp. 269–308.

ASTM – The American Society for Testing and Materials, *Annual Book of ASTM Standards*, 100 Barr Harbor Dr., West Conshohocken, PA. Website: http://www.astm.org.

ASTM D422 *Standard Test Method for Particle-size Analysis of Soils*, Vol. 04.08.

ASTM D638 *Standard Test Method for Tensile Properties of Plastics*, Vol. 08.01.

ASTM D698 *Standard Test Methods for Laboratory Compaction Characteristics of Soil using Standard Effort*, Vol. 04.08.

ASTM D792 *Standard Test Methods for Density and Specific Gravity (Relative Density) of Plastics by Displacement*, Vol. 08.01.

ASTM D883 *Standard Definitions of Terms Relating to Plastics*, Vol. 08.01.

ASTM D1004 *Standard Test Method for Initial Tear Resistance of Plastic Film and Sheeting*, Vol. 08.01.

ASTM D1505 *Standard Test Method for Density of Plastics by the Density-gradient Technique*, Vol. 08.01.

ASTM D1556 *Standard Test Method for Density and Unit Weight of Soil in Place by the Sand-cone Method*, Vol. 04.08.

ASTM D1557 *Standard Test Methods for Laboratory Compaction Characteristics of Soil using Modified Effort*, Vol. 04.08.

ASTM D1603 *Standard Test Method for Carbon Black in Olefin Plastics*, Vol. 08.01.

ASTM D2167 *Standard Test Method for Density and Unit Weight of Soil in Place by the Rubber Balloon Method*, Vol. 04.08.

ASTM D2216 *Standard Test Method for Laboratory Determination of Water (Moisture) Content of Soil and Rock*, Vol. 04.08.

ASTM D2657 *Standard Practice for Heat Fusion Joining of Polyolefin Pipe and Fittings*, Vol. 08.04.

ASTM D2922 *Standard Test Methods for Density of Soil and Soil-aggregate in Place by Nuclear Methods (Shallow Depth)*, Vol. 04.08.

ASTM D3017 *Standard Test Method for Water Content of Soil and Rock in Place by Nuclear Methods (Shallow Depth)*, Vol. 04.08.

ASTM D3895 *Standard Test Method for Oxidative-induction Time of Polyolefins by Differential Scanning Calorimetry*, Vol. 08.02.

ASTM D4318 *Standard Test Method for Liquid Limit, Plastic Limit, and Plasticity Index of Soils*, Vol. 04.08.

ASTM D4437 *Standard Practice for Determining the Integrity of Field Seams used in Joining Flexible Polymeric Sheet Geomembranes*, Vol. 04.09.

ASTM D4595 *Standard Test Method for Tensile Properties of Geotextiles by the Wide-width Strip Method*, Vol. 04.13.

ASTM D4632 *Standard Test Method for Grab Breaking Load and Elongation of Geotextiles*, Vol. 04.13.

ASTM D4643 *Standard Test Method for Determination of Water (Moisture) Content of Soil by the Microwave Oven Method*, Vol. 04.08.

ASTM D4646 *Standard Test Method for 24-h Batch-type Measurement of Contaminant Sorption by Soils and Sediments*, Vol. 11.04.

ASTM D4833 *Standard Test Method for Index Puncture Resistance of Geotextiles, Geomembranes, and Related Products*, Vol. 04.13.

ASTM D4885 *Standard Test Method for Determining Performance Strength of Geomembranes by the Wide Strip Tensile Method*, Vol. 04.13.

ASTM D5084 *Standard Test Method for Measurement of a Hydraulic Conductivity of Saturated Porous Materials Using a Flexible Wall Permeameter*, Vol. 04.08.

ASTM D5397 *Standard Test Method for Evaluation of Stress Crack Resistance of Polyolefin Geomembranes Using Notched Constant Tensile Load Test*, Vol. 04.09.

ASTM D5596 *Standard Test Method for Microscopic Evaluation of the Dispersion of Carbon Black in Polyolefin Geosynthetics*, Vol. 04.13.

ASTM D5617 *Standard Test Method for Multi-axial Tension Test for Geosynthetics*, Vol. 04.13.

ASTM D5721 *Standard Practice for Air-oven Aging of Polyolefin Geomembranes*, Vol. 04.13.

ASTM D5885 *Standard Test Method for Oxidative Induction Time of Polyolefin Geosynthetics by High-pressure Differential Scanning Calorimetry*, Vol. 04.09.

ASTM D5887 *Standard Test Method for Measurement of Index Flux through Saturated Geosynthetic Clay Liner Specimens Using a Flexible Wall Permeameter*, Vol. 04.13.

ASTM D5890 *Standard Test Method for Swell Index of Clay Mineral Component of Geosynthetic Clay Liners*, Vol. 04.13.

ASTM D5891 *Test Method for Fluid Loss of Clay Component of Geosynthetic Clay Liners*, Vol. 04.13.

ASTM D5993 *Standard Test Method for Measuring the Mass per Unit Area of GCL*, Vol. 04.13.

ASTM D6102 *Standard Guide for Installation of Geosynthetic Clay Liners*, Vol. 04.13.

ASTM D6243 *Standard Test Method for Determining the Internal and Interface Shear Resistance of Geosynthetic Clay Liner by the Direct Shear Method*, Vol. 04.13.

ASTM D6496 *Standard Test Method for Determining Average Bonding Peel Strength between the Top and Bottom Layers of Needle-punched Geosynthetic Clay Liners*, Vol. 04.13.

ASTM E1142 *Standard Terminology Relating to Thermophysical Properties*, Vol. 14.02.

ASTM F714 *Standard Specification for Polyethylene (PE) Plastic Pipe (SDR-PR) Based on Outside Diameter*, Vol. 08.04.

Aubertin, M., Aachib, M. and Authier, K. (2000) Evaluation of diffusive gas flux through covers with GCLs, *Geotextiles and Geomembranes*, **18**(2–4), 215–33.

August, H. and Tatzky, R. (1984) Permeability of commercially available polymeric liners for hazardous landfill leachate organic constituents, in *Proceedings of International Conference on Geomembranes*, June 20–24, Denver, CO, Vol. I, Industrial Fabrics Association International, St Paul, USA, pp. 163–8.

August, H., Tatzky-Gerth, R., Preuschmann, R. and Jakob, I. (1992) Permeationsverhalten von kombinationsdichtungen bei deponien und altlasten gegenueber wassergefaehrdenden stiffen, Report on the R&D Project 10203412 of the BMBF, published by BAM, Sub-Department IV.3, Landfill and Remedial Engineering, D-12200 Berlin, Germany.

Averesch, U. and Schicketanz, R. (1998) Installation procedure and welding of geomembranes in the construction of composite landfill liner systems: the fixing berm construction method, in *Proceedings of 6th International Conference on Geosynthetics*, Atlanta, Industrial Fabrics Association International, St Paul, MN, pp. 307–13.

Badu-Tweneboah, K., Giroud, J.P., Carlson, D.S. and Schmertmann, G.R. (1998) Evaluation of the effectiveness of HDPE geomembrane liner protection, in

Proceedings of 6th International Conference on Geosynthetics, Atlanta, Industrial Fabrics Association International, St Paul, MN, pp. 279–84.

Badv, K. and Rowe, R.K. (1996) Contaminant transport through a soil liner underlain by an unsaturated stone collection layer, *Canadian Geotechnical Journal*, **33**(2), 416–30.

Bailey, S.W. (1980) Summary of recommendations of AIPEA nomenclature committee, *Clay Minerals*, **5**, 85.

Baker, J.A. and Eith, A.W. (2000) The bioreactor landfill: perspectives from an owner/operator, in *Proceedings of 14th Geosynthetic Research Institute Conference*, pp. 21–39.

BAM – Bundesanstalt fuer Materialforschung und-pruefung (1995) *Anforderungen an die Scutzschicht fuer die Dichtungsbahnen in der Kombinationsdichtung, Zulassungsrichtlinie fuer Schutzschichten*, Federal Institute for Materials Research and Testing, Certification Guidelines for Protection Layers for Geomembranes in Composite Liners (in German).

Banerji, S.K., Piontek, K. and O'Connor, J.T. (1986) Pentachlorophenol adsorption on soils and its potential for migration into groundwater, *Hazardous and Industrial Solid Waste Testing and Disposal, ASTM Special Technical Publication*, **933**, 120–39.

Barbour, S.L. and Fredlund, D.G. (1989) Mechanisms of osmotic flow and volume change in clay soils, *Canadian Geotechnical Journal*, **26**, 551–62.

Barker, J.A. (1982) Laplace transform solutions for solute transport in fissured aquifers, *Advanced Water Research*, **5**(2), 98–104.

Barker, J.F. (1992) The persistence of aromatic hydrocarbons in various groundwater environments, Waterloo Centre for Groundwater Research, Research Paper.

Barker, J.A. and Foster, S.S.D. (1981) A diffusion exchange model for solute movement in fissured porous rock, *Quarterly Journal of Engineering Geology (London)*, **14**, 17–24.

Barker, J.F., Cherry, J.A., Carey, D.A. and Mattes, M.E. (1987) Hazardous organic chemicals in groundwater at Ontario landfills, in *Proceedings 1987 Technology Transfer Conference*, Part C, Paper C3, 29 pp.

Barlaz, M.A., Ham, R.K. and Schaefer, D.M. (1990) Methane production from municipal refuse: a review of enhancement techniques and microbial dynamics, *Critical Reviews in Environmental Control*, **19**, 557–84.

Barone, F.S. (1990) *Determination of diffusion and adsorption coefficients for some contaminants in clayey soil and rock: laboratory determination and field evaluation*, Ph.D. Thesis, University of Western Ontario, London, Ont., Canada, 325 pp.

Barone, F.S., Yanful, E.K., Quigley, R.M. and Rowe, R.K. (1989) Effect of multiple contaminant migration on diffusion and adsorption of some domestic waste contaminants in a natural clayey soil, *Canadian Geotechnical Journal*, **26**(2), 189–98.

Barone, F.S., Rowe, R.K. and Quigley, R.M. (1990) Laboratory determination of chloride diffusion coefficient in an intact shale, *Canadian Geotechnical Journal*, **27**, 177–84.

Barone, F.S., Rowe, R.K. and Quigley, R.M. (1992a) Estimation of chloride diffusion coefficient and tortuosity factor for mudstone, *ASCE Journal of Geotechnical Engineering*, **118**, 1031–46.

Barone, F.S., Rowe, R.K. and Quigley, R.M. (1992b) A laboratory estimation of diffusion and adsorption coefficients for several volatile organics in a natural clayey soil, *Journal of Contaminant Hydrogeology*, **10**, 225–50.

Barone, F.S., Costa, J.M.A., King, K.S., Edelenbos, M. and Quigley, R.M. (1993) Chemical and mineralogical assessment of in situ clay liner Keele Valley landfill, Maple, Ontario, in *Proceedings Joint CSCE–ASCE National Conference on Environmental Engineering* (eds R.N. Yong, J. Hadjinicolaou and A.M.O. Mohamed), The Geotechnical Research Centre of McGill University in conjunction with the CSCE and ASCE, pp. 1563–72.

Barone, F.S., Barbour, S.L., Bews, B.E. and Costa, J.M. (1999) Behavior of lysimeters installed within and below a compacted clay liner underlain by unsaturated sand, in *Proceedings of the Canadian Geotechnical Conference*, Regina, Saskatchewan, pp. 367–72.

Barone, F.S., Costa, J.M.A. and Ciardullo, L. (2000) Temperatures at the base of a municipal solid waste landfill, in *Proceedings of 6th Environmental Engineering Specialty Conference of the CSCE* (ed. R.K. Rowe), pp. 41–8.

Barsamyan, G. and Sokolov, V.B. (1999) Surface fluorination of polymer using xenon difluoride, in *Fluoropolymers. 1. Synthesis* (eds G. Hougham, P.E. Cassidy, K. Johns and T. Davidson), Kluwer Academic/Plenum Publishers, New York, pp. 223–40.

Basnett, C.R. and Bruner, R.J. (1993) Clay desiccation of a single-composite liner system, in *Proceedings of Geosynthetics 93*, Vancouver, Industrial Fabrics Association International, St Paul, MN, pp. 1329–40.

Basnett, C. and Brungard, M. (1992) The clay desiccation of a landfill composite lining system, *Geotechnical Fabrics Report*, **38**(1), 38–41.

Bear, J. (1979) *Hydraulics of Groundwater*, McGraw-Hill, New York.

Bedient, P.B., Rifai, H.S. and Newell, C.J. (1999) *Groundwater Contamination: Transport and Remediation*, 2nd edn, Prentice-Hall, Englewood Cliffs, N.J.

Belevi, H. and Baccini, P. (1989) Long-term behavior of municipal solid waste landfills, *Waste Management and Research*, **7**, 43–56.

Bennett, P.J. (1998) *The stable isotopic characterization of carbon and water cycling in municipal solid waste landfills*, M.Sc. thesis, The University of Western Ontario, London, Ontario.

Bennett, P.J., Longstaffe, F.J. and Rowe, R.K. (2000) The stability of dolomite in landfill leachate collection systems, *Canadian Geotechnical Journal*, **37**, 371–8.

Benson, C. and Boutwell, G. (1992) Compaction control and scale-dependent hydraulic conductivity of clay liners, in *Proceedings of 15th Annual Madison Waste Conference*, University of Wisconsin, Madison, pp. 62–83.

Benson, C.H. and Daniel, D.E. (1990) Influence of clods on hydraulic conductivity of compacted clay, *ASCE Journal of Geotechnical Engineering*, **116**(8), 1231–48.

Benson, C.H. and Khire, M.V. (1995) Earthen covers for arid and semi-arid sites, in *Landfill Closures: Environmental Protection and Land Recovery* (eds R.J. Dunn and R.P. Singh), Geotechnical Special Publication No. 53, American Society of Civil Engineers, Reston, VA, pp. 201–17.

Benson, C.H., Zhai, H. and Wang, X. (1994) Estimating hydraulic conductivity of compacted clay liners, *ASCE Journal of Geotechnical Engineering*, **120**(2), 366–87.

Benson, C.H., Daniel, D.E. and Boutwell, G.P. (1999) Field performance of compacted clay liners, *ASCE Journal of Geotechnical and Geoenvironmental Engineering*, **125**(5), 390–403.

Berens, A.R. and Hopfenberg, H.B. (1982) Diffusion of organic vapors at low concentration on glassy PVC, polystyrene and PMMA, *Journal of Membrane Science*, **10**, 283–303.

Biener, E. and Sasse, T. (1993) Construction and rehabilitation of landfill shafts, in *Proceedings 4th International Landfill Symposium*, Cagliari, Italy, pp. 451–460.

Biot, M.A. (1941) General theory of three-dimensional consolidation, *Journal of Applied Physics*, **12**, 155–64.

Bishop, D.J. (1996) Discussion of 'A comparison of puncture behavior of smooth and textured HDPE geomembranes' and 'Three levels of geomembrane puncture protection' by Narejo, D.B., *Geosynthetics International*, **3**(3), 441–3.

Blakey, N., Bradshaw, K., Reynolds, P. and Knox, K. (1997) Bio-reactor landfill – a field trial of accelerated waste stabilization, in *Proceedings 6th International Landfill Symposium*, Sardinia, Cagliari, Italy.

Blümel, W., Müller-Kirchenbauer, Reuter, E., Ehrenberg, H. and von Maubeuge, K.P. (2002) Performance of geosynthetic clay liners in lysimeters, in *Proceedings of International Symposium on Geosynthetic Clay Barriers*, Nuremberg, Germany, pp. 287–94.

Boardman, T. and Daniel, D. (1996) Hydraulic conductivity of desiccated geosynthetic clay liners, *ASCE Journal of Geotechnical and Geoenvironmental Engineering*, **122**, 204–8.

Bonaparte, R. and Yanful, E.K. (2001) Covers for waste, Chapter 27, *Geotechnical and Geoenvironmental Engineering Handbook* (ed. R.K. Rowe), Kluwer Academic Publishers, Norwell, pp. 825–77.

Bonaparte, R., Ah-Line, A.M., Charron, R. and Tisinger, L. (1988) Survivability and durability of a nonwoven geotextile, in *Proceedings Geosynthetics for Soil Improvement*, Nashville, pp. 68–91.

Bonaparte, R., Daniel, D. and Koerner, R.M. (2002) Assessment and recommendations for improving the performance of waste containment systems, EPA Report, Co-operative Agreement Number CR-821448–01–0.

Booker, J.R. and Rowe, R.K. (1987) One dimensional advective–dispersive transport into a deep layer having a variable surface concentration, *International Journal for Numerical and Analytical Methods in Geomechanics*, **11**(2), 131–42.

Bouazza, A. (2002) Geosynthetic clay liners, *Geotextiles and Geomembranes*, **20**(1), 1–17.

Bouazza, A. and Vangpaisal, T. (2000) Advective gas flux through partially saturated geosynthetic clay liners, *Advances in Geosynthetics Uses Transportation and Geoenvironmental Engineering*, Geotechnical Special Publication No. 101, 54–67.

Bouazza, A., Van Impe, W.F. and Van Den Broeck, M. (1996) Hydraulic conductivity of a geosynthetic clay liner under various conditions, in *Proceedings 2nd International Congress on Environmental Geotechnics*, Osaka, Japan, Vol. 1, pp. 453–8.

Bouazza, A., Vangpaisal, T. and Rahman, F. (2002a) Gas migration through needle punched geosynthetic clay liners, in *Proceedings International Symposium on Geosynthetics Clay Barriers, Nuremberg*, pp. 165–76.

Bouazza, A., Zornberg, J. and Adam, D. (2002b) Geosynthetics in waste containments: recent advances, in *Proceedings 7th International Geosynthetics Conference*, Nice, France, **2**, 445–507.

Bove, J.A. (1990) Direct shear friction testing for geosynthetics in waste containment, in *Geosynthetic Testing for Waste Containment Applications* (ed. R.M. Koerner), ASTM 1081, Philadelphia, PA.

Bowders, J.J. and Daniel, D.E. (1987) Hydraulic conductivity of compacted clay to dilute organic chemicals, *ASCE Journal of Geotechnical Engineering*, **113**(12), 1432–48.

Bowders, J.J., Daniel, D.E., Broderick, G.P. and Liljestrand, H.M. (1986) Methods for testing the compatibility of clay liners with landfill leachate, in *Hazardous and Industrial Solid Waste Testing: Fourth Symposium* (eds J.K. Petros, W.J. Lacy and R.A. Conway), ASTM STP 886, American Society for Testing and Materials, Philadelphia, PA, pp. 233–50.

Bowders, J.J., Daniel, D.E., Wellington, J. and Houssidas, V. (1997) Managing desiccation cracking in compacted clay liners beneath geomembranes, in *Proceedings of Geosynthetics'97*, Long Beach (eds L. Well and

R. Thiel), Industrial Fabrics Association International, St Paul, MN, pp. 527–40.

Boyd, S.A., Lee, J.F. and Mortland, M.M. (1988a) Attenuating organic contaminant mobility by soil modification, *Nature*, **33**, 345–7.

Boyd, S.A., Mortland, M.M. and Chiou, C.T. (1988b) Sorption characteristics of organic compounds on hexadecyltrimethylammonium-smectite, *Soil Science Society of America Journal*, **52**, 652–7.

Bracci, G., Giardi, M. and Paci, B. (1991) The problem of clay liners testing in landfills, in *Proceedings 3rd International Landfill Symposium*, Cagliari, Italy, pp. 679–89.

Brachman, R.W.I. (1999) *Mechanical performance of landfill leachate collection pipes*, Ph.D. thesis, Faculty of Engineering Science, The University of Western Ontario, London, Ont, Canada.

Brachman, R.W.I. (2001) Tensile stresses and the long term durability of PE pipes, in *Proceedings Tailings and Mine Waste'01*, Balkema, Rotterdam, pp. 201–10.

Brachman, R.W.I. and Gudina, S. (2002) A new laboratory apparatus for testing geomembranes under large earth pressures, in *Proceedings of 55th Canadian Geotechnical Conference*, Niagara Falls, Ont, pp. 993–1000.

Brachman, R.W.I. and Krushelnitzky, R.P. (2002) Stress concentrations around circular holes in perforated drainage pipes, *Geosynthetics International*, **9**(2), 189–213.

Brachman, R.W.I., Moore, I.D. and Rowe, R.K. (2000a) The design of a laboratory facility for evaluating the structural response of small diameter buried pipes, *Canadian Geotechnical Journal*, **37**(2), 281–95.

Brachman, R.W.I., Moore, I.D. and Rowe, R.K. (2000b) Local strain on a leachate collection pipe, *Canadian Journal of Civil Engineering*, **27**, 1273–85.

Brachman, R.W.I., Moore, I.D. and Rowe, R.K. (2001) The performance of a laboratory facility for evaluating the structural response of small diameter buried pipes, *Canadian Geotechnical Journal*, **38**(2), 260–75.

Brady, K.C., McMahon, W. and Lamming, G. (1994) *Thirty-year ageing of plastics*, Transport Research Laboratory, Project Report 11, E472A/BG, ISSN 0968–4093.

Brand, E.W. and Premchitt, J. (1982) Response characteristics of cylindrical piezometers, *Geotechnique*, **32**(3), 203–16.

Brandl, H. (1995) Stability of waste deposits, in *Proceedings of 10th Danube-European Conference on Soil Mechanics and Foundation Engineering*, Romania, pp. 1043–56.

Brebbia, C.A. and Dominguez, J. (1989) *Boundary Elements an Introductory Course*, Computational Mechanics Publications, McGraw-Hill, New York.

Brebbia, C.A. and Skerget, P. (1984) Diffusion–convection problems using boundary elements, in *Proceedings 5th International Conference on Finite Elements in Water Resources*, Vermont, pp. 747–68.

Bresler, E. (1973) Anion exclusion and coupling effects in non-steady transport through unsaturated soils. I. Theory, *Soil Science Society of America Proceedings*, **37**, 663–9.

Bressi, G., Zinessi, M., Montanelli, F. and Rimoldi, P. (1995) The slope stability of GCL layers in geosynthetic lining system, in *Proceedings 5th International Symposium on Landfills*, Cagliari, Italy, Vol. 1, pp. 595–610.

Bright, M.I., Thornton, S.F., Lerner, D.N. and Tellam, J.H. (2000) Attenuation of landfill leachate by clay liner materials in laboratory columns. 1. Experimental procedures and behaviour of organic contaminants, *Waste Management and Research*, **18**, 198–214.

Brink, D., Day, P.W. and Du Preez, L. (1999) Failure and remediation of Bulbul Drive landfill: KwaZulu-Natal, South Africa, in *Proceedings of 7th International Waste Management and Landfill Symposium*, Sardinia, Italy, October, 1999, Vol. 3, pp. 555–62.

Britton, L.N., Ashman, R.B., Aminabhavi, T.M. and Cassidy, P.E. (1989) Permeation and diffusion of environmental pollutants through flexible polymers, *Journal of Applied Polymer Science*, **38**(2), 227–35.

Broholm, M.M., Broholm, K. and Arvin, E. (1999) Sorption of heterocyclic compounds from a complex mixture of coal-tar compounds on natural clayey till, *Journal of Contaminant Hydrology*, **39**, 201–26.

Brown, K.W. and Anderson, D.C. (1983) *Effects of organic solvents on the permeability of clay soils*, US EPA, EPA-600/2–83–016, Environmental Protection Agency, Cincinnati, OH.

Brown, K.W., Green, J.W. and Thomas, J.C. (1983) The influence of selected organic liquids on the permeability of clay liners, in *Proceedings 9th Annual Research Symposium on Land Disposal, Incineration, and Treatment of Hazardous Waste*, EPA-600/9-83-018, Environmental Protection Agency, Cincinnati, OH, Ft. Mitchell, KY, May 2–4, pp. 114–25.

Brown, D.W., Thomas, J.C. and Green, J.W. (1984) Permeability of compacted soils to solvent mixtures and petroleum products, in *Proceedings 10th Annual Research Symposium on Land Disposal of Hazardous Waste*, EPA-600/9-84-007, Environmental Protection Agency, Cincinnati, OH, Ft. Mitchell, KY, April 3–5, pp. 124–37.

Brown, K.W., Thomas, J.C., Lytton, R.L., Jayawickrama, P. and Bahrt, S. (1987) *Quantification of leakage rates through holes in landfill liners*, US EPA Report CR810940, Cincinnati, OH, 147 pp.

Brummermann, K., Blümel, W. and Stoewahse, C. (1995) Protection layers for geomembranes: effectiveness and testing procedures, in *Proceedings of 5th International Conference on Geosynthetics*, Singapore, pp. 1003–6.

Brune, M., Ramke, H.G., Collins, H.J. and Hanert, H.H. (1991) Incrustation processes in drainage systems of sanitary landfills, in *Proceedings 3rd International Landfill Symposium*, Cagliari, Italy, pp. 999–1035.

Byrne, R.J., Kendall, J. and Brown, S. (1992) Cause and mechanism of failure Kettleman Hills Landfill B-19, Phase IA, in *Proceedings of ASCE Conference on Stability and Performance of Slopes and Embankments, Geotechnical Special Publication No. 31*, Vol. 2, pp. 1188–215.

Cancelli, A. and Cazzuffi, D. (1987) Permittivity of geotextiles in presence of water and pollutant fluids, in *Proceedings of Geosynthetics 87*, New Orleans, pp. 471–81.

Carslaw, H.S. and Jaegar, J.C. (1948) *Operational Methods in Applied Mathematics*, 2nd edn, Oxford University Press, London.

Carslaw, H.S. and Jaegar, J.C. (1959) *Conduction of Heat in Solids*, 2nd edn, Clarendon Press, Oxford.

Carstens, P.A.B., De Beer, J.A. and Le Roux, J.P. (1999) New surface-fluorinated products, in *Fluoropolymers. 1. Synthesis* (eds G. Hougham, P.E. Cassidy, K. Johns and T. Davidson), Kluwer Academic/Plenum Publishers, New York, pp. 241–59.

Cazzuffi, D., Cossu, R., Ferruti, L. and Lavagnolo, C. (1991) Efficiency of geotextiles and geocomposites in landfill drainage systems, in *Proceedings Third International Landfill Symposium*, Cagliari, Italy, pp. 759–80.

Cedergren, H.R. (1989) *Seepage, Drainage and Flow Nets*, 3rd edn, John Wiley & Sons, New York.

Cey, D.B., Barbour, S.L. and Hendry, J.M. (2001) Osmotic flow through a Cretaceous clay in southern Saskatchewan, Canada, *Canadian Geotechnical Journal*, **38**, 1025–33.

Chainey, M. (1990) Transport phenomena in polymer films, in *Handbook of Polymer Science and Technology*, Composites and Specialty Applications, Vol. 4 (ed. N.P. Cheremisinoff), Marcel Dekker, New York, pp. 499–540.

Chapuis, R.P. (1990) Sand–bentonite liners: predicting permeability from laboratory tests, *Canadian Geotechnical Journal*, **27**(1), 47–57.

Cheng, S.C.J., Corcoran, G.T., Miller, C.J. and Lee, J.Y. (1994) The use of a spray elastomer for landfill cover liner applications, in *Proceedings of 5th International Conference on Geotextiles, Geomembranes and Related Products*, Singapore, pp. 1037–40.

Chian, E.S.K. (1977) Stability of organic matter in landfill leachates, *Water Research*, **11**, 225–32.

Choi, J.W. and Oscarson, D.W. (1991) Mass transport through compacted bentonite: effect of exchangeable cation, in *Proceedings 1st Canadian Conference on Environmental Geotechnics*, pp. 643–51.

Christensen, T.H. and Kjeldsen, P. (1989) Basic biochemical processes in landfills, in *Sanitary Landfilling: Process, Technology and Environmental Impact* (eds T.H. Christensen, R. Cossu, and R. Stegmann), Academic Press, Toronto, pp. 29–49.

Christensen, T.H., Bjerg, P.L., Rugge, K., Albrechtensen, H.J., Heron, G., Pedersen, J.K., Foverskov, A., Skov, B., Wurtz, S. and Refstrop, M. (1993) Attenuation of organic leachate pollutants in groundwater, in *Proceedings, 4th International Waste Management and Landfill Symposium*, Sardinia'93, Cagliari, Italy, pp. 1105–16.

Christensen, T.H., Kjeldsen, P., Albrechtsen, H.-J., Heron, G., Nielsen, P.H., Bjerg, P.L. and Holm P.E. (1994) Attenuation of landfill leachate pollutants in aquifers, *Critical Reviews in Environmental Science and Technology*, **24**(2), 119–202.

Christensen, T.H., Bjerg, P.L., Banwart, S.A., Jakobsen, R., Heron, G. and Albrechtsen, H.-J. (2000) Characterization of redox conditions in groundwater contaminant plumes – review article, *Journal of Contaminant Hydrology*, **45**, 165–241.

Christensen, T.H., Kjeldsen, P., Bjerg, P.L., Jensen, D.L., Christensen, J.B., Baun, A., Albrechtsen, H.-J. and Heron, G. (2001) Biogeochemistry of landfill leachate plumes, *Applied Geochemistry*, **16**, 659–718.

Clark, M.M. (1996) *Transport Modeling for Environmental Engineers and Scientist*, John Wiley & Sons, New York.

Collins, H.J. (1991) Influences of recycling household refuse upon sanitary landfills, in *Proceedings of 3rd International Landfill Symposium*, Cagliari, Italy, pp. 1111–23.

Collins, H.J. (1993) Impact of temperature inside the landfill on the behaviour of the barrier system, in *Proceedings of 4th International Landfill Symposium*, Cagliari, Italy, pp. 417–32.

Colmanetti, J.P. and Palmeira, E.M. (2002) A study on geotextile–leachate interaction by large laboratory tests, in *Proceedings 7th International Conference on Geosynthetics* (eds Nice, Ph. Delmas, and J.P. Gourc), A.A. Balkema, Lisse, Vol. 2, pp. 749–52.

Colucci, P. and Lavagnolo, M.C. (1995) Three years field experience in electrical control of synthetic landfill liners, in *Proceedings of 5th International Landfill Symposium*, S. Margherita di Pula, Cagliari, Italy, pp. 437–52.

Cooke, A.J., Rowe, R.K., Rittmann, B.E. and Fleming, I.R. (1999) Modelling biochemically driven mineral precipitation in anerobic biofilms, *Water Science and Technology*, **39**(7), 57–64.

Cooke, A.J., Rowe, R.K., Rittmann, B.E., VanGulck, J. and Millward, S. (2001) Evaluating the relationship between biofilm growth and mineral precipitation in column experiments conducted using synthetic leachate, *ASCE Journal of Geotechnical and Geoenvironmental Engineering*, **127**(10), 849–56.

Cooper, H.H. and Jacob, C.E. (1946) A generalized graphical method for evaluating formation constants and summarizing well field history, *Transactions of American Geophysics Union*, **27**, 526–34.

Corser, P., Pellicer, J. and Cranston, M. (1992) Observations on the long term performance of composite clay liners and covers, *Geotechnical Fabrics Report*, **10**(8), 6–16.

Crooks, V.E. and Quigley, R.M. (1984) Saline leachate migration through clay: a comparative laboratory and field investigation, *Canadian Geotechnical Journal*, **21**(2), 349–62.

Cunha, R.C.A., Camargo, O.A. and Kinjo, T. (1993) Application of three isotherms on the adsorption of zinc in oxisoils, alfisoils and ultisoils, in *Proceedings of the 12th Latinoamerican Congress on Soil Science*, Salamanca, Sociedad Española de la Ciencia del Suelo, Vol. 1, pp. 222–9.

Cussler, E.L. (1997) *Diffusion – Mass Transfer in Fluid Systems*, 2nd edn, Cambridge University Press, Cambridge, 580 pp.

Daniel, D.E. (1984) Predicting hydraulic conductivity of clay liners, *ASCE Journal of Geotechnical Engineering*, **110**(2), 285–300.

Daniel, D.E. (1989) In situ hydraulic conductivity tests for compacted clay, *ASCE Journal of Geotechnical Engineering*, **115**(9), 1205–26.

Daniel, D.E. (1998) Landfills for solid and liquid wastes, in *Proceedings of 3rd International Congress on Environmental Geotechnics*, Lisbon, A.A. Balkema, Rotterdam.

Daniel, D.E. (2000) Hydraulic durability of geosynthetic clay liners, in *Proceedings 14th GRI Conference (Hot Topics in Geosynthetics)*, Las Vegas, USA, pp. 118–35.

Daniel, D.E. and Koerner, R.M. (1993a) *Quality assurance and quality control of waste containment facilities*, US EPA/600/R-93/182, US Environmental Protection Agency, Washington, DC.

Daniel, D.E. and Koerner, R.M. (1993b) Final cover systems, *Geotechnical Aspects of Waste Disposal*, Chapman & Hall, London, pp. 225–72.

Daniel, D.E. and Koerner, R.M. (1995) *Waste Containment Facilities: Guidance for Construction, Quality Assurance and Quality Control of Liner and Cover Systems*, ASCE, New York.

Daniel, D. and Liljestrand, H.M. (1984) Effects of landfill leachates on natural clay liner systems, Report to Chemical Manufacturers Association, Washington, DC.

Daniel, D.E. and Shan, H.Y. (1992) *Effects of partial wetting of Gundseal on strength and hydrocarbon permeability*, Report submitted to Gundle Lining Systems, Inc., 56 pp.

Daniel, D.E. and Trautwein, S.J. (1986) Field permeability test for earthen liners, in *Proceedings In Situ '86 ASCE Specialty Conference*, Blacksburg, VA, June 1986, pp. 146–60.

Daniel, D.E., Trautwein, S.J., Boynton, S.S. and Foreman, D.E. (1984) Permeability testing with flexible-wall permeameters, *Geotechnical Testing Journal*, GTJODJ, **7**(3), 113–22.

Daniel, D.E., Shan, H.Y. and Anderson, J.D. (1993) Effects of partial wetting on the performance of the bentonite component of a geosynthetic clay liner, in *Proceedings, Geosynthetics'93*, Vancouver, BC, Canada, IFAI, St Paul, MN pp. 1483–96.

Daniel, D.E., Koerner, R.M., Bonaparte, R., Landreth, R.E., Carson, D.A. and Scranton, H.B. (1998) Slope stability of geosynthetic clay liner test plots, *ASCE Journal of Geotechnical and Geoenvironmental Engineering*, **124**(7), 628–37.

Darwish, M.I.M., Rowe, R.K., van der Maarel, J.R.C., Pel, L., Huinink, H. and Zitha, P.L.J. (2004) Contaminant containment using polymer gel barriers, *Canadian Geotechnical Journal*, **41**(1):106–117.

D'Astous, A.Y., Ruland, W.W., Bruce, J.R.G., Cherry, J.A. and Gillham, R.W. (1989) Fracture effects in the shallow groundwater zone in weathered Sarnia-area clay, *Canadian Geotechnical Journal*, **26**(1), 43–56.

Day, M.J. (1977) *Analysis of movement and hydrochemistry of groundwater in the fractured clay and till deposits of the Winnipeg area, Manitoba*, M.Sc. thesis, University of Waterloo, 210 pp.

Day, S.R. and Daniel, D.E. (1985) Hydraulic conductivity of two prototype clay liners. *ASCE Journal of Geotechnical Engineering*, **111**(8), 957–70. See also discussion in *Journal of Geotechnical Engineering*, 1987, **113**(7), 796–819.

Deipser, A. and Stegmann, R. (1994) The origin and fate of volatile trace components in municipal solid waste landfills, *Waste Management and Research*, **12**, 129–39.

Demetrocopoulos, A.C., Korfiatis, G.P., Bourodimos, E.L. and Nawy, E.G. (1984) Modelling for design of landfill bottom liners, *ASCE Journal of Environmental Engineering*, **110**(6), 1084.

Demirekler, E., Rowe, R.K. and Ünlü, K. (1999) Modelling leachate production from municipal solid waste landfills, in *Proceedings of 7th International Landfill Symposium*, S. Margherita di Pula, Cagliari, Sardinia, October, II, pp. 17–24.

Desaulniers, D.D., Cherry, J.A. and Fritz, P. (1981) Origin, age and movement of pore water in argillaceous quaternary deposits at four sites in Southwestern Ontario, *Journal of Hydrology*, **50**, 231–57.

de Smedt (1981) Solute transfer through unsaturated porous media, *Quality of Groundwater – Studies in Environmental Science*, **17**, 1011–16.

Deusto, I.A., Lopez, J.I. and Rodriguez Frutos, J.L. (1998) Assessment and influence of specific parameters on a high density, aerobic landfill, *Waste Management and Research Journal*, **16**, 574–81.

Dhar, A.S. and Moore, I.D. (2001) Liner buckling in profiled polyethylene pipes, *Geosynthetics International*, **8**(4), 303–26.

Didier, G., Bouazza, A. and Cazaux, D. (2000a) Gas permeability of geosynthetic clay liners, *Geotextiles and Geomembranes*, **18**(2–4), 235–50.

Didier, G., Al Nassar, M., Plagne, V. and Cazaux, D. (2000b) Evaluation of self healing ability of geosynthetic clay liners, *International Conference on Geotechnical and Geological Engineering*, Melbourne (CDROM).

Dittrich, J.P., Rowe, R.K., Becker, D.E. and Lo, K.Y. (2002) Analysis of the 1990s excavation of the Sarnia approach for the new St. Clair tunnel project, in *Proceedings of 55th Canadian Geotechnical Conference*, Niagara, October, pp. 299–306.

Doedens, H. and Cord-Landwehr, K. (1989) Leachate recirculation, in *Sanitary Land Filling: Process, Technology and Environmental Impact* (eds T.H. Christensen, R. Cossu and R. Stegmann), Academic Press, New York.

Döll, P. (1996) *Modeling of moisture movement under the influence of temperature gradients: desiccation of mineral liners below landfills*, Ph.D. thesis, Technical University of Berlin, Germany.

Döll, P. (1997) Desiccation of mineral liners below landfills with heat generation, *ASCE Journal of Geotechnical and Geoenvironmental Engineering*, **123**, 1001–9.

Domenico, P.A. and Schwartz, F.W. (1998) *Physical and Chemical Hydrogeology*, John Wiley & Sons, New York, 506 pp.

Donahue, R.B., Barbour, S.L. and Headley, J.V. (1999) Diffusion and adsorption of benzene in Regina clay, *Canadian Geotechnical Journal*, **36**, 430–42.

Dunn, R.J. and Mitchell, J.K. (1984) Fluid conductivity testing of fine-grained soils, *ASCE Journal of Geotechnical Engineering*, **110**(10), 1648–65.

Durin, L., Touze, N. and Duquennoi, C. (1998) Water and organic solvent transport parameters in geomembranes, in *Proceedings of 6th International Conference on Geosynthetics*, Atlanta, Industrial Fabrics Association International, **1**, 249–56.

Dutt, G.R. and Low, P.F. (1962) Diffusion of alkali chlorides in clay-water systems, *Soil Science*, **93**, 233–40.

Eberle, M.A. and von Maubeuge, K. (1997) Measuring the in-situ moisture content of geosynthetic clay liners (GCLs) using time domain reflectometry (TDR), in *Proceedings of 6th International Conference on Geosynthetics*, Atlanta, (ed. R.K. Rowe), Industrial Fabrics Association International, vol. 1, pp. 205–10.

Egloffstein, T.A. (1997) Ion Exchange in Geosynthetic Clay Liners. *Geotechnical Fabrics Report*, **15**(5), June/July 1997, St Paul, USA.

Egloffstein, T.A. (2001) Natural bentonites – influence of the ion exchange and partial desiccation on permeability and self-healing capacity of bentonites used in GCLs, *Geotextiles and Geomembranes*, **19**(7), 427–44.

Ehrig, H.-J. (1989) Leachate quality. Chapter 4.2 in *Sanitary Landfilling* (eds T.H. Christensen, R. Cossu and R. Stegmann), Academic Press, London.

Ehrig, H.-J. and Scheelhaase, T. (1993) Pollution potential and long term behaviour of sanitary landfills, in *Proceedings of 4th International Landfill Symposium*, Cagliari, pp. 1204–25.

Eichenauer, T. and Reither, P. (2002) Comparison of different shear tests for GCLs and the use of these data in designs, in *Proceedings of the International Symposium on Geosynthetic Clay Liners, Nuremberg*, April, pp. 141–50.

Eid, H.T. and Stark, T.D. (1997) Shear behavior of an unreinforced geosynthetic clay liner, *Geosynthetics International*, **4**(6), 645–59.

Eid, H.T., Stark, T.D. and Doerfler, C.K. (1999) Effect of shear displacement rate on internal shear strength of a reinforced geosynthetic clay liners, *Geosynthetics International*, **6**(3), 219–39.

Eid, H.T., Stark, T.D., Evans, W.D. and Sherry, P.E. (2000) Municipal solid waste slope failure. I. Waste and foundation soil properties, *ASCE Journal of Geotechnical and Geoenvironmental Engineering*, **126**(5), 397–407.

Eith, A.W. and Koerner, G.R. (1997) Assessment of HDPE geomembrane performance in a municipal waste landfill double liner system after eight years of service, *Geotextiles and Geomembranes*, **15**(4–6), 277–87.

Eloy-Giorni, C., Pelte, T., Pierson, P. and Margarita, R. (1996) Water diffusion through geomembranes under hydraulic pressure, *Geosynthetics International*, **3**(6), 741–69.

Elrick, D.E., Smiles, D.E., Baumgartner, N. and Groenevelt, P.H. (1976) Coupling phenomena in saturated homo-ionic and montmorillonite: I, *Soil Science Society America Proceedings*, **40**, 490–1.

Elsbury, B.R., Daniel, D.E., Sraders, G.A. and Anderson, D.C. (1990) Lessons learned from compacted clay liners, *ASCE Journal of Geotechnical Engineering*, **116**(11), 1641–60.

Environmental Protection Agency (1983) *Lining of waste impoundment and disposal facilities*, SW-870, Office of Solid Waste and Emergency Response, Washington, DC.

Erdelyi, A., Magnus, W., Oberhetting, F. and Tucomi, F.G. (1954) *Tables of Integral Transforms*, Vol. 1 McGraw-Hill, New York.

Estornell, P. and Daniel, D.E. (1992) Hydraulic conductivity of three geosynthetic clay liners, *ASCE Journal of Geotechnical Engineering*, **118**(10), 1592–606.

Estrin, D. and Rowe, R.K. (1995) Landfill design and the regulatory system, in *Proceedings 5th International Landfill Symposium*, Sardinia'95, Cagliari, Italy, Vol. 3, pp. 15–26.

Estrin, D. and Rowe, R.K. (1997) Legal liabilities of landfill design engineers and regulators, in *Proceedings of 6th International Landfill Symposium*, Sardinia'97, Cagliari, Italy, pp. 65–76.

Fannin, R.J., Choy, H.W. and Atwater, J.W. (1998) Interpretation of transmissivity data for geonets, *Geosynthetics International*, **5**(3), 265–84.

Feki, N., Garcin, P., Faure, Y.H., Gourc, J.P. and Berroir, G. (1997) Shear strength tests on geosynthetic clay liner systems, in *Proceedings Geosynthetics 97*, Long Beach, USA, Vol. 2, pp. 899–912.

Felon, R., Wilson, P.E. and Janssens, J. (1992) Résistance Mécanique des Membranes d'Étanchéité. Membranes Armées en Théorie et en Pratique, in *Vade Mecum Pour la Réalisation des Systèmes d'Étanchéité-Drainage Artificiels Pour les Sites d'Enfouissement Technique en Wallonie*. Journées d'Études ULg 92 (eds A. Monjoie, J.M. Rigo, Cl. Polo-Chiapolini and R. Degeimbre), Liege University, Belgium.

Fernandez, F. and Quigley, R.M. (1985) Hydraulic conductivity of natural clays permeated with simple liquid hydrocarbons, *Canadian Geotechnical Journal*, **22**, 205–14.

Fernandez, F. and Quigley, R.M. (1988a) Viscosity and dielectric constant controls on the hydraulic conductivity of clayey soils permeated with water-soluble organics, *Canadian Geotechnical Journal*, **25**, 582–9.

Fernandez, F. and Quigley, R.M. (1988b) Effects of increasing amounts of non-polar organic liquids in domestic waste leachate on the hydraulic conductivity of clay liners in southwestern Ontario, in *Proceedings Technology. Transfer Conference No. 8*, Session C, Ontario MOE, pp. 55–79.

Fernandez, F. and Quigley, R.M. (1991) Controlling the destructive effects of clay–organic liquid interactions by application of effective stresses, *Canadian Geotechnical Journal*, **28**(3), 388–98.

Fleming, I.R. (1999) *Biogeochemical processes and clogging of landfill leachate collection systems*, Ph.D. thesis, Department of Civil and Environmental Engineering, The University of Western Ontario, London, Ontario.

Fleming, I.R. and Rowe, R.K. (2000) Toward sustainable leachate treatment in landfills, in *Proceedings 6th Canadian Environmental Engineering Conference*, London, June, pp. 15–19.

Fleming, I.R., Rowe, R.K. and Cullimore, D.R. (1999) Field observations of clogging in a landfill leachate collection system, *Canadian Geotechnical Journal*, **36**(4), 289–96.

Flyhammer, P. (1997) Estimation of heavy metals transformation in municipal solid waste, *Science Total Environment*, **198**, 123.

Fontes, M.P.F., Matos, A.T., Costa, L.M. and Neves, J.C.L. (2000) Competitive adsorption of zinc, cadmium, copper, and lead in three highly-weathered Brazilian soil, *Commun. Soil Sci. Plant. Anal.*, **31**(17–18), 2939–58.

Foose, G.J., Benson, C.H. and Edil, T.B. (2001) Predicting leakage through composite landfill liners, *ASCE Journal of Geotechnical and Geoenvironmental Engineering*, **122**(9), 760–7.

Forchheimer, P. (1930) *Hydraulik*, 3rd edn, B.G. Teubner, Leipzig, Germany, 596 pp.

Foreman, D. and Daniel, D.E. (1986) Permeation of compacted clay with organic chemicals, *ASCE Journal of Geotechnical Engineering*, **112**, 669–81.

Foster, S.S.D. (1975) The chalk groundwater tritium anomaly – a possible explanation, *Journal of Hydrology*, **25**, 159–65.

Fourie, A.B., Kuchena, S.M. and Blight, G.E. (1994) Effect of biological clogging on the filtration capacity of geotextiles, in *Proceedings of 5th International Conference on Geosynthetics*, Singapore, pp. 721–24.

Fox, P.J., De Battista, D.J. and Chen, S.H. (1996) Bearing capacity of GCLs for cover soils of varying particle size, *Geosynthetics International*, **3**(4), 447–61.

Fox, P.J., Rowland, M.G. and Scheithe, J.R. (1998a) Internal shear strength of three geosynthetic clay liners, *ASCE Journal of Geotechnical and Geoenvironmental Engineering*, **124**(10), 933–44.

Fox, P.J., Triplett, E.J., Kim, R.H. and Olsta, J.T. (1998b) Field study of installation damage for geosynthetic clay liners, *Geosynthetics International*, **5**(5), 491–520.

Fox, P.J., De Battista, D.J. and Mast, D.G. (2000) Hydraulic performance of geosynthetic clay liners under gravel cover soils, *Geotextiles and Geomembranes*, **18**(2–4), 179–201.

Fredlund, D.G. and Rahardjo, H. (1993) *Soil Mechanics for Unsaturated Soils*, John Wiley & Sons, New York, NY.

Fredlund, D.G., Wilson, G.W. and Barbour, S.L. (2001) Unsaturated soil mechanics and property assessment, Chapter 5 of *Geotechnical and Geoenvironmental Engineering Handbook*, Kluwer Academic Publishing, Norwell, USA, pp. 107–46.

Freeze, R.A. and Cherry, J.A. (1979) *Groundwater*, Prentice-Hall, Englewood Cliffs, NJ.

Fried, J.J. (1976) *Groundwater Pollution*, Elsevier, New York.

Frind, E.O. (1987) Modelling of contaminant transport in groundwater: an overview, *The Canadian Society for Civil Engineering Centennial Symposium on Management of Waste Contamination of Groundwater*, Montreal, May 1987, 30 pp.

Frind, E.O. and Hokkanen, G.E. (1987) Simulation of the Borden plume using the Alternating Direction Galerkin Technique, *Water Resource Research*, **23**(5), 918–30.

Frind, E.O., Sudicky, E.A. and Schellenberg, S.L. (1987) Micro-scale modelling in the study of plume evolution in heterogeneous media, *Stochastic Hydrol. Hydraul.*, **1**, 263–79.

Fritz, P., Matthess, G. and Brown, R.M. (1976) Deuterium and oxygen-18 as indicators of leachwater movement from a sanitary landfill, in *Interpretation of Environmental Isotope and Hydrochemical Data in Groundwater Hydrology*, International Atomic Energy Agency, Vienna, pp. 131–42.

Fukuoka, M. (1986) Large scale permeability test for geomembrane-subgrade system, in *Proceedings of 3rd International Conference on Geotextiles*, Vienna, Austria, pp. 917–22.

Fullerton, D.S. (1980) Preliminary correlation of post-Erie interstadial events (16,000–10,000 radiocarbon years before present), Central and Eastern Great Lakes Region, Hudson, Champlain and St. Lawrence Lowlands United States and Canada. US GS Professional Paper 1089, 52 pp. accompanied by 2 charts.

Gabr, M.A. and Valero, S.N. (1995) Geotechnical properties of municipal solid waste. *Geotechnical Testing Journal*, **18**, 241–51.

Gachet, C., Gotteland, Ph., Lemarechal, D. and Prudhomme, B. (1998) An in-situ household refuse density measurement protocol, in *Proceedings 3rd International Congress on Environmental Geotechnics*, Lisbon, Portugal, pp. 849–54.

Gallagher, E.M., Darbyshire, W. and Warwick, R.G. (1999) Performance testing of landfill geoprotectors: background, critique, development, and current UK practice, *Geosynthetics International*, **6**(4), 283–301.

Gambelin, D.J. and Cochrane, D. (1998) Life cycle analysis of a bioreactor landfill in California, in *Proceedings 3rd Annual Landfill Symposium*, SWANA, Palm Beach Gardens, FL.

Garcin, P., Faure, Y.H., Gourc, J.P. and Purwanto, E. (1995) Behaviour of geosynthetic clay liner (GCL): laboratory tests, in *Proceedings 5th International Symposium on Landfills*, Cagliari, Italy, Vol. 1, pp. 347–58.

Gardner, W.R. (1958) Some steady-state solutions of the unsaturated moisture flow equation with applications to evaporation from a water table, *Soil Science*, **85**(4), 228–32.

Gartner Lee and Associates Ltd (1986) *Burlington Regional Landfill Plume Delineation Report for the Regional Municipality of Halton*. See also: Halton Regional Landfill – Burlington. 1986 Monitoring Report.

Gartner Lee Ltd (1988) *Halton Regional Landfill – Burlington 1987 Monitoring Report* GLL87–224, Report for the Regional Municipality of Halton, March 1988.

Gartung, E., Prühs, H. and Hoch, A. (1989) Design of vertical shafts in landfills, in *Proceedings Sardinia 89 – 2nd International Landfill Symposium*, Cagliari, Italy, pp. C.XI-1–C.XI-8.

Gartung, E., Prühs, H. and Nowack, F. (1993) Measurements on vertical shafts in landfills, in *Proceedings 4th International Landfill Symposium*, Cagliari, pp. 461–8.

Gaudet, J.P., Jegat, H., vachand, G. and Wierenga, P.J. (1977) Solute transfer, with exchange between mobile and stagnant water, through unsaturated sand, *Soil Science Society of America Journal*, **41**, 665–71.

Gelhar, L.W. and Axness, C.L. (1983) Three-dimensional stochastic analysis of macrodispersion in aquifers, *Water Resources Research*, **15**(6), 1387–97.

Gelhar, L.W., Mantoglou, A., Welty, C. and Rehfeldt, K.R. (1985) *A review of field-scale physical solute transport processes in saturated and unsaturated porous media*, Electric Power Research Institute EPRI EA-4190 Project 2485–5, 116 pp.

Gelhar, L.W., Welty, C. and Rehfeldt, K.R. (1992) A critical review of data on field-scale dispersion in aquifers, *Water Resources Research*, **28**(7), 1955–74.

Gerhardt, R.A. (1984) Landfill leachate migration and attenuation in the unsaturated zone in layered and nonlayered coarse-grained soils, *Groundwater Monitoring Review*, **4**(2), 56–65.

Gilbert, R.B., Fernandez, F. and Horsfield, D.W. (1996) Shear strength of reinforced geosynthetic clay liner, *ASCE Journal of Geotechnical and Geoenvironmental Engineering*, **122**(4), 259–66.

Gillham, R.W. and Cherry, J.A. (1982) Contaminant migration in saturated unconsolidated geologic deposits, *Geophysical Society of America, Special Paper*, **189**, 31–62.

Gillham, R.W., Robin, M.J.L., Dytynyshyn, D.J. and Johnston, H.M. (1984) Diffusion of nonreactive and reactive solutes through fine-grained barrier materials, *Canadian Geotechnical Journal*, **21**, 541–50.

Giroud, J.P. (1995) Wrinkle management for polyethylene geomembranes requires active approach, *Geotechnical Fabrics Report*, **13**(3), 14–17.

Giroud, J.P. (1996a) Granular filters and geotextile filters, in *Proceedings of GeoFilters'96* (eds J. Lafleur and A.L. Rollin), Montréal, Canada, May 1996, pp. 565–680.

Giroud, J.P. (1996b) Workshop on testing of geosynthetic materials for landfill liners and covers, *IGS News*, **12**(2), 3–4.

Giroud, J.P. (1997) Equations for calculating the rate of liquid migration through composite liners due to geomembrane defects, *Geosynthetics International*, **4**(3–4), 335–48.

Giroud, J.P. and Beech, J.F. (1989) Stability of soil layers on geosynthetic lining systems, in *Proceedings of Geosynthetics'89*, pp. 35–46.

Giroud, J.-P. and Bonaparte, R. (1989a) Leakage through liners constructed with geomembranes. Part I. Geomembrane liners, *Geotextiles and Geomembranes*, **8**, 27–68.

Giroud, J.-P. and Bonaparte, R. (1989b) Leakage through liners constructed with geomembranes. Part

II. Composite liners, *Geotextiles and Geomembranes*, **8**, 71–112.

Giroud, J.P. and Bonaparte, R. (2001) Geosynthetics in liquid-containing structures, Chapter 26 of *Geotechnical and Geoenvironmental Engineering Handbook* (ed. R.K. Rowe), Kluwer Academic Publishers, Norwell, USA, pp. 789–824.

Giroud, J.P. and Houlihan, M.F. (1995) Design of leachate collection layers, in *Proceedings of 5th International Landfill Symposium*, S. Margherita di Pula, Cagliari, Italy, Vol. 2, pp. 613–40.

Giroud, J.P. and Morel, N. (1992) Analysis of geomembrane wrinkles, *Geotextiles and Geomembranes*, **11**(3), 255–76 (Erratum: **12**(4), 378).

Giroud, J.P. and Peggs, I.D. (1990) Geomembrane construction quality assurance, Waste Containment Systems: Construction, Regulation, and Performance, *ASCE Geotechnical Special Publication No. 26* (ed. R. Bonaparte), pp. 190–225.

Giroud, J.P. and Soderman, K.L. (1995) Design of structures connected to geomembranes, *Geosynthetics International*, **2**(2), 379–428.

Giroud, J.-P and Soderman, K.L. (2000) Criterion for acceptable bentonite loss from a GCL incorporated into a liner system, *Geosynthetics International, Special Issue on Liquid Collection Systems*, **7**(4–6), 529–81.

Giroud, J.P., Badu-Tweneboah, K. and Bonaparte, R. (1992) Rate of leakage through a composite liner due to geomembrane defects, *Geotextiles and Geomembranes*, **11**(1), 1–29.

Giroud, J.P., Bachus, R.C. and Bonaparte, R. (1995a) Influence of water flow on the stability of geosynthetic–soil layered systems on slopes, *Geosynthetics International*, **2**(6), 1149–80.

Giroud, J.P., Badu-Tweneboah, K. and Soderman, K.L. (1995b) Theoretical analysis of geomembrane puncture, *Geosynthetics International*, **2**(6), 1019–48.

Giroud, J.P., Badu-Tweneboah, K. and Soderman, K.L. (1997a) Comparison of leachate flow through compacted clay and geosynthetic clay liners in landfill liner systems, *Geosynthetics International*, **4**(3–4), 391–431.

Giroud, J.P., Khire, M.V. and Soderman, K.L. (1997b) Liquid migration through defects in a geomembrane overlain and underlain by permeable media, *Geosynthetics International*, **4**(3–4), 293–321.

Giroud, J.P., Palmer, B. and Dove, J.E. (2000) Calculation of flow velocity in pipes as a function of flow rate, *Geosynthetics International*, **7**(4–6), 583–600.

Giroud, J.P., Thiel, R.S., Kavazanjian, E. and Lauro, F.J. (2002) Hydrated area of bentonite layer encapsulated between two geomembranes, in *Proceedings of 7th International Conference on Geosynthetics*, Nice, September, pp. 827–32.

Gleason, M., Daniel, D. and Eykholt, G. (1997) Calcium and sodium bentonite for hydraulic containment applications, *ASCE Journal of Geotechnical and Geoenvironmental Engineering*, **123**(5), 438–45.

Goldman, L.J., Greenfield, L.I., Damle, A.S., Kingsbury, G.L., Northeim, C.M. and Truesdale, R.S. (1990) *Clay Liners for Waste Management Facilities, Design, Construction and Evaluation*, Noyes Data Corporation for US EPA, New Jersey, USA, 524 pp.

Gomes, M.P.F., Gomes, P.C., Silva, A.G., Mendonça, E.S. and Neto, A.R. (2001) Selectivity sequence and competitive adsorption of heavy metals by Brazilian soils, *Soil Science Society of America Journal*, **65**, 1115–21.

Goodall, D.E. and Quigley, R.M. (1977) Pollutant migration from two sanitary landfill sites near Sarnia, Ontario, *Canadian Geotechnical Journal*, **14**, 223–36.

Gordon, M.E., Huebner, P.M. and Miazga, T.J. (1989) Hydraulic conductivity of three landfill clay liners, *ASCE Journal of Geotechnical Engineering*, **115**(8), 1148–62.

Grant, B.L., McKenna, R.W. and von Maubeuge, K.P. (1997) Investigations of liner leakage rates and EPA guidelines, *Environmental Geotechnics*, A.A. Balkema, Rotterdam, pp. 315–24.

Grassie, N. and Scott, G. (1985) *Polymer Degradation and Stabilization*, Cambridge University Press, New York, 222 pp.

Green, W.J., Lee, G.F. and Jones, R.A. (1981) Clay–soils permeability and hazardous waste storage, *Journal of the Water Pollution Control Federation*, **53**(8), 1347–54.

Griffin, R.A. and Shimp, N.F. (1978) *Attenuation of pollutants in municipal landfill leachate by clay minerals. Part 2. Heavy metal adsorption*, Municipal Environmental Research Lab., US EPA, Report 600/14, Cincinnati, OH.

Griffin, R.A., Cartwright, K., Shimp, N.F., Steel, J.D., Ruch, R.R., White, W.A., Hughes, G.M. and Gilkeson, R.H. (1976) Attenuation of pollutants in municipal landfill leachate by clay minerals. Part I. Column leaching and field verification, Illinois State Geological Survey, Environmental Geology Notes, No. 78.

Griffin, R.A., Sack, W.A., Roy, W.R., Ainsworth, C.C. and Krapac, I.G. (1985) Batch-type 24-hour distribution ratio for contaminant adsorption by soil materials, in *Hazardous and Industrial Solid Waste Testing and Disposal* (eds D. Lorenzen *et al.*), ASTM, STP 933, Vol. 6, pp. 390–408.

Griffin, R.A., Sack, W.A., Roy, W.R., Ainsworth, C.C. and Krapac, I.G. (1986) Batch type 24-h distribution ratio for contaminant adsorption by soil materials, *Hazardous and Industrial Solid Waste Testing and Disposal*, ASTM Special Technical Publication 933, pp. 390–408.

Grim, R.E. (1953) *Clay Mineralogy*, McGraw-Hill, New York, 384 pp.

Grim, R.E. (1962) *Applied Clay Mineralogy*, McGraw-Hill, New York, 422 pp.

Grisak, G.E. and Pickens, J.F. (1980) Solute transport through fractured media. 1. The effect of matrix diffusion, *Water Resources Research*, **16**(4), 719–30.

Grisak, G.E., Pickens, J.F. and Cherry, J.A. (1980) Solute transport through fractured media. 2. Column study of fractured till, *Water Resources Research*, **16**, 731–9.

Gross, B.A., Bonaparte, R. and Giroud, J.P. (1990) Evaluation of flow from landfill leachate detection layers, in *Proceedings of 4th International Conference on Geotextiles and Geomembranes*, The Hague, pp. 481–6.

Grouch, S.L. and Starfield, A.M. (1983) *Boundary Element Methods in Solid Mechanics*, Allen and Unwin, London.

Gschwend, P.M. and Wu, S. (1985) On the constancy of sediment-water partition coefficients of hydrophobic pollutants, *Journal of Environmental Science and Technology*, **19**, 90–6.

Gullick, R.W. (1998) *Effects of sorbent addition on the transport of inorganic and organic chemicals in soil–bentonite cutoff wall containment barriers*, Ph.D. thesis, Department of Environmental Engineering, The University of Michigan, Ann Arbor, MI, 372 pp.

Gvirtzman, H. and Gorelick, S.M. (1991) Dispersion and advection in unsaturated porous media enhanced by anion exclusion, *Nature*, **352**(8), 793–5.

Ham, R.K. and Bookter, T.J. (1982) Decomposition of solid waste in test lysimeters, *ASCE Journal of Environmental Engineering*, **108**(6), 1147–70.

Ham, R.K., Hekimian, K., Ketter, S., Lockman, W.J., Lofty, R.J., McFaddin, D.E. and Daley, E.J. (1979) *Recovery, processing and utilization of gas from sanitary landfills*, EPA-600/2-79-001. US Environmental Protection Agency, Cincinnati, OH, 133 pp.

Hammoud, A. (1989) *A theoretical examination of containment in regularly fractured media*, M.E.Sc. thesis, The University of Western Ontario, London, Canada.

Hantush, M.S. (1956) Analysis of data from pumping tests in leaky aquifers, *Trans. Amer. Geophys. Union*, **37**, 702–14.

Hantush, M.S. (1960) Modification of the theory of leaky aquifers, *Journal of Geophysical Research*, **65**, 3713–25.

Harpur, W.A., Wilson-Fahmy, R.F. and Koerner, R.M. (1993) Evaluation of the contact between geosynthetic clay liners and geomembranes in terms of transmissivity, in *Proceedings GRI Seminar on Geosynthetic Liner Systems* (eds R.M. Koerner and R.F. Wilson-Fahmy), Philadelphia, PA, Industrial Fabrics Association International, pp. 143–54.

Harr, M.E. (1962) *Groundwater and Seepage*, McGraw-Hill, New York, pp. 43–4.

Haxo, H.E., Jr. and Haxo, P.D. (1988) *Consensus report of the ad hoc meeting on the service life in landfill environments of flexible membrane liners and other synthetic polymeric materials of construction*, Matrecon, Inc., Alameda, CA, Report to USEPA.

Haxo, H.E., Jr. and Lahey, T. (1988) Transport of dissolved organics from dilute aqueous solutions through flexible membrane liner, *Hazardous Waste and Hazardous Materials*, **5**, 275–94.

Headley, J.V., Boldt-Leppin, B.E.J., Haug, M.D. and Peng, J. (2001) Determination of diffusion and adsorption coefficients for volatile organics in an organophilic clay–sand–bentonite liner, *Canadian Geotechnical Journal*, **38**, 809–17.

Health Canada (2002) *Guidelines for Canadian Drinking Water Quality*, Federal–Provincial–Territorial Committee on Drinking Water.

Heerten, G. (2000) Geosynthetic clay liners in geotechnical application, *Workshop II–GCL Durability and Lifetime. GRI-14 Conference*, Las Vegas 2000.

Heerten, G., von Maubeuge, K., Simpson, M. and Mills, C. (1993) Manufacturing quality control of geosynthetic clay liners – a manufacturers perspective, in *Proceedings of 6th GRI Seminar, MQC/MQA and CQC/CQA of Geosynthetics*, St Paul, MN:IFAI, pp. 86–95.

Heerten, G., Saathoff, F., Scheu, C. and von Maubeuge, K.P. (1995) On the long-term shear behavior of geosynthetic clay liners (GCLs) in capping sealing systems, in *Proceedings of the International Symposium Geosynthetic Clay Liners*, Nuremberg, April, pp. 141–50.

Heibrock, G. (1997) Desiccation cracking of mineral sealing liners, in *Proceedings of 6th International Landfill Symposium*, Cagliari, Italy, Vol. 3, pp. 101–13.

Heilshorn, E.D. (1991) Removing VOCs from contaminated water, *Chemical Engineering*, **98**(2), 120–4.

Hendron, D.M., Fernandez, G., Prommer, P.J., Giroud, J.P. and Orozco, L.F. (1999) Investigation of the cause of the 27 September 1997 slope failure at the Dona Juana landfill, in *Proceedings of 7th International Landfill Symposium*, Cagliari, Italy, Vol. 3, pp. 545–54.

Herman, J.S. and White, W.B. (1985) Dissolution kinetics of dolomite: effects of lithology and fluid flow velocity, *Geochimica et Cosmochimica Acta*, **49**, 2017–26.

Herzog, B.L. and Morse, W.J. (1986) Hydraulic conductivity at a hazardous waste disposal site: comparison of laboratory and field determined values, *Waste Management and Research*, Vol. 4, Academic Press, New York, pp. 177–87.

Herzog, B.L., Griffin, R.A., Stohr, C.J., Follmer, L.R., Morse, W.J. and Su, W.J. (1989) Investigation of failure mechanisms and migration of organic contaminant at Wilsonville, Illinois, *Spring 1989 GWMR*, pp. 82–9.

Hewetson, J.P. (1985) *An investigation of the groundwater zone in fractured shale at a landfill*, M.E.Sc. thesis, The University of Waterloo.

Hewitt, R. and Daniel, D. (1997) Hydraulic conductivity of GCLs after freeze-thaw, *ASCE Journal of Geotechnical and Geoenvironmental Engineering*, **123**, 305–13.

Heyer, K.U. *et al.* (1997) Langfristiges Gefahrdungspontential und Deponieverhalten von Ablagerungen, Bericht zum Teilvorhaben, TV4 im *BMBF-Verbundvorhaben Deponiekorper, Forderkennzeichen BMBF*: 1460799 D3.

Hillaire-Marcel, C.M. (1979) *Les mers post-glacieres au Quebec: Quelques aspects*, Doctorat d'Etat thesis, Université Pierre et Marie Curie, Paris, France, 2 volumes.

Hoeg, K. (1968) Stresses against underground structural cylinders, *ASCE Journal of the Soil Mechanics and Foundations Division*, **94**(4), 833–58.

Hoek, E. and Bray, J.W. (1981) *Rock Slope Engineering*, 3rd edn, The Institute of Mining and Metallurgy, London, 133 pp.

Holzlöhner, U. (1989) Moisture behaviour of soil liners and subsoil beneath landfills, in *Proceedings of 2nd International Landfill Symposium*, Porto Conte, Vol. 1, pp. X1–X11.

Holzlöhner, U. (1995) Moisture balance, risk of desiccation in earthen liners, *State of the Art Report: Landfill Liner Systems* (eds U. Holzlöhner, H. August, T. Meggyes and M. Brune), Penshaw Press, pp. H1–22.

Howard, A.K. (1977) Modulus of soil reaction values for buried flexible pipe, *ASCE Journal of Geotechnical Engineering*, **103**(1), 33–43.

Howard, P.H. (1989) *Handbook of Environmental Fate and Exposure Data for Organic Chemicals*, Lewis Publishers, Michigan.

Hrapovic, L. (2001) *Laboratory study of intrinsic degradation of organic pollutants in compacted clayey soil*, Ph.D. thesis, The University of Western Ontario, 300 pp.

Hsuan, Y.G. (1999) Database of field incidents used to establish HDPE geomembrane stress crack resistance specifications, *Geotextiles and Geomembranes*, **18**(1), 1–22.

Hsuan, Y.G. (2000a) Laboratory index and performance tests of polyolefin geomembranes, in *Proceedings of EuroGeo 2000*, Bologna, Italy, pp. 132–47.

Hsuan, Y.G. (2000b) Database of field incidents used to establish HDPE geomembrane stress crack resistance specifications, *Geotextiles and Geomembranes*, **18**, 1–22.

Hsuan, Y.G. and Koerner, R.M. (1995) Long term durability of HDPE geomembranes. Part I. Depletion of antioxidants, *GRI Report* **16**, 35.

Hsuan, Y.G. and Koerner, R.M. (1997) GRI finalizes its first specification, *Geotechnical Fabrics Report* **15**(5), 17–20.

Hsuan, Y.G. and Koerner, R.M. (1998) Antioxidant depletion lifetime in high density polyethylene geomembranes, *ASCE Journal of Geotechnical and Geoenvironmental Engineering*, **124**(6), 532–41.

Hsuan, Y. and Koerner, R.M. (2002) Durability and lifetime of polymers fibres with respect to reinforced geosynthetic clay barriers – reinforced GCLs, in *Proceedings of the International Symposium Geosynthetic Clay Liners*, Nuremberg, April, pp. 73–86.

Hsuan, Y.G., Lord, A.E., Jr. and Koerner, R.M. (1991) Effects of outdoor exposure on high density polyethylene geomembrane, *Geosynthetics'91 Conference Proceedings*, Atlanta, USA, IFIA, pp. 287–302.

Hsuan, Y.G., Koerner, R.M. and Lord, A.E., Jr. (1993) Stress cracking resistance of high density polyethylene geomembranes, *ASCE Journal of Geotechnical Engineering*, **11**, 1840–55.

Hudson, A.P., Beaven, R. and Powrie, W. (1999) Measurement of the horizontal hydraulic conductivity of household waste in a large scale compression cell, in *Proceedings of 7th International Landfill Symposium*, Cagliari, Italy, Vol. 3, pp. 461–68.

Hughes, J.W. and Monteleone, M.J. (1987) Geomembrane/synthesized leachate compatibility testing, in *Geotechnical and Geohydrological Aspects of Waste Management* (eds Van Zyl *et al.*), Fort Collins, Co, pp. 35–50.

Hughes, G.M., Landon, R.A. and Farvolden, R.N. (1971) *Hydrogeology of solid waste disposal sites in northeastern Illinois*, Report SW-12d, US Environmental Protection Agency.

Huyakorn, P.S., Lester, B.H. and Mercer, J.W. (1983) An efficient finite element technique for modelling transport in fractured porous media. 1. Single species transport, *Water Resources Research*, **19**(3), 841–54.

Hvorslev, M.J. (1951) *Time lag and soil permeability in groundwater observations*, United States Army, Corps of Engineers, Waterways Experiment Station, Vicksburg, Miss., Bulletin 36.

Hwu, B., Koerner, R.M. and Sprague, C.J. (1990) Geotextile intrusion into geonets, in *Proceedings of 4th International Conference on Geotextiles*, The Hague, pp. 351–6.

James, A.N., Fullerton, D. and Eykolt, G.R. (1997) Field performance of GCL under ion exchange conditions, *ASCE Journal of Geotechnical and Geoenvironmental Engineering*, **123**(10), 897–902.

Jayawickrama, P., Brown, K.W., Thomas, J.C. and Lytton, R.L. (1988) Leakage rates through flaws in membrane liners, *ASCE Journal of Environmental Engineering*, **114**(6), 1401–19.

Jefferis, S.A. and Bath, A. (1999) Rationalising the debate on calcium carbonate clogging and dissolution in landfill drainage materials, in *Proceedings of the 7th International Landfill Symposium*, S. Margherita di Pula, Cagliari, Italy, pp. 277–84.

Jewell, R.A. (1991) Application of revised design charts for steep reinforced slopes, *Geotextiles and Geomembranes*, **10**, 203–33.

Ji, G.L. (1997) Electrostatic adsorption of anions, in *Chemistry of Variable Charge Soils* (ed. T.R. Yu), Oxford University Press, New York, pp. 112–37.

Jo, H.Y., Katsumi, T., Benson, C.H. and Edil, T.B. (2001) Hydraulic conductivity and swelling of non-prehydrated GCLs permeated with single species salt solutions, *ASCE Journal of Geotechnical and Geoenvironmental Engineering*, **127**(7), 557–67.

Jones-Lee, A. and Lee, G.F. (2000) Appropriate use of MSW leachate recycling in municipal solid waste landfilling, in *Proceedings Air and Waste Management Association 93rd National Annual Meeting*, CD-ROM Paper 00–455, Pittsburgh, PA.

Kalbe, U., Müller, W., Berger, W. and Eckardt, J. (2002) Transport of organic contaminants within composite liner systems, *Applied Clay Science*, **21**, 67–76.

Karickhoff, S.W., Brown, D.S. and Scott, T.A. (1979) Sorption on hydrophobic pollutants on natural sediments, *Water Resources Research*, **13**, 241–8.

Karimi, A.A., Farmer, W.J. and Cliath, J.C. (1987) Vapor-phase diffusion of benzene in soil, *Journal of Environmental Quality*, **16**(1), 38–43.

Kau, P.M.H., Smith, D.W. and Binning, P. (1997) Fluoride retention by kaolin clay liners in waste storage, *Journal of Contaminant Hydrology*, **28**, 267–88.

Kemper, W.D. and Van Schaik, J.C. (1966) Diffusion of salts in clay–water systems, *Soil Science Society American Proceedings*, **30**, 534–40.

Khandelwal, A., Rabideau, A.J. and Shen, P. (1998) Analysis of diffusion and sorption of organic solutes in soil–bentonite barrier materials, *Environmental Science and Technology*, **32**, 1333–9.

Kim, J.Y., Edil, T.B. and Park, J.K. (1997) Effective porosity and seepage velocity in column tests on compacted clay, *ASCE Journal of Geotechnical and Geoenvironmental Engineering*, **123**(12), 1135–42.

Kim, J.Y., Edil, T.B. and Park, J.K. (2001) Volatile organic compound (VOC) transport through compacted clay, *ASCE Journal of Geotechnical and Geoenvironmental Engineering*, **127**(2), 126–34.

Kimani Njoroge, B.N., Ball, W.P. and Cherry, R.S. (1998) Sorption of 1,2,4-trichlorobenzene and tetrachloroethylene within an authigenic soil profile: changes in K_{oc} with soil depth, *Journal of Contaminant Hydrology*, **29**(4), 347–77.

King, K.S., Quigley, R.M., Fernandez, F., Reades, D.W. and Bacopoulos, A. (1993) Hydraulic conductivity and diffusion monitoring of the Keele valley landfill liner, Maple, Ontario, *Canadian Geotechnical Journal*, **30**(1), 124–34.

Kjeldsen, P., Barlaz, M., Rooker, A.P., Baun, A., Ledin, A. and Christensen, H. (2002) Present and long-term composition of MSW landfill leachate: a review, *Critical Reviews in Environmental Science and Technology*, **32**(4), 297–336.

Klein, R., Baumann, T., Kahapka, E. and Niessner, R. (2001) Temperature development in a modern municipal solid waste incineration (MSWI) bottom ash landfill with regard to sustainable waste management, *Journal of Hazardous Materials*, **B83**, 265–80.

Klute, A. and Letey, J. (1958) The dependence of ionic diffusion on the moisture content of nonadsorbing porous media, *Soil Science Society American Proceedings*, **22**, 213–15.

Knochenmus, G., Wojnarowicz, M. and Van Impe, W.F. (1998) Stability of municipal solid wastes, in *Proceedings of 3rd International Congress on Environmental Geotechnics*, Lisboa, Portugal, pp. 977–1000.

Koch, R., Gaube, E., Hessel, J., Gondro, C. and Heil, H. (1988) Langzeitfestigkeit von deponiedichtungsbahnen aus polyethylene, *Mull und Abfall*, **8**, 348–61.

Kodikara, J. (2000) Analysis of tension developed in geomembranes placed on landfill slopes, *Geotextiles and Geomembranes*, **18**, 47–61.

Kodikara, J.K., Rahman, F. and Barbour, S.L. (2002) Towards a more rational approach to chemical compatibility testing of clay, *Geotechnical Journal*, **39**(3), 597–607.

Koerner, G. (2001) In situ temperature monitoring of geosynthetics used in a landfill, *Geotechnical Fabrics Report May 2001*, pp. 12–13.

Koerner, R.M. (1998) *Designing with Geosynthetics*, 4th edn, Prentice-Hall, Englewood Cliffs, NJ.

Koerner, R.M. and Daniel, D.E. (1997) *Final Covers for Solid Waste Landfills and Abandoned Dumps*, American Society of Civil Engineers, Reston, VA.

Koerner, R.M. and Hwu, B. (1989) Behaviour of double geonet drainage systems, *Geotechnical Fabrics Report*, IFAI Publ., pp. 39–44.

Koerner, G.R. and Koerner, R.M. (1989) Biological clogging of leachate collection systems, in *Durability and Aging of Geosynthetics* (ed. R.M. Koerner), Elsevier Applied Science, London, pp. 260–77.

Koerner, G.R. and Koerner, R.M. (1990) Biological activity and potential remediation involving geotextile landfill leachate filters, Geosynthetic Testing for Waste Containment Applications, ASTM STP 1081 (ed. R.M. Koerner), American Society for Testing and Materials, Philadelphia, PA, pp. 313–35.

Koerner, G.R. and Koerner, R.M. (1995a) Temperature behaviour of field deployed HDPE geomembranes, in *Proceedings Geosynthetics'95*, Nashville, pp. 921–37.

Koerner, R.M. and Koerner, G.R. (1995b) *Leachate clogging assessment of geotextile (and soil) landfill filters*, US EPA Report, CR-819371, March.

Koerner, R.M and Narejo, D. (1995) Bearing capacity of hydrated geosynthetic clay liner, ASCE *Journal of Geotechnical Engineering*, **121**(1), 82–5.

Koerner, R.M. and Soong, T.Y. (1998) Analysis and design of veneer cover soils, in *Proceedings of 6th International Conference on Geosynthetics*, Atlanta,

Industrial Fabric Association International, Vol. 1, pp. 1–24.

Koerner, R.M., Bove, J.A. and Martin, J.P. (1984) Water and air transmissivity of geotextiles, *Geotextiles and Geomembranes*, **1**, 57–73.

Koerner, R.M., Lucianai, V.A., Freese, J.S. and Carroll, R.G., Jr. (1986) Prefabricated drainage composites: evaluation and design guidelines, in *Proceedings of 3rd International Conference on Geotextiles*, Vienna, pp. 551–6.

Koerner, R.M., Halse, Y.H. and Lord, A.E., Jr. (1990) Long-term durability and aging of geomembranes, in *Waste Containment Systems: Construction, Regulation, and Performance* (ed. R. Bonaparte) ASCE Spec. Tech. Publ. No. 26, New York, pp. 106–34.

Koerner, G.R., Koerner, R.M. and Martin, J.P. (1993) Field performance of leachate collection systems and design implications, in *Proceedings of 31st Annual SWANA Conference*, San Jose, CA, pp. 365–80.

Koerner, G.R., Koerner, R.M. and Martin, J.P. (1994) Geotextile filters used for leachate collection system: testing, design and field behavior, *ASCE Journal of Geotechnical Engineering Division*, **120**(10), 1792–803.

Koerner, G.R., Yazdani, R. and Mackey, R.E. (1996a) Long term temperature monitoring of landfill geomembranes, in *Proceedings SWANA 1st Annual Landfill Symposium*, Publication, pp. 61–73.

Koerner, R.M., Wilson-Fahmy, R.F. and Narejo, D. (1996b) Puncture protection of geomembranes. Part III. Examples, *Geosynthetics International*, **3**(5), 655–75.

Koerner, R.M., Soong, T-Y., Koerner, G.R. and Gontar, A. (2001) Creep testing and data extrapolation of reinforced GCLs, *Geotextiles and Geomembranes*, **19**(7), 412–25.

Kraus, J.F., Benson, C.H., Erickson, A.E. and Chamberlain, E.J. (1997) Freeze thaw cycling and hydraulic conductivity of bentonitic barriers, *ASCE Journal of Geotechnical and Geoenvironmental Engineering*, **123**(3), 229–38.

Krol, M.M. and Rowe, R.K. (2004) Diffusion of TCE through soil–bentonite slurry walls, *Soil and Sediment Contamination* 13 (in press).

Kruempelbeck, I. and Ehrig, H.J. (1999) Long-term behaviour of municipal solid waste landfills in Germany, in *Proceedings of the 7th International Landfill Symposium*, S. Margherita di Pula, Cagliari, Italy, October 1999, pp. 27–36.

Kruse, K. (1994) *Langfristiges emissionsgeschehen von siedlungsabfalldeponiewn*, Institut fur Siedlungswasserwirtschaft der TU Braunschweig, Heft 54.

Krushelnitzky, R. and Brachman, R.W.I. (2002) Effect of backfill configuration on the structural response of leachate collection pipes, in *Proceedings of 55th Canadian Geotechnical Conference*, Niagara Falls, Ont., pp. 557–64.

Krushelnitzky, R. and Brachman, R.W.I. (2003) Large scale laboratory test of a leachate collection pipe in a trench, in *Proceedings of 56th Canadian Geotechnical Conference*, Winnipeg MB, pp. 536–41.

LaGatta, M.D., Boardman, B.T., Cooley, B.H. and Daniel, D.E. (1997) Geosynthetic clay liners subjected to differential settlement, *ASCE Journal of Geotechnical and Geoenvironmental Engineering*, **123**(5), 402–10.

Lai, T.M. and Mortland, M.M. (1962) Self-diffusion of exchangeable cations in bentonite. Clays and Clay Minerals, in *9th Conference*, Pergamon Press, New York, pp. 229–47.

Laine, D.L. and Darilek, G.T. (1993) Locating leaks in geomembrane liners of landfills covered with a protective soil, in *Proceedings Geosynthetics'93*, Vancouver, Canada, pp. 1403–12.

Lake, C.B. (2000) *Contaminant transport through geosynthetic clay liners and composite liner system*, Ph.D. thesis, The University of Western Ontario, Ontario, Canada, 408 pp.

Lake, C.B. and Rowe, R.K. (2000a) Swelling characteristics of needle-punched, thermally treated GCLs, *Geotextiles and Geomembranes*, **18**(2), 77–102.

Lake, C.B. and Rowe, R.K. (2000b) Diffusion of sodium and chloride through geosynthetic clay liners, *Geotextiles and Geomembranes*, **18**(2), 102–32.

Lake, C.B. and Rowe, R.K. (2004) Volatile organic compound migration through a GCL, *Geosynthetics International* (in press).

Lambe, T.W. (1956) The storage of oil in an earth reservoir, *Journal of Boston Society of Civil Engineers*, **43**, 179–241.

Lambe, T.W. (1958) The engineering behaviour of compacted clay, *Journal of Soil Mechanics and Foundation Division, ASCE*, **84**, (SM2) Paper 1655, 35 pp.

Lambe, T.W. (1960) Compacted clay – a symposium, *Trans. ASCE*, **125**(I), 681–756 (Two papers – Compacted clay: Structure; and Compacted Clay: Engineering Behaviour plus discussions).

Lambe, T.W. and Martin, R.T. (1953, 1954, 1955, 1956, 1957) Composition and engineering properties of soil I, II, III, IV, V, in *Proceedings 32, 33, 34, 35 and 36 Annual Meetings of the Highway Research Board*, USA.

Lambe, T.W. and Whitman, R.E. (1979) *Soil Mechanics*, SI Version. John Wiley & Sons, 553 pp.

Landva, A.O., Pelkey, S.G. and Valsangkar, A.J. (1998) Coefficient of permeability of municipal refuse, in *Proceedings of 3rd International Congress on Environmental Geotechnics*, Lisboa, Portugal, pp. 63–8.

Lapidus, L. and Amundson, N.R. (1952) Mathematics of adsorption in beds. VI. The effect of longitudinal diffusion in ion exchange and chromatographic columns, *Journal of Physical Chemistry*, **56**(8), 984–8.

Leite, A.L., Paraguassu, A.B. and Rowe, R.K. (2003) Sorption of Cd^{2+}, K^{+}, F^{-}, and Cl^{-} by some tropical soils, *Canadian Geotechnical Journal*, **40**(3), 629–42.

Lerman, A. (1979) *Geochemical Processes – Water and Sediment Environments*, Wiley–Interscience, New York.

Leroueil, S., Le Bihan, J.P. and Bouchard, R. (1992a) Remarks on the design of clay liners in lagoons as hydraulic barriers, *Canadian Geotechnical Journal*, **29**, 512–15.

Leroueil, S., Le Bihan, J.P. and Bouchard, R. (1992b) Discussion of water content–density criteria for compacted soil liners by Daniel, D. and Benson, C., *ASCE Journal of Geotechnical Engineering*, **118**(2), 963–5.

Leroueil, S., Locat, J., Sève, G., Picarelli, L. and Faure, R.M. (2001) Chapter 14, Slopes and Mass Movements, *Geotechnical and Geoenvironmental Engineering Handbook* (ed. R.K. Rowe), Kluwer Academic Publishers, Norwell, USA, pp. 397–428.

Lewis, C.F.M. (1969) Late quaternary history of lake levels in the Huron and Erie Basin, in *Proceedings 12th Conference on Great Lakes Res.*, Inter. Assoc. for Great Lakes Research, pp. 250–70.

Li, H.M., Bathurst, R.J. and Rowe, R.K. (2002) Use of GCLs to control migration of hydrocarbons in severe environmental conditions, in *Proceedings of International Symposium on Geosynthetic Clay Barriers*, Nuremberg, Germany, April, pp. 187–98.

Lin, L.C. and Benson, C.H. (2000) Effect of wet dry cycling on swelling and hydraulic conductivity of GCLs, *ASCE Journal of Geotechnical and Geoenvironmental Engineering*, **126**(1), 40–9.

Lindstrom, F.T., Haque, R., Freed, V.H. and Boersma, L. (1967) Theory of the movement of some herbicides in soils – linear diffusion and convection of chemicals in soils, *Environmental Science Technology*, **1**(7), 561–5.

Lo, I.M.C. (1992) *Development and evaluation of clay-liner materials for hazardous waste sites*, Ph.D. dissertation, The University of Texas at Austin.

Lo, I.M.C., Mak, R.K.M. and Lee, S.C.H. (1997) Modified clays for waste containment and pollutant attenuation, *ASCE Journal of Environmental Engineering*, **123**, 25–32.

Lord, A.E., Koerner, R.M. and Swan, J.R. (1988) Chemical mass transport measurement to determine flexible membrane liner lifetime, *Geotechnical Testing Journal, ASTM*, **11**(2), 83–91.

Lu, J.C.S., Morrison, R.D. and Stearns, R.J. (1981) Leachate production and management from municipal landfills: summary and assessment, in *Proceedings of the 7th Annual Research Symposium, Land Disposal: Municipal Solid Waste. EPA-600/9-81-002a*, US Environmental Protection Agency, Cincinnati, OH, pp. 1–17.

Lu, J.C.S., Eichenberger, B. and Stearns, R.J. (1985) Leachate from municipal landfills, production and management, *Pollution Technology Review*, Vol. 119, Noyes Publications, Park Ridge, NJ.

Luber, M. (1992) *Diffusion of chlorinated organic compounds through synthetic landfill liners*, Report, Department of Earth Sciences, University of Waterloo.

Lundell, C.M. and Menoff, S.D. (1989) The use of geosynthetics as drainage media at solid waste landfills, in *Proceedings Geosynthetics '89*, San Diego, pp. 10–17.

Lyngkilde, J. and Christensen, T.H. (1992) Fate of organic contaminants in the redox zones of a landfill leachate pollution plume (Vejen, Denmark) *Journal of Contaminant Hydrology*, **10**, 291–307.

MacKay, P. (1997) *Oxygen diffusion through clay covers*, M.E.Sc. thesis, The University of Western Ontario, London, Canada.

Madsen, F.T. and Kahr, G. (1993) Diffusion of ions in compacted bentonite, in *Proceedings of the International Conference on Nuclear Waste Management and Environmental Remediation*, Prague, September, pp. 229–46.

Manassero, M., Van Impe, W.F. and Bouazza, A. (1996) Waste disposal and containment, in *Proceedings of 2nd International Congress on Environmental Geotechnics*, Osaka, Japan, pp. 1425–74.

Manassero, M., Pasqualini, E. and Sani, D. (1998) Potassium sorption isotherms of a natural clayey-silt for pollutant containment, in *Proceedings of the 3rd International Congress on Environmental Geotechnics* (ed. P.S. Pinto), Balkema, Rotterdam.

Manassero, M., Benson, C. and Bouazza, A. (2000) Solid waste containment systems, in *Proceedings International Conference on Geological and Geotechnical Engineering, GeoEng.*, Melbourne, Australia, Vol. 1, pp. 520–642.

Manning, D.A.C. and Robinson, N. (1999) Leachate–mineral reactions: implications for drainage system stability and clogging, in *Proceedings of the 7th International Landfill Symposium*, S. Margherita di Pula, Cagliari, Italy, pp. 269–76.

Marques, A.C.M., Vilar, O.M. and Kaimoto, L.S.A. (1998) Urban solid waste – conception and design of a test fill, in *Proceedings, 3rd International Congress on Environmental Geotechnics*, Lisboa, Portugal, pp. 127–32.

Martensson, A.M., Aulin, C., Wahlberg, O. and Argen, S. (1999) Effect of humic substances on the mobility of toxic metals in a mature landfill, *Waste Management Research*, **17**, 296.

Matich, M.A.J. and Tao, W.F. (1984) A new concept of waste disposal, in *Proceedings of the CEO-SOS Seminar on Design and Construction of Municipal and Industrial Waste Disposal Facilities*, Toronto, pp. 43–60.

Matos, A.T., Costa, L.M., Fontes, M.P.F. and Martinez, M.A. (1999) Retardation factors and the dispersion/diffusion coefficients of Zn, Cd, Cu, and Pb in soils from Viçosa MG, Brazil, *Transactions of the American Society of Agricultural Engineers*, **42**(4), 903–10.

Matos, A.T., Fontes, M.P.F., Costa, L.M. and Martinez, M.A. (2000) Mobility of heavy metals as related to soil chemical and mineralogical characteristics of Brazilian soils, *Environmental Pollution*, **111**, 429–35.

Mazzieri, F. and Pasqualini, E. (1997) Field performance of GCLs: a case study, *Proceedings 1st ANZ Conference Environmental Geotechnics*, Melbourne, pp. 289–94.

Mazzieri, F. and Pasqualini, E. (2000) Permeability of damaged geosynthetic clay liners, *Geosynthetics International*, **7**(2), 101–18.

Mazzieri, F., Pasqualini, E. and Van Impe, W.F. (2000) Compatibility of GCLs with organic solutions, *International Conference on Geotechnical and Geological Engineering*, Melbourne, TechNo. mic Press (CDROM).

McBean, E.A., Poland, R., Rovers, F.A. and Crutcher, A.J. (1982) Leachate collection design for contaminant landfills, *ASCE Journal of Environmental Engineering Division*, **108**, 204.

McBean, E.A., Mosher, F.R. and Rovers, F.A. (1993) Reliability based design for leachate collection systems, in *Proceedings of 4th International Landfill Symposium*, Cagliari, pp. 431–41.

McBean, E.A., Rovers, R.A. and Farquhar, G.J. (1995) *Solid Waste Landfill Engineering and Design*, Prentice-Hall PTR, Toronto, Canada, 521 pp.

McCarthy, K.A. and Johnson, R.L. (1995) Measurement of trichloroethylene diffusion as a function of moisture content in sections of gravity-drained soil columns, *Journal of Environmental Quality*, **24**, 49–55.

McIsaac, R.S. and Rowe, R.K. (2003) *An experimental examination of biologically induced clogging of tire shreds*, Geotechnical Research Centre Report, The University of Western Ontario, London, Canada.

McIsaac, R.S., Rowe, R.K., Fleming, I.R. and Armstrong, M.D. (2000) Leachate collection system design and clog development, in *Proceedings of 6th Canadian Environmental Engineering Conference*, London, Ontario, pp. 66–73.

McKay, L.D. and Trudell, M.R. (1989) The sorption of trichloroethylene in clayey till, in *Proceedings of Symposium on Ground–Groundwater Contamination*, Saskatoon, Saskatchewan, June 14–15. Sponsored by GroundWater Division, National Hydrology Research Institute.

McQuade, S.J. and Needham, A.D. (1999) Geomembrane liner defects – causes, frequency and avoidance, in *Proceedings, Institution of Civil Engineers, Geotechnical Engineering*, Vol. 137, pp. 203–13.

Merry, S.M. and Bray, J.D. (1997) Time-dependent mechanical response of HDPE geomembranes, *ASCE Journal of Geotechnical and Geoenvironmental Engineering*, **123**(1), 57–65.

Mesri, G. and Olsen, R.E. (1970) Shear strength of montmorillonite, *Geotechnique*, **20**(3), 261–70.

Mesri, G. and Olson, R.E. (1971) Mechanisms controlling the permeability of clays, *Clays and Clay Minerals*, **19**, 151–8.

Michaels, A.S. and Lin, C.S. (1954) The permeability of kaolinite, *Industrial and Engineering Chemistry*, **46**, 1239–46.

Millington, R.J. (1959) Gas diffusion in porous media, *Science*, **30**(1), 49–55.

Millward, S.C. (2000) *Diffusion through composite liners and underlying unsaturated soils*, M.E.S.c Thesis, University of Western Ontario, 150 pp.

Mineralogical Society (1980) *Crystal Structures of Clay Minerals and Their X-ray Identification* (eds G.W. Brindley and G. Brown), Mineralogical Society Monograph, No. 5, 495 pp.

Mitchell, J.K. (1993) *Fundamentals of Soil Behaviour*, 2nd edn, John Wiley & Sons, New York.

Mitchell, J.K. and Madsen, F.T. (1987) Chemical effects on clay hydraulic conductivity, in: *Geotechnical Practice for Waste Disposal '87* (ed. R.D. Woods), American Society of Civil Engineers, Geotechnical Special Publication 13, pp. 87–116.

Mitchell, J.K., Hooper, D.R. and Campanella, R.G. (1965) Permeability of compacted clay, *Journal of the Geotechnical Engineering Division, ASCE*, **91**(SM4), 41–65.

Mitchell, J.K., Seed, R.B. and Seed, H.B. (1990a) Kettleman Hills Waste landfill slope failure. I. Liner system properties, *ASCE Journal of Geotechnical Engineering*, **116**(4), 647–68.

Mitchell, J.K., Seed, R.B. and Seed, H.B. (1990b) Stability considerations in the design and construction of lined waste repositories, ASTM STP1070, *Geotechnics of Waste Fills – Theory and Practice* (eds A.O. Landva and G. Knowles), pp. 207–24.

MoE (1998) *Landfill standards: a guideline on the regulatory and approval requirements for the new or expanding landfilling sites*, Ontario Ministry of the Environment, Ontario Regulation 232/98, Queen's Printer for Ontario, Toronto.

MoEE (1993a) *Ministry of the Environment, Ontario Policy, Incorporation of the reasonable use concept into MOE groundwater management activities*, Policy 15–08, March 1993.

MoEE (1993b) *Ministry of the Environment. Engineered facilities at landfills that receive municipal and non-hazardous wastes*, Policy 14–15, March 1993.

Moore, C.A. (1983) Landfill and surface impoundment performance evaluation manual, US EPA SW 869.

Moore, I.D. (2001) Chapter 18, Buried Pipes and Culverts, *Geotechnical and Geoenvironmental Engineering*

Handbook (ed. R.K. Rowe), Kluwer Academic Publishers, Norwell, USA, pp. 541–67.

Montogomery, J.H. and Welkom, L.M. (1990) *Groundwater Chemicals Desk Reference*, Lewis Publisher, Chelsea, MI.

Moreno, L. and Rasmuson, A. (1986) Contaminant transport through a fractured porous rock: impact of the inlet boundary condition on the concentration profile in the rock matrix, *Water Resources Research*, **22**(12), 1728–30.

Morrison, R.T. and Boyd, R.N. (1983) *Organic Chemistry*, Allyn and Bacon, Toronto, 1370 pp.

Mucklow, J.P. (1990) *Phenol migration from landfills by diffusion in natural clayey soils*, M.E.Sc. thesis, University of Western Ontario, London, Ont., Canada, 234 pp.

Müller, W., Jakob, R., Tatzky-Gerth and August, H. (1998) Solubilities, diffusion and partitioning coefficients of organic pollutants in HDPE geomembranes: experimental results and calculations, in *Proceedings of 6th International Conference on Geosynthetics*, Atlanta, Industrial Fabrics Association International, pp. 239–48.

Myrand, D., Gillham, R.W., Cherry, J.A. and Johnson, R.L. (1987) *Diffusion of Volatile Organic Compounds in Natural Clay Deposits*, Department of Earth Sciences, University of Waterloo, Waterloo, Ontario, Canada.

Narejo, D., Koerner, R.M. and Wilson-Fahmy, R.F. (1996) Puncture protection of geomembranes. Part II. Experimental, *Geosynthetics International*, **3**(5), 629–53.

Naylor, T. de V. (1989) Permeation properties, in *Comprehensive Polymer Science* (eds Colin Booth and Colin Price), Oxford, England, Pergamon Press, Vol. 2, pp. 643–68.

Neretnieks, I. (1980) Diffusion in the rock matrix: an important factor in radio nuclide retardation, *Journal of Geophysical Research*, **85**(B8), 4379–97.

Neuman, S.P. and Witherspoon, P.A. (1972) Field determination of the hydraulic properties of leaky multiple aquifer systems, *Water Resources Research*, **8**(5), 1284–98.

Nicholson, R.V., Gillham, R.W., Cherry, J.A. and Reardon, E.J. (1989) Reduction of acid generation in mine tailings through the use of moisture-retaining cover layers as oxygen barriers, *Canadian Geotechnical Journal*, **26**(1), 1–8.

Niemann, W.L. and Hatheway, A.W. (1997) Effect of variable-pH landfill leachate on a carbonate rock aggregate, *Environmental and Engineering Geoscience*, **3**, 423–30.

Nkedi-Kissa, P., Rao, P.S.C. and Hornsby, A.G. (1985) Influence of organic cosolvents on sorption of hydrophobic organic chemicals by soils, *Journal of Environmental Science Technology*, **19**, 975–9.

Nosko, V. and Touze-Foltz, N. (2000) Geomembrane liner failure: modeling of its influence on contaminant transfer, in *Proceedings of 2nd European Geosynthetics Conference*, Bologna, Italy, pp. 557–60.

Ogata, A. (1970) *Theory of dispersivity in a granular medium*, US Geological Survey, Professions Paper.

Ogata, A. and Banks, R.B. (1961) *A solution of the differential equation of longitudinal dispersion in porous media*, US Geol. Survey, Prof. Pap. 411-A.

Ogunbadejo, T.A. (1973) *Physico-chemistry of weathered clay crust formation*, Ph.D. thesis, The University of Western Ontario, London, Ontario.

O'Kane, M., Wilson, G.W., Barbour, S.L. and Swanson, D.A. (1995) Aspects of the performance of the till cover system at Equity Silver Mines Ltd, in *Proceedings of Sudbury 95 – Mining and the Environment*, Sudbury, Ontario, Canada, Vol. 2, pp. 565–73.

Olsen, H.W. (1966) Darcy's law in saturated kaolinite, *Water Resources Research*, **2**(2), 287–95.

Ontario Ministry of the Environment (2001) *Ontario Drinking Water Standards*, Revised January 2001, Queen's Printer for Ontario.

Orsini, C. and Rowe, R.K. (2001) Testing procedure and results for the study of internal erosion of geosynthetic clay liners, in *Proceedings of Geosynthetics 2001*, Portland, Oregon, pp. 189–201.

Othman, M.A. and Benson, C.H. (1993) Effect of freeze-thaw on the hydraulic conductivity and morphology of compacted clay, *Canadian Geotechnical Journal*, **30**(2), 236–46.

Ouvry, J.F., Gisbert, T. and Closset, L. (1995) Back analysis of a slide in a waste storage centre, *Recontres '95*, pp. 148–52.

Owen, J.A. and Manning, D.A.C. (1997) Silica in landfill leachates: implications for clay mineral stabilities, *Applied Geochemistry*, **12**, 267–80.

Pankow, J.F. and Cherry, J.A. (1996) *Dense Chlorinated Solvents*, Waterloo Press, Guelph, 522 pp.

Panu, U.S. and Filice, A. (1992) Techniques of flow rates into draintubes with circular perforations, *Journal of Hydrology*, **137**, 57–72.

Park, J.K. and Nibras, M. (1993) Mass flux of organic chemicals through polyethylene geomembranes, *Water Environment Research*, **65**(3), 227–37.

Park, J.K., Sakti, J.P. and Hooper, J.A. (1995) Effectiveness of geomembranes as barriers of organic compounds, in *Proceedings of Geosynthetics '95*, IFAI, Vol. 3, pp. 879–92.

Park, H., II, Lee, S.R. and Do, N.Y. (2002) Evaluation of decomposition effect on long-term settlement prediction for fresh municipal solid waste landfills, *ASCE Journal of Geotechnical and Geoenvironmental Engineering*, **128**(2), 107–18.

Parmar, R. and Bowman, J. (1989) Crack initiation and propagation paths for brittle failures in aligned and misaligned pipe butt fusion joints, *Polymer Engineering and Science*, **29**(19), 1396–405.

Parra, J.O. (1988) Electrical response of a leak in a geomembrane liner, *Geophysics*, **53**(11), 1445–52.

Peggs, I.D. and Carlson, D.S. (1989) Stress cracking of polyethylene geomembrane: field experience, in *Durability and Aging of Geosynthetics* (ed. R.M. Koerner), pp. 195–211.

Peggs, I.D. and Olsta, J.T. (1998) A GCL and incompatible soil case history: a design problem, in *Proceedings 12th GRI Conference, Philadelphia*, USA, pp. 117–38.

Peggs, I.D. and Thiel, R. (1998) Selecting a geomembrane material, in *Proceedings of 6th International Conference on Geosynthetics*, Atlanta, Industrial Fabrics Association International, St Paul, MN, pp. 381–88.

Pelte, T., Pierson, P. and Gourc, J.P. (1994) Thermal analysis of geomembranes exposed to solar radiation, *Geosynthetics International*, **1**(1), 21–44.

Perkins, T.K. and Johnston, D.C. (1963) A review of diffusion and dispersion in porous media, *Society of Petroleum Engineering Journal*, **3**(1), 70–84.

Peters, G. and Smith, D.W. (1999) Modelling of the dispersion–advection equation with spatially varying mechanical dispersion, in *Proceedings of the 8th Australian New Zealand Conference on Geomechanics*, Institution of Engineers Australia, Hobart, Vol. 1, pp. 969–74.

Petroff, L.J. (1994) *Design Methodology for High Density Polyethylene Manholes: Buried Plastic Pipe Technology*, Vol. 2, STP 1222 (ed. D. Eckstein), ASTM, Philadelphia, pp. 52–65.

Petrov, R.J. and Rowe, R.K. (1997) GCL – chemical compatibility by hydraulic conductivity testing and factors impacting its performance, *Canadian Geotechnical Journal*, **34**(6), 863–85.

Petrov, R.J., Rowe, R.K. and Quigley, R.M. (1997a) Comparison of laboratory measured GCL hydraulic conductivity based on three permeameter types, *Geotechnical Testing Journal*, ASTM, **20**(1), 49–62.

Petrov, R.J., Rowe, R.K. and Quigley, R.M. (1997b) Selected factors influencing GCL hydraulic conductivity, *ASCE Journal of Geotechnical and Geoenvironmental Engineering*, **123**, 683–95.

Phaneuf, R.J. (2000) Bioreactor landfills – regulatory issues, in *Proceedings 14th Geosynthetic Research Institute Conference*, pp. 9–26.

Plastic Pipes Institute (1996) *Plastic Pipes Institute Handbook of Polyethylene Pipe*, The Plastics Pipe Institute, Wayne, NJ.

Pohland, F.G. (1980) Leachate recycle as a management option, *Journal of the Environmental Engineering Division*, ASCE, **106**, 1057–69.

Porter, L.K., Kempter, W.D., Jackson, R.J. and Stewart, B.A. (1960) Chloride diffusion in soils as influenced by moisture content, *Soil Science Society of America, Proceedings*, **24**, 460–3.

Poulos, H.G. and Davis, E.H. (1980) *Pile Foundation Analysis and Design*, John Wiley & Sons, New York.

Powrie, W. and Beaven, R.P. (1999) Hydraulic properties of household waste and implications for landfills, in *Proceedings of the Institution of Civil Engineers, Geotechnical Engineering*, **137**, 235–47.

Powrie, W., Beaven, R. and Harkness, R.M. (1999) Applicability of soil mechanics principles to household wastes, in *Proceedings of 7th International Landfill Symposium*, S. Margherita di Pula, Cagliari, Italy, Vol. 3, pp. 429–36.

Prasad, T.V., Brown, K.W. and Thomas, J.C. (1994) Diffusion coefficients of organics in high density polyethylene (HDPE), *Waste Management and Research*, **12**, 61–71.

Press, W.H., Flannery, B.P., Teukolsky, S.A. and Vetterling, W.T. (1986) *Numerical Recipes – The Art of Scientific Computing*, Cambridge University Press, Cambridge.

Puls, R.W., Powel, R.M., Clark, D. and Eldred, C.J. (1991) Effects of pH, solid/solution ratio, ionic strength, and organic acids on Pb and Cd sorption on kaolinite, *Water, Air, and Soil Pollution*, **57–58**, 423–30.

Qian, X., Koerner, R.M. and Gray, D.H. (2002) *Geotechnical Aspects of Landfill Design and Construction*, Prentice-Hall, Upper Saddle River, NJ.

Quigley, R.M. (1991) *Chemical assessment of an in situ clayey liner – Keele Valley landfill, Maple, Ontario (Dec. 8, 1988 Exhumation)*, Final Report to Golder Associates Ltd, Mississauga, Ontario.

Quigley, R.M. and Crooks, V.E. (1983) Chemical profiles in soft clays and the role of long-term diffusion, in *Geological Environment and Soil Properties* (ed. R.N. Yong), ASCE Special Publication (unnumbered), pp. 5–18.

Quigley, R.M. and Fernandez, F. (1989) Clay/organic interactions and their effect on the hydraulic conductivity of barrier clays, in *Contaminant Transport in Groundwater* (eds H.E. Kobus and W. Kinzelbach), Balkema, Rotterdam, pp. 117–24.

Quigley, R.M. and Fernandez, F. (1992) Organic liquid interactions with water-wet barrier clays, in *Subsurface Contamination by Immiscible Fluids* (ed. K.U. Weyer), A.A. Balkema, Rotterdam, pp. 49–56 in *Proceedings International Conference on Subsurface Contamination by Immiscible Fluids*, Calgary, Alberta, April 1990.

Quigley, R.M. and Rowe, R.K. (1986) Leachate migration through clay below a domestic waste landfill, Sarnia, Ontario, Canada, in *Chemical Interpretation and Modelling Philosophies. Hazardous and Industrial Solid Waste Testing and Disposal, Vol. 6* (eds D. Lorenzen, R.A. Conway, L.P. Jackson, A. Hamza, C.L. Perket and W.J. Lacy), ASTM STP 933, American Society for Testing and Materials, Philadelphia, pp. 93–103.

Quigley, R.M., Gwyn, Q.H.J., White, O.L., Rowe, R.K., Haynes, J.E. and Bohdanowicz, A. (1983) Leda clay from deep boreholes at Hawkesbury, Ontario. Part I. Geology and geotechnique, *Canadian Geotechnical Journal*, **20**(2), 288–98.

Quigley, R.M., Crooks, V.E. and Yanful, E. (1984) Contaminant migration through clay below a domestic waste landfill site, Sarnia, Ontario, Canada, in *Proceedings International Groundwater Symposium on Groundwater Resources Utilization and Contaminant Hydrogeology*, Montreal, May 1984, Vol. II, pp. 499–506.

Quigley, R.M., Fernandez, F., Yanful, E., Helgason, T., Margaritis, A. and Whitby, J.L. (1987a) Hydraulic conductivity of contaminated natural clay directly beneath a domestic landfill, *Canadian Geotechnical Journal*, **24**(3), 377–83.

Quigley, R.M., Yanful, E.K. and Fernandez, F. (1987b) Ion transfer by diffusion through clayey barriers, in *Geotechnical Practice for Waste Disposal '87*, ASCE, Geot. Spec. Publ. No. 13, pp. 137–58.

Quigley, R.M., Fernandez, F. and Rowe, R.K. (1988) Clayey barrier assessment for impoundment of domestic waste leachate (southern Ontario) including clay–leachate compatibility by hydraulic conductivity testing, *Canadian Geotechnical Journal*, **25**(3), 574–81.

Quigley, R.M., Mucklow, J.P. and Yanful, E.K. (1990a) Contaminant migration by diffusion at the Confederation Road landfill, Sarnia, Ontario, in Engineering in our Environment, in *Proceedings of 1990 Conference of Canadian Society for Civil Engineering*, Hamilton, May, pp. 876–92.

Quigley, R.M., Yanful, E.K. and Fernandez, F. (1990b) Biological factors influencing laboratory and field diffusion, in *Microbiology in Civil Engineering, Proceedings Federation of European Microbiological Societies Symposium* (ed. P. Howsam), Cranfield Institute of Technology, UK, pp. 261–73.

Rahman, F. and Bouazza, A. (2002) *Oxygen diffusion through geosynthetic clay liners*, Progress Report, Australian Research Council, Monash University, Department of Civil Engineering, Melbourne, Australia.

Ramke, H.-G. and Brune, M. (1990) Untersuchungen zur Funktionsfähigkeit von Entwässerungsschichten, in *Deponiebasisabdichtungssystemen, Abschlu ßbericht Bundesminister für Forschung und Technologie*, FKZ BMFT 145 0457 3.

Ramsey, D.R. (1993) Diffusivities of organic contaminants in high density polyethylene geomembranes, in *Proceedings of Geosynthetics 93*, Vancouver, Canada, pp. 645–57.

Randolph, M.F. and Booker, J.R. (1982) Analysis of seepage into a cylindrical permeameter, in *Proceedings of 4th International Conference on Numerical Methods in Geomechanics*, Edmonton, pp. 349–58.

Rao, P.S.C., Green, R.E., Balasubramanian, V. and Kanehio, Y. (1974) Field study of solute movement in a highly aggregated oxisol with intermittent flooding. II. Picloram, *Journal Environmental Quality*, **3**, 197–202.

Rasmuson, A. and Eriksson, J.C. (1986) Capillary barriers in covers for mine tailings, *National Swedish Environmental Protection Board*, Report 3307.

Reades, D.W., King, K.S., Benda, E., Quigley, R.M., LeSarge, K. and Heathwood, C. (1989) The results of ongoing monitoring of the performance of a low permeability clay liner, Keele Valley landfill, Maple, Ontario, in *Proceedings Focus Conference on Eastern Regional Ground Water Issues*, National Water Well Assoc., Kitchener, Ontario, pp. 79–91.

Reddy, K.R., Bandi, S.R., Rohr, J.J., Finy, M. and Siebken, J. (1996) Field evaluation of protective covers for landfill geomembrane liners under construction loading, *Geosynthetics International*, **3**(6), 679–700.

Reinhart, D.R. (1996) Full-scale experiences with leachate recirculation landfills: case studies, *Waste Management Research Journal*, **14**, 347–65.

Reinhart, D.R. and Al-Yousfi, A.G. (1996) The impact of leachate recirculation on municipal solid waste landfill operating characteristics, *Waste Management and Research Journal*, **14**, 337–46.

Reinhart, D.R. and Townsend, T.G. (1998) *Landfill Bioreactor Design and Operation*, Lewis Publishers, Boca Raton, FL.

Reinhart, D.R., McCreanor, P.T. and Townsend, T. (2000) The bioreactor landfill: research and development needs, in *Proceedings of the 14th GRI Conference*, pp. 1–8.

Reitzel, S.F. (1990) *The temporal characterization of municipal solid waste*, M.A.Sc. Thesis, University of Waterloo, Waterloo, Ontario, Canada.

Renaud, R. (2001) Steam injection landfill bioreactors, *Geotechnical Fabrics Report*, **19**(1), 26–28.

Reuter, E. and Egloffstein, T. (2000) Self healing of geosynthetic clay liners (GCLs) after extreme drying periods, in *Proceedings of 2nd European Geosynthetics Conference*, Bologna, pp. 745–54.

Revens, A., Ross, D., Gregory, B., Meadows, M., Harries, C. and Gronow, J. (1999) Long-term fate of metals in landfill, in *Proceedings of 7th International Waste Management and Landfill Symposium*, Sardinai'99, CISA, Cagliari, Italy, Vol. 1, p. 199.

Reynolds, R.T. (1991) Geotechnical field techniques used in monitoring slope stability at a landfill, in *Proceedings of Field Measurements in Geotechniques* (ed. G. Sorum), Rotterdam, Balkema, pp. 833–91.

Richardson, G.N. (1997a) GCL internal shear strength requirements, *Geotechnical Fabrics Report*, March, pp. 20–5.

Richardson, G.N. (1997b) GCLs: alternative subtitle D liner systems, *Geotechnical Fabrics Report*, May, pp. 36–42.

Richardson, G.N. and Koerner, R.M. (1988) *Geosynthetic design guidance for hazardous waste landfill cells and surface impoundments*, EPA/600/52–87/097, Cincinnati, OH.

Rittmann, B.E., Fleming, I.R. and Rowe, R.K. (1996) Leachate chemistry: its implications for clogging, in *Proceedings of the North American Water and Environment Congress'96*, June 1996, American Society of Civil Engineers, Anaheim, CA (CD-ROM).

Rittmann, B.E., Banaszak, J.E., Cooke, A. and Rowe, R.K. (2003) Biogeochemical evaluation of mechanisms controlling $CaCO_3$ precipitation in landfill leachate-collection systems, *ASCE Journal of Environmental Engineering*, **129**(8), 730–3.

Robinson, H.D. and Maris, P.J. (1985) The treatment of leachates from domestic waste in landfill sites, *Journal of the Water Pollution Control Federation*, **57**, 30–8.

Rogers, C.D.F. (1999) The structural performance of flexible pipe for landfill drainage, *Geotechnical Engineering*, **137**, 249–60.

Rogers, C.E. (1985) Permeation of gases and vapors in polymers, Chapter 2, *Polymer Permeability* (ed. J. Comyn), Elsevier Applied Science Publisher, London, pp. 11–73.

Rollin, A.L., Mlynarek, J., Lafleur, J. and Zanescu, A. (1994) Performance changes in aged in-situ HDPE geomembrane, *Landfilling of Wastes: Barriers* (eds T.H. Christensen, R. Cossu and R. Stegmann), E & FN Spon, pp. 431–43.

Rollin, A.L., Marcotte, M., Jacquelin, T. and Chaput, L. (1999) Leak location in exposed geomembrane liners using an electrical leak detection technique, in *Proceedings of Geosynthetics '99*, Boston, MA, pp. 615–26.

Rowe, R.K. (1987) Pollutant transport through barriers, in *Proceedings of ASCE Specialty Conference, Geotechnical Practice for Waste Disposal'87*, Ann Arbor, June, pp. 159–81.

Rowe, R.K. (1988) Contaminant migration through groundwater: the role of modelling in the design of barriers, *Canadian Geotechnical Journal*, **25**(4), 778–98.

Rowe, R.K. (1991a) Contaminant impact assessment and the contaminating lifespan of landfills, *Canadian Journal of Civil Engineering*, **18**(2), 244–53.

Rowe, R.K. (1991b) Environmental geotechnology: some pertinent considerations, in *Proceedings of 7th International Conference of the International Association for Computer Methods and Advances in Geomechanics*, Cairns, Vol. 1, pp. 35–48.

Rowe, R.K. (1992) Integration of hydrogeology and engineering in the design of waste management sites,in *Proceedings of the International Association of Hydrogeologists Conference on Modern Trends in Hydrogeology*, Hamilton, Ontario, Canada, pp. 7–21.

Rowe, R.K. (1993) Some challenging applications of geotextiles in filtration and drainage, *Geotextiles in Filtration and Drainage*, Thomas Telford, London, pp. 1–12.

Rowe, R.K. (1994) *Leachate characterization*, Report available from Geotechnical Research Centre, The University of Western Ontario 126 pp.

Rowe, R.K. (1995) Leachate characteristics for MSW landfills, in *Proceedings of 5th International Landfill Symposium*, Cagliari, Italy, October 1995, pp. 327–44.

Rowe, R.K. (1998a) Geosynthetics and the minimization of contaminant migration through barrier systems beneath solid waste, in *Proceedings of 6th International Conference on Geosynthetics*, Atlanta, Industrial Fabrics Association International, St Paul, USA, pp. 27–103.

Rowe, R.K. (1998b) From the past to the future of landfill engineering through case histories, in *Proceedings of 4th International Conference on Case Histories in Geotechnical Engineering*, St. Louis, USA, pp. 145–66.

Rowe, R.K. (1999) Solid waste disposal facilities for urban environments, in *Proceedings of 11th Pan American Conference on Soil Mechanics and Geotechnical Engineering*, Foz do Iguassu, Brazil, August, pp. 89–111.

Rowe, R.K. (2001) Barrier Systems, Chapter 25 of *Geotechnical and Geoenvironmental Engineering Handbook*, Kluwer Academic Publishing, Norwell, USA, pp. 739–88.

Rowe, R.K. and Badv, K. (1996a) Chloride migration through clayey silt underlain by fine sand or silt, *ASCE Journal of Geotechnical Engineering*, **122**(1), 60–8.

Rowe, R.K. and Badv, K. (1996b) Advective–diffusive contaminant migration in unsaturated sand and gravel, *ASCE Journal of Geotechnical Engineering*, **122**(12), 965–75.

Rowe, R.K. and Badv, K. (1996c) Use of a geotextile separator to minimize intrusion of clay into a coarse stone layer, *Geotextiles and Geomembranes*, **14**(2), 73–94.

Rowe, R.K. and Booker, J.R. (1983) *SFIN – a finite element analysis program for single contaminant migration under 1D conditions*, Geotechnical Research Centre, The University of Western Ontario, London, Ontario.

Rowe, R.K. and Booker, J.R. (1985a) 1-D pollutant migration in soils of finite depth, *ASCE Journal of Geotechnical Engineering*, **111**(GT4), 479–99.

Rowe, R.K. and Booker, J.R. (1985b) 2D pollutant migration in soils of finite depth, *Canadian Geotechnical Journal*, **22**(4), 429–36.

Rowe, R.K. and Booker, J.R. (1986) A finite layer technique for calculating three-dimensional pollutant migration in soil, *Geotechnique*, **36**(2), 205–14.

Rowe, R.K. and Booker, J.R. (1987) An efficient analysis of pollutant migration through soil, Chapter 2 in the Book, *Numerical Methods in Transient and Coupled Systems* (eds Lewis, Hinton, Bettess and Schrefler), John Wiley & Sons, pp. 13–42.

Rowe, R.K. and Booker, J.R. (1988a) Modelling of contaminant movement through fractured or jointed media with parallel fractures, in *Proceedings of 6th International Conference on Numerical Methods in Geomechanics*, Innsbruck, April, pp. 855–62.

Rowe, R.K. and Booker, J.R. (1988b) *MIGRATE – Analysis of 2D Pollutant Migration in a Non-homogeneous Soil System: Users Manual*, Report Number GEOP-1–88, Geotechnical Research Centre, The University of Western Ontario, London, Ontario.

Rowe, R.K. and Booker, J.R. (1989) A semi-analytic model for contaminant migration in a regular two or three dimensional fractured network: conservative contaminants, *International Journal for Numerical and Analytical Methods in Geomechanics*, **13**, 531–50.

Rowe, R.K. and Booker, J.R. (1990a) Contaminant migration through fractured till into an underlying aquifer, *Canadian Geotechnical Journal*, **27**, 484–95.

Rowe, R.K. and Booker, J.R. (1990b) A semi-analytic model for contaminant migration in a regular two or three dimensional fractured network: reactive contaminants, *International Journal for Numerical and Analytical Methods in Geomechanics*, **14**, 401–25.

Rowe, R.K. and Booker, J.R. (1991a) Modelling of 2D contaminant migration in a layered and fractured zone beneath landfills, *Canadian Geotechnical Journal*, **28**(3), 338–52.

Rowe, R.K. and Booker, J.R. (1991b) Pollutant migration through a liner underlain by fractured soil, *ASCE Journal of Geotechnical Engineering*, **118**(7), 1031–46.

Rowe, R.K. and Booker, J.R. (1998) Modelling impacts due to multiple landfill cells and clogging of leachate collection systems, *Canadian Geotechnical Journal*, **35**, 1–14.

Rowe, R.K. and Booker, J.R. (1983, 1994, 1997, 1999, 2004) POLLUTE v.7 – 1-D pollutant migration through a non homogeneous soil, Distributed by GAEA Technologies Ltd. www.gaea.ca.

Rowe, R.K. and Booker, J.R. (2000) Theoretical solutions for calculating leakage through composite liner systems, in *Developments in Theoretical Geomechanics– The John Booker Memorial Symposium*, Sydney, November, pp. 580–602.

Rowe, R.K. and Fleming, I.R. (1998) Estimating the time for clogging of leachate collection systems, in *Proceedings of the 3rd International Congress on Environmental Geotechnics*, Lisbon, September, **1**, 23–8.

Rowe, R.K. and Fraser, M.J. (1993a) Long-term behaviour of engineered barrier systems, in *Proceedings of 4th International Landfill Symposium*, Cagliari, Italy, pp. 397–406.

Rowe, R.K. and Fraser, M.J. (1993b) Service life of barrier systems in the assessment of contaminant impact, in *Proceedings of Joint CSCE–ASCE National Conference on Environmental Engineering*, Montreal, July, Vol. 2, pp. 1217–24.

Rowe, R.K. and Giroud, J.-P. (1994) Quality assurance of barrier systems for landfills, *IGS News*, **10**(1), 6–8.

Rowe, R.K. and Lake, C.B. (1997) *Program LEAK – a program for calculating leakage through holes in a geomembrane in composite liners.*

Rowe, R.K. and Lake, C.B. (2000) Geosynthetic clay liners (GCLs) for municipal solid waste landfills, in *Environmental Mineralogy: Microbial Interactions, Anthropogenic Influences, Contaminated Land and Waste Management* (eds J.D. Cotter-Howells, L.S. Campbell, E. Valsami-Jones and M. Batchelder), pp. 395–406.

Rowe, R.K. and Nadarajah, P. (1993) Evaluation of the hydraulic conductivity of aquitards, *Canadian Geotechnical Journal*, **30**(5), 781–800.

Rowe, R. and Nadarajah, P. (1995) Transport modelling under transient flow conditions, in *Proceedings of 5th International Symposium on Numerical Models in Geomechanics*, NUMOG V, Davos, Switzerland, pp. 337–42.

Rowe, R.K. and Nadarajah, P. (1996a) An analytical method for predicting the velocity field beneath landfills, *Canadian Geotechnical Journal*, **34**, 264–82.

Rowe, R.K. and Nadarajah, P. (1996b) Effects of temporal fluctuation of leachate concentration in landfills, in *Proceedings of 2nd International Congress on Environmental Geotechnics*, Osaka, Japan (ed. Kamon), pp. 305–10.

Rowe, R.K. and Nadarajah, P. (1996c) Estimating leachate drawdown due to pumping wells in landfills, *Canadian Geotechnical Journal*, **33**(1), 1–10.

Rowe, R.K. and Orsini, C. (2003) Effect of GCL and subgrade type on internal erosion in GCLs, *Geotextiles and Geomembrane*, **21**(1), 1–24.

Rowe, R.K. and San, K.W. (1992) Effect of source characteristics of landfills on environmental impact, in *Proceedings of International Association of Hydrogeologists Conference on Modern Trends in Hydrogeology*, Hamilton, Canada, pp. 249–61.

Rowe, R.K. and Sangam, H.P. (2002) Durability of HDPE geomembranes, *Geotextiles and Geomembrane*, **20**(2), 77–95.

Rowe, R.K. and Sawicki, D.W. (1992) The modelling of a natural diffusion profile and the implications for landfill design, in *Proceedings of 4th International Symposium on Numerical Methods in Geomechanics*, Swansea, August, pp. 481–9.

Rowe, R.K. and VanGulck, J. (2001) Clogging of leachate collection systems: from laboratory and field study to modelling and prediction, in *Proceedings of 2nd Australian–New Zealand Conference on Environmental Geotechnics*, Newcastle, November, pp. 1–22.

Rowe, R.K. and Weaver, T.R. (1997) Contaminant transport in groundwater, in *Proceedings of 1st Australian–New Zealand Conference on Environmental Geotechnics*, Melbourne, November, pp. 97–114.

Rowe, R.K., Caers, C.J. and Barone, F. (1988) Laboratory determination of diffusion and distribution coefficients of contaminants using undisturbed soil, *Canadian Geotechnical Journal*, **25**, 108–18.

Rowe, R.K., Caers, C.J. and Chan, C. (1993) Evaluation of a compacted till liner test pad constructed over a granular subliner contingency layer, *Canadian Geotechnical Journal*, **30**(4), 667–89.

Rowe, R.K., Golder Associates, Fenco MacLaren and M.M. Dillon (1994) *Evaluation of Service Life of the Engineered Component of Landfills*, Report available from Geotechnical Research Centre, The University of Western Ontario, 153 pp.

Rowe, R.K., Fleming, I., Cullimore, R., Kosaric, N. and Quigley, R.M. (1995a) *A research study of clogging and encrustation in leachate collection systems in municipal solid waste landfills*, Report available from Interim Waste Authority Ltd, Geotechnical Research Centre, The University of Western Ontario.

Rowe, R.K., Hrapovic, L. and Kosaric, N. (1995b) Diffusion of chloride and dichloromethane through an HDPE geomembrane, *Geosynthetics International*, **2**(3), 507–36.

Rowe, R.K., Hrapovic, L. and Armstrong, M.D. (1996) Diffusion of organic pollutants through HDPE geomembrane and composite liners and its influence on groundwater quality, in *Proceedings 1st European Geosynthetics Conference*, Maastricht, October, pp. 737–42.

Rowe, R.K., Barone, F.S. and Hrapovic, L. (1997a) Laboratory and field studies of salt diffusion through a composite liner, in *Proceedings of 6th International Landfill Symposium*, S. Margherita di Pula, Cagliari, Italy, October, Vol. 3, pp. 241–50.

Rowe, R.K., Lake, C., von Maubeuge, K. and Stewart, D. (1997b) Implications of diffusion of chloride through geosynthetic clay liners, *Geoenvironment'97*, Melbourne, Australia, November, pp. 295–300.

Rowe, R.K., Hsuan, Y.G., Lake, C.B., Sangam, P. and Usher, S. (1998) Evaluation of a composite (geomembrane/clay) liner for a lagoon after 14 years of use, in *Proceedings of 6th International Conference on Geosynthetics*, Atlanta, Industrial Fabrics Association International, St Paul, MN, pp. 191–6.

Rowe, R.K., Armstrong, M.D. and Cullimore, D.R. (2000a) Mass loading and the rate of clogging due to municipal solid waste leachate, *Canadian Geotechnical Journal*, **37**, 355–70.

Rowe, R.K., Armstrong, M.D. and Cullimore, D.R. (2000b) Particle size and clogging of granular media permeated with leachate, *ASCE Journal of Geotechnical and Geoenvironmental Engineering*, **126**(9), 775–86.

Rowe, R.K., Caers, C.J., Reynolds, G. and Chan, C. (2000c) Design and construction of barrier system for the Halton Landfill, *Canadian Geotechnical Journal*, **37**(3), 662–75.

Rowe, R.K., Lake, C.B. and Petrov, R.J. (2000d) Apparatus and procedures for assessing inorganic diffusion coefficients through geosynthetic clay liners, *Geotechnical Testing Journal*, **23**(2), 206–14.

Rowe, R.K., Goveas, L. and Dittrich, J.P. (2002a) Excavations in gassy soils, *Geotechnical Engineering*, **155**(3), 159–61.

Rowe, R.K., VanGulck, J. and Millward, S. (2002b) Biologically induced clogging of a granular media permeated with synthetic leachate, *Canadian Journal of Environmental Engineering and Science*, **1**(2), 135–56.

Rowe, R.K., Sangam, H.P. and Lake, C.B. (2003a) Evaluation of an HDPE geomembrane after 14 years as a leachate lagoon liner, *Canadian Geotechnical Journal*, **40**(3), 536–50.

Rowe, R.K., Mukunoki, T. and Sangam, P.H. (2003b) *Effect of temperature BTEX diffusion and sorption through geosynthetic clay liners*, GeoEngineering Centre at Queen's-RMC Report.

Roy, W.R., Krapac, I.G., Chou, S.F.J. and Griffin, R.A. (1992) *Batch-type procedures for estimating soil adsorption of chemicals*, Technical Resource Document, EPA/530-SW-87–006-F, EPA, Cincinnati, USA.

Rugge, K., Bjerg, P.L. and Christensen, T.H. (1995) Distribution of organic compounds from municipal solid waste in the groundwater downgradient of a landfill (Grindsted, Denmark), *Environmental Science and Technology*, **29**(5), 1395–400.

Ruhl, J.L. and Daniel, D.E. (1997) Geosynthetic clay liners permeated with chemical solutions and leachates, *ASCE Journal of Geotechnical and Geoenvironmental Engineering*, **123**(4), 369–81.

Ruland, W.W. (1988) *Fracture depth and active groundwater flow in clayey till in Lambton County*, Ontario, M.Sc. thesis, The University of Waterloo.

Saathoff, F. and Sehrbrock, U. (1995) Indicators for selection of protection layers for geomembranes, in *Proceedings of 5th International Conference on Geosynthetics*, Singapore, pp. 1019–22.

Sakti, P.J., Park, K.J. and Hoopes, J.A. (1992) Permeation of organic chemicals through HDPE geomembranes, *Water Forum 92*, ASCE, Baltimore, MD, pp. 201–6.

Salame, M. (1961) An empirical method for predicting of liquid permeation in polyethylene and related polymers, *SPE Transactions*, October 1961, pp. 153–63.

Saleem, M., Asfour, A.A., De Kee, D. and Harrison, B. (1989) Diffusion of organic penetrant through low density polyethylene (LDPE) films: effect of size and shape of the penetrant molecules, *Journal of Applied Polymer Science*, **137**, 617–25.

Samson, E. and Marchand, J. (1999) Numerical solution of the extended Nernst–Planck model, *Journal of Colloid and Interface Science*, **215**, 1–8.

Samson, E., Marchand, J., Robert, J.-L. and Bournazel, J.-P. (1999) Modelling ion diffusion mechanisms in porous media, *International Journal for Numerical Methods in Engineering*, **46**, 2043–60.

Sangam, H.P. (2001) *Performance of HDPE geomembrane liners in landfill applications*, Ph.D. dissertation, The University of Western Ontario, Ontario, Canada, 394 pp.

Sangam, H.P. and Rowe, R.K. (2001) Migration of dilute aqueous organic pollutants through HDPE geomembranes, *Geotextiles and Geomembranes*, **19**(6), 329–57.

Sangam, H.P. and Rowe, R.K. (2002) Effects of exposure conditions on the depletion of antioxidants from HDPE geomembranes, *Canadian Geotechnical Journal*, **39**(6), 1221–30.

Sangam, H.P. and Rowe, R.K. (2003) *Effect of surface fluorination on diffusion through an HDPE geomembrane*, GeoEngineering Centre at Queen's – RMC Research Report, June 2003.

Santamaria, C. (2003) *Circumferential buckling of polymer pipes under earth and fluid pressures*, M.Sc. thesis, University of Alberta, 171 pp.

Santayana, P.D. and Pinto, A.A.V. (1998) The Beirolas landfill eastern expansion landslide, *Environmental Geotechnics* (ed. P. Seco e Pinto), Balkema, Rotterdam, pp. 905–10.

Sarkar, D., Essington, M.E. and Misra, K.C. (2000) Adsorption of mercury(II) by kaolinite, *Soil Science Society of America Journal*, **64**, 1968–75.

Sauvé, S., Martínez, C.E., Mcbride, M. and Hendershot, W. (2000) Adsorption of free lead (Pb^{2+}) by pedogenic oxides, ferrihydrite, and leaf compost, *Soil Science Society of America Journal*, **64**, 595–9.

Schmidt, R.K., Young, C. and Helwitt, J. (1984) Long term field performance of geomembranes – fifteen years experience, in *Proceedings of International Conference on Geomembranes*, Denver, CO, IFIA, Vol. 2, pp. 173–87.

Schnoor, L.J. (1996) *Environmental Modeling – Fate and Transport of Pollutants in Water, Air, and Soil*, John Wiley and Sons, New York.

Schubert, W.R., Harrington, T.J. and Finno, R.J. (1984) Glacial clay liners in waste disposal practice, *Environmental Engineering Specialty Conference*, American Society of Civil Engineers, pp. 36–41.

Schwarzenbach, R.P. and Westall, J. (1981) Transport of non-polar organic compounds from surface water to groundwater, Laboratory sorption studies, *Environmental Science and Technology*, **15**(11), 1360–7.

Seed, R.B., Mitchell, J.K. and Seed, H.B. (1990) Kettleman Hills waste landfill slope failure. II. Stability analyses, *ASCE Journal of Geotechnical Engineering*, **116**(4), 669–89.

Seeger, S. and Müller, W. (1996) Requirements and testing of protective layer systems for geomembranes, *Geotextiles and Geomembranes*, **14**, 365–76.

Selim, H.M. and Mansell, R.S. (1976) Analytical solution of the equation for transport of reactive solute, *Water Resources Research*, **12**(3), 528–32.

Shackelford, C.D. and Daniel, D.E. (1991) Diffusion in saturated soil. II. Results for compacted clay, *ASCE Journal of Geotechnical Engineering*, **117**(3), 485–506.

Shackelford, C.D., Cotten, T.E., Rohal, K.M. and Strauss, S.H. (1997) Acid buffering a high pH soil for zinc diffusion, *ASCE Journal of Geotechnical and Geoenvironmental Engineering*, **123**(3), 260–71.

Shackelford, C.D., Benson, C.H., Katsumi, T., Edil, T.B. and Lin, L. (2000) Evaluating the hydraulic conductivity of GCLs permeated with non-standard liquids, *Geotextiles and Geomembranes*, **18**, 133–61.

Shackelford, C.D., Malusis, M.A. and Olsen, H.W. (2001) Clay membrane barriers for waste containment, *Geotechnical News*, **19**(2), 39–43.

Shan, H.S. and Daniel, D.E. (1991) Results of laboratory tests on geotextile/bentonite liner material, in *Proceedings of Geosynthetics '91 Conference*, Atlanta, USA, pp. 517–32.

Shaw, P.A. and Knight, A.J. (2000) Dollars and sense of implementing a bioreactor landfill, in *Proceedings of 14th Geosynthetic Research Institute Conference*, pp. 40–58.

Sherard, J.L., Decker, R.S. and Ryker, N.L. (1972) Piping – earth dams of dispersive clay, in *Proceedings of ASCE Specialty Conference on the Performance of Earth and Earth Supported Structures*, Purdue University, West Lafayette, IN, pp. 589–626.

Shultz, D.W., Duff, B.M. and Peters, W.R. (1984) *Electrical resistivity technique to assess the integrity of geomembrane liners*, Final Technical Report, US Environmental Protection Agency, Contract 68–03–30331, SwRI Project 14–6289.

Siebken, J. and Cunningham, D. (1999) Environmental stress cracking in geonets manufactured from high density polyethylene, in *Proceedings of Geosynthetics 99*, Portland, pp. 1071–81.

Singh, U. and Uehara, G. (1986) Eletrochemistry of the double layer: principles and applications to soils, in *Soil Physical Chemistry* (ed. D.L. Sparks), CRC Press, Boca Raton, FL.

Sivakumar Babu, G.L., Sporer, H., Zanzinger, H. and Gartung, E. (2001) Self-healing properties of geosynthetic clay liners, *Geosynthetics International*, **8**(5), 461–70.

Sleep, B.E. and McClure, P.D. (2001) The effect of temperature on adsorption of organic compounds to soils, *Canadian Geotechnical Journal*, **38**, 46–52.

Smith, D.W. (2000) One-dimensional contaminant transport through a deforming porous medium: theory and a solution for a quasi-steady-state problem, *International Journal for Analytical and Numerical Methods in Geomechanics*, **24**, 693–722.

Smith, J.A. and Jaffe, P.R. (1994) Benzene transport through landfill liners containing organophilic bentonite, *ASCE Journal of Environmental Engineering*, ASCE, **120**(6), 1577.

Smith, D.W., Rowe, R.K. and Booker, J.R. (1993) The analysis of pollutant migration through soil with linear hereditary time dependent sorption, *International Journal for Analytical and Numerical Methods in Geomechanics*, **17**(4), 255–74.

Soltani, F. (1997) *Etude de l'écoulement de gaz a travers les geosynthetiques bentonitiques utilizés en couverture des centres de stockage de déchets*, These de Doctorat, INSA-Lyon, France.

Soong, T.Y. and Koerner, R.M. (1998) Laboratory study of high density polyethylene geomembrane waves, in *Proceedings of 6th International Conference on Geosynthetics*, Atlanta, Industrial Fabrics Association International, St. Paul, MN, pp. 301–6.

Southen, J.M. and Rowe, R.K. (2002) Desiccation behaviour of composite landfill lining systems under thermal gradients, in *Proceeding of International Symposium on Geosynthetic Clay Barriers*, Nuremberg, Germany, pp. 311–20.

Sposito, G. (1984) *The Surface Chemistry of Soils*. Oxford University Press, New York.

Sposito, G. (1989) *The Chemistry of Soils*, Oxford University Press, New York.

Srinivasan, K.R. and Fogler, H.S. (1990) Use of organophilic-clays in the removal of priority pollutants for industrial waste waters, *Structural Aspects*, **38**, 277–86.

Stack, T.M., Potter, S.T. and Suthersan, S.S. (1999) Putting down roots, *Civil Engineering*, Vol. 69, April, pp. 46–9.

Stam, T.G. (2000) Geosynthetic clay liner field performance, in *Proceedings 14th GRI Conference (Hot Topics in Geosynthetics)*, Las Vegas, USA, pp. 242–54.

Stark, T.D. (1998) Bentonite Migration in Geosynthetic Clay Liners, in *Proceedings of 6th International Conference on Geosynthetics* (ed. R.K. Rowe), Industrial Fabrics Association International, Atlanta.

Stark, T.D. and Eid, H.T. (1996) Shear behavior of reinforced geosynthetic clay liners, *Geosynthetics International*, **3**(6), 771–86.

Stark, T.D. and Poeppel, A.R. (1994) Landfill liner interface strengths from torsional-ring shear tests, *ASCE Journal of Geotechnical Engineering*, **120**(3), 597–615.

Stark, T.D., Eid, H.T., Evans, W.D. and Sherry, P.E. (2000) Municipal solid waste slope failure. II. Stability analyses, *ASCE Journal of Geotechnical and Geoenvironmental Engineering*, **126**(5), 408–19.

Stegmann, R. (1983) New aspects on enhancing biological process in sanitary landfills, *Waste Management and Research Journal*, **1**, 201–11.

Stone, J.L. (1984) Leakage monitoring of the geomembrane for proton decay experiment, in *Proceedings International Conference on Geomembranes*, Denver, USA, pp. 475–80.

Sudicky, E.A. (1986) A natural gradient experiment on solute transport in a sand aquifer: spatial variability of hydraulic conductivity and its role in the dispersion process, *Water Resources Research*, **22**(13), 2069–82.

Sudicky, E.A. and Frind, E.O. (1982) Contaminant transport in fractured porous media: analytical solutions for a system of parallel fractures, *Water Resources Research*, **18**(6), 1634–42.

Talbot, A. (1979) The accurate numerical integration of Laplace transforms, *Journal of Institute of Mathematics Applications*, **23**, 97–120.

Tanays, E., Le Tellier, I., Bernhard, C. and Gourc, J.P. (1994) Behaviour of lining systems on waste landfill slopes: an experimental approach, in *Proceedings of 5th International Conference on Geosynthetics*, Singapore, Vol. 3, pp. 977–80.

Tang, D.H., Frind, E.O. and Sudicky, E.A. (1981) Contaminant transport in fractured porous media: analytical solution for a single fracture, *Water Resources Research*, **17**(3), 555–64.

Tavenas, F., Tremblay, M. and Leroueil, S. (1983) Mesure in situ de la perméabilité des argiles, Symposium international sur la reconnaissance des sols et des roches par essais en place, Paris, *Bulletin of the International Association of Engineering Geology*, **26–27**, 509–15.

Tavenas, F., Tremblay, M., Larouche, G. and Leroueil, S. (1986) In situ measurement of permeability in soft clays, *ASCE Specialty Conference, In Situ '86*, Blacksburg, pp. 1034–48.

Tavenas, F., Diene, M. and Leroueil, S. (1990) Analysis of the in situ constant-head permeability test in clays, *Canadian Geotechnical Journal*, **27**, 305–14.

Thomas, R.W. (2002) Thermal oxidation of a polypropylene geotextile used in a geosynthetic clay liner, in *Proceedings of the International Symposium Geosynthetic Clay Liners*, Nuremberg, April, pp. 87–96.

Thomas, H.R. and Missoum, H. (1999) Three-dimensional coupled heat, moisture and air transfer in a deformable unsaturated soil, *International Journal for Numerical Methods in Engineering*, **44**, 919–43.

Thomas, G.W. and Swoboda, A.R. (1970) Anion exclusion effects on chloride movement in soils, *Soil Science*, **110**, 163–66.

Thornton, S.F., Bright, M.I., Lerner, D.N. and Tellam, J.H. (2000) Attenuation of landfill leachate by UK triassic sandstone aquifer materials. 2. Sorption and degradation of organic pollutants in laboratory columns, *Journal of Contaminant Hydrology*, **43**, 355–83.

Tisinger, L.G. and Giroud, J.P. (1993) The durability of HDPE geomembranes, *Geotechnical Fabrics Report*, **11**(6), 4–8.

Tognon, A.R. (1999) *Laboratory testing of geosynthetic landfill components*, M.E.Sc. Thesis, Faculty of Engineering Science, The University of Western Ontario, London, Ont., Canada.

Tognon, A.R., Rowe, R.K. and Brachman, R.W.I. (1999) Evaluation of side wall friction for a buried pipe test facility, *Geotextiles and Geomembranes*, **17**, 193–212.

Tognon, A.R., Rowe, R.K. and Moore, I.D. (2000) Geomembrane strain observed in large-scale testing of protection layers, *ASCE Journal of Geotechnical and Geoenvironmental Engineering*, **126**(12), 1194–208.

Touze-Foltz, N. (2001) Large scale tests for the evaluation of composite liners hydraulic performance, in *Proceedings of 8th International Landfill Symposium*, S. Margherita di Pula, Cagliari, Italy, pp. 133–42.

Touze-Foltz, N., Schmittbuhl, J. and Memier, M. (2001a) Geometric and spatial parameters of geomembrane wrinkles on large scale model tests, in *Proceedings of Geosynthetics 2001*, Portland, USA, pp. 715–28.

Touze-Foltz, N., Rowe, R.K. and Navarro, N. (2001b) Liquid flow through composite liners due to geomembrane defects: nonuniform hydraulic transmissivity at the liner interface, *Geosynthetics International*, **8**(1), 1–26.

Trast, J. and Benson, C.H. (1995) Estimating field hydraulic conductivity at various effective stresses, *ASCE Journal of Geotechnical Engineering*, **121**(10), 736–40.

Triplett, E.J. and Fox, P.J. (2001) Shear strength of HDPE geomembrane/geosynthetic clay liner interface, *ASCE Journal of Geotechnical and Geoenvironmental Engineering*, **127**(6), 543–52.

Uchrin, C.G. and Kataz, J. (1986) Sorption kinetics of competing organic substances on New Jersey coastal plain aquifer solids, *Hazardous and Industrial Solid Waste Testing and Disposal*, ASTM Special Technical Publication **933**, 140–50.

US EPA (2002) *National Primary Drinking Water Standards*. EPA-816-F-02-013. Office of Water (4101), United States Environmental Protection Agency, Washington, DC.

Usher, S.J. and Cherry, J.A. (1988) Evaluation of a two layer aquitard through the use of a one-dimensional analytical solution of the transient head distribution, in *41st Canadian Geotechnical Conference*, Waterloo, Ontario.

van Bavel, C.H.M. (1952) Gaseous diffusion and porosity in porous media, *Soil Sciences*, **73**, 91–104.

van Genuchten, M.Th. (1978) *Calculating the unsaturated hydraulic conductivity with a new closed form analytical model*, Research Report 78-WR-08, Department of Civil Engineering, Princeton University, Princeton, NJ.

Vangpaisal, T. and Bouazza, A. (2001) Gas permeability of three needle punched geosynthetic clay liners, in *Proceedings of 2nd ANZ Conference on Environmental Geotechnics*, Newcastle, Australia, pp. 373–8.

VanGulck, J. (1998) *Determination of the calcium carbonate yield coefficient*, Undergraduate Thesis, The University of Western Ontario, London, Ontario, Canada.

VanGulck, J.F. (2003) *Biodegradation and clogging in gravel size material*, Ph.D. Thesis, Queen's University, Kingston, Ont., 536 pp.

VanGulck, J.F., Rowe, R.K., Rittmann, B.E. and Cooke, A.J. (2003) Predicting biogeochemical calcium precipitation in landfill leachate collection systems, *Biodegradation*, **14**, 331–46.

Van Olphen, H. (1977) *An Introduction to Clay Colloid Chemistry*, 2nd edn, John Wiley and Sons, 318 pp.

Vasko, S.M., Jo, H.Y., Benson, C.G., Edil, T.B. and Katsumi, T. (2001) Hydraulic conductivity of partially prehydrated geosynthetic clay liners permeated with aqueous calcium chloride solutions, in *Proceedings of Geosynthetics 2001*, Portland, pp. 685–99.

Viebke, J., Elble, E., Ifwarson, M. and Gedde, U.W. (1994) Degradation of unstabilized medium-density polyethylene pipes in hot-water applications, *Polymer Engineering and Science*, **34**(17), 1354–61.

Viste, D.R. (1997) Waste processing and biosolids incorporation to enhance landfill gas, in *Proceedings of 6th International Landfill Symposium*, Sardinia '97, Cagliari, Italy, Vol. 1, pp. 369–74.

Vogel, T.M., Criddle, C.S. and McCarty, P.L. (1987) Transformations of halogenated diphatic compounds, *Environmental Science and Technology*, **2**(8), 722–36.

Voice, T.C. (1988) Activated carbon adsorption, in *Standard – Handbook of Hazardous Waste Treatment and Disposal* (ed. H.H. Freeman), McGraw-Hill, New York, pp. 6.3–6.21.

Voice, T.C., Rice, C.P. and Weber, W.J. (1983) Effect of solids concentration on the sorptive partitioning of hydrophobic pollutants in aquatic systems, *Journal of Environment Science Technology*, **17**, 513–18.

von Maubeuge, K.P. and Eberle, M.A. (1998) Can geosynthetic clay liners be used on slopes to achieve long-term stability, in *Proceedings of 3rd International Congress on Environmental Geotechnics*, Lisboa, Portugal, September, pp. 375–80.

von Maubeuge, K.P. and Ehrenberg, H. (2000) Long-term resistance to oxidation of PP and PE geotextiles, in *Proceedings of 2nd European Geosynthetics Conference*, Bologna, October, pp. 465–71.

von Maubeuge, K. and Heerten, G. (1994) Needle-punched geosynthetic clay liners (GCLs), in *Proceedings of 8th GRI Conference Geosynthetic Resins, Formulations and Manufacturing*, December 1994, Philadelphia, PA, pp. 199–207.

Walker, A.N., Beaven, R.P. and Powrie, W. (1997) Overcoming problems in the development of a high rate flushing bioreactor landfill, in *Proceedings of 6th International Landfill Symposium*, Sardinia'97, Cagliari, Italy.

Walton, J.C., Rahman, M., Casey, D., Picornell, M. and Johnson, F. (1997) Leakage through flaws in geomembrane liners, *ASCE Journal of Geotechnical and Geoenvironmental Engineering*, **123**(6), 534–9.

Warith, M.A. and Sharma, R. (1998) Technical review of methods to enhance biological degradation in sanitary landfills, *Water Quality Research Journal of Canada*, **33**, 417–37.

Warith, M.A., Zekry, W. and Gawri, N. (1999) Effect of leachate recirculation on municipal solid waste biodegradation, *Water Quality Research Journal of Canada*, **34**, 267–80.

Wiedemeier, T.H., Rifai, H.S., Wilson, J.T. and Newell, C. (1999) *Natural Attenuation of Fuels and Chlorinated Solvents in the Subsurface*, John Wiley & Sons, New York.

Wigh, R.J. (1979) *Boone County Field Site Interim Report*, EPA-600/2–79–058.

Williams, N., Giroud, J.P. and Bonaparte, R. (1984) Properties of plastic nets for liquid and gas drainage associated with geomembranes, in *Proceedings of International Conference on Geomembranes*, Denver, pp. 399–404.

Wilson-Fahmy, R.F. and Koerner, R.M. (1995) Leakage rates through holes in geomembranes overlying geosynthetic clay liners, in *Proceedings of Geosynthetics'95*, Industrial Fabrics Association International, pp. 655–68.

Wilson-Fahmy, R.F., Narejo, D. and Koerner, R.M. (1996) Puncture protection of geomembranes. Part I. Theory, *Geosynthetics International*, **3**(5), 605–28.

Wong, J. (1977) The design of a system for collecting leachate from a lined landfill site, *Water Resources Research*, **13**(2), 404–10.

Wong, L.C. and Haug, M.D. (1991) Cyclical closed-system freeze-thaw permeability testing of soil liner and cover materials, *Canadian Geotechnical Journal*, **28**(6), 784–93.

Workman, J.P. (1993) Interpretation of leakage rates in double-lined systems, in *Proceedings of 7th GRI Conference*, pp. 91–108.

Woyshner, M.R. and Yanful, E.K. (1995) Modelling and field measurements of water percolation through an experimental soil cover on mine tailings, *Canadian Geotechnical Journal*, **32**(4), 601–9.

Xu, S., Sheng, G. and Boyd, S.A. (1997) Use of organo-clays in pollution abatement, *Advances in Agronomy*, **59**, 25–62.

Yanful, E.K. (1993) Oxygen diffusion through soil covers on sulphidic mill tailings, *ASCE Journal of Geotechnical Engineering*, **119**(8), 1207–28.

Yanful, E.K. and Aubé, B. (1993) Modelling moisture-retaining soil covers, in *Proceedings of Joint Canadian Society of Civil Engineers – American Society of Civil Engineers National Conference on Environmental Engineering*, Montreal, Canada, pp. 273–80.

Yanful, E.K. and Quigley, R.M. (1986) Heavy metal deposition at the clay/waste interface of a landfill site, Sarnia, Ontario, in *Proceedings of 3rd Canadian Hydrogeological Conference*, Saskatoon, April 1986, pp. 35–42.

Yanful, E.K. and Quigley, R.M. (1990) Tritium, oxygen-18 and deuterium diffusion at the Confederation Road landfill site, Sarnia, Ontario, Canada, *Canadian Geotechnical Journal*, **27**(3), 271–5.

Yanful, E.K., Nesbitt, H.W. and Quigley, R.M. (1988a) Heavy metal migration at a landfill site, Sarnia, Ontario, Canada. I. Thermodynamic assessment and chemical interpretations, *Applied Geochemistry*, **3**, 523–33.

Yanful, E.K., Quigley, R.M. and Nesbitt, W. (1988b) Heavy metal migration at a landfill site. Part II. Metal partitioning and geotechnical implications, *Applied Geochemistry*, **3**, 623–9.

Yanful, E.K., Bell, A.V. and Woyshner, M.R. (1993a) Design of a composite soil cover for an experimental waste rock pile near Newcastle, New Brunswick, Canada, *Canadian Geotechnical Journal*, **30**(4), 578–87.

Yanful, E.K., Riley, M.D., Woyshner, M.R. and Duncan, J. (1993b) Construction and monitoring of a composite soil cover on an experimental waste rock pile near Newcastle, New Brunswick, Canada, *Canadian Geotechnical Journal*, **30**(4), 588–99.

Yazdani, R., Augenstein, D. and Pacey, J. (2000) US EPA Project XL: Yolo County's accelerated anaerobic and aerobic composting (full-scale controlled landfill bioreactor) project, in *Proceedings of 14th Geosynthetic Research Institute Conference*, pp. 77–105.

Yeh, G.T. (1984) Solution of contaminant transport equations using an orthogonal upstream weighting finite element scheme, in *Proceedings of 5th International Conference on Finite Elements in Water Resources*, Vermont, pp. 285–93.

Yong, R.N. and Sheremata, T.W. (1991) Effect of chloride ions on adsorption of cadmium from a landfill leachate, *Canadian Geotechnical Journal*, **28**(3), 378–87.

Yong, R.N. and Warkentin, B.P. (1975) *Soil Properties and Behaviour*, 2nd edn, Elsevier Scientific Publ. Col., 449 pp.

Yoshida, H. and Rowe, R.K. (2003) Consideration of landfill liner temperature, in *Proceedings of 8th International Waste Management and Landfill Symposium*, S. Margherita di Pula, Cagliari, Sardinia, Italy.

You, S.-J., Yin, Y. and Allen, H.E. (1999) Partitioning of organic matter in soils: effects of pH and water/soil ratio, *The Science of the Total Environment*, 227, 155–60.

Yuen, S.T.S., Styles, J.R., Wang, Q.J. and McMahon, T.A. (1999) Findings from a full-scale bioreactor landfill study in Australia, in *Proceedings of 7th International Landfill Symposium*, Sardinia'99, Cagliari, Italy, pp. 53–58.

Zagorski, G.A. and Wayne, M.H. (1990) Geonet seams, *Geotextiles and Geomembranes*, **9**, 487–99.

Zanzinger, H. (1999) Efficiency of geosynthetic protection layers for geomembrane liners: performance in a large-scale model test, *Geosynthetics International*, **6**(4), 303–17.

Zanzinger, H. and Gartung, E. (1995) Large-scale model test of leachate pipes in landfills under heavy load, in *Proceedings of an International Conference on Advances in Underground Pipeline Engineering* 2, Belvue, WA, pp. 114–25.

Zehnder, A.J.B. (1978) Ecology of methane formation, *Water Pollution Microbiology* (ed. R. Mitchell), Vol. 2, John Wiley & Sons, New York, pp. 349–76.

Zhang, X.N. and Zhao, A.Z. (1997) Surface charge, in *Chemistry of Variable Charge Soils* (ed. T.R. Yu), Oxford University Press, New York, pp. 17–60.

Zhang, X., Barbour, S.L. and Headley, J.V. (1998) A diffusion batch method for determination of the adsorption coefficient of benzene on clay soils, *Canadian Geotechnical Journal*, **35**, 622–9.

Zhou, Y. and Rowe, R.K. (2003) Development of a technique for modelling clay liner desiccation, *International Journal for Numerical and Analytical Methods in Geomechanics*, **27**(6), 473–93.

Zienkiewicz, O.C. (1977) *The Finite Element Method*, McGraw-Hill, UK.

Zornberg, J.G. and Christopher, B.R. (1999) Geosynthetics, Chapter 27 in *Handbook of Groundwater Engineering* (ed. J.W. Delleur), CRC Press, Boca Raton, FL.

Zornberg, J.G., LaFountain, L. and Caldwell, J.A. (2003) Analysis and design of evapotranspirative cover for hazardous waste landfill, *ASCE Journal of Geotechnical and Geoenvironmental Engineering*, **129**(6), 427–38.

Index

Barrier Systems for Waste Disposal